Springer Texts in Statistics

Series Editors:
Richard DeVeaux
Stephen E. Fienberg
Ingram Olkin

More information about this series at http://www.springer.com/series/417

Also by Richard M. Heiberger

R through Excel:
A Spreadsheet Interface for Statistics,
Data Analysis, and Graphics,
with Erich Neuwirth, Springer 2009

Computation for the Analysis of Designed Experiments, Wiley 1989

Richard M. Heiberger • Burt Holland

Statistical Analysis and Data Display

An Intermediate Course with Examples in R

Second Edition

 Springer

Richard M. Heiberger
Department of Statistics
Temple University
Philadelphia, PA, USA

Burt Holland
Department of Statistics
Temple University
Philadelphia, PA, USA

ISSN 1431-875X   ISSN 2197-4136   (electronic)
Springer Texts in Statistics
ISBN 978-1-4939-2121-8   ISBN 978-1-4939-2122-5   (eBook)
DOI 10.1007/978-1-4939-2122-5

Library of Congress Control Number: 2015945945

Springer New York Heidelberg Dordrecht London

Printed on acid-free paper

Springer Science+Business Media LLC New York is part of Springer Science+Business Media (www.springer.com)

In loving memory of Mary Morris Heiberger

To my family: Margaret, Irene, Andrew, and Ben

Preface

1 Audience

Students seeking master's degrees in applied statistics in the late 1960s and 1970s typically took a year-long sequence in statistical methods. Popular choices of the course textbook in that period prior to the availability of high-speed computing and graphics capability were those authored by Snedecor and Cochran (1980) and Steel and Torrie (1960).

By 1980, the topical coverage in these classics failed to include a great many new and important elementary techniques in the data analyst's toolkit. In order to teach the statistical methods sequence with adequate coverage of topics, it became necessary to draw material from each of four or five text sources. Obviously, such a situation makes life difficult for both students and instructors. In addition, statistics students need to become proficient with at least one high-quality statistical software package.

This book *Statistical Analysis and Data Display* can serve as a standalone text for a contemporary year-long course in statistical methods at a level appropriate for statistics majors at the master's level and for other quantitatively oriented disciplines at the doctoral level. The topics include concepts and techniques developed many years ago and also a variety of newer tools.

This text requires some previous studies of mathematics and statistics. We suggest some basic understanding of calculus including maximization or minimization of functions of one or two variables, and the ability to undertake definite integrations of elementary functions. We recommend acquired knowledge from an earlier statistics course, including a basic understanding of statistical measures, probability distributions, interval estimation, hypothesis testing, and simple linear regression.

2 Motivation

The Second Edition in 2015 has four major changes since the First Edition in 2004 Heiberger and Holland (2004). The changes are summarized here and described in detail in Section 5.

- The computation for the Second Edition is entirely in R (R Core Team, 2015). R is a free open-source publicly licensed software environment for statistical computing and graphics. The computation for the First Edition is mostly in S-Plus, with some R and some SAS. R uses a dialect of the S language developed at Bell Labs. The R dialect is closely related to the dialect of S used by S-Plus. R is much more powerful now than it was when the First Edition was written.

- All graphs from the First Edition have been redrawn in color. There are many additional graphs new to the Second Edition. The graphs are easier to specify because they are built with the much more powerful graphical primitives that exist now and didn't exist 12 years ago. Most graphs are constructed with **lattice**, the R implementation of **trellis** graphics pioneered by S-Plus. Some, particularly in Chapter 15, are drawn using `mosaic` and related functions in the **vcd** package. Functions for the graphic displays designed for this book are included in the **HH** package available at CRAN (Heiberger, 2015).

- Most chapters in the Second Edition are similar in content to the chapters in the First Edition. There are several revised and expanded chapters and several additional appendices.

- The new appendices respond to shifts in the software landscape and/or in the assumed knowledge of computing by the intended audience since 2004.

3 Structure

The book is organized around statistical topics. Each chapter introduces concepts and terminology, develops the rationale for its methods, presents the mathematics and calculations for its methods, and gives examples supported by graphics and computer output, culminating in a writeup of conclusions. Some chapters have greater detail of presentation than others, based on our personal interests and expertise.

Our emphasis on graphical display of data is a distinguishing characteristic of this book. Many of our graphical displays appeared here for the first time. We show graphs, how to construct and interpret them, and how they relate to the tabular outputs that appear automatically when a statistical program "analyzes" a data set. The graphs are not automatic and so must be requested. Gaining an understanding of a data set is always more easily accomplished by looking at appropriately drawn graphs than by examining tabular summaries. In our opinion, graphs are the heart of most statistical analyses; the corresponding tabular results are formal confirmations of our visual impressions.

We believe that a firm control of the language gives the analyst the tools to think about the ideal way to detect and display the information in the data. We focus our presentation on the written command languages, the most flexible descriptors of the statistical techniques. The written languages provide the opportunity for growth and understanding of the underlying techniques. The point-and-click technology of icons and menus is sometimes convenient for routine tasks. However, many interesting data analyses are not routine and therefore cannot be accomplished by pointing and clicking the icons provided by the program developers.

4 Computation

In the First Edition, and again in the Second Edition, the code and data for all examples and figures in the book is available for download.

For the Second Edition, the datasets and R code will be distributed as the R package **HH** through CRAN (Heiberger, 2015).

For the First Edition, the download containing S-Plus, R, and SAS code was initially (in 2004) available from my web site. In 2007, the R code was placed on CRAN (the Comprehensive R Archive Network) as the R package **HH**. In 2009, the S-Plus code was placed on CSAN (the Comprehensive S Archive Network) as the S-Plus package **HH** (Heiberger, 2009).

All datasets in the **HH** package are documented in the book.

4.1 R

R (R Core Team, 2015) is free, publicly licensed, extensible, open-source software. The R language is a dialect of the S language (Becker et al., 1988), similar to that used by S-Plus (Insightful Corp., 2002; TIBCO Software Inc., 2010). Much code (both functions and examples) written for one will also work in the other. R has been increasing its reach—within academia, industry, government, and internationally. Please see Appendix A for information on downloading and using R.

The S language was originally developed at Bell Labs in the 1970s. The Association for Computing Machinery (ACM) awarded John M. Chambers of Bell Labs the 1998 Software System Award for developing the S system.

The R language is an exceptionally well-developed tool for statistical research and analysis, that is for exploring and designing new techniques of analysis, as well as for analysis. The trellis graphics implementation in R's **lattice** package is especially strong for statistical graphics, the output of data analysis through which both the raw data and the results are displayed for the analyst and the client.

R is available by download. The developers are The R Development Core Team, an international group that includes John Chambers and other former Bell Labs researchers.

4.2 The HH Package in R

An important feature of this book is its graphical displays of statistical analyses. For the Second Edition, the **HH** functions for graphing have been rewritten using the more powerful graphing infrastructure that is now available in the **lattice** package in R. The package version number has been changed from the **HH_2.3.x** series to the **HH_3.1-x** series to reflect the redesign. The First Edition had black-and-white figures in print, even though the software at that time produced color figures. In the Second Edition all figures, both in print and in the eBook edition, are in color.

Please see Appendix B for information on working with the **HH** package.

R graphics have much improved since the time of the First Edition. The **lattice** graphics package for plotting coordinated sets of displays was in its infancy when we wrote the First Edition, not yet as capable as the equivalent **trellis** graphics system in S-Plus, and specifically not capable of all the figures in the book. Now **lattice** is much more powerful than **trellis**, and can be even further extended with the capabilities since encoded in the **latticeExtra** package (Sarkar and Andrews, 2013).

The R package system was also not as extensive at that time, and the S-Plus package system did not yet exist. The code and examples for the First Edition of the book were distributed as a zip file on my website and accessible through the Springer website. The code and examples were revised and distributed as an R package **HH** beginning in 2007, and as an S-Plus package in 2009, when S-Plus created their package system. I have continually maintained and extended the software.

4.3 S-Plus, now called S+

S+ is still available, but less commonly used. TIBCO, the owner of S+ is now distributing a Developer's Edition of R called TERR (TIBCO Enterprise Runtime for R) based on their new enterprise-grade, high-performance statistical engine (TIBCO

Software Inc., 2014). The design goal of TERR is to be able to install all R packages. As of July 2014, TERR had not yet implemented their graphics system. Once their graphics system is implemented, **HH_3.1-x** will work with TERR.

The older version of **IIII** (Heiberger, 2009), designed for the First Edition of this book, continues to work with S+.

4.4 SAS

SAS is an important statistical computing system in industry. All the code from our First Edition still works. My own personal work has become more highly R-focused. I have chosen to drop most of the SAS discussion and examples from the body of the Second Edition.

Some SAS material is still in the body of the Second Edition. Now-standard terminology introduced by SAS, primarily the notation for "Types" of Sums of Squares described in Section 13.6, is referenced and described. The notation of the SAS MODEL statement is similar to the notation of the R model formula. Comparisons of the two notations are in Sections 9.4.1, 12.13.1, 12.15, 12.A, 13.4, and 13.5.

All datasets in the Second Edition can be used with SAS. See Appendix H for details.

5 Chapters in the Second Edition

5.1 Revised Chapters

All graphs from the First Edition have been redrawn in color and with the use of much more powerful graphical primitives that didn't exist 12 years ago.

There are many additional graphs new to the Second Edition.

Chapters 3 and 5 have many new figures, most built with the NTplot function. The graphs, showing significance and power of hypothesis tests for the normal and t distributions, produced by this single function cover most of the standard first semester introductory Statistics course.

Chapter 11 "Multiple Regression—Regression Diagnostics" has a new section 11.3.7 "Residuals vs Leverage" to discuss one of the panels produced by R's plot.lm function that was not in the similar S-Plus function.

Chapter 15 "Bivariate Statistics—Discrete Data" has undergone major revision. The examples are now centered on `mosaic` graphics, using the **vcd** package that was not available when the First Edition was written.

Section 15.8 "Example—Adverse Experiences" is new. The discussion focuses on the Adverse Effects dotplot, and shows how multi-panel plots graphical displays can replace pages of tabular data. The discussion is based on the work in which I participated while at research leave at GSK (Amit et al., 2008).

Section 15.9 "Likert Scale Data" is new. This section is based on my recent work with Naomi Robbins (Heiberger and Robbins, 2014). Rating scales, such as Likert scales and semantic differential scales, are very common in marketing research, customer satisfaction studies, psychometrics, opinion surveys, population studies, and numerous other fields. We recommend diverging stacked bar charts as the primary graphical display technique for Likert and related scales. We discuss the perceptual issues in constructing the graphs. Many examples of plots of Likert scales are given.

5.2 Revised Appendices

We have made major changes to the Appendices. There are more appendices now and the previous appendices have been restructured and expanded. The description of the Second Edition appendices is in Section 1.3.5.

6 Exercises

Learning requires that the student work a fair selection of the exercises provided, using, where appropriate, one of the statistical software packages we discuss. Beginning with the exercises in Chapter 5, even when not specifically asked to do so, the student should routinely plot the data in a way that illuminates its structure, and state all assumptions made and discuss their reasonableness.

Acknowledgments: First Edition

We are indebted to many people for providing us advice, comments, and assistance with this project. Among them are our editor John Kimmel and the production staff at Springer, our colleagues Francis Hsuan and Byron Jones, our current and former students (particularly Paolo Teles who coauthored the paper on which Chapter 18 is based, Kenneth Swartz, and Yuo Guo), and Sara R. Heiberger. Each of us gratefully acknowledges the support of a study leave from Temple University. We are also grateful to Insightful Corp. for providing us with current copies of S-Plus software for ourselves and our student, and to the many professionals who reviewed portions of early drafts of this manuscript.

Philadelphia, PA, USA   Richard M. Heiberger
Philadelphia, PA, USA   Burt Holland
July 2004

Acknowledgments

We are indebted to many additional people for support in preparing the Second Edition. Our editors at Springer Jon Gurstelle (now at Wiley), Hannah Bracken, and Michael Penn encouraged the preparation of this Second Edition. Alicia Strandberg at Villanova University used a preliminary version of this edition with two of her classes. She and her students provided excellent feedback and suggestions for the preparation of this material. I also used drafts of this edition in my own courses at Temple University and incorporated the classes' feedback into the revision.

We are grateful to the R Core and the many R users and contributors who have provided the software we use so heavily in our graphical and tabular analyses.

The material in the new section on Adverse Effects is based on the work with the GSK team investigating graphics for safety data in clinical trials, particularly coauthors Ohad Amit and Peter W. Lane.

The material in the new section on Likert scale plots is based on the work with Naomi Robbins.

The First Edition was coauthored by Burt Holland. Even though Burt died in 2010, I am writing this second preface mostly in the plural. Burt's voice is present in much of the text of the Second Edition. Most of the numbered chapters have essentially the same content as in the First Edition.

The new sections and the Appendices in the Second Edition are entirely by me. All graphs in this edition are newly drawn by me using the more powerful graphics infrastructure that is now available in R.

I had several discussions with Kenneth Swartz when I was initially considering writing this edition and at various points along the way.

Barbara Bloomfield provided me overall support in everything. She also responded to my many queries on stylistic and appearance issues in the revised manuscript and graphs.

Philadelphia, PA, USA   Richard M. Heiberger
October 2015

Contents

Author Bios

Richard M. Heiberger is Professor Emeritus in the Department of Statistics of Temple University, an elected Fellow of the American Statistical Association, and a former Chair of the Section on Statistical Computing of the American Statistical Association. He was Graduate Chair for the Department of Statistics and Acting Associate Vice Provost for the University. He participated in the design of the linear model and analysis of variance functions while on research leave at Bell Labs. He has taught short courses at the Joint Statistics Meetings, the American Statistical Association Conference on Statistical Practice, the R Users Conference, and the Deming Conference on Applied Statistics. He has consulted with several pharmaceutical companies.

Burt Holland was Professor in the Department of Statistics of Temple University, an elected Fellow of the American Statistical Association, Chair of the Department of Statistics of Temple University, and Chair of Collegial Assembly of the Fox School. He has taught short courses at the Joint Statistics Meetings and the Deming Conference on Applied Statistics. He has made many contributions to linear modeling and simultaneous statistical inference. He frequently served as consultant to medical investigators. He developed a very popular General Education course on Statistics and the News.

Chapter 1

Introduction and Motivation

Statistics is the science and art of making decisions based on quantitative evidence. This introductory chapter motivates the study of statistics by describing where and how it is used in all endeavors. It gives examples of applications, a little history of the subject, and a brief overview of the structure and content of the remaining chapters.

Almost all fields of study (including but not limited to physical science, social science, business, and economics) collect and interpret numerical data. Statistical techniques are the standard ways of summarizing and presenting the data, of turning data from an accumulation of numbers into usable information. Not all numbers are the same. No group of people are all the same height, no group has an identical income, not all cars get the same gas mileage, not all manufactured parts are absolutely identical. How much do they differ? Variability is the key concept that statistics offers. It is possible to measure how much things are not alike. We use standard deviation, variance, range, interquartile range, and MAD (median absolute deviation from the median) as measures of not-the-sameness. When we compare groups we compare their variability as well as their range.

Statistics uses many mathematical tools. The primary tools—algebra, calculus, matrix algebra, analytic geometry—are reviewed in Appendix I. Statistics is not purely mathematics. Mathematics problems are usually well specified and have a single correct answer on which all can agree. Data interpretation problems calling for statistics are not yet well specified. Part of the data analyst's task is to specify the problem clearly enough that a mathematical tool may be used. Different answers to the same initial decision problem may be valid because a statistical analysis requires assumptions about the data and its manner of collection, and analysts can reasonably disagree about the plausibility of such assumptions.

Statistics uses many computational tools. In this book, we use R (R Core Team, 2015) as our primary tool for statistical analysis. R is an exceptionally well-developed tool for statistical research and analysis, that is for exploring and

© Springer Science+Business Media New York 2015
R.M. Heiberger, B. Holland, *Statistical Analysis and Data Display*,
Springer Texts in Statistics, DOI 10.1007/978-1-4939-2122-5_1

designing new techniques of analysis, as well as for analysis. We discuss installation and use of R in Appendix A.

We make liberal use of graphs in our presentations. Data analysts are responsible for the *display* of data with graphs and tables that summarize and represent the data and the analysis. Graphs are often the output of data analysis that provide the best means of communication between the data analyst and the client. We study a variety of display techniques.

While producing this book, we designed many innovative graphical displays of data and analyses. We introduce our displays in Section 1.3.4. We discuss the displays throughout the book in the context of their associated statistical techniques. These discussions are indexed under the term *graphical design*. In the appendix to Chapter 4, we summarize the large class of newly created graphs that are based on Cartesian products.

The R code for all the graphs and tables in this book is included in the **HH** package for R (Heiberger, 2015). See Appendix B for a summary of the **HH** package. We consider the **HH** package to be an integral part of the book.

Statistics is an art. Skilled use of the mathematical tools is necessary but not sufficient. The data analyst must also know the subject area under study (or must work closely with a specialist in the subject area) to ensure an appropriate choice of statistical techniques for solving a problem. Experience, good judgment, and considerable creativity on the part of the statistical analyst are frequently needed.

Statistics is "the science of doing science" and is perhaps the only discipline that interfaces with all other sciences. Most statisticians have training or considerable knowledge in one or more areas other than statistics. The statistical analyst needs to communicate successfully both orally and in writing with the client for the analysis.

Statistics uses many communications skills, both written and oral. Results must be presented to the client and to the client's management. We discuss some of the mechanics of writing programs and technical reports in Appendices K, L, M, N, and O.

A common statistical problem is to discover the characteristics of an unobservable population by examining the corresponding characteristics of a sample *randomly* selected from the population and then (inductively) inferring the population characteristics (parameters) from the corresponding sample characteristics (statistics). The task of selecting a random sample is not trivial. The discipline of statistics has developed a vast array of techniques for inferring from samples to populations, and for using probabilities to quantify the quality of such inferences.

Most statistical problems involve simultaneous consideration of several related measurements. Part of the statistician's task is to determine the interdependence among such measures, and then to account for it in the analysis.

The word "statistics" derives from the political science collections of numerical data describing demographics, business, politics that are useful for management of the "state". The development of statistics as a scientific discipline dates from the end of the 19th century with the design and analysis of agricultural experiments aimed at finding the best combination of fertilization, irrigation, and variety to maximize crop yield. Early in the 20th century, these ideas began to take hold in industry, with experiments designed to maximize output or minimize cost. Techniques for statistical analysis are developed in response to the needs of specific subject areas. Most of the techniques developed in one subject field can be applied unchanged to other subjects.

1.1 Statistics in Context

We write as if the statistician and the client are two separate people. In reality they are two separate roles and the same person often plays both roles. The client has a problem associated with the collection and interpretation of numerical data. The statistician is the expert in designing the data collection procedures and in calculating and displaying the results of statistical analyses.

The statistician's contribution to a research project typically includes the following steps:

1. Help the client phrase the question(s) to be answered in a manner that leads to sensible data collection and that is amenable to statistical analysis.

2. Design the experiment, survey, or other plan to approach the problem.

3. Gather the data.

4. Analyze the data.

5. Communicate the results.

In most statistics courses, including the one for which this book is designed, much of the time is spent learning how to perform step 4, the science of statistics. However, step 2, the art of statistics, is very important. If step 2 is poorly executed, the end results in step 5 will be misleading, disappointing, or useless. On the other hand, if step 4 is performed poorly following an excellent plan from step 2 and a correct execution of step 3, a reanalysis of the data (a new step 4) can "save the day".

Today (2015) there are more than 18,000 statisticians practicing in the United States. Most fields in the biological, physical, and social sciences require training in statistics as educational background. Over 100 U.S. universities offer graduate degrees in statistics. Most firms of any size and most government agencies employ statisticians to assist in decision making. The profession of *statistician* is highly placed in the *Jobs Rated Almanac* Krantz (1999). A shortage of qualified statisticians to fill open positions is expected to persist for some time American Statistical Association (2015).

1.2 Examples of Uses of Statistics

Below are a few examples of the countless situations and problems for which statistics plays an important part in the solution.

1.2.1 Investigation of Salary Discrimination

When a group of workers believes that their employer is illegally discriminating against the group, legal remedies are often available. Usually such groups are minorities consisting of a racial, ethnic, gender, or age group. The discrimination may deal with salary, benefits, an aspect of working conditions, mandatory retirement, etc. The statistical evidence is often crucial to the development of the legal case.

To illustrate the statistician's approach, we consider the case of claimed salary discrimination against female employees. The legal team and statistician begin by developing a defensible list of criteria that the defendant may legally use to determine a worker's salary. Suppose such a list includes years of experience (yrsexp), years of education (yrsed), a measure of current job responsibility or complexity (respon), and a measure of the worker's current productivity (product). The statistician then obtains from a sample of employees, possibly following a subpoena by the legal team, data on these four criteria and a fifth criterion that distinguishes between male and female employees (gender). Using regression analysis techniques we introduce in Chapter 9, the statistician considers two statistical models, one that explains salary as a function of the four stipulated permissible criteria, and another that explains salary as a function of these four criteria plus gender. If the model containing the predictor gender predicts salary appreciably better than does the model excluding gender and if, according to the model with gender included, females receive significantly less salary than males, then this may be regarded as statistical evidence of discrimination against females. Tables and graphs based on techniques discussed in Chapters 15, 17, and 4 (and other chapters) are often used in legal proceedings.

In the previous section it is pointed out that two statisticians can provide different analyses because of different assumptions made at the outset. In this discrimination context, the two legal teams may disagree over the completeness or relevance of the list of permissible salary determinants. For example, the defense team may claim that females are "less ambitious" than males, or that women who take maternity or child care leaves have less continuous or current experience than men. If the court accepts such arguments, this will undermine the plaintiff statistician's finding of the superiority of the model with the extra predictor.

1.2.2 Measuring Body Fat

In Chapters 8, 9, and 13 we discuss an experiment designed to develop a way to estimate the percentage of fat in a human body based only on body measurements that can be made with a simple tape measure. The motivation for this investigation is that measurement of body fat is difficult and expensive (it requires an underwater weighing technique), but tape measurements are easy and inexpensive to obtain. At the outset of this investigation, the client offered data consisting of 15 inexpensive measurements and the expensive body fat measurement on each of 252 males of various shapes and sizes. Our analysis in Chapter 9 demonstrates that essentially all of the body fat information in the 15 other measurements can be captured by just three of these other measurements. We develop a regression model of body fat as a function of these three measurements, and then we examine how closely these three inexpensive measurements alone can estimate body fat.

1.2.3 Minimizing Film Thickness

In Section 13.3.1 we discuss an experiment that seeks to find combinations of temperature and pressure that minimize the thickness of a film deposited on a substrate. Each of these factors can affect thickness, and the complication here is the possibility that the optimum amount of one of these factors may well depend on the chosen amount of another factor. Modeling such *interaction* between factors is key to a proper analysis. The statistician is also expected to advise on the extent of sensitivity of thickness to small changes in the optimum mix of factors.

1.2.4 Surveys

Political candidates and news organizations routinely sample potential voters for their opinions on candidates and issues. Results gleaned from samples selected by contemporary methods are often regarded as sufficiently reliable to influence candidate behavior or public policy.

The marketing departments of retail firms often sample potential customers to learn their opinions on product composition or packaging, and to help find the best matches between specialized products and locales for their sale.

Manufacturers sample production to determine if the proportion of output that is defective is excessive. If so, this may lead to the decision the output should be scrapped, or at least that the production process be inspected and corrected for problems.

All three of these examples share statistical features. The data are collected using techniques discussed in Section 3.11. The initial analysis is usually based on techniques of Chapter 5.

1.2.5 Bringing Pharmaceutical Products to Market

The successful launching of a new pharmaceutical drug is a huge undertaking in which statisticians are key members of the investigative team. After candidate drugs are found to be effective for alleviation of a condition, experiments must be run to check them for toxicity, safety, side effects, and interactions with other drugs. Once these tests are passed, statisticians help to determine the optimum quantity and spacing of dosages. Much of the testing is done on lab animals; only at the later stages are human subjects involved. The entire process is performed in a manner mandated by government regulatory agencies (such as the Food and Drug Administration (FDA) in the United States, The European Medicines Agency (EMA) in the European Union, or the Ministry of Health, Labour and Welfare (MHLW) in Japan). Techniques are based on material developed in all chapters of this book.

1.3 The Rest of the Book

1.3.1 Fundamentals

Chapters 2 through 5 discuss data, types of data analysis, and graphical display of data and of analyses.

Chapter 2 describes data acquisition and how to get the data ready for its analysis. We emphasize that an important early step in any data analysis is graphical display of the data.

Chapter 3 provides an overview of basic concepts—probability, distributions, estimation, testing, principles of inference, and sampling—that are background material for the remainder of the book. Several common distributions are discussed and illustrated here. Others appear in Appendix J. Two important fitting criteria— least squares and maximum likelihood—are introduced. Random sampling is a well-defined technique for collecting data on a subset of the population of interest. Random sampling provides a basis for making inferences that a haphazard collection of data cannot provide.

A variety of graphical displays are discussed and illustrated in Chapter 4. The graphs themselves are critically important analysis tools, and we show examples where different display techniques help in the interpretation of the data. On occasion we display graphs that are intermediate steps leading to other graphs. For example, Figure 14.17 belongs in a final report, but Figure 14.15, which suggests the improved and expanded Figure 14.17, should not be shown to the client.

Chapter 5 introduces some of the elementary inference techniques that are used throughout the rest of the book. We focus on tests on data from one or two normal distributions. We show the algebra and graphics for finding the center and spread of the distributions. These algebraic and graphical techniques are used in all remaining chapters.

1.3.2 Linear Models

Chapters 6 through 13 build on the techniques developed in Chapter 5. The word "linear" means that the equations are all linear functions of the model parameters and that graphs of the analyses are all straight lines or planes.

In Chapter 6 we extend the t-tests of Chapter 5 to the comparison of the means of several (more than two) populations.

With $k > 2$ populations, there are only $k - 1$ independent comparisons possible, yet we often wish to make $\binom{k}{2}$ comparisons. In Chapter 7 we discuss the concept of *multiple comparisons*, the way to make valid inferences when there are more comparisons of interest than there are degrees of freedom. We introduce the fundamental concept of "contrasts", direct comparisons of linear combinations of the means of these populations, and show several potentially sensible ways to choose $k - 1$ independent contrasts. We introduce the *MMC* plot, the mean–mean plot for displaying arbitrary multiple comparisons.

Chapters 8 through 11 cover regression analysis, the process of modeling a continuous response variable as a linear function of one or more predictor variables.

In Chapter 8 we plot a continuous response variable against a single continuous predictor variable and develop the least-squares procedure for fitting a straight line to the points in the plot. We cast the algebra of least squares in matrix notation (relevant matrix material is in Appendix I) and apply it to more than one predictor variable. We introduce the statistical assumptions of a normally distributed error term and show how that leads to estimation and testing procedures similar to those introduced in Chapter 5.

Chapter 9 builds on Chapter 8 by allowing for more than one predictor for a response variable and introducing additional structure, such as interactions, among the predictor variables. We show techniques for studying the relationships of the predictors to each other as well as to the response.

Chapter 10 shows how dummy variables are used to incorporate categorical predictors into multiple regression models. We begin to use dummy variables to encode the contrasts introduced in Chapter 6, and we continue using dummy variables and contrasts in Chapters 12, 13, and 14. We show how the use of continuous (concomitant) variables (also known as covariates) can enhance the modeling of designed experiments.

Chapter 11 evaluates the models, introduces diagnostic techniques for checking assumptions and detecting outliers, and uses tools such as transformation of the variables to respond to the problems detected.

In Chapter 12 we extend the analysis of one-way classifications of continuous data to several types of two-way classifications. We cast the analysis of variance into the regression framework with dummy variables that code the classification factors with sets of contrasts.

In Chapters 13 and 14 we consider the principles of experimental design and their application to more complex classifications of continuous data. We discuss the analysis of data resulting from designed experiments.

1.3.3 Other Techniques

The analysis of tabular categorical data is considered in Chapter 15. We discuss contingency tables, tables in which frequencies are classified by two or more factors. For 2×2 tables or sets of 2×2 tables we use odds ratios or the Mantel–Haenszel test. For larger tables we use χ^2 analysis. We discuss several situations in which contingency tables arise, including sample surveys and case–control studies.

In Chapter 16 we briefly survey nonparametric testing methods that don't require the assumption of an underlying normal distribution.

Chapter 17 is concerned with logistic regression, the statistical modeling of a response variable which is either dichotomous or which represents a probability. We place logistic regression in the setting of generalized linear models (although we do not fully discuss generalized linear models in this volume). We extend the graphical and algebraic analysis of linear regression to this case.

We conclude in Chapter 18 with an introduction to ARIMA modeling of time series. Time series analysis makes the explicit assumption that the observations are *not* independent and models the structure of the dependence.

1.3.4 New Graphical Display Techniques

This book presents many new graphical display techniques for statistical analysis. Most of our new displays are based on defining the panels of a multipanel graphical display by a Cartesian product of sets of variables, of transformations of a variable, of functions of a fit, of models for a fit, of numbers of parameters, or of levels of a factor. The appendix to Chapter 4 summarizes how we use the Cartesian products to design these new displays and gives a reference to an example in the book for each. The displays, introduced throughout this book's 18 chapters, cover a wide variety of statistical methods. The construction and interpretation of each display are provided in the chapter where it appears.

We produced these displays with the functions that are included in the **HH** package available at CRAN (Heiberger, 2015) and CSAN (Heiberger, 2009). We use R because it is especially strong for designing and programming statistical graphics. We encourage readers and software developers to write and publish functions and macros for these displays in other software systems that have a similarly rich graphics environment.

1.3.5 Appendices on Software

Appendix A discusses the installation and use of R. Some of its material was in the First Edition Appendix B.

Appendix B discusses the **HH** package. The scripts for all examples in both the First and Second Editions of the book are included in the **HH** package. The Appendix shows how to use the scripts to duplicate the figures and tables in the book. Some of its materials were in the First Edition Appendix B.

Appendix C "**Rcmdr**" is new. It discusses and illustrates menu-driven access to the functions and graphics in the book. It is based on my R package **RmcdrPlugin.HH**, an add-in for the R package **Rcmdr** that provides the menu system.

Appendix D "RExcel" is new. It discusses the RExcel interface described in my book with Erich Neuwirth (Heiberger and Neuwirth, 2009) describing his RExcel software (Neuwirth, 2014). RExcel provides a seamless integration of R and Excel. RExcel both places R inside the Excel automatic recalculation model and makes the Rcmdr menu system available on the Excel menu bar.

Appendix E "**Shiny**" is new. It discusses and illustrates web-based access to R functions using the **shiny** package written by R-Studio and distributed on CRAN. **shiny** provides an R language interface for writing interactive web pages.

Appendix F "R Packages" gives a very brief discussion of software design. It includes references to the R documentation.

Appendix G "Computational Precision and Floating Point Arithmetic" is new. Computers use *floating point* arithmetic. The floating point system is not identical to the real-number system that we (teachers and students) know well, having studied it from kindergarten onward. In this appendix we show several examples to illustrate and emphasize the distinction.

Appendix H "Other Statistical Software" is new. It tells how to use the datasets for this book with software other than R.

1.3.6 Appendices on Mathematics and Probability

Appendix I "Mathematics Preliminaries" has been expanded from First Edition Appendix F with many more graphs and tables.

Appendix J "Probability Distributions" has been expanded from First Edition Appendix D to include additional probability distributions. It now covers all probability distributions in the R **stats** package, and it now includes a density graph for each distribution.

1.3.7 Appendices on Statistical Analysis and Writing

Appendix K "Working Style" has been split off and expanded from First Edition Appendix E. It includes a discussion of the importance of a good R-aware text editor and defines what that means. It includes a discussion of our process in writing this book and my process in writing and maintaining the **HH** package.

Appendix L "Writing Style" has been split off and expanded from First Edition Appendix E. It discusses some of the basics of clear writing—including typography, presentation of graphs, and alignment in tables, and programming style.

Appendix M "Accessing R through a Powerful Editor—with Emacs and **ESS** as the Example" has been split off and expanded from First Edition Appendix E. A good editor is one of the most important programs on your computer. It is the direct contact with all the documents, including R scripts and R functions, that you write. A good editor will understand the syntax of your programming language (R specifically) and will simplify the running and testing of code. We write in the terminology of Emacs because it is our working environment. Most of what we illustrate applies to other high-quality editors.

Appendix N "LATEX" has been split off and expanded from First Edition Appendix E. It provides basic information about LATEX, the document preparation system in which we wrote this book.

Appendix O "Word Processors and Spreadsheets" has been split off and expanded from First Edition Appendix E. Unless there are specific add-ins that understand R, we do not recommend word processing software for working with R. We can recommend spreadsheet software for use as a small-scale database management system and as a way of organizing calculations. Unless you are working with RExcel (discussed in Appendix D) we do not recommend the use of spreadsheets for the actual statistical calculations.

Chapter 2

Data and Statistics

Statistics is the field of study whose objective is the transformation of data (usually sets of numbers along with identifying characteristics) into information (usually in the form of tables, graphs, and written and verbal summaries) that can inform sound policy decisions. We give examples of applications of statistics to many fields in Chapter 1. Here we focus on the general concepts describing the collection and arrangement of the numbers themselves.

2.1 Types of Data

Traditionally, we refer to five different *types* of data: count, categorical, ordered, interval, and ratio.

count data: The observational unit either has, or does not have, a particular property. For example, tossed coins can come up heads or tails. We count the number n of heads when a total of N coins are tossed.

categorical data: The data values are distinct from each other. Categorical variables are also referred to as *nominal* variables, *class* variables, or *factors*. The various categories or classes of a categorical variable are called its *levels*. An example of a factor, from the introductory paragraph of Chapter 6, is `factory` having six levels. That is, the investigation takes place at six factories. If we *code* `factory` as $\{1, 2, 3, 4, 5, 6\}$, meaning that we arbitrarily assign these six numbers to the six factories, we must be careful not to interpret these codes as ratio data. Coding the `factory` levels as integers doesn't give us the right to do arithmetic on the code numbers.

© Springer Science+Business Media New York 2015
R.M. Heiberger, B. Holland, *Statistical Analysis and Data Display*,
Springer Texts in Statistics, DOI 10.1007/978-1-4939-2122-5_2

ordered data: The data values can be placed in a rank ordering. For any two observations, the analyst knows which of the two is larger, but not necessarily the magnitude of the difference between them. There is a distinct concept of *first*, *second*, ..., *last*. There is no way to measure the distance between values. An example is military ranks: A general is higher-ranked than a colonel, which in turn is higher than a major. There is no easy way to say something like, "A general is twice as far above a colonel as a colonel is above a major."

interval data: The data values have well-defined distances between them, but there is not a ratio relationship. School grades are an example. Students in 10th grade have studied one year longer than students in 9th grade; similarly, students in 9th grade have studied one year longer than students in 8th grade. It is not meaningful to say a 10th-grade student is twice as knowledgeable as a 5th-grade student.

ratio data: The data values are measured by real numbers: There are a well-defined origin and a well-defined unit. Height of people is an example. There is a well-defined 0 height. We can speak of one person being 1 inch taller than another or of being 10% taller than another.

We also have another categorization of data as *discrete* or *continuous*. Discrete data have a finite or countably infinite number of possible values the data can take. Continuous data take any real number value in an interval; the interval may be either closed or open.

Many of the datasets we will study, both in this book and in the data analysis situations that this book prepares you for, have several variables. Frequently, there are one or more ratio-scaled numeric variables associated with each value of a categorical variable. When only one numeric variable is measured for each observational unit, the dataset is said to be *univariate*. When there are k ($k > 1$) variables measured on each observational unit, the dataset is said to be *multivariate*. Multivariate datasets require additional techniques to identify and respond to correlations among the observed variables.

2.2 Data Display and Calculation

Data are often collected and presented as tables of numbers. Analysis reports are also frequently presented as numbers. Tables of numbers can be presented on a page in ways that make them easier to read or harder to read. We illustrate some of each here and will identify some of the formatting decisions that affect the legibility of numerical tables.

2.2.1 Presentation

There are two general principles:

alignment of decimal points: Units digits of each number are in the same vertical column. Larger numbers extend farther to the left.

display comparable numbers with common precision: Numbers to be compared are displayed so the positions to be compared are in the same column.

Table 2.1 shows two tables with identical numerical information. The first is legible because it follows both principles; the second is not because it doesn't.

Table 2.1 Legible and illegible tabular displays of the same numerical data: In panel a the numbers are aligned on the decimal point and are displayed to the same precision (the same number of decimal digits). In panel b the numbers are centered or left justified—with the effect of hiding the comparability, and displayed with different precisions—which further hides comparability.

a. Legible			b. Illegible		
109.209	133.502	112.219	109.209	133.50234	112.21
153.917	78.971	109.311	153.9	78	109.31152
80.269	83.762	77.036	80.26	83.76253	77.036
74.813	112.720	119.719	74.81323	112.72001	119.7
84.228	103.849	85.586	84.2	103.	85.58
80.558	100.944	115.134	80.55801	100.94474	115.13436
85.519	89.280	109.247	85.51940	89.28095	109.24788

2.2.2 Rounding

The number of decimal digits in a number indicates the precision with which the number was observed. Any analysis normally retains the same precision. Any changes in the number of decimal digits that are not justified by the method of analysis implicitly suggests that the data are either more or less precise than they actually are. This can lead to misleading inferences and wrong policy decisions.

Please see Appendix G for an illustration of the potential problems and references to more detailed discussion. Be sure to read FAQ 7.31 in file
```
        system.file("../../doc/FAQ")
```
The help menus in **Rgui** in Windows and **R.app** on Macintosh have direct links to the FAQ file.

There are simple rules:

1. DO NOT ROUND intermediate values! Keep all 16 digits of double precision arithmetic in a computer program and all 12 digits on pocket calculators. For example, if a correct calculation $7.1449/3.6451 = 1.9601$ is rounded to $7.14/3.65$, the quotient is less than 1.96 and a decision based on whether or not the result exceeds 1.96 will reach an incorrect conclusion.

2. Final answers may be rounded to the SAME number of significant digits as the original data. You may never have final answers with more digits than any intermediate value or with more digits than the original data.

3. Standard deviations can give a hint as to the number of believable digits. For example, if $\bar{x} = 1.23456$ and $s_{\bar{x}} = .0789$, then we can justifiably round to $\bar{x} \approx 1.234$ (using the logic that $t = 1.23456/.0789 = 15.64715 \approx 15.647$ is good to three decimal positions).

2.3 Importing Data

R, and other statistical software systems, have functions that can read data in a variety of formats. All the datasets used in this book are included in the **HH** package. Section 2.3.1 tells how to access them.

Access to datasets in other formats and in other locations (anywhere on the internet) is described in Section 2.3.2.

2.3.1 Datasets for This Book

We have many datasets that we analyze in examples or make available for analysis in exercises. Most datasets are real, taken from journal articles; data repositories of governments, corporations and organizations; data libraries; or our own consulting experience. Citations to these datasets are included in the text. As befits a text, most data we present are structured for the techniques of the chapter in which we present it. Our datasets are frequently used in more than one chapter. We have an Index of Datasets with which you can locate all references to a specific dataset across chapters.

The datasets discussed in this book are available for readers in two different formats.

For use with R, all datasets mentioned in the book are available in the **HH** package for R. The **HH** package can be downloaded from CRAN (Comprehensive R

Archive Network) for use on any computer on which R is installed. Details on installing R are in Appendix A. Once the **HH** is loaded into an R session, the ABCD dataset is made accessible with the statement

    data(ABCD)

Additional information on the **HH** package is in Appendix B.

For use with any other software system, the datasets mentioned in the book are available in ASCII format as csv files. These are text files in which each row of data appears on one row of the file. Within a row, the items are separated by commas. Further discussion of the ASCII files, including the url where they are available, is in Appendix H.

2.3.2 Other Data sources

In consulting environments data is often collected and stored in a database management system. R has packages that can read directly from database management systems that may be housed anywhere on the internet.

Datasets are often stored in MS Excel xls files. These can be directly read into R on any operating system using the **XLConnect** package. See Section A.1.4 for further discussion. On MS Windows machines, the RExcel software is available for direct interaction between R and Excel. See Appendix D for further information.

Datasets stored as datafiles in the internal format of other statistical software systems may be migrated to an R analysis. R can read and write most of them with the aid of the **foreign** package. See the help file help(package="foreign") for further information.

2.4 Analysis with Missing Values

Statisticians frequently encounter situations where an analysis cannot be completed in a straightforward fashion because some portion of the data is missing. For some analyses, missing or unbalanced data cause no difficulty beyond the need to calculate results with a revised formula. Examples include the two-sample t-test of Section 5.4.3 and the one-way analysis of variance of Chapter 6. In other circumstances, such as multiple regression analyses discussed in Chapters 9 to 11, the analyst must either discard the observations carrying incomplete information or use sophisticated techniques beyond the scope of this book to impute, or estimate, the missing portions of the existing data. If the reasons for "missingness" are related to the problem being addressed, abandoning observations is likely to lead to incorrect inferences. If the data are missing at random, discarding a few observations may be a satisfactory solution, but the smaller the ultimate sample size, the less likely the analysis will produce useful and correct results. Imputing the values of missing data is usually preferable to discarding cases with incomplete information. We recommend Little and Rubin (2002) as a comprehensive reference on how to handle missing data, particularly techniques of imputation.

A discussion of how missing values are handled is in R is in Section 2.A.

2.5 Data Rearrangement

Datasets are not necessarily arranged in the most convenient way for the analysis you have in mind. Rearrangement is usually easy. Frequently the functions in the **reshape2** package will be helpful.

We usually work with one of the two data arrangements. Table 2.2 shows both arrangements and the use of the **reshape2** functions `melt` and `dcast` to convert between them. One arrangement (the `data.frame wide` in Table 2.2) is a set of multiple columns (x and y), one per variable, with factor levels `Names` explicitly indicated by data values in the appropriate column. Each observation has all its values listed in the same row of all columns.

The other (the `data.frame long` in Table 2.2) contains all the numeric values in a single column (`value`), with levels of factors explicitly identified in their own columns (`Names` and `variable`). Note that the two different variables in the `wide` arrangement are represented by two levels of the `variable` factor in the `long` arrangement.

2.6 Tables and Graphs

Graphs constructed from data arranged in a table are generally more useful and informative than the table. The human eye and brain can quickly discern patterns from a well-constructed picture that would be far from obvious from the underlying tabular data. Excellent examples are contained in Tufte (2001) and Wainer (1997).

Characteristics that we wish to reveal with our graphs are location, variability, scale, shape, correlation, interaction, clustering, and outliers. In Chapter 4 we illustrate many of these characteristics, primarily through our discussion of scatterplots and scatterplot matrices. Additional types of displays are presented in many subsequent chapters. We discuss both the information about the data that we obtain from the graphs and the structure of the graphs. We introduce many new types of graphs throughout the book. In the appendix to Chapter 4 we provide a summary on those new graphs that are based on Cartesian products.

2.7 R Code Files for *Statistical Analysis and Data Display* (HH)

The **HH** package is available for R from CRAN. See Appendix A for details on installing R with our recommended packages on your computer. The **HH** package includes all datasets used in the book. R scripts for all figures and tables in the book are included in files in the **HH** package. See Appendix B for details. Many of the graphs were produced with functions that are included and fully documented in the **HH** package.

Table 2.2 Define `wide`, a `data.frame` in the wide arrangement. Convert it to the long arrangement with the `melt` function, and convert it back with the `dcast` function.

```
> library(reshape2)

> wide <- data.frame(Names=LETTERS[1:5], x=1:5, y=6:10)

> wide
  Names x  y
1     A 1  6
2     B 2  7
3     C 3  8
4     D 4  9
5     E 5 10

> long <- melt(wide, id="Names")

> long
   Names variable value
1      A        x     1
2      B        x     2
3      C        x     3
4      D        x     4
5      E        x     5
6      A        y     6
7      B        y     7
8      C        y     8
9      D        y     9
10     E        y    10

> wideagain <- dcast(Names ~ variable, value="value", data=long)

> wideagain
  Names x  y
1     A 1  6
2     B 2  7
3     C 3  8
4     D 4  9
5     E 5 10
```

The R code for all *examples*, and for occasional *exercises* is included with the **HH** package from CRAN. Thus you can duplicate all tables and figures in the book and you can use these as templates for analyzing other datasets. The R code for the examples in each chapter of the Second Edition is in a file named after the chapter. For example, the code file for Chapter 6, the one-way analysis of variance chapter, is in file hh2/oway.R. The full path to the file on your computer is found by entering

```
HHscriptnames(6)
```

at the R prompt. The content of the tables and figures is not available as files. They may all be reproduced by running the code.

The R code for the First Edition is also available. Enter

```
HHscriptnames(6, edition=1)
```

at the R prompt.

The Second edition code is identical to the code that actually produced the tables and figures. The book was written using the Sweave and Stangle functions from the **utils** package with the LaTeX document preparation system. See Appendix N for links to LaTeX. All code is included within the LaTeX source for the book. See help(Sweave, package="utils") for details on writing using Sweave.

For the reader of this book, all you need to know is how to find the code for a chapter (HHscriptnames(6) as indicated above), and the structure of the files. Each file starts with a line that tells the name of my LaTeX source file for that chapter. It then has code *chunks*, with each chunk being the code associated with a table or figure. The first chunk in all files is the line

```
library(HH)
```

Each file is independent of all other chapters and assumes only that the HH package is loaded. Multiple chunks associated with the same dataset in the same file assume the previous chunks have already been run.

The chunks begin with function calls to the hhpdf or hhcapture functions. When I was writing the book, these calls were defined to capture the figure or table as a file. For the reader, these calls are defined in the **HH** package as *noops* (NO OPeration)—that is, they don't do anything. All output goes directly to your console window or your graphics window.

The best way to use the files is to pick up their lines and paste them in the R console window. It will often be helpful to study how the lines are constructed.

It is possible to source the entire file. While it works, all it does is produce all the tables and figures that you already have in the book. Sourcing the files won't help in learning.

2.A Appendix: Missing Values in R

The R convention for missing values is NA (a standard abbreviation for "Not Available" or "No Answer"). When R knows that a value is missing it prints "NA" (without the quotes). When R is reading an ASCII data file, it will recognize by default the character sequence "NA" as a missing observation.

If the ASCII data file uses some other convention (such as the "." that SAS uses by default), then we must tell R to use a different convention for reading missing values either with an argument to the read.table function or, after the reading, by some logical investigation of the data values.

R has several conventions for working with datasets containing NA values.

Data Input: See Tables 2.3, 2.4, and 2.5 for an example. We use the default missing value indicator in Table 2.3, an explicitly defined missing value indicator in Table 2.4, and a non-default missing value indicator in Table 2.5 without telling the read.table function that we were doing so.

Table 2.3 The data are read from a text argument, which is equivalent to reading from a text file. In the AA example, the default missing value is NA. In the BB example in Table 2.4, the argument na.strings defines strings "999" and "." to indicate missing values. The internal representation is the R value NA. In the CC example in Table 2.5, where we didn't use the argument na.strings, the y variable has been coerced to be a factor. Read the help file

```
        help("read.table", package="utils")
```
for more information

```
> AA <- read.table(text="
+ x y
+ 1 2
+ 3 NA
+ 5 6
+ ", header=TRUE)

> AA
  x  y
1 1  2
2 3 NA
3 5  6

> sapply(AA, class)
        x         y
"integer" "integer"
```

Table 2.4 The argument na.strings defines strings "999" and "." to indicate missing values.

```
> BB <- read.table(text="
+ x y
+ 1 2
+ 3 999
+ 5 6
+ 7 .
+ 9 10
+ ", header=TRUE, na.strings=c("999", "."))

> BB
  x  y
1 1  2
2 3 NA
3 5  6
4 7 NA
5 9 10

> sapply(BB, class)
        x         y
"integer" "integer"
```

Table 2.5 We neglected to use the argument na.strings. The y variable has become a factor.

```
> CC <- read.table(text="
+ x y
+ 1 2
+ 3 999
+ 5 6
+ 7 .
+ 9 10
+ ", header=TRUE)

> CC
  x   y
1 1   2
2 3 999
3 5   6
4 7   .
5 9  10

> sapply(CC, class)
        x         y
"integer"  "factor"

> CC$y
[1] 2   999 6   .   10
Levels: . 10 2 6 999
```

Printing: Missing numerical values are displayed as NA. Missing character and
factor items are displayed as <NA>. See Table 2.6

Table 2.6 Missing numerical values are displayed as NA. Missing character and factor items are
displayed as <NA>.

```
> abcd <- data.frame(x=c(1, 2, NA, 4, 5, 6, 7, 8),
+                    y=c(6, 5, 8, NA, 10, 9, 12, 11),
+                    ch=c(NA, "N", "O", "P", "Q", "R", "S", "T"),
+                    stringsAsFactors=FALSE)

> abcd
   x  y   ch
1  1  6 <NA>
2  2  5    N
3 NA  8    O
4  4 NA    P
5  5 10    Q
6  6  9    R
7  7 12    S
8  8 11    T

> sapply(abcd, class)
          x          y          ch
  "numeric"  "numeric" "character"
```

Graphs: Points whose coordinates are not known (points "O" and "P") are not
   printed. Points with known coordinates and unknown value (the first point whose
   value should have been "M") are displayed as NA in the known position. See
   Figure 2.1.

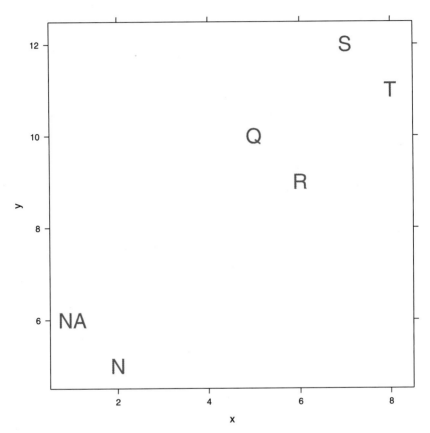

Fig. 2.1   The dataset abcd is defined in Table 2.6. The plot was drawn with

```
> xyplot(y ~ x, data=abcd, labels=abcd$ch, panel=panel.text,
+        col=c("red", "blue"))
```

Arithmetic: Arithmetic with missing values returns a missing value. Many functions, sum is illustrated in Table 2.7, can be told to remove the missing values and sum the non-missing values.

Table 2.7 Arithmetic with missing values returns a missing value. Many functions, sum and mean are illustrated in Table 2.7, can be told to remove the missing values and sum the non-missing values.

```
> 3 + NA
[1] NA

> sum(3, NA)
[1] NA

> sum(3, NA, na.rm=TRUE)
[1] 3

> abcd$x
[1]  1  2 NA  4  5  6  7  8

> mean(abcd$x)
[1] NA

> mean(abcd$x, na.rm=TRUE)
[1] 4.714
```

Linear Models: Default (na.action=na.omit) behavior is to remove the row.
Table 2.8 shows the default behavior of the lm and related modeling functions.
The entire row containing the missing values is removed from the analysis and
subsequent processing. Table 2.9 shows an optional better behavior.

Table 2.8 The default behavior of the lm and related modeling functions. The entire row containing the missing values is removed from the analysis and subsequent processing. See Table 2.9 for
an optional better behavior.

```
> a.lm <- lm(y ~ x, data=abcd)

> summary(a.lm)

Call:
lm.default(formula = y ~ x, data = abcd)

Residuals:
    1       2       5       6       7       8
0.704  -1.219   1.013  -0.910   1.167  -0.755

Coefficients:
            Estimate Std. Error t value Pr(>|t|)
(Intercept)    4.373      1.053    4.16   0.0142 *
x              0.923      0.193    4.79   0.0087 **
---
Signif. codes:
0 '***' 0.001 '**' 0.01 '*' 0.05 '.' 0.1 ' ' 1

Residual standard error: 1.2 on 4 degrees of freedom
  (2 observations deleted due to missingness)
Multiple R-squared:  0.851,Adjusted R-squared:  0.814
F-statistic: 22.9 on 1 and 4 DF,  p-value: 0.00872


> predict(a.lm)
     1      2      5      6      7      8
 5.296  6.219  8.987  9.910 10.833 11.755
```

Linear Models: Better behavior (na.action=na.exclude) is to keep track of which rows have been omitted. Table 2.9 shows an optional better behavior. The rows with missing values are still removed from the calculations of the "lm" object, but information on which rows were suppressed is retained.

Table 2.9 With na.action=na.exclude, the rows with missing values are still removed from the calculations of the "lm" object, but information on which rows were suppressed is retained.

```
> b.lm <- lm(y ~ x, data=abcd, na.action=na.exclude)

> summary(b.lm)

Call:
lm.default(formula = y ~ x, data = abcd, na.action = na.exclude)

Residuals:
      1      2      5      6      7      8
  0.704 -1.219  1.013 -0.910  1.167 -0.755

Coefficients:
            Estimate Std. Error t value Pr(>|t|)
(Intercept)    4.373      1.053    4.16   0.0142 *
x              0.923      0.193    4.79   0.0087 **
---
Signif. codes:
0 '***' 0.001 '**' 0.01 '*' 0.05 '.' 0.1 ' ' 1

Residual standard error: 1.2 on 4 degrees of freedom
  (2 observations deleted due to missingness)
Multiple R-squared:  0.851,Adjusted R-squared:  0.814
F-statistic: 22.9 on 1 and 4 DF,  p-value: 0.00872


> predict(b.lm)
      1      2      3      4      5      6      7      8
  5.296  6.219     NA     NA  8.987  9.910 10.833 11.755
```

Chapter 3

Statistics Concepts

In this chapter we discuss selected topics on probability. We define and graph several basic probability distributions. We review estimation, testing, and sampling from populations. The discussion here is at an intermediate technical level and at a speed appropriate for review of material learned in the prerequisite course.

3.1 A Brief Introduction to Probability

The quality of inferences are commonly conveyed by probabilities. Therefore, before discussing inferential techniques later in this chapter, we briefly digress to discuss *probability* in this section and *random variables* in Section 3.2.

If A is any *event*, $P(A)$ represents the probability of occurrence of A. Always, $0 \leq P(A) \leq 1$. The odds in favor of the occurrence of event A are

$$\frac{P(A)}{1 - P(A)} \tag{3.1}$$

and the odds against the occurrence of event A are

$$\frac{1 - P(A)}{P(A)} \tag{3.2}$$

Thus, if $P(A) = \frac{3}{4}$, then the odds in favor of A are 3, also referred to as 3 to 1, and the odds against A are $\frac{1}{3}$.

If B is a second event, $A \cup B$ represents the event that "either A or B occurs", that is, the *union* of A and B, then

$$P(A \cup B) = P(A) + P(B) - P(A \cap B) \tag{3.3}$$

© Springer Science+Business Media New York 2015

R.M. Heiberger, B. Holland, *Statistical Analysis and Data Display*,

Springer Texts in Statistics, DOI 10.1007/978-1-4939-2122-5_3

where $A \cap B$ is the event that "both A and B occur", that is the *intersection* of A and B. Events A and B are said to be *mutually exclusive* events if they cannot both occur; in this case, $A \cap B = \emptyset$ (the impossible event) and so $P(A \cap B) = 0$. Events A and B are said to be *independent* events if the occurrence or nonoccurrence of one of them does not affect the probability of occurrence of the other one; for independent events,

$$P(A \cap B) = P(A)\,P(B)$$

The *conditional probability* of B given A, written $P(B \mid A)$, is the probability of occurrence of B given that A occurs. If $P(A) \neq 0$,

$$P(B \mid A) = \frac{P(A \cap B)}{P(A)}$$

Note that $P(B \mid A) = P(B)$ if A and B are independent events, but not otherwise.

To illustrate these ideas, imagine a box containing six white and four red billiard balls, identical to the touch. Suppose we select two balls from the box and let $A =$ "the first ball is white" and $B =$ "the second ball is white". A and B are independent events if the first ball is replaced in the box prior to drawing the second ball, but not otherwise. Let us assume that the first ball is not replaced so that the two events are dependent. Various sets of events are listed with their probabilities in Table 3.1.

In this table we demonstrate two ways to calculate the probability $\frac{78}{90}$ that we get a white ball in either the first selection or second selection or both selections. One way is with the formula for $P(A \cup B)$ in Equation (3.3). Another method is to recognize that the event "at least one white" can be partitioned into three mutually exclusive events: First draw white and second draw red; first draw red and second draw white; and both draws white. The probability of "at least one white" is seen to be the sum of the probabilities of the events comprising this partitioning.

3.2 Random Variables and Probability Distributions

A *random variable*, abbreviated as r.v., is a function that associates events with real numbers. For example, if we toss a coin 10 times, we can define an r.v. X to be the number of heads observed in these 10 tosses. This r.v. has *possible values* $x = 0, 1, 2, \ldots, 10$. Observing 7 heads among the 10 tosses is an event, and "7" is the number that this r.v. X associates with it.

A closely related concept is the r.v.'s *probability distribution*, which indicates how the total probability, 1, is distributed or allocated to the possible values of the r.v. It is usual to denote an r.v. with a capital letter and a possible value of this r.v. with the corresponding lowercase letter.

Table 3.1 Probability of intersection events, conditional events, union events in the setting of a box containing six white and four red billiard balls. We select two balls from the box. The A event is "the first ball is white" and the B event is "the second ball is white". See Figure 3.1 for an illustration of this distribution.

Event	Position		Probability		Probability of event
	1	2	1	2	
A	W	?	$\frac{6}{10}$	1	$\frac{6}{10}$
B	?	W	1	$\frac{6}{10}$	$\frac{6}{10}$
$B \cap A$	W	W	$\frac{6}{10}$	$\frac{5}{9}$	$\frac{30}{90}$
$\bar{B} \cap \bar{A}$	R	R	$\frac{4}{10}$	$\frac{3}{9}$	$\frac{12}{90}$
$B \mid A$	$[W]$	W	$\frac{(\frac{6}{10})}{(\frac{6}{10})}$	$\frac{5}{9}$	$\frac{5}{9}$

$$B \cup A \begin{cases} \begin{array}{cccc} W & R & \frac{6}{10} & \frac{4}{9} \\ R & W & \frac{4}{10} & \frac{6}{9} \\ W & W & \frac{6}{10} & \frac{5}{9} \end{array} \end{cases}$$

$$P(WR) + P(RW) + P(WW) = P(A) + P(B) - P(B \cap A) = P(B \cup A)$$
$$\frac{24}{90} + \frac{24}{90} + \frac{30}{90} = \frac{6}{10} + \frac{6}{10} - \frac{30}{90} = \frac{78}{90}$$

3.2.1 Discrete Versus Continuous Probability Distributions

There are essentially two distinct types of probability distribution of a quantitative variable: discrete and continuous. (Random variables are also classified as discrete or continuous according to the classification of their probability distributions.) It is important to distinguish between the two types because they differ in their methods of display and calculation.

The key distinction between these two types relates to the spacings between adjacent possible values of the data. For discrete data, the distance separating consecutive possible values of the variable does not depend on a measurement device; indeed it may be completely arbitrary. For continuous data, the distances may (theoretically) assume all possible values in some interval.

For example, the number of times an archer hits a target in 10 attempts is a discrete variable because the answer is a count of the number of occurrences. It is impossible for there to be 3.5 hits. A discrete variable need not be integer-valued.

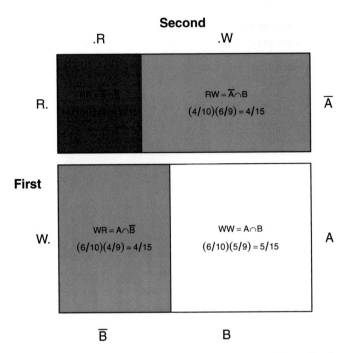

Fig. 3.1 Mosaic plot corresponding to Table 3.1. The area of each panel is equal to the probability of the event identified in that panel. The bottom row representing the event $A =$ "W is selected first" consists of the two panels WR and WW. The bottom row has height .6 $= P(A)$. The right-hand column represents the event $B =$ "W is selected second" consists of the two panels RW and WW. The event "$B \cap A$" is the white region WW in the lower right corner. The event WW has height .6 and width 5/9, hence area .6 × 5/9 = 1/3. The event $B \mid A$ is also the white area WW, but now thought of as the proportion of the A area that is also B. The probability of $B \mid A$ is the ratio of the area of $B \mid A$ to the A area (1/3)/.6 = 5/9. The event $B \mid \bar{A}$ is the pink region RW in the upper right corner. The probability of $B \mid \bar{A}$ is the ratio of the pink area RW to the $\bar{A}$ area (4/15)/.4 = 2/3. The event $\bar{B} \cap \bar{A}$ is the red region RR in the upper left corner. The event RR has height .4 and width 3/9, hence area .4 × 3/9 = 2/15.

The proportion of hits in 10 attempts is also discrete. It is impossible for this proportion to be .35. It is possible for a discrete variable to have a *countably infinite* number of possible values. An example would be the number of attempts needed for the archer to achieve her ninth hit. This variable can assume any positive integer value; it is possible but unlikely that the archer will need 100 attempts.

On the other hand, the archer's height in inches is a continuous variable because it can be anything between perhaps 3 feet and 8 feet (90–240 cm). While as a practical matter it would be difficult to measure height to within $\frac{1}{4}$-inch (6 mm) accuracy, it is not theoretically impossible for someone to be $68\frac{3}{4}$ inches (174.6 cm) tall.

In summary, it is possible to make a list of the possible values of a discrete random variable, but this is not true for a continuous random variable.

For completeness, we also point out that it is possible for data to be a mixture of discrete and continuous types. Let Y = the total measurable daily precipitation measured at Philadelphia International Airport. On some fraction of all days, roughly 70% of them, there is no precipitation. So $P(Y = 0) \approx .7$. But considering only those dates with measurable precipitation, Y is continuous, i.e., the distribution of $(Y \mid Y > 0)$ is continuous.

3.2.2 Displaying Probability Distributions—Discrete Distributions

The display of a probability distribution varies according to whether the r.v. is discrete or continuous. We can make an ordered list of the possible values of a discrete r.v. For example, if X denotes the number of heads in two tosses of a fair coin, then X has three possible values $\{0,1,2\}$. We will see later that for this coin, the probabilities are as given in Table 3.2.

Table 3.2   The total probability 1.0 has been *distributed* to the three possible values: 0, 1, 2.

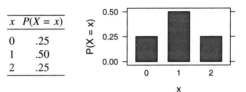

x	$P(X = x)$
0	.25
1	.50
2	.25

Sometimes we choose to study several interdependent random variables at the same time. In such instances, we require their bivariate or multivariate probability distribution.

In Table 3.3 we consider an example of a discrete bivariate and conditional distribution. Here p.m.f. stands for *probability mass function*.

Here X and Y are dependent r.v.'s because, e.g., $f(1,0) = .10$, which differs from $f(1) \times g(0) = .60 \times .15 = .09$. Alternatively, $f(1 \mid 0) = \frac{2}{3}$, which differs from $f(1) = .6$. In general, if U and V are discrete random variables, then U and V are independent r.v.'s if

$$P\big((U = u) \cap (V = v)\big) = P(U = u) \times P(V = v)$$

for all possible values u of U and v of V, i.e., the distribution of U doesn't depend on the value of V.

Table 3.3 Example of Discrete Bivariate and Conditional Distributions. The top panel shows the probabilities of each of the six events in the distribution. The area of the six events adds up to 1. The center panel shows conditioning of x on y. Within each column, the area adds up to 1. The bottom panel shows conditioning of y on x. Within each row, the area adds up to 1.

Joint p.m.f. $f(x, y)$

x	y 0	1	2	$y(x) = x$-margin
1	.10	.20	.30	.60
2	.05	.10	.25	.40
$g(y) = y$-margin	.15	.30	.55	1.00

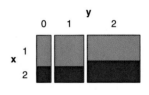

Conditional p.m.f. $f(x \mid y)$

x	y 0	1	2
1	$\frac{2}{3}$	$\frac{2}{3}$	$\frac{6}{11}$
2	$\frac{1}{3}$	$\frac{1}{3}$	$\frac{5}{11}$
all	1	1	1

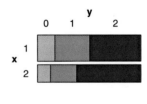

Conditional p.m.f. $g(y \mid x)$

x	y 0	1	2	all
1	$\frac{1}{6}$	$\frac{2}{6}$	$\frac{3}{6}$	1
2	$\frac{1}{8}$	$\frac{2}{8}$	$\frac{5}{8}$	1

The cumulative distribution $\mathcal{F}$ of a discrete random variable is calculated as

$$\mathcal{F}(x) = P(X \le x) = \sum_{t \le x} f(t)$$

where the sum is taken over all possible values t of X that are less than or equal to x.

3.2.3 Displaying Probability Distributions—Continuous Distributions

The probability distribution of a continuous random variable cannot be described in the manner of Table 3.2 or 3.3 (listing its possible values alongside their associated probabilities) because a continuous r.v. has an *uncountably infinite* number of possible values. Instead the probability distribution of a continuous r.v. X is described by its probability density function (p.d.f.), say $f(x)$. This function has the properties that

1. $f(x) \geq 0$

2. the probability that X lies in any interval is given by the area under $f(x)$ above this interval.

In the p.d.f. in Figure 3.2, the shaded area under the density and above the horizontal axis represents the probability that the random variable lies between 2 and 4.

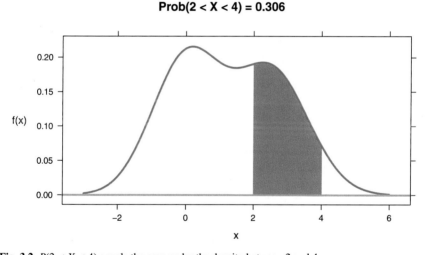

Prob(2 < X < 4) = 0.306

Fig. 3.2 $P(2 < X < 4)$ equals the area under the density between 2 and 4.

The cumulative distribution $\mathcal{F}$ of a continuous random variable is calculated as

$$\mathcal{F}(x) = P(X \leq x) = \int_{-\infty}^{x} f(t)\, dt$$

Continuous r.v.'s U and V are also independent if the distribution of U doesn't depend on the value of V or, equivalently, if the distribution of V doesn't depend on the value of U. In this case, we can express the independence condition as

$$P\big((U \le u) \cap (V \le v)\big) = P(U \le u) \times P(V \le v) \qquad (3.4)$$

for all u and v.

Appendix J catalogs frequently encountered probability distributions, illustrates their density functions, and includes function names in R for calculations with the distributions.

3.3 Concepts That Are Used When Discussing Distributions

Understanding the distribution of observations is critical to interpreting data. In this section we introduce several concepts that are used to describe distributions: mean, variance, median, symmetry, correlation; and types of graphs that are used to display these concepts: histogram, stem-and-leaf, density, scatterplot.

3.3.1 Expectation and Variance of Random Variables

The expectation of an r.v. X, denoted $E(X)$, is its expected or long-run average value; alternatively it is the mean of the probability distribution of X and so we write $E(X) = \mu$. If X is discrete with p.m.f. $p(x)$, then $E(X) = \sum x\, p(x)$. If X is continuous, then $E(X) = \int x f(x)\, dx$, where the range of integration extends over the set of real numbers that X may assume. The variance of X is defined by $\sigma^2 = \text{var}(X) = E(X - \mu)^2 = E(X^2) - \mu^2$. The square root σ of the variance is called the *standard deviation*, abbreviated s.d. It is a more useful measure of variability than the variance because it is measured in the same units as X, rather than in artificial squared units.

If $x_1, x_2, \ldots, x_n$ is a random sample of n items selected from some population, the sample mean

$$\bar{x} = \frac{1}{n} \sum_{i=1}^{n} x_i \qquad (3.5)$$

estimates the population mean μ, and the sample variance

$$s^2 = \frac{1}{n-1} \sum_{i=1}^{n} (x_i - \bar{x})^2 \qquad (3.6)$$

estimates the population variance σ^2. In addition, the sample standard deviation $s = \sqrt{s^2}$ estimates the population standard deviation σ. Please see Section G.12 for a discussion on the importance of using the two-pass algorithm based on the definition in Equation 3.6, and not the alternative one-pass algorithm based on Equation 3.7,

$$s^2 = \frac{1}{n-1} \sum_{i=1}^{n} \left(x_i^2 - n\bar{x}^2 \right) \tag{3.7}$$

when doing arithmetic by computer. The short explanation is that you will always get the right answer with Equation 3.6 and may sometimes get a very wrong answer with Equation 3.7.

It can be shown that if a_1 and a_2 are constants and x_1 and x_2 are any two random variables, then

$$E(a_1 x_1 \pm a_2 x_2) = a_1 E(x_1) \pm a_2 E(x_2) \tag{3.8}$$

If, in addition, x_1 and x_2 are uncorrelated random variables, then

$$\text{var}(a_1 x_1 \pm a_2 x_2) = a_1^2 \, \text{var}(x_1) + a_2^2 \, \text{var}(x_2) \tag{3.9}$$

When x_1 and x_2 are correlated, then the variance of the sum is given by

$$\text{var}(a_1 x_1 \pm a_2 x_2) = a_1^2 \, \text{var}(x_1) + a_2^2 \, \text{var}(x_2) \pm 2 a_1 a_2 \, \text{cov}(x_1, x_2) \tag{3.10}$$

These three formulas (Equations 3.8, 3.9, and 3.10) generalize to the multivariate situation in Equations 3.16 and 3.17.

3.3.2 Median of Random Variables

The median of an r.v. X, denoted median$(X) = \eta$, is the middle value of the distribution. The population median is defined as the value η such that

$$\int_{-\infty}^{\eta} f(x)\, dx = .5 \qquad \text{for continuous distributions} \tag{3.11}$$

or

$$\sum_{x \le \eta} p(x) \ge .5 \text{ and } \sum_{x < \eta} p(x) \le .5 \qquad \text{for discrete distributions.} \tag{3.12}$$

We show an example of the median of a distribution in Figure 3.6.

The order statistics $X_{(i)}$ are the values of the observed X_i ordered from smallest to largest. The middle order statistic $\overset{+}{X}$ is called the sample median and is defined as

$$\overset{+}{X} = \begin{cases} X_{\left(\frac{n+1}{2}\right)} & \text{odd } n \\ \left(X_{\left(\frac{n}{2}\right)} + X_{\left(\frac{n+1}{2}\right)} \right) / 2 & \text{even } n \end{cases} \tag{3.13}$$

The notation $\overset{\perp}{X}$ for the sample median used here is intended to be self-descriptive, with an overbar split in the middle into two equal halves. We believe the notation is due to Tukey. There is no standard notation for the median.

3.3.3 Symmetric and Skewed Distributions

Symmetry and skewness are classifications applicable to both continuous and discrete distributions. The mean of a symmetric distribution coincides with its median. A continuous distribution example is the normal distribution having a density function such as that plotted in Figure 3.13. A symmetric distribution has equivalent behavior on either side of its mean. In particular, its *tails*, the values of the density function away from the center, are mirror images.

A skewed distribution is one that is not symmetric. Unimodal distributions (ones having a single point where the probability mass is higher than at adjacent points) that are skewed are further classified as being positively or negatively skewed. A positively skewed distribution has a long, thin tail on its right side and a short, fat tail on its left side. Its mean exceeds its median. A negatively skewed distribution has a long, thin tail on its left side and a short, fat tail on its right side. Its median exceeds its mean. Note that the left/right naming convention for skewed distributions is based on the side containing the long, thin tail. We illustrate a negatively skewed, symmetric, and positively skewed distribution in Figure 3.3. We show boxplots of negatively skewed, symmetric, and positively skewed data in Figure 3.7.

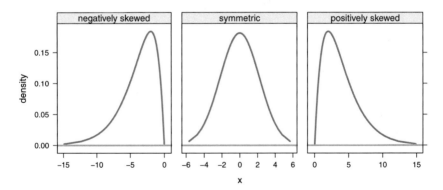

Fig. 3.3 Negatively skewed, symmetric, and positively skewed distributions.

The χ^2 distribution described in Section J.1.3 is an example of a continuous positively skewed distribution. The (discrete) binomial distribution to be described in Section 3.4.1 is negatively skewed, symmetric, or positively skewed according to whether its parameter p is less than, equal to, or greater than 0.5.

The skewness terminology often comes into play because many statistics procedures work best when underlying distributions are symmetric, and tactics that move the distribution toward symmetry (for example, with data transformations such as the power transformations described in Section 4.8) are frequently used in the analysis of skewed distributions.

Each of the densities in Figure 3.3 has a single mode. Some densities have more than one mode. Figure 3.2 is an example of a bimodal density, with one mode between 0 and 1 and another mode between 2 and 3. Multimodal distributions, ones having more than two modes, are occasionally encountered. Sometimes bimodality and multimodality arise as a result of interpreting samples coming from two or more populations with different locations as having arisen from a single population. Therefore, bimodality or multimodality may suggest a need for disaggregation of samples.

3.3.4 Displays of Univariate Data

It is difficult to gain an understanding of data presented as a table of numbers. Summary statistics such as those presented in the preceding sections are helpful for this purpose but may fail to capture some important features. In this section we present three displays (Histogram, Stem-and-leaf, and Boxplots) for univariate data that are basic tools for studying both the distributional shape and unusual data values. We illustrate these displays with the variable male.life.exp (1990 male life expectancy) in each of 40 countries, part of the datafile data(tv) to be examined in more detail in Section 4.6. We summarize the variable in Table 3.4 as a frequency table, a partitioning of the data into k evenly spaced nonoverlapping categories, and a tally of the number or proportion of items in each category.

3.3.4.1 Histogram

The construction of a histogram begins with the frequency table. Usually the number of categories is between 6 and 12—the use of fewer than 6 categories tends to under-summarize the data while the use of more than 12 categories tends to oversummarize the data. For male.life.exp we chose 6 age-range categories that encompass the ages from all 40 countries.

The corresponding histogram in Figure 3.4 is a graph consisting of rectangles with width covering the breadth of the classes and heights equal to the class frequencies. This plot is also called a *relative frequency* histogram, particularly when the vertical axis is labeled to show the *proportion* of countries in each category, for example $\frac{6}{40} = 0.15$ in the first category for ages 50–54. We show both axis labelings in Figure 3.4 with the proportion axis on the right.

Table 3.4 Frequency Distribution of Male Life Expectancy

```
> data(tv)

> tmp <- as.matrix(table(cut(tv$male.life.exp, breaks=seq(49.5,79.5,5))))

> dimnames(tmp) <-
+     list("Male Life Expectancy"=
+               c("50--54","55--59","60--64","65--69","70--74","75--79"),
+          " "="Frequency")

> tmp

Male Life Expectancy Frequency
               50--54         6
               55--59         4
               60--64         9
               65--69        11
               70--74         7
               75--79         3
```

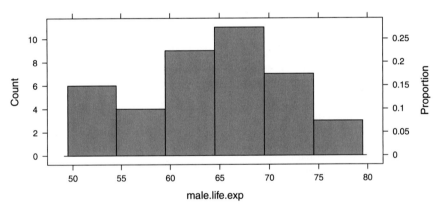

Fig. 3.4 Life Expectancy for Males. The count axis is on the left and the proportion axis is on the right.

Figure 3.4 is an example of a bimodal distribution, one having two peaks. In this example, the lower peak may correspond to economically poorer countries and the upper peak to wealthier countries, with relatively few countries falling between these extremes. In general, bimodal distributions sometimes suggest an amalgamation of samples from two separate populations that perhaps should be investigated separately. An advantage of histograms is that they can be constructed from huge datasets with no more effort than from small data sets. A disadvantage is that the data used to construct a histogram cannot be recovered from the histogram itself.

3.3.4.2 Stem-and-Leaf Display

Stem-and-leaf displays, designed by John Tukey, resemble histograms in that they portray the shape of a distribution. The stem-and-leaf display is usually preferable because it is possible to recover the data used to construct a stem-and-leaf display (at least to some degree of precision). Unlike histograms, stem-and-leaf displays are limited to data sets of not more than a few hundred observations in order that the display fits entirely on one page or one computer monitor.

A stem-and-leaf display for male life expectancy is in Table 3.5. This is a table, not a figure, because stem-and-leaf is a text-based graphic display.

Table 3.5 Stem-and-Leaf Display of Male Life Expectancy

```
> stem(tv$male.life.exp)

  The decimal point is 1 digit(s) to the right of the |

  5 | 002234
  5 | 6799
  6 | 012223344
  6 | 66777888899
  7 | 1223334
  7 | 556
```

The column of numbers in this display to the left of the vertical bars represent the tens digit of each of the life expectancies. This column is the *stem*. The numbers to the right of the vertical bars, one digit for each country, are the leaves, the unit digits of the life expectancies for the 40 countries. The stem-and-leaf display, following Tukey, rounds down, to maintain the same digit as appears in the data table. A 90° counterclockwise rotation of the stem and leaves gives a picture that closely resembles Figure 3.4. The legend locating the decimal point tells the reader that "5 | 0" in the display stands for 50, rather than .05 or 500.

Stem-and-leaf displays can accommodate measurements containing more than two significant digits. This is accomplished either by suppressing the values of trailing digits or by allowing more than a single digit for each leaf. For example, suppose in a different problem the measurement is 564. This can be represented as "5 | 6", with the stem indicating the hundreds, rounding the units digit down to a multiple of 10, and with a legend locating the decimal point 2 places to the right of the vertical bar. Alternatively, it can be represented with a stem indicating the hundreds and with two-digit leaves as "5 | 64,", again locating the decimal point two places to the right of the vertical bar, and with the "," indicating that the leaf is two digits wide. Or, another option, as "56 | 4" with a stem of 56 tens (representing 560) and with a single-digit leaf of 4.

3.3.4.3 Boxplots

Boxplots, also known as box-and-whisker plots, are among the many inventions of John Tukey. Their main use is as a compact, simultaneous display to compare several related data sets. Many examples of side-by-side boxplots appear in this book. Boxplots may be arranged along either a vertical or horizontal scale. This book contains examples illustrating both options.

Boxplots make use of the sample first quartile Q_1, median $\overset{\scriptscriptstyle\downarrow}{x} = Q_2$, and third quartile Q_3. The statistics $Q_1, \overset{\scriptscriptstyle\downarrow}{x}, Q_3$ divide the sample into four equal parts. Q_1 is the median of the sample values that are less than or equal to $\overset{\scriptscriptstyle\downarrow}{x}$ and Q_3 is the median of the sample values that are greater than or equal to $\overset{\scriptscriptstyle\downarrow}{x}$. Approximately 25% of the sample lies within each of the four intervals (all finite intervals are closed, so double counting is possible)

$$(-\infty, Q_1], \qquad [Q_1, \overset{\scriptscriptstyle\downarrow}{x}], \qquad [\overset{\scriptscriptstyle\downarrow}{x}, Q_3], \qquad [Q_3, \infty)$$

A rectangle (box) is drawn so that when placed against a numerical scale its edges occur at Q_1 and Q_3. A line is drawn, parallel to the edges, through the inside of the box at the median $\overset{\scriptscriptstyle\downarrow}{x}$. Lines perpendicular to the edges of the box extend outward from the midpoints of the edges. These lines are sometimes called "whiskers". The lower whisker extends to the lowest sample item not more than $1.5 \times$ IQR below Q_1. The upper whisker extends to the largest sample item not more than $1.5 \times$ IQR above Q_3. Points outside the range of the whiskers are plotted as filled-in circles. Such points are deemed extreme or outlying values ("outliers"). In general, outliers should be carefully scrutinized. Sometimes they are due to transcription errors and are not legitimately part of the data under consideration (in which case you should attempt to correct the data). Other times, they are the critical data points that provide the key to an explanation of the study. One example of a critically important outlier is the Gulf of Mexico oil spill. On most days very little oil is released into the ocean. If we ignored the large spill detected on 20 April 2010, we would be missing the important information. In astronomy, "transient" events are very important. That is how supernovas are detected (Table 3.6).

Table 3.6 shows the quartiles for the `male.life.exp` variable. Figure 3.5 shows the boxplot for the `male.life.exp` variable.

Table 3.6 Quartiles of Life Expectancy for Males

```
> quantile(tv$male.life.exp)
    0%    25%    50%    75%   100%
 50.00  59.75  66.00  69.50  76.00
```

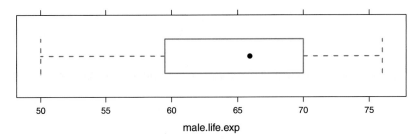

Fig. 3.5 Boxplot of Life Expectancy for Males

See the illustration in Figure 3.6 for the quartiles of a continuous distribution. The interquartile range

$$IQR = Q_3 - Q_1$$

is a measure of dispersion of the central portion of a distribution. When X is normally distributed $X \sim N(\mu, \sigma^2)$, we have IQR = 1.34898σ.

Figure 3.7 contains parallel boxplots depicting three samples on a common scale, illustrating the distinctions between boxplots for negatively skewed, symmetric, and positively skewed distributions. This parallels the density presentations in Figure 3.3. Asymmetry is nicely displayed in this figure.

Several more elaborate versions of the boxplot exist. For example, adding a *notch* to the sides of a box provides information on the variability of the sample median. For details, see Hoaglin et al. (1983).

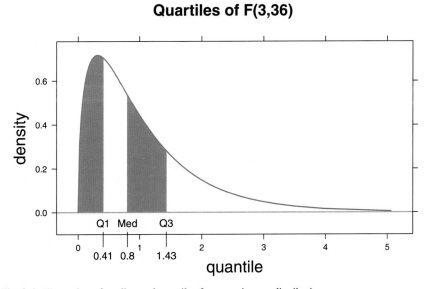

Fig. 3.6 Illustration of median and quartiles for a continuous distribution.

Boxplots are generally unsuccessful in conveying the existence of multiple modes. For such data, histograms and stem-and-leaf displays are often preferred choices.

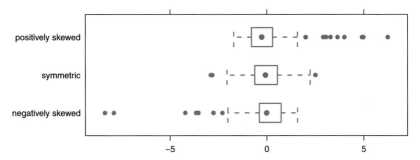

Fig. 3.7 Boxplots illustrating negatively skewed, symmetric, and positively skewed distributions.

3.3.5 Multivariate Distributions—Covariance and Correlation

In Section 3.2.2 we give an example of a discrete multivariate (actually bivariate) probability distribution. We now touch on the notion of the continuous multivariate distribution of a continuous random vector $X = (X_1, X_2, \ldots, X_p)'$. For example, variable X_1 could be height and variable X_2 weight, all measured on the same set of people. The mean or expectation of X is $\mu = (\mu_1, \mu_2, \ldots, \mu_p)'$, the vector of means of the univariate distribution of the $X_i's$. The variance–covariance matrix of X, say V, also called the covariance matrix or dispersion matrix, is the symmetric $p \times p$ matrix having the variances of the $X_i's$ on its main diagonal, and the covariances of different $X_i's$ elsewhere. The covariance of X_i and X_j is

$$V_{ij} = \sigma_{ij} = \text{cov}(X_i, X_j) = E\big((X_i - \mu_i)(X_j - \mu_j)\big)$$

is the element in the row i column j position of V. If we denote the standard deviations of X_i and X_j by σ_i and σ_j, respectively, then the *correlation* between X_i and X_j is

$$\rho_{ij} = \frac{\text{cov}(X_i, X_j)}{\sigma_i \sigma_j} = \frac{V_{ij}}{\sqrt{V_{ii}V_{jj}}} \tag{3.14}$$

This is a rescaling of the covariance, interpreted as a measure of the strength of the (straight line) linear relationship between X_i and X_j. It can be shown that $-1 \le \rho_{ij} \le 1$. If this correlation is close to ± 1, X_i and X_j are closely linearly associated; the association is direct if $\rho_{ij} > 0$ and inverse if $\rho_{ij} < 0$. If $\rho_{ij} = 0$, then X_i and X_j are said to be *uncorrelated*, i.e., the X's are not linearly related. It is easy to construct an example of correlated variables for any specified correlation. Figure 3.8 gives a static view of a sequence of related variables with specified correlation coefficient. A dynamic illustration of the effect of the correlation coefficient can be constructed by plotting a sequence of panels similar to those in Figure 3.8 and cycling through them. We do so in a **shiny** app in the **HH** package with the statement

```
shiny::runApp(system.file("shiny/bivariateNormalScatterplot",
                          package="HH"))
```

at the R prompt. See Figure E.3 for a screenshot. In both the static and dynamic illustrations the formula is very simple. Define x and e as independent realizations from the $N(0, 1)$ distribution. Then

$$y = \rho x + (1 - \rho^2)^{1/2} e \tag{3.15}$$

has correlation ρ with x.

Matrix algebra plays an important role in the study of multivariate distributions. For example, in matrix notation, the covariance matrix is

$$V = E\big((X - \mu)(X - \mu)'\big)$$

and the correlation matrix P (uppercase ρ) is given by

$$P = \big(\text{diag}(V)\big)^{-\frac{1}{2}} V \big(\text{diag}(V)\big)^{-\frac{1}{2}}$$

When the individual x_i are normally distributed, their joint distribution is called the multivariate normal and is notated $x \sim N(\mu, V)$. The bivariate ($p = 2$) normal distribution with means $\mu_i = 0$, variances $\sigma_i^2 = 1$, and correlation $\rho = .7$ $\left[\text{hence } V = \left(\begin{smallmatrix} 1.0 & 0.7 \\ 0.7 & 1.0 \end{smallmatrix}\right)\right]$ is plotted as a three-dimensional object in Figure 3.9. This is actually one panel of the set of rotated views of the density shown in Figure 3.10. A rotating version (see the **shiny** screenshot in Figure E.2) of the bivariate normal density example runs in R with the statement

```
shiny::runApp(system.file("shiny/bivariateNormal",
                          package="HH"))
```

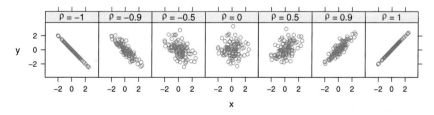

Fig. 3.8 Bivariate Normal distribution—scatterplot at various correlations. The distributions in the panels are related. The x-variable in all panels is the same. The y is generated from a common e-variable by the formula $y = \rho x + (1 - \rho^2)^{1/2} e$ for a sequence of values for ρ. The x- and e-variables were independently generated from the N(0,1) distribution. We provide a **shiny** app `bivariateNormalScatterplot` for a dynamic version of this set of panels. See Figure E.3 for a screenshot.

Bivariate Normal, $\rho = 0.7$

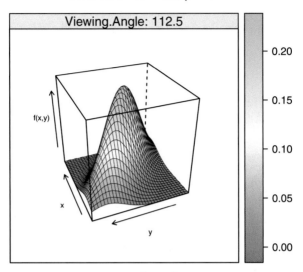

Fig. 3.9 Bivariate Normal density with $(\mu_1, \sigma_1^2, \mu_2, \sigma_2^2, \rho) = (0, 1, 0, 1, .7)$ in 3D space with viewing angle $= 112.5°$. A set of eight viewing angles is shown in Figure 3.10

If X and Y are random vectors with

$$Y = B + CX$$

for some vector B and some matrix C, then

$$E(Y) = B + C\,E(X) \tag{3.16}$$

and

$$\mathrm{var}(Y) = C\,\mathrm{var}(X)\,C' \tag{3.17}$$

If, moreover, X has a multivariate normal distribution, then so does Y. In other words, linear functions of normal r.v.'s are normal. Equations 3.16 and 3.17 generalize the scalar versions in Equations 3.8, 3.9, and 3.10.

It follows from Equation 3.17 that if X_1, X_2, X_3, X_4 are univariate random variables, then

$$\mathrm{var}(X_1 + X_2) = \mathrm{var}(X_1) + \mathrm{var}(X_2) + 2\,\mathrm{cov}(X_1, X_2)$$

and

$$\mathrm{cov}(X_1 + X_3, X_2 + X_4) = \mathrm{cov}(X_1, X_2) + \mathrm{cov}(X_1, X_4) + \mathrm{cov}(X_3, X_2) + \mathrm{cov}(X_3, X_4)$$

If Y has a k-dimensional multivariate normal distribution with mean μ and covariance matrix V, then

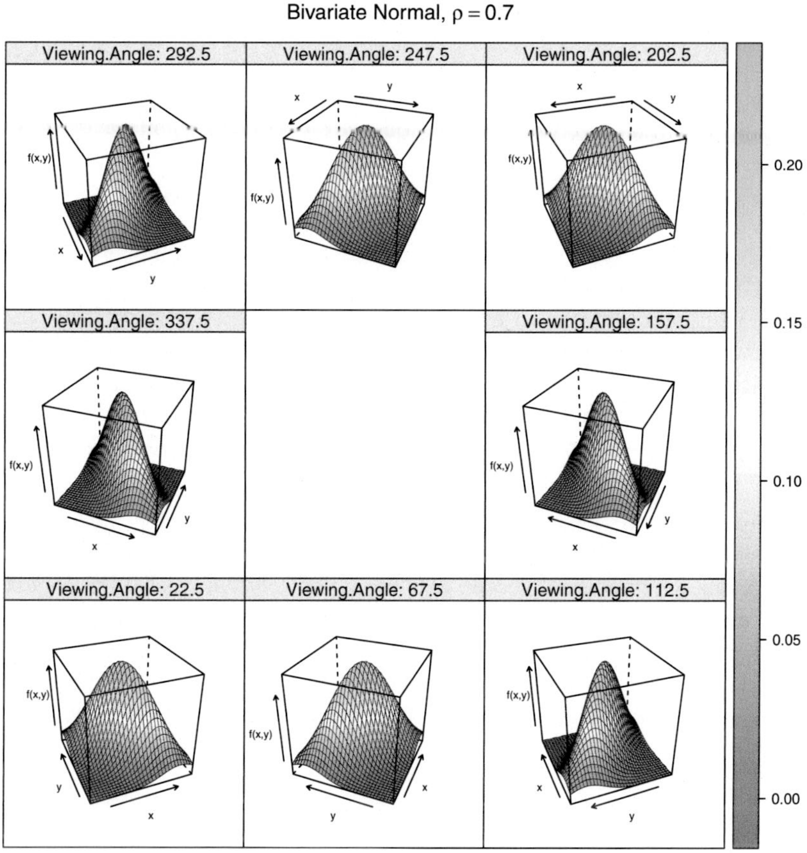

Bivariate Normal, $\rho = 0.7$

Fig. 3.10 Bivariate Normal density in 3D space with various viewpoints. Figure 3.9 shows a higher resolution view of the 112.5° panel. The reader can view an interactive version of this plot with the **shiny** app shiny::runApp(system.file("shiny/bivariateNormal", package="HH")). See Figure E.2 for a screenshot.

$$Q = (Y - \mu)'V^{-1}(Y - \mu)$$

has a χ^2 distribution with k degrees of freedom (See Appendix J).

3.4 Three Probability Distributions

In this section we introduce three probability distributions, the (discrete) binomial distribution and the (continuous) Normal and t distributions, that frequently arise in practice. Details of how to perform probability-related calculations for these and other frequently encountered distributions are discussed in Appendix J.

3.4.1 The Binomial Distribution

The binomial distribution is perhaps the most commonly encountered discrete distribution in statistics. Consider a sequence of n independent trials, or mini-experiments, each of which can result in one of just two possible outcomes. For convenience these outcomes are labeled *success* and *failure* although in context the success outcome may not connote a favorable event. Further assume that the probability of success, p, is the same for each trial. Let X denote the number of successes observed in the n trials. Then X has a binomial distribution with parameters n and p. This distribution has mean $\mu = np$ and standard deviation $\sigma = \sqrt{np(1-p)}$. We show an illustration of the discrete density for the binomial with $n = 15$ and $p = .4$ in Section J.3.2. In Figure 3.11 we show the discrete density for the binomial with $n = 15$ and $p = .4$, underlaid with the normal approximation with $\mu = np = 15 \times .4 = 6$ and $\sigma = \sqrt{np(1-p)} = \sqrt{15 \times .4 \times .6} = \sqrt{3.6} = 1.897$.

The above scenario is widely applicable. If one randomly samples with replacement from a population with a proportion p of successes, then the number of successes in the sample is binomially distributed. Even if the sampling is *without* replacement, the number of successes is approximately binomial if the population size is much greater than the sample size; in this case the first two assumptions above are only mildly violated. Applications include the number of voters favoring a candidate in a political poll, the number of patients in a population that suffer from a particular illness, and the number of defective items in one day's output from an assembly line.

However, it is not unusual for one or more of the binomial assumptions to be violated. For example, suppose we sample *without* replacement from a population of successes and failures and the population size is not much greater than the sample size, say less than 20 times as large as the sample. Then the trials are not independent and the *success* probability is not constant from trial to trial. (In this situation the correct distribution to use for X is the *hypergeometric* distribution. See Appendix J.)

Similarly, the binomial model is unlikely to apply to the number of hits by the archer in Section 3.2.1 because her shots (trials) may not be independent and may not have the same probability of a hit.

Usually in practice, we need to calculate not just $P(X = x)$, the probability of achieving *exactly* x successes, but probabilities of an interval of successes such as $P(X \leq x)$, the probability of *at most* x successes, or $P(a \leq X \leq b)$, the probability of observing between a and b successes inclusive.

A table of binomial probabilities can be used when n and p appear in the table. Otherwise, as illustrated in Appendix J, R functions can easily be used to produce accurate results.

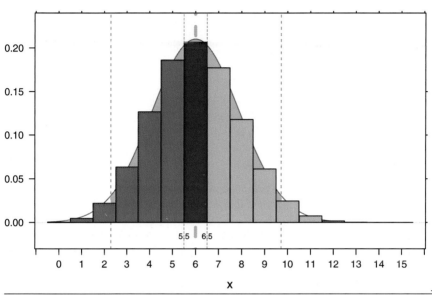

dbinom(x, size = 15, prob = 0.4)

```
> pbinom(size=15, prob=.4, q=6)
[1] 0.6098

> pnorm(q=6.5, mean=15*.4, sd=sqrt(15*.4*(1-.4)))
[1] 0.6039

> dbinom(size=15, prob=.4, x=6)
[1] 0.2066

> diff(pnorm(q=c(5.5, 6.5), mean=15*.4, sd=sqrt(15*.4*(1-.4))))
[1] 0.2079
```

Fig. 3.11 We show the discrete density for the binomial with $n = 15$ and $p = .4$, underlaid with the normal approximation with $\mu = np = 15 \times .4 = 6$ and $\sigma = \sqrt{np(1-p)} = \sqrt{15 \times .4 \times .6} = \sqrt{3.6} = 1.897$. The dark bar at $x = 6$ has probability $P(x = 6) = .2066$ from the binomial and $P(5.5 < x < 6.5) = .2079$ from the normal. The dark bar at $x = 6$ and all bars to its left together have probability $P(x \leq 6) = .6098$ from the binomial and $P(x < 6.5) = .6039$ from the normal approximation. The normal approximations are calculated with the correction for continuity (the interval $[6-.5, 6+.5]$ is the full width of the dark bar at $x = 6$).

3.4.2 The Normal Distribution

Many natural phenomena follow the normal distribution, whose probability density function is the familiar "bell-shaped" curve, symmetric about the mean μ. In addition, a celebrated theoretical result called the Central Limit Theorem says that the sampling distributions of sample means (see Section 3.5), sample proportions, and sample totals each are approximately normally distributed if the sample size is "sufficiently large." Since this theorem applies to almost all possible probability distributions from which a sample might be selected, including discrete distributions, the theorem brings the normal distribution into play in a wide variety of circumstances.

If X has a normal distribution with mean μ and standard deviation σ, and we define the standardization of X as $Z = \frac{X-\mu}{\sigma}$, then Z is normally distributed with mean 0 and standard deviation 1, *i.e.*, the *standard normal* distribution. We write $X \sim N(\mu, \sigma^2)$ to indicate that X has a normal distribution with mean μ and variance σ^2 (or we could say with standard deviation σ). In this notation, the standard normal distribution is $N(0, 1)$. The density function $\phi(z)$ and cumulative distribution function $\Phi(z)$ are defined in Section J.1.9 and illustrated in Figure 3.12.

The normal distribution is "bell-shaped" and symmetrically distributed about μ, which is also this distribution's median and mode. Almost all of the probability is concentrated in the interval $\mu \pm 3\sigma$. We use z_α to be the solution to the equation $P(Z > z_\alpha) = \alpha$. This is the value on the horizontal axis that has area α under the curve and to its right. For example, $z_{.05} = 1.645$. Figure 3.13 shows the normal density function for a $N(100, 25)$ distribution. If X has this distribution, the left shaded area in Figure 3.13 represents 95% of the area under the density function. That is,

$$P(Z < 1.645) = P\left(\frac{X - \mu}{\sigma} < 1.645\right) = P(X < 108.225) = .95$$

after substituting $\mu = 100$ and $\sigma = 5$. The right shaded area is

$$\alpha = .05 = P\left((X - \mu)/\sigma \geq \Phi^{-1}(1 - \alpha) = 1.645\right)$$

A dynamic version of any call to the NTplot function is available as a **shiny** app in the **HH** package with the argument shiny=TRUE included as an additional argument, for example

```
NTplot(shiny=TRUE)
```
A dynamic version of Figure 3.13 is initialized with the call

```
NTplot(mean0=100, mean1=NA, xbar=NA, xlim=c(75, 125),
       sd=5, digits=6, zaxis=TRUE, cex.z=0.6,
       cex.prob=.9, shiny=TRUE)
```
A screenshot of a dynamic NTplot example is in Figure E.1.

3.4.3 The (Student's) t Distribution

The t distribution is similar to the standard normal distribution in that its density is a bell-shaped curve symmetric about 0. However, as we see in Figure 3.14, where

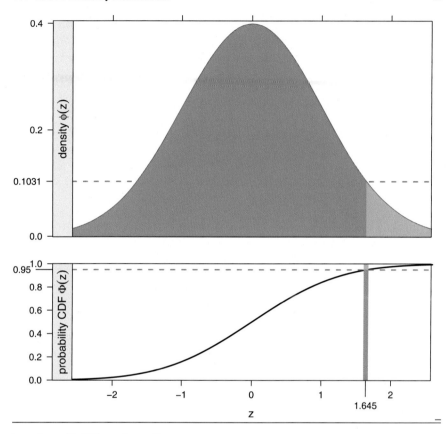

```
>    dnorm(1.645, m=0, s=1)
[1] 0.1031

>    pnorm(1.645, m=0, s=1)
[1] 0.95

>    qnorm(0.95, m=0, s=1)
[1] 1.645
```

Fig. 3.12 The standard normal density $N(0, 1)$ is shown in the top panel. The darker colored area is $\Phi(1.645) = P(Z \leq 1.645) = .95$. The lighter colored area is $1 - \Phi(1.645) = P(Z > 1.645) = .05$. The height of the density function in the top panel at $z = 1.645$ is $\phi(1.645) = .1031$. The cumulative distribution is shown in the bottom panel. The height of the darker line segment (below the curve) at $z = 1.645$ is $P(Z \leq 1.645) = .95$. The height of the lighter line segment (above the curve) at $z = 1.645$ is $P(Z > 1.645) = .05$.

we compare several t distributions to the normal distribution, the probability density function for the t is lower in the center and "heavier" in the tails. If the mean of a sample of size n is standardized with a sample standard deviation s rather than with a population standard deviation σ, then the resulting standardization, $\frac{\bar{X}-\mu}{s/\sqrt{n}}$, has a Student's t distribution with *degrees of freedom* parameter $n-1$. The t distribution is used for inference on population means and regression coefficients.

That $\frac{\bar{X}-\mu}{s/\sqrt{n}}$ has a t distribution rests on the fact that $\bar{X}$ and s are independent random variables when sampling from a normal population.

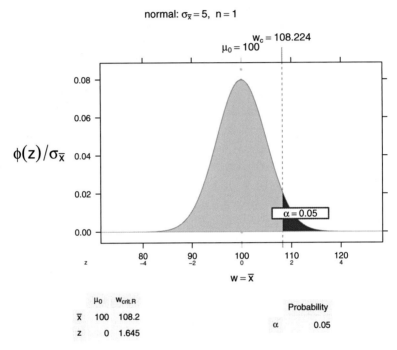

Fig. 3.13 A normal curve centered on the assumed true mean $\mu = 100$. We assume $\sigma = 5$ and $\alpha = .05$. The left lightly shaded area is $.95 = P\left(z = (X-\mu)/\sigma \leq \Phi^{-1}(1-\alpha) = 1.645\right)$. The right darkly shaded area is $\alpha = .05 = P\left(z = (X-\mu)/\sigma \geq \Phi^{-1}(1-\alpha) = 1.645\right)$. The plot shows both the $\bar{x}$ scale and (in smaller font) the z scale. The table below the plot shows μ_0 and the right critical value $\bar{x}_{crit.R}$ in both scales. The critical value in the z scale is directly from the normal table.

As the sample size n and hence the degrees of freedom get large, the sample standard deviation s increasingly approximates σ so that $\frac{\bar{X}-\mu}{s/\sqrt{n}}$ increasingly approximates $\frac{\bar{X}-\mu}{\sigma/\sqrt{n}}$. In other words, as the degrees of freedom increases, a t distribution increasingly resembles a standard normal distribution.

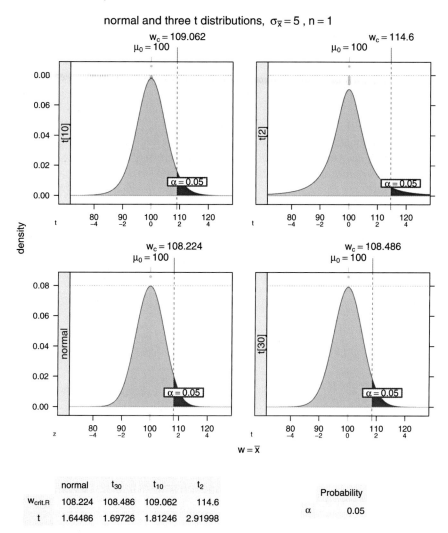

normal and three t distributions, $\sigma_{\bar{x}} = 5$, $n = 1$

	normal	t_{30}	t_{10}	t_2
$w_{crit.R}$	108.224	108.486	109.062	114.6
t	1.64486	1.69726	1.81246	2.91998

	Probability
α	0.05

Fig. 3.14 These panels are similar to Figure 3.13, the first panel is identical to Figure 3.13. The remaining panels show t-distributions with 30 df, 10 df, and 2 df. Each panel has less area in the center and more area in the tails. Use the reference line at $y = .08$ to see the drop in central area, use the thickness of the tails at $\bar{x} = 120$ to see the increase in the probability in the tails. Use the location of the critical value $w_c = \bar{x}_c$ on the graph and in the table below the graph to see that the critical value for the $\alpha = .05$ test is moving away from the null hypothesis value μ_0 as the df gets larger.

3.5 Sampling Distributions

In Chapter 1 we learn that knowledge about characteristics of populations can be gleaned from analogous characteristics of random samples from these populations. Also recall that population characteristics are called parameters and sample characteristics are called statistics. In the next two sections we discuss the two main techniques for using statistics to infer about parameters: estimation, and hypothesis testing. Implementation of these techniques requires that we use knowledge about the likely values of statistics. Such information about statistics is contained in their *sampling distribution*. The sampling distribution of a statistic depends on our assumed knowledge of the distribution of values in the population to which we are inferring. The term *standard error* is used to refer to the standard deviation of a sampling distribution.

Consider first the mean $\bar{X}$ of a sample of n items randomly selected from a normal population, $N(\mu, \sigma^2)$. It can be shown that the sampling distribution of $\bar{X}$ is also normally distributed with this same mean but with a much smaller variance:

$$\bar{X} \sim N(\mu, \sigma^2/n)$$

We illustrate this phenomenon in Figures 3.15 and 3.16. Figure 3.15 shows the individual observations and their means. Figure 3.16 shows the distribution of the means.

In the more likely situation where σ^2 is unknown, analogous probability statements are made with reference to the Student's t distribution.

Next suppose that the population is not necessarily normal. Then under fairly general conditions, a statistical theory result called the Central Limit Theorem states that $\bar{X}$ has "approximately" a $N(\mu, \sigma^2/n)$ distribution if the sample size n is "sufficiently large". Thus, the inferential statements concerning μ made in the normal distribution case are also approximately valid if the population is not normal.

What is meant here by "approximately" and "sufficiently large"? We mean that the closer the population is to a normal population, the smaller the sample size needs to be for the approximation to be acceptably accurate. Unless the population is multimodal or severely skewed, a sample size of 30 to 50 is usually sufficient for the approximation to hold.

Another application of the Central Limit Theorem implies that the sampling distribution of the proportion $\hat{p} = X/n$ of successes in n binomial trials is approximately normally distributed with mean $\mu = np$ and variance $\sigma^2 = npq$, where $q = 1 - p$. This result is used for inferences concerning the proportion of successes in a dichotomous population where the binomial assumptions apply.

If S^2 is the variance of a random sample of size n from a normal population having variance σ^2, then the sampling distribution of $(n-1)S^2/\sigma^2$ is χ^2 with $n-1$ degrees of freedom. We use this result for inferences concerning the population standard deviation σ.

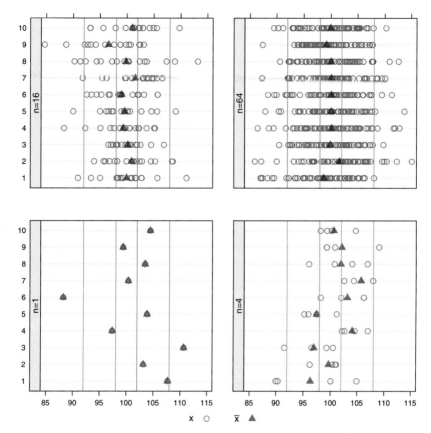

Fig. 3.15 Each panel shows 10 sets of n observations from the $N(\mu = 100,\ \sigma^2 = 5^2)$ distribution. The number n (from the set $\{1, 4, 16, 64\}$) differs by panel. The open circles show each individual observation. The semi-transparent triangle overlay shows the mean of each set of observations. As n gets larger, the set of 10 means are closer together. In the $n = 1$ panel, the means are identical to the individual observations and they occupy the full width of the panel. More precisely the variance of the means in the $n = 1$ panel is $\sigma^2 = 5^2$. In the $n = 4$ panel, the means are the average of 4 observations and they spread over only the central half of the panel with $\sigma_{\bar{x}}^2 = 5^2/4$. In the $n = 16$ panel, the means are the average of 16 observations and they spread over only the central quarter of the panel with $\sigma_{\bar{x}}^2 = 5^2/16$. In the $n = 64$ panel, the means are the average of 64 observations and they spread over only the central eighth of the panel with $\sigma_{\bar{x}}^2 = 5^2/64$.

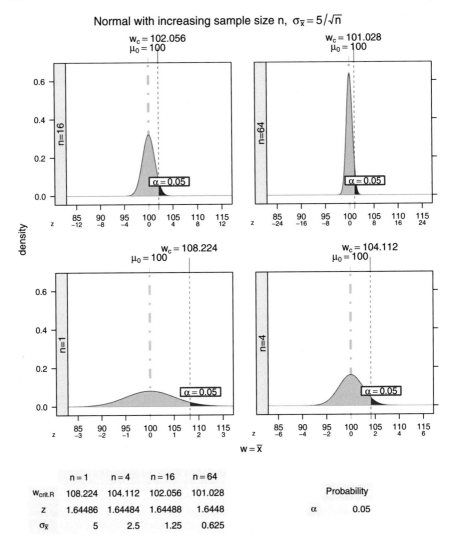

Fig. 3.16 These panels are also similar to Figure 3.13, with both the $\bar{x}$-scale and the z-scale shown in each panel of the graph. Again the panel with $n = 1$ is identical to Figure 3.13. The remaining panels show the sampling distribution of $\bar{x}$ as n increases. Each time the sample size goes up by a multiple of 4, the distance on the $\bar{x}$-scale from the critical value $w_c = \bar{x}_c$ to μ_o is halved, and the height of the density is doubled. On the z-scale, the distance from $w_c = \bar{x}_c$ to μ_o is always exactly $z_\alpha = 1.645$.

3.6 Estimation

A fundamental task of statistical analysis is inference of the characteristics of a large population from a sample of n items or individuals selected at random from the population. Sampling is commonly undertaken because it is

a. cheaper and

b. less prone to error

than examining the entire population. Estimation is one of the two broad categories of statistical techniques used for this purpose. The other is hypothesis testing, discussed in Section 3.7.

An *estimator* is a formula that can be evaluated with numbers from the sample. When the sample values are plugged into the formula, the result becomes an *estimate*. An estimator is a particular example of a statistic.

3.6.1 Statistical Models

A key component of statistical analysis involves proposing a statistical model. A statistical model is a relatively simple approximation to account for complex phenomena that generate data. A statistical model consists of one or more equations involving both random variables and parameters. The random variables have stated or assumed distributions. The parameters are unknown fixed quantities. The random components of statistical models account for the inherent variability in most observed phenomena. Subsequent chapters of this book contain numerous examples of statistical models.

The term *estimation* is used to describe the process of determining specific values for the parameters by fitting the model to the data. This is followed by determinations of the quality of the fit, often via hypothesis testing or evaluation of an index of goodness-of-fit.

Model equations are often of the form

$$\texttt{data} = \texttt{model} + \texttt{residual}$$

where `model` is an equation that explains most of the variation in the data, and `residual`, or lack-of-fit, represents the portion of the data that is not accounted for by the model. A good-quality model is one where `model` accounts for most of the variability in the data, that is, the data are well fitted by the model.

A proposed model provides a framework for the statistical analysis. Experienced analysts know how to match models to data and the method of data collection. They are also prepared to work with a wide variety of models, some of which are discussed in subsequent chapters of this book. Statistical analysis then proceeds by estimating the model and then providing figures and tables to support a discussion of the model fit.

3.6.2 Point and Interval Estimators

There are essentially two types of estimation: point estimation and interval estimation.

A typical example begins with a sample of n observations collected from a normal distribution with unknown mean μ and unknown standard deviation σ. We calculate the sample statistics

$$\bar{x} = \left(\sum_{i=1}^{n} x_i \right) / n$$

$$s^2 = \left(\sum_{i=1}^{n} (x - \bar{x})^2 \right) / (n - 1)$$

Then $\bar{x}$ is a point estimator for μ. Define the standard error of the mean $s_{\bar{x}}$ as $s_{\bar{x}} = s/\sqrt{n}$. We then have

$$\bar{x} \pm t_{\alpha/2,\nu}\, s_{\bar{x}} = (\bar{x} - t_{\alpha/2,\nu}\, s_{\bar{x}},\ \bar{x} + t_{\alpha/2,\nu}\, s_{\bar{x}})$$

as a two-sided $100(1 - \alpha)\%$ confidence interval for μ.

For specificity, let us look in Figure 3.17 at the situation with $n = 25$, $\bar{x} = 8.5$, $\nu = 24$, $s^2 = 4$, $\alpha = .05$. From the t-table, the critical value $t_{\alpha/2,24} = 2.064$. We get $s_{\bar{x}} = s/\sqrt{n} = 2/\sqrt{25} = .4$ as the standard error of the mean.

Point estimators are single numbers calculated from the sample, in this example $\hat{\mu} = 8.5$. Interval estimators are intervals within which the parameter is expected to fall, with a certain degree of confidence, in this example $95\% \,\mathrm{CI}(\mu) = 8.5 \pm 2.064 \times 0.4 = (7.6744, 9.3256)$. Interval estimators are generally more useful than point estimators because they indicate the precision of the estimate. Often, as here, interval estimators are of the form:

$$\text{point estimate} \pm \text{constant} \times \text{standard error}$$

where "standard error" is the observed standard deviation of the statistic used as the point estimate. The constant is a percentile of the standardized sampling distribution of the point estimator. We summarize the calculations in Table 3.7.

3.6.3 Criteria for Point Estimators

There are a number of criteria for what constitutes "good" point estimators. Here is a heuristic description of some of these.

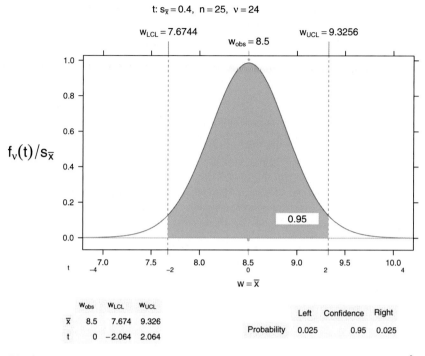

t: $s_{\bar{x}} = 0.4$, $n = 25$, $v = 24$

w_{obs}	w_{LCL}	w_{UCL}			Left	Confidence	Right
$\bar{x}$	8.5	7.674	9.326				
t	0	-2.064	2.064	Probability	0.025	0.95	0.025

Fig. 3.17 Confidence interval plot for the t distribution with $n = 25$, $\bar{x} = 8.5$, $v = 24$, $s^2 = 4$, $\alpha = .05$. We calculate $t_{\alpha/2,24} = 2.064$ and the two-sided 95% confidence interval (7.674, 9.326). The algebra and R notation for the estimators are shown in Table 3.7.

Table 3.7 Algebra and R notation for the example in Figure 3.17.

$\bar{x}$	`> xbar <- 8.5`
s	`> s <- sqrt(4)`
n	`> n <- 25`
$s_{\bar{x}}$	`> s.xbar <- s/sqrt(n)`
	`> s.xbar`
	`[1] 0.4`
$t_{\alpha/2,24}$	`> qt(.975, df=24)`
	`[1] 2.063899`
$\bar{x} \pm t_{\alpha/2,24}\, s_{\bar{x}}$	`8.5 + c(-1,1) * 2.064 * 0.4`
	`[1] 7.6744 9.3256`

unbiasedness: The expected value of the sampling distribution of the estimator is the parameter being estimated. The bias is defined as:

$$\text{bias} = \text{expected value of sampling distribution} - \text{parameter}$$

Unbiasedness is not too crucial if the bias is small and if the bias decreases with increasing n. The sample mean $\bar{x}$ is an unbiased estimator of the population mean μ and the sample variance s^2 is an unbiased estimate of the population variance

σ^2. The sample standard deviation s is a biased estimator of the population standard deviation σ. However, the bias of s decreases toward zero as the sample size increases; we say that s is an *asymptotically unbiased* estimator of σ.

small variance: Higher precision. For example, for estimating the mean μ of a normal population, the variance $s_{\bar{x}} = s/\sqrt{n}$ of the sample mean $\bar{x}$ is less than the variance $s_{\overset{+}{x}} = \sqrt{\frac{\pi}{2}}\, s/\sqrt{n}$ of the sample median $\overset{+}{x}$.

consistency: The quality of the estimator improves as n increases.

sufficiency: the estimator fully uses all the sample information. Example: If X is distributed as continuous uniform on $[0, a]$, how would you estimate a? Since the population mean is $a/2$, you might think that $2\bar{x}$ is a "good" estimator for a. The largest item in the sample of size n, denoted $x_{(n)}$, is a better and *sufficient* estimator of a. This estimator cannot overestimate a while $2\bar{x}$ can either underestimate or overestimate a. If $x_{(n)}$ exceeds $2\bar{x}$, then it must be closer to a than is $2\bar{x}$.

3.6.4 Confidence Interval Estimation

A confidence interval estimate of a parameter is an interval that has a certain probability, called its *confidence coefficient*, of containing the parameter. The confidence coefficient is usually denoted $1-\alpha$ or as a percentage, $100(1-\alpha)\%$. Common values for the confidence coefficient are 95% and 99%, corresponding to $\alpha = .05$ or $.01$, respectively. Figure 3.17 illustrates a 95% confidence interval for the mean of a normal distribution.

If we construct a 95% confidence interval (CI), what is the meaning of 95%? It is easy to incorrectly believe that 95% is the probability that the CI contains the parameter. This is false because the statement *"CI contains the parameter"* is not an event, but rather a situation that is certainly either true or false. The correct interpretation refers to the *process used to construct the CI:* If, hypothetically, many people were to use this same formula to construct this CI, plugging in the results of their individual random samples, about 95% of the CI's of these many people would contain the parameter and about 5% of the CI's would exclude the parameter.

It is important to appreciate the tradeoff between three quantities:

- confidence coefficient (the closer to 1 the better)
- interval width (the narrower the better)
- sample size (the smaller the better)

In practice it is impossible to optimize all three quantities simultaneously. There is an interrelationship among the three so that specification of two of them uniquely determines the third. A common practical problem is to seek the sample size required to attain a given interval width and confidence. Examples of such formulas appear in Section 5.6.

3.6.5 Example—Confidence Interval on the Mean μ of a Population Having Known Standard Deviation

The interpretation of the confidence coefficient may be further clarified by the following illustration of the construction of a $100(1-\alpha)\%$ confidence interval on an unknown mean μ of a normal population having known standard deviation σ, using a random sample of size n from this population. If $\bar{X}$ denotes the sample mean, then $\frac{\bar{X}-\mu}{\sigma/\sqrt{n}}$ has a standard normal distribution. Let $z_{\frac{\alpha}{2}}$ denote the $100(1-\frac{\alpha}{2})^{\text{th}}$ percentile of this distribution. Then

$$P\left(-z_{\frac{\alpha}{2}} < \frac{\bar{X}-\mu}{\sigma/\sqrt{n}} < z_{\frac{\alpha}{2}}\right) = 1 - \alpha$$

After a bit of algebraic rearrangement, this becomes

$$P\left(\bar{X} - z_{\frac{\alpha}{2}} \frac{\sigma}{\sqrt{n}} < \mu < \bar{X} + z_{\frac{\alpha}{2}} \frac{\sigma}{\sqrt{n}}\right) = 1 - \alpha$$

The endpoints of the interval $\left(\bar{X} - z_{\frac{\alpha}{2}} \frac{\sigma}{\sqrt{n}}, \ \bar{X} + z_{\frac{\alpha}{2}} \frac{\sigma}{\sqrt{n}}\right)$ are random variables, so the probability statement refers to the probability that the interval contains the parameter, not the probability that the parameter is contained in the interval.

In practice, we replace the random variable $\bar{X}$ with $\bar{x}$, the realized value from the sample, and wind up with the $100(1-\alpha)\%$ confidence interval for μ:

$$\left(\bar{x} - z_{\frac{\alpha}{2}} \frac{\sigma}{\sqrt{n}}, \ \bar{x} + z_{\frac{\alpha}{2}} \frac{\sigma}{\sqrt{n}}\right) \tag{3.18}$$

Figure 3.18 shows an example from the situation with known variance σ^2.

3.6.6 Example—One-Sided Confidence Intervals

One-sided confidence intervals correspond to one-sided tests of hypotheses. Such intervals have infinite width and therefore are much less commonly used in practice than two-sided confidence intervals, which have finite width. The rationale for using

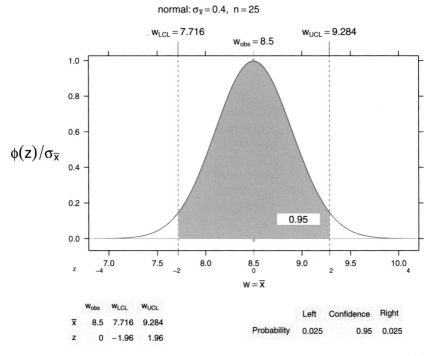

Fig. 3.18 Confidence interval plot for the normal distribution with $n = 25$, $\bar{x} = 8.5$, $\sigma^2 = 4$, $\alpha = .05$. We calculate $z_{\alpha/2} = 1.96$ and the two-sided 95% confidence interval (7.716, 9.284). Compare this to the t-based confidence interval in Figure 3.17 and note that the width of the interval is narrower here because we have more information, that is, because we know the variance, we don't have to estimate the variance.

one-sided intervals matches that for one-sided tests—sometimes the analyst believes the value of a parameter is at least or at most some value rather than on either side. One-sided confidence intervals on the mean of a population having known standard deviation are shown in Table 5.1. Other examples of one-sided confidence intervals appear in Tables 5.2 and 5.3. Figure 3.19 shows a one-sided example from the situation with known variance σ^2.

3.7 Hypothesis Testing

The statistician sets up two competing hypotheses, the null hypothesis H_0 and the alternative hypothesis H_1, for example in Figure 3.21 in Section 3.8,

$H_0: \mu = 32$ vs $H_1: \mu \neq 32$. The task is to decide whether the sample evidence better supports H_0 (decision to "retain H_0") or H_1 (decision to "reject H_0").

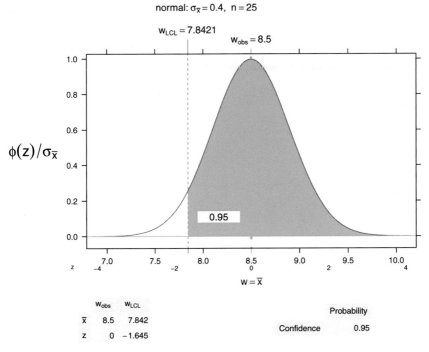

normal: $\sigma_{\bar{x}} = 0.4$, $n = 25$

$W_{LCL} = 7.8421$

$W_{obs} = 8.5$

	W_{obs}	W_{LCL}
$\bar{x}$	8.5	7.842
z	0	−1.645

	Probability
Confidence	0.95

Fig. 3.19 One-sided confidence interval plot for the normal distribution with $n = 25$, $\bar{x} = 8.5$, $\sigma^2 = 4$, $\alpha = .05$. We are confident that the true mean is larger than the calculated value. We calculate $z_\alpha = 1.645$ and the one-sided 95% confidence interval $(7.842, \infty)$.

There are two types of errors: the Type I error of *rejecting* H_0 when H_0 is true, and the Type II error of *retaining* H_0 when H_1 is true. In the classical hypothesis setup, the statistician prespecifies α—the maximum probability of committing a Type I error. Subject to this constraint, we select a testing procedure that gives good control over β—the probability of committing a Type II error. This probability is a function of the unknown parameter being tested. A plot of the probability against the parameter is called an *operating characteristic curve* (O.C. curve) of the test.

The *power* of a hypothesis test is the probability of correctly rejecting a false null hypothesis, equivalently, the probability $1 - \beta$. A *power curve* is a plot of the proba- bility of rejecting H_0 against the true value of the parameter. It contains information identical to that conveyed by an O.C. curve. It is a convention in various scientific fields whether the power or the O.C. curve is used. We illustrate both in Figure 3.20. Statisticians can determine the sample size needed to conduct a test that has a high probability of detecting a departure from H_0 by studying O.C. or power curves for a variety of proposed sample sizes. Examination of these curves displays the tradeoffs between Type I error, Type II error, and sample size. See Figure 3.20 for a static example. The reader can explore these options dynamically with the `shiny=TRUE` argument to the `NTplot` function. Figure E.1 shows a screenshot of our **shiny** app duplicating Figure 3.20. Further discussion of Operating Characteristic and power curves is in Section 3.9.

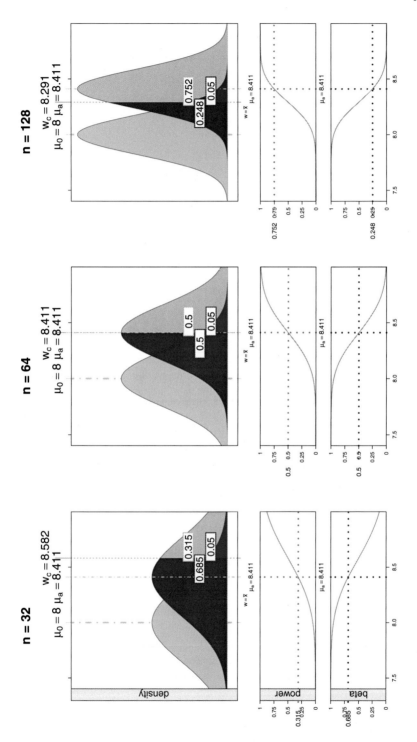

Fig. 3.20 (continued)

The three sets of panels show the same null hypothesis ($\mu_0 = 8$) and alternative hypothesis ($\mu_1 = 8.411$) with three different sample sizes ($n = 32, 64, 128$) and their corresponding powers (.315, .500, .752). Each set contains a normal plot in the top panel, the corresponding power curve in the middle panel, and the beta curve (operating characteristic curve) in the bottom panel.

The pink area in each top panel shows the power, the probability that an observed $\bar{x}$ will be to the right of the critical value $\bar{x}_C = 8.411$ when the true mean is $\mu_1 = 8.411$. The gray curve in each middle panel is the power curve, showing the power for all possible values of the alternate mean μ_1. The crosshairs in the middle panel are at $\mu_1 = 8.411$ and power($\mu_1 = 8.411$). The red area in the top panels shows $\beta = 1 -$ power, the probability of the Type II Error. The gray curve in each bottom panel is the beta curve (the Operating Characteristic

curve) showing the β for all possible values of the alternate mean μ_1. The crosshairs in the bottom panel are at $\mu_1 = 8.411$ and beta($\mu_1 = 8.411$). As we increase the sample size (move from the left set of panels toward the right set of panels), the density functions get taller and thinner while maintaining a constant area of 1, the power and beta curves get steeper, and the power increases (hence beta decreases) for any specified value of μ_1. The reader can duplicate these panels by running the R code in file HHscriptnames(3). The reader can set up a dynamic version of this plot from the same code with NTplot(tmp64, shiny=TRUE) and then clicking the animate icon for the n-slider. See Figure E.1 for a screenshot. The screenshot initially doesn't show the Power and Beta curves. They can be included by checking the Power and Beta checkboxes on the Display Options tab.

Do not confuse the decision to retain H_0 with the statement that H_0 is true. We might be committing a Type II error. Similarly, the decision to reject H_0 is not the same as saying that H_0 is false because we might be committing a Type I error.

Commonly selected values of α are .05 or .01. The choice is sometimes governed by what is traditional in a research area.

With the prespecification of α, the statistician maintains better control over Type I error than Type II error. When we have a choice, the names H_0 and H_1 should be assigned such that the hypothesis with the more serious error is called H_0 and its more serious error is the Type I error. The hypothesis with the less serious error is called H_1 and its less serious error is the Type II error. In many applications, H_0 is essentially the statement that the status quo is better, while H_1 is the statement that an innovation is better. The Type I error of incorrectly deciding in favor of an innovation is typically more serious than the error of incorrectly maintaining the status quo because innovation is usually costly. As a result, classical testing puts the burden of proof on the innovation H_1; H_0 is retained unless there is compelling evidence not to do so.

The preceding rules for deciding which hypothesis is H_0 are based on the fact that classical hypothesis testing places more control over Type I error at the cost of reduced control over Type II error. The logic for this approach is seen by comparing in Table 3.8 the definitions of these two errors in the hypothesis testing context with the potential errors in a U.S. courtroom.

Table 3.8 Comparison of Hypothesis Testing with the Decision Options in a Court of Law

	Hypothesis Testing			Court of Law	
	True situation			True situation	
Decision	H_0 true	H_0 false	Decision	Innocent	Guilty
Reject H_0	Type I error	correct	Convict	greater error	correct
Retain H_0	correct	Type II error	Acquit	correct	lesser error

In the United States, the error of convicting an innocent defendant is viewed as far more serious than the error of acquitting a guilty defendant. Accordingly, the U.S. legal system places the burden of proof on the prosecution to establish guilt beyond a reasonable doubt. If sufficient evidence is not presented to the court, the defendant is acquitted. Similarly, in hypothesis testing, the burden is placed on the analyst to provide convincing evidence that H_0 is false; in the absence of such evidence, H_0 is accepted. Continuing the analogy, in the hypothesis testing framework, the way to reduce the probability of committing a Type II error without compromising control of Type I error is to seek an increased sample size. In the legal framework, courts can best reduce the probability of acquitting guilty defendants by obtaining as much relevant evidence as possible.

Table 3.8 also demonstrates that if we modify a hypothesis testing procedure to less readily reject a null hypothesis, this results in both greater control of Type I error and reduced control of Type II error.

Tests of hypotheses are conducted by determining what sample results would be likely if H_0 is true. If then a sufficiently unlikely sample statistic is observed, doubt is cast on the truth of H_0; i.e., H_0 is rejected.

Most tests are constructed by calculating a test statistic from a random sample. This is compared to a critical value, or values. If the test statistic is on one side of the critical value(s), H_0 is retained; if on the other side, H_0 is rejected. If the value of the test statistic leads to rejection of H_0, the test statistic is said to be (statistically) *significant*.

A criticism of classical hypothesis testing is the requirement that α be prespecified. One way around this is to calculate the *p*-value of the test.

The *p*-value is the probability of observing, in hypothetical repeated samples from the null distribution (that is, when H_0 is true), a value of the test statistic at least as extreme in the direction of H_1 as the test statistic calculated from the present sample.

For most testing procedures, calculating the *p*-value requires the use of the computer. We reject H_0 (that is, we make the decision to act as if H_0 does not describe the world) if $\alpha > p$-value; we retain H_0 (that is, we make the decision to act as if H_0 does describe the world) otherwise. Then the analyst needs only to know how α compares with the *p*-value, and does not have to commit to a particular value of α. Most software provides *p*-values as part of the output rather than requesting α as part of the input.

Another criticism of classical hypothesis testing is that if H_0 is barely false, it is always possible to reject H_0 simply by taking a large enough sample size. For example, if we test $H_0: \mu = 32$, where μ is the mean amount of soda a bottling plant puts into 32-ounce (0.946 liter) bottles, and if in reality, $\mu = 32.001$ ounces, H_0 can be rejected even though as a practical matter it makes no sense to act as though anything is wrong with the filling mechanism. This would be an instance of a statistically significant result that is not of practical significance. Because of this criticism, many statisticians are much more comfortable using CIs than tests.

In practice, a very small *p*-value may be regarded as sufficiently strong evidence against H_0 to convince us to act as though H_0 is false (that is, as though H_0 does not describe the world). However, even in this situation and especially if the sample size is large, we should be mindful of the possibility that one is making a Type I error. Also, we should always be alert to the possibility that an underlying assumption about the population is incorrect; if so, the *p*-value calculation may be distorted.

3.8 Examples of Statistical Tests

Suppose in the example of the previous section, the standard deviation of fill volume is known to be 0.3 ounces, and that a sample of 100 bottles yields a mean of 31.94 ounces. If the alternative hypothesis is $H_1: \mu \neq 32$, then we should reject H_0 if $\bar{x}$ is sufficiently above or below 32. We illustrate this example in Figure 3.21. In this example, in order to maintain Type I error probability at $\alpha = .01$, we should reject H_0 if

$$\bar{x} < 32 - z_{.005} \; \sigma \; / \sqrt{n}$$
$$= 32 - 2.576\,(0.3)/ \; 10 = 31.923$$

or

$$\bar{x} > 32 + z_{.005} \; \sigma \; / \sqrt{n}$$
$$= 32 + 2.576\,(0.3)/ \; 10 = 32.077$$

Since $\bar{x}$ meets neither condition, we should retain H_0 when testing at $\alpha = .01$. This is an example of a "two-tailed" (or "two-sided") test because we reject H_0 if $\bar{x}$ lies sufficiently far on either tail of the Z distribution with the null hypothesized mean.

At this point we might ask whether a larger choice of α would have led to the "retain H_0" decision. This is answered by finding the p-value, here equal to $2P(Z > |z_{calc}|)$ for $z_{calc} = (\bar{x} - \mu_0)/(\sigma/ \sqrt{n}) = -2$. Thus p-value$= 2P(Z > 2.00) = 0.046$. Then any choice of $\alpha \leq 0.046$ requires retention of H_0; i.e., the decision to act as if the filling machine is in control.

A two-tailed test can be conducted as follows. Reject the null hypothesis at level α if the null hypothesized value of the parameter lies outside the $100(1 - \alpha)\%$ confidence interval for the parameter.

Sometimes analysts prefer to conduct a "one-tailed" (or "one-sided") test where the alternative hypothesis statement is a one-sided inequality. Suppose in the soda bottling example it was felt that the error of incorrectly claiming bottles are being underfilled is much more serious than an error of incorrectly claiming bottles are being overfilled. We illustrate the one-tailed test in Figure 3.22. Then we might test $H_0: \mu \geq 32$ vs $H_1: \mu < 32$, because this way the more serious error is the better controlled Type I error. Now H_0 will be rejected only when $\bar{x}$ is sufficiently below 32. If once again we take $\alpha = .01$, we reject H_0 if

$$\bar{x} < 32- z_{.01} \; \sigma \; / \sqrt{n}$$
$$= 32-2.326\,(0.3)/ \; 10 = 31.93$$

As with the two-tailed test, H_0 is retained.

Note that, if instead we had observed $\bar{x} = 31.925$ ounces, we would have rejected H_0 with the one-tailed alternative but retained it with the two-tailed alternative. The explanation for this distinction is that the portion of the left side of the parameter space where H_1 is true is larger under the one-tailed setup than under the analogous two-tailed setup.

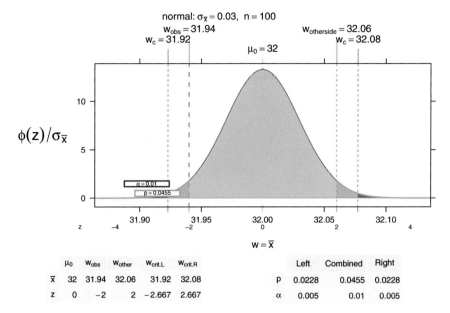

Fig. 3.21 Test whether the bottle production is within bounds. The figure shows a two-sided rejection region—anything in the deep blue region outside the critical bounds ($\bar{x}_{\text{critLeft}}$ = 31.92, $\bar{x}_{\text{critRight}}$ = 32.08). The observed value $\bar{x}$ = 31.94 is within the central light-blue do-not-reject region. The p-value is the green shaded area outside the bounds ($\bar{x}$ = 31.94, $\bar{x}_{\text{otherside}}$ = 32.06) where $\bar{x}_{\text{otherside}}$ = $\mu_0 + (\mu_0 - \bar{x})$ = 32.06 is the value equally far from the null value $\mu_0 = 32$ in the other direction.

3.9 Power and Operating Characteristic (O.C.) (Beta) Curves

These two types of curves are used to assess the degree of Type II error control of a proposed test. The O.C. curve is a plot of the probability of retaining H_0 under the condition of a specified value of the parameter vs the specified value of the parameter being tested, and the power curve is a plot of the probability of rejecting H_0 vs the parameter being tested. These two plots give equivalent information, and the choice of which to use is a matter of taste or tradition in one's discipline.

Power and O.C. curves are used to display the menu of competing choices of sample size, α, and Type II error probability. One desires that all three of these quantities be as small as possible, but fixing any two of them uniquely determines the third. Analysts commonly use one of these curves to assess the needed sample size to achieve desired control over the two errors. If the required sample size is infeasibly large, the analyst can see what combinations of diminished control over the two errors are possible with the maximum attainable sample size. Note that $\beta = P(\text{Type II error})$ is a function of the true value of the unknown parameter being tested and that α is the *maximum* probability of committing a Type I error.

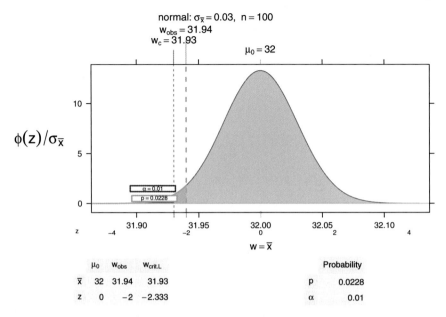

Fig. 3.22 Test whether the bottle production is within bounds. The figure shows a one-sided rejection region—anything in the deep blue region below the limit $\bar{x}_c = 31.93$. The observed value $\bar{x} = 31.94$ is in the right light-blue do-not-reject region. The p-value is the green shaded area to the left of $\bar{x} = 31.94$).

In the case discussed above, $\beta = P(\text{Type II error}|\mu_a)$ is a function of the true (and unknown) value μ_a of the parameter.

We illustrate the formulation of an O.C. curve and its construction using R. The pnorm function calculates the normal c.d.f. Φ. The qnorm function calculated the inverse normal c.d.f. Φ^{-1}.

Consider a situation where we have a normal population with unknown mean μ and known s.d. $\sigma = 2.0$. Suppose we wish to test $H_0: \mu \leq 8$ vs $H_1: \mu > 8$, using $\alpha = .05$ and a sample of $n = 64$ items. Here we retain H_0 if

$$\bar{X} \leq \mu_0 + \Phi^{-1}(.95)\,\sigma/\sqrt{n}$$
$$= 8 + 1.645 \quad 2/8$$
$$= 8.411$$

i.e., H_0 is retained if $\bar{X} \leq 8.411$. Since the true μ is unknown, the probability that H_0 is retained is a function of this μ:

$$P(\bar{X} \leq 8.411 \mid \mu) = P\left[\frac{\bar{X}-\mu}{\sigma/\sqrt{n}} \leq \frac{8.411-\mu}{(2/8)}\right]$$
$$= P\left[Z \leq 4(8.411 - \mu)\right]$$
$$= \Phi(33.644 - 4\mu)$$

where Z is $N(0, 1)$. The power curve for this problem is the plot of $1 - \Phi(33.644 - 4\mu)$ vs μ. Figure 3.23 shows the normal plot under both the null and alternative hypotheses for several values of μ_1, and the associated power plot and beta (Operating Characteristic) plots. The power and beta curves in all three columns of Figure 3.23 are identical. The crosshairs identify the location on the curves of the power and probability of the Type II error for the specified value μ_a of the alternative.

For most distributions, tests of hypotheses, calculation of Type II error probabilities, and construction of O.C. and power curves involves the use of a *noncentral* probability distribution. Noncentral distributions are discussed in Section J.2 in Appendix J. Noncentrality is not an issue for tests using the normal distribution, as the normal does not have a noncentral form.

We illustrate a noncentral alternative distribution in Figure 3.24.

The `power.t.test` function is essentially the same as the right panel in Figure 3.24, the only difference is that `power.t.test` assumes $\mu_0 = 0$.

```
PowerT <- power.t.test(n=12, sd=2, delta=1.4,
                       type="one.sample",
                       alternative="one.sided")
NTplot(PowerT, beta=TRUE, power=TRUE)
```

3.10 Efficiency

Efficiency is a measure of value (usually information in Statistics) per unit cost. We wish to maximize efficiency. We want small sample sizes because each observation has a cost, and fewer observations cost less than more observations. We want larger sample sizes because that gives us a better estimate of the precision of our study. A larger sample size increases the degrees of freedom for the error term. When we look at a table of t- or F- or χ^2-values we see that the critical value of the test statistics for a specified significance level is smaller as the sample size increases. We can see this in many of the figures in this chapter. Figure 3.20 shows that the critical value for a normal test goes down as the sample size goes up. Figure 3.14 shows that the critical value is smaller as the degrees of freedom increase. Choosing the right sample size is therefore important. It needs to be large enough that there is information about the population, and small enough that the client is willing to pay for the observations.

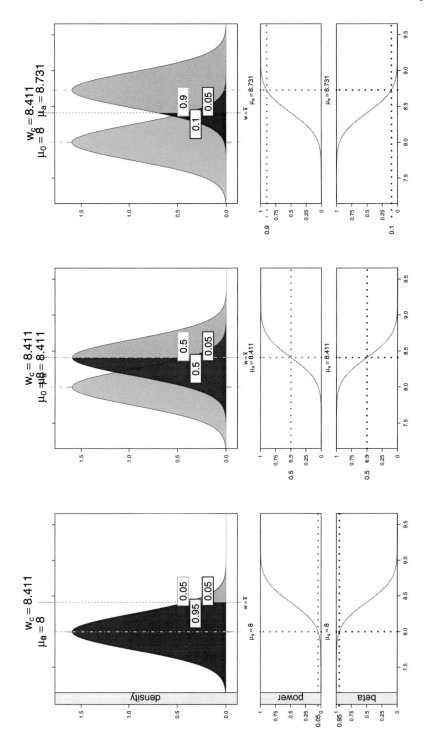

Fig. 3.23 (continued)

The three sets of panels show the same null hypothesis ($\mu_0 = 8$) with a sequence of alternative hypothesis values ($\mu_1 = 8$, $\mu_1 = 8.411 = \mu_c$, $\mu_1 = 8.7314$) and their corresponding powers (.05, .5, .90). Each set contains a normal plot in the top panel, the corresponding power curve in the middle panel, and the beta curve in the bottom panel.

Start on the right with $\mu_1 = 8.7314$, power = .90, and $\beta = .10$. In this rightmost set, the μ_1 and μ_0 are far enough apart that there is less overprinting in the top axis. The pink area in the top panel shows the power, the probability that an observed $\bar{x}$ will be to the right of the critical value $\bar{x}_C = \mu_1 = 8.411$ when the true mean is $\mu_1 = 8.7314$. The gray curve in the middle panel is the power curve, showing the power for all possible values of the alternate mean μ_1. The crosshairs in the middle panel on the right are at $\mu_1 = 8.7314$ and power($\mu_1 = 8.7314$) = .9. The red area in the top panel shows $\beta = 1 -$ power, the probability of the Type II Error. The gray curve in the bottom panel is the beta curve (the Operating Characteristic curve) showing the β for all possible values of the alternate mean μ_1. The crosshairs in the bottom panel on the right are at $\mu_1 = 8.7314$ and beta($\mu_1 = 8.7314$) = .1.

Move to the center set of panels. Now the $\mu_0 = \bar{x}_c = 8.411$ and $\mu_1 = 8.411$ and the power is exactly 1/2. In the top axis, we see that the labels μ_0 and μ_1 are close together and partially obscure each other. In the bottom panel, the $\beta = 1 - 1/2 = 1/2$.

Move to the left panels where $\mu_0 = \mu_1 = 8$ and note that the power is equal to $\alpha = .05$ and $\beta = 1 - \alpha = .95$. The value and labels μ_0 and μ_1 are now identical and the coloring for the alternative hypothesis regions completely masks the coloring for the null hypothesis regions.

The reader can duplicate these panels by running the R code in file HHscriptnames (3). The user can set up a dynamic version of this plot from the same code with NTplot(tmp8411, shiny=TRUE) and then clicking the animate icon for the mu[a]-slider. The user can also get there from the example in Figure E.1. On the Display Options tab, check Power and Beta. On the Normal and t tab, click the ▶ button on mu[a].

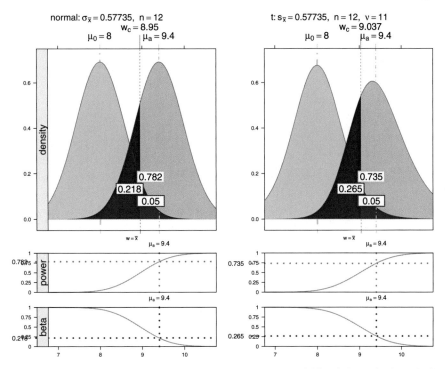

Fig. 3.24 The *t*-test of the same null hypothesis ($\mu_0 = 8$) as Figure 3.23 and alternative hypothesis values ($\mu_1 = 9.4$). On the left, under the assumption of known variance which implies that the density curve for the alternative is also normal with the same variance as under the null, the power is .782. On the right, under the assumption of unknown variance which requires that *s* must be estimated from the data, the alternative distribution has a noncentral *t* distribution. The null has a smaller central peak value and larger critical value. The alternative is no longer symmetric and has an even smaller peak value. See further discussion of the noncentral *t* distribution in Section J.2.2.

3.11 Sampling

Whenever we wish to learn the characteristics of a large *population* or *universe* that is unwieldy or expensive to completely examine, we may instead select a *sample* from the population. If the sample has been selected by a *random* mechanism, it is usually possible to infer population characteristics from the analogous characteristics in the sample. Much of the remainder of this volume deals with methods for conducting such inferences. In this section we discuss methods for selecting random samples. Only rarely is it practical to sample the entire population; such a sample is called a *census* of the population.

Here are some examples of situations where we would learn about a population by choosing a random sample from it.

- A factory wishes to know if the proportion of today's output that is defective is sufficiently small that the output may be shipped for sale rather than scrapped. Examining the entire output stream is likely to be impractical and expensive, and clearly impossible if examining an item results in its destruction. Instead, a quality-control worker may suggest a random sample of the output, with a size of sample that is sufficient to accurately estimate the proportion of defectives without being excessively costly. [Formula (5.17) may be used for determining the sample size in this situation.]

- A candidate for statewide political office wants to assess whether more than half of the electorate will vote for her. An accurate estimate of the proportion favoring her would greatly influence her future campaign strategy. She obviously must contract for a sample because her campaign cannot afford to contact all potential voters. A complication in this situation is that the population of voters and their opinions are apt to be somewhat different on election day from what they are at the time the sample is selected.

- A timber company wishes to estimate the average height of the trees in a forest under its control. Such measurements are expensive to obtain because they involve sighting a tree's top at a fixed ground distance from the tree. Therefore, a census of the forest would be prohibitively expensive and some type of random sample of trees is preferred.

If an arbitrary sample (essentially any procedure that isn't based on a specified probability distribution) is used, there is no guarantee that it will truly represent the population. To ensure that the sample adequately reflects the population, a randomization mechanism must be used. The techniques for inferring from sample to population discussed in the following chapters rest on the assumption that samples are randomly selected. If this assumption is unjustified, the probability-based statements that accompany the inferences will be incorrect.

For a given sample size n the analyst seeks to maximize the likely precision of the inference from sample to population while minimizing the cost of selecting and using the sample information. The most straightforward random sampling plan is termed *simple random sampling*. Sometimes, however, a different sampling plan can afford greater precision, or lower cost, or be easier to administer. We discuss simple random sampling and several commonly used alternatives.

3.11.1 Simple Random Sampling

A simple random sample of size n from a population of size N is one selected according to a mechanism guaranteeing that each of the $\binom{N}{n}$ potential samples have the same probability, $1/\binom{N}{n}$, of being the sample actually selected.

If, as is usually the case, the population is already identified with a numbering from 1 to N, or if it is easy to set up such a numbering, then statistical software can be used to select n distinct integers in the range 1 to N so that all potential selections are equally likely to occur.

Such a sample is easily produced in R with the statement sample(N, n). If the population is not numbered but exists as a character vector x [where $n \leq$ length(x)], then sample(x, n) produces the required sample from x.

3.11.2 Stratified Random Sampling

Sometimes the population of interest is meaningfully partitioned into groups, called *strata* in the sampling literature. For example, in a school situation the strata could be individual classrooms. In addition to making inferences about the entire population, it is also desired to learn about each *stratum* (the singular of *strata*). When this is the case, we may wish to select a random sample within each stratum. Then sample estimates are available for each stratum, and these can be combined into estimates for the entire population.

Suppose there are k strata and the number of population items in stratum i is $N_i, i = 1, \ldots, k$, where $\sum_{i=1}^{k} N_i = N$. The analyst then needs to decide how many of the n total sample items should be selected from stratum i. One popular possibility, called *proportional allocation*, stipulates sampling $n_i = \left(\frac{N_i}{N}\right)n$ items from the ith stratum. Since n_i need not be an integer, it is customary to round this calculation to the nearest integer. The mean estimated from the stratified random sample is $\bar{x}_{\text{ST}} = \frac{1}{N}\sum_i N_i\bar{x}_i$, i.e., a weighted average of the stratum sample means using the relative strata sizes as weights.

As an example, suppose it is desired to estimate the average annual malpractice premium paid by physicians licensed to practice in Pennsylvania. Since the risk of malpractice differs across medical specialties, it is likely also to be of interest to determine such estimates for each medical specialty. A physician considering relocation to Pennsylvania from elsewhere will be more interested in the estimated premium for her own medical specialty than the average premium of all Pennsylvania physicians. Accordingly, an investigator first decides the size n of a statewide sample she can afford. Then she obtains a directory of Pennsylvania physicians classified according to specialty and notes the number N_i of Pennsylvania physicians in each specialty $i, i = 1, \ldots, k$, where k is the number of distinct medical specialties. (Such a directory may be available for purchase from the American Medical Association.) Then a sample of approximately $n_i = \left(\frac{N_i}{N}\right)n$ physicians is selected from among the Pennsylvania practitioners of specialty i.

Stratified sampling has the virtue of avoiding an undersampling of any stratum and so guarantees some minimum degree of precision for estimates from each stratum. When the population exhibits minimal variability within strata but considerable variability between units in different strata, estimates based on stratified random sampling are likely to be more precise than ones based on simple random samples of comparable total size. This fact will be demonstrated in Section 3.11.5.

3.11.3 Cluster Random Sampling

This technique is designed to control the cost of sampling in exchange for some decrease in precision of estimation. It is most frequently used when it is necessary to make personal contact with the *sampling units* (entity that is to be sampled), and the sampling units are physically dispersed to the extent that traveling from one unit to another is an appreciable cost.

As with stratified sampling, cluster sampling involves two stages. Assume that the population is partitioned into c clusters. A cluster is typically formed from geographically contiguous units so that sampling units within the same cluster are much closer to one another than two units in different clusters. In stage 1 the analyst selects c_0 of these clusters, where c_0 is considerably less than c. Then in stage 2 the analyst randomly samples n_i items from each selected cluster i, where $\sum_{i=1}^{c_0} n_i = n$. The samples within each cluster can be simple random samples, stratified random samples, etc. As in the case of stratified random sampling, we must decide on a rule for allocating the total sample size n to the clusters.

If T_i is the total for all observations in cluster i, then the mean estimated from the cluster random sample is $\bar{y}_{\mathrm{CRS}} = \left(\sum_i T_i \right) / \left(\sum_i N_i \right)$, where both sums extend from 1 to c_0.

Cluster random sampling saves costs because it involves much less travel from one cluster to another than other sampling methods. But precision is sacrificed because this method prevents a large part of the population from appearing in the sample. In contrast to stratified sampling of strata, cluster sampling of clusters is most efficient when the variation within clusters is large compared to the variation between clusters.

When it is required to personally interview persons sampled from a city's population of eligible voters, a good strategy would be to identify voting districts as clusters and use cluster sampling. If, instead, we wanted to interview city residents as to their product preferences, an analyst might prefer to use zip codes as clusters because geography-based marketing strategies are more likely to be segmented by zip code than by voting district.

3.11.4 Systematic Random Sampling

This method may be considered when simplicity of the sampling design and administration is of prime importance.

Order the population from 1 to N and initially assume that N is an integral multiple of n, say $N = mn$. Then randomly select an integer i, $1 \le i \le m$. Then sample population item i and every m^{th} item thereafter. For example, if $N = 120$, $n = 20$, $m = 6$, we might randomly sample items $4, 10, 16, \ldots, 118$.

Suppose instead that $N = mn + l$, $1 \le l < n$. The analyst may then seek to move toward the N proportional to n situation. Suppose we modify the preceding illustration to $N = 132$. A possibility is to accept a larger $n = 22$. Another option that maintains $n = 20$ is to randomly remove $l = 12$ observations from sampling consideration and then proceed as before with the mn remaining observations.

This method should not be used if the population displays a periodic characteristic with the same period as m. For example, if we wish to randomly sample 20 houses in a subdivision consisting of 120 houses where each block has exactly 6 houses, then the preceding plan would either contain, or avoid, sampling houses on the end of blocks. Such houses tend to be on larger lots than ones in the middle of blocks and the plan would either include them exclusively or miss them entirely.

3.11.5 Standard Errors of Sample Means

In this section we provide standard errors for the means of random samples selected by various methods. Then according to the Central Limit Theorem, an approximate large-sample $100(1-\alpha)\%$ confidence interval for the population mean is of the form

$$\text{sample mean} \pm \text{standard error} \cdot z_{(1-\frac{\alpha}{2})}$$

For a simple random sample, the standard error is

$$s_{\text{SRS}} = \sqrt{\frac{s^2}{n}\left(\frac{N-n}{N-1}\right)}$$

For a stratified random sample with sample variance s_i^2 from stratum i, the standard error is

$$s_{\text{ST}} = \frac{1}{N}\sqrt{\sum_i N_i^2\left(\frac{N_i-n_i}{N_i-1}\right)\frac{s_i^2}{n_i}}$$

If the $\{s_i^2\}$ tend to be smaller than s, then s_{ST} will tend to be smaller than s_{SRS} with the conclusion that stratification was worthwhile.

To present the standard error for the mean of a cluster random sample, define $\bar{N} = N/c$ to be the average cluster size. The standard error is

$$s_{CRS} = \sqrt{\left(\frac{c - c_0}{c_0 c \bar{N}^2}\right) \frac{\sum_i (T_i - \bar{y}_{CRS} N_i)^2}{c_0 - 1}}$$

The summation extends from 1 to c_0, where as before, c_0 is the number of clusters that were sampled.

3.11.6 Sources of Bias in Samples

Sampling error is the discrepancy between the estimate and parameter being estimated. This error decreases as the sample size increases. Nonsampling errors are more serious than sampling errors because they can't be minimized by increasing the sample size. Continuing the example discussed in Section 3.11.2, we discuss two such sources of bias in the context of randomly sampling physicians who practice in Pennsylvania. *Selection bias* occurs when it is impossible to sample some members of the population. *Nonresponse bias* occurs if responses are not obtained from some members of the sample.

In order to randomly sample from the population consisting of all physicians licensed to practice medicine in Pennsylvania, we must obtain a list or computer file of such physicians. Even if we could obtain a list of physicians licensed to practice, there is no way to know which physicians on such a list are in fact practicing medicine (as opposed to performing medical research or administrative tasks). Therefore, use of such a list would introduce selection bias. A better approach might be to obtain a list of the Pennsylvania membership of the American Medical Association (AMA). This list does indicate the nature of the physician's practice, if any, so nonpractitioners on the list can be ignored. However, not all physicians practicing in Pennsylvania are AMA members; such membership is not legally required in order to practice medicine. Thus some selection bias would still be present with this approach. Selection bias would be eliminated if the client can be persuaded to amend the target population to AMA members practicing in Pennsylvania.

Next suppose that this amendment is accepted and that a random sample of n practicing physicians is selected from the list. How should the physicians be contacted? Since physicians are busy individuals; visiting them in person or contacting them by telephone is unlikely to yield a response. Ignoring nonrespondents is likely to result in nonresponse bias because busier physicians are less likely to respond, and busyness may be associated with the survey questions.

Mail contact of the sampled physicians is preferred for several reasons. Since a written questionnaire can be answered at the physician's convenience, the physician is more likely to respond. Second, the questionnaire can be placed under a cover letter that encourages participation, written by a person respected by the respondents. Third, it is possible to keep track of who does not initially respond so that such individuals can be contacted again. This is accomplished by asking respondents to mail in a signed postcard indicating that they have participated, and to return the anonymous questionnaire in an envelope mailed separately.

Even this elaborate mail questionnaire approach does not eliminate the possibility of nonresponse bias. The extent of any remaining bias can be judged by comparing characteristics of the sampled physicians with those of the physician population reported in the AMA membership directory.

3.12 Exercises

3.1. Refer to the discrete bivariate distribution considered in Table 3.3.

a. Let $Z = X + 1$. Find the distribution of Z.

b. Find $E(2X + 1)$ and $2E(X) + 1$. Then find $E(X^2)$ and $[E(X)]^2$.

c. Find $P(X < Y)$.

d. Let X_1 and X_2 be independent and identically distributed as X. Make a table of the joint distribution of X_1 and X_2, and use this to find $P(X_1 < X_2 + 1)$.

3.2. How large a random sample is required for there to be a 92% probability of sampling at least one defective from a lot of 100,000 items which contains 100 defectives? (Hints: What is the random variable here? Consider the event that is the complement of "at least one defective".)

3.3. Suppose X is binomial(50, .10), and Y is binomial(20, .25). Draw the distribution functions of X and Y. Which one has a bigger mean? Which one has a bigger standard deviation?

3.4. If X, Y are each standard normal random variables, and they are independent of one another, what is the distribution of $Z = 3X + 2Y$?

3.5. Suppose that Y is a 2×1 random vector such that

$$W = \begin{pmatrix} 80 \\ 40 \end{pmatrix} + \begin{pmatrix} 10 & 7 \\ 7 & 5 \end{pmatrix} Y$$

has a bivariate normal distribution with mean $\begin{pmatrix} 60 \\ 70 \end{pmatrix}$ and covariance matrix $\begin{pmatrix} 100 & 40 \\ 40 & 50 \end{pmatrix}$. Find the probability distribution of Y, including its mean vector and covariance matrix.

3.6. In class #1, 32 out of 40 students earned fewer than 70 points on the final exam. In class #2, 40 out of 50 students earned fewer than 75 points on the same exam. Restate the given class information in terms of percentiles. Is it possible to tell which class had a higher average score?

3.7. Somebody tells you that a 95% confidence interval for the mean number of customers per day is (74.2, 78.5), and that this indicates that 95% is the probability that the mean is between 74.2 and 78.5. Criticize this statement and replace it with one correct sentence.

3.8. Acme, Inc. thinks it has a new way of manufacturing a key product. It is trying to choose between A = "new way is better than old way" or B = "old way is better than new way". Acme plans to reach its tentative conclusion by sampling some of the product produced the new way and conducting a statistical test. The new way is much more expensive than the old way. Which statement, A or B, should be the null hypothesis? Justify your answer.

3.9. The probability that a project succeeds in New York is .4, the probability that it succeeds in Chicago is .5, and the probability that it succeeds in at least one of these cities is .6. Find the probability that this project succeeds in Chicago given that it succeeds in New York.

3.10. You are considering two projects, A and B. With A you estimate a payoff of $60,000 with probability .6 and $30,000 with probability .4. With B you estimate a payoff of $80,000 with probability .5 or $30,000 with probability .5. Answer the following questions after performing appropriate calculations.

a. Which project is better in terms of expected payoff?

b. Which project is better in terms of variability of payoff?

3.11. If X has a mean of 15 and a standard deviation of 4, and if $Y = 5 - 3X$, what are the mean and standard deviation of Y?

3.12. State the two ways in which a data analyst can modify a statistical test in order to decrease its Type II error probability.

3.13. An analyst makes three independent inferences. For each of these inferences, the probability is .05 that it is *in*correct. Find the probability that *all three* inferences are *correct*.

3.14. Let A = "a McDonald's franchise in Kansas is profitable" and let B = "the Philadelphia Eagles will have a winning season next year". If $P(A)$ = .8 and $P(B)$ = .6, find the probability that *either A or B* occurs.

3.15. Use statistical software commands to do this problem. A new medicine has probability .70 of curing gout. If a random sample of 10 people with gout are to be given this medicine, what is the probability that among the 10 people in the sample, between 5 and 8 people will be cured?

3.16. Use statistical software commands to do this problem. The daily output of a production line is normally distributed with a mean of 163 units and a standard deviation of 4 units.

a. Find the probability that a particular day's output will be 160 units or less.

b. The production manager wants to tell her supervisor, "80% of the time our production is at least x units". What number should she use for x?

3.17. Find the expected value and standard deviation of a random variable U if its probability distribution is as follows:

u	$P(U = u)$
1	.6
2	.3
3	.1

3.18. A random variable W has probability density function $f(w) = 2 - 2w$, $0 < w < 1$, and $f(w) = 0$ for all other values of w.

a. Verify that $f(w)$ is indeed a probability density function.

b. Find the corresponding cumulative distribution function, $\mathcal{F}(w)$.

c. Find the expectation of W.

d. Find the standard deviation of W.

e. Find the median of this distribution, i.e., the number w_m such that $P(W < w_m)$ = .5.

3.19. Use a statistical software command to approximate the value of $z_{.08}$.

3.20. State the two things that a data analyst can do in order to make a confidence interval *narrower*.

3.21. A data analyst tentatively decides on values for α and n for a statistical test. Before performing the test she investigates its Type II error control and finds this to be unsatisfactory. What two options does she have to improve Type II error control?

3.22. In the discussion of *sufficiency* of a point estimator in Section 3.6.3, we indicated that $2\bar{x}$ is not a good estimator of a from a sample of n items from a continuous uniform distribution on $[0, a]$. Can you suggest a better estimator of a and explain why it is better than $2\bar{x}$?

3.23. The dataset data(salary), from Forbes Magazine (1993), contains the ages and salaries of the chief executives of the 60 most highly ranked firms among *Forbes Magazine*'s "Best small firms in 1993." Consider the variable age.

a. Produce a boxplot and a stem-and-leaf plot for age.

b. Construct a 95% confidence interval for the mean age. What assumptions were made in your construction?

c. Test $H_0: \mu \leq 50$ against $H_1: \mu > 50$, reporting and interpreting the *p*-value for this test.

d. Approximate the power of this test for the alternative $\mu_1 = 53$ by using the normal distribution as an approximation for the test statistic in part c, assuming $\alpha = .05$.

3.24. The dataset data(cereals) contains various nutritional measurements for 77 breakfast cereals. We are concerned here with the variable carbo (carbohydrates) measured in grams per serving. Be aware that the cereal Quaker Oatmeal shows a negative value for carbohydrates, probably indicating a missing value for that observation. Be sure that you inform your data analysis package of this anomaly and that the package does something sensible with that information. Elimination of the observation is one possible response to missingness.

a. Produce boxplots and stem-and-leaf plot for carbo. Do these plots suggest that this variable comes from a normal population?

b. Construct at 99% confidence interval for the mean carbohydrate content.

c. Test $H_0: \mu \geq 16$ against $H_1: \mu < 16$, reporting and interpreting the *p*-value for this test.

d. Approximate the probability of committing a Type II error for the alternative $\mu_1 = 15$. Use the normal distribution to approximate the test statistic in part c, assuming $\alpha = .05$.

3.25. The sampling bias in the December 1969 U.S. Draft Lottery, with data in file data(draft70mn), is described in Exercise 4.1. Suppose you had been the administrator of that lottery. Explain how you would have performed the sampling without incurring such bias.

3.26. Royalties paid to authors of novels have sometimes been based on the number of words contained in the novel. Recommend to an old-fashioned author how to estimate the number of words in a handwritten manuscript she is planning to give to her publisher.

3.27. Samples are taken from two strata. Suppose the variance of the two samples combined is $s^2 = 7.6$ and the following within-stratum information is known:

Stratum	N_i	n_i	s_i^2
1	100	30	1.2
2	120	40	1.4

Observe that there is far less variability within the two strata than between the two strata. Calculate s_{SRS} and s_{ST} to verify that for estimating the common population mean in this situation, $\bar{x}_{SRS}$ is much preferred to $\bar{x}_{ST}$.

3.28. The organization of a candidate for a city political office wishes to poll the electorate. For this purpose, discuss the relative advantages and disadvantages of personal interview polling vs telephone polling.

3.29. Explain how it is possible for a census to yield less accurate results than a random sample from the same population.

3.30. A student claims that a random sample of n items from a population of N items is one selected so that each item in the population has the same probability $\frac{n}{N}$ of appearing in the sample. Demonstrate that this definition is inadequate.

3.31. A four-drawer file cabinet contains several thousand sheets of paper, each containing a statement of the dollar amount due to be paid to your company. The sheets are arranged in the order that the debt was incurred. You are asked to spend not more than one hour to estimate the average dollar amount on all sheets in the file cabinet. Propose a plan for accomplishing this.

Chapter 4

Graphs

Graphs are used to inspect and display patterns in data. Appropriately drawn graphs are, in our opinion, the best way to gain an understanding of what data have to say. In this chapter we present several of the types of graphs and plots we will be using throughout. We discuss the visual impact of the graphs and relate them to the tabular presentation of the same material.

Statistical techniques have underlying assumptions. An important use of graphs is to aid in the checking of a list of assumptions a technique requires in order for an analysis using the technique to be correct. For example, regression analysis, discussed in Chapters 8 to 11, requires that model residuals are randomly distributed. Residual plots, discussed in these chapters, must show random scatter rather than a systematic pattern.

We discuss the construction of graphs and pay attention to each of the components of graphs that can aid (or hinder) the interpretation of the data. We show good (and some bad) examples of plots and discuss why we make those value judgments. Appendix 4.A gives an overview of R Graphics with an emphasis on the design and use of the **lattice** package.

The appendix to this chapter summarizes many graphs that are introduced in this book.

We see graphs as the heart of most statistical analyses; the corresponding tabular results are formal confirmations of our visual impressions. The graphs are not automatically produced by most software; instead it is up to the analyst to request them.

© Springer Science+Business Media New York 2015
R.M. Heiberger, B. Holland, *Statistical Analysis and Data Display*,
Springer Texts in Statistics, DOI 10.1007/978-1-4939-2122-5_4

4.1 What Is a Graph?

A graph is a geometrical representation of the information in a table of numbers. Our prototype graph is the traditional scatterplot—a two-dimensional plot of two variables (x on the abscissa or horizontal axis and y on the ordinate or vertical axis). For each observation (x, y) in the data we locate a point on the graphing surface with coordinates (x, y). For a dataset with n observations we mark n points on the graphing surface. Figure 4.1 is an example of such a plot.

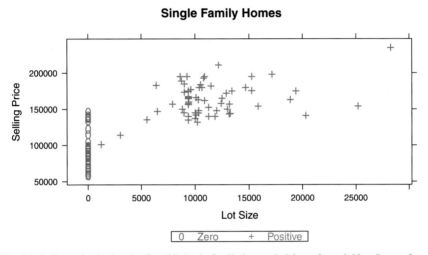

Fig. 4.1 Selling price by lot size for 105 single-family homes in Mount Laurel, New Jersey, from March 1992 through September 1994. What is the meaning of a lot size of zero?

Note the graphic features of the plot in Figure 4.1 (the interpretation of this graph will be discussed in Section 4.3):

x-axis tick marks, tick labels, label: Information on the variable that defines the horizontal direction of the graph. The range of the scale is large enough to show all points.

y-axis tick marks, tick labels, label: Information on the variable that defines the vertical direction of the graph. The range of the scale is large enough to show all points.

main title: Information on the subject matter of the graph.

plotting character: A plotting character, in this case a blue '0' or a red '+', is placed at each x–y coordinate. The color and character represent some aspect of the data. In this example they are redundant and represent whether the reported Lot Size has a zero or nonzero value.

color: The characters are color coded to represent a factor within the data. Please see Section 4.A.4 for more information on color.

legend: There are several different plotting characters and colors used in this graph. The reader needs identification of each. For this plot, they indicate levels of a factor. In other examples, they might be used to indicate different response variables.

caption: A short paragraph describing the structure of the graph and the message that the graph is designed to illustrate.

The goal of statistical graphics is to make evident the characteristics of the data such as location, variability, scale, shape, correlation, interaction, and clustering. Once we have a visual understanding of the data, we usually attempt to model it with formal algebraic procedures. We will normally translate our algebraic understanding back to a graphical presentation and to a verbal discussion of our findings.

4.2 Example—Ecological Correlation

Examination of plots of the data at early stages of the analysis, before requesting and examining tabular output, is an essential part of data analysis. This point is demonstrated in Figure 4.2, which illustrates what is known as the Ecological Fallacy. If without examining a plot of these (simulated) data we perform a simple regression of y on x, we find that y and x are directly related. The plot strongly suggests that what we have is the amalgamation of three disparate groups. Within each of the groups it is clear that y and x are inversely related, the opposite conclusion from the amalgamated result. In practice it is likely that the existence of the groups is meaningful information that must be accounted for by the analyst. In this case the individual within-group results are what should be reported.

Robinson (1950) introduced the idea by showing that the correlation between percentage illiterate and percentage black racial group for the United States as a whole, based on the 1930 U.S. Census, is different from this correlation within various subgroups of the U.S. population. The terms Ecological Fallacy and Ecological Correlation were coined by Selvin (1965). Human ecology is a branch of sociology dealing with the relationship between human groups and their environments. The fallacy is that we cannot necessarily use a finding from an entire population to reach conclusions about various subsets of the population.

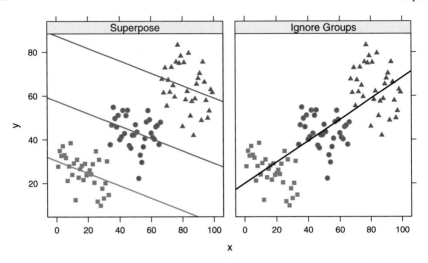

Fig. 4.2 Ecological Correlation. The overall slope ignoring groups is strongly positive. The slope within each group is strongly negative. These are simulated data.

4.3 Scatterplots

Figure 4.1 shows the selling price by lot size for 105 single-family homes in Mount Laurel, New Jersey, from March 1992 through September 1994. The data, from Asabere and Huffman (1996), are accessed with `data(njgolf)`.

There is much information in Figure 4.1. We start by listing the most obvious items, and then we will look at the less obvious and more puzzling items. The range of lot sizes is 0–30,000 square feet, with most of the lots in the 8,000–15,000-square-foot range. But what is that large cluster of lot sizes at 0 square feet? The range of sale prices is $50,000–$250,000, with most of the 0-size lots selling for under $130,000 and most of the nonzero lots selling above $130,000. Within the 5,000–25,000-square-foot range price seems independent of size of lot, that is, for any lot size in that range the best estimate of sales price is the same, about $165,000.

The scatterplot is an ordinary 2-dimensional plot with one variable `lotsize` on the *x*-axis (horizontal axis or abscissa) and the other variable `sprice` on the *y*-axis (vertical axis or ordinate). The plotting routine automatically determines the appropriate scale and tick locations for both axes and prints the variable names as the default labels for the axes.

We raised many questions in our perusal of Figure 4.1. Answering them requires us to look carefully at the definitions of the variables we displayed in the figure. We find that the variable labeled `lotsize` is actually a conflation of two distinct concepts. If the property is a condominium (a form of ownership of an apartment that combines single ownership of the residence unit with joint ownership of the building and associated grounds), the variable `lotsize` was arbitrarily coded to 0. If the property is a single-family house, then the variable `lotsize` contains the actual lot size in square feet. This explains the numerous observations having `lotsize=0` in Figure 4.1.

We must also look at additional variables. We will start with three measures of the size of the dwelling unit, rather than of the lot on which it is built. In Figure 4.3 we look at selling price against the number of bedrooms, the dining room area, and the kitchen area. All three plots show a rise in selling price as the *x*-variable increases. We can also see a hint in Figure 4.3 that selling price increases with *x* for both the lower-priced properties (the condominiums) and the higher-priced ones (the single-family houses).

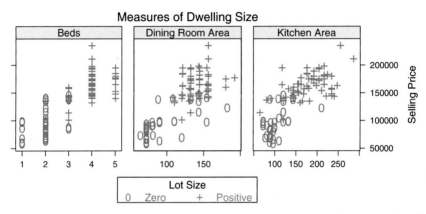

Fig. 4.3 Selling price by number of bedrooms, by dining room area, and by kitchen area for 105 single-family homes in Mount Laurel, New Jersey, from March 1992 through September 1994.

We investigate that possibility in Figure 4.4 where we show all three plots conditioned on whether the property is a condominium or house. Now we see very clear uphill trends of price on the measures of size within each of the panels of the figure.

4.4 Scatterplot Matrix

We looked at five variables in Figure 4.4 and nominally two, but actually three, variables in Figure 4.1. In both figures we used selling price as the *y*-variable and the others as either *x*-variables or as conditioning variables. In Figure 4.5 we look at all six variables together. This display shows all the individual panels that we looked at in the previous graphs in the `sprice` row and also shows the relationships among the other variables.

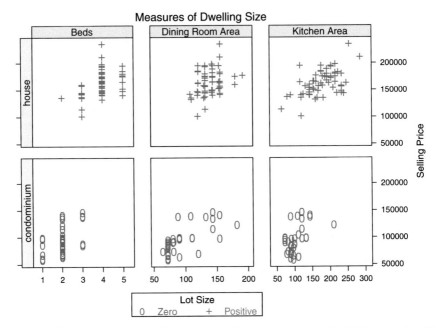

Fig. 4.4 Selling price by number of bedrooms, by dining room area, and by kitchen area for 105 single-family homes, conditioned on whether the property is a condominium or house, in Mount Laurel, New Jersey, from March 1992 through September 1994.

The display type is a *scatterplot matrix* or *splom* (Scatter PLOt Matrix), a matrix of scatterplots with each of the six variables taking the role of x-variable and y-variable against all the others. Thus there are $_6P_2 = 30$ distinct plots in Figure 4.5. Each of these 30 plots is a plot of a pair of variables comparable to Figure 4.1. Since each of the six variables appears in both the x- and y-position, there are only $_6P_2/2 = 30/2 = \binom{6}{2} = 15$ distinct pairs of variables in the plots. We see that the (i, j) panel of the splom (counting from the lower-left corner) is the reflection of the (j, i) panel.

A defining property of the scatterplot matrix is that all panels in the same row have identical y-scaling and all panels in the same column have identical x-scaling. It is therefore easy to trace an interesting point in one panel across to the other panels. For example, the single point visible in the condominium position of the `lotsize ~ cond.house` panel is recognized as an overplotting of many condominium points when we trace it in the other panels to the left and see that the dining area of condominiums runs the full range of dining areas for the entire dataset.

Unfortunately, Figure 4.5 has also lost (although we partially retain it by different colors) the distinction between the condominiums and houses that we worked so hard to find. We recover that distinction in Figure 4.6 where we now show the five numeric variables separately for condominiums and houses. We can look across the subpanels in each main panel of Figure 4.6 and see relationships among multiple variables. On the condominium panel of Figure 4.6 we see that the condominium with largest kitchen and dining room is one of the higher-priced properties (but not the highest) and it has only two bedrooms. On the house panel of Figure 4.6 we see that the highest-priced house has the largest lot size, but not the largest dining area and only four bedrooms.

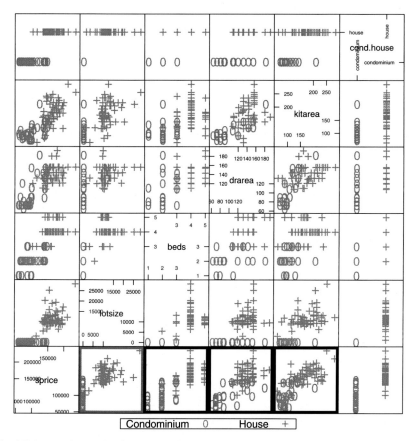

Fig. 4.5 Scatterplot matrix of six variables for the 105 single-family homes in Mount Laurel, New Jersey, from March 1992 through September 1994. The `sprice ~ lotsize` panel outlined in gray (bottom row, second column) is the same as Figure 4.1. The three panels in the bottom row outlined in black (third, fourth, and fifth columns) are the same as Figure 4.3. The same three panels, separated by color, are in Figure 4.4.

Additional discussion of scatterplot matrices appears in Section 4.7.

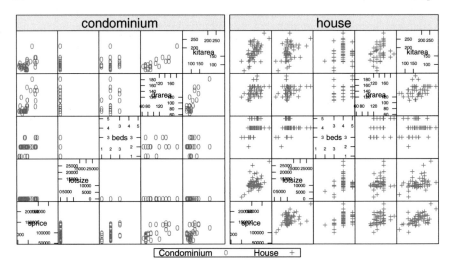

Fig. 4.6 Scatterplot matrix of five variables for the 105 single-family homes, conditioned on whether the property is a condominium or house, in Mount Laurel, New Jersey, from March 1992 through September 1994.

4.5 Array of Scatterplots

Let us step away from data analysis for a moment and look at the structure of the graphs. Figures 4.2 and 4.3 contain graphs with multiple panels. The panels are clearly labeled with a *strip label* that shows the level of the factor on which the panels are conditioned. In Figure 4.2 the factor levels in the strip labels describe the model by which the slopes were calculated. The *x*- and *y*-axes are identical in both panels.

In Figure 4.3 the factor levels in the strip labels name the room of the house for which the measurements are shown. Here, the three panels represent different variables (number of bedrooms) or value ranges (square feet), hence each panel has its own *x*-scale. All three panels have the same response variable Selling Price and are therefore shown on the same *y*-scale.

Figure 4.4 is more elaborate, with conditioning on two factors. The horizontal factor, the room of the house, is the same as the factor in Figure 4.3 and therefore the top strips and the *x*-scales are the same as in Figure 4.3. The vertical factor is the form of ownership (condominium or house), which in this case makes the same distinction as the Lot Size factor. We distinguish the panels for the ownership/Lot Size factor with *left strip labels* for each row. The response variable Selling Price is the same in all six panels, and all are shown on the same *y*-scale. The panels are defined by the 3 × 2 crossing of the factor levels. Within each panel we show the data points for only the specified factor levels.

A set of scatterplots can be conditioned on more than two factors. Figure 4.7 has three factors. Figure 4.7 is a preview of Figure 17.10 and its content will be discussed in Chapter 17. In this figure we use two rows of top strip labels, one for the pairs of columns representing stage, and the other for levels of grade nested within stage. We have a single column of left strip labels for levels of X.ray.

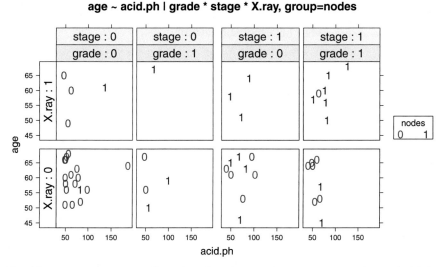

Fig. 4.7 A scatterplot array conditioned on three factors. There are two rows of top strip labels and one column of left strip labels. The upper strip label distinguishes pairs of columns for levels of the stage factor. We also have additional horizontal space between the pairs of columns. The lower strip label distinguishes levels of grade within each level of stage. The left strip label distinguishes levels of X.ray. The plotting symbol and color represent a fourth factor. Within each panel, the points show a plot of age ~ acid.ph.

4.6 Example—Life Expectancy

4.6.1 Study Objectives

For each of the 40 largest countries in the world (according to 1990 population figures), the dataset data(tv) gives the country's life expectancy at birth partitioned by gender, number of people per television set, and number of people per physician Rossman (1994).

4.6.2 Data Description

`life.exp:`   Life expectancy at birth

`ppl.per.tv:`   Number of people per television set

`ppl.per.phys:`   Number of people per physician

`fem.life.exp:`   Female life expectancy at birth

`male.life.exp:`   Male life expectancy at birth

4.6.3 Initial Graphs

We initially focus on the male and female life expectancies in Table 4.1 and Figure 4.8.

Figure 4.8 shows each data row of Table 4.1 as a distinct point. Since both x and y are the same variable in the same units for two different subsets of the population, it is important to use a common range and ticks on both axes and maintain an aspect ratio of 1. In a good graphical system we have control of the plotting symbols. We plotted the points with a solid dot • and labeled one point (Japan) with text to show its coordinates $(x, y) = (82, 76)$.

The first impression we get from reading Figure 4.8 is that most of the points are below the 45° line. This is such an important part of the interpretation of this graph that we drew the 45° line. Once the line is there for reference we immediately note that one country's point is above the 45° line. Which one? The easiest way to find out is to plot the abbreviated country names instead of dots (Figure 4.9a). Bangladesh is the country that has a longer life expectancy for males than females. On an interactive graphics system we merely click on the point and the system will label it [see file (`code/grap.identify.s`)]. We have simulated the interactive appearance in Figure 4.9b.

We see from the figures that life expectancy for males and females is related; as one goes up the other tends to go up as well. We have done several other fine tunings on Figure 4.8. Life expectancy is measured on the same numerical scale for both male and female; therefore, we forced both scales to have the same range and we forced the graph to be square. By default, most plotting systems independently determine the x- and y-scales and use the maximum available area for the graph. Figure 4.9c releases the constraint on the ranges and we see that the male and female ranges are different (the female range is offset from the male range by 5 years). Since the graph goes from the lower-left corner to the upper-right corner, it falsely gives the visual impression that the two ranges are the same. When we plot the 45° line in Figure 4.9d we get much of the correct impression back. In Figure 4.9e, where we no longer constrain the graph to be square, we lose the visual effect of forcing the ranges to be the same on both axes. In Figure 4.9e we have plotted the least-squares line through the points in addition to the 45° line. Least squares will be discussed in detail in Chapter 8. For now we note that this line attempts to get close to most of the points. It is used as an indicator of the linear relationship between the two variables male and female life expectancy.

Table 4.1 Life expectancy. The country abbreviations used here are from the R function call `abbreviate(row.names(tv))`.

Abbrev	Country	Female	Male		Abbrev	Country	Female	Male
Argn	Argentina	74	67		M(B)	Myanmar (Burma)	56	53
Bngl	Bangladesh	53	54		Pkst	Pakistan	57	56
Brzl	Brazil	68	62		Peru	Peru	67	62
Cand	Canada	80	73		Phlp	Philippines	67	62
Chin	China	72	68		Plnd	Poland	77	69
Clmb	Colombia	74	68		Romn	Romania	75	69
Egyp	Egypt	61	60		Russ	Russia	74	64
Ethp	Ethiopia	53	50		StAf	South Africa	67	61
Frnc	France	82	74		Span	Spain	82	75
Grmn	Germany	79	73		Sudn	Sudan	54	52
Indi	India	58	57		Tawn	Taiwan	78	72
Indn	Indonesia	63	59		Tnzn	Tanzania	55	50
Iran	Iran	65	64		Thln	Thailand	71	66
Itly	Italy	82	75		Trky	Turkey	72	68
Japn	Japan	82	76		Ukrn	Ukraine	75	66
Keny	Kenya	63	59		UnKn	United Kingdom	79	73
K,Nr	Korea, North	73	67		UnSt	United States	79	72
K,St	Korea, South	73	67		Vnzl	Venezuela	78	71
Mexc	Mexico	76	68		Vtnm	Vietnam	67	63
Mrcc	Morocco	66	63		Zair	Zaire	56	52

4.7 Scatterplot Matrices—Continued

There are five variables in the `tv` dataset. Figure 4.10 plots them all in a scatterplot matrix.

Continuing with the discussion begun in Section 4.4, the scatterplot matrix is a coordinated set of scatterplots, one for each pair of variables in the dataset. We refer to the individual scatterplots comprising the matrix as *panels*. The panels are labeled by their $Y \sim X$, that is *RowName by ColumnName*, variable names. Thus, in Figure 4.10, the panel in the upper-left-hand corner (also called the NW or Northwest corner) is called the `ppl.per.phys ~ fem.life.exp` panel. Variable names are unambiguous and are constant across multiple views of the data: The `male.life.exp ~ fem.life.exp` panel refers to the same data values all of Figures 4.8, 4.9, and 4.10. We would NOT say "row 1 by column 4" because the sequencing of variables and the direction of ordering the rows and columns (is row 1 at the top or bottom?) are unclear.

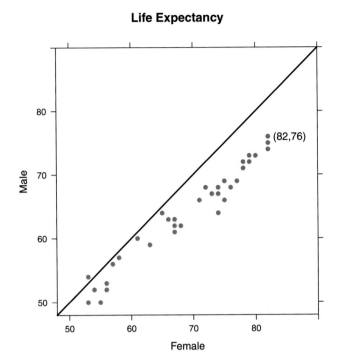

Fig. 4.8 Life Expectancy. In most countries, female life expectancy is longer than male life expectancy.

There are several possible orientations of the panels; we display the best in Figure 4.10 and will discuss other orientations in Figures 4.11 and 4.12. There are five variables; hence the matrix consists of a 5×5 array of scatterplots. Indexing for the set of plots is sorted in the same way as the axes in each individual panel. Indexing begins at the lower left and proceeds from left to right and from bottom to top. The main diagonal runs from southwest to northeast (SW–NE). Each panel containing one scatterplot is square. Each pair of variables appears twice, once below the main diagonal and again as a mirror image above the main diagonal. There is a single axis of symmetry for the entire *splom*.

The variables in Figure 4.10 are all continuous measurements. When using a *splom* to display data with categorical variables, we recommend avoiding inclusion of categorical variables among the variables comprising the *splom* itself, particularly for categorical variables having few categories, as they will usually appear as a noninformative regular lattice (see, for example, the `customf×cornerf` panel of Figure 9.3). It is usually more informative to produce two or more adjacent *splom*s, by conditioning on the categorical variables, or to use different plotting symbols for the different levels of one of the factors. We use both strategies in Figure 9.4, conditioning on the levels of `corner` and using different plotting symbols for the

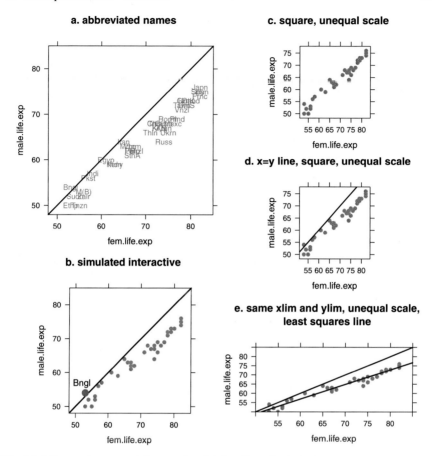

Fig. 4.9 Life expectancy—variations on the plot. See discussion in text.

levels of custom. Another example is Figure 11.1, which contains two adjacent *splom*s conditioned on the two levels of the categorical variable lime.

We have presented what we consider to be the best orientation of the splom in Figure 4.10. Two other orientations are commonly used. When scatterplot matrices were first invented, the importance of a single axis of symmetry was not yet realized. Older scatterplot matrix programs (and some current ones) default, or are limited, to the more difficult main diagonal from northwest to southeast (NW–SE). The R splom function defaults to the optimal SW–NE main diagonal. It can be told to use the nonoptimal alternate diagonal with the argument as.matrix=TRUE. The older R function pairs defaults to the nonoptimal NW–SE diagonal but provides the option to change it with the row1attop=FALSE argument. pairs also defaults to rectangular panels (the goal is maximal use of the plotting surface) but fortunately provides an option to force square panels (with a previous use of par(pty="s").

Televisions, Physicians, and Life Expectancy

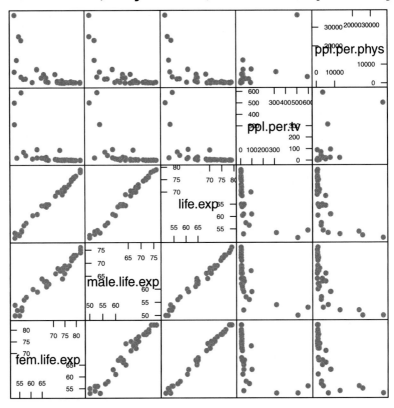

Fig. 4.10 Televisions, physicians, and life expectancy. Variable ppl.per.tv has two missing values. We notice this immediately in panel ppl.per.tv ˜ ppl.per.phys, where the two points at ppl.per.phys = 25000 in the bottom three rows of the ppl.per.phys column do not appear. Similarly these points are missing in Figures 4.11, 4.14, and 4.15.

We recommend also using a square plotting region.). We show the pairs plot with both suboptimal choices in Figure 4.11.

The major difficulty with Figure 4.11 is that the multiple axes of symmetry are hard to find. The axes of symmetry are illustrated in Figure 4.12. The confusion in Figure 4.12a occurs because pairs of plots with the same variable names appear to the lower left and upper right of the NW–SE main axis of the matrix of plots. Within each pair, the upper plot needs to be reflected about its own SW–NE axis to match the lower plot. By comparison, Figure 4.12b has a single axis of symmetry for the entire plot. All pairs of plots and reflections within each pair occur around a single SW–NE axis of symmetry. Note also that the individual panels of the display are square to help ease the eye's task of seeing the symmetry.

Earlier versions of S (Becker et al., 1988) defaulted to printing just one triangle of the two mirror image triangles in `pairs` and had an option `full=TRUE` to print the full matrix. From the manual, "By default, only the lower triangle is produced, saving space and plot time, but making interpretation harder." That option made sense with typewriter terminals at 10 characters per second. It no longer makes sense with desktop workstations, windowing terminals, and laser printers. The single triangle of a scatterplot matrix can be created in R by suppressing one of the triangles, for example with a call similar to

```
splom(iris, lower.panel=function(...){}).
```

Older programs sometimes display a very confusing subset of the lower triangle in which the rows and columns of the display show different sets of variables. The intent is to save space by suppressing a presumably non-informative main diagonal. The effect on the reader is to add confusion by breaking symmetry. A symbolic version of this form of the plot is in Figure 4.13.

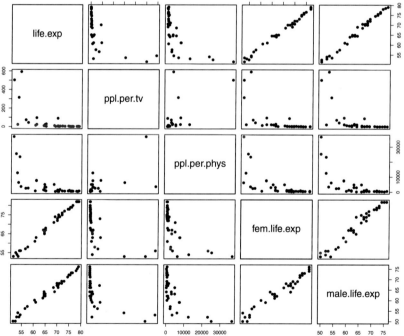

Fig. 4.11 Nonoptimal alternate orientation with rectangular panels for splom. The downhill diagonal is harder to read (see Figure 4.12). The rectangular panels make it hard to compare each panel with its transpose.

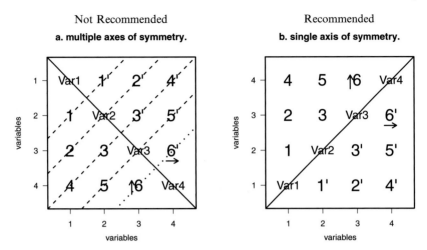

Fig. 4.12 Axes of symmetry for splom. Figure 4.12a has six axes of symmetry. We focus on panel 6, which appears in positions reflected about the main NW–SE axis. The individual points within panels 6 and 6' are reflected about the dashed SW–NE line, as indicated by the position of the arrow. The other four axes, which reflect respectively panel 5, panels 3 and 4, panel 2, and panel 1 are indicated with dotted lines. Figure 4.12b has only one axis of symmetry. The arrow for panel 6 is reflected by the same SW–NE axis that reflects panels 6 and 6'.

Not Recommended

2	**21**		
variables 3	**31**	**32**	
4	**41**	**42**	**43**
	1	2	3
		variables	

Fig. 4.13 Symbolic form of very confusing subset of panels for the scatterplot matrix. This form has different variables along the rows and columns and has very little symmetry that might aid the reader. Note, for example, that panels 31 and 42 are positioned such that the eye wants to treat them as symmetric. This form is mostly obsolete and is strongly not recommended.

4.8 Data Transformations

Since the three life expectancy variables are similar, let us look at the simplified splom in Figure 4.14. The bottom row of the splom, with life.exp as the *y*-coordinate, shows an L-shaped pattern against both ppl.per.tv and ppl.per.phys as the *x*-variables. We have learned (or will learn in this chapter and again in Chapter 8) that straight lines are often helpful in understanding a plot. There is no sensible way to draw a straight line here. The plot of the two potential *x*-variables against each other is bunched up in the lower-left corner. The bunching

Televisions, Physicians, and Life Expectancy

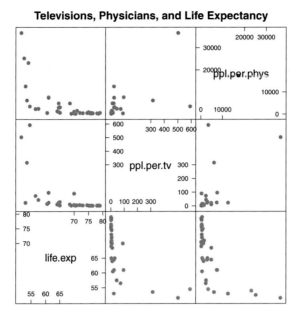

Fig. 4.14 Televisions, physicians, and life expectancy.

suggests that a log transformation of the ppl.* variables will straighten out the plot. We see in Figure 4.15 that it has done so.

We also see that the log transformation has *stabilized the variance*. By this we mean that the ppl.per.phys ~ life.exp panel of Figure 4.14 has a range that fills the vertical dimension of the panel for values of life.exp near 50 and that is almost constant for values of life.exp larger than 65. After the log transformation of ppl.per.phys shown in Figure 4.15, for any given value of life.exp we observe that the vertical range of the response is about $\frac{1}{3}$ of the vertical dimension of the panel.

There are several issues associated with data transformations. In the life expectancy example the natural logarithm ln was helpful in straightening out the plots. In other examples other transformations may be helpful. We will take a first look at a family of power transformations. We recommend Emerson and Stoto (1983) for a more complete discussion. We identify some of the issues here and then focus on the use of graphics to help determine which transformation in the family of power transformation would be most helpful in any given situation.

- Stabilize variance. This chapter and also Chapters 6 and 14.

- Remove curvature. This chapter.

- Remove asymmetry. This chapter.

- Respond to systematic residuals. Chapters 8 and 11.

log(Televisions, Physicians), and Life Expectancy

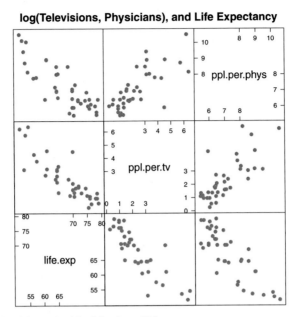

Fig. 4.15 log(televisions), log(physicians), and life expectancy.

The family of power transformations $T_p(x)$, often called the *Box–Cox transformations* Box and Cox (1964), are given by

$$T_p(x) = \begin{cases} x^p & (p > 0) \\ \ln(x) & (p = 0) \\ -x^p & (p < 0) \end{cases} \tag{4.1}$$

Notice that the family includes both positive and negative powers, with the logarithm taking the place of the 0 power. The negative powers have a negative sign to maintain the same direction of the monotonicity; if $x_1 < x_2$, then $T_p(x_1) < T_p(x_2)$ for all p. When the data are nonnegative but contain zero values, logarithms and negative powers are not defined. In this case we often add a "start" value, frequently $\frac{1}{2}$, to the data values before taking the log or power transformation.

When we wish to study the mathematical properties of these transformations, we use the related family of scaled power transformations $T_p^*(x)$ given by

$$T_p^*(x) = \begin{cases} \frac{x^p-1}{p} & (p \neq 0) \\ \ln(x) & (p = 0) \end{cases} \tag{4.2}$$

The scaling in $T_p^*(x)$ gives the same value $T_p^*(1) = 0$ and derivative $\frac{d}{dx}T_p^*(1) = 1$ for all p.

There is also a third family of power transformations $W_p(x)$ given by

$$W_p(x) = \begin{cases} x^p & (p \neq 0) & \text{Do not use this form,} \\ \ln(x) & (p = 0) & \text{the reciprocal is not negated.} \end{cases} \qquad (4.3)$$

that is occasionally (and incorrectly) used. This family does not negate the reciprocals; hence, as we see in Figure 4.16b, it is very difficult to read.

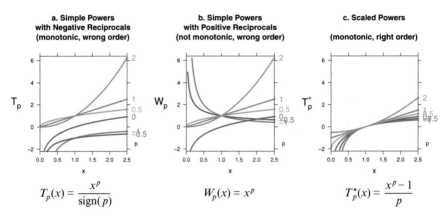

$$T_p(x) = \frac{x^p}{\text{sign}(p)} \qquad\qquad W_p(x) = x^p \qquad\qquad T_p^*(x) = \frac{x^p - 1}{p}$$

Fig. 4.16 Power Transformations. The smooth transitions between the scaled curves in Figure 4.16c is the justification for using the family of power transformations $T_p^*(x)$ in Equation (4.2). This is the only one of three panels in which both (a) the monotonicity of the individual powers is visible and (b) the simple relation between the curves and the sequence of powers in the ladder of powers $p = -1, -\frac{1}{2}, 0, \frac{1}{2}, 1, 2$ is retained over the entire x domain. Figure 4.16a keeps the monotonicity but loses the sequencing. Figure 4.16b, which doesn't negate the reciprocals, is very hard to read because two of the curves are monotone decreasing and four are monotone increasing. Figure 4.16 is based on Figures 4-2 and 4-3 of Emerson and Stoto (1983).

Figure 4.16 shows the plots of all three families: the two parameterizations of the Box–Cox power transformations $T_p(x)$ and $T_p^*(x)$, and the third, poorly parameterized power family $W_p(x)$. There are several things to note in these graphs.

1. Figure 4.16a, the plots of $T_p(x)$, correctly negates the reciprocals, thereby maintaining a positive slope for all curves and permitting the perception that these are all monotone transformations.

2. In Figure 4.16b, the plots of $W_p(x)$, we see that the plots of the two reciprocal transformations have negative slope and that all the others have positive slope. This reversal interferes with the perception of the monotonicity of the transformations.

3. Figure 4.16c, the plots of $T_p^*(x)$, is used to study the mathematical and geometric properties of the family of transformations. The individual formulas in Equations (4.1) and (4.2) are linear functions of each other; hence the properties and appearance of the individual lines in the graphs based on them are equiva-

lent. Equation (4.1) is simpler for hand arithmetic. Equation (4.2) makes evident that the powers (including 0 and negative) are simply and systematically related. Taking the negative of the reciprocal explains how the negative powers fits in. Showing how the 0 power or logarithm fits in is trickier; we use l'Hôpital's rule:

$$\lim_{p \to 0} \frac{x^p - 1}{p} = \lim_{p \to 0} \frac{\frac{d}{dp}(x^p - 1)}{\frac{d}{dp}p} = \lim_{p \to 0} x^p \ln x = \ln x$$

The *ladder of powers* is the sequential set of power transformations with $p = -1, -\frac{1}{2}, 0, \frac{1}{2}, 1, 2$.

4.9 Life Expectancy Example—Continued

We look again at the plot of `life.exp` vs `ppl.per.phys` from Figures 4.14 and 4.15 where we see that taking the logarithm of `ppl.per.phys` straightened out the graph. In Figure 4.17 we use the `ladder.fstar` function in the **HH** package to take the full set of scaled powers (in the ladder of powers T_p^*) of each of these two variables and plot them against each other. It is apparent from these plots that any power of `life.exp` plots as a straight line against the log of the number of physicians (power = 0). This is unusual behavior. More typically the shape of the plot shifts as the power of either variable shifts. This calls for further investigation.

We plot in Figure 4.18 the various scaled powers using equation (4.2) against `life.exp`. This equation is plotted for each of the values $p = -2, -1, 0, .5, 1, 2$, where $p = 0$ represents the log transformation. We see that within the observed range of values (51, 79) of `life exp`, all the simple power transformations are essentially linear. This explains why all panels in the ppp^0 column of Figure 4.17 are almost identical.

The columns of Figure 4.17 look different from each other. We look (for convenience) at row `life.exp^1`, with the original scaling of life expectancy, and note that the shape of the graphs shifts from concave-SW through diagonal to concave-NE as the power of ppp (people per physician) increases. We need to look at just this single variable as it moves through the series of powers. We do so in Figure 4.19. Panel 4.19a shows the boxplots, panel 4.19b shows the dotplots, and panel 4.19c shows the stem-and-leaf plots. All three panels show the same information. At the positive powers, the data for ppp are extremely asymmetric; they are bunched up at the low end of the scale. As the power moves from positive to negative, the center moves toward the higher end and the distribution becomes more symmetric. At the negative powers the data become asymmetric again; this time they are bunched up at the high end. If symmetry for just one variable was the only objective, we might try the $-.3$ power $-x^{-.3}$ (`-(x^-.3)`).

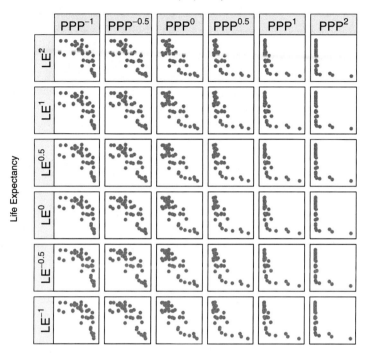

Fig. 4.17 Ladder of powers for "Life Expectancy" and "People per Physician".

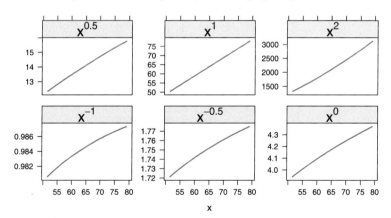

Fig. 4.18 Powers of "Life Expectancy". All powers in the range we are looking at yield a graph very close to straight line. Read the panels starting on the bottom-left position, then the rest of the bottom row, then the top row.

The boxplots show the shift in center as the dot for the median moves from one side to the other. They show the shift in symmetry as the center box increases from a small portion to a large fraction of the total width and as the whisker and outliers shift from one side to the other. The dotplots show the same information by density and spread of dots. Both dotplots and boxplots have the same scale. The stem-and-leaf is essentially a density plot. It shows the points bunched up at the low values for positive powers, centered and symmetric for 0 power, and bunched at the high values for the negative powers.

a. boxplots

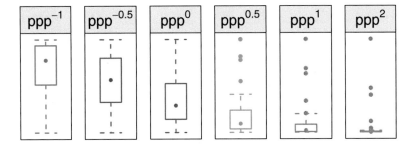

b. stripplots

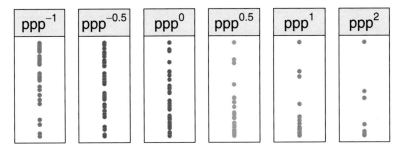

Fig. 4.19 Powers of "People per Physician": boxplots, strip plots, and stem-and-leaf. Stem-and-leaf appears in continuation of figure.

c. stem-and-leaf

```
> stem(-'-1', scale=2)        > stem(-'-0.5', scale=3)      > stem(-'0', scale=3)

  The decimal point is          The decimal point is          The decimal point is
4 digit(s) to the left        2 digit(s) to the left        1 digit(s) to the left
of the |                      of the |                      of the |

   -9998 | 76627740              -199 | 0                     -104 | 1
   -9996 | 91870                 -198 | 77                    -102 |
   -9994 | 8                     -198 | 2                     -100 | 45
   -9992 | 65                    -197 | 775                    -98 |
   -9990 | 6662                  -197 | 11                     -96 |
   -9988 | 6                     -196 | 6                      -94 | 4
   -9986 |                       -196 | 430                    -92 |
   -9984 | 44                    -195 | 9                      -90 |
   -9982 | 866361                -195 |                        -88 | 41
   -9980 |                       -194 | 95                     -86 | 3
   -9978 | 2                     -194 | 2                      -84 | 99
   -9976 | 7                     -193 | 9976                   -82 |
   -9974 | 22                    -193 |                        -80 | 640
   -9972 | 00                    -192 |                        -78 | 1
   -9970 | 1                     -192 | 41                     -76 | 7
   -9968 |                       -191 | 999875                 -74 |
   -9966 |                       -191 |                        -72 | 50
   -9964 |                       -190 | 96                     -70 | 8
   -9962 | 6                     -190 | 00                     -68 | 7727
   -9960 | 4                     -189 | 66                     -66 |
   -9958 |                       -189 | 2                      -64 | 372210
   -9956 | 1                     -188 |                        -62 | 63
   -9954 | 8                     -188 |                        -60 | 7100
                                 -187 | 96                     -58 | 115
                                 -187 |                        -56 | 2
                                 -186 | 97                     -54 | 652


> stem(-'0.5', scale=2)       > stem(-'1', scale=2)         > stem(-'2', scale=2)

  The decimal point is          The decimal point is          The decimal point is
1 digit(s) to the right       3 digit(s) to the right       8 digit(s) to the right
of the |                      of the |                      of the |

    -38 | 1                       -36 | 7                       -6 | 7
    -36 |                         -34 |                         -6 |
    -34 |                         -32 |                         -5 |
    -32 |                         -30 |                         -5 |
    -30 | 63                      -28 |                         -4 |
    -28 |                         -26 |                         -4 |
    -26 |                         -24 | 2                        -3 |
    -24 |                         -22 | 2                        -3 | 2
    -22 | 2                       -20 |                         -2 | 7
    -20 |                         -18 |                         -2 |
    -18 |                         -16 |                         -1 |
    -16 | 30                      -14 |                         -1 |
    -14 | 5                       -12 | 5                       -0 | 8
    -12 | 88                      -10 |                         -0 | 33211000000000000000000000000000000000
    -10 | 697                      -8 |
     -8 | 75                       -6 | 642
     -6 | 7173320                  -4 | 99
     -4 | 0987776520               -2 | 51054
     -2 | 886651098                -0 | 63211007666666654444433322
```

Fig. 4.19 continued. Powers of "People per Physician": stem-and-leaf. We are using non-syntactic names for the variables taken to a power ('-1' for the ppp^-1) in this display. The negative sign in the function call is a response to the difference in display conventions between graphics (with small numbers at the bottom of the response axis) and the stem and leaf (with small numbers at the top).

4.10 Color Vision

About 10% of the population have color deficient vision. Your job is make your graphs legible to everyone. Download ImageJ (Rasband, 2015) and VischeckJ (Dougherty and Wade, 2006) and follow the instructions in those sites. This program will allow you to simulate color deficient vision on your computer.

Figure 4.20 shows the first six colors of the **lattice** default colors and the color scheme col3x2, designed to work well for use with a 3 × 2 classification, and used here in many of the figures. We constructed the color scheme col3x2 using the Dark2 and Set2 palettes in the **RColorBrewer** package (Brewer, 2002; Neuwirth, 2011).

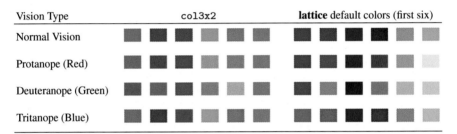

Fig. 4.20 Four visual appearances of two color schemes. The four vision types are crossed with the two color schemes to create eight cells. The three color deficiency simulations were made with the vischeck software. Six color choices which are intended to be distinct are in each of the eight cells. It is very easy to track each color scheme across the four vision types. The six colors on the left, consisting of three dark colors followed by lighter versions of the same three colors, are the Dark2 and Set2 palettes from the **RColorBrewer** package. For all four vision types, the six colors on the left are perceived as a set of darker three and lighter three colors. The six colors on the right are the first six of the standard **lattice** colors. They are not clearly distinct in any of the color deficiency simulations.

4.11 Exercises

We recommend that you begin all exercises by examining a scatterplot matrix of the variables. Based on the scatterplot matrix, you might wish to consider transforming some of the variables.

4.1. The U.S. Draft Lottery held in December 1969 was meant to prioritize the order in which young men would be drafted during 1970 for service in the Vietnam War. Each of the 366 dates was written on a small piece of paper and placed in a capsule. In chronological order the capsules were placed in a vessel and the vessel was stirred. The capsules were then drawn one at a time, thereby assigning ranks 1 to 366 to the dates. But because of inadequate stirring, men with birthdays toward the end of the year tended to have higher rank and thus greater vulnerability to the draft than men born early in the year. The dataset data(draft70mn), originally from Data Archive (1997), contains 12 columns for the months January through December. For each month, the mth entry represents the rank between 1 and 366 for the mth day of that month. Produce parallel boxplots for the months arranged chronologically, and draw the line segments connecting the medians of adjacent months. This illustrates the claim that the drawing was not random.

4.2. Sokal and Rohlf (1981), later in Hand et al. (1994), examined factors contributing to air pollution in 41 U.S. cities, as assessed by sulfur dioxide content. The dataset appears in data(usair). The variables are

SO2: SO_2 content of air in mcg per cubic meter

temp: average annual temperature in degrees Fahrenheit

mfgfirms: number of manufacturing firms employing at least 20 workers

popn: 1970 census population size, in thousands

wind: average annual wind speed in mph

precip: average annual precipitation in inches

raindays: average number of days per year having precipitation

Produce a scatterplot matrix for these data both before and after log-transforming all 7 variables. Compare the splom and explain why the log transformation is appropriate for these data. Which of the 6 predictor variables are most highly correlated with the logged response SO2? Which of the 15 pairs of logged predictors appear to be highly correlated?

4.3. Vandaele (1978), also in Hand et al. (1994), contains data on the reported 1960 crime rate per million population and 13 potential explanatory variables for each of 47 states. The data appear in the file data(uscrime). The variables are

R: reported crime rate per million population

Age: the number of males aged 14 to 24

S: 1 if Southern state, 0 otherwise

Ed: 10 times mean years of schooling of population age 25 or older

Ex0:   1960 per capital expenditures by state and local government on police protection

Ex1:   same as Ex0 but for 1959

LF: number of employed urban males aged 14–24 per 1000 such individuals

M: number of males per 1000 females

N: state population size in hundred thousands

NW: number of nonwhites per 1000 population

U1:   unemployment rate per 1000 among urban males aged 14–24

U2:   unemployment rate per 1000 among urban males aged 25–39

W: a measure of wealth, units = 10 dollars

X: number of families per 1000 earning below one half of the median income
      (a measure of income inequality)

Construct a scatterplot matrix for these data. The variables other than R will be referred to as predictors in Exercise 9.6. Based on this plot, which pairs of predictors are highly correlated? Which predictors are most closely linearly associated with R?

4.4. Hand et al. (1994) contains data on the average `mortality` rate for males per 100,000 and the `calcium` concentration (ppm) in the public drinking water in 61 large towns in England and Wales, averaged over the years 1958 to 1964. Each town was also identified as being at least as far north as the town Derby (`derbynor=1`) or south of Derby (`derbynor=0`). The data are accessed as `data(water)`. Exercise 10.4 will request investigation of the relationship between water hardness (`calcium`) and `mortality`. The sampling units are towns in two regions. Produce two separate but adjacent plots of `mortality` vs `calcium` for the two regions specified by `derbynor`. Discuss the differences you see in the two plots.

4.5. Williams (1959), also in Hand et al. (1994), presents data on the `density` and hardness of 36 Australian eucalyptus trees. The dataset is accessed as `data(hardness)`. Determine a transformation from the Box–Cox family that will make `hardness` as close as possible to normally distributed. The result will be useful for Exercise 11.2, which requests a model of `hardness` as a function of `density`.

4.6. Following a severe water shortage in Concord, New Hampshire, during the late 1970s, conservation measures were instituted there in 1980. The shortage became especially acute during the summer of 1981. Hamilton (1983) and Hamilton (1992) discuss models of the 1981 household water consumption in Concord, New Hampshire, in terms of several other variables. The dataset, accessed as `data(concord)`, contains information on the following variables from each of 496 households:

water81:   cubic feet of household water use in 1981

water80:   cubic feet of household water use in 1980

income: 1981 household income in $1000s

educat: education of head of household, in years

peop81:   number of people living in household in summer 1981

retired: 1 if head of household is retired, otherwise 0

Exercise 11.3 requests the modeling of household water use in 1981 in Concord as a function of 5 predictors. To assist with this task, investigate which transformation from the ladder of powers family will bring the response variable, `water81`, as close as possible to normality.

4.A Appendix: R Graphics

R has three major tool sets for graphics specification: base, lattice/trellis, and ggplot. R has a fourth tool set in the **vcd** package for Visualizing Categorical Data.

Base graphics, in the **graphics** package, is the oldest, going back to the beginning of S (Becker et al., 1988). It provides functions for drawing plots and components of plots directly on the graphics device (computer screen or paper).

Lattice graphics with the **lattice** package (Sarkar, 2008, 2014) dates back to the Trellis system (Becker et al., 1996a,b) of S and S-Plus. **lattice** functions construct R objects which represent the graph. The objects can be stored and updated with additional labeling or other annotation. When the objects are printed, they produce a visible plot on the graphics device.

Grammar of Graphics (Wilkinson, 1999), implemented in package **ggplot2** (Wickham, 2009), is the newest system. While **ggplot2** functions also construct R objects which represent the graph, they do so with a completely different partitioning of the components of a graph.

Packages **lattice**, **ggplot2**, and **vcd** have all been implemented in the **grid** package (R Core Team, 2015; Murrell, 2011).

Most graphics in this book were constructed using **lattice**. Many were drawn by direct use of the functions provided in the **lattice** package. Others were drawn by first constructing new functions, distributed in the **HH** package, and then using the new functions. The **vcd** graphics package (Meyer et al., 2012, 2006) is used for mosaic plots and related plots in Chapter 15.

The R code for all graphs in this book is available in the **HH** package. To see the code for any chapter, say Chapter 7, enter at the R prompt the line:

```
HHscriptnames(7)
```

and discover the pathname for the script file. Open that file in your favorite R-aware editor. See Appendix B for more details on the R scripts distributed with the **HH** package.

4.A.1 Cartesian Products

A feature common to many of the displays in this book is the Cartesian product principle behind their construction.

The Cartesian product of two sets A and B is the set consisting of all possible ordered pairs (a, b) where a is a member of the set A and b is a member of the set B. Many of our graphs are formed as a rectangular set of panels, or subgraphs, where each panel is based on one pair from a Cartesian product. The sets defining the

Cartesian product differ for each graph type. For example, a set can be a collection of variables, functions of a single variable, levels of a single factor, functions of a fitted model, different models, etc.

When constructing a graph that can be envisioned as a Cartesian product, it is necessary that the code writer be aware of the Cartesian product relationship. The **lattice** code for such a graph includes a command that explicitly states the Cartesian product.

4.A.2 Trellis Paradigm

Most of the graphs in this book have been constructed using the trellis paradigm as implemented in **lattice**. The trellis system of graphics is based on the paradigm of repeating the same graphical specifications for each element in a Cartesian product of levels of one or more factors.

The majority of the methods supplied in the R **lattice** package are based on a typical formula having the structure

$$y \sim x \mid a * b \tag{4.4}$$

where

y is either continuous or factor
x is continuous
a is factor
b is factor

and each panel, as defined by the Cartesian product of the levels of a and b, is a plot of y ~ x for the subset of the data with the stated values of a and b.

4.A.3 Implementation of Trellis Graphics

The concept of trellis plots can be implemented in any graphics system. In the S family of languages (**S-Plus** and **R**), selection of the set of panels, assignment of individual observations to one panel in the set, and coordinated scaling across all panels are automated in response to a formula specification in the user level.

The term *trellis* comes from gardening, where it describes an open structure used as a support for vines. In graphics, a trellis provides a framework in which related graphs can be placed. The term *lattice* has a similar meaning.

4.A.4 Coordinating Sets of Related Graphs

There are several graphical issues that needed attention in any multipanel graph. See Figure 10.8 for an example illustrating these issues.

positioning: The panels containing marginal displays (if any) need to be clearly delineated as distinct from the panels containing data from just a single set of levels of the factors. We do this by placing extra space between the set of panels for the individual factor values and the panels containing marginal displays.

scaling: All panels need to be on exactly the same scale to enhance the reader's ability to compare the panels visually. We use the automatic scaling feature of trellis plots to scale simultaneously both the individual panels and the marginal panels.

labeling: We indicate the marginal panels by use of the strip labels.

shape of plotting characters: We used three distinct plotting characters for the three-level factor.

color of plotting characters: We used three contrasting colors for the three-level factor. The choice to use both distinct plotting characters and distinct colors is redundant (reemphasizing the difference between levels), accessible (making the graph work for people with color vision deficiencies), and defensive (protecting the interpretability of the graph from black-and-white copying by a reader).

There are several packages in R that address color selection. The **RColorBrewer** package (Neuwirth, 2011), based on the ColorBrewer website (Brewer, 2002), gives a discussion on the principles of color choice and gives a series of palettes for distinguishing nominal sets of items or sequences of items. The **colorspace** package (Ihaka et al., 2013) provides qualitative, sequential, and diverging color palettes based on HCL colors.

4.A.5 Cartesian Product of Model Parameters

Figure 10.12 displays four different models of a response variable as a function of a factor and a continuous covariate. The model in the center row and right column is the same model shown in Figure 10.8. The models are shown as a Cartesian product of model parameters. The models in the columns of Figure 10.12 are distinguished by the absence or presence of a parameter for Type—forcing a common intercept in the left column and allowing different intercepts by Type in the right column.

The three rows are distinguished by how the covariate Calories is handled: separate slopes by Type in the top row, constant slope for all Types in the middle row, or identically zero slope (horizontal line) in the bottom row.

Figure 10.12 is structured as a set of *small multiples*, a term introduced by Tufte Tufte (2001) to indicate repetition of the same graphical design structure. "Small multiples are economical: once viewers understand the design of one slice, they have immediate access to the data in all other slices. Thus, as the eye moves from one slice to the next, the constancy of the design allows the viewer to focus on changes in the data rather than on changes in graphical design (Tufte (2001), page 48)." Figure 10.12 may be interpreted as a four-way Cartesian product: slope (α vs α_i), intercept ($\beta = 0$, β, β_j), individual panels vs superpose, hotdog Type (Beef, Meat, Poultry) with a an ordinary two-way scatterplot with a fitted line inside each element of the four-way product.

4.A.6 Examples of Cartesian Products

1. In the plots illustrating lack of homogeneity of variance (Figure 6.6), one of the sets in the Cartesian product is the function of the data represented (observed data, median-centered data, absolute value of the median-centered data). The other set is the levels of the catalyst factor. We discuss in Section 6.10 the Brown–Forsyth test for variance homogeneity.

2. In the logistic regression plots (Figure 17.12) there are several sets used to define the Cartesian products. The rows of the array are functions of the fitted probability. The columns of the array are the levels of one of the factors (X-ray) with a marginal value of X-ray in the left-most column. The individual lines within the panels, as identified in the legend, are levels of the X.ray $\times$ stage $\times$ grade interaction. This is an ordinary xyplot of the predicted response variable displayed on three scales—the logit scale, the odds scale, and the probability scale—against one of the predictor variables acid.ph.

3. In the ladder-of-power plots (Figure 4.17) the rows of the array are powers of y and the columns are powers of x. This plot is useful in a regression context for determining the optimal power transformations of both the response and predictor variables.

4. Figure 4.21 shows the ability to control the position and color of boxplots. This simulated example shows the results of a clinical trial where the patients' followup visits were scheduled with nonconstant intervals between visits. Here, the boxes for both treatment levels are grouped by week and the weeks are correctly spaced. The default positioning for bwplot places the boxes evenly spaced, honoring neither the week nor the treatment factor.

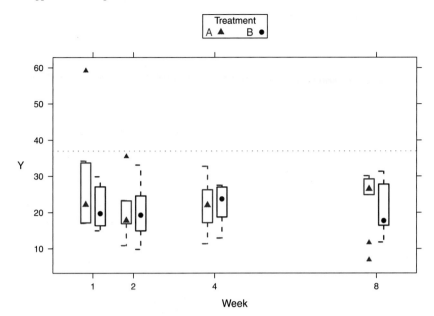

Fig. 4.21 The response to treatments A and B was measured at weeks 1, 2, 4, and 8. The boxplots have been positioned at distances illustrating the time difference and with A and B adjacent at each time point.

5. Mosaic plots (Figure 15.11 and other figures in Chapter 15) as constructed as Cartesian products of several factors.

6. Diverging stacked bar charts as used in displays of Likert scale data (Figure 15.14 and others in Section 15.9 are a crossing of a set of questions (possibly nested in another factor) with a set of potential responses.

4.A.7 latticeExtra—Extra Graphical Utilities Based on Lattice

The **latticeExtra** provides many functions for combining independently constructed **lattice** plots and for controlling the size and placement of arrays of **lattice** plots. We use these functions in many of our graphs. The `mmcplot` (Figure 7.18 and elsewhere in the book) is built by constructing the two panels independently and then combining them with the `latticeExtra:::c.trellis` function. Many of our plots are constructed by overlaying two independently drawn graphs with the `layer` function or with the `latticeExtra:::'+.trellis'` as illustrated in Figure 4.22.

4.B Appendix: Graphs Used in This Book

We emphasize throughout this book that graphical display is an integral part of data analysis. Superior data analysis almost always benefits from high-quality graphics. Appropriately drawn graphs are, in our opinion, the best way to gain an understanding of what data have to say, and also to convey results of the analysis to others, both other statisticians and persons with minimal training in statistics.

We illustrate many standard graphs. We also illustrate many graphical displays that are not currently standard and some of which are new. The software for our displays is included in the **HH** package.

Analysts occasionally require a graph unlike any readily available elsewhere. We recommend that serious data analysts invest time in becoming proficient in writing code rather than using the graphical user interface (GUI). Very few of the graphs in this book can be produced using a standard GUI. Some of them can be produced using the menus in our package **RcmdrPlugin.HH**. Users of a GUI are limited to the current capabilities of the GUI. While the design of GUIs will continually improve, their capabilities will always remain far behind what skilled programmers can produce. Even less-skilled analysts can take advantage of cutting-edge graphics by accessing libraries of graphing functions such as those included in the **HH** package and other packages available on CRAN.

4.B.1 Structured Sets of Graphs

Several of our examples extend the concept of a structured presentation of plots of different sets of variables, or of different parametric transformations of the same set of variables. Several of our examples extend the interpretation of the model formula, that is, the semantics of the formula, to allow easier exposition of standard statistical techniques.

In this appendix we list these displays in order to comment on their construction. We provide a reference to an example in the book for each type of display. Discussion of the interpretation of the graphs appears in the indicated chapters.

4.B.2 Combining Panels

1. Scatterplot Matrices: splom A scatterplot matrix (splom) is a trellis display in which the panels are defined by a Cartesian product of variables. In the standard scatterplot matrix constructed by splom, Figure 4.5 for example, the same set of variables define both the rows and columns of the matrix.

A scatterplot matrix (splom) does not follow the semantic paradigm of Equation (4.4). It differs from the majority of trellis-based methods in two ways. First, each of the panels is a plot of a different set of variables. Second, each of the panels is based on the entire set of observations.

Subsections 4.4 and 4.7 contain extensive discussions of scatterplot matrices. We strongly recommend the use of a splom, sometimes conditioned on values of relevant categorical variables, as an initial step in analyzing a set of data.

2. xyplot can be used to construct more general matrices of panels, for example with different of sets of variables for the rows and columns of the scatterplot matrix. Figure 4.4, for example, shows that xyplot can be used to specify a set of variables to define the columns of the matrix and subsets of the observations (specified as different levels of a factor) to define the rows. The formula is essentially

> sprice ~ beds + drarea + kitarea | CondoHouse

Sets of xyplots with coordinated subsets of variables can be useful in situations where the number of variables under study is too large to produce a legible splom containing all variables on a single page. In such a circumstance we recommend the use of two or more pages of xyplots to display pairwise relationships among variables.

3. Figure 4.22 shows several ways to combine multiple variables in one or more panels. The figure shows overlaying plots, concatenating plots, and conditioning panels on the levels of a factor.

4.B.3 Regression Diagnostics

In the regression diagnostics plots (Figure 11.6), the panels are defined by conditioning on a set of functions (one for each statistic). This plot displays all common regression diagnostics on a single page. Included are thresholds for flagging cases as unusual along with identification of such cases.

4.B.4 Graphs Requiring Multiple Calls to xyplot

When one of the sets in the Cartesian product is a set of functions, the easiest way to construct the product is to make several xyplot calls, one for each function in the set.

1. Partial residual plots (Figure 9.10) — [functions of fitted values and residual] × [variables]. Response against predictors, residuals against predictors, partial

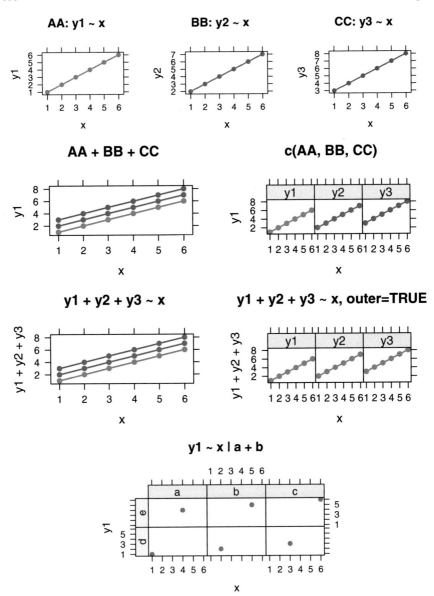

Fig. 4.22 Several ways to plot multiple variables simultaneously. The top row shows the "trellis" objects from three separate calls to the xyplot function. The second row shows two ways of combining the "trellis" objects in the top row. On the left they are overlaid into the same panel using the **latticeExtra** +.trellis function. On the right they are concatenated into a multi-panel "trellis" object by using the **latticeExtra** c.trellis function. The third row shows two ways of specifying similar displays with a single xyplot command. On the left there are three response variables in the model formula with the default setting that places them into the same panel. On the right the outer=TRUE argument places them into three adjacent panels. The bottom row shows placement of the points into separate panels by specifying the Cartesian product of the levels of the factors a and b in the conditioning section (following the "|" symbol) of the model formula. The code for these plots is included in the file identified by HHscriptnames(4).

residuals against predictors (partial residual plots), and partial residuals of Y against partial residuals of X (added variable plots). Each row of Figure 9.10 is a different function of fitted values or residual. Each column is either one of the predictor variables or a function of the predictor variables. See the discussion in Section 9.13.

2. Analysis of covariance plots (One example is in the set of Figures 10.6, 10.7, 10.8, and 10.9. Another example is in Figure 14.6) — [models] × [levels]. A key feature of this set of plots is its presentation of all points both superposed into one panel and also segregated into individual panels defined by the levels of a factor. In this framework, the superposition of all levels of the factor is itself considered a level.

3. ODOFFNA plots (Figure 14.17) — [transformation power] × [factors] × [factors], a 3-dimensional Cartesian product. This is a series of interaction plots indexed by a third variable, the transformation power, all on a single page. Figure 14.17 is intended to find a satisfactory power transformation to achieve homogeneity of variance and then assess interaction among the two factors for the chosen power transformation.

4.B.5 Asymmetric Roles for the Row and Column Sets

1. Interaction plots (Figure 12.1) — [factors] × [factors]. Each off-diagonal panel is a standard interaction plot. Panels in transpose positions interchange the trace- and x-factors. Rows are labeled by the trace factor. Columns are labeled by the x-factor. The main diagonal is used for boxplots of the main effects.

2. ARIMA-trellis plots (Figure 18.8) — [number of AR parameters] × [number of MA parameters] × [type of display]. Each of the 3×3 displays contains diagnostic information about each of the 9 models indexed by the numbers of autoregressive and moving average parameters p and q. In addition we group several types of display on a single page. This plot displays most commonly used diagnostics for identifying the number of AR and MA parameters in time series models of the ARIMA class.

4.B.6 Rotated Plots

Mean–mean multiple comparisons plots (MMC plots) (Figure 7.19) — [means at levels] × [means at levels]. The plot is designed as a crossing of the means of a response variable at the levels of one factor with itself. It is then rotated 45° so the horizontal axis can be interpreted as the differences in mean levels and the vertical axis can be interpreted as the weighted averages of the means comprising each comparison. This class of plots is used to display the results of a multiple comparison procedure.

4.B.7 Squared Residual Plots

The fundamental concept of "least squares" is difficult to present to introductory classes. Here, we illustrate the squares. The sum of their areas is the "sum of squares" that is minimized according to the "least-squares" principle.

Illustrations of 2D and 3D least-squares fits (Figures 8.2, 9.1, and 9.5)—[fitted models] × [methods of displaying residuals]. The rows of Figure 8.2 are ways of displaying residuals; the first row shows the residuals as vertical lines, the second as squares. The columns show different models: none, least-squares, and a too-shallow fit.

4.B.8 Adverse Events Dotplot

There are two primary panels in Figure 15.13 — [factor] × [functions of percents]. The first panel shows the observed percentages on the x-axis. The second panel shows the relative risk with its confidence interval on the x axis. Both panels have the same y axis showing the event names.

4.B.9 Microplots

Microplots (as in Table 13.2) are small plots embedded into a table of numbers. The plot often carries as much or more information as the numbers.

4.B.10 Alternate Presentations

We have alternate presentations of existing ideas.

1. Transposed trellis plots are sometimes helpful. In Figure 13.13 we show a set of boxplots with the response variable on the vertical axis. The vertical orientation places the response variable in the vertical direction and accords with how we have been trained to think of functions—levels of the independent variable along the abscissa and the response variable along the ordinate. In Section 13.A we show in Figure 13.17 the same graphs with the response variable on the horizontal axis.

2. Odds-ratio CI plot (Figure 15.10). The odds ratio

$$\left(\frac{p_1}{q_1}\right) \bigg/ \left(\frac{p_2}{q_2}\right)$$

does not, by construction, give information on both underlying p_1- and p_2-values. It is necessary to specify one of them to estimate the other. We backtransform the CI on the odds ratio to a CI on the probability scale and plot the CI of p_2 for all possible values of p_1. The two axes have the same $(0, 1)$ probability scale.

Chapter 5

Introductory Inference

In this chapter we discuss selected topics and issues dealing with statistical inferences from samples to populations, building upon the brief introduction to these ideas in Chapter 3. The discussion here is at an intermediate technical level and at a speed appropriate for review of material learned in the prerequisite course.

We provide procedures for constructing confidence intervals and conducting hypothesis tests for several frequently encountered situations.

5.1 Normal (z) Intervals and Tests

A confidence interval and test concerning a population mean were briefly described in Chapter 3. This is a more extensive presentation.

The confidence interval on the mean μ of a normal population when the standard deviation is known was given in Equation (3.18). The development there assumed that the population was normal. However, since the Central Limit Theorem discussed in Section 3.4.2 guarantees that $\frac{\bar{Y}-\mu}{\sigma/\sqrt{n}}$ is approximately normally distributed if n is "sufficiently large", the interval

$$\left(\bar{y} - z_{\frac{\alpha}{2}} \frac{\sigma}{\sqrt{n}}, \ \bar{y} + z_{\frac{\alpha}{2}} \frac{\sigma}{\sqrt{n}}\right) \tag{5.1}$$

is an approximate two-sided $100(1 - \alpha)\%$ confidence interval when the population is not normal. The closer the population is to a normal population, the closer will be this interval's coverage probability to $1 - \alpha$. Thus, in the nonnormal case, this interval is an approximate CI for μ.

© Springer Science+Business Media New York 2015
R.M. Heiberger, B. Holland, *Statistical Analysis and Data Display*,
Springer Texts in Statistics, DOI 10.1007/978-1-4939-2122-5_5

Also shown in the rightmost column of Table 5.1 are one-sided confidence intervals for μ. These are less commonly used than two-sided intervals because they have infinite width. But they are sometimes encountered in contexts where an upper or lower bound for μ is required.

5.1.1 Test of a Hypothesis Concerning the Mean of a Population Having Known Standard Deviation

We consider three pairs of null and alternative hypotheses in Table 5.1 and Figure 5.1.

Table 5.1 Confidence intervals and tests with known standard deviation σ, where $\sigma_{\bar{y}} = \dfrac{\sigma}{\sqrt{n}}$ and $z_{\text{calc}} = \dfrac{\bar{y} - \mu_0}{\sigma_{\bar{y}}}$. The six situations are shown graphically in Figure 5.1

		Tests			Confidence Interval	
H_0	H_1	Rejection Region		p-value		
		z-scale	y-scale		Lower	Upper
$\mu \leq \mu_0$	$\mu > \mu_0$	$z_{\text{calc}} > z_\alpha$	$\bar{y} > \mu_0 + z_\alpha \sigma_{\bar{y}}$	$P(Z > z_{\text{calc}})$	$(\bar{y} - z_\alpha \sigma_{\bar{y}},$	∞ $)$
$\mu \geq \mu_0$	$\mu < \mu_0$	$z_{\text{calc}} < -z_\alpha$	$\bar{y} < \mu_0 - z_\alpha \sigma_{\bar{y}}$	$P(Z < z_{\text{calc}})$	$(\qquad -\infty,$	$\bar{y} + z_\alpha \sigma_{\bar{y}})$
$\mu = \mu_0$	$\mu \neq \mu_0$	$\lvert z_{\text{calc}} \rvert > z_{\frac{\alpha}{2}}$	$\lvert \bar{y} - \mu_0 \rvert > z_{\frac{\alpha}{2}} \sigma_{\bar{y}}$	$2P(Z > \lvert z_{\text{calc}} \rvert)$	$(\bar{y} - z_{\frac{\alpha}{2}} \sigma_{\bar{y}},$	$\bar{y} + z_{\frac{\alpha}{2}} \sigma_{\bar{y}})$

The first two pairs are called *one-tailed* or *one-sided* tests because their rejection regions lie on one side of the normal distribution. The third pair has a two-sided rejection region and hence is termed a two-tailed or two-sided test. In any given problem, only one of these three is applicable. For expository purposes, it is convenient to discuss them together.

Some authors formulate the one-sided tests with sharp null hypotheses

H_0	H_1
$\mu = \mu_0$	$\mu > \mu_0$
$\mu = \mu_0$	$\mu < \mu_0$

However, with the sharp formulation it can happen that neither the null nor alternative hypothesis is true, in which case the action of rejecting the null hypothesis has an uncertain interpretation.

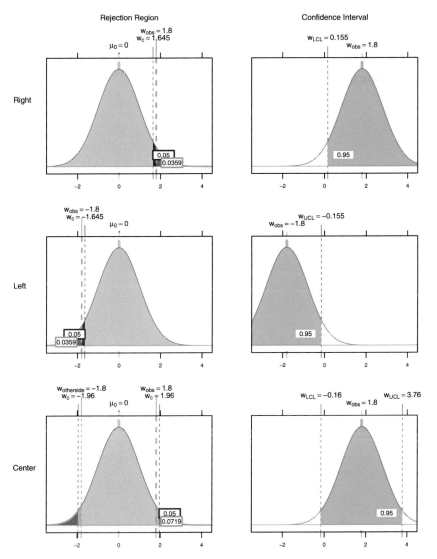

Fig. 5.1 Graphical display of the six situations described in Table 5.1: Confidence intervals and tests with known standard deviation σ. See Section 5.1.1 for full discussion.

For the first pair of hypotheses, we reject H_0 if the sample mean is sufficiently greater than μ_0, specifically, if $\bar{y} > (\mu_0 + z_\alpha \sigma / \sqrt{n})$. Otherwise, H_0 is retained. Equivalently, if we define the calculated Z statistic under the null hypothesis,

$$z_{\text{calc}} = \frac{\bar{y} - \mu_0}{\sigma / \sqrt{n}} \tag{5.2}$$

then we reject H_0 if $z_{\text{calc}} > z_\alpha$; otherwise H_0 is retained. The p-value of this test is $P(Z > z_{\text{calc}})$.

The testing procedure for the second pair of hypotheses is the mirror image of the first pair. H_0 is rejected if $\bar{y} < (\mu_0 - z_\alpha \sigma / \sqrt{n})$ and retained otherwise. Equivalently, we reject H_0 if $z_{\text{calc}} < -z_\alpha$. The p-value of this test is $P(Z < z_{\text{calc}})$.

For the third pair, the two-sided test, we reject H_0 if either

$$\bar{y} < (\mu_0 - z_{\frac{\alpha}{2}} \sigma / \sqrt{n}) \quad \text{or} \quad \bar{y} > (\mu_0 + z_{\frac{\alpha}{2}} \sigma / \sqrt{n});$$

equivalently, if $|z_{\text{calc}}| > z_{\frac{\alpha}{2}}$. The p-value of this two-sided test is $2P(Z > |z_{\text{calc}}|)$. Hence H_0 is rejected if $\bar{y}$ is sufficiently above or sufficiently below μ_0. Another equivalent rule is to reject H_0 if and only if μ_0 falls outside the $100(1 - \alpha)\%$ confidence interval for μ.

The rejection region for all three pairs is included in Table 5.1.

5.1.2 Confidence Intervals for Unknown Population Proportion p

We consider a confidence interval on the unknown proportion p of successes in a population consisting of items or people labeled as successes and failures. Such populations are very frequently encountered in practice. For example, we might wish to estimate the proportion p of voters who will ultimately vote for a particular candidate, based on a random sample from a population of likely voters. Inspectors of industrial output may wish to estimate the proportion p of a day's output that is defective based on a random sampling of this output.

Suppose the sample size is n, of which Y items are successes and that $\hat{p} = \frac{Y}{n}$, a point estimator of p, is the proportion of sampled items that fall into the success category. Until recently, the usual $100(1 - \alpha)\%$ confidence interval for p suggested in the statistics literature was

$$\hat{p} \pm z_{\frac{\alpha}{2}} \sqrt{\frac{\hat{p}(1 - \hat{p})}{n}}$$

This interval is satisfactory when $n \geq 100$ unless p is close to either 0 or 1. The large sample is needed for the Central Limit Theorem to assure us that the discrete probability distribution of $\hat{p}$ is adequately approximated by the continuous normal distribution.

Agresti and Caffo (2000) suggest the following alternative confidence interval for p, where $\tilde{p} = \frac{Y+2}{n+4}$ and $\tilde{n} = n + 4$:

$$\tilde{p} \pm z_{\frac{\alpha}{2}} \sqrt{\frac{\tilde{p}(1 - \tilde{p})}{\tilde{n}}} \tag{5.3}$$

Agresti and Caffo show that their interval has coverage probability that typically is much closer to the nominal $1 - \alpha$ than the usual confidence interval. It differs from the usual interval in that we artificially add two successes and two failures to the original sample. For p near 0 or 1, the usual interval, which is symmetric about $\hat{p}$, may extend beyond one of these extremes and hence not make sense, while the alternative interval is likely to remain entirely between 0 and 1.

Conventional one-sided confidence intervals for p are shown in Table 5.2. Comparable to Agresti and Caffo's proposed two-sided interval, Cai (2003) proposes improved one-sided confidence intervals for p having coverage probabilities closer to $1 - \alpha$ than the conventional intervals. These lower and upper intervals, respectively, are

$$\left[0, \mathcal{F}_{\mathrm{Be}}^{-1}(1 - \alpha \mid Y + .5, n - Y + .5)\right] \tag{5.4}$$

and

$$\left[\mathcal{F}_{\mathrm{Be}}^{-1}(\alpha \mid Y + .5, n - Y + .5), 1\right] \tag{5.5}$$

where $\mathcal{F}_{\mathrm{Be}}^{-1}(\alpha \mid a, b)$ denotes the value x of a random variable corresponding to the 100α percentile of the beta distribution with parameters a and b. See Section J.1.1 for a brief discussion of the beta distribution.

5.1.3 Tests on an Unknown Population Proportion p

Assume we have a sample of $n \geq 100$ items from a population of successes and failures, and we wish to test a hypothesis about the proportion p of successes. Paralleling the previous discussion of tests on a population mean, there are two one-tailed tests and one two-tailed test as detailed in Table 5.2 and Figure 5.2. As in the discussion of the confidence interval on p, the normal approximation to the distribution of $\hat{p}$ requires that n not be too small. Note that the confidence intervals are based on densities centered on the observed proportion $\hat{p} = x/n$. They therefore have a different standard deviation $\sqrt{\hat{p}(1 - \hat{p})/n}$, and therefore height at the center of the density, than the densities centered at the null hypothesis p_0 with standard deviation $\sqrt{p_0(1 - p_0)/n}$.

5.1.4 Example—One-Sided Hypothesis Test Concerning a Population Proportion

As an illustration, suppose a pollster wishes to test the hypothesis that at least 50% of a city's voting population favors a certain bond issue. The pollster observed only

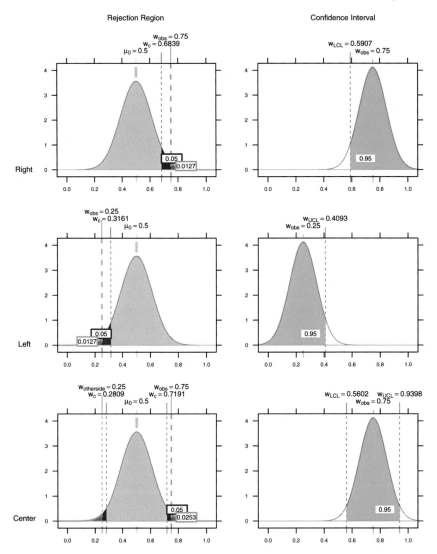

Fig. 5.2 Graphical display of the six situations described in Table 5.2: Confidence intervals and tests for population proportions. Note that the confidence intervals, centered on the observed $\hat{p}$, have differently scaled density functions than the null hypothesis distributions, centered on the hypothesized p_0. For the tests, the standard deviation is $\sigma_{p_0} = \sqrt{\big(p_0/(1 - p_0)\big)/n}$. For the confidence intervals, the standard deviation is $s_{\hat{p}} = \sqrt{\big(\hat{p}/(1 - \hat{p})\big)/n}$. In this example the densities for the confidence interval are taller and narrower. See Section 5.1.3 for full discussion.

Table 5.2 Conventional confidence intervals and tests with unknown population proportion p, where

$$\sigma_{p_0} = \sqrt{\frac{p_0(1-p_0)}{n}} \text{ and } z_{calc} = \frac{\hat{p} - p_0}{\sigma_{p_0}} \text{ for tests, and } s_{\hat{p}} = \sqrt{\frac{\hat{p}(1-\hat{p})}{n}} \text{ for confidence intervals.}$$

		Tests			Confidence Interval	
H_0	H_1	Rejection Region		p-value	Lower	Upper
		z-scale	p-scale			
$p \leq p_0$	$p > p_0$	$z_{calc} > z_\alpha$	$\hat{p} > p_0 + z_\alpha \sigma_{p_0}$	$P(Z > z_{calc})$	$(\hat{p} - z_\alpha s_{\hat{p}}, 1$	$)$
$p \geq p_0$	$p < p_0$	$z_{calc} < -z_\alpha$	$\hat{p} < p_0 - z_\alpha \sigma_{p_0}$	$P(Z < z_{calc})$	$($	$0, \hat{p} + z_\alpha s_{\hat{p}})$
$p = p_0$	$p \neq p_0$	$|z_{calc}| > z_{\frac{\alpha}{2}}$	$|\hat{p} - p_0| > z_{\frac{\alpha}{2}} \sigma_{p_0}$	$2P(Z > |z_{calc}|)$	$(\hat{p} - z_{\frac{\alpha}{2}} s_{\hat{p}},$	$\hat{p} + z_{\frac{\alpha}{2}} s_{\hat{p}})$

222 of a random sample of 500 persons in the population favors this bond issue. Let us conduct this test at $\alpha = 0.01$.

Here H_1 is of the form $H_1: p < .50$. We reject H_0 if

$$\hat{p} < p_0 - z_{.01} \sqrt{\frac{p_0(1-p_0)}{n}} \tag{5.6}$$

With $p_0 = .50$, $\hat{p} = 222/500 = 0.444$, $z_{.01} = 2.326$, and

$$\sqrt{p_0(1-p_0)/n} = .0224 \tag{5.7}$$

we find that the right side of (5.6) is 0.448 so that H_0 is (barely) rejected. In this example, $z_{calc} = -2.500$ so that the p-value $= P(Z < -2.500) = .0062$. Hence we reject H_0 because $\alpha = .01 > p = .0062$.

5.2 *t*-Intervals and Tests for the Mean of a Population Having Unknown Standard Deviation

When we wish to construct a confidence interval or test a hypothesis about an unknown population mean μ, more often than not the population standard deviation σ is also unknown. Then we must use the sample standard deviation $s = \sum((x - \bar{x})^2)/(n - 1)$ from Equation 3.9 in place of σ when standardizing $\bar{y}$. But while $(\bar{y} - \mu)/(\sigma/\sqrt{n})$ has an approximate normal distribution if n is sufficiently large, $(\bar{y}-\mu)/(s/\sqrt{n})$ has an approximate t distribution with $n-1$ degrees-of-freedom. The latter standardization with s in the denominator has more variability than the former standardization with σ in the denominator. The t distribution reflects this increased variability because it has less probability concentrated near zero than does the standard normal distribution.

The confidence interval and tests for μ using the t distribution are similar to those using the normal (Z) distribution (that is, Table 5.1 is applicable), with t_{calc} replacing z_{calc} and t_α replacing z_α. For this problem, the degrees-of-freedom parameter for the t distribution is always $n - 1$.

For example, to test $H_0: \mu \geq \mu_0$ vs $H_1: \mu < \mu_0$, we reject H_0 if

$$t_{calc} = \frac{\bar{y} - \mu}{s/\sqrt{n}} < -t_\alpha \tag{5.8}$$

Here the p-value $= P(t < t_{calc})$ is calculated from the t distribution with $n-1$ degrees of freedom.

Calculating the power associated with t-tests is more difficult than for the normal tests because the alternative distribution is not the same as the null distribution. With the normal tests, both distributions have the same shape. With the t-tests, the alternative distribution has the *noncentral* t distribution with noncentrality parameter $(\mu_1 - \mu_0)/(\sigma/\sqrt{n})$. We postpone further discussion of the noncentral t distribution to Section 5.6.2 and Figure 5.10 in the context of sample size calculations. Also see the illustration in Section J.2.2.

The approximate confidence interval on μ is $\bar{y} \pm t_{\frac{\alpha}{2}} \dfrac{s}{\sqrt{n}}$.

5.2.1 Example—Inference on a Population Mean μ

Hand et al. (1994) presents a data set, reproduced in data(vocab), containing the scores on a vocabulary test of a sample of 54 students from a study population. Assume that the test was constructed to have a mean score of 10 in the general population. We desire to assess whether the mean score of the study population is also $\mu = 10$. Assuming that standard deviation for the study population is not known, we wish to calculate a 95% confidence interval for μ and to test $H_0: \mu = 10$ vs $H_1: \mu \neq 10$.

We begin by looking at a stem-and-leaf display of the sample data to see if the underlying assumption of normality is tenable. We observe in Figure 5.3 that the sample is slightly positively skewed with one high value that may be considered an outlier. Based on the Central Limit Theorem, the t-based procedures in Figure 5.4 are justified here. The small p-value ($p \approx 3_{10}^{-14}$) is a strong evidence that μ is not 10. The 95% confidence interval (12.30, 13.44) suggests that the mean score is close to 12.9 in the study population.

We examine a nonparametric approach to this problem in Section 16.2.

```
> stem(vocab$score, scale=2)

  The decimal point is at the |

   9 | 0
  10 | 0000
  11 | 0000000000000
  12 | 0000000
  13 | 000000000
  14 | 000000000
  15 | 0000
  16 | 00000
  17 | 0
  18 |
  19 | 0
```

Fig. 5.3 Stem-and-leaf display of vocabulary scores.

5.3 Confidence Interval on the Variance or Standard Deviation of a Normal Population

Let the (unbiased) estimator of σ^2 based on a sample of size n be denoted s^2. Then $(n-1)s^2/\sigma^2$ has a χ^2 distribution with df $= n-1$. Thus

$$P\left(\chi^2_{\frac{\alpha}{2},n-1} < (n-1)\,s^2/\sigma^2 < \chi^2_{1-\frac{\alpha}{2},n-1}\right) = 1 - \alpha$$

Inverting this statement leads to the $100(1-\alpha)\%$ confidence interval for σ^2:

$$\left(\frac{(n-1)s^2}{\chi^2_{1-\frac{\alpha}{2},n-1}}, \frac{(n-1)s^2}{\chi^2_{\frac{\alpha}{2},n-1}}\right)$$

If instead a CI on σ is desired, take the square roots of both the lower and upper limits in the above. We graph the estimation of a confidence interval in Figure 5.5.

The distribution of $(n-1)s^2/\sigma^2$ can also be used to conduct a test about σ^2 (or σ). For example, to test $H_0: \sigma^2 \leq \sigma_0^2$ vs $H_1: \sigma^2 > \sigma_0^2$, the p-value is $1 - \mathcal{F}_{\chi^2_{n-1}}\left((n-1)s^2/\sigma_0^2\right)$. Tests of the equality of two or more variances are addressed in Section 6.10.

```
> vocab.t <- t.test(vocab$score, mu=10)

> vocab.t

One Sample t-test

data:  vocab$score
t = 10.08, df = 53, p-value = 6.372e-14
alternative hypothesis: true mean is not equal to 10
95 percent confidence interval:
 12.30 13.44
sample estimates:
mean of x
    12.87
```

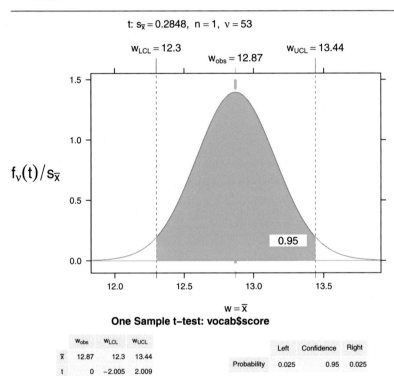

$$t: s_{\bar{x}} = 0.2848, \ n = 1, \ \nu = 53$$

$w_{LCL} = 12.3$ $w_{obs} = 12.87$ $w_{UCL} = 13.44$

$f_\nu(t)/s_{\bar{x}}$

0.95

$w = \bar{x}$

One Sample t–test: vocab$score

	w_{obs}	w_{LCL}	w_{UCL}
$\bar{x}$	12.87	12.3	13.44
t	0	−2.005	2.009

	Left	Confidence	Right
Probability	0.025	0.95	0.025

Fig. 5.4 *t*-test and *t*-based confidence interval of vocabulary scores.

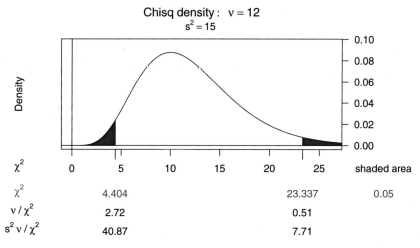

Fig. 5.5 Confidence interval for variance assuming a chi-square distribution with $v = 12$ degrees of freedom and an observed $s^2 = 15$. The estimated 95% confidence interval on σ^2 is $(7.71, 40.87)$. By taking the square root, we find the estimated 95% confidence interval on σ is $(2.777, 6.393)$.

5.4 Comparisons of Two Populations Based on Independent Samples

Two populations are often compared by constructing confidence intervals on the difference of the population means or proportions. In this discussion it is assumed that random samples are independently selected from each population.

5.4.1 Confidence Intervals on the Difference Between Two Population Proportions

The need for confidence intervals on the difference of two proportions is frequently encountered. We might wish to estimate the difference in the proportions of voters in two populations who favor a particular candidate, or the difference in the proportions of defectives produced by two company locations.

Labeling the populations as 1 and 2, the traditional confidence interval, assuming that both populations are large and that neither proportion is close to either 0 or 1, is

$$(\hat{p}_1 - \hat{p}_2) \pm z_{\frac{\alpha}{2}} \sqrt{\frac{\hat{p}_1(1 - \hat{p}_1)}{n_1} + \frac{\hat{p}_2(1 - \hat{p}_2)}{n_2}} \tag{5.9}$$

Agresti and Caffo (2000) also provided an improved confidence interval for this situation, which again provides confidence closer to $100(1-\alpha)\%$ than the preceding interval. For $i = 1, 2$, let $\tilde{p}_i = \frac{Y_i+1}{n_i+2}$, i.e., revise the estimate of p_i by adding one success and one failure to both samples. Then the improved interval is

$$(\tilde{p}_1 - \tilde{p}_2) \pm z_{\frac{\alpha}{2}} \sqrt{\frac{\tilde{p}_1(1 - \tilde{p}_1)}{n_1 + 2} + \frac{\tilde{p}_2(1 - \tilde{p}_2)}{n_2 + 2}} \qquad (5.10)$$

To test the null hypothesis H_0: $p_1 - p_2$ the appropriate statistic is

$$z = \frac{\hat{p}_1 - \hat{p}_2}{\sqrt{\hat{p}(1 - \hat{p})\left(\dfrac{1}{n_1} + \dfrac{1}{n_2}\right)}} \qquad (5.11)$$

where $\hat{p} = \dfrac{n_1 p_1 + n_2 p_2}{n_1 + n_2}$.

Notice the distinction between the standard error portions of Equations 5.9 and 5.10. The standard error in the test statistic 5.11 is calculated under the assumption that the null hypothesis is true. The larger standard error in 5.9 cannot utilize this assumption.

5.4.2 Confidence Interval on the Difference Between Two Means

For a CI on a difference of two means under the assumption that the population variances are unknown, there are two cases. If the variances can be assumed to be equal, their common value is estimated as a weighted average of the two individual sample variances. In general, the process of calculating such meaningfully weighted averages is referred to as *pooling*, and the result in this context is called a pooled variance:

$$s_p^2 = \frac{(n_1 - 1)s_1^2 + (n_2 - 1)s_2^2}{n_1 + n_2 - 2} \qquad (5.12)$$

The pooled estimator s_p^2 has more degrees of freedom (uses more information) than either s_1^2 or s_2^2 for the estimation of the common population variance. When the pooled variance is used as the denominator of F-tests it provides a more powerful test than either of the components, and therefore it is preferred for this purpose. Then the CI is

$$(\bar{y}_1 - \bar{y}_2) \pm t_{\frac{\alpha}{2},n_1+n_2-2} \, s_p \sqrt{\frac{1}{n_1} + \frac{1}{n_2}}$$

In the case where the variances cannot be assumed equal, there are two procedures. The Satterthwaite option is

$$(\bar{y}_1 - \bar{y}_2) \pm t_{\frac{\alpha}{2},\mathrm{df}} \sqrt{\frac{s_1^2}{n_1} + \frac{s_2^2}{n_2}}$$

where df is the integer part of

$$\frac{\left(\dfrac{s_1^2}{n_1} + \dfrac{s_2^2}{n_2}\right)^2}{\dfrac{\left(\dfrac{s_1^2}{n_1}\right)^2}{n_1 - 1} + \dfrac{\left(\dfrac{s_2^2}{n_2}\right)^2}{n_2 - 1}}$$

The Satterthwaite option is sometimes referred to as the Welch option.

The Cochran option is

$$(\bar{y}_1 - \bar{y}_2) \pm t \sqrt{\frac{s_1^2}{n_1} + \frac{s_2^2}{n_2}}$$

where $t = \dfrac{w_1 t_1 + w_2 t_2}{w_1 + w_2}$, $\quad w_i = s_i^2/n_i$, $\quad$ and $\quad t_i$ is $t_{\frac{\alpha}{2},(n_i-1)}$.

The Satterthwaite option is more commonly used than the Cochran option. In practice, they lead to similar results.

5.4.3 Tests Comparing Two Population Means When the Samples Are Independent

There are two situations to consider with independent samples. When the populations may be assumed to have a common unknown variance σ, the calculated t statistic is

$$t_{\mathrm{calc}} = \frac{\bar{y}_1 - \bar{y}_2}{s_p \sqrt{\frac{1}{n_1} + \frac{1}{n_2}}} \tag{5.13}$$

where s_p was defined in Equation (5.12) and t_{calc} has $n_1 + n_2 - 2$ degrees of freedom.

When the two samples might have different unknown variances, then the test is based on

$$s_{(\bar{y}_1 - \bar{y}_2)} = \sqrt{\mathrm{var}\,(\bar{y}_1 - \bar{y}_2)} = \sqrt{\frac{s_1^2}{n_1} + \frac{s_2^2}{n_2}} \quad \text{and} \quad t_{\mathrm{calc}} = \frac{\bar{y}_1 - \bar{y}_2}{s_{(\bar{y}_1 - \bar{y}_2)}} \quad (5.14)$$

In either case, we consider one of the three tests in Table 5.3.

Table 5.3 Confidence intervals and tests for two population means. When the samples are independent and we can assume a common unknown variance, use $s_{\Delta\bar{y}} = s_p \sqrt{\frac{1}{n_1} + \frac{1}{n_2}}$ and t_{calc} as given by Equation (5.13). When the samples are independent and we assume different unknown variances, use $s_{\Delta\bar{y}} = s_{(\bar{y}_1 - \bar{y}_2)}$ and t_{calc} as given by Equation (5.14). When the samples are paired, use $s_{\Delta\bar{y}} = s_{\bar{d}}$ and t_{calc} as given by Equation (5.15).

| | | Tests | | Confidence Interval | |
| | | Rejection | | | |
H_0	H_1	Region	p-value	Lower	Upper
$\mu_1 \le \mu_2$	$\mu_1 > \mu_2$	$t_{\mathrm{calc}} > t_\alpha$	$P(t > t_{\mathrm{calc}})$	$\big((\bar{y}_1 - \bar{y}_2) - t_\alpha s_{\Delta\bar{y}},$	$\infty \quad\big)$
$\mu_1 \ge \mu_2$	$\mu_1 < \mu_2$	$t_{\mathrm{calc}} < -t_\alpha$	$P(t < t_{\mathrm{calc}})$	$\big(\qquad\qquad -\infty,$	$(\bar{y}_1 - \bar{y}_2) + t_\alpha s_{\Delta\bar{y}}\big)$
$\mu_1 = \mu_2$	$\mu_1 \ne \mu_2$	$\lvert t_{\mathrm{calc}}\rvert > t_{\frac{\alpha}{2}}$	$2P(t > \lvert t_{\mathrm{calc}}\rvert)$	$\big((\bar{y}_1 - \bar{y}_2) - t_{\frac{\alpha}{2}} s_{\Delta\bar{y}},$	$(\bar{y}_1 - \bar{y}_2) + t_{\frac{\alpha}{2}} s_{\Delta\bar{y}}\big)$

R uses the `t.test` function which calculates a one-sample, two-sample, or paired t-test, or a Welch modified two-sample t-test. The Welch modification is synonymous with the Satterthwaite method.

The example in Tables 5.4 and 5.5 and Figure 5.6 compares two means where the samples are independent and assumed to have a common unknown variance. Table 5.4 shows the t-test calculated with the `t.test` function. Table 5.5 calculates the t-value manually using the definitions in Equations 5.12 and 5.13. Figure 5.6 plots the result of the `t.test` with the `NTplot` function.

Table 5.4 Select the subset of the `cereals` dataset for "Cold cereal" and manufacturers "G" and "K". Use `t.test` to compare their mean carbohydrate values assuming independent samples with a common unknown variance. The result from the `t.test` is plotted in Figure 5.6.

```
> data(cereals)

> table(cereals[,c("mfr","type")])
   type
mfr  C  H
  A  0  1
  G 22  0
  K 23  0
  N  5  1
  P  9  0
  Q  7  1
  R  8  0

> C.KG <- cereals$type=="C" & cereals$mfr %in% c("K","G")

> cerealsC <- cereals[C.KG, c("mfr", "carbo") ]

> cerealsC$mfr <- factor(cerealsC$mfr)

> bwplot(carbo ~ mfr, data=cerealsC) +
+ dotplot(carbo ~ mfr, data=cerealsC)

> t.t <- t.test(carbo ~ mfr, data=cerealsC, var.equal=TRUE)

> t.t

    Two Sample t-test

data:  carbo by mfr
t = -0.3415, df = 43, p-value = 0.7344
alternative hypothesis: true difference in means is not equal to 0
95 percent confidence interval:
 -2.784  1.978
sample estimates:
mean in group G mean in group K
          14.73           15.13
```

Table 5.5 The t-value $-.3415$ in Table 5.4 is calculated manually.

```
> mm <- tapply(cerealsC$carbo, cerealsC$mfr, mean)

> vv <- tapply(cerealsC$carbo, cerealsC$mfr, var)

> ll <- tapply(cerealsC$carbo, cerealsC$mfr, length)

> s2p <- ((ll-1) %*% vv) / sum(ll-1)

> tt <- -diff(mm) / (sqrt(s2p) * sqrt(sum(1/ll)))

> tt
          [,1]
[1,] -0.3415
```

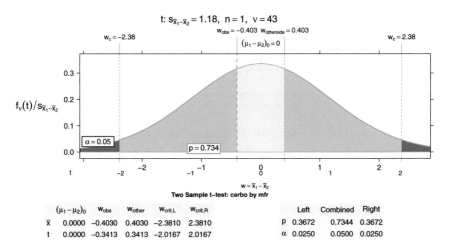

Fig. 5.6 Show the NTplot(t.t, zaxis=TRUE) of the t-test in Table 5.4. There are two horizontal scales on the bottom axis of the plot. The $w = \bar{x}_1 - \bar{x}_2$ scale is the top scale and the t scale is the bottom scale. Specific interesting values in the w scale are identified on the top axis. $w_{obs} = -.403$ and its symmetrically placed $w_{otherside} = .403$ are very close to the center of the graph, illustrating that the observation is not anywhere near the rejection region $|W| > 2.38$.

5.4.4 Comparing the Variances of Two Normal Populations

We assume here that independent random samples are available from both populations. The F distribution is used to compare the variances σ_1^2 and σ_2^2 of two normal populations. Let s_1^2 and s_2^2 be the variances of independent random samples of size $n_i, i = 1, 2$ from these populations.

To test

$$H_0: \sigma_1^2 \leq \sigma_2^2$$

vs

$$H_1: \sigma_1^2 > \sigma_2^2$$

define $F = s_1^2/s_2^2$ and reject H_0 if F is sufficiently large. The p-value of the test is $1 - \mathcal{F}_{F_{(n_1-1, n_2-1)}}(F)$. The power of this and other F-tests is sensitive to the second (denominator) df parameter and is usually not adequate unless this df ≥ 20.

A $100(1 - \alpha)\%$ confidence interval for a ratio of variances of two normal populations, σ_1^2/σ_2^2, is

$$\left(\frac{s_1^2}{s_2^2} \frac{1}{F_{\text{low}}}, \frac{s_1^2}{s_2^2} F_{\text{high}} \right)$$

where

F_{low} is $F_{1-\frac{\alpha}{2}, n_1-1, n_2-1}$, the upper $100(1 - \frac{\alpha}{2})$ percentage point of an F distribution with $n_1 - 1$ and $n_2 - 1$ degrees of freedom, and

F_{high} is $F_{1-\frac{\alpha}{2}, n_2-1, n_1-1}$, the upper $100(1 - \frac{\alpha}{2})$ percentage point of an F distribution with $n_2 - 1$ and $n_1 - 1$ degrees of freedom.

An extension to testing the homogeneity of more than two population variances will be presented in Section 6.10.

5.5 Paired Data

Sometimes we wish to compare the mean change in a measurement observed on an experimental unit under two different conditions. For example:

1. Compare the subject knowledge of students before and after they receive instruction on the subject.

2. Compare the yield per acre of a population of farms for a crop grown with two different fertilizers.

3. Compare the responses of patients to both an active drug and a placebo, when they are administered each of them in sequential random order with a suitable "washout" period between the administrations.

This "matched pairs" design is superior to a design of the same total size using independent samples because (in illustrations 1 and 3 above) the person to person variation is removed from the comparison of the two administrations, thereby improving the precision of this comparison. The principles of designing experiments to account for and remove extraneous sources of variation are discussed in more detail in Chapter 13.

It is assumed that the populations have a common variance and are approximately normal. Let $y_{11}, y_{12}, \ldots, y_{1n}$ be the sample of n items from the population under the first condition, having mean μ_1, and similarly let $y_{21}, y_{22}, \ldots, y_{2n}$ be the sample from the population under the second condition, having mean μ_2.

Define the n differences $d_1 = y_{11} - y_{21}$, $d_2 = y_{12} - y_{22}, \ldots, d_n = y_{1n} - y_{2n}$. Let $\bar{d}$ and s_d be the mean and standard deviation, respectively, of the sample of n d_i's. Then an approximate $100(1-\alpha)\%$ confidence interval on the mean difference $\mu_1 - \mu_2$ is $\bar{d} \pm t_{\frac{\alpha}{2}, n-1} s_{\bar{d}}$ where $s_{\bar{d}} = s_d / \sqrt{n}$. Tests of hypotheses proceed similarly to t-tests for two independent samples. Table 5.3 can still be used, but with

$$s_{\bar{d}} = s_d / \sqrt{n}, \quad \text{and} \quad t_{\text{calc}} = \frac{\bar{d}}{s_{\bar{d}}} \tag{5.15}$$

with degrees of freedom $n - 1$.

5.5.1 Example—t-test on Matched Pairs of Means

Woods et al. (1986), later in Hand et al. (1994), investigate whether native English speakers find it easier to learn Greek than native Greek speakers learning English. Thirty-two sentences are written in both languages. Each sentence is scored according to the quantity of errors made by an English speaker learning Greek and by a Greek speaker learning English. It is desired to compare the mean scores of the two groups. The data are available as data(teachers); the first column is the error score on the English version of the sentence and the second column is the error score on the Greek version of the sentence.

These are 32 pairs of observations because the same sentence is evaluated in both languages. It would be incorrect to regard these as independent samples. The dotplot in Figure 5.7 reveals that for most sentences the English version shows fewer errors. The stem-and-leaf of the differences in Figure 5.8a shows the difference variable is positively skewed so that a transformation, as discussed in Section 4.8, is required. Care must be used with a power transformation because many of the differences are negative. The smallest difference is -16. Therefore, we investigate a

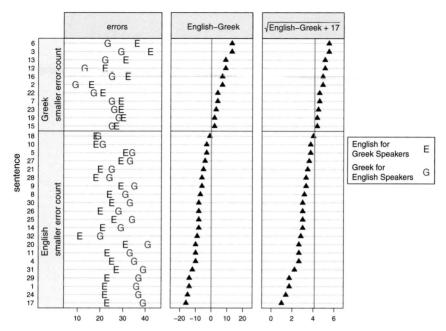

Fig. 5.7 Dotplot of language difficulty scores. The difficulty in learning each of 32 sentences written in English for Greek speakers (marked English) and written in Greek for English speakers (marked Greek) is noted. The panels are defined by placing the sentences in which the English version showed fewer errors on the bottom and the sentences in which the Greek version showed fewer errors on the top. The sentences have been ordered by the difference in the English and Greek error scores. The left panels show the observed error scores. The center panels show the differences, English–Greek, of the error scores. The right panels show the square root transformed differences, $\sqrt{\text{English–Greek} + 17}$. The t-tests in Table 5.8 will be based on the differences and the transformed differences.

square root transformation following the addition of 17 to each value. The second stem-and-leaf in Figure 5.8b illustrates that this transformation succeeds in bringing the data closer to symmetry. Since a difference of zero in the original scale corresponds to a transformed difference of $\sqrt{17} \approx 4.123$, the null hypothesis of equal difficulty corresponds to a comparison of the sample means in the transformed scale to 4.123, not to 0. The observed p-value is .0073, showing a very clear difference in difficulty of learning the two languages. For comparison, the t-test on the untransformed differences show a p-value of only .0346.

```
> stem(teachers$"English-Greek")          > stem(sqrt(teachers$"English-Greek" + 17),
                                           +      scale=.5)
  The decimal point is 1 digit(s) to the
  right of the |                             The decimal point is at the |

   -1 | 65                                    1 | 0477
   -1 | 442000                                2 | 26668
   -0 | 988887665                             3 | 00002335677
   -0 | 4331                                  4 | 04456699
    0 | 22344                                 5 | 1155
    0 | 7799
    1 | 33
```

```
> t.test(teachers$"English-Greek")         > t.test(sqrt(teachers$"English-Greek" + 17),
                                           +        mu=sqrt(17))

One Sample t-test                          One Sample t-test

data:  teachers$"English-Greek"            data:  sqrt(teachers$"English-Greek" + 17)
t = -2.211, df = 31, p-value = 0.03457     t = -2.871, df = 31, p-value = 0.00731
alternative hypothesis:                    alternative hypothesis:
    true mean is not equal to 0                true mean is not equal to 4.123
95 percent confidence interval:            95 percent confidence interval:
 -6.2484 -0.2516                            3.086 3.947
sample estimates:                          sample estimates:
mean of x                                  mean of x
    -3.25                                      3.517
```

a. Original Scale   b. Transformed Scale

Fig. 5.8 Stem-and-leaf display and *t*-test of sentence difference scores from Figure 5.7 in the original scale and in the offset square-root transformed scale.

5.6 Sample Size Determination

Deciding on an appropriate sample size is a fundamental aspect of experimental design. In this section we provide discussions of the minimum required sample size for some situations of inference about population means:

- A confidence interval on μ with specified width W and confidence coefficient $100(1 - \alpha)\%$.

- A test about μ having specified Type I error α, and power $1 - \beta$ at a specified distance δ from the null hypothesized parameter.

These are key design objectives for many experiments with modest inferential goals. Specialized software exists for the purpose of determining sample sizes in a vast array of inferential situations. But our discussion here is limited to a few commonly encountered situations for which the formulas are sometimes mentioned in elementary statistics texts.

We assume throughout this discussion that the sample size will be large enough to guarantee that the standardized test statistic is approximately normally distributed. If, as is usual, a sample size calculation does not yield an integer, it is conservative to take n as the next-higher integer. The sample size formulas here are all the result of explicitly solving a certain equation for n. In situations not discussed here, an explicit solution for n may not exist, and the software may produce an iterative solution for n.

5.6.1 Sample Size for Estimation

Since the width of a confidence interval can be expressed as a function of the sample size, the solution of the problem of sample size for a confidence interval is straightforward in the case of a single sample.

For a CI on a single mean, assuming a known population variance σ^2,

$$n = \frac{4\sigma^2\left(\Phi^{-1}(1 - \frac{\alpha}{2})\right)^2}{W^2} \tag{5.16}$$

where Φ^{-1} is the inverse cumulative distribution of a standard normal distribution defined in Section J.1.9. Equation 5.16 is found by solving Equation 5.1 for n when we want the width of the confidence interval to be $W = 2\,z_{\frac{\alpha}{2}}\frac{\sigma}{\sqrt{n}}$. If σ^2 is unknown, a reasonable guess may be made in its place. (Note that the sample variance is not known prior to selecting the sample.) If we are unable to make a reasonable guess, an ad hoc strategy would be to take a small pilot sample of n_0 items and replace σ in the formula with the standard deviation of the pilot sample. Then if the calculation results in a recommended n greater than n_0, one samples $n - n_0$ additional items.

The required sample size for the Agresti and Caffo CI on a single proportion, Equation (5.3), is

$$n = \frac{\left(\Phi^{-1}(1 - \frac{\alpha}{2})\right)^2}{W^2} - 4 \tag{5.17}$$

This formula is based on the normal approximation to the binomial distribution. Many statistics texts contain charts for finding the required sample size based on the exact binomial distribution.

5.6.2 Sample Size for Hypothesis Testing

For hypothesis testing we are interested in controlling the specified Type II error probability β when the unknown parameter being tested is a distance δ from the null hypothesized value. For a one-tailed test on the mean of a population with known variance σ^2, use

$$n = \sigma^2 \left(\Phi^{-1}(1 - \alpha) + \Phi^{-1}(1 - \beta) \right)^2 / \delta^2 \tag{5.18}$$

We illustrate Equation 5.18 in Figure 5.9.

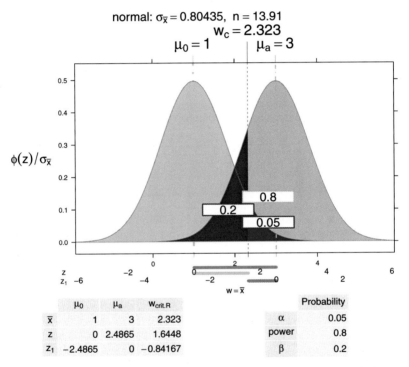

	μ_0	μ_a	$w_{crit.R}$		Probability
$\bar{x}$	1	3	2.323	α	0.05
z	0	2.4865	1.6448	power	0.8
z_1	-2.4865	0	-0.84167	β	0.2

Fig. 5.9 Sample size and power for the one-sample, one-sided normal test. This figure illustrates Equation 5.18. Both the null and alternative distributions are normal with the same standard error $\sigma = 3$. There are three colored line segments in the horizontal axis region. The top line segment (light gray) on the $w = \bar{x}$ scale is $\delta = \mu_a - \mu_0 = 2$ w-units wide, going from $w = \mu_0 = 1$ to $w = \mu_a = 3$. The middle line segment (light blue) and the bottom line segment (pink) together also are $\delta = 2 w$ units wide. The middle segment is $2.323 - 1 = 1.323$ w units wide which is equal to 1.6448 z units. The bottom segment is $3 - 2.323 = 0.677$ w units wide which is equal to $.84167$ z units We know $\sigma = 3$ and we know that $\sigma_{\bar{x}} = \sigma / \sqrt{n}$. We need to solve for $n = 3^2(1.6488 + .84167)^2/2^2 = 13.96$. We round up to use $n = 14$.

For a two-tailed test, use

$$n = \sigma^2 \left(\Phi^{-1}(1 - \tfrac{\alpha}{2}) + \Phi^{-1}(1 - \beta) \right)^2 / \delta^2 \qquad (5.19)$$

For testing the equality of the means of two populations with a common variance, with δ now equal to the mean difference under the alternative hypothesis, use

$$n = 2\sigma^2 \left(\Phi^{-1}(1 - \alpha) + \Phi^{-1}(1 - \beta) \right)^2 / \delta^2 \qquad (5.20)$$

for the one-tailed test, and

$$n = 2\sigma^2 \left(\Phi^{-1}(1 - \tfrac{\alpha}{2}) + \Phi^{-1}(1 - \beta) \right)^2 / \delta^2 \qquad (5.21)$$

for the two-tailed test.

When the variance σ^2 is unknown and has to be estimated with s^2 from the sample, the formulas are more difficult because the inverse t cumulative function for the alternative depends on the standard deviation through the noncentrality parameter $(\mu_1 - \mu_0)/(\sigma/\sqrt{n})$. The t formulas might require several iterations as the degrees of freedom, hence the critical values are a function of the sample size.

Tables 5.6, 5.7, and 5.8 show sample size calculations for the one-sample, one-sided test. The example is done three times. Table 5.6 shows the calculation using Equation 5.18 when σ^2 is assumed and the normal equations apply. Table 5.7 uses the R function power.t.test which solves the t equations efficiently. Table 5.8 iterates the definitions for the t distribution. Figure 5.10 shows the power plot for the n value in Table 5.7 and one of the n values in Table 5.8.

Lastly, consider attempting to detect a difference between a proportion p_1 and a proportion p_2. The required common sample size for the one-tailed test is

$$n = \frac{\left(p_1(1 - p_1) + p_2(1 - p_2)\right)\left(\Phi^{-1}(1 - \alpha) + \Phi^{-1}(1 - \beta)\right)^2}{(p_1 - p_2)^2} \qquad (5.22)$$

From the preceding pattern, you should be able to deduce the modification for the two-tailed test (see Exercise 5.16).

Table 5.6 We calculate the sample size for a one-sided, one-sample test for the normal distribution with the assumption that the variance is known. In Tables 5.7 and 5.8 we show the same calculations for the *t*-test under the assumption that the variance has been estimated from the sample.

```
> ## one sided
> alpha <- .05

> power <- .80

> beta <- 1-power

> delta <- 1

> sd <- 2

> ## Approximation using formula assuming normal is appropriate
> sd^2*(qnorm(1-alpha) + qnorm(1-beta))^2 / delta^2
[1] 24.73

> ## [1] 24.73
> ## n is slightly smaller with the normal assumption.
>
```

Table 5.7 We calculate the sample size for a one-sided, one-sample *t*-test using the power.t.test function. We show the same calculation manually in Table 5.8. We show a static plot of the result in the left column of Figure 5.10. We also show the **shiny** code to specify a dynamic plot.

```
> ## solve using power.t.test
> PTT <-
+ power.t.test(delta=delta, sd=sd, sig.level=alpha, power=power,
+              type="one.sample", alternative="one.sided")

> PTT

     One-sample t test power calculation

              n = 26.14
          delta = 1
             sd = 2
      sig.level = 0.05
          power = 0.8
    alternative = one.sided


> NTplot(PTT, zaxis=TRUE)  ## static plot

> ## NTplot(PTT, zaxis=TRUE, shiny=TRUE)  ## dynamic plot
>
```

Table 5.8 We manually calculate the sample size for a one-sided, one-sample t-test to illustrate the iterative process directly. The power.t.test function does this much more efficiently (see Table 5.7). The iterative process starts with an initial sample size n_0 and calculates the critical value $t_{c,0}$ using the central t distribution for that sample size. The second step in the process is to evaluate the power associated with that critical value assuming fixed δ and a series of sample sizes and their associated df and ncp. For the next iterate choose as the new sample size n_1 the sample size whose power is closest to the target power. Calculate a new critical value $t_{c,1}$ and then a new set of powers associated with that critical value. Continue until convergence, meaning the new sample size is the same as the previous one.

```
> ## solve manually with t distribution.  Use ncp for alternative.
> n0 <- 30 ## pick an n0 for starting value

> t.critical <- qt(1-alpha, df=n0-1)

> t.critical
[1] 1.699

> ## [1] 1.699
>
> ## a series of n values
> nn <- 23:30

> names(nn) <- nn

> nn
23 24 25 26 27 28 29 30
23 24 25 26 27 28 29 30

> ## find the power for a series of n values for the specified critical value
> pt(t.critical, df=nn-1, ncp=delta/(sd/sqrt(nn)), lower=FALSE)
    23     24     25     26     27     28     29     30
0.7568 0.7722 0.7868 0.8006 0.8136 0.8258 0.8374 0.8483

> ##     23     24     25     26     27     28     29     30
> ## 0.7568 0.7722 0.7868 0.8006 0.8136 0.8258 0.8374 0.8483
>
> ## recalculate critical value with new n=26
> t.critical <- qt(1-alpha, df=26-1)

> t.critical
[1] 1.708

> ## find the power for a series of n values for the new critical value
> pt(t.critical, df=nn-1, ncp=delta/(sd/sqrt(nn)), lower=FALSE)
    23     24     25     26     27     28     29     30
0.7540 0.7695 0.7842 0.7981 0.8112 0.8235 0.8352 0.8461

> ##     23     24     25     26     27     28     29     30
> ## 0.7540 0.7695 0.7842 0.7981 0.8112 0.8235 0.8352 0.8461
> ## conclude n between 26 and 27
>
```

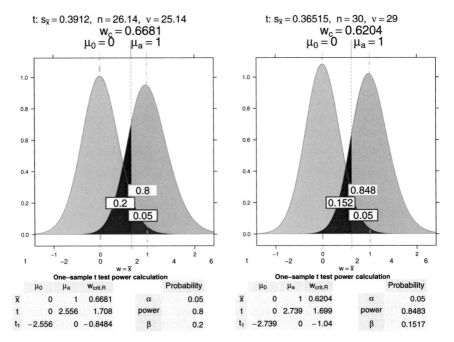

Fig. 5.10 Sample size and power figures for the one-sample, one-sided *t*-test. The left figure shows the sample size *n*=26.14 calculated in Table 5.7. The right figure shows the starting position with *n*=30 from Table 5.8. When the sample size *n* is larger (on the right), the df goes up, the height of the densities (both null and alternative) go up, the densities become thinner, the critical value in the *t* scale and in the $\bar{x}$ scale goes down. The alternative distribution is noncentral *t*, has a different maximum height, and is not symmetric.

5.7 Goodness of Fit

Goodness-of-fit tests are used to assess whether a dataset is consistent with having been sampled from a designated hypothesized distribution. In this section we discuss two general goodness-of-fit tests, the Chi-Square Goodness-of-Fit Test and the Kolmogorov–Smirnov Goodness-of-Fit Test. For testing goodness of fit to specific distributions, there may be better (more powerful) specialized tests than these. For example, the Shapiro–Wilk test of normality (`shapiro.test`) is more powerful than either general test.

Since many statistics procedures assume an underlying normal distribution, a test of goodness of fit to normal, either before or after transformation, is frequently performed. Occasionally, analysts need to check for fit to other distributions. For example, it is often the case that the distribution of a test statistics is known asymptotically (i.e., if the sample is "large"), but not if the sample is of modest size. It is therefore of interest to investigate how large a sample is needed for the asymptotic distribution to be an adequate approximation. This requires a series of goodness-of-fit tests to the asymptotic distribution. In Chapter 15, we will learn in our discussion of the analysis of contingency table data that the distribution of $\chi^2 = \sum \frac{(O-E)^2}{E}$ is approximately chi-square provided that no cell sizes are too small. A determination of the ground rule for "too small" required tests of goodness of fit to chi-square distributions with appropriate degrees of freedom.

This class of tests assesses whether a sample may be assumed to be taken from a null hypothesized distribution.

5.7.1 Chi-Square Goodness-of-Fit Test

The chi-square distribution may be used to conduct goodness-of-fit tests, i.e., ones of the form

H_0: the data are from a [specified population]

vs

H_1: the data are from some other population

For certain specific populations, including normal ones, other specialized tests are more powerful.

The test begins by partitioning the population into k classes or categories. For a discrete population the categories are the possible values; for a continuous population the choice of a decomposition is rather arbitrary, and the ultimate conclusion may well depend on the selected size of k and the selected partition.

The test statistic is the same as that used for contingency tables. For each category, calculate from the probability distribution the theoretical or expected frequency E. If over all k categories, there is a substantial discrepancy between the k observed frequencies O and the k E's, then H_0 is rejected. The measure of discrepancy is the test statistic $\chi^2 = \sum \frac{(O-E)^2}{E}$. A "large" value of χ^2 is evidence against H_0. If the total sample size, $n = \sum O = \sum E$, is sufficiently "large", χ^2 is approximately chi-square distributed and the p-value is approximately the chi-square tail probability associated with χ^2 with $k - 1$ degrees of freedom.

For adequacy of the chi-square approximation it is suggested that all expected frequencies be at least 5. If this is not the case, the analyst may consider combining adjacent categories after which this condition is met. Then k represents the number of categories following such combining.

Sometimes, the statement of the null hypothesis is so vague that calculation of expected frequencies requires that some parameters be estimated from the data. In such instances, the df is further reduced by the number of such parameters estimated. This possibility is illustrated in Example 5.7.3.

5.7.2 Example—Test of Goodness-of-Fit to a Discrete Uniform Distribution

A six-sided die (singular of the word *dice*) is rolled 30 times with the following outcomes: 1, 3 times; 2, 7 times; 3, 5 times; 4, 8 times; 5, 1 time; and 6, 6 times. Test whether the die is fair.

A fair die is one that has a discrete uniform distribution on 1, 2, 3, 4, 5, 6. Each of these six possibilities has $\frac{1}{6}$ chance of occurring, and all six E's are $30(\frac{1}{6}) = 5$. Then

$$\chi^2 = \frac{(3-5)^2}{5} + \ldots + \frac{(6-5)^2}{5} = 6.8$$

and the *p*-value from χ_5^2 is 0.236. Hence these 30 observations do not provide evidence to refute the fairness of the die. We show the calculations in Table 5.9 and the plot of the test in Figure 5.11.

Table 5.9 Test of Goodness-of-Fit to a Discrete Uniform Distribution. The test is plotted in Figure 5.11.

```
> dice <- sample(rep(1:6, c(3,7,5,8,1,6)))

> dice
 [1] 4 6 4 2 3 2 4 4 6 3 6 4 3 2 3 4 6 2 6 2 1 4 3 5 1 2 1 6
[29] 4 2

> table(dice)
dice
1 2 3 4 5 6
3 7 5 8 1 6

> chisq.test(table(dice))

Chi-squared test for given probabilities

data:  table(dice)
X-squared = 6.8, df = 5, p-value = 0.2359
```

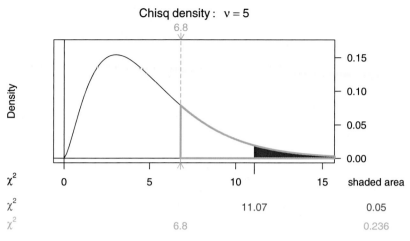

Fig. 5.11 Plot of the hypothesis test of Table 5.9. The observed value $\chi^2 = 6.8$ shows $p = 0.236$ and is in the middle of the do-not-reject region,

Table 5.10 Observed and expected frequencies for the goodness-of-fit example in Section 5.7.3.

Y	O	E	$\dfrac{(O-E)^2}{E}$
0	13	6.221	7.388
1	18	20.736	0.361
2	20	27.648	2.116
3	18	18.432	0.010
4	6	6.144	0.003
5	5	0.819	21.337
			31.215

5.7.3 Example—Test of Goodness-of-Fit to a Binomial Distribution

In a certain community, there were 80 families containing exactly five children. It was noticed that there was an excess of boys among these. It was desired to test whether Y = "number of girls in family" is a binomial r.v. with $n = 5$ and $p = .4$. The expected frequencies calculated from this binomial distribution are shown in Table 5.10 along with the observed frequencies and the calculated χ^2_5 statistic. Then the p-value is, $8.5 \cdot 10^{-6}$, calculated as the tail probability at 31.215 for a chi-square distribution with 5 df. We conclude that the sample data contain more dispersion than does binomial(5, .4). The excess dispersion is visible in the left panel of Figure 5.12.

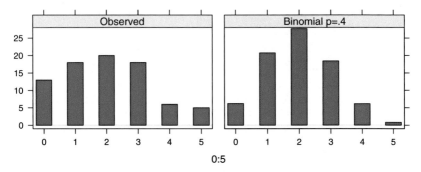

Fig. 5.12 Plot of family size data from Table 5.11. The Observed data is more spread out than the Expected (binomial) data. The sample variance for the Observed is `var(rep(0:5,` `times=Observed)) == 1.987` and the sample variance for the Expected is `var(rep(0:5,` `times=Expected)) == 1.131`.

In this example, the value of the binomial proportion parameter, p, was specified. If instead it had to be estimated, the df would decrease from 5 to 4. We illustrate the calculation of both tests in R in Table 5.11.

5.8 Normal Probability Plots and Quantile Plots

Quantile plots (Q-Q plots) are visual diagnostics used to assess whether (a) a dataset may reasonably be treated as if it were a sample from a designated probability distribution, or (b) whether two datasets show evidence of coming from a common unspecified distribution.

The normal probability plot, an important special case of the more general quantile plot, is used to assess whether data are consistent with a normal distribution. The normal probability plot is a standard diagnostic plot in regression analysis (Chapters 8–11) used to check the assumption of normally distributed residuals. This condition is required for the validity of many of the usual inferences in a regression analysis. If the normality assumption appears to be violated, it is often possible to retain a simple analysis by transforming the data scale, for example by a power transformation, and then reanalyzing and replotting to see if the residuals from the transformed data are close to normal. The choice of transformation may be guided by the interpretation of the normal probability plot.

In R, a normal probability plot is produced with the qqmath function (in **lattice**) or the qqnorm function (in base graphics). function. Normal probability plots are included in the default plots for the results of linear model analyses.

A quantile plot to assess consistency of observed data y_i with a designated distribution is easily constructed. We sort the observed data to get $y_{[i]}$, find the quantiles of the distribution by looking up the fractions $(i - \frac{1}{2})/n$ in the inverse cumulative

Table 5.11 Calculation of p-value for chi-square test with known p and with estimated $\hat{p}$. The Observed and Expected frequencies are plotted in Figure 5.12.

```
> Observed <- c(13, 18, 20, 18, 6, 5)

> names(Observed) <- 0:5

> ## binomial proportion p=.4 is specified
> Expected <- dbinom(0:5, size=5, p=.4)*80

> names(Expected) <- 0:5

> chisq.test(Observed, p=Expected, rescale.p=TRUE)

Chi-squared test for given probabilities

data:  Observed
X-squared = 31.21, df = 5, p-value = 8.496e-06

Warning message:
In chisq.test(Observed, p = Expected, rescale.p = TRUE) :
  Chi-squared approximation may be incorrect

> ## binomial proportion p is calculated from the observations
> p <- sum(Observed * (0:5)/5)/sum(Observed)

> p
[1] 0.4025

> Expected <- dbinom(0:5, size=5, p=p)*80

> names(Expected) <- 0:5

> WrongDF <- chisq.test(Observed, p=Expected, rescale.p=TRUE)
Warning message:
In chisq.test(Observed, p = Expected, rescale.p = TRUE) :
  Chi-squared approximation may be incorrect

> WrongDF

Chi-squared test for given probabilities

data:  Observed
X-squared = 30.72, df = 5, p-value = 1.066e-05


> c(WrongDF$statistic, WrongDF$parameter)
X-squared        df
    30.72      5.00

> ## correct df and p-value
> pchisq(WrongDF$statistic, df=WrongDF$parameter - 1, lower=FALSE)
X-squared
3.498e-06
```

distribution function to get $q_i = F^{-1}((i - \frac{1}{2})/n)$, and then plotting the sorted data $y_{[i]}$ against the quantiles q_i. Consistency is suggested if the points tend to fall along a straight line. A pattern of a departure from a straight-line quantile plot usually suggests the nature of the departure from the assumed distribution. The R one-sample quantile plots (both the **lattice** qqmath and the base graphics qqnorm) default to the usual convention of plotting the data against the theoretical values. Other software and a number of references reverse the axes. Readers of presentations containing quantile plots should be alert to which convention is used, and writers must be sure to label the axes to indicate the convention, because the choice matters considerably for interpretation of departures from compatibility.

A general Q-Q (or quantile-quantile) plot is invoked in R with the base graphics command qqplot(x, y, plot=TRUE), whereby the quantiles of two samples, x and y, are compared. As with a normal probability case, the straightness of the Q-Q plot indicates the degree of agreement of the distributions of x and y, and departure from a well-fitting straight line on an end of the plot indicates the presence of outlier(s). Quoting from the S-Plus online help for qqplot:

> A Q-Q plot with a "U" shape means that one distribution is skewed relative to the other. An "S" shape implies that one distribution has longer tails than the other. In the default configuration (data on the y-axis) a plot from qqnorm that is bent down on the left and bent up on the right means that the data have longer tails than the Gaussian [normal].

For a normal probability plot with default configuration, a plot that is bent up on the left and bent down on the right indicates that the data have shorter tails than the normal. A curved plot that opens upward suggests positive skewness and curvature opening downward suggests negative skewness.

It is possible to construct a Q-Q plot comparing a sample with any designated distribution, not just the normal distribution. In R and S-Plus this is accomplished with the function ppoints(y), which returns a vector of n=length(y) fractions uniformly spaced between 0 and 1 which will be used as input to the quantile (inverse cumulative distribution) function. For example, all three R statements

```
plot(sort(y) ~ qlnorm(ppoints(y)))
qqplot(qlnorm(ppoints(y)), y)
qqmath(y, distribution=qlnorm)
```

produce a lognormal Q-Q plot of the data in y. See Appendix J for the lognormal distribution.

If it is unclear from a normal probability plot whether the data are in fact normal, the issue may be further addressed by a specialized goodness-of-fit test to the normal distribution, the Shapiro–Wilk test. This test works by comparing

$S(y)$ the empirical distribution function of the data, the fraction of the data that is less than or equal to y

with

$\Phi\big((y - \bar{y})/s\big)$ the probability that a normal r.v. Y (with mean $\bar{y}$ and s.d. s) is less than or equal to y

Over the observed sample, $S(y)$ and $\Phi\big((y - \bar{y})/s\big)$ should be highly correlated if the data are normal, but not otherwise. The Shapiro–Wilk statistic W is closely related to the square of this correlation. If the normal probability plot is nearly a straight line, W will be close to 1. A small value of W is evidence of nonnormality. The Shapiro–Wilk test is available in R with the shapiro.test function. For this specific purpose the Shapiro–Wilk test is more powerful than a general goodness-of-fit test such as the Kolmogorov–Smirnov procedure discussed in Section 5.9.

5.8.1 Normal Probability Plots

Figure 5.13 contrasts the appearance of normal probability plots for the normal distribution and various departures from normality. Typically, the plot has these appearances:

- An "S" shape for distributions with thinner tails than the normal.
- An inverted "S" shape for distribution with heavier tails than the normal.

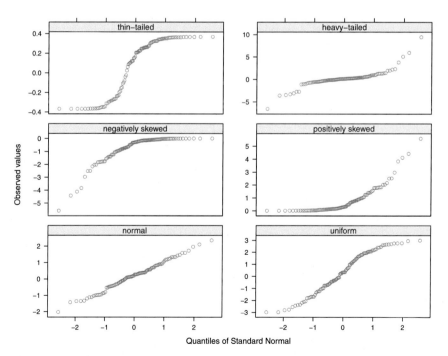

Fig. 5.13 Normal probability plots of data randomly selected from normal and other distributions. The density plots of these variables are in Figure 5.14.

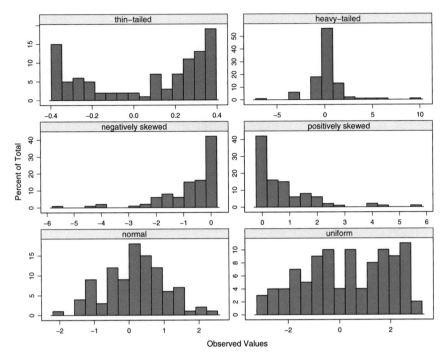

Fig. 5.14 Density plots of the data randomly selected from normal and other distributions. This is the same data whose normal probability plots are shown in Figure 5.13
.

- A "J" shape for positively skewed distributions.
- An inverted "J" shape for negatively skewed distributions.
- Isolated points at the extremes of a plot for distributions having outliers.

5.8.2 Example—Comparing t-Distributions

We compare a random sample of 100 from a t distribution with 5 df to quantiles from a longer-tailed t_3 distribution and from shorter-tailed t_7 and normal distributions. The four superimposed Q-Q plots and a reference 45° line are shown in Figure 5.15.

Note that the picture we get will vary according to the particular random sample selected. In this example the plot against the quantiles of t_5, the same distribution from which the sample was drawn, is close to the 45° line. The longer-tailed t_3 quantiles show a reflected "S" shape. The shorter-tailed t_7 and normal distributions show an "S" shape.

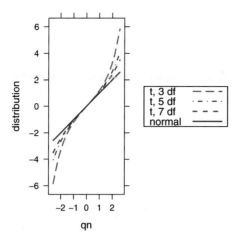

Fig. 5.15 Q-Q plots for the *t* distribution with several different degrees of freedom. The normal is the same as the *t* with infinite degrees of freedom. The scaling here is isometric, the same number of inches per unit on the *x* and *y* scales. The aspect ratio is chosen to place the normal QQ plot exactly on the 45° line.

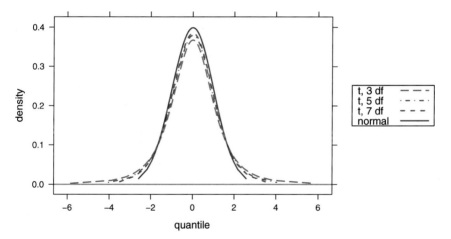

Fig. 5.16 *t* densities. The normal (*t* with infinite degrees of freedom) is the tallest and thinnest. As the degrees of freedom decrease, the center gets less high and the tails have noticeable weight farther away from the center.

Long and short tails refer to the appearance of plots of the density functions. Note that the normal has almost no probability (area) outside of ±2.5. The *t* distributions have more and more probability in the tails of the distribution (larger $|q|$) as the degrees of freedom decrease. The superimposed densities are displayed in Figure 5.16.

5.9 Kolmogorov–Smirnov Goodness-of-Fit Tests

The Kolmogorov–Smirnov goodness-of-fit tests are used to formally assess hypothesis statements concerning probability distributions. The K–S one-sample test tests whether a random sample comes from a hypothesized null distribution. The K–S two-sample test tests whether two independent random samples are coming from the same but unspecified probability distribution. The alternative hypothesis can be either one-sided or two-sided.

The K–S one-sample test involves comparing the maximum discrepancy between the empirical cumulative distribution of the data, defined as $S(y)$ = fraction of the data that is less than or equal to y, and the cumulative distribution function of the hypothesized population being sampled. The K–S two-sample test statistic is the maximum discrepancy between the empirical distribution functions of the two samples.

The Shapiro–Wilk test of normality is more powerful than K–S for assessing normality. The Shapiro–Wilk test statistic W more fully uses the sample than does K–S. If we have data that are close to normal except for one very unusual point, K–S will be more sensitive to this point than W. In general, the K–S procedure focuses on the most extreme departure from the hypothesized distribution while Shapiro–Wilk's assessment based on Q-Q focuses on the average departure.

The K–S tests are performed in R with the function ks.test. See the R help file for ks.test for details. This function can handle both one- and two-sample tests. For the one-sample test, a long list of probability distributions can be specified as the null hypothesis. The parameters of the null distribution can be estimated from the data or left unspecified. With some exceptions, the alternative hypothesis can be "greater" or "less" as well as "two-sided". The interpretation of a one-sided hypothesis is that one c.d.f. is uniformly and appreciably shifted to one side of the other c.d.f.

5.9.1 Example—Kolmogorov–Smirnov Goodness-of-Fit Test

We illustrate the One-Sample Kolmogorov–Smirnov Test in Table 5.12 and Figure 5.17. We illustrate the Two-Sample Kolmogorov–Smirnov Test in Table 5.13 and Figure 5.18.

We selected two random samples of 300 items, the first from a t distribution with 5 df, and the second from a standard normal distribution. Table 5.12 shows the K–S tests and Figure 5.17 the plot of the tests.

Table 5.12 Kolmogorov–Smirnov One-Sample Test. The first test corresponds to the left panels of Figure 5.17. We see a *p*-value of 0.2982 and do not reject the null. The second test corresponds to the right panels of Figure 5.17. We see a *p*-value of 0.003808 and reject the null.

```
> rt5 <- rt(300, df=5)

> rnn <- rnorm(300)

> ks.test(rt5, function(x)pt(x, df=2))

One-sample Kolmogorov-Smirnov test

data:  rt5
D = 0.0563, p-value = 0.2982
alternative hypothesis: two-sided


> ks.test(rnn, function(x)pt(x, df=2))

One-sample Kolmogorov-Smirnov test

data:  rnn
D = 0.1022, p-value = 0.003808
alternative hypothesis: two-sided
```

Table 5.13 Kolmogorov–Smirnov Two-Sample Test. The test corresponds to Figure 5.18. We see a *p*-value of 0.09956 and we do not reject the null.

```
> ks.test(rt5, rnn)

Two-sample Kolmogorov-Smirnov test

data:  rt5 and rnn
D = 0.1, p-value = 0.09956
alternative hypothesis: two-sided
```

In the table we test to see if these sample datasets are consistent with a t distribution with 2 df. The 5-df dataset is consistent with the 2-df null distribution. The normal dataset is not. The top panel in both columns of Figure 5.17 shows the distribution for the hypothesized t_2 distribution, and the vertical deviations of

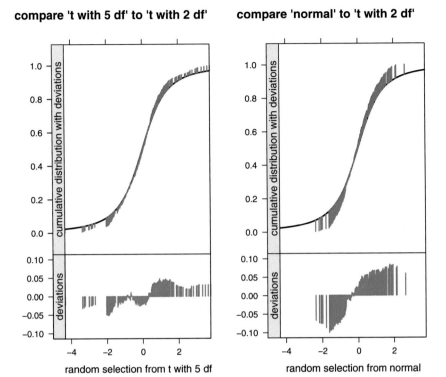

Fig. 5.17 Kolmogorov–Smirnov plots. Kolmogorov–Smirnov One-Sample Test. On the left we compare a random selection from the *t* distribution with 5 df to a null hypothesis distribution of *t* with 2 df. The ks.test in Table 5.12 shows a *p*-value of 0.2982 and does not reject the null. On the right we compare a random selection from the standard normal distribution to a null hypothesis distribution of *t* with 2 df. The ks.test in Table 5.12 shows a *p*-value of 0.003808 and rejects the null. The solid line in the top panels is the CDF for null distribution, in this example the *t* with 2 df. The deviation lines connect the observed *y*-values from the dataset under test to the hypothesized *y*-values from the null distribution. The deviation lines are magnified and centered in the bottom panels. The largest |vertical deviation| is the value of the K–S statistic in Table 5.12.

the data from the hypothesized distribution. The largest absolute value of these vertical deviations is the Kolmogorov–Smirnov statistic. The lower panel shows the deviations.

In Table 5.13 and Figure 5.18 we directly compare two different samples to see if the Two-Sample ks.test can distinguish between them. In this example the null hypothesis is retained. The plot shows both empirical distribution functions.

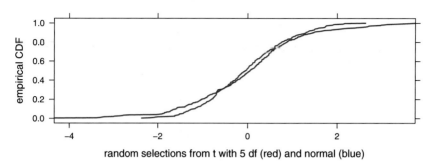

Fig. 5.18 Kolmogorov–Smirnov two-sample plot. We plotted the two empirical CDF on the same axes. The largest absolute vertical deviation is the value of the K–S statistic. Interpolation to calculate the vertical deviations is messier in the two-sample case, therefore we didn't do it for the figure. The `ks.test` function in Table 5.13 does do the interpolation.

5.10 Maximum Likelihood

Maximum likelihood is a general method of constructing "good" point estimators. *Likelihood ratio* is a general method of constructing tests with favorable properties. We briefly consider both of these ideas.

5.10.1 Maximum Likelihood Estimation

We start from the joint distribution of the sample statistics. The maximum likelihood estimator (MLE) is the value of the parameter that maximizes this expression of the joint distribution, which is called the likelihood function L. In practice it is usually easier to solve the equivalent problem of maximizing $\ln(L)$, equivalent since $\ln(\cdot)$ is an increasing function.

As a simple example, we derive the MLE of the mean μ of a normal population with known standard deviation σ, based on a random sample of n from this population.

The likelihood function $L(\mu)$ is a function of the parameter μ. $L(\mu)$ is constructed as the product of the individual density functions for the observed data values y_i.

$$L(\mu) = \prod \phi\left(\frac{y_i - \mu}{\sqrt{2}\,\sigma}\right) = (2\pi\sigma^2)^{-\frac{n}{2}} \exp\left(-\frac{\sum(y_i - \mu)^2}{2\sigma^2}\right) \qquad (5.23)$$

Apart from an additive constant that does not depend on μ, we find

$$\ln(L) = -\frac{\Sigma(y_i - \mu)^2}{2\sigma^2}$$

The value of μ that maximizes this expression is the value of μ that minimizes

$$\sum(y_i - \mu)^2 = \sum(y_i - \bar{y})^2 + n(\bar{y} - \mu)^2$$

The answer, $\hat{\mu} = \bar{y}$, is both the "least-squares" and maximum likelihood estimator of μ. (The least-squares and maximum likelihood estimators do not necessarily coincide for other estimands than μ.)

5.10.2 Likelihood Ratio Tests

Let $y_1, y_2, \ldots, y_n$ denote a random sample of some population and let $L = L(y_1, y_2, \ldots, y_n)$ denote the *likelihood* of this sample, i.e., the joint probability distribution of the sample values. Let H_0 be a null hypothesis about the parameter(s) of this population. A *likelihood ratio (LR) test* of H_0 uses the likelihood ratio

$$\lambda = \frac{\text{maximum of } L \text{ over only those parameter values for which } H_0 \text{ is true}}{\text{maximum of } L \text{ over all possible parameter values}}$$

$$(5.24)$$

or some random variable that is a strictly increasing or strictly decreasing function of only λ. H_0 is rejected if λ is sufficiently small, where "sufficiently small" depends on α.

While likelihood ratio tests do not, in general, have optimal properties, experience has taught that they frequently are competitive. One reason for their popularity is that they have a known asymptotic (i.e., large sample size n) distribution: $-2 \ln(\lambda)$ is approximately a χ^2 r.v. with d.f. equal to the number of parameters constrained by H_0. This fact can be used to construct a large sample test.

For example, to test $H_0: \mu = 0$ vs $H_1: \mu \neq 0$, where μ is the mean of a normal population with unknown variance, it is not difficult to show that the likelihood ratio test procedure gives $\lambda = \frac{1}{(1+t^2)^{n/2}}$, where $|t|$ is the usual absolute t statistic used for this purpose. Here $|t|$ arises as the appropriate test statistic because it is a strictly decreasing function of λ.

5.11 Exercises

5.1. Suppose that hourly wages in the petroleum industry in Texas are normally distributed with a mean of $17.60 and a standard deviation of $1.30. A large company in this industry randomly sampled 50 of its workers, determining that their hourly wage was $17.30. Stating your assumptions, can we conclude that this company's average hourly wage is below that of the entire industry?

5.2. The mean age of accounts payable has been 22 days. During the past several months, the firm has tried a new method to reduce this mean age. A simple random sample of 200 accounts payable last month had mean age 20.2 days and standard deviation 7.2 days. Use a confidence interval to determine if the new method has made a difference.

5.3. The Security and Exchange Commission (SEC) requires companies to file annual reports concerning their financial status. Firms cannot audit every account receivable, so the SEC allows firms to estimate the true mean. They require that a reported mean must be within $5 of the true mean with 98% confidence. In a small sample of 20 from firm Y, the sample standard deviation was $40. What must the total sample size be so that the audit meets the standard of the SEC?

5.4. The Kansas City division of a company produced 982 units last week. Of these, 135 were defective. During this same time period, the Detroit division produced 104 defectives out of 1,088 units. Test whether the two divisions differed significantly in their tendency to produce defectives.

5.5. A human resources manager is interested in the proportion of firms in the United States having on-site day-care facilities. What is the required sample size to be 90% certain that the sample proportion will be within 5% of the unknown population proportion?

5.6. A health insurance company now offers a discount on group policies to companies having a sufficiently high percentage of nonsmoking employees. Suppose a company with several thousand workers randomly samples 200 workers and finds that 186 are nonsmokers. Find a 95% confidence interval for the proportion of this company's employees who do not smoke.

5.7. Out of 750 people chosen at random, 150 were unable to identify your product. Find a 90% confidence interval for the proportion of all people in the population who will be unable to identify your product.

5.8. A national poll, based on interviews with a random sample of 1,000 voters, gave one candidate 56% of the vote. Set up a 98% confidence interval for the proportion of voters supporting this candidate in the population. You need not complete the calculations.

5.9. Two hundred people were randomly selected from the adult population of each of two cities. Fifty percent of the city #1 sample and 40% of the city #2 sample were opposed to legalization of marijuana. Test the two-sided hypothesis that the two cities have equal proportions of citizens who favor legalization of marijuana. (Calculate and interpret the p-value.)

5.10. A random sample of 200 people revealed that 80 oppose a certain bond issue. Find a 90% confidence interval for the proportion in the population who oppose this bond issue. Work the arithmetic down to a final numerical answer.

5.11. The confidence interval answer to the previous question is rather wide. How large a sample would have been required to reduce the confidence interval error margin to 0.02?

5.12. Random samples of 400 voters were selected in both New Jersey and Pennsylvania. There were 210 New Jersey respondents and 190 Pennsylvania respondents who stated that they were leaning toward supporting the Democratic nominee for President. Test the claim (alternative hypothesis) that the proportion of all New Jersey voters who lean Democratic exceeds the proportion of all Pennsylvania voters who lean Democratic.

a. Set up H_0 and H_1.

b. Calculate $\hat{p}_1$, $\hat{p}_2$, and $\hat{p}$.

c. Calculate z_{calc}.

d. Approximate the p-value.

e. State your conclusion concerning the claim.

5.13. The relative rotation angle between the L2 and L3 lumbar vertebrae is defined as the acute angle between posterior tangents drawn to each vertebra on a spinal X-ray. See Figure 7.20 for an illustration with different vertebrae. When this angle is too large the patient experiences discomfort or pain. Chiropractic treatment of this condition involves decreasing this angle by applying (nonsurgical) manipulation or pressure. Harrison et al. (2002) propose a particular such treatment. They measured the angle on both pre- and post-treatment X-rays from a random sample of 48 patients. The data are available as data(har1).

a. Test whether the mean post-treatment angle is less than the mean angle prior to treatment.

b. Construct a quantile plot to assess whether the post-treatment sample is compatible with a t distribution with 5 degrees of freedom.

5.14. The Harrison et al. (2002) study also measured the weights in pounds of the sample of 48 treated patients and a random sample of 30 untreated volunteer controls.

a. Use the data available as data(har2) to compare the mean weights of the treatment and control populations.

b. Use these data to compare the standard deviation of weights of the treatment and control populations.

c. Construct and interpret a normal probability plot for the weights of the treated patients.

5.15. The *Poisson* probability distribution is defined on the set of nonnegative integers. The Poisson is often used to model the number of occurrences of some event per unit time or unit space. Examples are the number of phone calls reaching a switchboard in a given minute (with the implication that the number of operators scheduled to answer the phones will be determined from the model) or the number of amoeba counted in a 1 ml. specimen of pond water. The probability that a Poisson r.v. Y has a particular (nonnegative integer) value y is given by

$$P(Y = y) = \frac{e^{-\mu}\mu^y}{y!}, \quad y = 0, 1, 2, \ldots$$

(While the value of y may be arbitrarily large, the probability of obtaining a very large y is infinitesimally small.) The parameter μ is the mean number of occurrences per unit. The mean μ of the Poisson distribution is either known in advance or must be estimated from the data. Poisson probabilities may be calculated with R as noted in Section J.3.6.

You are asked to perform a chi-square goodness-of-fit test of the Poisson distribution to the following data, which concern the number of specimens per microscope field in a sample of lake water.

y:	0	1	2	3	4	5	6	7
O:	21	30	54	26	11	3	3	2

The observed value O_y is the number of fields in which exactly y specimens were observed. In this example, $\sum O_y = 150$ fields were examined and, for example, exactly $O_2 = 54$ of the fields showed $y = 2$ specimens. The Poisson parameter μ is unknown and should be estimated as a weighted average of the possible values y, i.e.,

$$\hat{\mu} = \frac{\sum\limits_{y=0}^{7} y\,O_y}{\sum\limits_{y=0}^{7} O_y}$$

5.16. Extend the one-tailed sample size formula for comparing two proportions, Equation (5.22), to the two-tailed case.

Chapter 6

One-Way Analysis of Variance

In Chapter 5 we consider ways to compare the means of two populations. Now we extend these procedures to comparisons of means from several populations. For example, we may wish to compare the average hourly production of a company's six factories. We say that the investigation has a *factor* `factory` that has six *levels*, namely the six identifiers distinguishing the factories from one another. Or we may wish to compare the yields per acre of five different varieties of wheat. Here, the factor is `wheat`, and the levels of `wheat` are `variety1` through `variety5`. This chapter discusses investigations having a single factor. Experiments having two factors are discussed in Chapter 12, while situations with two or more factors are discussed in Chapters 13 and 14.

One-way analysis of variance (ANOVA) is the natural generalization of the two-sample t-test to more than two groups. Suppose that we have a factor A with a levels. We select independent samples from each of these a populations, where n_i is the size of the sample from population i. We distinguish between two possible assumptions about these populations comprising the single factor. We discuss *fixed effects* beginning in Section 6.1 and *random effects* beginning in Section 6.4.

6.1 Example—Catalyst Data

With the catalyst data from Montgomery (1997) we are interested in comparing the concentrations of one component of a liquid mixture in the presence of each of four catalysts. We investigate whether the catalysts provide for equal mean concentrations, and then since this does not appear to be true, we study the extent of differences among the mean concentrations. We access the dataset with `data(catalystm)` and plot it in Figure 6.1. We see that group D does not overlap groups A and B and that group C has a wider spread than the others.

© Springer Science+Business Media New York 2015
R.M. Heiberger, B. Holland, *Statistical Analysis and Data Display*,
Springer Texts in Statistics, DOI 10.1007/978-1-4939-2122-5_6

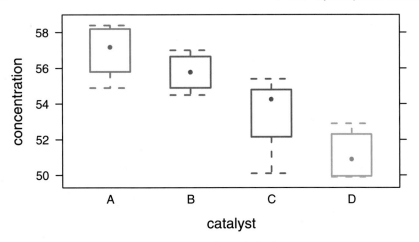

catalyst

Fig. 6.1 Boxplots Comparing the Concentrations for each Catalyst

The ANOVA (analysis of variance) table and the table of means are in Table 6.1. The F-test in the ANOVA table addresses the null hypothesis that the four catalysts have equal mean concentrations. We see immediately, from the small p-value ($p = .0014$), that these four catalysts do not provide the same average concentrations.

Table 6.1 ANOVA Table for Catalyst Data

```
> catalystm1.aov <- aov(concent ~ catalyst, data=catalystm)

> anova(catalystm1.aov)
Analysis of Variance Table

Response: concent
           Df Sum Sq Mean Sq F value Pr(>F)
catalyst    3   85.7   28.56    9.92 0.0014 **
Residuals  12   34.6    2.88
---
Signif. codes:  0 '***' 0.001 '**' 0.01 '*' 0.05 '.' 0.1 ' ' 1

> model.tables(catalystm1.aov, "means")
Tables of means
Grand mean

54.49

 catalyst
       A     B     C     D
     56.9 55.77 53.23 51.12
rep  5.0  4.00  3.00  4.00
```

6.2 Fixed Effects

Initially we assume that the a stated levels of A are the totality of all levels of interest to us. We call A a *fixed* factor. We model the jth observation from population i as

$$y_{ij} = \mu + \alpha_i + \epsilon_{ij} \qquad \text{for} \quad i = 1, \ldots, a \quad \text{and} \quad j = 1, \ldots, n_i \qquad (6.1)$$

where μ and the α_i are fixed quantities with the constraint

$$\sum_i \alpha_i = 0 \qquad (6.2)$$

and the ϵ_{ij} are assumed to be normally and independently distributed (NID) with common mean 0 and common variance σ^2, which we denote by

$$\epsilon_{ij} \sim \text{NID}(0, \sigma^2) \qquad (6.3)$$

We interpret μ as the grand mean of all a populations, α_i as the deviation of the mean of population i from μ, and assume that the responses from all a populations have a normal distribution with a common variance. If the normality assumption is more than mildly violated, we must either transform the response variable to one for which this assumption is satisfied, perhaps with a power transformation such as those discussed in Section 4.8, or use a nonparametric procedure as described in Chapter 16. The common variance assumption may be examined with the hypothesis test described in Section 6.10. If the variances are not homogeneous, a transformation such as those discussed in Section 4.8 sometimes can fix the inhomogeneity of variance problem as well as the nonnormality problem by changing to a scale in which the transformed observations show homogeneity of variance.

We discuss in Appendix 6.A the correspondence between the notation of Equation (6.1) and the software notation in Table 6.1.

The initial question of interest is the equality of the a population means, which we investigate with the test of

$$H_0: \alpha_1 = \alpha_2 = \ldots = \alpha_a$$

$$\text{vs} \qquad (6.4)$$

$$H_a: \text{ the } \alpha_i \text{ are not all equal.}$$

When $a = 2$, the test is the familiar

$$t_{n_1+n_2-2} = \frac{\bar{y}_1 - \bar{y}_2}{s_p \sqrt{(\frac{1}{n_1} + \frac{1}{n_2})}}$$

where

$$s_p^2 = \left((n_1 - 1)s_1^2 + (n_2 - 1)s_2^2\right)/(n_1 + n_2 - 2)$$

from Equations (5.13) and (5.12). By squaring both sides, we can show

$$F_{1,n_1+n_2-2} = t_{n_1+n_2-2}^2 = \frac{n_1(\bar{y}_1 - \bar{\bar{y}})^2 + n_2(\bar{y}_2 - \bar{\bar{y}})^2}{s_p^2} \tag{6.5}$$

where

$$\bar{\bar{y}} = \frac{n_1\bar{y}_1 + n_2\bar{y}_2}{n_1 + n_2}$$

In the special case where $n_1 = n_2$, Equation (6.5) is easily proved by using these hints:

1. $\bar{\bar{y}} = \dfrac{\bar{y}_1 + \bar{y}_2}{2}$

2. $(\bar{y}_1 - \bar{\bar{y}}) = -(\bar{y}_2 - \bar{\bar{y}})$

3. $\dfrac{1}{n_1} + \dfrac{1}{n_2} = \dfrac{2}{n_1}$

The equality (6.5) is also true for unequal n_i, but the proof is messier.

When $a \geq 2$, we generalize formula (6.5) to

$$F_{a-1,\,(\Sigma n_i)-a} = \frac{\left(\sum n_i(\bar{y}_i - \bar{\bar{y}})^2\right)/(a-1)}{s_p^2} \tag{6.6}$$

where $\bar{\bar{y}}$ and s_p^2 are the weighted mean

$$\bar{\bar{y}} = \frac{\sum n_i\bar{y}_i}{\sum n_i} \tag{6.7}$$

and pooled variance

$$s^2 = s_p^2 = \frac{\sum(n_i - 1)s_i^2}{\sum(n_i - 1)} = MS_{residual} \tag{6.8}$$

over all a samples.

The usual display of this formula is in the analysis of variance table and the notation is

$$F_{(a-1),\,(\Sigma n_i)-a} = \frac{SS_{treatment}/df_{treatment}}{SS_{residual}/df_{residual}} = \frac{MS_{treatment}}{MS_{residual}} = \frac{MS_{Tr}}{MS_{Res}} \tag{6.9}$$

The sample ANOVA table in Table 6.2 illustrates the structure. Note that Figure 6.2 includes a section "Total" that is missing in Table 6.1. The Total Sum of Squares is the sum of the Treatment and Residual Sums of Squares, and the Total Degrees of Freedom is the sum of the Treatment and Residual Degrees of Freedom. R does not print the Total line in its ANOVA tables.

Table 6.2 Sample Table to Illustrate Structure of the ANOVA Table

Analysis of Variance of Dependent Variable y					
Source	Degrees of Freedom	Sum of Squares	Mean Square	F	p-value
Treatment	df_{Tr}	SS_{Tr}	MS_{Tr}	F_{Tr}	p_{Tr}
Residual	df_{Res}	SS_{Res}	MS_{Res}		
Total	df_{Total}	SS_{Total}			

The terms of the table are defined by

Treatment	
df_{Tr}	$a - 1$
SS_{Tr}	$\sum_{i=1}^{a} n_i (\bar{y}_i - \bar{\bar{y}})^2$
MS_{Tr}	SS_{Tr}/df_{Tr}
F_{Tr}	MS_{Tr}/MS_{Res}
p_{Tr}	$1 - \mathcal{F}_F(F_{Tr} \mid df_{Tr}, df_{Res})$

Residual	
df_{Res}	$\left(\sum_{i=1}^{a} n_i \right) - a$
SS_{Res}	$\sum_{i=1}^{a} \sum_{j=1}^{n_i} (y_{ij} - \bar{y}_i)^2$
MS_{Res}	SS_{Res}/df_{Res}

Total		
df_{Total}	$\left(\sum_{i=1}^{a} n_i \right) - 1$	$= df_{Tr} + df_{Res}$
SS_{Total}	$\sum_{i=1}^{a} \sum_{j=1}^{n_i} (y_{ij} - \bar{\bar{y}})^2$	$= SS_{Tr} + SS_{Res}$

As in Section 5.4.4, this F-test of the pair of hypotheses in Equation (6.4) compares two estimates of the population variance σ^2. MS_{Res} is an unbiased estimator of σ^2 whether or not H_0 is true. MS_{Tr} is unbiased for σ^2 when H_0 is true but an overestimate of σ^2 when H_a is true. Hence, the larger the variance ratio $F = MS_{Tr}/MS_{Res}$, the stronger the evidence in support of H_a. Comparing two variances facilitates the comparison of a means. For this reason, the foregoing procedure is called *analysis of variance*. It involves decomposing the total sum of squares SS_{Total} into the variances used to conduct this F-test. The p-value in this table is calculated as the probability that a central F random variable with df_{Tr} and df_{Res} degrees of freedom exceeds the calculated F_{Tr}.

6.3 Multiple Comparisons—Tukey Procedure for Comparing All Pairs of Means

Multiple comparisons refer to procedures for simultaneously conducting all inferences in a family of related inferences, while keeping control of a Type I error concept that relates to the entire family. This class of inferential procedures is discussed in detail in Chapter 7. In the present chapter, we introduce the Tukey procedure, used for the family of all $\binom{a}{2}$ pairwise comparisons involving a population means.

We illustrate the Tukey procedure with a continuation of the analysis of the catalyst data. We seek to determine which catalyst mean differences are responsible for the overall conclusion that the catalyst means are not identical.

Under the assumption that `catalyst` is a fixed factor, we investigate the nature of the differences among the four catalysts. There are $\binom{4}{2} = 6$ pairs of catalysts, and for each of these pairs we wish to determine whether there is a significant difference between the concentrations associated with the two catalysts comprising the pair. (If the levels of `catalyst` had instead been quantitative or bore a structural relationship to one another, a different follow-up to the analysis of variance table would have been more appropriate. An example of such a situation is the analysis of the turkey data presented in Section 6.8.)

We seek to control at a designated level α the familywise error rate, FWE, defined as the probability of incorrectly rejecting at least one true null hypothesis under any configuration of true and false null hypotheses. For the family consisting of all pairs of means, the Tukey procedure maximizes, in various senses, the probability of detecting truly differing pairs of means while controlling the FWE at α.

The Tukey procedure uses a critical value q_α from the Studentized range distribution (see Section J.1.10), i.e., the distribution of standardized difference between the maximum sample mean and the minimum sample mean, rather than an ordinary t distribution for comparing two means discussed in Section 5.4.3. The Tukey output may be presented in the form of simultaneous confidence intervals on each of the mean differences rather than, or in addition to, tests on each difference. The interpretation is that the confidence coefficient $1 - \alpha$ is the probability that all of the $\binom{a}{2}$ pairwise confidence intervals among a sample means contain their respective true values of the difference between the two population means:

$$
\begin{aligned}
1 - \alpha \leq P\big(\mathrm{CI}_{12} \cap \mathrm{CI}_{13} \cap \ldots \cap \mathrm{CI}_{1a} \\
\cap\, \mathrm{CI}_{23} \cap \ldots \cap \mathrm{CI}_{2a} \\
\cap \ldots \cap \mathrm{CI}_{(a-1)a}\big)
\end{aligned}
\tag{6.10}
$$

where

$$
\mathrm{CI}_{ii'} : (\bar{y}_i - \bar{y}_{i'}) - \frac{q_\alpha}{\sqrt{2}}\, s_{(\bar{y}_i - \bar{y}_{i'})} \leq (\mu_i - \mu_{i'}) \leq (\bar{y}_i - \bar{y}_{i'}) + \frac{q_\alpha}{\sqrt{2}}\, s_{(\bar{y}_i - \bar{y}_{i'})}
$$

$$
\tag{6.11}
$$

and

$$S_{(\bar{y}_i - \bar{y}_{i'})} = s \sqrt{\frac{1}{n_i} + \frac{1}{n_{i'}}}$$

$$s = \sqrt{MS_{residual}}$$

If the sample sizes are unequal, the confidence intervals (6.11) are conservative in the sense that the coverage probability in Equation (6.10) exceeds $1 - \alpha$. If the sample sizes are equal, the inequality in Equation (6.10) is instead an equality and the simultaneous $1 - \alpha$ confidence for the set of intervals in (6.11) is exact.

We show the listing for the Tukey test of the catalyst data in Table 6.3 and the *MMC* multiple comparisons plot in Figure 6.2. The *Mean–mean Multiple Comparisons* display is discussed in Section 7.2. Denoting the mean concentration associated with catalyst i as μ_i, since the confidence intervals on $\mu_A - \mu_D$ and $\mu_B - \mu_D$ lie entirely above 0 while all other confidence intervals include 0, we conclude that both catalysts A and B provide, on average, a significantly greater concentration than catalyst D; no other significant differences between catalysts were uncovered. We continue with this example in Section 7.2.4

In view of this finding one might be tempted to focus on the differences demonstrated to be significant in Table 6.3, and construct hypothesis tests or confidence intervals using a method from Section 5.4.2. A more general framing of this temptation is to ask, "Is it permissible to use preliminary examinations of the data to develop subsequent hypotheses about the data" (a practice referred to as *data snooping*)? With few exceptions, the answer is *no* because the two-stage nature of the procedure distorts the claimed significance levels or confidence coefficients of the analyses in the second stage. Inferential strategies should be developed before the data are collected—based entirely on the structure of the data and the sampling method used. Strategies should not depend on the observed data. Here one should be content with the analyses in Table 6.3 and supporting graphical displays such as Figure 6.2, assuming the correctness of the assumptions underlying their construction.

Although in this example there were equal sample sizes from the levels of catalyst, neither the basic analysis of variance nor the Tukey multiple comparison procedure requires that the factor levels have the same sample size. Analyses of one-way data having unequal sample sizes are requested in the Exercises.

6.4 Random Effects

We could assume that the a observed levels of A are a random sample from a large or conceptually infinite population of levels. We call A a *random* factor. For example, in a study to compare the daily productivity of assembly line workers in a large

Table 6.3 Tukey Multiple Comparisons for Catalyst Data

```
> catalystm.mmc <-
+     mmc(catalystm1.aov, linfct = mcp(catalyst = "Tukey"))

> catalystm.mmc
Tukey contrasts
Fit: aov(formula = concent ~ catalyst, data = catalystm)
Estimated Quantile = 2.966
95% family-wise confidence level
$mca
      estimate stderr    lower upper height
A-B      1.125  1.138 -2.25211 4.502  56.34
A-C      3.667  1.239 -0.00986 7.343  55.07
B-C      2.542  1.296 -1.30334 6.387  54.50
A-D      5.775  1.138  2.39789 9.152  54.01
B-D      4.650  1.200  1.09022 8.210  53.45
C-D      2.108  1.296 -1.73667 5.953  52.18
$none
    estimate stderr lower upper height
A      56.90 0.7590 54.65 59.15  56.90
B      55.77 0.8485 53.26 58.29  55.77
C      53.23 0.9798 50.33 56.14  53.23
D      51.12 0.8485 48.61 53.64  51.12
```

firm, the workers in the study may be a random sample of a employees from among thousands of employees performing identical tasks.

We still work with Equation (6.1), and still maintain the same assumptions about μ and the ϵ_{ij}'s. We have a different interpretation of the α_i. Now the term α_i in Equation (6.1) is assumed to be a $N(0, \sigma_A^2)$ random variable and the restriction $\sum_i \alpha_i$ no longer applies. Instead we work with the hypotheses

$$H_0: \sigma_A^2 = 0$$

vs   (6.12)

$$H_a: \sigma_A^2 > 0$$

The sample ANOVA table in Table 6.2 still applies. The F statistic now compares the hypotheses in Equation 6.12. In the context of the worker productivity example, the factor worker is referred to as a *random* factor. We are using the a sampled workers to assess whether the entire population of workers has identical or nonidentical productivity.

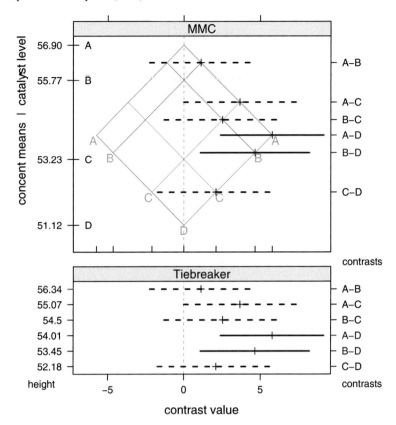

Fig. 6.2 Tukey Multiple Comparisons of All Pairwise Comparisons of Catalyst Means with the MMC plot. The MMC plot is fully developed in Section 7.2. The top panel is the MMC display. The cell means are on the vertical axis and the confidence intervals for the contrasts are on the horizontal axis. The isomeans grid in the MMC panel displays the cell means on both diagonal axes. The bottom panel, labeled "Tiebreaker" even though there are no ties in the top panel for this example, shows the contrasts in the same order as the MMC panel, but evenly spaced in the vertical direction. The left tick labels in the Tiebreaker panel are the heights of the confidence lines in the MMC panel. For example, the A–B line has height 56.34, halfway between the height of the A mean at 56.90 and the height of the B mean at 55.77. The bottom panel does not have any information on the values of the means themselves.

6.5 Expected Mean Squares (EMS)

To better understand the distinction between the F-test in the fixed and random factor cases, it is useful to compare the expected mean squares (EMS) for the ANOVA table under the two assumptions. The EMS are algebraically displayed in Table 6.4. Calculation of the EMS values in this table are outlined in Exercise 6.14.

In the case of factor A fixed, the F statistic is testing whether $\sum_i n_i(\alpha_i - \bar{\alpha})^2 = 0$, where $\bar{\alpha} = (\sum_i n_i\alpha_i) / (\sum_i n_i)$. This statement is true if and only if the α_i are identical.

Table 6.4 Expected Mean Squares in One-Way Analysis of Variance. Similar tables for Two-Way models and Three-Way models are in Tables 12.8 and 13.11.

Source	df	$E(MS)$	EMS, factor A fixed	EMS, factor A random
Treatment A	$a - 1$	$E(MS_{Tr})$	$\sigma^2 + \left(\frac{1}{a-1}\right)\sum_i n_i(\alpha_i - \bar{\alpha})^2$	$\sigma^2 + \frac{1}{a-1}\left(\sum_i n_i - \frac{\sum_i n_i^2}{\sum_i n_i}\right)\sigma_A^2$
Residual	$\sum_i(n_i - 1)$	$E(MS_{Res})$	σ^2	σ^2
Total	$(\sum_i n_i) - 1$			

In the case of factor A random, the F statistic tests whether $\sigma_A^2 = 0$ because the coefficient of σ_A^2 is positive whether or not H_0 is true.

For fixed effects the power of the F-test is an increasing function of the non-centrality parameter of the F statistic, which in turn is an increasing function of $EMS_{Treatment}/EMS_{Residual}$. When factor A is random, it follows that the power is an increasing function of $\sum_i n_i - \sum_i n_i^2 / \sum_i n_i$. For fixed total sample size $\sum_i n_i$, this quantity and hence power is maximized when $n_i = \sum_i n_i/a$, that is, when the sample is equally allocated to the levels, or nearly equal allocation if $\sum_i n_i/a$ is not an integer.

In general in Analysis of Variance tables, examination of expected mean squares suggests the appropriate numerator and denominator mean squares for conducting tests of interest. We look for $EMS_{Treatment}/EMS_{Residual}$ that exceeds 1 if and only if the null hypothesis of interest is false. This idea is especially useful in analyzing mixed models (i.e., ones containing both fixed and random factors) as is discussed in Section 12.10.

6.6 Example—Catalyst Data—Continued

In Section 6.1 the four levels of the factor catalyst were assumed to be qualitative rather than quantitative. It was also assumed that these are the only catalysts of interest. In this situation catalyst is a *fixed* factor since the four catalyst levels we study are the only levels of interest.

If instead these four catalysts had been regarded as a random sample from a large population of catalysts, then catalyst would have been considered a *random* factor. Figure 6.1 provides a tentative answer to the question of whether the four distributions are homogeneous. This figure also addresses the reasonableness of the assumption that the data come from normal homoskedastic populations, that is, populations having equal variances. The boxplots hint at the possibility that catalyst 3 has a more variable concentration than the others, but the evidence is not substantial in view of the small sample sizes (5,4,3,4). We look more formally at the homogeneity of the variances of these four catalysts in Section 6.10.

The F-test in Table 6.1 addresses the null hypothesis that the four catalysts have equal mean concentrations. The small p-value suggests that these four catalysts provide different average concentrations.

If instead, the factor catalyst in this experiment had been a random factor rather than a fixed factor, the F-test would be addressing the hypothesis that there is no variability in concentration over the population of catalysts from which these four catalysts are a random sample.

6.7 Example—Batch Data

In the batch data data(batch) taken from Montgomery (1997), the 5 sampled batches constitute a random sample from a large population of batches. Thus batch is a random factor, not a fixed factor. The response variable is calcium content. The ANOVA is in Table 6.5. The small p-value, .0036, leads us to conclude that the population of batches, from which these 5 batches were a random sample, had non-homogeneous calcium content. We must investigate whether the variances within batches are the same. We do so in Table 6.6 and Figure 6.3 and 6.4.

Table 6.5 ANOVA of Batch data. The batches are a random effect, therefore means are not meaningful. Instead the test compares the variability between groups with the variability of within groups.

```
> data(batch)

> bwplot(Calcium ~ Batch, data=batch, groups=Batch,
+         panel=panel.bwplot.superpose, xlab="Batch")

> batch1.aov <- aov(Calcium ~ Batch, data=batch)

> anova(batch1.aov)
Analysis of Variance Table

Response: Calcium
          Df Sum Sq Mean Sq F value Pr(>F)
Batch      4 0.0970 0.02424    5.54 0.0036 **
Residuals 20 0.0876 0.00438
---
Signif. codes:  0 '***' 0.001 '**' 0.01 '*' 0.05 '.' 0.1 ' ' 1
```

Table 6.6 Homogeneity of Variance test for Batch data. With p = .9978, the conclusion is to retain the null hypothesis and act as if all the group variances are equal. See caption of Figure 6.3 for discussion of this test.

```
> hovBF(Calcium ~ Batch, data=batch)

hov: Brown-Forsyth

data:  Calcium
F = 0.03219, df:Batch = 4, df:Residuals = 20, p-value = 0.998
alternative hypothesis: variances are not identical
```

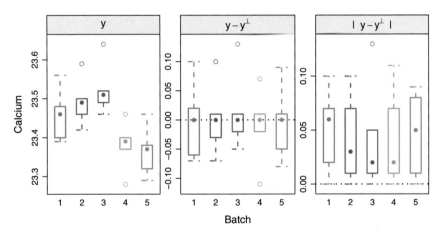

Fig. 6.3 Homogeneity of Variance plot for Batch. The left panel shows the data within each group. The center panel shows the same variabilities centered on the group medians ($y_{ij} - \overset{+}{y}_j$). The right panel shows the boxplot of the absolute deviations from the median ($|y_{ij} - \overset{+}{y}_j|$). The Brown–Forsyth test is an ordinary analysis of variance of these absolute deviations from the median. If the means of each group of the absolute deviations differ, then the test says the variances of the groups of original data differ.

6.8 Example—Turkey Data

6.8.1 Study Objectives

The goal in many agricultural experiments is to increase yield. In the Turkey experiment (data from Ott (1993)) data(turkey) the response is weight gain (in pounds) of turkeys and the treatments are diet supplements.

Variability of Groups, Centers of Groups, and all Data

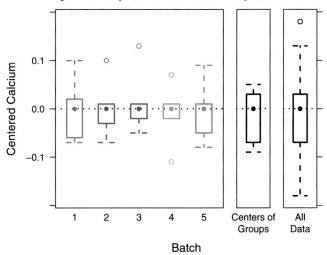

Fig. 6.4 Centered variability for the Batch data. The data has been centered on the group medians. The left panel is identical to the middle panel of Figure 6.3. The center panel shows the variability of the set of group means. The right panel shows the variability of the entire set of $y =$ `Calcium` values. The central box in each glyph shows the interquartile range, which is approximately proportional to the standard deviation. The variability in the three panels corresponds to the square root of the values in the Mean Square column of the ANOVA table in Table 6.5. The variability of the right panel represents the Total line of the ANOVA table (the one that R doesn't show) which is the variance of the response variable ignoring all the predictor variables. When the center panel variability is larger than the left panel variabilities, then the F value will be large (use the p-value from the F table to see whether it is significantly large).

6.8.2 Data Description

Six turkeys were randomly assigned to each of 5 diet groups and fed for the same length of time. The diets have a structure such that it is possible and desirable to undertake an *orthogonal contrast analysis*, a systematic set of comparisons among their mean responses. A contrast is a comparison of two or more means such that the expected value of the comparison is zero when the null hypothesis is true. (Contrasts and orthogonal contrasts are discussed in Section 6.9.) The diets are

`control:`   control

`A1:`   control + amount 1 of additive A

`A2:`   control + amount 2 of additive A

`B1:`   control + amount 1 of additive B

`B2:`   control + amount 2 of additive B

The data are accessed as `data(turkey)` and plotted in Figure 6.5.

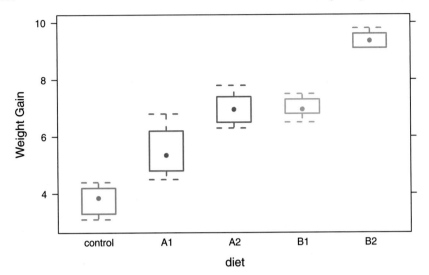

Fig. 6.5 Turkey Data: Boxplots of Weight Gain for each Diet

Table 6.7 ANOVA Table for Turkey Data

```
> turkey.aov <- aov(wt.gain ~ diet, data=turkey)

> summary(turkey.aov)
            Df Sum Sq Mean Sq F value  Pr(>F)
diet         4  103.0   25.76    81.7 5.6e-14 ***
Residuals   25    7.9    0.32
---
Signif. codes:  0 '***' 0.001 '**' 0.01 '*' 0.05 '.' 0.1 ' ' 1

> model.tables(turkey.aov, type="means", se=TRUE)
Tables of means
Grand mean

6.53

 diet
diet
control      A1      A2      B1      B2
  3.783   5.500   6.983   7.000   9.383

Standard errors for differences of means
          diet
        0.3242
replic.      6
```

6.8.3 Analysis

The ANOVA table and table of means are in Table 6.7. The first thing we notice is that the diets differ significantly in their promotion of weight gain ($F_{4,25}$ = 81.7, p-value ≈ 0). Then we observe that the diets are structured so that particular comparisons among them are of special interest. We make these comparisons by partitioning the sum of squares to reflect several well-defined contrasts. The contrasts are displayed in Table 6.8. The ANOVA table using them is in Table 6.9.

The interaction line `diet: A.vs.B.by.amount` in Table 6.8 asks the question, "Does the increase from amount 1 to amount 2 of additive A have the same effect as the increase from amount 1 to amount 2 of additive B?" This question may equivalently be stated as: "Does the change from amount 1 of additive A to amount 1 of additive B have the same effect as the change from amount 2 of additive A to amount 2 of additive B?" (We interpret the description of the experiment to mean that the amounts 1 and 2 of the additives are measured in the same units). The concept of interaction is discussed in detail in Chapter 12.

These contrasts decompose the 4-df sum of squares for diet into four single-df sums of squares, one for each of the four contrasts. This set of contrast sums of squares is additive because we have defined the contrasts in such a way that they are *mutually orthogonal*. In essence this means that the information contained in one of the contrasts is independent of the information contained in any of the other contrasts. The independence of information makes each of the contrasts more readily interpretable than they would be if the contrasts had been defined without the property of orthogonality.

6.8.4 Interpretation

We tentatively interpret the contrast analysis as follows:

1. trt.vs.control: averaged over the 4 treatments, turkeys receiving a dietary additive gain significantly more weight than ones not receiving an additive.

2. additive: turkeys receiving additive B gain significantly more weight than turkeys receiving additive A.

3. amount: turkeys receiving amount 2 gain significantly more weight than turkeys receiving amount 1.

4. interaction between additive and amount: the extent of increased weight gain as a result of receiving amount 2 rather than amount 1 is not significantly different for the two additives.

Table 6.8 Specification of contrasts for turkey data. This set of contrasts has been constructed to reflect the intent of the experiment: to compare A vs B, amount 1 vs amount 2, and control vs treatment. We first show the default treatment contrasts, then replace them with our constructed contrasts.

```
> contrasts(turkey$diet)
        A1 A2 B1 B2
control  0  0  0  0
A1       1  0  0  0
A2       0  1  0  0
B1       0  0  1  0
B2       0  0  0  1

> contrasts(turkey$diet) <-
+    cbind(control.vs.treatment=c(1,-.25,-.25,-.25,-.25),
+             A.vs.B              =c(0, .5,  .5, -.5, -.5 ),
+             amount              =c(0, .5, -.5,  .5, -.5 ),
+             A.vs.B.by.amount    =c(0, .5, -.5, -.5,  .5 ))

> contrasts(turkey$diet)
        control.vs.treatment A.vs.B amount A.vs.B.by.amount
control                1.00    0.0    0.0              0.0
A1                    -0.25    0.5    0.5              0.5
A2                    -0.25    0.5   -0.5             -0.5
B1                    -0.25   -0.5    0.5             -0.5
B2                    -0.25   -0.5   -0.5              0.5

> tapply(turkey$wt.gain, turkey$diet, mean) %*%
+      contrasts(turkey$diet)
        control.vs.treatment A.vs.B amount A.vs.B.by.amount
[1,]                  -3.433  -1.95 -1.933              0.45
```

Our conclusions derive from the definitions of the contrasts, the signs of their estimates in Table 6.7, and the results of the tests that each contrast is 0, shown in Table 6.8. We give further discussion of appropriate techniques for simultaneously testing the point estimates of the contrasts in Section 7.1.4.1. We illustrate the conclusions in Figure 7.4. In general, conclusions such as these are tentative because we are making several simultaneous inferences. Therefore, it may be appropriate to use a form of Type I error control that accounts for the simultaneity. See the discussion of multiple comparisons in Chapter 7. In this example, with very small p-values, the use of simultaneous error control will not lead to different conclusions.

Table 6.9 ANOVA Table for Turkey Data with Contrasts. The sum of the individual sums of squares from each "`diet:`" line is the sum of squares for `diet`. The point estimates of the contrasts are in Table 7.4. Development of this table's F statistic for `diet: A.vs.B` is explained in the discussion surrounding Equations (6.15)–(6.17).

```
> turkey2.aov <- aov(wt.gain ~ diet, data=turkey)

> summary(turkey2.aov)
            Df Sum Sq Mean Sq F value  Pr(>F)
diet         4  103.0   25.76    81.7 5.6e-14 ***
Residuals   25    7.9    0.32
---
Signif. codes:  0 '***' 0.001 '**' 0.01 '*' 0.05 '.' 0.1 ' ' 1

> old.width <- options(width=67)

> summary(turkey2.aov,
+           split=list(diet=list(
+                     control.vs.treatment=1,
+                     A.vs.B=2,
+                     amount=3,
+                     A.vs.B.by.amount=4)))
                            Df Sum Sq Mean Sq F value  Pr(>F)
diet                         4  103.0    25.8   81.67 5.6e-14 ***
  diet: control.vs.treatment 1   56.6    56.6  179.40 6.6e-13 ***
  diet: A.vs.B               1   22.8    22.8   72.34 7.6e-09 ***
  diet: amount               1   22.4    22.4   71.11 8.9e-09 ***
  diet: A.vs.B.by.amount     1    1.2     1.2    3.85   0.061 .
Residuals                   25    7.9     0.3
---
Signif. codes:  0 '***' 0.001 '**' 0.01 '*' 0.05 '.' 0.1 ' ' 1

> options(old.width)
```

6.8.5 Specification of Analysis

The partitioned ANOVA table in Table 6.8 is constructed and displayed in two separate steps in Tables 6.8 and 6.9.

We specify the contrasts in several steps: We display the default contrasts, we define new contrasts, we display the new contrasts. Sometimes several iterations are needed until we get it right. Table 6.8 displays both the default and the new contrasts.

Once the contrasts are defined, we use them in the `aov()` command in Table 6.9. The `aov()` command uses the contrasts that are in the `data.frame` when it is called. Redefining the contrasts after the `aov()` command has been used has no effect on the `aov` object that has already been created.

The split argument to the summary command indexes the columns of the contrasts that are in the aov object. The index numbers in the list argument are necessary. The names of the items in the list are optional. They are used to provide pretty labels in the ANOVA table.

6.9 Contrasts

Once we have determined that there are differences among the means of the groups, that is, that the null hypothesis is rejected, we must follow through by determining the pattern of differences. Is one specific group responsible for the differences? Are there subsets of groups that behave differently than other subsets? We make this determination by partitioning the treatment sum of squares $SS_{treatment}$ into single degree-of-freedom components, each associated with a *contrast* among the group means. There are many possible contrasts that might be chosen. In this section we discuss the algebra of a single contrast vector c. In Chapter 10 we discuss sets of contrast vectors collected into a contrast matrix, and the relation between different possible sets.

Contrasts are associated with precisely formulated hypotheses. In the turkey example of Section 6.8 the initial null hypothesis was

$$H_0: \ \mu_{control} = \mu_{A1} = \mu_{A2} = \mu_{B1} = \mu_{A2}$$
$$H_1: \ \text{Not all } \mu_i \text{ are the same}$$

That initial null hypothesis was rejected when we observed the overall p-value 5.6×10^{-14}.

For the next step in the analysis we must refine the hypotheses we are testing. In this example there are 5 levels of diet, hence $5 - 1 = 4$ degrees of freedom for the diet effect. That means there are 4 statements that can be tested independently. For this example we will specify a set of 4 null hypothesis statements:

$$\mu_{control} = (\mu_{A1} + \mu_{A2} + \mu_{B1} + \mu_{A2})/4$$
$$(\mu_{A1} + \mu_{A2})/2 = (\mu_{B1} + \mu_{B2})/2 \tag{6.13}$$
$$(\mu_{A1} + \mu_{B1})/2 = (\mu_{A2} + \mu_{B2})/2$$
$$(\mu_{A1} - \mu_{A2})/2 = (\mu_{B1} - \mu_{B2})/2$$

These statements are usually written as inner products of the vector of group means with a *contrast vector C*. We can rewrite Equations 6.13 by moving all terms to the left-hand side and then using the inner product notation:

$$
\begin{aligned}
(\mu_{\text{control}} \quad \mu_{A1} \quad \mu_{A2} \quad \mu_{B1} \quad \mu_{A2}) \cdot (1 \;\; -.25 \;\; -.25 \;\; -.25 \;\; -.25) &= 0 \\
(\mu_{\text{control}} \quad \mu_{A1} \quad \mu_{A2} \quad \mu_{B1} \quad \mu_{A2}) \cdot (0 \quad .50 \quad .50 \;\; -.50 \;\; -.50) &= 0 \\
(\mu_{\text{control}} \quad \mu_{A1} \quad \mu_{A2} \quad \mu_{B1} \quad \mu_{A2}) \cdot (0 \quad .50 \;\; -.50 \quad .50 \;\; -.50) &= 0 \\
(\mu_{\text{control}} \quad \mu_{A1} \quad \mu_{A2} \quad \mu_{B1} \quad \mu_{A2}) \cdot (0 \quad .50 \;\; -.50 \;\; -.50 \quad .50) &= 0
\end{aligned}
\tag{6.14}
$$

The rest of this section discusses the properties of the various choices for the contrasts vector C.

The concept of a contrast among group means was first encountered in Section 6.8. Contrasts are chosen primarily from the structure of the levels, for example, the average effect of Treatment A at several levels compared to the average effect of Treatment B at several levels (the A.vs.B contrast in Tables 6.8 and 7.4 and in Figure 7.4). Or, for another example, a linear effect of the response to a linear increase in speed (the .L contrast in Section 10.4).

6.9.1 Mathematics of Contrasts

The mathematics of contrasts follows directly from the mathematics of the independent two-sample t-test:

$$
t_{\text{calc}} = \frac{\bar{y}_1 - \bar{y}_2}{s_p \sqrt{\frac{1}{n_1} + \frac{1}{n_2}}}
\tag{5.13}
$$

The residual mean square s_{Resid}^2 from the ANOVA table takes the place of s_p^2.

We will look closely at the A.vs.B contrast in Table 6.8 comparing the average of the A treatments $\bar{Y}_1 = (\bar{Y}_{A1} + \bar{Y}_{A2})/2$ to the average of the B treatments $\bar{Y}_2 = (\bar{Y}_{B1} + \bar{Y}_{B2})/2$ with $n_1 = n_{A1} + n_{A2} = n_2 = n_{B1} + n_{B2}$.

Direct substitution of these values into Equation (5.13) with $n \overset{\text{def}}{=} n_{\text{control}} = n_{A1} = n_{A2} = n_{B1} = n_{B2}$, followed by simplification (see Exercise 6.12) leads to

$$
t_{\text{calc}} = \frac{(\bar{Y}_{A1} + \bar{Y}_{A2})/2 - (\bar{Y}_{B1} + \bar{Y}_{B2})/2}{\frac{1}{2} s_{\text{Resid}} \sqrt{\frac{1}{n} + \frac{1}{n} + \frac{1}{n} + \frac{1}{n}}} \overset{\text{def}}{=} t_{\text{A.vs.B}}
\tag{6.15}
$$

We can write the numerator of Equation (6.15) as the dot product

$$
\begin{aligned}
C_{\text{A.vs.B}} &= (\bar{Y}_{\text{control}} \;\; \bar{Y}_{A1} \;\; \bar{Y}_{A2} \;\; \bar{Y}_{B1} \;\; \bar{Y}_{B2}) \cdot (0 \;\; \tfrac{1}{2} \;\; \tfrac{1}{2} \;\; -\tfrac{1}{2} \;\; -\tfrac{1}{2}) \\
&= (\bar{Y}_j) \cdot (c_j)
\end{aligned}
\tag{6.16}
$$

and then recognize the denominator of Equation (6.15) as the square root of the estimator of the variance of the numerator when the null hypothesis is true

$$\widehat{\text{var}}(C_{\text{A.vs.B}}) = \tfrac{1}{4} s_{\text{Resid}}^2 \left(\frac{1}{n} + \frac{1}{n} + \frac{1}{n} + \frac{1}{n} \right) \tag{6.17}$$

When we do the arithmetic, the value

$$t_{\text{A.vs.B}} = \frac{C_{\text{A.vs.B}}}{\sqrt{\widehat{\text{var}}(C_{\text{A.vs.B}})}} = 8.5051 = \sqrt{72.3367} = \sqrt{F_{\text{A.vs.B}}}$$

is recognized as the square root of the F-statistic for the diet: A.vs.B line of the ANOVA table in Table 6.8.

The vector $c = (c_j) = (0\ \tfrac{1}{2}\ \tfrac{1}{2}\ -\tfrac{1}{2}\ -\tfrac{1}{2})$ is called a contrast vector and the product $C_{\text{A.vs.B}}$ is called a contrast. The numbers c_j in a contrast vector satisfy the constraint

$$\sum_j c_j = 0 \tag{6.18}$$

Under the null hypothesis that $\mu_1 = \mu_2 = \ldots = \mu_5$, we have

$$E(C_{\text{A.vs.B}}) = E\left(\sum_j c_j \mu_j \right) = 0$$

Under both the null and alternative hypotheses, assuming that all σ_j^2 are identical and equal to σ^2, we see that

$$\text{var}(C_{\text{A.vs.B}}) = \frac{\sigma^2}{n} \sum c_j^2$$

A similar argument shows that each of the columns listed under contrasts(turkey\$diet) in Table 6.8 can be used to construct the correspondingly named row of the ANOVA table (Exercise 6.13).

This set of contrasts has an additional property. They are orthogonal. This means that the dot product of each column with any of the others is 0, for example,

$$c_{\text{A.vs.B}} \cdot c_{\text{amount}} = (0\ \tfrac{1}{2}\ \tfrac{1}{2}\ -\tfrac{1}{2}\ -\tfrac{1}{2}) \cdot (0\ \tfrac{1}{2}\ -\tfrac{1}{2}\ \tfrac{1}{2}\ -\tfrac{1}{2}) = 0 \tag{6.19}$$

This implies that the covariance of the contrasts is zero, for example

$$\text{cov}(C_{\text{A.vs.B}}, C_{\text{amount}}) = 0$$

that is, the contrasts are uncorrelated. As a consequence, the sum of sums of squares for each of the four contrasts in Table 6.8 is the same as the sum of squares for diet given by the SS$_{\text{treatment}}$ term in Equations (6.6) and (6.9).

The SS$_{\text{treatment}}$, and the sum of squares for each of the single degree-of-freedom contrasts comprising it, is independent of the MS$_{\text{residual}} = s_{\text{Resid}}^2$. The F-tests for each of the orthogonal contrasts are not independent of each other because all use the same denominator term.

In general, the n_j are not required to be identical. The general statement for a contrast vector

$$(c_j) = (c_1, \ldots, c_J) \tag{6.20}$$

is that the contrast $C = \sum c_j \bar{Y}_j$ has variance

$$\text{var}(C) = \sigma^2 \sum \frac{c_j^2}{n_j} \tag{6.21}$$

6.9.2 Scaling

The contrasts displayed here were scaled to make the sum of the positive values and the sum of the negative values each equal to 1. This scaling is consistent with the phrasing that a contrast is a comparison of the average response over several levels of a factor to the average response over several different levels of the factor. Any alternate scaling is equally valid and will give the same sum of squares.

6.9.2.1 Absolute-Sum-2 Scaling

We recommend the *absolute-sum-2* scaling where the sum of the absolute values of the coefficients equals 2,

$$\sum_j |c_j| = 2 \tag{6.22}$$

Equivalently, the sum of the positive coefficients equals 1 and the sum of the negative coefficients also equals 1. The *absolute-sum-2* scaling makes it easy to extend the mean–mean multiple comparisons plots to arbitrary sets of contrasts. See Section 7.2.3 for details on the mean–mean multiple comparisons plots.

6.9.2.2 Normalized Scaling

The normalized scaling, with $c_j^* = c_j / \sqrt{\sum c_j^2}$, is frequently used because the corresponding dot product

$$
\begin{aligned}
C^*_{A.vs.B} &= (\bar{Y}_{\text{control}} \ \bar{Y}_{A1} \ \bar{Y}_{A2} \ \bar{Y}_{B1} \ \bar{Y}_{B2}) \cdot (0 \ \tfrac{1}{2} \ \tfrac{1}{2} \ -\tfrac{1}{2} \ -\tfrac{1}{2}) \\
&= (\bar{Y}_j) \cdot (c_j^*)
\end{aligned} \tag{6.23}
$$

is simply related to the A.vs.B sum of squares by

$$SS_{A.vs.B} = n(C^*_{A.vs.B})^2 = 22.815. \qquad (6.24)$$

Under the null hypothesis

$$\text{var}(C^*_{A.vs.B}) = \sigma^2_{Resid} \qquad (6.25)$$

and under the alternate hypothesis

$$\text{var}(C^*_{A.vs.B}) \gg \sigma^2_{Resid} \qquad (6.26)$$

This provides the justification for the F-test.

In this example, the normalized scaling in Equation (6.23) is identical to the scaling in Equation (6.16) that makes the positive and negative sums each equal to 1. That is not always the case. The control.vs.treatment contrast with positive and negative values each summing to 1 as displayed in Table 6.8 is

$$(1 - .25 - .25 - .25 - .25)$$

The same control.vs.treatment contrast with normalized scaling is

$$\sqrt{.8}\,(1 - .25 - .25 - .25 - .25)$$

6.9.2.3 Integer Scaling

Another frequently used scaling makes each individual value c_j an integer. For the examples shown here, this gives

```
A.vs.B                   (0  1  1 −1 −1)
control.vs.treatment (4 −1 −1 −1 −1)
```

Because this scaling eases hand arithmetic, it was very important prior to digital computers. This scaling is included in tables of orthogonal contrasts in many texts, see for example, Cochran and Cox (1957), Table 3.4, page 64.

6.10 Tests of Homogeneity of Variance

In Sections 5.3, 5.4.4, 6.2, and 6.6 we mention that the assumption that several populations have a common variance can be checked via a statistical test. Assuming there are a populations having variances σ^2_i for $i = 1, 2, \ldots, a$, the test is of the form

$$H_0: \ \sigma_1^2 = \sigma_2^2 = \ldots = \sigma_a^2$$

vs   (6.27)

$$H_1: \ \text{not all the } \sigma_i^2 \text{ are identical to each other.}$$

For this purpose, Brown and Forsyth (1974) present the recommended test. Intensive simulation investigations, including Conover et al. (1981), have found that this test performs favorably compared with all competitors in terms of Type I error control and power for a wide variety of departures from Normality.

The Brown and Forsyth test statistic is the F statistic resulting from an ordinary one-way analysis of variance on the absolute deviations from the median

$$Z_{ij} = |Y_{ij} - \overset{\perp}{Y}_i| \tag{6.28}$$

where $\overset{\perp}{Y}_i$ is the median of $\{Y_{i1}, \ldots, Y_{i,n_i}\}$.

The test is available as the hovBF function in the **HH** package with the form

hovBF(y ~ A)

where A is a factor. The plot illustrating the test is available as the hovplotBF function in the **HH** package.

We continue the data(catalystm) example of Sections 6.1 and 6.6. Our impression from Figure 6.1 is that catalyst 3 has a larger variance than the other three catalysts. We formally test this possibility with the Brown–Forsyth test, illustrated in Figure 6.6. Because of the large p-value, .74, we are unable to conclude that the variances of the concentrations are not identical.

6.11 Exercises

6.1. Till (1974), also cited in Hand et al. (1994), compared the salinity (in parts per 1000) for three distinct bodies of water in the Bimini Lagoon, Bahamas. The data are available as data(salinity). Analyze the data under the assumption that the 3 bodies of water constitute a random sample from a large number of such bodies of water.

6.2. Milliken and Johnson (1984) report on an experiment to compare the rates of workers' pulses during 20-second intervals while performing one of 6 particular physical tasks. Here 68 workers were randomly assigned to one of these tasks. The data are available as data(pulse). Investigate differences between the mean pulse rates associated with the various tasks.

Brown–Forsyth Homogeneity of Variance

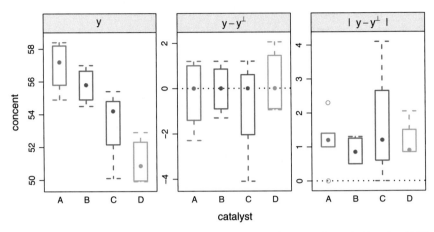

Fig. 6.6 Catalyst data: Brown–Forsyth test of the hypothesis of equal variances in Equation (6.27). The left panel shows the original data. The middle panel shows the deviations from the median for each group, hence is a recentering of the left panel. The right panel shows absolute deviations from the median. The central dots for each catalyst in the right panel, the MAD (median absolute deviation from the median), are approximately equal, reflecting the null distribution of the Brown–Forsyth test statistic. The Brown–Forsyth test shows $F = .42$ with 3 and 12 df and $p = .74$. Hence we do not reject this test's null hypothesis. We conclude that the variances of the concentrations in the four catalyst groups are approximately equal.

6.3. Johnson and Leone (1967) provide `data(operator)`. Five operators randomly selected from all company operators are each given four equal time slots in which their production is measured. Perform an appropriate analysis. Does it appear as if the population of operators has homogeneous productivity?

6.4. Anionwu et al. (1981), also reprinted in Hand et al. (1994), examine whether hemoglobin levels for patients with sickle cell anemia differ across three particular types of sickle cell disease. Here `type` is a fixed factor and its three qualitative levels are *"HB SS"*, *"HB S/thalassaemia"*, and *"HB SC"*. The data are available as `data(sickle)`. Perform an analysis of variance and multiple comparison with the Tukey procedure to compare the patients' hemoglobin for the three types.

6.5. Cameron and Pauling (1978), also reprinted in Hand et al. (1994), compare the survival times in days of persons treated with supplemental ascorbate following a diagnosis of cancer at five organ sites: *Stomach, Bronchus, Colon, Ovary,* and *Breast.* The dataset is available as `data(patient)`.

a. Perform a log transformation of the response `days` for each of the five levels of the factor `site` in order to improve conformity with the required assumption that the data be approximately normally distributed with equal within-site variance. Produce and compare boxplots to compare the response before and after the transformation.

b. Perform an analysis to assess differences in mean survival between the different cancer sites.

6.6. NIST (2002) reports the result of an experiment comparing the absorbed energy produced by each of four machines. The machines are labeled *Tinius1*, *Tinius2*, *Satec*, and *Tokyo*. The data are available as data(notch). Assuming that these were the only machines of interest, compare the responses on the four machines and use the Tukey procedure to assess significant differences among them.

6.7. An experiment was designed to examine the effect of storage temperature on the potency of an antibiotic. Fifteen antibiotic samples were obtained and three samples were stored at each of the five indicated temperatures (degrees F). The potencies of each sample were checked after 30 days. The dataset, taken from Peterson (1985), are available as data(potency).

a. Perform an analysis of variance to confirm that potency changes with storage temperature.

b. Set up two orthogonal contrasts to assess the nature of the dependence of potency on temperature. You may use the contrast $(-2, -1, 0, 1, 2)$ to assess linearity and the contrast $(2, -1, -2, -1, 2)$ to assess whether there is a quadratic response. (Further discussion of polynomial contrasts is in Section 10.4.)

c. Test whether each of the contrasts you proposed in part b) is significantly different from 0.

d. Report your recommendations for the temperature at which this antibiotic should be stored.

6.8. Anderson and McLean (1974) report the results of an experiment to compare the disintegration times in seconds of four types of pharmaceutical tablets labeled A, B, C, D. These were the only tablet types of interest. The data are available as data(tablet1). Perform an analysis of variance to see if the tablets have equivalent disintegration times. The time to disintegration determines when the medication begins to work. Shorter times mean the tablet will begin disintegrating in the stomach. Longer times mean the table will disintegrate in the small intestines where it is more easily absorbed and less susceptible to degradation from the digestive enzymes. Assuming that longer times to disintegration are desirable, use the Tukey procedure to prepare a recommendation to the tablet manufacturer.

6.9. The dataset data(blood) contain the results of an experiment reported by Box et al. (1978) to compare the coagulation times in seconds of blood drawn from animals fed four different diets labeled A, B, C, D. Assuming that these were the only diets of interest, set up an analysis of variance to compare the effects of the diets on coagulation. Use the Tukey procedure to investigate whether any pairs of the diets can be considered to provide essentially equivalent coagulation times.

6.10. Reconsider data(draft70mn) from Data Archive (1997), previously visited in Exercises 4.1 and 3.25. Assuming that the ranks were randomly assigned to the dates of the year, construct a one-way analysis of variance with the ranks as response and the months as groups. Isolate the linear effect of month.

6.11. Westfall and Rom (1990) considered the nonbirth litter weights of mice whose mothers were previously subjected to one of three treatments or a control, with the objectives of relating weight differences to treatment dosages. (It is conjectured that "nonbirth weight" refers to weight at some definite time following birth.) The data are available as data(mice). Perform a Brown–Forsyth homogeneity of variance test on these data and carefully state your conclusions.

6.12. Derive Equation (6.15) from Equation (5.13) by substitution and simplification as outlined in Section 6.9.

6.13. Verify that the four single degree-of-freedom lines in the ANOVA table in Table 6.8 can be obtained from the contrasts in the contrasts(turkey$dist) section of Table 6.8.

6.14. Calculate the EMS values in Table 6.4. MS_{Tr} is defined in Table 6.2. Define $y_{ij} = \mu + \alpha_i + \epsilon_{ij}$ from Equation (6.1) and substitute into the formula for MS_{Tr} to get the EMS. When α_i is fixed ($\sum_i \alpha_i = 0$), then $E\left((\alpha_i - \bar{\alpha})^2\right)$ is σ^2 plus a function of the α_i values. When α_i is random $\left(\alpha_i \sim N(0, \sigma_A^2)\right)$, then $E\left((\alpha_i - \bar{\alpha})^2\right)$ is σ^2 plus a constant times σ_A^2.

6.A Appendix: Computation for the Analysis of Variance

Model formulas are expressed in R with a symbolic notation which is a simplification of the more extended traditional notation

$$y_{ij} = \mu + \alpha_i + \epsilon_{ij} \quad \text{for} \quad i = 1, \dots, a \quad \text{and} \quad j = 1, \dots, n_i \quad (6.1)$$

The intercept term μ and the error term ϵ_{ij} are usually assumed. The existence of the subscripts is implied and the actual values are specified by the data values.

With R we will be using aov for the calculations and anova and related commands for the display of the results. aov can be used with equal or unequal cell sizes n_i. Model (6.1) is denoted in R by the formula

```
Y ~ A
```

The operator ~ is read as "is modeled by".

Two different algorithms are used to calculate the analysis of variance for data with one factor: sums of squared differences of cell means and regression on dummy variables. Both give identical results.

The intuition of the analysis is most easily developed with the sums of squared differences algorithm. We began there in Equation 6.6 and the definitions in the notes to Table 6.2. We show in Table 6.10 the partitioning of the observed values for the response variable concent in catalystm example into columns associated with the terms in the model. The sum of each row reproduces the response variable. This is called the linear identity. The sum of the squares in each column is the ANOVA table. This is called the quadratic identity. In the notation of Table 6.2 the numbers in the (Intercept) column are $\bar{\bar{y}}$, the numbers in the catalyst column are the treatment effects $\bar{y}_i - \bar{\bar{y}}$, and the numbers in the Residuals column are $y_{ij} - \bar{y}_i$. The numbers in the result of the apply statement are the sums of squares: $\sum_{ij} \bar{\bar{y}}^2$, $SS_{Tr} = \sum_{i=1}^{a} n_i (\bar{y}_i - \bar{\bar{y}})^2$, $SS_{Res} = \sum_{i=1}^{a} \sum_{j=1}^{n_i} (y_{ij} - \bar{y}_i)^2$, and $\sum_{ij} y_{ij}^2$. We come back to the linear and quadratic identities in Table 8.6.

The regression formulation is easier to work with and generalizes better. Once we have developed our intuition we will usually work with the regression formulation. The discussion of contrasts in Section 6.9 leads in to the regression formulation in Chapter 10. For the moment, In Table 6.11 we step forward into the notation of Chapter 10 and express the catalystm example in regression notation.

Table 6.10 Linear and quadratic identities for the one way Analysis of Variance. The column labeled Sum is the sum of the three columns of the projection matrix onto the space of the Grand Mean (labeled (Intercept)), the effects due to the factor catalyst, and the Residuals. The Sum column is identical to the observed response variable concent. The sums of squares of each column of the projection matrix are the numbers in the similarly labeled row in the "Sum of Squares" column of the ANOVA table.

```
> data(catalystm)

> catalystm.aov <- aov(concent ~ catalyst, data=catalystm)

> anova(catalystm.aov)
Analysis of Variance Table

Response: concent
          Df Sum Sq Mean Sq F value Pr(>F)
catalyst   3   85.7   28.56    9.92 0.0014 **
Residuals 12   34.6    2.88
---
Signif. codes:
0 '***' 0.001 '**' 0.01 '*' 0.05 '.' 0.1 ' ' 1

> model.tables(catalystm.aov)
Tables of effects

  catalyst
        A     B      C      D
    2.412 1.287 -1.254 -3.362
rep 5.000 4.000  3.000  4.000

> Proj <- proj(catalystm.aov)

> Proj <-cbind(Proj, Sum=apply(Proj, 1, sum))

> Proj
   (Intercept) catalyst Residuals  Sum
1        54.49    2.412    1.3000 58.2
2        54.49    2.412    0.3000 57.2
3        54.49    2.412    1.5000 58.4
4        54.49    2.412   -1.1000 55.8
5        54.49    2.412   -2.0000 54.9
6        54.49    1.287    0.5250 56.3
7        54.49    1.287   -1.2750 54.5
8        54.49    1.287    1.2250 57.0
9        54.49    1.287   -0.4750 55.3
10       54.49   -1.254   -3.1333 50.1
11       54.49   -1.254    0.9667 54.2
12       54.49   -1.254    2.1667 55.4
13       54.49   -3.362    1.7750 52.9
14       54.49   -3.362   -1.2250 49.9
15       54.49   -3.362   -1.1250 50.0
16       54.49   -3.362    0.5750 51.7

> apply(Proj, 2, function(x) sum(x^2))
(Intercept)    catalyst    Residuals         Sum
   47502.20       85.68        34.56    47622.44
```

Table 6.11 The aov by the factor `catalyst` in Table 6.10 is identical to the `lm` shown here by the three dummy variables generated from the `catalyst` factor. The degrees of freedom (1+1+1=3) and the Sums of Squares (8.8+2.7+741.1=85.7) are both the same.

```
> contrasts(catalystm$catalyst)
  B C D
A 0 0 0
B 1 0 0
C 0 1 0
D 0 0 1

> X <- model.matrix(catalystm.aov)[,2:4]

> X
   catalystB catalystC catalystD
1          0         0         0
2          0         0         0
3          0         0         0
4          0         0         0
5          0         0         0
6          1         0         0
7          1         0         0
8          1         0         0
9          1         0         0
10         0         1         0
11         0         1         0
12         0         1         0
13         0         0         1
14         0         0         1
15         0         0         1
16         0         0         1

> catalystm.lm <-
+      lm(concent ~ X[,"catalystB"] + X[,"catalystC"] + X[,"catalystD"],
+          data=catalystm)

> anova(catalystm.lm)
Analysis of Variance Table

Response: concent
                  Df Sum Sq Mean Sq F value  Pr(>F)
X[, "catalystB"]   1    8.8     8.8    3.07 0.10526
X[, "catalystC"]   1    2.7     2.7    0.95 0.35012
X[, "catalystD"]   1   74.1    74.1   25.73 0.00027 ***
Residuals         12   34.6     2.9
---
Signif. codes:
0 '***' 0.001 '**' 0.01 '*' 0.05 '.' 0.1 ' ' 1
```

6.B Object Oriented Programming

Many of R's functions are designed to be sensitive to the class of object to which they are applied. Figure 6.7 shows that the same syntax plot(x) produces a different form of plot depending on the class of the argument x.

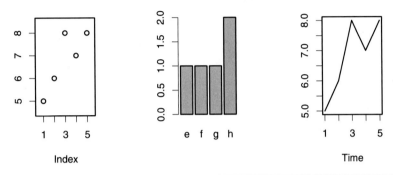

```
> tmp <- data.frame(AA=c(5,6,8,7,8),
+                   BB=factor(letters[c(5,6,8,7,8)]),
+                   CC=ts(c(5,6,8,7,8)),
+                   stringsAsFactors=FALSE)

> tmp
  AA BB CC
1  5  e  5
2  6  f  6
3  8  h  8
4  7  g  7
5  8  h  8

> sapply(tmp, class)
        AA          BB          CC
 "numeric"    "factor"        "ts"

> is.numeric(tmp$A)
[1] TRUE

> plot(tmp$AA)

> plot(tmp$BB)

> plot(tmp$CC)
```

Fig. 6.7 The three columns of the data.frame tmp have three different classes. The plot function is sensitive to the class of its argument and draws a different style plot for each of these classes. The integer object (more generally numeric object) is plotted as a scatterplot with an index on the horizontal axis. The factor object is plotted as a barchart with the level names on the horizontal axis. The time series object is plotted as a line graph with the time value on the horizontal axis.

```
> class(catalystm.aov)
[1] "aov" "lm"

> summary(catalystm.aov)
            Df Sum Sq Mean Sq F value Pr(>F)
catalyst     3   85.7   28.56    9.92 0.0014 **
Residuals   12   34.6    2.88
---
Signif. codes:
0 '***' 0.001 '**' 0.01 '*' 0.05 '.' 0.1 ' ' 1

> old.par <- par(mfrow=c(1,4))

> plot(catalystm.aov)

> par(old.par)
```

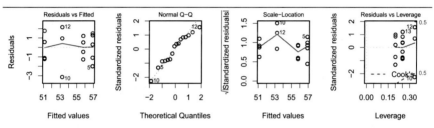

Fig. 6.8 The two accessor functions summary and plot are sensitive to the class of their argument
and produce a form of output appropriate to the argument, in this case an "aov" object. Note that
"aov" objects are a special case of "lm" objects. The summary function for an "aov" object
produces an ANOVA table. The plot function for an "lm" object is a set of four diagnostic plots
of the residuals from the fitted model. The contents of the panels of the plot are discussed in
Sections 8.4 and 11.3.7.

The result of a function call (aov for example) is an object with a class ("aov").
Accessor functions such as summary or plot are sensitive to the class of their argument and produce an appropriate form of output as shown in Figures 6.7 and
6.8.

Chapter 7

Multiple Comparisons

In Exercise 3.13 we discover that the probability of simultaneously making three correct inferences, when each of the three individually has P(correct inference) = $1 - \alpha$ = 0.95, is only $(1 - \alpha)^3$ = $.95^3$ = 0.857. Alternatively, the probability of making at least one incorrect inference is $1 - -0.857 = 0.143 \approx 3\alpha$. In general, the more simultaneous inferences we make at one time, the smaller the probability that all are correct. In this chapter we learn how to control the probability that all inferences are simultaneously correct. We usually phrase the goal as controlling the probability of making at least one incorrect inference.

We consider all inferences in a related *family* of inferences. Such a family is typically a natural and coherent collection; for example, all inferences resulting from a single experiment. The inferences can be individual tests of hypotheses or confidence intervals. In the context of a family of hypothesis tests, if we control the Type I error probability for each test at level α, the probability of committing at least one Type I error in the family will be much larger than α. For example, if the tests are independent and α = .05, then the probability of at least one Type I error is $1 - (1 - .05)^6 \approx .26$, which seems an unacceptably large error threshold. The way to control the probability of at least one Type I error in the family is to choose a smaller α for each individual test. For example, with a single two-sided test with α = .05 from a standard normal the critical value is 1.96. The intention of all multiple comparison procedures is to provide a larger critical value than the default value.

A way to avoid such errors when conducting many related inferences simultaneously is to employ a *multiple comparison procedure*. Such a procedure for *simultaneous hypothesis testing* may seek to (strongly) control the familywise error rate (FWE), defined as P(reject at least one true hypothesis under any configuration of true and false hypotheses). A procedure for *simultaneous confidence intervals* should control the probability that at least one member of the family of confidence

© Springer Science+Business Media New York 2015
R.M. Heiberger, B. Holland, *Statistical Analysis and Data Display*,
Springer Texts in Statistics, DOI 10.1007/978-1-4939-2122-5_7

intervals does not contain the parameter being estimated by the interval. When a multiple comparison procedure is used, it is said that the analyst is *controlling for multiplicity*.

In order to exert FWE control over a family of related hypothesis tests, it is necessary to have a reduced probability of rejecting any particular null hypothesis in the family. As explained in Section 3.7, reducing the probability of rejecting particular hypotheses results in an increased probability of retaining them, and therefore reduced power for tests of these hypotheses. This implies that, as compared with testing hypotheses in isolation from one another, a multiple comparison procedure has a diminished ability to reject false null hypotheses. In other words, a test of a particular hypothesis using a multiple comparison procedure will be less powerful than the test of the same hypothesis in isolation. In deciding whether to use a multiple comparison procedure, the protection against the possibility of an excessive number of incorrect hypothesis rejections must be weighted against this loss of power. An analogous statement holds for simultaneous versus isolated confidence intervals.

In general, the choice of multiple comparison procedure to be used depends on the structure of the *family* of related inferences and the nature of the collection of statistics from which the confidence intervals or tests will be calculated.

Section 7.1 summarizes the most frequently used multiple comparisons procedures. Section 7.2 presents a graphical procedure for looking at the results of the multiple comparisons procedures.

7.1 Multiple Comparison Procedures

7.1.1 Bonferroni Method

A very general way to control the FWE is based on the Bonferroni inequality, $P(\bigcup E_i) \leq \sum_i P(E_i)$, where the E_i are arbitrary events. If the family consists of m related tests, conducting each test at level $\frac{\alpha}{m}$ ensures that FWE $\leq \alpha$. If the family consists of m related confidence intervals, maintaining confidence $100(1 - \frac{\alpha}{m})\%$ for each interval will ensure that the overall confidence of all m intervals will be at least $100(1 - \alpha)\%$. The Bonferroni method should be considered for use when the family of related inferences is unstructured (e.g., not like the structured families required for the procedures discussed in Sections 7.1.2–7.1.4), or when the statistics used for inference about each family member have nonidentical probability distributions.

The Bonferroni inequality is very blunt in the sense that its right side is typically much larger than its left. One reason for this is that it does not seek to take into account information about the intersections of the events E_i. As a result, the Bonferroni approach is very conservative in the sense of typically guaranteeing an FWE substantially less than its nominal value of α, and the extent of this conservativeness increases with m. The value of this approach is that it is very generally applicable, for example, when the pivotal statistics associated with the m inferences have nonidentical probability distributions. Hochberg (1988) provides an easy-to-understand improvement to the Bonferroni approach for hypothesis testing that tends to reject more false null hypotheses than Bonferroni. Hochberg's procedure has been proven to be applicable to a wide variety of testing situations; see Sarkar (1998).

7.1.2 Tukey Procedure for All Pairwise Comparisons

Often a family of inferences has a special structure that allows us to use available information about the joint distributions of the pivotal statistics, thus enabling the use of a less conservative approach than Bonferroni. An example of this, discussed in Section 6.3, is the family consisting of all $m = \binom{k}{2}$ comparisons among all pairs of means of k populations. For this family, Tukey's Studentized range test is usually recommended.

7.1.3 The Dunnett Procedure for Comparing One Mean with All Others

The Dunnett procedure is used when the family of inferences of interest is the comparisons of the mean of one designated population with each of the means of the remaining populations, all populations being at least approximately normal with approximately the same variance. Often in practice the designated population is a control and the others are active treatments. The Dunnett procedure uses the percentiles of a multivariate t distribution rather than a univariate t distribution discussed in Section 5.4.3.

For purposes of illustration of the Dunnett procedure, we use weightloss data. A random sample of 50 men who were matched for pounds overweight was randomly separated into 5 equal groups. Each group was given exactly one of the weight loss agents A, B, C, D, or E. After a fixed period of time, each man's weight loss was recorded. The data, taken from Ott (1993), are accessed as data(weightloss) and shown in Figure 7.1.

The F-statistic tests the null hypothesis that the five groups have identical mean weight loss vs the alternative that the groups do not have identical mean weight loss. The small p-value from the F test in the basic ANOVA in Table 7.1 suggests that the agents have differing impacts on weight loss.

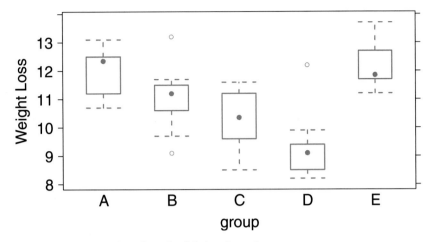

Fig. 7.1 Weightloss data: Boxplots of weight loss for each group.

Table 7.1 Weightloss ANOVA

```
> weightloss.aov <- aov(loss ~ group, data=weightloss)

> summary(weightloss.aov)
            Df Sum Sq Mean Sq F value  Pr(>F)
group        4   59.9   14.97    15.1 6.9e-08 ***
Residuals   45   44.7    0.99
---
Signif. codes:  0 '***' 0.001 '**' 0.01 '*' 0.05 '.' 0.1 ' ' 1
```

When we regard agent D as the control, we seek to investigate whether any of the other four agents appear to promote significantly greater weight loss than agent D. From Figure 7.1 we see that the five populations are approximately normal with approximately the same variance. Therefore, we may proceed with the Dunnett procedure. Since we are investigating whether the other agents *improve* on D, we display infinite upper one-sided confidence intervals against D in Table 7.2 and Figures 7.2 and 7.3.

The (default) 95% confidence level in Table 7.2 applies simultaneously to all four confidence statements. The fact that all four confidence intervals lie entirely above zero suggests that D is significantly inferior to the other four weightloss agents.

Figure 7.3 is a mean–mean display of Dunnett's multiple comparison procedure applied to the weightloss data. Tabular results are shown in Table 7.3. Figure 7.3 is analogous to Figure 6.2 in Section 6.3. The mean–mean display technique is discussed in detail in Section 7.2. In Figure 7.3, reflecting the results for upper one-sided Dunnett confidence intervals, all horizontal lines except that for comparing

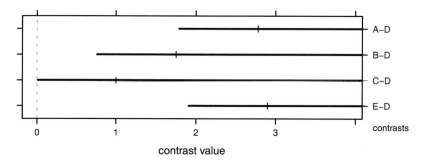

Fig. 7.2 Weightloss data: Standard display of one-sided multiple comparisons using the Dunnett method against the control treatment D.

groups D and C fall to the right of zero. The line for C-D sits on the boundary with lower limit .009. Consistent with the boxplots in Figure 7.1, we conclude that all weightloss agents (except possibly C) provide superior mean weight loss to that provided by agent D.

The Dunnett procedure is used in Exercises 7.7 and 12.4.

7.1.3.1 Computing Note—Specifying the Alternative Hypothesis

There are at least three conventions for indicating the alternative hypothesis. Be very clear which you are using.

As shown here, the `glht` function in R uses the argument `alternative` to indicate the alternative hypothesis. `glht` uses `alternative="greater"` to indicate an infinite upper bound, `alternative="less"` for an infinite lower bound, and defaults to `alternative="two-sided"`.

The S-Plus function `multicomp` uses the argument `bounds="lower"` to indicate a finite lower bound, implying an infinite upper bound. `multicomp` uses `bounds="upper"` for a finite upper bound, implying an infinite lower bound. For two-sided intervals `multicomp` defaults to `bounds="both"`.

SAS `PROC ANOVA` specifies the alternative hypothesis by using a different option name for each. SAS uses the option `dunnettu`, with the suffix "u" to indicate an infinite upper interval, the option `dunnettl` with the suffix "l" to indicate an infinite lower bound, and the option `dunnett` with no suffix for two-sided intervals.

Table 7.2 Weight loss using the Dunnett procedure.

```
> weightloss.dunnett <-
+ glht(weightloss.aov,
+     linfct=mcp(group=
+                contrMat(table(weightloss$group), base=4)),
+     alternative = "greater")

> confint(weightloss.dunnett)

 Simultaneous Confidence Intervals

Multiple Comparisons of Means: User-defined Contrasts


Fit: aov(formula = loss ~ group, data = weightloss)

Quantile = -2.222
95% family-wise confidence level


Linear Hypotheses:
           Estimate lwr       upr
A - D <= 0 2.78000  1.78949      Inf
B - D <= 0 1.75000  0.75949      Inf
C - D <= 0 1.00000  0.00949      Inf
E - D <= 0 2.90000  1.90949      Inf
```

Table 7.3 MMC calculations for weightloss using the Dunnett procedure.

```
> weightloss.mmc <-
+    mmc(weightloss.aov,
+        linfct=mcp(group=
+                   contrMat(table(weightloss$group), base=4)),
+        alternative = "greater")

> weightloss.mmc
Dunnett contrasts
Fit: aov(formula = loss ~ group, data = weightloss)
Estimated Quantile = -2.222
95% family-wise confidence level
$mca
     estimate stderr   lower upper height
E-D     2.90    -Inf 1.909477  Inf  10.72
A-D     2.78    -Inf 1.789477  Inf  10.66
B-D     1.75    -Inf 0.759477  Inf  10.14
C-D     1.00    -Inf 0.009477  Inf   9.77
$none
   estimate stderr lower upper height
E    12.17    -Inf 11.47   Inf  12.17
A    12.05    -Inf 11.35   Inf  12.05
B    11.02    -Inf 10.32   Inf  11.02
C    10.27    -Inf  9.57   Inf  10.27
D     9.27    -Inf  8.57   Inf   9.27
```

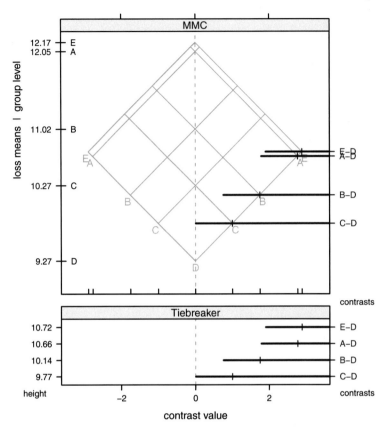

Fig. 7.3 Weightloss data: Mean–mean display of one-sided multiple comparisons using the Dunnett method against the control treatment D. The Tiebreaker panel is needed in this example because the E–D and A–D contrasts are at almost the same height in the top panel and are therefore overprinted. The similar heights for these two contrasts follow from the similar means for the E and A levels of the loss factor. Please see the discussion of the mean–mean display in Section 7.2.

7.1.4 Simultaneously Comparing All Possible Contrasts Scheffé and Extended Tukey

7.1.4.1 The Scheffé Procedure

In the context of comparing the means of a populations, the Scheffé multiple comparison procedure controls the familywise error rate over the infinite-sized family consisting of all possible contrasts $\sum_{j=1}^{a} c_j \mu_j$ involving the population means. The Scheffé procedure is therefore appropriate for exerting simultaneous error control over the set of four contrasts in our analysis of the turkey data data(turkey) from Section 6.8. In exchange for maintaining familywise error control over so large a family, the Scheffé method gives rise to wide confidence limits and relatively unpowerful tests. Therefore, we recommend its use only in the narrowly defined situation of simultaneously inferring about mean contrasts more complex than a comparison of two means. The Scheffé procedure uses a percentile of an F distribution, derived as the distribution of the most significant standardized contrast among the sample means.

The confidence interval formula by the Scheffé procedure is

$$\text{CI}\left(\sum_{j=1}^{a} c_j \mu_j \right) = \sum_{j=1}^{a} c_j \bar{y}_j \pm \sqrt{(a-1)F_{.05,a-1,N-a}} \ s \sqrt{\sum_{j=1}^{a} \frac{c_j^2}{n_j}} \tag{7.1}$$

This provides the set of $100(1 - \alpha)\%$ simultaneous confidence intervals for all possible contrasts among the population means. In this equation $N = \sum_{j=1}^{a} n_j$.

For R glht, we must manually calculate the critical value with, for example in the turkey data,

```
scheffe.quantile <- sqrt(4*qf(.95, 4, 25))
```

The Scheffé test is one of the methods available in the S-Plus multicomp function and is one of the options for the MEANS statement in SAS PROC ANOVA.

7.1.4.2 Scheffé Intervals with the Turkey Data

Table 6.8 provides F-tests of the hypotheses that the members of a basis set of four contrasts are zero. These four tests do not control for multiplicity. The finding in Table 6.8 is that three of these contrasts differ significantly from zero. We do not declare the fourth contrast significantly different from zero because its p-value exceeds 0.05.

The Scheffé procedure allows us to make inferences about these same contrasts while controlling for multiplicity. The confidence interval and testing results are shown in Table 7.4 and in Figure 7.4. An additional advantage of the Scheffé analysis is that the results specify the direction of contrasts' significant difference from zero. For example, in Table 7.4, the fact that the confidence interval on A.vs.B lies entirely below zero implies that, on average, the mean weight gain from diet B exceeds that from diet A. The F-statistics in Table 6.8 are essentially squared t-statistics, and this obscures information on directionality unless the definitions of the contrasts being tested are carefully examined alongside the test results.

We may use the results of the Scheffé analysis to assess the extent to which, if any, of the Scheffé simultaneous confidence intervals cause us to modify our previous conclusions about the contrasts. When doing so it is important to observe the contrast codings, that is, the numerical values defining the contrast. Observing that

Table 7.4 Scheffé Test for Turkey Data Contrasts. See also Figure 7.4.

```
> data(turkey)

> turkey.aov <- aov(wt.gain ~ diet, data=turkey)

> scheffe.quantile <- sqrt(4*qf(.95, 4, 25))

> turkey.lmat <-
+    cbind(control.vs.treatment=c(1,-.25,-.25,-.25,-.25),
+          A.vs.B              =c(0, .5,   .5,  -.5,  -.5 ),
+          amount              =c(0, .5,  -.5,   .5,  -.5 ),
+          A.vs.B.by.amount    =c(0, .5,  -.5,  -.5,   .5 ))

> row.names(turkey.lmat) <- row.names(contrasts(turkey$diet))

> turkey.mmc <- mmc(turkey.aov, calpha=scheffe.quantile, focus="diet",
+                   focus.lmat=turkey.lmat,
+                   estimate.sign=0, order.contrasts=FALSE)

> turkey.mmc$lmat
                     estimate stderr   lower  upper height
control.vs.treatment   -3.433 0.2563 -4.2849 -2.582  5.500
A.vs.B                 -1.950 0.2293 -2.7116 -1.188  7.217
amount                 -1.933 0.2293 -2.6950 -1.172  7.217
A.vs.B.by.amount        0.450 0.2293 -0.3116  1.212  7.217
```

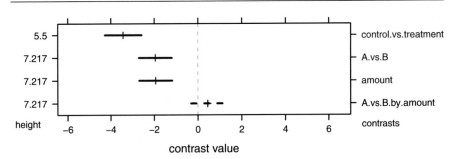

Fig. 7.4 Scheffé plot for turkey data. See also Table 7.4.

the first three of the four Scheffé intervals exclude 0 while the last one includes 0, the Scheffé results reinforce our original impressions from the nonsimultaneous F-tests of these contrasts in Table 6.8.

In this example, examination of the Scheffé results did not cause us to revise our earlier results ignoring multiplicity. In general, use of a multiple comparison procedure is an appropriately conservative approach that may not declare a difference found by nonsimultaneous tests or confidence intervals.

Figure 7.5 is a graphic presentation of the Scheffé procedure applied to comparisons of all pairs of means. We use Scheffé intervals here because these pairwise comparisons are part of a larger family of contrasts that includes those displayed in Figure 7.6. There are $10 = \binom{5}{2}$ pairwise differences among the means of the 5 diet combinations studied. Figure 7.5 is a mean–mean display of Scheffé simultaneous confidence intervals on these mean differences.

Figures 7.5 and 7.6 contain overprinting of the confidence lines and labels for several of their comparisons of level means. The overprinting in Figure 7.5 is due to almost identical mean values for levels B1 and A2. The overprinting in Figure 7.6 is a consequence of the same almost identical mean values, now reflected as identical heights for the contrasts because the interaction of the A.vs.B and the amount comparisons is not significant. In situations with such overprinting, we augment the mean–mean display with a traditional display of these same confidence intervals. This *Tiebreaker* plot lists the contrasts in the same vertical order as in the mean–mean plot. The conclusions here, based on the fact that 9 of the 10 intervals lie entirely above zero, are

- For both amount 1 and 2, the mean weight gain from additive B is significantly greater than the mean weight gain from additive A.

- For both additive A or B, the mean weight gain from amount 2 significantly exceeds the mean weight gain from amount 1.

- The weight gain from the control diet is significantly below that from any of the other 4 diets.

We graphically summarize these conclusions with the orthogonal contrasts in Figure 7.6. The 3 contrasts that differ significantly from zero do not cross the vertical $d = 0$ axis. The nonsignificant contrast does cross the $d = 0$ axis.

Table 7.4 and Figure 7.4 show three of the user-defined contrasts to have negative estimates. Figure 7.6 shows those contrasts to be reversed to have positive contrasts. We believe that multiple comparisons are most easily interpreted when the means are sequenced in numerical order (not lexicographic order), and consequently that all displayed contrasts should compare the larger value to the smaller value. That is, all displayed contrast values should be positive. Such reversal of the direction of a contrast creates no problem when assessing how contrasts relate to zero so long as the reversal is noted. We note the reversal by appending a "−" to the names of the reversed contrasts.

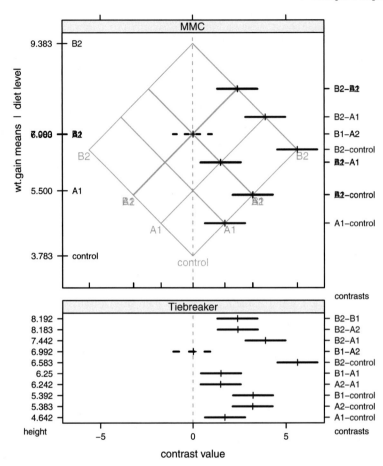

Fig. 7.5 MMC: mca plot for Turkey data. Overprinting of contrasts at the same height in the MMC panel are separated in the Tiebreaker panel by a standard multiple comparisons plot ordered to match the order of the MMC plot.

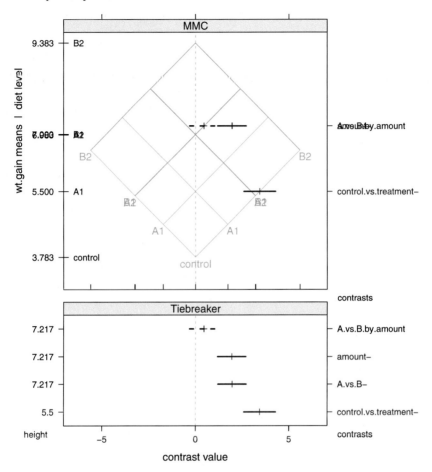

Fig. 7.6 MMC: Orthogonal basis set of contrasts for Turkey data. Overprinting of the confidence lines (for contrasts `A.vs.B-`, `amount-`, and `A.vs.B.by.amount`, in this example) and their labels in the right-axis labels of the MMC panel is a consequence of almost identical values for the group means in the left-axis labels (`A2` and `B1`). The overprinting is resolved in the Tiebreaker panel, a standard multiple comparisons plot (without information on the group means) ordered to match the order of the MMC plot. The contrasts in these panels are the same contrasts that appear in Figure 7.4, but negative estimates there have been reversed here. During the reversal a "−" was appended to contrast names for which it was not possible to figure out how to reverse the contrast name.

7.1.4.3 The Extended Tukey Procedure

The Tukey procedure can be extended to cover the family of all possible contrasts when the samples are of the same size n. Generalizing Equation (6.11) to any contrast vector (c_j) in the equal n case, we get

$$\text{CI}\left(\sum_{j=1}^{a} c_j \mu_j\right) = \sum_{j=1}^{a} c_j \bar{y}_j \pm \frac{q_\alpha}{2} \frac{s}{\sqrt{n}} \sum_{j=1}^{a} |c_j| \tag{7.2}$$

as the set of $100(1-\alpha)\%$ simultaneous confidence intervals for all possible contrasts among the population means.

The q_α here is the same value used in Equation (6.11). Except for very simple contrasts, such as between pairs of means, these generalized Tukey intervals will be even wider than the analogous Scheffé intervals, Hochberg and Tamhane (1987). The generalized Tukey intervals (7.2) may be considered for use when interest lies in a family consisting of the union of all pairwise contrasts with a small number of more complicated contrasts.

As discussed in Hochberg and Tamhane (1987), the family encompassed by the generalized Tukey intervals also includes the set of individual intervals on each population mean,

$$\text{CI}(\mu_j) = \bar{y}_j \pm q_\alpha \frac{s}{\sqrt{n_j}} \tag{7.3}$$

These intervals are illustrated for the artificial data in Figure 7.11.

7.2 The Mean–Mean Multiple Comparisons Display (MMC Plot)

7.2.1 Difficulties with Standard Displays

The conclusions from the application of the Tukey procedure to the catalyst data are not well conveyed by the standard tabular and graphical output shown in Table 7.5 and Figure 7.7. In both displays, the magnitudes of the sample means themselves are not shown. These displays are therefore not capable of depicting the relative distances between adjacent sorted sample means. Indeed, the standard display ignores the sample means entirely and instead sorts the contrasts alphabetically. Compare Table 7.5 to the $mca section of Table 6.3, and Figure 7.7 to the bottom panel of Figure 6.2.

Another standard display of results of a Tukey test, shown here in Figure 7.8, is often used to communicate results when sample sizes are equal. The sample means are listed in ascending magnitude. Straight-line segments are used to indicate significance according to the following rules. If two sample means are not covered by the same line segment, the corresponding population means are declared significantly different. If two sample means are covered by a common line segment, the corresponding population means are declared not significantly different.

Table 7.5 Tukey Multiple Comparisons for Catalyst Data—Standard Display (not showing means)

```
> catalystm.glht <-
+    glht(catalystm1.aov, linfct = mcp(catalyst = "Tukey"))

> confint(catalystm.glht)

 Simultaneous Confidence Intervals

Multiple Comparisons of Means: Tukey Contrasts


Fit: aov(formula = concent ~ catalyst, data = catalystm)

Quantile = 2.966
95% family-wise confidence level


Linear Hypotheses:
           Estimate lwr     upr
B - A == 0 -1.125   -4.501  2.251
C - A == 0 -3.667   -7.342  0.009
D - A == 0 -5.775   -9.151 -2.399
C - B == 0 -2.542   -6.386  1.302
D - B == 0 -4.650   -8.209 -1.091
D - C == 0 -2.108   -5.952  1.736
```

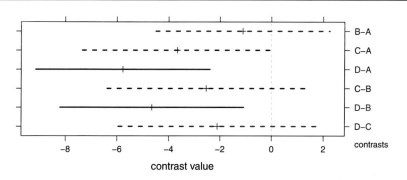

Fig. 7.7 All Pairwise Comparisons of Catalyst Means. In this standard display, the group means are not displayed. The contrasts are sorted alphabetically. Compare this figure to the bottom panel of Figure 6.2 where the contrasts are sorted by the values of the means being compared.

With this procedure it is difficult to depict correctly the relative distances between adjacent sorted sample means because the table is constrained by the limited resolution of a fixed-width typewriter font rather than the high resolution of a graphical display.

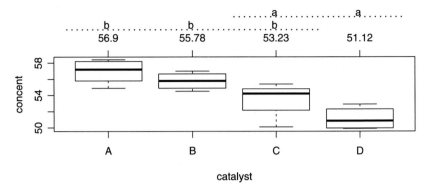

Fig. 7.8 This example is constructed from the `cld` function in the **multcomp** package. The function call `plot(cld(catalystm.glht))` draws the boxplots and the letter values. We manually (by supplementary code) placed the numerical values of the means and the underlines connecting the letters.

Further, the procedure cannot be used when sample sizes are unequal. Table 7.6 and Figure 7.9 illustrate this limitation using artificial data:

Group	N	Mean
A	5	2.0
B	100	2.1
C	100	2.8
D	5	3.0

The Tukey procedure shown in Table 7.6 uncovers a significant difference between the means of populations B and C, for which the sample sizes are large, but no significant difference between the means of populations A and D, for which the sample sizes are small. With the lines-type graph in Figure 7.9 the nonsignificant difference between the means of A and D requires that a common line covers the range from 2.0 to 3.0, including the location of the means of groups B and C. The presence of this line contradicts the finding of a significant difference between the means of groups B and C that is seen in the standard displays in Figures 7.10 and 7.11 and in the mean–mean display (described in Section 7.2.2) in Figure 7.12.

Table 7.6 Simultaneous confidence intervals on all pairs of mean differences. The means of samples B and C both lie between the means of samples A and D. This example is based on highly unbalanced artificial data. Sample sizes were 100 from populations B and C and 5 from populations A and D. The Tukey procedure finds a significant difference between the means of populations B and C but no significant difference between the means of populations A and D.

```
> group <- factor(LETTERS[1:4])

> n <- c(5,100,100,5)

> ybar <- c(2, 2.1, 2.8, 3)

> inconsistent.aov <- aovSufficient(ybar ~ group, weights=n, sd=.8)

> anova(inconsistent.aov)
Analysis of Variance Table

Response: ybar
          Df Sum Sq Mean Sq F value  Pr(>F)
group      3     27    9.01    14.1 2.2e-08 ***
Residuals 206    132    0.64
---
Signif. codes:  0 '***' 0.001 '**' 0.01 '*' 0.05 '.' 0.1 ' ' 1

> inconsistent.glht <-
+    glht(inconsistent.aov, linfct=mcp(group="Tukey"),
+         vcov.=vcovSufficient, df=inconsistent.aov$df.residual)

> crit.point <- qtukey(.95, 4, 206)/sqrt(2)

> confint(inconsistent.glht, calpha=crit.point)

 Simultaneous Confidence Intervals

Multiple Comparisons of Means: Tukey Contrasts


Fit: aov(formula = formula, data = data, weights = weights, x = TRUE)

Quantile = 2.59
95% confidence level


Linear Hypotheses:
           Estimate lwr      upr
B - A == 0   0.1000 -0.8496  1.0496
C - A == 0   0.8000 -0.1496  1.7496
D - A == 0   1.0000 -0.3105  2.3105
C - B == 0   0.7000  0.4070  0.9930
D - B == 0   0.9000 -0.0496  1.8496
D - C == 0   0.2000 -0.7496  1.1496
```

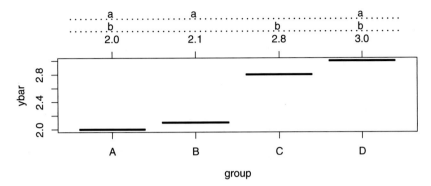

Fig. 7.9 Underlining of means that are not significantly different. Both the a and b lines, which are valid for the comparison of catalysts A and D (in this example based on the low precision test for small sample sizes), mask the significant difference between catalysts B and C (based on a much higher precision test for much larger sample sizes).

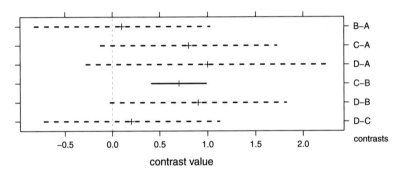

Fig. 7.10 Simultaneous confidence intervals on all pairs of mean differences. The means of samples B and C both lie between the means of samples A and D. Sample sizes were 100 from populations B and C and 5 from populations A and D. The Tukey procedure finds a significant difference between the means of populations B and C but no significant difference between the means of populations A and D. The short confidence interval for the C-A contrast reflects the higher precision of the contrasts based on the larger sample sizes.

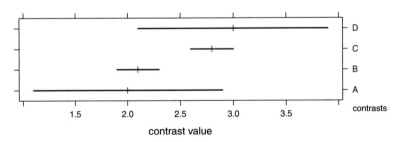

Fig. 7.11 The means of samples B and C both lie between the means of samples A and D. Sample sizes were 100 from populations B and C and 5 from populations A and D. The Tukey procedure finds a significant difference between the means of populations B and C but no significant difference between the means of populations A and D. The underlying formula for these intervals appears in Equation (7.3).

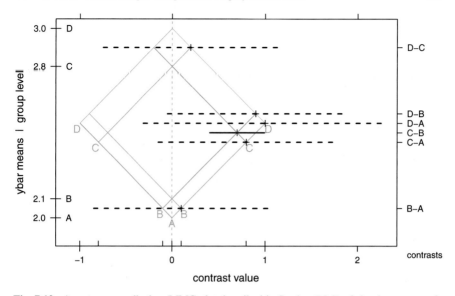

Fig. 7.12 A mean–mean display (MMC plot described in Section 7.2.2) of simultaneous confidences on the means from populations A, B, C, D in the artificial data. Each confidence interval on a mean difference is represented by a horizontal line. If and only if an horizontal line crosses the vertical "contrast value = 0" line, the corresponding population mean difference is declared nonsignificant. In this display we use dashed black lines for nonsignificant comparisons and solid red lines for significant comparisons. This display shows the relative differences between sample means and allows for unequal sample sizes. The short confidence interval for the `C-A` contrast reflects the higher precision of the contrasts based on the larger sample sizes.

7.2.2 Hsu and Peruggia's Mean–Mean Scatterplot

Hsu and Peruggia (1994) address the deficiencies in standard displays of multiple comparison procedures with their innovative graphical display of the Tukey procedure for all pairwise comparisons. In Section 7.2.2.1 we show the details of the construction of the MMC plot displayed in Figure 7.13. We postpone interpretation of Figure 7.13 until Section 7.2.2.2.

In Section 7.2.3 we extend their display to show other multiple comparison procedures for arbitrary sets of contrasts. Software for our extension is included in the **HH** package as function mmc and its related functions.

7.2.2.1 Construction of the Mean–Mean Scatterplot

We begin with data-oriented orthogonal h- and v-axes in Figures 7.14 and 7.15 and then move to rotated difference $(h - v)$ and mean $(h + v)/2$ axes in Figure 7.16. The rotations by $45°$ introduce factors of $\sqrt{2}$ that are there to maintain the orthogonality of h and v in the rotated coordinates.

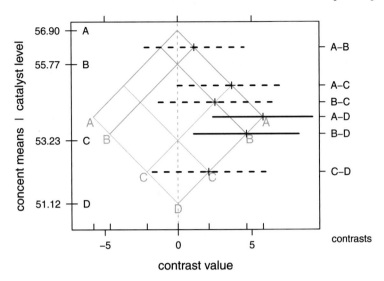

Fig. 7.13 Multiple comparisons of all pairwise comparisons of catalyst means with the MMC display. This is a repeat of the top panel of Figure 6.2.

Construction of MMC plot: concent ~ catalyst

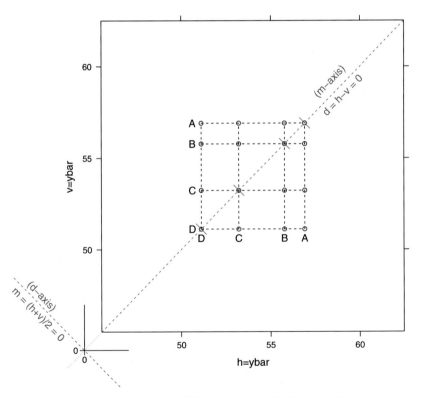

Fig. 7.14 Construction of mean–mean multiple comparisons plot for the catalyst data. Data-oriented axes and isomeans grid, steps 1–6 in the discussion in Section 7.2.

Construction of MMC plot: concent ~ catalyst

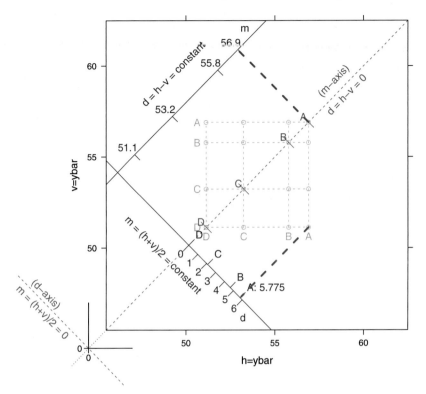

Fig. 7.15 Construction of mean–mean multiple comparisons plot for the catalyst data. Data-oriented axes, steps 7–9 in the discussion in Section 7.2.

1. Draw a square plot in Figure 7.14 on (h, v)-axes. Define (d, m)-axes at $\pm 45°$.

2. Plot each $\bar{y}_i$ against $\bar{y}_j$.

3. Connect the points with $h = \bar{y}_i$ and $v = \bar{y}_j$ lines. The lines are labeled with the level names of the group means. We call this the *isomeans grid*. It is used as the background reference for the MMC plot.

4. Draw the 45° line $h = v$. Define the value $d = h - v$, where the letter d indicates differences between group means. The line we just drew corresponds to $d = 0$. We will call this line the m-axis, where the name $m = (h + v)/2$ indicates the group means.

5. Place tick marks on the m-axis at the points $(\bar{y}_i, \bar{y}_i)$.

6. Draw the $-45°$ line through the origin ($h = 0, v = 0$). The line we just drew corresponds to $m = 0$. We will call this line the d-axis, where the name d indicates the differences.

7. Copy Figure 7.14 to Figure 7.15.

8. Draw another m-axis parallel to the $d = 0$ line. Drop a perpendicular from the $(\bar{y}_A, \bar{y}_A)$ intersection on the $d = 0$ line to the new m-axis. Place a tick at that point and label it with the $m = \bar{y}_A$ value. Place similar tick marks at the heights $m = \bar{y}_i$. (The actual distances from the $m = 0$ line to the tick marks are $\bar{y}_i \sqrt{2}$.)

9. Draw another d-axis parallel to the line $m = 0$. We will place two sets of tick marks on the new d-axis: at the projections of the observed differences $(h, v) = (\bar{y}_i, \min_i(\bar{y}_i))$, and at unit intervals on the difference scale. Drop a perpendicular from the $(\bar{y}_A, \bar{y}_D)$ intersection to the new d-axis. Place a tick at that point and label it with the level name A and the value $\bar{y}_A - \bar{y}_D$. Place similar ticks at the distances $\bar{y}_i - \bar{y}_D$. (The actual distances from the $d = 0$ line to the tick marks are $(\bar{y}_i - \bar{y}_D)/\sqrt{2}$.) Place ticks below the d-axis at the distances $(0, 1, 2, 3, 4, 5, 6)/\sqrt{2}$ and label them $(0, 1, 2, 3, 4, 5, 6)$.

10. Rotate Figure 7.15 counterclockwise by $45°$ to get Figure 7.16.

11. Construct the confidence intervals. We show just one pairwise interval, the one centered on the point $\left(d = \bar{y}_B - \bar{y}_D, \ m = (\bar{y}_B + \bar{y}_D)/2\right)$. The confidence interval line is parallel to the d-axis at a height equal to the average of the two observed means. The interval is on the d-scale and covers all points $(\bar{y}_B - \bar{y}_D) \pm \hat{\sigma} q \sqrt{1/n_B + 1/n_D}$, where $\hat{\sigma}$ is the standard deviation from the ANOVA table and q is the critical value used for the comparison. In this example we use the critical value $q = q_{.05,4,12}/\sqrt{2} = 2.969141$ from the Studentized range distribution.

12. We show all $\binom{4}{2} = 6$ pairwise differences $\bar{y}_i - \bar{y}_j$ with their confidence intervals in Figure 7.13.

Figures 7.14, 7.15, and two additional intermediate figures were drawn with function HH:::mmc.explain, an unexported but accessible function in the **HH** package. Figure 7.16 is an ordinary MMC plot with an lmat matrix indicating exactly one contrast. The code for all figures is included in file HHscriptnames(7).

7.2.2.2 Interpretation of the Mean–Mean Scatterplot

We construct the background of Figures 7.16 and 7.13 by rotating Figure 7.15 counterclockwise by $45°$ and suppressing the h- and v-axes. The horizontal d-axis shows the values of the contrasts and the vertical m-axis shows the average values of the two means being contrasted.

In Figure 7.13, each mean pair $(\bar{y}_i, \bar{y}_j)$ is plotted on the now-diagonal (h, v)-axes and can also be identified with its (d, m)-coordinates. In Figure 7.16, we focus on the pair of means $\bar{y}_B$ and $\bar{y}_D$. We begin with the (h, v)-system and identify the point as

$$(h, v) = (\bar{y}_B, \bar{y}_D) = (55.8, \ 51.1)$$

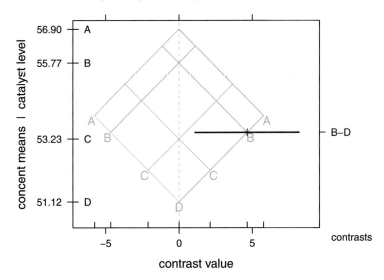

Fig. 7.16 Construction of mean–mean multiple comparisons plot for the catalyst data. Difference and mean-oriented axes. This figure shows steps 1–11 in the discussion in Section 7.2. This figure is essentially the same as Figure 7.15 with a single contrast and rotated 45° counter-clockwise. This figure shows only one of the six pairwise contrasts. All six contrasts and the result of all 12 construction steps are shown in Figure 7.13.

The coordinates of the same pair of means $(\bar{y}_B, \bar{y}_D)$ in the (d, m)-system are

$$(d, m) = \left(\bar{y}_B - \bar{y}_D, (\bar{y}_B + \bar{y}_D)/2\right)$$
$$= \left((55.8 - 51.1), (55.8 + 51.1)/2\right) = (4.65, 53.45)$$

We choose to label the ticks on the m-axis by the means because they are more easily interpreted: The confidence interval on $\bar{y}_B - \bar{y}_D$ is at the mean height $m = (\bar{y}_B + \bar{y}_D)/2$ in Figure 7.16. Hsu and Peruggia label the ticks on the m-axis by the sum $\bar{y}_B + \bar{y}_D = 2m$ because one unit on the $2m$-scale takes exactly the same number of inches as one unit on the d-scale.

Figure 7.13 is constructed from Figure 7.16 by including all of the $\binom{4}{2} = 6$ pairwise differences $\bar{y}_i - \bar{y}_j$, not just the single difference we use for the illustration.

Each of the confidence intervals for the $\binom{4}{2} = 6$ pairwise differences $\bar{y}_i - \bar{y}_j$ in Figure 7.13 is centered at a point whose height on the vertical m-axis is equal to the average of the corresponding means $\bar{y}_i$ and $\bar{y}_j$ and whose location along the horizontal d-axis is at distance $\bar{y}_i - \bar{y}_j$ from the vertical line $d = 0$. Horizontal lines are drawn at these heights so that the midpoints of these lines intersect their $(h = \bar{y}_i, v = \bar{y}_j)$ intersection. The width of each horizontal line is the width of a confidence interval estimating the difference $\bar{y}_i - \bar{y}_j$. By default the endpoints of the line are chosen to be the endpoints of the 95% two-sided confidence interval chosen by the Tukey procedure for all $\binom{4}{2}$ possible pairs.

If a horizontal confidence interval line crosses the vertical $d = 0$ line, the mean difference is declared not significant. Otherwise the mean difference is declared significant. If an end of a horizontal line is close to the vertical $d = 0$, this says that the declaration of significance was a close call.

When the critical value q is chosen by one of the standard multiple comparisons procedures (we illustrate with and default to the Tukey procedure), the widths of the horizontal confidence interval lines are the simultaneous confidence intervals for the six pairs of population mean differences. This depiction is not restricted to the case of equal sample sizes and hence equal interval widths.

The display in Figure 7.13 has several advantages over traditional presentations of Tukey procedure results. In a single graph we see

1. The means themselves, with correct relative distances,

2. The point and interval estimates of the $\binom{4}{2}$ pairwise differences,

3. The point and interval estimates for arbitrary contrasts of the level means,

4. Declarations of significance,

5. Confidence interval widths that are correct when the sample sizes are unequal.

7.2.3 Extensions of the Mean–Mean Display to Arbitrary Contrasts

Heiberger and Holland (2006) extend the mean–mean multiple comparisons plot to arbitrary contrasts, that is, contrasts that are not limited to the set of pairwise comparisons.

Two critical issues needed to be addressed. The first is the scaling of the contrast and the second is the set of contrasts selected for consideration.

7.2.3.1 Scaling

The standard definition of a contrast in Equation (6.20) requires that it satisfy the zero-sum constraint Equation (6.18). The variance of the contrast is calculated with Equation (6.21).

When we calculate sums of squares and F-tests, this definition is sufficient. When we wish to plot arbitrary contrasts on the mean–mean multiple comparisons plot described in Section 7.2.2, the contrasts must be comparably scaled. The heights must be in the range of the observed $\bar{y}_j$, and all confidence intervals must fall inside the range of the d-axis. To satisfy this additional requirement, we need to require the absolute-sum-2 scaling introduced in Section 6.9.2.1 and made explicit in Equation (6.22). Any other scaling makes it impossible to fit these values on the mean–mean plot.

With the absolute-sum-2 scaling we can think of any contrast as the comparison of two weighted averages of $\bar{y}_j$. Let us call them $\bar{y}_+ = \sum c_j^+ \bar{y}_j$ and $\bar{y}_- = \sum c_j^- \bar{y}_j$, where we use the superscript notation $a^+ = \max(a, 0)$ and $a^- = \max(-a, 0)$. We illustrate with the contrast comparing the average of means $\bar{y}_A$ and $\bar{y}_B$ with the mean $\bar{y}_D$.

	A	B	C	D	$\bar{y}_+$	$\bar{y}_-$
absolute-sum-2	.5	.5	0	−1	$(\bar{y}_A + \bar{y}_B)/2$	$\bar{y}_D$
integer	1	1	0	−2		
normalized	$1/\sqrt{6}$	$1/\sqrt{6}$	0	$-2/\sqrt{6}$		

We plot the contrast centered at the (h, v)-location $(\bar{y}_-, \bar{y}_+)$, where each term is at the correctly weighted average of the observed $\bar{y}_j$-values. The height on the m-axis of the MMC plot is $(\bar{y}_+ + \bar{y}_-)/2$ and the difference on the d-axis is $\bar{y}_+ - \bar{y}_-$. The confidence interval widths are proportional to the standard error of $\bar{y}_+ - \bar{y}_-$, which, from (6.21), is proportional to $\sqrt{\sum c_j^2/n_j}$.

7.2.3.2 Contrasts

The simplest set of contrasts is the set of all pairwise comparisons $\bar{y}_i - \bar{y}_j$ (as in Figure 6.2). Others sets include comparisons $\bar{y}_j - \bar{y}_0$ of all treatment values to a control (as in Figure 7.3) and a basis set of orthogonal contrasts that span all possible contrasts (as will be seen in Figure 7.17).

7.2.3.3 Labeling

Our presentation of the MMC plot, for example in Figure 6.2, has improved labeling compared to the Hsu and Peruggia presentation.

The left-axis ticks are the $\bar{y}_i$-values themselves, at the heights of the intersections of the 45° h- and v-lines with the vertical $d = 0$ line. The labels on the outside of the left axis are the $\bar{y}_i$-values. The labels on the inside of the left axis are the names of the factor levels.

The right-axis labels belong to the horizontal CI lines for the contrasts. The labels outside the right axis are the automatically generated contrasts, either pairwise $\bar{y}_i - \bar{y}_j$ or comparisons $\bar{y}_j - \bar{y}_0$ of all treatment values to a control. The labels inside the right axis are the requested contrasts from the explicitly specified lmat matrix. Each CI line is at the height corresponding to the average of the two $\bar{y}_* \left((\bar{y}_i + \bar{y}_j)/2 \quad \text{or} \quad (\bar{y}_+ + \bar{y}_-)/2\right)$ values they are comparing. Each CI line is centered at the observed difference

$\big((\bar{y}_i - \bar{y}_j)$ or $(\bar{y}_+ - \bar{y}_-)\big)$. The half-width of the (two-sided) CI line is $qs_{\bar{y}_i - \bar{y}_j}$, where q is calculated according to the specified multiple comparisons criterion.

The bottom axis is in the difference $\bar{y}_i - \bar{y}_j$ d-scale. The ticks and labels outside the bottom axis are regularly spaced values on the difference scale. The ticks inside the bottom axis, at distances $\pm|\bar{y}_j - \min_j \bar{y}_j|$, correspond to the horizontal d-axis positions of the foot of the 45° h- and v-lines. The names of the factor levels appear at the foot of each 45° line.

7.2.3.4 q Multipliers

Hypothesis test and confidence interval formulas, introduced in Chapter 3, depend on a multiple of the standard deviation. The multiplier is a quantile chosen from an appropriate distribution. When only one hypothesis is tested or only one interval is constructed, the multiplier is denoted z when the test statistic is normally distributed and t when the test statistic is from a t distribution. Multipliers denoted q, sometimes with a subscript, are used in many of this chapter's formulas for confidence intervals and rules for rejecting null hypotheses. In both Sections 7.1.2 and 7.1.4.3 discussing Tukey procedures, and in plots in Section 7.2 displaying results from these procedures, q refers to the Studentized range distribution. The multiplier used in the Dunnett procedure of Section 7.1.3 is a percentile of a marginal distribution of a multivariate t distribution. The multiplier for the Scheffé procedure is the square root of a percentile of an F distribution. For details, see Hochberg and Tamhane (1987).

7.2.4 Display of an Orthogonal Basis Set of Contrasts

The sum of squares associated with the factor A with a levels has $a - 1$ degrees of freedom. The missing degree of freedom is associated with the grand mean and is normally suppressed from the ANOVA table.

In Section 6.8.3 we note that it is always possible to construct an orthogonal set of contrasts that decompose the $a - 1$ df sum of squares for an effect into $a - 1$ independent single-df sums of squares. In this section we illustrate the mathematics for constructing an orthogonal basis set by constructing one from the set of pairwise contrasts. From this basis set, we show that we can construct any other set of contrasts. We also show that an orthogonal basis set, augmented with an additional contrast for the grand mean (not actually a contrast since it doesn't sum to 0), can be used to construct any linear combination of the group means.

This discussion uses all the matrix algebra results summarized in Appendix Section I.4. This section is placed here in Chapter 7 because it belongs to the discussion of the MMC plots. It might be more easily read after Section 10.3 where contrast matrices and their relation to dummy variables are discussed.

We illustrate the discussion with the catalyst data in data(catalystm). We begin with the set of pairwise contrasts behind the construction of Figure 7.13. We isolate the contrasts implicit in the "mmc" object with the lmatPairwise function in Table 7.7.

Table 7.7 Contrast matrix for pairwise comparisons. There are three matrices displayed here. The first is the contrasts for the catalyst factor as used by the aov function. We show the default contrasts as defined by the contr.treatment function. The first level is omitted. Note that these are not 'contrasts' as defined in the standard theory for linear models as they are not orthogonal to the intercept. Then the contrast matrix for pairwise comparisons is displayed in two different structures. The glht function uses the linfct (linear function) format. Each row is the difference of two columns of the contr.treatment matrix. The last matrix, structured to be used in the focus.lmat argument to the mmc function, shows columns which are standard contrasts (each column sums to zero).

```
> ## aov contrast matrix for catalyst factor.  The columns are
> ## constructed by contr.treatment with the default base=1
> contrasts(catalystm$catalyst)
  B C D
A 0 0 0
B 1 0 0
C 0 1 0
D 0 0 1


> ## Linear function used internally by glht for pairwise contrasts.
> ## The rows of linfct are the differences of the columns
> ## of the contrast matrix.
> catalystm.mmc$mca$glht$linfct
    (Intercept) catalystB catalystC catalystD
A-B           0        -1         0         0
A-C           0         0        -1         0
B-C           0         1        -1         0
A-D           0         0         0        -1
B-D           0         1         0        -1
C-D           0         0         1        -1


> ## Contrasts in lmat format, each column sums to zero.
> ## The last three rows are the transpose of the last three columns
> ## of the linfct matrix.
> ## The first row is prepended to make the column sum be zero.
> catalyst.pairwise <- lmatPairwise(catalystm.mmc)

> catalyst.pairwise
   A-B A-C B-C A-D B-D C-D
A    1   1   0   1   0   0
B   -1   0   1   0   1   0
C    0  -1  -1   0   0   1
D    0   0   0  -1  -1  -1
```

Table 7.8 illustrates an orthogonal basis set of contrasts for the catalyst data. This examination of 3 linearly independent contrasts succinctly summarizes the information contained in the 3 degrees of freedom for comparing the means of the 4 levels of the fixed factor `catalyst`. For completeness we show that `catalystm.lmat` and `catalyst.pairwise` span the same subspace.

Table 7.8 The orthogonal contrast matrix `catalysm.lmat` contains three columns that decompose the 3-df `catalyst` sum of squares term into three single-df sums of squares. The `crossprod` shows that `catalystm.lmat` is an orthogonal rank-3 matrix. The zero residuals from the regression of `catalystm.lmat` on `catalyst.pairwise` shows that they span the same subspace.

```
> ## An orthogonal set of ($4-1$) contrasts for the catalyst factor.
> ## user-specified contrasts       A  B  C  D
> catalystm.lmat <- cbind("AB-D" =c(1, 1, 0,-2),
+                         "A-B"  =c(1,-1, 0, 0),
+                         "ABD-C"=c(1, 1,-3, 1))

> dimnames(catalystm.lmat)[[1]] <- levels(catalystm$catalyst)

> catalystm.lmat
  AB-D A-B ABD-C
A    1   1     1
B    1  -1     1
C    0   0    -3
D   -2   0     1

> crossprod(catalystm.lmat)
        AB-D A-B ABD-C
AB-D       6   0     0
A-B        0   2     0
ABD-C      0   0    12

> catalyst.pairwise
   A-B A-C B-C A-D B-D C-D
A    1   1   0   1   0   0
B   -1   0   1   0   1   0
C    0  -1  -1   0   0   1
D    0   0   0  -1  -1  -1

> resid(lm(catalystm.lmat ~ catalyst.pairwise))
  AB-D A-B ABD-C
A    0   0     0
B    0   0     0
C    0   0     0
D    0   0     0
```

In Table 7.9 and Figure 7.17 we use the orthogonal basis to construct an easily interpretable MMC plot on the `catalyst` levels. The principal conclusion from Figure 7.13 is that the means of both catalysts A and B significantly exceed the mean of catalyst D. Figure 7.17 reforces this conclusion with the finding that the average of the means of catalysts A and B significantly exceeds the mean of catalyst D because the confidence interval for this contrast lies entirely above 0. A second new conclusion from Figure 7.17 is that the average of the means of catalysts A, B, and D is not significantly different from the mean of catalyst C because the confidence interval for this contrast includes 0.

Table 7.9 We use `catalysm.lmat` as the `focus.lmat` argument to `mmc` leading to Figure 7.17.

```
> catalystm.mmc <-
+     mmc(catalystm1.aov,
+         linfct = mcp(catalyst = "Tukey"),
+         focus.lmat=catalystm.lmat)

> catalystm.mmc
Tukey contrasts
Fit: aov(formula = concent ~ catalyst, data = catalystm)
Estimated Quantile = 2.966
95% family-wise confidence level
$mca
      estimate stderr     lower upper height
A-B      1.125  1.138 -2.251228 4.501  56.34
A-C      3.667  1.239 -0.008905 7.342  55.07
B-C      2.542  1.296 -1.302338 6.386  54.50
A-D      5.775  1.138  2.398772 9.151  54.01
B-D      4.650  1.200  1.091143 8.209  53.45
C-D      2.108  1.296 -1.735671 5.952  52.18
$none
   estimate stderr lower upper height
A     56.90 0.7590 54.65 59.15  56.90
B     55.77 0.8485 53.26 58.29  55.77
C     53.23 0.9798 50.33 56.14  53.23
D     51.12 0.8485 48.61 53.64  51.12
$lmat
       estimate stderr  lower upper height
A-B       1.125  1.138 -2.251 4.501  56.34
ABD-C     1.367  1.088 -1.860 4.594  53.92
AB-D      5.212  1.022  2.182 8.243  53.73
```

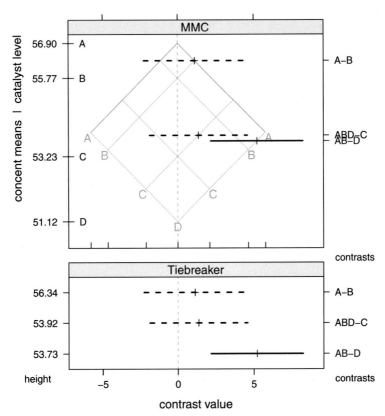

Fig. 7.17 MMC plot constructed with
```
mmcplot(catalystm.mmc, type="lmat", style="both")
```
using the orthogonal set of contrasts defined in Table 7.8 based on the pairwise set in Figures 7.13
and 6.2. The comparison between the average of $\bar{y}_A$ and $\bar{y}_B$ with the mean $\bar{y}_D$ is the only significant comparison. The other two confidence intervals include 0. The Tiebreaker panel is needed to
respond to the overprinting of labels in the right axis of the MMC panel.

7.2.5 Hsu and Peruggia's Pulmonary Example

This is the example that Hsu and Peruggia (1994) use to introduce the mean–mean
multiple comparisons plots. The response variable is FVC, forced vital capacity. The
groups are levels of the smoker factor

Table 7.10 ANOVA table for pulmonary data.

```
> data(pulmonary)

> pulmonary
    smoker   n  FVC    s
NS      NS 200 3.35 0.63
PS      PS 200 3.23 0.46
NI      NI  50 3.19 0.52
LS      LS 200 3.15 0.39
MS      MS 200 2.80 0.38
HS      HS 200 2.55 0.38

> pulmonary.aov <-
+   aovSufficient(FVC ~ smoker, data=pulmonary,
+                 weights=pulmonary$n, sd=pulmonary$s)

> summary(pulmonary.aov)
              Df Sum Sq Mean Sq F value Pr(>F)
smoker         5   89.3   17.85    83.9 <2e-16 ***
Residuals   1044  222.1    0.21
---
Signif. codes:  0 '***' 0.001 '**' 0.01 '*' 0.05 '.' 0.1 ' ' 1
```

NS nonsmokers
PS passive smokers
NI noninhaling smokers
LS light smokers (1–10 cigarettes per day for at least the last 20 years)
MS moderate smokers (11–39 cigarettes per day for at least the last 20 years)
HS heavy smokers (≥40 cigarettes per day for at least the last 20 years)

There are six levels of the smoker factor, hence 5 df for comparing them. The means for the six groups are accessed as data(pulmonary). The ANOVA table is in Table 7.10. The MMC plot is in Figure 7.18. The MMC plot of a set of orthogonal contrasts is in Figure 7.19.

Figure 7.18 shows that the three levels {PS, NI, and LS} are indistinguishable; we call this the low-smoker cluster. This comparison of three levels uses 2 df. There are only 3 df left. From the SW–NE HS line, we see that the MS-HS contrast is significant, that the comparisons between each of the three levels in the low-smoker cluster with MS is significant, and that the comparison of NS with HS and with MS are each significant. All three comparisons of NS with the low-smoker cluster have lower bounds close to zero, and one of the three comparisons is significant.

We can summarize these visual impressions by constructing an orthogonal set of contrasts that reflect them exactly. Figure 7.19 shows a basis set of five orthogonal contrasts. In the center, the p-nl and n-l contrasts show that the three levels in

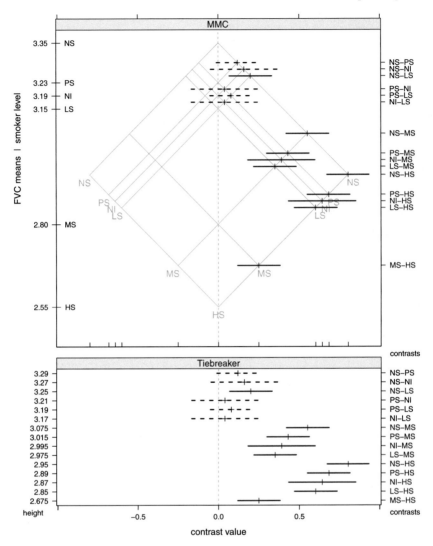

Fig. 7.18 Hsu and Peruggia's pulmonary example. The apparent clustering of the three groups PS, NI, LS suggests the set of contrasts we show in Figure 7.19.

the low-smoker cluster are indistinguishable. The other three lines show that the nonsmoker group is significantly different from the low-smoker cluster (n-pnl), that the moderate- and heavy-smoker groups are significantly different (m-h), and that the combined nonsmoker group and low-smoker cluster are significantly different from the combined moderate- and heavy-smoker groups (npnl-mh).

The center of the interval for each of the contrasts in Figure 7.19 is constructed by the linear combination of the means for the levels. For example, the n-pnl interval is on the NW–SE NS line and on the average of the NE–SW PS, NI, and LS

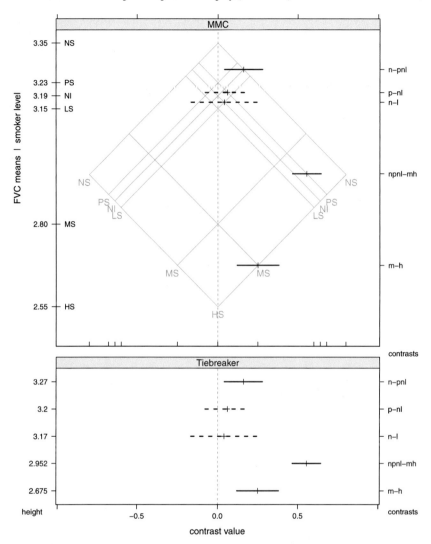

Fig. 7.19 Hsu and Peruggia's pulmonary example: An orthogonal set of contrasts. There are three significant contrasts and two not significant contrasts. The means for the three groups we discovered in Figure 7.18 are indistinguishable. The other differences are significant. The ability to display an arbitrary orthogonal set of contrasts is one of our enhancements to the mean–mean plot.

lines. The width of the interval is calculated from the algebra of the contrast. A simultaneous 95% coverage probability applies to the five confidence intervals in Figure 7.19 because they are constructed using the extended Tukey procedure. This procedure guarantees the coverage probability over the set of all possible contrasts. In exchange for this guarantee, these extended Tukey intervals are fairly wide. Having used the Tukey procedure to construct the intervals in Figure 7.18, it would be incorrect to switch to the narrower Scheffé procedure simultaneous intervals for the basis set of contrasts. With such a switch we would have two competing analyses, and this would distort the claimed coverage probabilities for the now distinct analyses in the two figures.

7.3 Exercises

7.1. Use an MMC plot to display the results of the Tukey procedure in Exercise 6.2.

7.2. Use an MMC plot to display the results of the Tukey procedure in Exercise 6.4.

7.3. Use an MMC plot to display the results of the Tukey procedure applied to the log-transformed data discussed in Exercise 6.5.

7.4. Use an MMC plot to display the results of the Tukey procedure in Exercise 6.6.

7.5. Use an MMC plot to display the results of the Tukey procedure in Exercise 6.8.

7.6. Use an MMC plot to display the results of the Tukey procedure in Exercise 6.9.

7.7. The relative rotation angle between tangents to cervical vertebrae C3 and C4 is a standard musculoskeletal measurement. Figure 7.20 illustrates the measurement of relative rotation angles. Harrison et al. (2004) hypothesize that the value

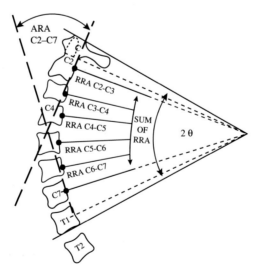

Fig. 7.20 Illustration of the relative rotation angles between the cervical vertebrae (neck area). Exercise 7.7 uses the C3–C4 angle.

of this angle, C3–C4, in persons complaining of neck pain tends to differ from that in healthy individuals. The dataset, accessed as data(c3c4), contains the C3–C4 measurements of a random sample of 194 patients of which 72 had no complaints of neck pain, 52 complained of acute neck pain of recent origin, and 70 have had chronic neck pain. The pain condition is coded 0 for none, 1 for acute, and 2 for chronic. There is no implied ordering in this coding scheme. Perform an analysis of variance followed by Dunnett's procedure to determine if the mean C3–C4 value of persons with acute or chronic neck pain differs from the mean C3–C4 value of persons without neck pain.

Chapter 8

Linear Regression by Least Squares

8.1 Introduction

We usually study more than one variable at a time. When the variables are continuous, and one is clearly a response variable and the others are predictor variables, we usually plot the variables and then attempt to fit a model to the plotted points. With one continuous predictor, the first model we attempt is a straight line; with two or more continuous predictors, we attempt a plane. We plot the model, the residuals from the model, and various diagnostics of the quality of the fit.

In this chapter we are primarily concerned with modeling a straight-line relationship between two variables using n pairs of observations on these variables, a common and fundamental task. One of these variables, conventionally denoted y, is a response or output variable. The other variable, often denoted x, is known as an explanatory or input or predictor variable. Usually, but not always, it is clear from the context which of the two variables is the response and which is the predictor. For example, if the two variables are personal income and consumption spending, then consumption is the response variable because the amount that is spent depends on how much income is available to be spent.

The relationship between y and x is almost never perfectly linear. When the n points are plotted in two dimensions, they appear as a random scatter about some unknown straight line. We model this line as

$$y_i = \beta_0 + \beta_1 x_i + \epsilon_i \quad \text{for } i = 1, \ldots, n \tag{8.1}$$

where

$$\epsilon_i \sim N(0, \sigma^2) \tag{8.2}$$

that is, the ϵ_i are assumed normally independently distributed with constant mean 0 and common variance σ^2 [abbreviated as $\epsilon_i \sim NID(0, \sigma^2)$]. In other words, we assume that the response variable is linearly related to the predictor variables, plus

© Springer Science+Business Media New York 2015
R.M. Heiberger, B. Holland, *Statistical Analysis and Data Display*,
Springer Texts in Statistics, DOI 10.1007/978-1-4939-2122-5_8

a normally distributed random component. Here the intercept β_0 and slope β_1 are unknown *regression coefficients* that must be estimated from the data. The variance σ^2 is a third unknown parameter, introduced along with the assumption of a normally distributed error term, which must also be estimated.

A commonly used procedure for estimating β_0 and β_1 is the method of *least squares* because, as we will see in Section 8.3.2, this mathematical criterion leads to simple "closed-form" formulas for the estimates. Under the stated normality assumptions in Equation (8.2) about the residuals ϵ_i of Model (8.1), the least-squares estimates of the regression coefficients are also the maximum likelihood estimates of these coefficients.

8.2 Example—Body Fat Data

8.2.1 Study Objectives

The example is taken from Johnson (1996). A group of subjects is gathered, and various body measurements and an accurate estimate of the percentage of body fat are recorded for each. Then body fat can be fit to the other body measurements using multiple regression, giving, we hope, a useful predictive equation for people similar to the subjects. The various measurements other than body fat recorded on the subjects are, implicitly, ones that are easy to obtain and serve as proxies for body fat, which is not so easily obtained.

Percentage of body fat, age, weight, height, and ten body circumference measurements (e.g., abdomen) are recorded for 252 men. Body fat, a measure of health, is estimated through an underwater weighing technique. Fitting body fat to the other measurements using multiple regression provides a convenient way of estimating body fat for men using only a scale and a measuring tape.

8.2.2 Data Description

We will initially use only 47 observations and only five of the measurements that have been recorded.

bodyfat: Percent body fat using Siri's equation, 495/Density − 450

abdomin: Abdomen circumference (cm) "at the umbilicus and level with the iliac crest"

biceps: Extended biceps circumference (cm)

wrist: Wrist circumference (cm) "distal to the styloid processes"

forearm: Forearm circumference (cm)

8.2.3 Data Input

We access the data from data(fat) and then look at the data with the scatterplot matrix in Figure 8.1.

Fat data

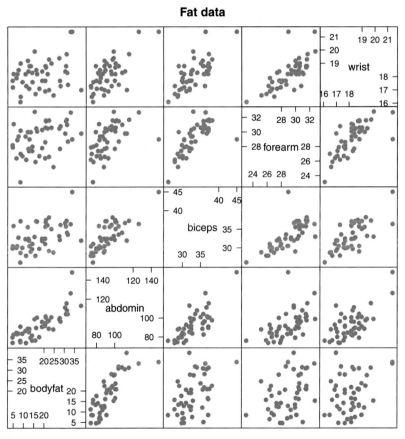

Fig. 8.1 Body Fat Data

The response variable bodyfat is in the bottom row of the plot. We can see that a linear fit makes sense against abdomin. A linear relationship between bodyfat and the other predictor variables is also visible in the plot, but is weaker. All the predictor variables show correlation with each other.

8.2.4 One-X Analysis

The initial analysis will look at just bodyfat and abdomin. We will come back to the other variables later. We expand the bodyfat ˜ abdomin panel of Figure 8.1 in the left column of Figure 8.2 and place two straight lines on the graph in the two rightmost columns. The line in column 3 is visibly not a good fit. It is too shallow and is far above the points in the lower left. The line in column 2, labeled "least-squares fit", is just right. The criterion we use is *least squares*, which means that the sum of the squared differences from the fitted to observed points is to be minimized. The *least-squares* line is the straight line that achieves the minimum.

The top row of Figure 8.2 displays the vertical differences from the fitted to observed points. The bottom row displays the squares of the differences from the fitted to observed points. The least-squares line minimizes the sum of the areas of these squares. It is evident that the sum of the squared areas in column 2 is smaller than the sum of squared areas for the badly fitting line in column 3.

From any of these panels it is apparent that on average, body fat is directly related to abdominal circumference. As will be explained in Section 8.3.5, the least-squares line in Figure 8.2 can be used to predict bodyfat from abdomin. Note that although it is mathematically correct to say that abdomin increases with bodyfat, this is a misleading statement because it implies an unlikely direction of causality among these variables.

8.3 Simple Linear Regression

8.3.1 Algebra

Figure 8.2 illustrates the least-squares line that best fits bodyfat to abdomin. Now that we see from the bottom row of the figure that the least-squares line actually does minimize the sum of squares, let us review the mathematics behind the calculation of the least-squares line. The standard notation we use for the least-squares straight line is

$$\hat{y} = \hat{\beta}_0 + \hat{\beta}_1 x \tag{8.3}$$

where $\hat{\beta}_0$ and $\hat{\beta}_1$ are called the *regression coefficients*. We define the residuals by

$$e_i = y_i - \hat{y}_i \tag{8.4}$$

We wish to find $\hat{\beta}_0$ and $\hat{\beta}_1$ that minimize the expression for the sum of squares of the calculated residuals:

$$\sum_{i=1}^{n} e_i^2 = \sum_{i=1}^{n} (y_i - \hat{y}_i)^2 = \sum_{i=1}^{n} \left(y_i - (\beta_0 + \beta_1 x_i)\right)^2 \tag{8.5}$$

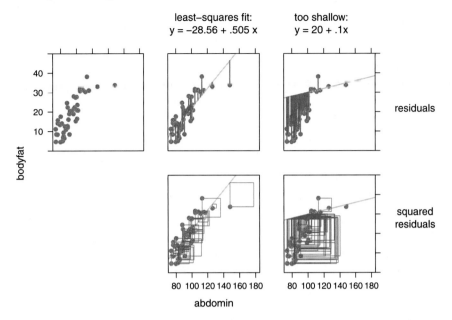

least–squares fit:
y = –28.56 + .505 x

too shallow:
y = 20 + .1x

Fig. 8.2 One *X*-variable and two straight lines. The second column is the least-squares line, the third is too shallow. Row 1 shows the residuals. Row 2 shows the squared residuals. The least-squares line minimizes the sum of the squared residuals.

We minimize by differentiation with respect to the parameters β_0 and β_1, setting the derivatives to 0 (thus getting what are called the *normal equations*)

$$\frac{\partial}{\partial \beta_0} \sum_{i=1}^{n} \left(y_i - (\beta_0 + \beta_1 x_i)\right)^2 = \sum_{i=1}^{n} 2\left(y_i - (\beta_0 + \beta_1 x_i)\right)(-1) = 0$$

$$\frac{\partial}{\partial \beta_1} \sum_{i=1}^{n} \left(y_i - (\beta_0 + \beta_1 x_i)\right)^2 = \sum_{i=1}^{n} 2\left(y_i - (\beta_0 + \beta_1 x_i)\right)(-x_i) = 0$$

(8.6)

and then solving simultaneously for the regression coefficients

$$\hat{\beta}_1 = \frac{\sum(y_i - \bar{y})(x_i - \bar{x})}{\sum(x_i - \bar{x})^2}$$

$$\hat{\beta}_0 = \bar{y} - \hat{\beta}_1 \bar{x}$$

(8.7)

In addition to minimizing the sum of squares of the calculated residuals, $\hat{\beta}_0$ and $\hat{\beta}_1$ have the property that the sum of the calculated residuals is zero, i.e.,

$$\sum_{i=1}^{n} e_i = 0$$

(8.8)

We request a proof of this assertion in Exercise 8.9.

For two or more predictor variables, the procedure (equating derivatives to zero) is identical but the algebra is more complex. We postpone details until Section 9.3.

8.3.2 Normal Distribution Theory

Under the normality assumption (8.2) for the residuals of Model (8.1), the least-squares estimates are also maximum likelihood estimates. This is true because if the residuals are normally distributed, their likelihood function is maximized when Equation (8.5) is minimized.

In Model (8.1), the unknown population variance of the ϵ_i, σ^2, is estimated by the sample variance

$$\hat{\sigma}^2 = s^2 = \frac{\sum_{i=1}^{n}(y_i - \hat{y}_i)^2}{n-2} \tag{8.9}$$

Because the sample variance is proportional to the residual sum of squares in Equation (8.5), minimizing the sample variance also leads us to the least-squares estimates $\hat{\beta}_0$ and $\hat{\beta}_1$ in Equations (8.7). The square root s of the sample variance in Equation (8.9), variously termed the *standard error of estimate*, the *standard error*, or the *root mean square error*, indicates the size of a typical vertical deviation of a point from the calculated regression line.

8.3.3 Calculations

The results of the statistical analysis are displayed in several tables, primarily the *ANOVA* (analysis of variance) table, the table of regression coefficients, and the table of other statistics shown in Table 8.1. These tables are fundamental to our interpretation of the analysis. The formulas for each number in these tables appear in Tables 8.2, 8.3, and 8.4. As with Tables 6.2 and 6.2, the ANOVA table in Section 8.1 does not include the "Total" line and the interpretation in Table 8.2 does include the "Total" line. R does not print the Total line in its ANOVA tables.

For one-x regression (this example), there is usually only one null and alternative hypothesis of interest:

$$H_0 : \beta_1 = 0 \quad \text{vs} \quad H_1 : \beta_1 \neq 0 \tag{8.10}$$

Both $t = 9.297$ in the table of coefficients and $F = 86.427 = 9.297^2 = t^2$ in the ANOVA table are tests between those hypotheses. The associated p-value ($p = .5_{10}^{-12}$, which we report as < 0.0001), is smaller than any reasonable α (the traditional .05 or .01, for example). Therefore, we are justified in rejecting the null

Table 8.1 ANOVA table and table of regression coefficients for the simple linear regression model with y=bodyfat and x=abdomin.

```
> data(fat)

> fat.lm <- lm(bodyfat ~ abdomin, data=fat)

> anova(fat.lm)
Analysis of Variance Table

Response: bodyfat
          Df Sum Sq Mean Sq F value  Pr(>F)
abdomin    1   2440    2440    86.4 4.9e-12 ***
Residuals 45   1271      28
---
Signif. codes:
0 '***' 0.001 '**' 0.01 '*' 0.05 '.' 0.1 ' ' 1

> summary(fat.lm)

Call:
lm(formula = bodyfat ~ abdomin, data = fat)

Residuals:
   Min     1Q Median     3Q    Max
-12.42  -4.11   1.21   3.52   9.65

Coefficients:
            Estimate Std. Error t value Pr(>|t|)
(Intercept) -28.5601     5.1100   -5.59  1.3e-06 ***
abdomin       0.5049     0.0543    9.30  4.9e-12 ***
---
Signif. codes:
0 '***' 0.001 '**' 0.01 '*' 0.05 '.' 0.1 ' ' 1

Residual standard error: 5.31 on 45 degrees of freedom
Multiple R-squared:  0.658, Adjusted R-squared:  0.65
F-statistic: 86.4 on 1 and 45 DF,  p-value: 4.85e-12
```

hypothesis in favor of the alternative. Inference on β_0 frequently makes no sense. In this example, for example, β_0 is the expected bodyfat of an individual having the impossible abdomin with zero circumference.

The Total line in the ANOVA table shows the sum of squares and degrees of freedom for the response variable bodyfat around its mean. When we divide these two numbers we recognize the formula $\sum_{i=1}^{n}(y_i - \bar{y})^2/(n - 1) = 80.678$ as Equation (3.6) for the sample variance of the response variable. The goal of the analysis is to *explain* as much of the variance in the response variable as possible with a model

Table 8.2 Interpretation of items in "ANOVA Table" from Table 8.1. The symbols in the `abdomin` section are subscripted `Reg`, short for "Regression". In this setting, "Regression" refers to the group of all model predictors. In this example there is only one predictor, `abdomin`.

Name	Notation	Formula	Value in Table 8.1
Total			
Sum of Squares	$\mathsf{SS_{Total}}$	$\sum_{i=1}^{n}(y_i - \bar{y})^2 = \mathsf{SS_{Reg}} + \mathsf{SS_{Res}}$	3711.199
Degrees of Freedom	$\mathsf{df_{Total}}$	$n-1$	46
Variance about Mean		$\mathsf{SS_{Total}}/\mathsf{df_{Total}}$	80.678
Residual			
Sum of Squares	$\mathsf{SS_{Res}}$	$\sum_{i=1}^{n}(y_i - \hat{y}_i)^2$	1270.699
Degrees of Freedom	$\mathsf{df_{Res}}$	$n-2$	45
Mean Square	$\mathsf{MS_{Res}}$	$\hat{\sigma}^2 = s^2 = \dfrac{\sum_{i=1}^{n}(y_i - \hat{y}_i)^2}{n-2}$	28.238
abdomin			
Sum of Squares	$\mathsf{SS_{Reg}}$	$\sum_{i=1}^{n}(\hat{y}_i - \bar{y})^2$	2440.500
Degrees of Freedom	$\mathsf{df_{Reg}}$	number of predictor variables	1
Mean Square	$\mathsf{MS_{Reg}}$	variability in $\hat{y}$ attributable to $\hat{\beta}_1$	
		$\left(\dfrac{\text{abdomin Sum of Squares}}{\text{abdomin Degrees of Freedom}}\right)$	2440.500
F-Value	F_{Reg}	$\left(\dfrac{\text{abdomin Mean Square}}{\text{Residual Mean Square}}\right)$	86.427
$\Pr(>F)$	p_{Reg}	$P(F_{1,45} > 86.427) = 1 - \mathcal{F}_{1,45}(86.427)$	< 0.0001

that relates the response to the predictors. When we have explained the variance, the *residual* (or leftover) mean square s^2 is much smaller than the sample variance of the response variable.

The *coefficient of determination*, also known as *Multiple R^2*, usually accompanies ANOVA tables. This measure, generally denoted R^2, is the proportion of variation in the response variable that is accounted for by the predictor variable(s). It is desirable that R^2 be as close to 1 as possible. Models with R^2 considerably below 1 may be acceptable in some disciplines. The defining formula for R^2 is

$$R^2 = \frac{\mathsf{SS_{Reg}}}{\mathsf{SS_{Total}}} \tag{8.11}$$

In regression models with only one predictor, an alternative notation is r^2. This notation is motivated by the fact that r^2 is the square of the sample correlation

Table 8.3 Interpretation of items in "Table of Regression Coefficients" from Table 8.1.

Name	Notation	Formula	Value in Table 8.1
(Intercept)			
Value	$\hat{\beta}_0$	$\bar{y} - \hat{\beta}_1 \bar{x}$	-28.560
Standard Error	$\hat{\sigma}_{\beta_0}$	$\hat{\sigma} \sqrt{\dfrac{1}{n} + \dfrac{\bar{x}^2}{\sum (x_i - \bar{x})^2}}$	5.110
t-value	t_{β_0}	$\dfrac{\hat{\beta}_0}{\hat{\sigma}_{\beta_0}}$	-5.589
Pr($> |t|$)	$p_{\hat{\beta}_0}$	$P(t_{45} > |-5.589|)$	< 0.0001
abdomin			
Value	$\hat{\beta}_1$	$\dfrac{\sum (y_i - \bar{y})(x_i - \bar{x})}{\sum (x_i - \bar{x})^2}$	0.505
Standard Error	$\hat{\sigma}_{\beta_1}$	$\hat{\sigma} / \sqrt{\sum (x_i - \bar{x})^2}$	0.054
t-value	t_{β_1}	$\hat{\beta}_1 / \hat{\sigma}_{\beta_1}$	9.297
Pr($> |t|$)	$p_{\hat{\beta}_1}$	$P(t_{45} > |9.297|)$	< 0.0001

coefficient r between the response and predictor variable. r is the usual estimate of the population correlation coefficient defined and interpreted in Equation (3.14). A formula for the sample correlation r is

$$r = \frac{\sum (y_i - \bar{y})(x_i - \bar{x})}{\sqrt{\sum (y_i - \bar{y})^2 \sum (x_i - \bar{x})}} \tag{8.12}$$

It can be shown that $-1 \leq r \leq 1$. If $r = \pm 1$, then x and y are perfectly linearly related, directly so if $r = 1$ and inversely so if $r = -1$. The arithmetic sign of r matches the arithmetic sign of $\hat{\beta}_1$.

In the present body fat example, we find $r = 0.811$ and $r^2 = 0.658$. This value of r is consistent with the moderately strong positive linear relationship between bodyfat and abdomin in the least-squares fit shown in Figure 8.2. Continuing with this example, the estimated response variance ignoring the predictor is 80.678 and the estimated response variance paying attention to the predictor abdomin, the **Residuals Mean Square**, is 28.238. Graphically, we see in Figure 8.3 that the variance estimate 80.678 about the mean belongs to Figure 8.3a and the variance estimate 28.238 about the regression line belongs to Figure 8.3b.

Table 8.4 Interpretation of additional items, some of which are shown in Table 8.1.

Name	Notation	Formula	Value based on Table 8.1
Coefficient of Determination Multiple R^2	R^2	$\left(\dfrac{\text{abdomin Sum of Squares}}{\text{Total Sum of Squares}}\right)$	0.6576
	p	Number of predictor x variables in the model in the model	1.
Adjusted R^2	R^2_{adj}	$1 - \left(\dfrac{n-1}{n-p-1}\right)(1 - R^2)$	0.6500
Dependent Mean	$\bar{Y}$	$\dfrac{\sum Y_i}{n}$	18.3957
Residual Standard Error	$\hat{\sigma} = s$	$\sqrt{s^2}$	5.3139
Coefficient of Variation	cv	$s/\bar{Y}$	28.8867

While these two estimates of response variance are intuitive, they are not actually the statistically correct numbers to compare because they are not independent. The Total Sum of Squares is the sum of the Residuals Sum of Squares and the abdomin Sum of Squares. These two components of the Total Sum of Squares are independent and are therefore the base for the correct quantities to compare. The abdomin mean square is an unbiased estimate of σ^2 if H_0 is true but an overestimate of σ^2 if H_0 is false. The Residuals Mean Square is unbiased for σ^2 in either case. Therefore, the ratio of these two mean squares will tend to be close to 1 if H_0 is true but greater than 1 otherwise. With the assumption of independent normally distributed ϵ_i, the ratio, given as the F-Value = 86.427 in the table, follows a (central) F distribution with 1 and 45 degrees of freedom if H_0 is true, but not otherwise. Appeal to this distribution tells us whether the ratio is significantly greater than 1. When the observed $\Pr(> F)$ value in the table (in this case < 0.0001) is small, we interpret that as evidence that H_0 is false.

The formal statement of the test is: Under the null hypothesis that $\beta_1 = 0$ (that is, that information about x=abdomin gives no information about y=bodyfat), the probability of observing an F-value as large as the one we actually saw (in this case 86.427) is very small (in this case the probability is less than 0.0001). This very small p-value (assuming H_0 is true) is very strong evidence that H_0 is not true, that is, it is evidence that $\beta_1 \neq 0$. We will therefore act as if H_0 is false and take further actions as if the relationship of the fitted regression model actually explains what is going on.

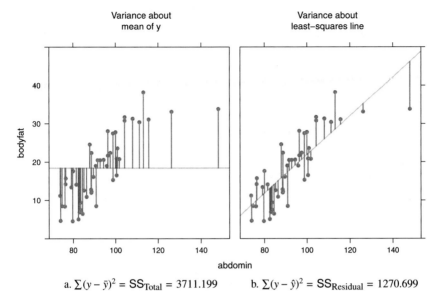

a. $\sum (y - \bar{y})^2 = SS_{\text{Total}} = 3711.199$ b. $\sum (y - \hat{y})^2 = SS_{\text{Residual}} = 1270.699$

Fig. 8.3 Variance about mean and about least-squares line.

The estimate $\hat{\beta}_1$ from Equation (8.7) can be rewritten as a weighted sum of y_i-values or of single-point slopes $\hat{\beta}_{1i} = (y_i - \bar{y})/(x_i - \bar{x})$

$$\hat{\beta}_1 = \sum_i (y_i - \bar{y}) \left(\frac{(x_i - \bar{x})}{\sum (x_i - \bar{x})^2} \right) \tag{8.13}$$

$$= \sum_i \left(\frac{y_i - \bar{y}}{x_i - \bar{x}} \right) \left(\frac{(x_i - \bar{x})^2}{\sum (x_i - \bar{x})^2} \right) \tag{8.14}$$

Figure 8.4 illustrates equation 8.14 with the R command
`demo("betaWeightedAverage", ask=FALSE)`.

The variance of $\hat{\beta}_1$

$$\sigma_{\hat{\beta}_1}^2 = \text{var}(\hat{\beta}_1) = \frac{\sigma^2}{\sum (x_i - \bar{x})^2} \tag{8.15}$$

is constructed from the sum in Equation (8.13) with formulas based on Equation (3.9) (see Exercise 8.7). The sample estimate of the standard error of $\hat{\beta}_1$ is

$$\hat{\sigma}_{\hat{\beta}_1} = \frac{\hat{\sigma}}{\sqrt{\sum (x_i - \bar{x})^2}} \tag{8.16}$$

Under H_0, and with the assumption of independent normally distributed ϵ_i, the t-ratio $t_{\hat{\beta}_1} = \hat{\beta}_1 / \hat{\sigma}_{\hat{\beta}_1}$ has a t_{45} distribution allowing us to use the t table in our tests. It

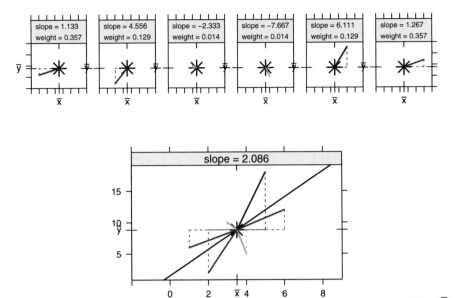

```
> bWA
     color x   y
1      red 1   6
2     blue 2   2
3    green 3  10
4   orange 4   5
5    brown 5  18
6   purple 6  12
```

Fig. 8.4 Equation 8.14 shows that the slope $\hat{\beta}_1$ can be written as the weighted sum of the single-point slopes $\hat{\beta}_{1i} = (y_i - \bar{y})/(x_i - \bar{x})$. The top set of panels shows the set of single-point slopes. The bottom panel shows all six single-point slopes and the regression line whose slope is the weighted sum of the individual slopes. The dataset for this example is displayed.

follows from this that a $100(1 - \alpha)\%$ confidence interval on β_1 is

$$\hat{\beta}_1 \pm t_{df,\frac{\alpha}{2}} \, \hat{\sigma}_{\hat{\beta}_1}$$

where $df = df_{\text{Res}}$ degrees of freedom.

Similarly, we can show (see Exercise 8.8)

$$\sigma_{\hat{\beta}_0}^2 = \text{var}(\hat{\beta}_0) = \sigma^2 \left(\frac{1}{n} + \frac{\bar{x}^2}{\sum(x_i - \bar{x})^2} \right) \tag{8.17}$$

8.3.4 Residual Mean Square in Regression Printout

The *residual mean square* is also called the *error mean square*. It is called *residual* because it is the variability in the response variable left over after fitting the model. It is called *error* because it is a measure of the difference between the model and the data. We prefer the term "residual" and discourage the term "error" because the term "error" suggests a mistake, and that is not the intent of this component of the analysis. Nevertheless, on occasion we use the term "error" as a synonym for "residual" to match the continued use by SAS of "Error Mean Square" rather than our preferred "Residual Mean Square". See Table 8.5 for a comparison of several notations. See Tables 8.5 and 8.6 for illustrations of how the fitted values and the residuals are related to the various sums of squares used in the ANOVA table. These tables show the linear and quadratic identities introduced in Section 6.A.

8.3.5 New Observations

One of the uses of a fitted regression equation is to make inferences about new observations. A new observation y_0 at x_0 has the model

$$y_0 = \beta_0 + \beta_1 x_0 + \epsilon_0 = \mu_0 + \epsilon_0$$

where

- y_0 is a single unobserved value

- x_0 is a the value of the predictor x at the new observation

- β_0 and β_1 are the regression coefficients.

The concepts that we introduce here extend, almost without change, to the multiple regression setting of Chapter 9. We therefore preview the slightly more elaborate notation of Chapter 9. The model in Equation (8.1) can be rewritten in matrix notation as

$$\begin{pmatrix} y_1 \\ \vdots \\ y_n \end{pmatrix} = \begin{pmatrix} 1 & x_1 \\ \vdots & \vdots \\ 1 & x_n \end{pmatrix} \begin{pmatrix} \beta_0 \\ \beta_1 \end{pmatrix} + \begin{pmatrix} \epsilon_1 \\ \vdots \\ \epsilon_n \end{pmatrix} \tag{8.18}$$

$$Y_{n\times 1} = X_{n\times(1+p)}\, \beta_{(1+p)\times 1} + \epsilon_{n\times 1}$$

We restrict $p = 1$ in Chapter 8. More generally, beginning in Chapter 9, p is a positive integer.

Table 8.5 Residual Mean Square in Regression Printout. The "Residual Mean Square" and "Error Mean Square" are two names for the same concept. Note that the (Std Err Residual)$_i$ is different for each i. It is smallest for x values closest to $\bar{x}$ and increases as the x values move away from $\bar{x}$. This is the reason that the confidence bounds for the regression line (see Figure 8.5) show curvature.

For each observation i the standard regression printout shows

$$
\begin{array}{ccccc}
\widehat{\mathrm{var}}(\hat{\mu}_i) & + & \widehat{\mathrm{var}}(e_i) & = \widehat{\mathrm{var}}(y_i) = & \hat{\sigma}^2 \\
h_i\hat{\sigma}^2 & + & (1-h_i)\hat{\sigma}^2 & = & \hat{\sigma}^2 \\
(\text{Std Err Predict})_i^2 & + & (\text{Std Err Residual})_i^2 & = \text{Residual Mean Square} & \\
& & & = \text{Error Mean Square} &
\end{array}
$$

```
> h <- hat(model.matrix(fat.lm))

> pred <- predict(fat.lm, se.fit=TRUE)

> res <- resid(fat.lm)

> sigma.hat.square <- anova(fat.lm)["Residuals", "Mean Sq"]

> fat.predvalues <-
+ data.frame("y=bodyfat"=fat$bodyfat,   "x=abdomin"=fat$abdomin,
+            h=h,                        mu.hat=pred$fit,
+            e=res,                      var.mu.hat=h*sigma.hat.square,
+            var.resid=(1-h)*sigma.hat.square,
+            sigma.hat.square=sigma.hat.square,
+            se.fit=sqrt(h*sigma.hat.square),
+            se.resid=sqrt((1-h)*sigma.hat.square))

> fat.predvalues[1:3, 1:7]
  y.bodyfat x.abdomin      h mu.hat      e var.mu.hat var.resid
1      12.6      85.2 0.02762  14.46 -1.860     0.7800     27.46
2       6.9      83.0 0.03171  13.35 -6.450     0.8954     27.34
3      24.6      87.9 0.02399  15.82  8.776     0.6773     27.56

> ## fat.predvalues
>
> ## linear identity
> all.equal(rowSums(fat.predvalues[,c("mu.hat", "e")]),
+           fat$bodyfat,
+           check.names=FALSE)
[1] TRUE

> ## quadratic identity
> (SSqReg <- sum((fat.predvalues$mu.hat - mean(fat$bodyfat))^2))
[1] 2440

> (SSqRes <- sum(res^2))
[1] 1271

> (SSqTot <- sum((fat$bodyfat - mean(fat$bodyfat))^2))
[1] 3711

> all.equal(SSqReg + SSqRes, SSqTot)
[1] TRUE
```

Table 8.6 We show the linear identity $y_i = \bar{y} + \hat{\beta}_1(x_i - \bar{x}) + e_i$ and the quadratic identity $\sum (y_i - \bar{y})^2 = \sum (\beta_1 x_i)^2 + \sum \epsilon_i^2$ for least squares regression. The linear identity is the partitioning of the column of y_i into columns for the grand mean, the product of the regression coefficient and the difference of x_i from $\bar{x}$, and the column of residuals e_i. The quadratic identity is the arithmetic behind the sums of squares in the ANOVA table.

$$
\begin{aligned}
y_i &= \hat{\beta}_0 \qquad\;\; + \hat{\beta}_1 x_i + e_i \quad \text{for } i = 1,\dots,n \;\; \text{from Equation (8.1)} \\
&= (\bar{y} - \hat{\beta}_1 \bar{x}) + \hat{\beta}_1 x_i + e_i \\
&= \bar{y} + \hat{\beta}_1(x_i - \bar{x}) \;\; + e_i \qquad\qquad\qquad \text{linear identity}
\end{aligned}
$$

i	y_i	$\bar{y}$	$\beta_1(x_i - \bar{x})$	e_i	i	y_i	$\bar{y}$	$\beta_1(x_i - \bar{x})$	e_i
1	12.6	18.4	−3.935	−1.860	25	14.2	18.4	−8.429	4.233
2	6.9	18.4	−5.046	−6.450	26	4.6	18.4	−6.712	−7.083
3	24.6	18.4	−2.572	8.776	27	8.5	18.4	−9.288	−0.608
4	10.9	18.4	−3.329	−4.166	28	22.4	18.4	−2.168	6.172
5	27.8	18.4	3.538	5.866	29	4.7	18.4	−9.641	−4.055
6	20.6	18.4	0.710	1.494	30	9.4	18.4	−4.794	−4.202
7	19.0	18.4	−1.158	1.762	31	12.3	18.4	−2.168	−3.928
8	12.8	18.4	−2.269	−3.327	32	6.5	18.4	−4.289	−7.607
9	5.1	18.4	−5.299	−7.997	33	13.4	18.4	−7.015	2.020
10	12.0	18.4	−2.218	−4.177	34	20.9	18.4	3.790	−1.286
11	7.5	18.4	−4.743	−6.153	35	31.1	18.4	11.415	1.289
12	8.5	18.4	−1.057	−8.839	36	38.2	18.4	10.152	9.652
13	20.5	18.4	−0.704	2.808	37	23.6	18.4	3.992	1.212
14	20.8	18.4	4.447	−2.042	38	27.5	18.4	2.932	6.172
15	21.7	18.4	1.720	1.584	39	33.8	18.4	27.825	−12.421
16	20.5	18.4	−0.098	2.202	40	31.3	18.4	7.628	5.276
17	28.1	18.4	1.720	7.984	41	33.1	18.4	16.767	−2.063
18	22.4	18.4	2.275	1.729	42	31.7	18.4	5.709	7.595
19	16.1	18.4	−1.714	−0.582	43	30.4	18.4	9.193	2.811
20	16.5	18.4	3.790	−5.686	44	30.8	18.4	5.709	6.695
21	19.0	18.4	1.468	−0.863	45	8.4	18.4	−8.581	−1.415
22	15.3	18.4	2.932	−6.028	46	14.1	18.4	−5.804	1.508
23	15.7	18.4	−8.379	5.683	47	11.2	18.4	−9.742	2.546
24	17.6	18.4	−6.561	5.765					

$\sum \text{columns}_i^2$ 19616 15905   2440   1271

$$
\begin{aligned}
\text{Total Sum of Squares} &= \sum y_i^2 - \sum \bar{y}^2 \\
&= \sum (y_i - \bar{y})^2 \\
&= 19616 - 15905 \\
&= 3711 \\
&= \qquad\qquad \sum (\beta_1 x_i)^2 + \sum \epsilon_i^2 \\
&= \qquad\qquad 2440 + 1271 \quad \text{quadratic identity}
\end{aligned}
$$

In the extended notation, a new observation y_0 at x_{0+} has the model

$$y_0 = x_{0+}\beta + \epsilon_0 = \mu_0 + \epsilon_0$$

where

- y_0 is a single unobserved value
- x_{0+} is a $1 \times (1 + p)$ row of predictors [$(1\ x_0)$ in Chapter 8]
- β is a $(1 + p)$-vector of regression coefficients [$(\beta_0\ \beta_1)'$ in Chapter 8].

There are two related questions to ask about the new observation:

1. Estimate the parameter $\mu_0 = E(y_0) = x_{0+}\beta$.
2. Predict a specific observation $y_0 = \mu_0 + \epsilon_0$.

Estimation intervals for new μ_0 and prediction intervals for new y_0 based on a new value x_{0+} depend on the quantity h_0 defined as

$$h_0 = \frac{1}{n} + \frac{(x_0 - \bar{x})^2}{\sum\limits_{i=1}^{n}(x_i - \bar{x})^2} \tag{8.19}$$

The formula for h_0 is similar to the leverage formula for h_i to be introduced in Equations (9.14) or (9.15), where the new value x_{0+} replaces one of the observed values X_{i+}. The notation i specifically means one of the original n observations and the notation 0 means an additional observation that need not be one of the original ones. Equation (8.19) is specifically for simple linear regression ($p = 1$). The more complex formula in Equations (9.14) or (9.15) is needed when $p > 1$.

Answering the questions requires information about estimated variances:

1. Estimate the

    a. parameter $\mu_0 = E(y_0) = x_{0+}\beta$ with
    b. estimator $\hat{\mu}_0 = x_{0+}\hat{\beta}$,
    c. variance of the estimator $\text{var}(\hat{\mu}_0) = h_0\sigma^2$, and
    d. estimated variance of the estimator $\widehat{\text{var}}(\hat{\mu}_0) = h_0\hat{\sigma}^2$.

2. Predict

    a. a specific observation $y_0 = \mu_0 + \epsilon_0$ with
    b. predictor $\hat{y}_0 = \hat{\mu}_0 = x_{0+}\hat{\beta}$ (the same as the parameter estimate),

c. variance of the predictor $\text{var}(\hat{y}_0) = \text{var}(\hat{\mu}_0 + \epsilon_0) = \text{var}(\hat{\mu}_0) + \text{var}(\epsilon_0)$, and

d. estimated variance of the predictor $\widehat{\text{var}}(\hat{y}_0) = \widehat{\text{var}}(\hat{\mu}_0) + \widehat{\text{var}}(\epsilon_0) = h_0\hat{\sigma}^2 + \hat{\sigma}^2 = \hat{\sigma}^2(h_0 + 1)$.

In the special case that $x_{0+} = x_{i+}$ (one of the observed points), we have

$$\widehat{\text{var}}(\hat{y}_i) = (1 + h_i)\hat{\sigma}^2 = \hat{\sigma}^2 + \widehat{\text{var}}(\hat{\mu}_i)$$

Note that the (standard error)^2 for prediction $\hat{\sigma}^2(h_0 + 1)$ is larger than the (standard error)^2 for estimation $\hat{\sigma}^2 h_0$. A prediction interval for individual observations $\hat{y}_0$ estimates the range of observations that we might see. A confidence interval for the estimated mean of the new observations estimates the center point of the predicted range.

Most regression programs print the standard error for estimation of the mean: $\hat{\sigma}\sqrt{h_0}$, the confidence interval for estimating $\mu_0 = E(y_0|x)$: $\hat{y}_0 \pm t_{df,\frac{\alpha}{2}}\hat{\sigma}\sqrt{h_0}$, [also shown in Equation (9.24)], and the prediction interval for a new observation $(y_0|x)$: $\hat{y}_0 \pm t_{df,\frac{\alpha}{2}}\hat{\sigma}\sqrt{1 + h_0}$ [also shown in Equation (9.25)]. These items are discussed in detail in Section 9.9.

The commands that construct the confidence and prediction intervals in R, and their interpretation, are shown in Table 8.7. To see the standard error for prediction of a new observation, we must manually do the arithmetic

$$\hat{\sigma}^2 h_0 + \hat{\sigma}^2 = (1 + h_0)\hat{\sigma}^2 \tag{8.20}$$

The two questions about a new observation are actually familiar questions in a new guise. They are the same questions addressed in Section 3.6 about the location parameter μ of a sample from a single variable. We elaborate on the comparison in Table 8.8.

In both the confidence interval and the prediction interval of the regression problem in Table 8.8, the magnitude of (Standard Deviation)^2 increases as the new value x moves further from the mean $\bar{x}$ of the existing x_i's. This indicates that we have more confidence in a prediction for an x in the vicinity of the x_i's of the existing data than in an x far from the x_i's of the existing data. The lesson is that extrapolations of the fitted regression relationship for remote values of x are likely to be unreliable.

Confidence and prediction intervals for a particular new observation at x_0 are shown in Table 8.7. These intervals can be extended to confidence and prediction *bands* by letting x_0 vary over the entire range of x. Figure 8.5 illustrates such 95% bands for fat.lm, the modeling of bodyfat as a function of abdomin, displayed in Table 8.1. The 0.95 probability statement applies to each particular value of $x = x_0$. It does not apply to statements that the bands enclose the infinite set of all possible means or predictions as x varies over its range.

Table 8.7 Construction of the confidence and prediction intervals for new observations in R. See also the discussion surrounding Equations (9.24) and (9.25).

```
> old.data <-
+      data.frame(y=rnorm(50), x1=rnorm(50), x2=rnorm(50), x3=rnorm(50))

> example.lm <- lm(y ~ x1 + x2 + x3, data=old.data)

> (example.coef <- coef(example.lm))
(Intercept)          x1          x2          x3
   -0.09670     0.11571    -0.12581    -0.09652

> (new.data <- data.frame(x1=3, x2=2, x3=45))
  x1 x2 x3
1  3  2 45

> predict(example.lm, newdata=new.data, se.fit=TRUE,
+          interval="confidence")
$fit
     fit    lwr   upr
1 -4.344 -18.03 9.337

$se.fit
[1] 6.797

$df
[1] 46

$residual.scale
[1] 0.9492


> predict(example.lm, newdata=new.data, se.fit=TRUE,
+          interval="prediction")
$fit
     fit    lwr   upr
1 -4.344 -18.16 9.47

$se.fit
[1] 6.797

$df
[1] 46

$residual.scale
[1] 0.9492


> c(1, data.matrix(new.data)) %*% example.coef
        [,1]
[1,] -4.344
```

Table 8.8 Comparison of confidence and prediction intervals in the one-sample problem (t-test) and in the regression problem.

	One Sample	Regression
Model Parameters:		
Model	$y = \mu_Y + \epsilon$	$y_x = \beta_0 + \beta_1 x + \epsilon$
Parameter	μ_Y	$\mu_{YX} = \beta_0 + \beta_1 x$
Variance of ϵ	$\mathrm{var}(\epsilon) = \sigma_Y^2$	$\mathrm{var}(\epsilon) = \sigma_{YX}^2$
Sample Statistics:		
Estimate	$\hat{\mu}_Y = \bar{y}$	$\hat{\mu}_{yx} = b_0 + b_1 x$
		$\hat{y}_i = b_0 + b_1 x_i$
Variance	$s_Y^2 = \sum_{i=1}^{n}(y_i - \bar{y})^2/(n-1)$	$s_{YX}^2 = \sum_{i=1}^{n}(y_i - \hat{y}_i)^2/(n-2)$

Estimate Parameter:

(Standard Deviation)^2 for Confidence Interval Estimate

What is the average height μ_Y of everyone?

What is the average height μ_{YX} of those people who are $x = 10$ years old?

$$s_{\hat{\mu}_Y}^2 = s_{\bar{y}}^2 = \frac{s_Y^2}{n} = s_Y^2\left(\frac{1}{n}\right)$$

$$s_{\hat{\mu}_{yx}}^2 = s_{YX}^2 h_x = s_{YX}^2\left(\frac{1}{n} + \frac{(x - \bar{x})^2}{\sum_{i=1}^{n}(x_i - \bar{x})^2}\right)$$

Prediction Interval:

(Standard Deviation)^2 for Prediction Interval for an Individual Response

How tall is the next person?

$$\hat{y} = \hat{\mu}_Y + \epsilon = \bar{y} + \epsilon$$

$$s_{\hat{y}}^2 = \frac{s_Y^2}{n} + s_Y^2 = s_Y^2\left(\frac{1}{n} + 1\right)$$

How tall is the next 10-year-old?

$$\hat{y}_x = \hat{\mu}_{yx} + \epsilon = (b_0 + b_1 x) + \epsilon$$

$$s_{\hat{y}_x}^2 = s_{YX}^2 h_x + s_{YX}^2 = s_{YX}^2(1 + h_x)$$

$$= s_{YX}^2\left(1 + \frac{1}{n} + \frac{(x - \bar{x})^2}{\sum_{i=1}^{n}(x_i - \bar{x})^2}\right)$$

95% confidence and prediction intervals for fat.lm

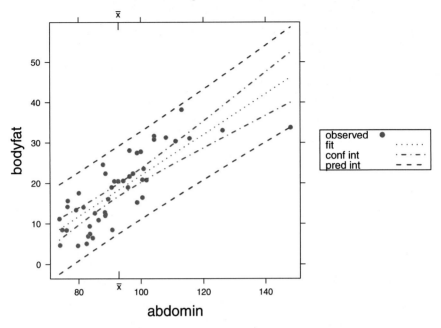

Fig. 8.5 Confidence and prediction bands for modeling `bodyfat ~ abdomin`, body fat data. The widths of these bands are minimized at $x = \bar{x}$ because h_0 is minimized at $x = \bar{x}$.

8.4 Diagnostics

There are two steps to a statistical analysis. The first step is to construct a model and estimate its parameters. Sections 8.3.2 and 9.2 discuss estimation of the parameters of linear models with one and two predictor variables. The second step is to study the quality of the fit of the data to that model and determine if the model adequately describes the data. This section introduces the diagnostics. They are investigated more thoroughly in Section 11.3.

The choice of diagnostic techniques is connected directly to the model and assumptions. If the assumption (8.2) that the error terms ϵ_i are normally independently distributed with constant mean 0 and variance σ^2 is valid, then the residuals $e_i = (y_i - \hat{y}_i)$ will be approximately normally distributed. More precisely, the n values e_i will behave exactly like n numbers independently chosen from the normal distribution and subjected to $p + 1$ linear constraints. In the simplest case, when $p = 0$ (one-sample t-test in Chapter 5), the residuals e_i behave like n independent normals centered on their observed mean $\bar{x}$. For simple linear regression ($p = 1$), the residuals behave like n independent normals vertically centered on a straight line specified by the two estimated parameters $\hat{\beta}_1$ and $\hat{\beta}_0$.

The diagnostic techniques are various procedures for looking at approximately normal numbers and seeing if they display any systematic behavior. If we see systematic behavior, then we conclude that the model did not capture all the interesting features of the data. We iterate the analysis steps by trying to model the systematic behavior we just detected and then looking at the residuals from the newer model.

Figure 8.6 shows several diagnostic plots from the simple regression model of Section 8.1. These are our versions of standard plots of the Fitted Values and Residuals from the regression analysis. The first three panels are based on the output of the R statement `plot(fat.lm)` (using the `plot.lm` method in the **stats** package). The fourth is based on an S-Plus plot. All four as displayed here were drawn with the statement `lmplot(fat.lm)` using the `lmplot` function in the **HH** package.

(We show plots from `plot(fat.lm)` in Figure 11.18. We prefer the **lattice**-based appearance of our first three plots to the **base** graphics of `plot(fat.lm)`. We believe the fourth panel of Figure 11.18 (enlarged in Figure 11.19) can't be described until Chapter 11. We believe the fourth panel of Figure 8.6 is highly informative and wish that R had included it as part of their standard display.)

We discuss each panel in turn, with the numbering sequence $\left(\begin{smallmatrix} 1 & 3 \\ 2 & 4 \end{smallmatrix}\right)$.

1. Panels 1 and 2 are coordinated. Panel 1 is a plot of the Residuals $e = y - \hat{y}$ against the Fitted Values $\hat{y}$ along with a horizontal line at $e = 0$. The horizontal line corresponds to the least-squares fit of the Residuals against the Fitted Values. There is, by construction, no linear effect in this panel. There may be quadratic (or higher-order polynomial) effects visible. The marginal distribution of the Fitted Values $\hat{y}$ may show patterns that need further investigation. When there is only one x-variable, as in the example in Figure 8.6, the Fitted Values are a linear transformation of the x-variable. In this example, we see that the x-value of the point with the largest absolute residual is noticeably larger than any of the other x-values.

2. Panel 2 plots $\sqrt{|e|} = \sqrt{|\text{Residuals}|}$ against the Fitted Values $\hat{y}$. It shows much of the same information as Panel 1. The absolute value folds the negative residuals onto the positive direction in order to emphasize magnitude of departure from the model at the expense of not showing direction. The square root transformation brings in the larger residuals and spreads out the smaller ones. See the discussion of the ladder of powers in Section 4.9 for more information on the effects of transformations. In this display we chose to retain the original directionality by choice of plotting symbol and color.

3. Panel 3 is a normal probability plot with the Residuals on the vertical axis and the normal quantiles on the horizontal axis. The diagonal line has the standard deviation s for its slope. When the residuals are approximately normal, the points will be close to the diagonal line. Asymmetries in the residuals will be visible. Short tails in the distribution of the residuals will be visible as an "S"-shaped

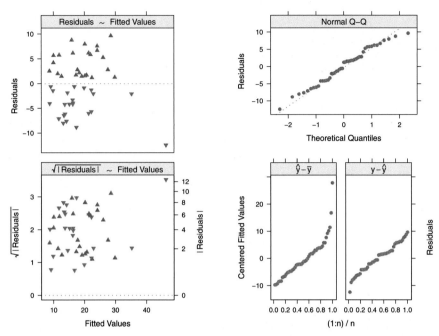

Fig. 8.6 Diagnostics for `lm(bodyfat ~ abdomin, data=fat)`. Diagnostic Plots of the Residuals and Fitted Values from a regression analysis. See Section 8.4 for an extensive discussion of each of the four panels in this display. On the left we show two views of the Residuals plotted against the Fitted Values, with the Residuals themselves on the top, and the square root of the absolute values of the Residuals on the bottom. On the top right, we show the QQ plot of the Residuals against the Normal quantiles. On the right bottom, we show the *r-f spread plot*—a two-panel display of the transposed empirical distributions of the Centered Fitted Values and of the Residuals (see Section 8.5).

display, and long tails in the distribution of the residuals (seen as vertical outliers in panels 1 and 2) will be visible as a mirror-image "2" shape. See Section 5.8 for further discussion of probability plots.

4. Panel 4 is subdivided into two transposed empirical distributions. The left panel shows the Centered Fitted Values $\hat{y} - \bar{y}$ and the right panel shows the Residuals $y - \hat{y}$. The relative vertical ranges of these two panels gives some information on the multiple correlation coefficient R^2. We develop the construction and interpretation of panel 4 in Section 8.5 and Figure 8.7.

8.5 ECDF of Centered Fitted Values and Residuals

The ECDF plot of Centered Fitted Values and Residuals is the *r-f spread plot* defined by Cleveland (1993). The empirical distribution of $S(x)$ is defined in Section 5.7 as the fraction of the data that is less than or equal to x. The empirical distribution is defined analogously to the cumulative distribution $F(x) = P(X \leq x)$ of a theoretical distribution.

We discuss each of the panels of Figure 8.7.

a. The plot of the cumulative distribution is a plot of $F(x)$ against x.

b. The empirical cumulative distribution of an observed set of data is a plot of proportion$(X \leq x)$ against x. If there are n observations in the dataset, we plot i/n against $x_{[i]}$. We use the convention here that subscripts in square brackets mean that the data have been sorted. For example, let us look at the fitted values $\hat{y}$ and residuals $e = y - \hat{y}$ from the regression analysis in Table 8.1. The left side of Figure 8.7b is the cumulative distribution of the fitted values. The right side is the cumulative distribution of the residuals. Note that these plots are on very different scales for the abscissa and therefore cannot easily be compared visually.

c. We construct Figure 8.7c by making two adjustments to Figure 8.7b. First, we center the fitted values on their mean. Second, we plot both graphs on the same abscissa scale by forcing them to have the same x-axis constructed as the range of the union of their individual abscissas.

d. Figure 8.7d is the transpose of the pair of graphs in Figure 8.7c. We interchange the axes, putting the proportions on the abscissa and the data (centered fitted values in the left panel and residuals in the right panel) on the ordinate. We therefore force the y-axes to have a common limits. S-Plus uses Figure 8.7d as the fifth diagnostic plot of their analog of Figure 8.6. The vertical axis now uses the same y units as panels 1 and 3 of Figure 8.6.

If our model explains the data well, then we would anticipate that the residuals have less variability than the fitted values.

The multiple correlation R^2 can be written as

$$R^2 = \frac{SS_{Reg}}{SS_{Total}} = \frac{SS_{Reg}}{SS_{Reg} + SS_{Res}} \tag{8.21}$$

We can use the squared range of the fitted values as a surrogate for the SS_{Reg} and the squared range of the residuals as a surrogate for the SS_{Res}. This leads to the interpretation of panel 4 of Figure 8.6 as an indicator of R^2. We show a series of illustrations of this interpretation in Figure 8.8. If the ranges of the $\hat{y} - \bar{y}$ and $y - \hat{y}$ panels are similar, then $R^2 \approx \frac{1}{2}$. If the range of the fitted values is larger, then the R^2 is closer to 1, and if the range of the fitted values is smaller, then the R^2 is closer to 0.

a. Cumulative distribution of the standard normal $\Phi(x)$ for $x \sim N(0, 1)$.

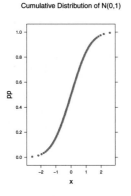

b. Empirical distributions of fitted values and residuals with independent ranges for the abscissa.

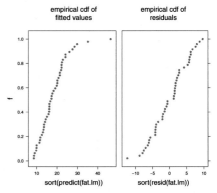

c. Empirical distributions of fitted values and residuals with common range for the abscissa.

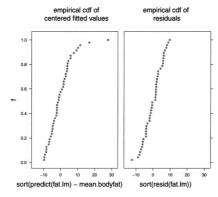

d. Transposed empirical distributions of fitted values and residuals with common range for the abscissa. This is panel 5 of Figure 8.6.

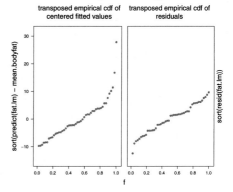

Fig. 8.7 Explanation of panel 4 of Figure 8.6. Panels a,b,c are empirical distribution plots and panel d is the transposed empirical distribution plot of the fitted values and residuals from the linear regression `fat.lm <- lm(bodyfat ~ abdomin, data=fat)`. Please see the discussion in the text of Section 8.5 for more detailed description of the panels in this figure.

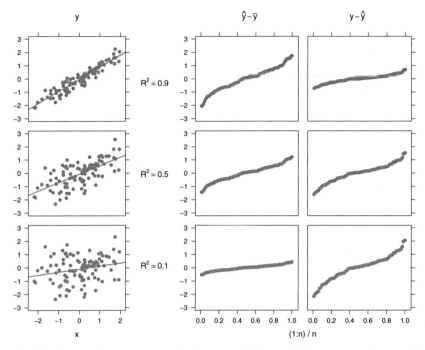

Fig. 8.8 There are three columns y, $\hat{y} - \bar{y}$, and $y - \hat{y}$. The rows of the y column shows a plot of y against x along with the fitted regression line for each of three levels of R^2 (.1, .5, .9). The $\hat{y} - \bar{y}$ and $y - \hat{y}$ columns show the transposed ECDF of the Fitted Values and Residuals for those situations. For $R^2 = .1$, the Residuals $y - \hat{y}$ has a wider range than the Centered Fitted Values $\hat{y} - \bar{y}$. For $R^2 = .5$, the two ranges are equal. For $R^2 = .9$, the Residuals $y - \hat{y}$ has a narrower range than the Centered Fitted Values $\hat{y} - \bar{y}$.

8.6 Graphics

The figures in this chapter represent several different types of plots.

Figure 8.1 is a scatterplot matrix, constructed in R with `splom()`.

Figures 8.2 and 8.3 use `regrresidplot`, a function in the **HH** package in R. Our function `panel.residSquare`, used by `regrresidplot`, constructs the squares that represent the squared residuals with real squares on the plotting surface. The heights of the squares are in y-coordinates. The widths of the squares are the same number of inches (or cm) on the plotting surface as the heights. Each of the figures has been placed into a lattice structure which enforces the same x- and y-ranges for comparability. Our function `regrresidplot` is based on the explanation of least-squares regression in Smith and Gonick (1993).

Figure 8.5 is a scatterplot drawn with **HH** function `ci.plot` with superimposed lines for the fitted regression line and the confidence and prediction intervals.

Figures 8.6, 8.7, and 8.8 use functions in the **HH** package that are based on the R function plot.lm to display the standard plots of Residuals and Fitted Values from a regression analysis. The ECDF plots of Centered Fitted Values and Residuals are drawn by the **HH** function diagplot5new, which is based on the S-Plus function rfplot, which in turn is based on a plot by Cleveland (1993).

8.7 Exercises

8.1. Hand et al. (1994) report on a study by Lea (1965) that investigated the relationship between mean annual temperature (degrees F) in regions of Britain, Norway, and Sweden, and the rate of mortality from a type of breast cancer in women. The data are accessed as data(breast).

a. Plot the data. Does it appear that the relationship can be adequately modeled by a linear function?

b. Estimate the regression line and add this to your plot.

c. Calculate and interpret R^2.

d. Calculate and interpret the standard error of estimate.

e. Interpret the estimated slope coefficient in terms of the variables mortality and temperature.

f. Find a 95% confidence interval on the population slope coefficient.

g. Find a 95% prediction interval for a region having mean annual temperature 45.

h. One of these 16 data points is unusual compared to the others. Describe how.

8.2. Shaw (1942), later in Mosteller and Tukey (1977), shows the level of Lake Victoria Nyanza relative to a standard level and the number of sunspots in each of 20 consecutive years. The data are accessed as data(lake). Use linear regression to model the lake level as a function of the number of sunspots in the same year.

8.3. Does muscle mass decrease with age? The age in years and muscle mass were obtained from 16 women. The data come from Neter et al. (1996) and are accessed as data(muscle).

a. Plot mass vs age and overlay the fitted regression line.

b. Interpret the slope coefficient in terms of the model variables.

c. Predict with 90% confidence the muscle mass of a 66-year-old woman.

d. Interpret the calculated standard error of estimate.

e. Interpret R^2 in terms of the model variables.

8.4. The dataset data(girlht) contains the heights (in cm) at ages 2, 9, and 18 of 70 girls born in Berkeley, California in 1928 or 1929. The variables are named h2, h9, and h18, respectively. The data come from a larger file of physical information on these girls in Cook and Weisberg (1999).

a. Regress h18 on h9 and also h18 on h2.

b. Discuss the comparative strengths of these two regression relationships.

c. Interpret the slope coefficients of both regressions.

8.5. We would expect that the price of a diamond ring would be closely related to the size of the diamond the ring contains. Chu (1996) presents data on the price (Singapore dollars) of ladies' diamond rings and the number of carats in the ring's diamond. The data are accessed as data(diamond).

a. Regress price on carats.

b. Notice that the estimated intercept coefficient is significantly less than 0. Therefore, this model is questionable, although the range of the predictor variables excludes 0. Instead fit a model without an intercept term.

c. Compare the goodness of fits of the two models. Which is preferable?

8.6. The data data(income), from Bureau of the Census (2001), contains year 2000 data on the percentage of college graduates and per capita personal income for each of the 50 states and District of Columbia. Regress income on college. Interpret the meaning of R^2 for these data. Discuss which states have unusually low or high per capita income in relation to their percentage of college graduates.

8.7. Prove Equation (8.15)

$$\sigma^2_{\hat{\beta}_1} = \text{var}(\hat{\beta}_1) = \frac{\sigma^2}{\sum(x_i - \bar{x})^2}$$

The proof is primarily algebraic manipulation. Rewrite (8.13) as a weighted sum of the independent y_i, that is as

$$\hat{\beta}_1 = \sum(y_i - \bar{y})\left(\frac{(x_i - \bar{x})}{\sum(x_i - \bar{x})^2}\right) = \sum y_i k_i \qquad (8.22)$$

then write

$$\text{var}(\hat{\beta}_1) = \sigma^2 \sum k_i^2 \qquad (8.23)$$

and simplify.

8.8. Prove Equation (8.17) that the variance of the estimate of the intercept $\hat{\beta}_0$ has variance

$$\sigma_{\hat{\beta}_0}^2 = \text{var}(\hat{\beta}_0) = \sigma^2 \left(\frac{1}{n} + \frac{\bar{x}^2}{\sum(x_i - \bar{x})^2} \right)$$

8.9. Algebraically prove the assertion in Equation (8.8) that in simple regression, the sum of the calculated residuals is zero.

8.10. In Figure 8.2 we construct the actual squares of the residuals and show that the sum of the areas of the squared residuals is smallest for the least-squares line. We do the construction in the simplest way, placing the other three sides on the side that is already there representing the residual. Other possibilities are

a. Place the left–right center of the square on the residual line. Use the function `panel.residSquare` as the model for your function.

b. Place a circle (a real circle in inches of graph surface) on the points. Base your function on the functions `panel.residSquare` and the descriptions of the R points function (`?points`). The value `pch=1` provides a circle. You can use the `cex` argument to control the size of the circles.

Option 1: Keep the existing residual line and center the circle on the observed point.

Option 2: Use the existing residual line as the diameter of the circle.

Chapter 9

Multiple Regression—More Than One Predictor

In Chapter 8 we introduce the algebra and geometry behind the fitting of a linear model relating a single response variable to one or more explanatory (predictor) variables using the criterion of least squares. In this chapter we consider in more detail situations where there are two or more predictors.

The two linear modeling techniques we have studied so far, regression in Chapter 8 and analysis of variance in Chapter 6, have much of their mathematics interpretation in common. In this chapter we explore the common mathematical features, with some examples of how they apply. In the following chapters we use this common structure.

We begin by extending the Chapter 8 discussion of regression with a single predictor (simple regression) to allow for two or more predictors. *Multiple* regression refers to regression analysis with at least two predictors. There is another term *multivariate regression* which refers to situations with more than one response variable. We do not discuss multivariate regression in this book.

9.1 Regression with Two Predictors—Least-Squares Geometry

The graphics for least squares with two x-variables, and in general for more than two x-variables, are similar to the graphics in Figure 8.2. We will work with two x-variables, abdomin and biceps, from the data(fat) dataset we used in Chapter 8. In the three snapshots of the basic 3-dimensional plot in Figure 9.1, bodyfat is plotted as y against the other two variables as x_1 and x_2.

The response variable is placed on the vertical dimension and the two x-variables biceps and abdomin define the horizontal plane. The red and green dots at the observations show the three-dimensional location of the observed points. Positive residuals are shown as green dots above the least-squares plane and are connected

© Springer Science+Business Media New York 2015
R.M. Heiberger, B. Holland, *Statistical Analysis and Data Display*,
Springer Texts in Statistics, DOI 10.1007/978-1-4939-2122-5_9

to the fitted value on the plane by a green residual line. The green residual line forms one edge of the square. Negative residuals are shown as red dots below the least-squares plane and are connected to the fitted value on the plane by a red residual line. The red residual line forms one edge of the square.

The least-squares plane minimizes the sum of the squared areas. The displayed squares are the squares whose sum has been minimized by the least-squares process. The view in the right panel is from above the plane. It shows biceps coming out of the page and abdomin going into the page. The view in the center panel is from a point that is on the least-squares plane. The view in the left panel is from below the least-squares plane. Variable biceps is coming out of the page and variable abdomin is almost along the page. The code in file HHscriptnames(9) constructs an interactive 3-d version of this plot. We selected these specific static snapshots from the interactive plot.

We think of this plot as a point cloud in 3-space floating over the surface defined by the x-variables. Any plane other than the least-squares plane will show a larger sum of squared areas than the least-squares plane illustrated here.

9.2 Multiple Regression—Two-X Analysis

The specification of the analysis for two x-variables is similar to that for one x-variable. The sequential ANOVA table and the table of coefficients for a two x-variable analysis of the body fat data data(fat) are in Table 9.1.

Since both predictors are significantly different from 0, the arithmetic justifies the illustration in Figure 9.1, where we see from the regression plane that $\hat{y}$ changes linearly with changes in either x_1 and x_2. The table of coefficients tells us that on average for this population, percent body fat increases by 0.683 if abdomen circumference increases by one cm and biceps is unchanged, and percent body fat decreases by .922 if biceps increases by one cm while abdomin is unchanged.

The t-value for biceps (the second variable in the ANOVA table) is related to the F-value for biceps: $t^2 = (-2.946)^2 = 8.677 = F$. The t-value (8.693) for abdomin (the first variable in the ANOVA table) is not simply related to the correspondingly labeled F-value (101.172). We investigate this relationship in the discussion of Table 13.27.

Figure 9.2 shows the diagnostics from the two-X regression model of Section 9.2. Compare this to the similar plot for one-X regression in Figure 8.6.

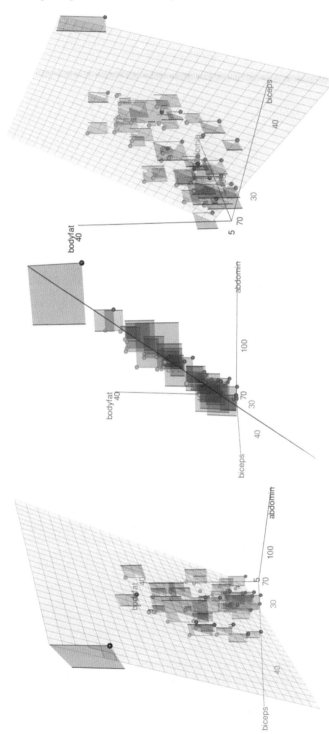

Fig. 9.1 Bodyfat data with one response variable bodyfat and two predictor variables biceps and abdomin. This figure is three static screenshots of an interactive 3-d plot. The plot is the two-*x* extension of the display of squared residuals we introduced in Figure 8.2. The center panel shows a viewpoint aligned with the least-squares plane. The left panel shows a rotation to the left and we see the negative residuals under the plane and the partially occluded positive residuals through the plane. The right panel shows a rotation to the right and we see the positive residuals above the plane and the partially occluded negative residuals through the plane. See the text for the detailed description of the three panels. See Table 9.1 for the corresponding ANOVA table.

Table 9.1 Sequential ANOVA table and table of regression coefficients from the two-x model with y=bodyfat, x_1=abdomin, and x_2=biceps. See Figure 9.1.

```
> fat2.lm <- lm(bodyfat ~ abdomin + biceps, data=fat)

> anova(fat2.lm)
Analysis of Variance Table

Response: bodyfat
          Df Sum Sq Mean Sq F value  Pr(>F)
abdomin    1   2440    2440  101.17 5.6e-13 ***
biceps     1    209     209    8.68  0.0051 **
Residuals 44   1061      24
---
Signif. codes:  0 '***' 0.001 '**' 0.01 '*' 0.05 '.' 0.1 ' ' 1

> summary(fat2.lm)

Call:
lm(formula = bodyfat ~ abdomin + biceps, data = fat)

Residuals:
    Min      1Q  Median      3Q     Max
-11.252  -3.674   0.716   3.771  10.241

Coefficients:
            Estimate Std. Error t value Pr(>|t|)
(Intercept) -14.5937     6.6922   -2.18   0.0346 *
abdomin       0.6829     0.0786    8.69  4.2e-11 ***
biceps       -0.9222     0.3130   -2.95   0.0051 **
---
Signif. codes:  0 '***' 0.001 '**' 0.01 '*' 0.05 '.' 0.1 ' ' 1

Residual standard error: 4.91 on 44 degrees of freedom
Multiple R-squared:  0.714,Adjusted R-squared:  0.701
F-statistic: 54.9 on 2 and 44 DF,  p-value: 1.1e-12
```

9.3 Multiple Regression—Algebra

Everything in simple regression analysis carries over to multiple regression. There are additional issues that arise because we must also study the relations among the predictor variables. The algebra for multiple regression is most easily expressed in matrix form. (A brief introduction to matrix algebra appears in Appendix I.) The formulas for simple regression can be derived as the special case of multiple regression with $p = 1$.

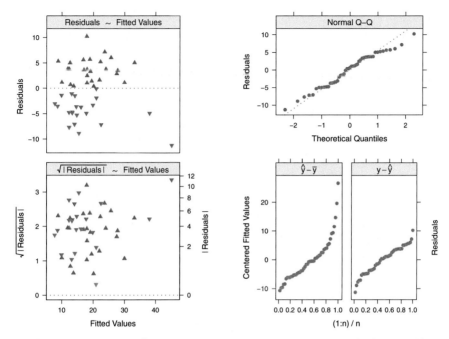

Fig. 9.2 Diagnostics for `lm(bodyfat ~ abdomin + biceps, data=fat)`. Compare this to the similar plot for one-X regression in Figure 8.6.

Assume

$$Y = X \beta + \epsilon \qquad (9.1)$$
$$\underset{n\times1}{} \ \underset{n\times(1+p)}{} \underset{(1+p)\times1}{} \ \underset{n\times1}{}$$

or equivalently

$$y_i = \beta_0 + \beta_1 X_{i1} + \cdots + \beta_p X_{ip} + \epsilon_i \quad \text{for } i = 1, \ldots, n \qquad (9.2)$$

where

- $\underset{n\times1}{Y}$ are observed values,

- $\underset{n\times(1+p)}{X} = [\mathbf{1}\, X_1 X_2 \ldots X_p]$ are observed values with $\underset{n\times1}{\mathbf{1}}$ representing the constant column with 1 in each row and X_j indicating the column with X_{ij} in the ith row, $\underset{n\times1}{}$

- $\underset{(1+p)\times1}{\beta}$ are unknown constants,

- $\underset{n\times1}{\epsilon} \sim N(0, \sigma^2 I)$ are independent.

Then the least-squares estimate $\widehat{\beta}$ is obtained by minimizing the sum of squared deviations

$$S = (Y - X\beta)'(Y - X\beta) = \sum_{i=1}^{n} \Big(y_i - (\beta_0 + \beta_1 X_{i1} + \cdots + \beta_p X_{ip})\Big)^2$$

by taking the derivatives $(\partial S/\partial \beta_j)$ with respect to all the β_j and setting them to 0. The resulting set of equations, called the *Normal Equations* and generalizing Equation 8.6,

$$(X'X)\widehat{\beta} = (X'Y) \tag{9.3}$$

are solved for $\widehat{\beta}$. The solution [equivalent to Equation 8.7] is equal to

$$\widehat{\beta} = (X'X)^{-1}(X'Y) = \left((X'X)^{-1}X'\right)Y = (X^+)Y \tag{9.4}$$

The symbol $X^+ \overset{\text{def}}{=} (X'X)^{-1}X'$ is the notation for the Moore–Penrose *generalized inverse* of any rectangular matrix. In the special case of square invertible matrices the generalized inverse becomes the familiar matrix inverse. We introduce this notation here because it simplifies the appearance of the equations. We start with the model $Y = X\beta + \epsilon$ in Equation (9.1) and conclude with the estimate $\hat{\beta} = X^+Y$ in Equation (9.4). We effectively moved the X to the other side and replaced the ϵ with the hat on the β. Note that Equation (9.4) is an identity, but neither efficient nor numerically stable as a computing algorithm. An efficient algorithm uses Gaussian elimination to solve the equations directly. See Section I.4.7 for further discussion on efficient computation.

We construct the fitted values with

$$\widehat{Y} = X\widehat{\beta} = \left(X(X'X)^{-1}X'\right)Y = HY \tag{9.5}$$

where the matrix

$$H \overset{\text{def}}{=} X(X'X)^{-1}X' \tag{9.6}$$

is a projection matrix. The sum of squares (SS) for the regression is $SS_{Reg} = Y'HY$. The projection matrix H is called the *hat matrix* because multiplying H by Y places a hat $\frown$ on Y. We can see that $H_{ij} = \partial \widehat{Y}_i/\partial Y_j$. We discuss the hat matrix in Section 9.3.1.

The *residuals* are defined as the difference

$$e = Y - \widehat{Y} = (I - H)Y \tag{9.7}$$

between the observed values Y and the fitted values $\widehat{Y}$. With least-squares fitting, the residuals are orthogonal to the observed x-values

$$e'X = 0 \tag{9.8}$$

and therefore to the fitted values

$$e'\widehat{Y} = e'X\hat{\beta} = 0 \tag{9.9}$$

The variance–covariance matrix of the residuals e is $\sigma^2(I-H)$. Note in particular that $\text{var}(e_i) = \sigma^2(1 - H_{ii})$ is not constant for all i. As a consequence the confidence bands in Figure 8.5 are not parallel to the regression line, but instead have a minimum width at the mean of the x values.

An unbiased estimator of σ^2 is

$$s^2 = \frac{Y'(I - H)Y}{n - p - 1} = MS_{Res} = SS_{Res}/df_{Res} \tag{9.10}$$

Its square root, s, sometimes called the standard error of estimate, is an asymptotically unbiased estimator of σ. As in the case of simple regression, the sum of the residuals is zero, that is,

$$\sum_{i=1}^{n} e_i = 1'e = 0 \tag{9.11}$$

where $1'$ is a row vector of ones. The proof of this assertion is requested in Exercise 9.1.

Both $\widehat{\beta}$ and $\widehat{Y}$ are linear combinations of y_i. The y_i are independent because the ϵ_i are independent. Hence the elementary theorems

$$E(a_1y_1 \pm a_2y_2) = a_1E(y_1) \pm a_2E(y_2) \tag{3.8}$$

and

$$var(a_1y_1 \pm a_2y_2) = a_1^2 \, var(y_1) + a_2^2 \, var(y_2) \tag{3.9}$$

are applicable. These are where we get Equation (8.15), the standard error for β_1, the corresponding formula

$$var(\hat{\beta}) = \sigma^2(X'X)^{-1} \tag{9.12}$$

for the estimator of β in Equation (9.4), and formulas (9.24) and (9.25) for tests and confidence intervals about $E(Y|X)$ and for prediction intervals about Y for new values of X.

9.3.1 The Hat Matrix and Leverage

The hat matrix in Equation (9.6) is called that because premultiplication by H places a hat '$\frown$' on Y: $\widehat{Y} = HY$. The ith diagonal of H is called the leverage of the ith case because it tells how changes in Y_i affect the location of the fitted regression line, specifically:

$$\frac{\partial \widehat{Y_i}}{\partial Y_i} = H_{ii} \tag{9.13}$$

If $\left(H_{ii} > 2(p + 1)/n\right)$, then the ith point is called a high leverage point. See Section 11.3.1. Equation (9.13) shows that changes in the observed Y_i-value of high leverage points have a large effect on the predicted value $\hat{Y}_i$, that is, they have a large effect on the location of the fitted regression plane.

The hat matrix is used in regression diagnostics, that is, techniques for evaluating how the individual data points affect the regression analysis. Many diagnostics are discussed in Section 11.3.

Frequently these diagonals of H are denoted by $h_i = H_{ii}$. They are calculated in R with the command hat(X).

A specific formula for the leverage h_i itself is almost simple:

$$h_i = X_{i.}(X'X)^{-1}X'_{i.} \qquad \text{where } X_{i.} \text{ is the } i^{\text{th}} \text{ row of } X \qquad (9.14)$$

In an alternate but common notation, the predictor matrix does not include the column **1**. To avoid excessive confusion, define Z to be all the columns of X except the initial column **1**:

$$\underset{n \times p}{Z} = [X_1 X_2 \dots X_p]$$

and let

$$\bar{Z} = (\bar{X}_1 \bar{X}_2 \dots \bar{X}_p)$$

In this notation the formula for leverage looks worse:

$$h_i = \frac{1}{n} + (Z_{i.} - \bar{Z})\left((Z - \mathbf{1}\bar{Z})'(Z - \mathbf{1}\bar{Z})\right)^{-1}(Z_{i.} - \bar{Z})' \qquad (9.15)$$

The term $\frac{1}{n}$ in Equation (9.15), with the Z matrix which excludes the column **1**, is not needed in Equation (9.14), with the X matrix which includes the column **1**. In simple regression, with $Z = X_1 = x$, formula (9.15) simplifies to Equation (8.19)

$$h_i = \frac{1}{n} + \frac{(x_i - \bar{x})^2}{\sum\limits_{i=1}^{n}(x_i - \bar{x})^2} \qquad (8.19)$$

9.3.2 Geometry of Multiple Regression

Several types of pictures go along with multiple regression. We have already looked at the scatterplot matrix, drawn with the R command splom(data.frame); for example, see Figure 8.1 for the splom of the body fat dataset fat.

The picture that goes best with the defining least-squares equations is the multi-dimensional point cloud. It is easiest to illustrate this with Y and two X-variables. See Figures 8.2 and 9.1 for one-X and two-X examples.

A similar construction is in principle possible for more X-variables. Illustrating the projection of four or more dimensions onto a two-dimensional graph is difficult at best.

9.4 Programming

9.4.1 Model Specification

We use several notations for the specification of a regression model to a computer program. How are the statements constructed in each notation, and what are their syntax and their semantics?

For specificity, let us look at a linear regression model with a response variable y and two predictor variables x_1 and x_2. We express this model in several equivalent notations. In the algebraic notation of Section 9.3, we have

$$\underset{n\times 1}{Y} = \underset{n\times(1+2)}{X}\ \underset{(1+2)\times 1}{\beta} + \underset{n\times 1}{\epsilon} \tag{9.16}$$

or equivalently

$$y_i = \beta_0 + \beta_1 X_{i1} + \beta_2 X_{i2} + \epsilon_i \quad \text{for } i = 1, \ldots, n \tag{9.17}$$

In R model formula notation, we have

$$y \ \tilde{} \ x1 + x2 \tag{9.18}$$

In SAS model statement notation (with the space character indicating the formulaic sum), we have

$$y = x1 \quad x2 \tag{9.19}$$

In both computer languages the statement is read, "y is modeled as a linear function of x_1 and x_2."

The four statements (9.16)–(9.19) are equivalent. Both computational specifications remove the redundancy in notation used by the traditional scalar algebra notation. The program knows that the variables (y, x1, and x2) have length n; there is no need to repeat that information. All linear model specifications have regression coefficients, and most have a constant term (we discuss models without a constant term in Section 9.8); there is no need to specify the obvious. There is always an error term because the model does not fit the data exactly; there is no need to specify the error term explicitly. The two pieces of information unknown to the program are

- Which variable is the response and which are the predictors. This is indicated positionally—the response is on the left, and notationally—the "~" or "=" separates the response from the predictors. A separation symbol is needed because the same notation can be generalized to express multiple response variables.

- The relationship between the predictors. R indicates summation explicitly with the "+" and SAS indicates it implicitly by leaving a space between the predictor variable names. Other relationships, for example crossing or nesting (to be discussed beginning in Section 13.5), are indicated by other algebraic symbols as indicated in Table 13.18.

The interpretation of operator symbols in the model specification notation is related to, but not identical to, the interpretation of the same symbols in an ordinary algebra statement. The model formulas (9.18) and (9.19) mean:

find the coefficients $\hat{\beta}_0, \hat{\beta}_1, \hat{\beta}_2$ that best fit

$$y_i = \hat{\beta}_0 + \hat{\beta}_1 x_{i1} + \hat{\beta}_2 x_{i2} + \hat{\epsilon}_i \qquad (9.20)$$

for the observed values (y_i, x_{i1}, x_{i2}) for all $i: 1 \leq i \leq n$.

The "+" and space " " in formulas (9.18) and (9.19) do not have the ordinary arithmetic sense of $x_{i1} + x_{i2}$.

9.4.2 Printout Idiosyncrasies

The R summary and anova functions do not print the Total line in their ANOVA tables.

SAS PROC GLM uses the name "Type I Sum of Squares" for the sequential ANOVA table. See the discussion of sums of squares types in Section 13.6.1.

9.5 Example—Albuquerque Home Price Data

9.5.1 Study Objectives

Realtors can use a multiple regression model to justify a house selling price based on a list of desirable features the house possesses. Such data are commonly compiled by local boards of realtors. We consider a data file containing a random sample of 117 home sales in Albuquerque, New Mexico during the period February 15 through April 30, 1993, taken from Albuquerque Board of Realtors (1993).

9.5.2 Data Description

We use a subset of five of the eight variables for which data are provided, and 107 of the 117 houses that have information on all five of these variables.

price: Selling price in $100's

sqft: Square feet of living space

custom: Whether the house was built with custom features (1) or not (0)

corner: Whether the house sits on a corner lot (1) or not (0)

taxes: Annual taxes in $

We investigate models of price as a function of some or all of the candidate predictors sqft, custom, corner, and taxes. This example assumes that taxes potentially determine price. In some real estate contexts the causality could work in the opposite direction: selling prices can affect subsequent home appraisals and hence tax burden.

9.5.3 Data Input

The data are accessed with data(houseprice) and looked at initially with the scatterplot matrices in Figures 9.3 and 9.4. Two of the four candidate predictors, custom and corner, are dichotomous variables, and the panels involving them in Figure 9.3 are wasteful of space and not very informative. Figure 9.4, with separate superpanels for the two values of corner and separate plot symbols for the two values of custom, displays the information much more efficiently. We learn from these figures that custom houses tend to have higher prices than regular houses, and corner houses have different patterns of relationships between price and the continuous predictors than middle houses.

Figure 9.4 suggests that price is directly related to all four candidate predictors. We proceed with the analysis by regressing price on the four variables in Table 9.2. In this Table we examine the signs of the regression coefficients and the magnitudes of their p-values. We see that price is strongly positively associated with sqft, taxes and custom (as opposed to regular) houses. Such conclusions are consistent with common knowledge of house valuation. The predictor corner has a marginally significant negative coefficient. Hence there is moderate evidence that, on average, corner houses tend to be lower priced than middle houses.

The magnitudes of the regression coefficients also convey useful information. For example, on average, each additional square foot of living space corresponds to a $0.2076 \times \$100 = \$20.76$ increase in price, and on average custom houses sell

houseprice

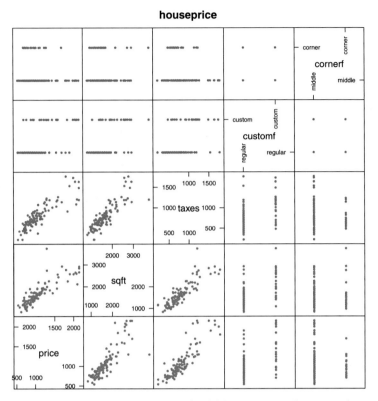

Fig. 9.3 House-price data. The discreteness of variables `customf` and `cornerf` decreases the informativeness of this splom, particularly the panel for this pair of variables. Figure 9.4 is a preferred splom presentation of these data.

for $156.81481 \times \$100 = \$15,681.48$ more than regular houses. The $R^2 = 0.8280$ says that in the population of houses from which `data(houseprice)` is a random sample, 82.8% of the variability in price is accounted for by these four predictors.

9.6 Partial F-Tests

Sometimes we wish to examine whether two or more predictor variables *acting together* have a significant impact on the response variable. For example, suppose we consider the house-price data of Section 9.5 with four candidate predictors, `sqft`, `custom`, `corner`, and `taxes`, and wish to examine if `custom` and `corner` together have a significant impact on `price`, above and beyond the impacts of `sqft` and `taxes`. R (in Table 9.3) approaches this by direct comparison of two models. The *full model* contains all predictors under consideration. The *reduced*

Table 9.2 Analysis of variance table for house-price data.

```
> houseprice.lm2 <- lm(price ~ sqft + taxes + custom + corner,
+                      data=houseprice)

> anova(houseprice.lm2)
Analysis of Variance Table

Response: price
          Df    Sum Sq   Mean Sq F value  Pr(>F)
sqft       1 11102445 11102445  421.34 < 2e-16 ***
taxes      1  1374474  1374474   52.16 9.5e-11 ***
custom     1   350716   350716   13.31 0.00042 ***
corner     1   114215   114215    4.33 0.03985 *
Residuals 102  2687729     26350
---
Signif. codes:  0 '***' 0.001 '**' 0.01 '*' 0.05 '.' 0.1 ' ' 1

> summary(houseprice.lm2)

Call:
lm(formula = price ~ sqft + taxes + custom + corner,
   data = houseprice)

Residuals:
   Min     1Q Median     3Q    Max
-544.6  -99.5   -4.8   64.8  510.2

Coefficients:
            Estimate Std. Error t value Pr(>|t|)
(Intercept) 175.166     56.312    3.11  0.00242 **
sqft          0.208      0.061    3.40  0.00096 ***
taxes         0.677      0.101    6.70  1.2e-09 ***
custom      156.815     44.495    3.52  0.00064 ***
corner      -83.401     40.059   -2.08  0.03985 *
---
Signif. codes:  0 '***' 0.001 '**' 0.01 '*' 0.05 '.' 0.1 ' ' 1

Residual standard error: 162 on 102 degrees of freedom
Multiple R-squared:  0.828,Adjusted R-squared:  0.821
F-statistic:  123 on 4 and 102 DF,  p-value: <2e-16
```

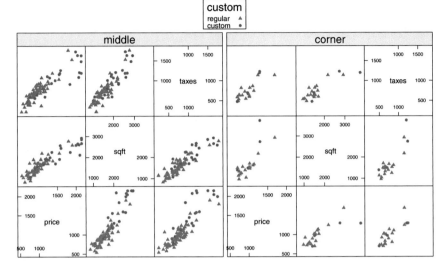

Fig. 9.4 Albuquerque house-price data. Custom houses go for higher prices than regular houses. Corner houses have a different pattern than middle houses.

Table 9.3 Partial F-tests of H_0: $\beta_{custom} = \beta_{corner} = 0$ using the R anova() function with two linear models as arguments.

```
> houseprice.lm1 <- lm(price ~ sqft + taxes, data=houseprice)

> anova(houseprice.lm1, houseprice.lm2)
Analysis of Variance Table

Model 1: price ~ sqft + taxes
Model 2: price ~ sqft + taxes + custom + corner
  Res.Df      RSS Df Sum of Sq      F  Pr(>F)
1     104 3152660
2     102 2687729  2    464931  8.82 0.00029 ***
---
Signif. codes:  0 '***' 0.001 '**' 0.01 '*' 0.05 '.' 0.1 ' ' 1
```

model contains all predictors apart from the ones we test in order to see if then can be eliminated from the model. *Partial F* refers to the fact that we are simultaneously testing *part* of the model's predictors, not all predictors but perhaps more than just one of them. The idea behind this test is apparent from Table 9.3. The F-test examines whether the reduction in residual sum of squares as a result of fitting the more elaborate model is a significant reduction. This assessment is performed by measuring the *extra sum of squares*, defined as

$$\text{(residual SS from reduced model)} - \text{(residual SS from full model)} \quad (9.21)$$

against the residual sum of squares from the full model. The degrees of freedom associated with the extra sum of squares equals the number of parameters being tested for possible elimination.

The general form of the test is

$$F = \frac{(\text{extra SS })/(\text{df associated with extra SS})}{(\text{full model residual SS})/(\text{df associated with full model residual SS})} \quad (9.22)$$

The strategy of this approach is used whenever one wishes to compare the fits of two linear models, one of which has the same terms as the other plus at least one more term.

For testing the hypothesis that the population regression coefficients of custom and corner are both equal to 0, we see that the F-statistic is 8.82 on 2 and 102 degrees of freedom. There are two numerator degrees of freedom because the null hypothesis involves constraints on two model parameters. The very small p-value strongly suggests that this null hypothesis is false. We conclude that at least one of custom and corner is needed in the model.

The preceding discussion assumes that sqft and taxes were already in the model. It is also possible to test the combined effect on price of custom and corner compared with no other predictors, or exactly one of the predictors sqft and taxes. However, we do not pursue these possibilities here.

9.7 Polynomial Models

If the relationship between a response Y and an explanatory variable X is believed to be nonlinear, it is sometimes possible to model the relationship by adding an X^2-term to the model in addition to an X-term. For example, if Y is product demand and X is advertising expenditure on the product, an analyst might feel that beyond some value of X there is "diminishing marginal returns" on this expenditure. Then the analyst would model Y as a function of X, X^2, and possibly other predictors, and anticipate a significant negative coefficient for X^2. Occasionally a need is encountered for higher-order polynomial terms.

An example from Hand et al. (1994), original reference Williams (1959), is data(hardness) which we first encountered in Exercise 4.5. In this section we investigate the modeling of hardness as a quadratic function of density. We pursue this analysis in Exercise 11.2 from another angle, a transformation of the response variable hardness.

Hardness of wood is more difficult to measure than density. Modeling hardness in terms of density is therefore desirable. These data come from a sample of Australian Janka timbers. The Janka hardness test measures the resistance of a sample of wood to denting and wear. A quadratic model fits these data better than a linear model. An additional virtue of the quadratic model is that its intercept term differs insignificantly from zero; this is not true of a model for these data containing only a linear term. (If wood has zero hardness, it certainly has zero density.)

The fitted quadratic model in Table 9.4 is

$$\texttt{density} = -118.007 + 9.4340\,\texttt{hardness} + 0.5091\,\texttt{hardness}^2$$

The regression coefficient for the quadratic term is significantly greater than zero, indicating that the plot is a parabola opening upwards as shown in Figure 9.5. The p-value for the quadratic regression coefficient is identical to the p-value for the quadratic term in the ANOVA table because both tests are for the marginal effect of the quadratic term assuming the linear term is already in the model. The two p-values for the linear term differ because they are testing the linear coefficient in two different models. The p-value for linear regression coefficient assumes the presence of a quadratic term in the model, but the linear p-value in the sequential ANOVA table addresses a model with only a linear component.

When fitting a truly quadratic model, it is necessary to include the linear term in the model even if its coefficient does not significantly differ from zero unless there is subject area theory stating that the relationship between the response and predictor lacks a linear component.

The regression coefficients of the x^2 term are difficult to interpret. An interpretation should be done with the coefficients of the orthogonal polynomials shown in Table 9.5, not the simple polynomials of Table 9.4. See Section 10.4 for further discussion.

Table 9.4 Quadratic regression of hardness data. The quadratic term, with $p=.0027$, is very important in explaining the curvature of the observations. See Figure 9.5 to compare this fit with the linear fit. Compare the regression coefficients here with the regression coefficients in Table 9.5 where we use the orthogonal quadratic polynomial, rather than the simple square, for the quadratic regressor.

```
> data(hardness)

> hardness.lin.lm  <- lm(hardness ~ density,
+                        data=hardness)

> anova(hardness.lin.lm)
Analysis of Variance Table

Response: hardness
          Df   Sum Sq  Mean Sq F value Pr(>F)
density    1 21345674 21345674     637 <2e-16 ***
Residuals 34  1139366    33511
---
Signif. codes:  0 '***' 0.001 '**' 0.01 '*' 0.05 '.' 0.1 ' ' 1

> hardness.quad.lm <- lm(hardness ~ density + I(density^2),
+                        data=hardness)

> anova(hardness.quad.lm)
Analysis of Variance Table

Response: hardness
             Df   Sum Sq  Mean Sq F value Pr(>F)
density       1 21345674 21345674   815.9 <2e-16 ***
I(density^2)  1   276041   276041    10.6 0.0027 **
Residuals    33   863325    26161
---
Signif. codes:  0 '***' 0.001 '**' 0.01 '*' 0.05 '.' 0.1 ' ' 1

> coef(summary.lm(hardness.quad.lm))
             Estimate Std. Error t value Pr(>|t|)
(Intercept) -118.0074   334.9669 -0.3523 0.726857
density        9.4340    14.9356  0.6316 0.531970
I(density^2)   0.5091     0.1567  3.2483 0.002669
```

Table 9.5 Quadratic regression of hardness data with orthogonal polynomials. The quadratic term, with $p=.0027$, is very important in explaining the curvature of the observations. See Figure 9.5 to compare this fit with the linear fit. In this fit with the orthogonal polynomial for the quadratic term, the regression coefficient for the linear term is identical to the regression coefficient in the simple linear regression. Compare to the very different regression coefficients in Table 9.4. The ANOVA tables are identical.

```
> data(hardness)

> hardness.lin.lm <- lm(hardness ~ density,
+                       data=hardness)

> anova(hardness.lin.lm)
Analysis of Variance Table

Response: hardness
          Df   Sum Sq  Mean Sq F value Pr(>F)
density    1 21345674 21345674     637 <2e-16 ***
Residuals 34  1139366    33511
---
Signif. codes:  0 '***' 0.001 '**' 0.01 '*' 0.05 '.' 0.1 ' ' 1

> coef(summary.lm(hardness.lin.lm))
            Estimate Std. Error t value  Pr(>|t|)
(Intercept) -1160.50    108.580  -10.69 2.066e-12
density        57.51      2.279   25.24 1.333e-23

> h2 <- data.frame(density=hardness$density, poly(hardness$density, 2))

> xyplot(X1 + X2 ~ density, data=h2)  ## graph not shown in book

> hardness.quad.orth.lm <- lm(hardness ~ density + h2$X2,
+                             data=hardness)

> anova(hardness.quad.orth.lm)
Analysis of Variance Table

Response: hardness
          Df   Sum Sq  Mean Sq F value Pr(>F)
density    1 21345674 21345674   815.9 <2e-16 ***
h2$X2      1   276041   276041    10.6 0.0027 **
Residuals 33   863325    26161
---
Signif. codes:  0 '***' 0.001 '**' 0.01 '*' 0.05 '.' 0.1 ' ' 1

> coef(summary.lm(hardness.quad.orth.lm))
            Estimate Std. Error t value  Pr(>|t|)
(Intercept) -1160.50     95.937 -12.096 1.125e-13
density        57.51      2.013  28.564 7.528e-25
h2$X2         525.40    161.745   3.248 2.669e-03
```

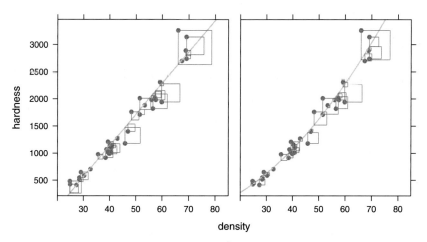

Fig. 9.5 Linear $y \sim x$ and quadratic $y \sim x + x^2$ fits of y=hardness to x=density. The quadratic curve fits much better as can be seen from the much smaller squares (leading to smaller residual sum of squares) at the left and right ends of the `density` range in the quadratic fit. See Table 9.4 for the numerical comparison.

9.8 Models Without a Constant Term

Sometimes it is desired that the statistical model for a response not contain a constant (i.e., vertical intercept) term because the response is necessarily equal to zero if all predictors are zero. An example is the modeling of the body fat data discussed in Section 9.1. Obviously, if a "subject" has zero measurements for `abdomin` and `biceps`, then the response `bodyfat` is necessarily zero also. Similarly, if we wish to model the volume of trees in a forest as a function of trees' diameters and heights, a "tree" having zero diameter and height must have no volume.

An advantage to explicitly recognizing the zero intercept constraint is that a degree of freedom that would be used to estimate the intercept is instead used to estimate the model residual. This results in slightly increased power of tests and decreased sizes of interval estimates of model parameters.

Figure 9.6 and Table 9.6 are for regressions of `bodyfat` on `biceps`, both with and without a constraint that the regression pass through the origin. Note the appreciably smaller slope of the no-intercept regression and that the no-intercept model has 46 df for residual as compared with 45 df for the unconstrained model.

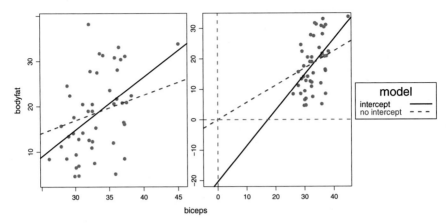

Fig. 9.6 Regressions with and without a constant term for a portion of the body fat data. See Table 9.6. The left panel is limited to the range of the data. The right panel extends the range to include both intercepts. The dotted line through the origin at (0,0) makes an unwarranted extrapolation outside the range of the data.

Table 9.6 Body fat data: Regressions of bodyfat on biceps, with an intercept term (here) and without an intercept term (in Table 9.7). See Figure 9.6. As compared with the intercept model, the no-intercept model has larger values of both the regression sum of squares and the total sum of squares, and hence also a larger value of R^2.

```
> data(fat)

> ## usual model with intercept
> xy.int.lm <- lm(bodyfat ~ biceps, data=fat)

> summary(xy.int.lm)

Call:
lm(formula = bodyfat ~ biceps, data = fat)

Residuals:
    Min      1Q  Median      3Q     Max
-16.580  -5.443  -0.846   5.255  21.088

Coefficients:
            Estimate Std. Error t value Pr(>|t|)
(Intercept)  -20.364     10.855   -1.88  0.06715 .
biceps         1.171      0.326    3.59  0.00081 ***
---
Signif. codes:  0 '***' 0.001 '**' 0.01 '*' 0.05 '.' 0.1 ' ' 1

Residual standard error: 8.01 on 45 degrees of freedom
Multiple R-squared:  0.223,Adjusted R-squared:  0.206
F-statistic: 12.9 on 1 and 45 DF,  p-value: 0.00081


> anova(xy.int.lm)
Analysis of Variance Table

Response: bodyfat
          Df Sum Sq Mean Sq F value  Pr(>F)
biceps     1    827     827    12.9 0.00081 ***
Residuals 45   2884      64
---
Signif. codes:  0 '***' 0.001 '**' 0.01 '*' 0.05 '.' 0.1 ' ' 1
```

Table 9.7 Body fat data: Regressions of bodyfat on biceps, without an intercept term. See Table 9.6 for the model with an intercept term. See Figure 9.6. R uses the notation − 1 in the formula to indicate that the column of **1** is to be suppressed from the dummy variable matrix. As compared with the intercept model, the no-intercept model has larger values of both the regression sum of squares and the total sum of squares, and hence also a larger value of R^2. The no-intercept model has a very high regression sum of squares and corresponding F-value because it includes the contribution from the constant term.

```
> data(fat)

> ## model without a constant term
> xy.noint.lm <- lm(bodyfat ~ biceps - 1, data=fat)

> summary(xy.noint.lm)

Call:
lm(formula = bodyfat ~ biceps - 1, data = fat)

Residuals:
    Min     1Q  Median      3Q     Max
-15.110  -6.145  -0.006   6.841  20.185

Coefficients:
       Estimate Std. Error t value Pr(>|t|)
biceps    0.563      0.036    15.6   <2e-16 ***
---
Signif. codes:  0 '***' 0.001 '**' 0.01 '*' 0.05 '.' 0.1 ' ' 1

Residual standard error: 8.22 on 46 degrees of freedom
Multiple R-squared:  0.841, Adjusted R-squared:  0.838
F-statistic:  244 on 1 and 46 DF,  p-value: <2e-16


> anova(xy.noint.lm)
Analysis of Variance Table

Response: bodyfat
          Df Sum Sq Mean Sq F value Pr(>F)
biceps     1  16506   16506     244 <2e-16 ***
Residuals 46   3110      68
---
Signif. codes:  0 '***' 0.001 '**' 0.01 '*' 0.05 '.' 0.1 ' ' 1
```

9.9 Prediction

Generalizing the discussion in Section 8.3.5 for simple regression, the multiple regression model equation, with regression coefficients estimated by the least-squares analysis, is commonly used for two distinct but related problems.

1. Find a confidence interval on the conditional mean of the population of $Y|x$. That is, estimate a range of mean $E(Y|x)$-values that (with high confidence) bracket the true mean for the specified values of the predictors x.

2. Find a prediction interval for a new observed response Y_0 from these values of the predictors x; i.e., an interval within which a particular new observation will fall with a certain probability.

We continue the analysis of data(fat) to illustrate the distinction between these two problems. Using R we continue with fat2.lm displayed in Table 9.1. For specificity, we work with $x_1 = $ abdomin $= 93$ and $x_2 = $ biceps $= 33$.

The algebraic setup begins from the model in Equation (9.1), from which it follows that

$$s_e^2 = \frac{Y'Y - \hat{\beta}'X'Y}{n - p - 1} = \frac{(Y - \hat{Y})'(Y - \hat{Y})}{n - p - 1}$$

Let $x_0 = (93\ 33)$ denote the vector of predictor values for which we wish to construct these two intervals. Define

$$h_0 = x_0(X'X)^{-1}x_0' \tag{9.23}$$

Let $t_{\frac{\alpha}{2}, n-p-1}$ denote the $100(1 - \frac{\alpha}{2})$ percentage point of the t distribution with $n - p - 1$ degrees of freedom. The expected response $E(y|x_0)$ (the center of the confidence interval) and the predicted response $\hat{y}_{x_0}$ for a new observation (the center of the prediction interval) are both equal to $x_0'\hat{\beta}$. Then the $100(1 - \alpha)\%$ confidence interval is

$$x_0'\hat{\beta} \pm t_{\frac{\alpha}{2}, n-p-1}\, s_e\, \sqrt{h_0} \tag{9.24}$$

and the $100(1 - \alpha)\%$ prediction interval is

$$x_0'\hat{\beta} \pm t_{\frac{\alpha}{2}, n-p-1}\, s_e\, \sqrt{1 + h_0} \tag{9.25}$$

The prediction interval is wider than the confidence interval because we are predicting one particular y corresponding to x_0, but estimating with confidence the mean $E(y|x_0)$ of all possible y's that could arise from x_0. A particular y could be much smaller or larger than the mean, and hence there is more uncertainty about y than about the mean. This is captured in the distinction between the two preceding formulas: the "1+" inside the square root. The "1+" arises from the fact that we must predict the ϵ_0 part of the model, but in the estimation problem, we estimate that ϵ_0

Table 9.8 95% Confidence and prediction intervals for the body-fat example. See Tables 9.1 and 13.27 for the ANOVA table and the regression coefficients. The `predict` function produces $s_e \sqrt{h_0}$=se.fit, s_e=residual.scale and the confidence and prediction intervals.

```
> fat2.lm <- lm(bodyfat ~ abdomin + biceps, data=fat)

> pi.fit <- predict(fat2.lm,
+                        newdata=data.frame(abdomin=93:94, biceps=33:34),
+                        se.fit=TRUE, interval="prediction")

> ci.fit <- predict(fat2.lm,
+                        newdata=data.frame(abdomin=93:94,
+                        biceps=33:34),
+                        se.fit=TRUE, interval="confidence")

> pi.fit
$fit
     fit   lwr   upr
1 18.49 8.485 28.49
2 18.25 8.236 28.26

$se.fit
     1      2
0.7171 0.7518

$df
[1] 44

$residual.scale
[1] 4.911


> ci.fit$fit
     fit   lwr   upr
1 18.49 17.04 19.93
2 18.25 16.73 19.76
```

is zero. As a result, the prediction interval for a given set of explanatory variables is always wider than the corresponding confidence interval.

The confidence and prediction intervals for this example are shown in Table 9.8. The confidence interval (17.0, 19.9) is for the mean percentage body fat of a population of individuals each having abdomin circumference 93 cm and biceps circumference 33 cm. The prediction interval (8.5, 28.5) is for one particular individual with this combination of abdomin and biceps. Observe that the prediction interval is wider than the confidence interval. This is because a single person can have atypically low or high body fat, but "many" people includes those with both atypically low and high body-fat percentages in comparison to their abdomin and biceps, and the lows and highs tend to cancel out when averaging. See Table 8.8 for an illustration of this in the more familiar setting of estimation of a sample mean.

9.10 Example—Longley Data

9.10.1 Study Objectives

The Longley data is a classic small set containing 16 years of annual macroeconomic data that Longley (1967) used to illustrate difficulties arising in computations involving highly intercorrelated variables. R does accurately calculate the regression coefficients for these data. Less numerically sophisticated statistical software packages, including most in existence at the time Longley wrote his article, produce incorrect analyses because the high intercorrelation, or ill-conditioning of the data, is a computational challenge for the numerical solution of linear equations and related matrix operations. Please see the computational discussion in Section I.4.7 for details.

We use data(longley), distributed with R, a subset of all variables in Longley's original data set. Our intent here is to develop a parsimonious model to explain the response variable Employed as a function of the remaining variables as candidate predictors. The extreme collinearity arises in this data set because all of its economic variables tend to increase as time progresses. We acknowledge that these are really time series data, and if more than 16 years were involved, it would be appropriate to use time series techniques such as those in Chapter 18 for a proper analysis. We use this example because it is now a classical dataset for investigating a set of poorly conditioned linear equations. Our intention in this section is to analyze these data using multiple regression, demonstrating ways to bypass or confront the difficulties collinearity presents for regression modeling. In contrast, time series analyses specifically seek to model the interdependence caused by time.

9.10.2 Data Description

GNP.deflator: GNP adjusted for inflation based on year 1954 = 100

GNP: Gross National Product, 1964 Economic Report of the President

Unemployed: 1964 Economic Report of the President

Armed.Forces: Number serving in the U.S. Armed Forces

Population: Noninstitutional, aged at least 14

Year: 1947 through 1962

Employed: Total employment, U.S. Department of Labor, March 1963

9.10.3 Discussion

Figure 9.7 contains a scatterplot matrix of the Longley data. Here the response variable Employed appears in the top (last) row and last column. (In general, for ease of interpretation, response variables should appear in this way or in the bottom (first) row and first column. Remember from Section 4.7 and Figure 4.12 that we strongly recommend that sploms have the main diagonal in the SW–NE direction.)

We see that Employed is highly positively correlated with four of the six predictors and mildly positively correlated with the others. In addition, the predictors (including Year) that are highly correlated with Employed are also highly correlated with one another. This suggests that these four predictors carry redundant information and therefore some of them are unnecessary for modeling the response.

Consider the listing in Table 9.9 for a model containing all six candidate predictors. The proportion of variability in the response Employed that is collectively explained by all six predictors is given by R^2, the proportion of the Sum of Squares

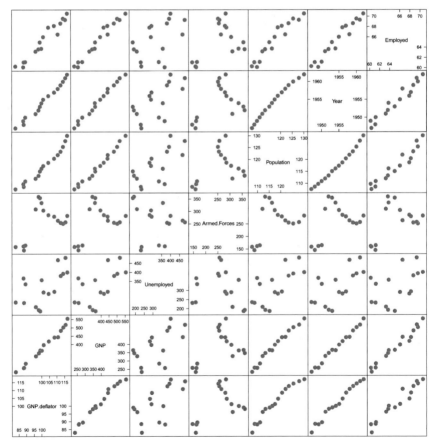

Fig. 9.7 Longley data splom. Notice the high positive correlations of four predictors (including Year) with one another and with the response variable Employed.

Table 9.9 Longley data regression using all six original predictors.

```
> longley.lm <- lm( Employed ~ . , data=longley)

> summary(longley.lm)

Call:
lm(formula = Employed ~ ., data = longley)

Residuals:
    Min      1Q  Median      3Q     Max
-0.4101 -0.1577 -0.0282  0.1016  0.4554

Coefficients:
               Estimate Std. Error t value Pr(>|t|)
(Intercept)   -3.48e+03   8.90e+02   -3.91  0.00356 **
GNP.deflator   1.51e-02   8.49e-02    0.18  0.86314
GNP           -3.58e-02   3.35e-02   -1.07  0.31268
Unemployed    -2.02e-02   4.88e-03   -4.14  0.00254 **
Armed.Forces  -1.03e-02   2.14e-03   -4.82  0.00094 ***
Population    -5.11e-02   2.26e-01   -0.23  0.82621
Year           1.83e+00   4.55e-01    4.02  0.00304 **
---
Signif. codes:  0 '***' 0.001 '**' 0.01 '*' 0.05 '.' 0.1 ' ' 1

Residual standard error: 0.305 on 9 degrees of freedom
Multiple R-squared:  0.995,Adjusted R-squared:  0.992
F-statistic:  330 on 6 and 9 DF,  p-value: 4.98e-10


> anova(longley.lm)
Analysis of Variance Table

Response: Employed
             Df Sum Sq Mean Sq F value  Pr(>F)
GNP.deflator  1  174.4   174.4 1876.53 9.3e-12 ***
GNP           1    4.8     4.8   51.51 5.2e-05 ***
Unemployed    1    2.3     2.3   24.36 0.00081 ***
Armed.Forces  1    0.9     0.9    9.43 0.01334 *
Population    1    0.3     0.3    3.75 0.08476 .
Year          1    1.5     1.5   16.13 0.00304 **
Residuals     9    0.8     0.1
---
Signif. codes:  0 '***' 0.001 '**' 0.01 '*' 0.05 '.' 0.1 ' ' 1

> vif(longley.lm)
GNP.deflator          GNP    Unemployed Armed.Forces
     135.532     1788.513        33.619        3.589
  Population         Year
     399.151      758.981
```

column *not* in the Residuals row: more than 0.99. So the predictors can be used to adequately explain Employed. In this model, three predictors that seem to be closely correlated with the response Employed in Figure 9.7, Population, GNP, and GNP.deflator, are not statistically significant in Table 9.9. We continue to discuss the Longley data, focusing on the selection of an appropriate subset of the predictors, in Sections 9.11 and 9.12.

9.11 Collinearity

Collinearity, also called multicollinearity, is a condition where the model's predictors variables are highly intercorrelated. A consequence of this situation is the inability to estimate the model's regression coefficients with acceptable precision. Therefore, models with this problem are not considered useful. It is unacceptable to reach a final model that has this condition to an appreciable extent.

Collinearity arises when investigators include predictors carrying redundant information in the model. A symptom is a model with a high R^2, showing that collectively the predictors bear heavily on the response, but paradoxically, few or none of the predictors have regression coefficients significantly different from zero.

Consider the case of a single response Y and two predictors X_1 and X_2. The fitted model plots as a plane in the 3-dimensional space of (Y, X_1, X_2). A near-collinear situation exists if the correlation between X_1 and X_2 is close to ± 1. Geometrically, this occurs when the data points congregate close to a (2-dimensional) straight line when plotted in the 3-dimensional space. When this happens, the points can be fitted fairly well by any plane containing this straight line. Since each of these many planes is a candidate for the best model, the model decided upon as being *the* best will be similar to other model candidates. Therefore, declaring any model to be best will be a tentative decision. This tentativeness is expressed by large standard errors of the estimated regression coefficients that comprise the coefficients of the plane corresponding to the best model.

Figure 9.8, based on a portion of the Longley data introduced in Section 9.10, illustrates these ideas. Here the variables GNP and Year are almost perfectly correlated and so the scattering of points falls close to a line in 3-dimensional space. Many planes fit this line approximately equally well. The uncertainty about the best fitting of these many planes causes the coefficients of the estimated plane, the regression coefficients, to have large standard errors.

When there are more than two predictors, the geometric argument extends to discussions of hyperplanes. The consequence is again unacceptably large standard errors of regression coefficients.

Although collinearity limits our ability to model the relationship between the predictors and the response accurately, it does not necessarily impede our ability to use the predictors to predict the response. In the context of the example associated

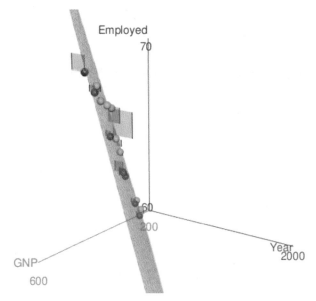

Fig. 9.8 The two *X*-variables, Year and GNP, are highly collinear. See model
        longley2.lm <- lm(Employed ~ Year + GNP, data=longley)
in file HHscriptnames(9). The response variable Employed is essentially on a straight line in the
three-dimensional space of the figure. The specific plane displayed is almost arbitrary. Any plane
that goes through the straight line of the observed points on the plane we see would work just as
well.

with Figure 9.8, if we want to predict the response for values of the predictors near
the straight line in 3-dimensional space, many planes that are good fits to this straight
line will yield roughly the same prediction.

A simple diagnostic of collinearity is the *variance inflation factor*, VIF, one
for each regression coefficient (other than the intercept). Since the condition of
collinearity involves the predictors but not the response, this measure is a function
of the *X*'s but not of *Y*. The VIF for predictor i is

$$\text{VIF}_i = 1/(1 - R_i^2) \tag{9.26}$$

where R_i^2 is the R^2 from a regression of predictor i against the remaining predictors.
If R_i^2 is close to 1, this means that predictor i is well explained by a linear function
of the remaining predictors, and, therefore, the presence of predictor i in the model
is redundant. Values of VIF exceeding 5 are considered evidence of collinearity:
The information carried by a predictor having such a VIF is contained in a subset of
the remaining predictors. If, however, all of a model's regression coefficients differ
significantly from 0 (p-value $< .05$), a somewhat larger VIF may be tolerable.

VIF is an imperfect measure of collinearity. Occasionally the condition can be
attributable to more complicated relationships among the predictors than VIF can
detect.

The best approach for alleviating collinearity is to reduce the set of predictors to a noncollinear subset. Methods for accomplishing this are presented in Section 9.12. An ad hoc (manual) procedure, presented in Section 9.12.1, involves eliminating predictors one at a time, at each stage deleting the predictor having the highest VIF. If two predictors are almost tied for highest, then subject area information should be used to choose between them. Proceed until all remaining predictors have VIF $\leq$ 5. Other approaches (not discussed in this book) include ridge regression and regression on principal components Gunst and Mason (1980).

For the regression analysis of the Longley data, evidence of collinearity appears in Table 9.9 in the variance inflation factors (VIF) for the six predictors. Five of these exceed 33. The next section discusses an approach for dealing with multicollinearity.

Collinearity often arises in polynomial regression models discussed in Section 9.7 because polynomials can be approximated by linear functions within a restricted domain. To avoid both collinearity in polynomial models and numerical instability caused by working with variables of greatly differing orders of magnitude, it is recommended to recenter the response variable to have mean = 0 prior to initiating a polynomial modeling.

9.12 Variable Selection

In building a regression model the analyst should consider for use any explanatory variable that is likely to bear upon the response while avoiding the use of two explanatory variables that carry essentially the same information. For example, in modeling the monthly cost of energy needed to heat a 2000-square-foot home, one should avoid using both the mean monthly exterior temperature and the heating degree days (a measure used by heating fuel suppliers) in the same model. The use of redundant explanatory variables is likely to lead to a model with unacceptable collinearity having large standard errors for the predictor regression coefficients.

When subject area theory does not suggest a parsimonious model (i.e., one with relatively few predictors), it is tempting to construct a model using all possibly relevant predictors for which data are available. However, doing so is again likely to result in a collinearity problem. In such circumstances, how can the analyst decide on an appropriate subset of the candidate predictors for a regression model?

Stepwise regression is a tool for answering this question. But this mechanical technique should not be used in order to avoid careful thought about potentially useful predictor variables. Careless use of stepwise regression can, to some extent, distort the significance and confidence levels of inferences in the ultimately specified model, potentially leading to erroneous conclusions. In addition, a model that makes reasonable subject area sense to the client is much preferred to an equally well fitting one that is less intuitive and harder to understand and explain.

In our experience, a careful systematic approach can often be used to develop a more interpretable model than one produced by a mechanical stepwise algorithm. The starting point is a scatterplot matrix that, along with examination of variance inflation factors, can be used to identify redundant predictors. If two predictors are seen to be highly correlated, we prefer to avoid using the one that has a less obvious subject matter connection to the response variable. An algorithm cannot make such a judgment. Inspection of sploms invite the analyst to consider whether an original variable should be transformed before inclusion in the model. Nevertheless, stepwise approaches to model selection continue to be commonly used, particularly when there are a large number of potential predictors and the analyst has minimal feel for which variables should be or need not be included in the model.

We discuss in turn two systematic methods for model selection, a manual approach and an automated approach, and apply both methods to the Longley data.

9.12.1 Manual Use of the Stepwise Philosophy

The first approach involves manual inspections of the VIFs, the p-values associated with the t-tests on the regression coefficients, and any available subject matter information to eliminate variables one at a time until a final model is reached with all predictors significant and all VIFs under 5. This approach is viable if the number of predictors is small as in this example. It would be too cumbersome in a situation with more than 12 to 15 predictors.

The three largest VIFs belong to GNP, Year, and Population. The splom implies that they carry almost identical information. We begin by removing one of them from the model. We choose to eliminate Population because the t-test that its regression coefficient is zero has a larger p-value than the tests for either GNP or Year.

The analysis with all predictors except population appears in Table 9.10.

The outstanding feature of this model is the high p-value associated with variable GNP.deflator. Its VIF is well in excess of 5. We proceed with an analysis eliminating GNP.deflator in Table 9.11.

All four predictors in this model have significant regression coefficients. However, two of the VIFs are still large, and one of the predictors corresponding to them must be eliminated. We choose to eliminate GNP because its p-value, while small, is larger than those of the three other remaining predictors.

The results of the analysis with the remaining predictors Unemployed, Year, and Armed.Forces are in Table 9.12.

This is our tentative final model. The collinearity has been eliminated (all VIFs are below 5), and all regression coefficients differ significantly from zero. In addition, $R^2 = 0.993$, so these three predictors account for virtually all of the variability in Employed.

Table 9.10 Longley data regression. Best five-predictor model after eliminating one predictor using the manual stepwise approach.

```
> longley3.lm <- lm( Employed ~
+            GNP.deflator + GNP + Unemployed + Armed.Forces + Year,
+            data=longley)

> summary(longley3.lm)

Call:
lm(formula = Employed ~ GNP.deflator + GNP + Unemployed +
    Armed.Forces + Year, data = longley)

Residuals:
    Min      1Q  Median      3Q     Max
-0.3901 -0.1434 -0.0356  0.0973  0.4614

Coefficients:
               Estimate Std. Error t value Pr(>|t|)
(Intercept)  -3.56e+03   7.72e+02   -4.62  0.00096 ***
GNP.deflator  2.77e-02   6.07e-02    0.46  0.65798
GNP          -4.21e-02   1.76e-02   -2.39  0.03789 *
Unemployed   -2.10e-02   3.03e-03   -6.95     4e-05 ***
Armed.Forces -1.04e-02   2.00e-03   -5.21  0.00040 ***
Year          1.87e+00   3.99e-01    4.68  0.00087 ***
---
Signif. codes:  0 '***' 0.001 '**' 0.01 '*' 0.05 '.' 0.1 ' ' 1

Residual standard error: 0.29 on 10 degrees of freedom
Multiple R-squared:  0.995, Adjusted R-squared:  0.993
F-statistic:  438 on 5 and 10 DF,  p-value: 2.27e-11


> anova(longley3.lm)
Analysis of Variance Table

Response: Employed
             Df Sum Sq Mean Sq F value   Pr(>F)
GNP.deflator  1  174.4   174.4 2073.3 6.3e-13 ***
GNP           1    4.8     4.8   56.9 2.0e-05 ***
Unemployed    1    2.3     2.3   26.9 0.00041 ***
Armed.Forces  1    0.9     0.9   10.4 0.00905 **
Year          1    1.8     1.8   21.9 0.00087 ***
Residuals    10    0.8     0.1
---
Signif. codes:  0 '***' 0.001 '**' 0.01 '*' 0.05 '.' 0.1 ' ' 1

> vif(longley3.lm)
GNP.deflator          GNP  Unemployed Armed.Forces         Year
      76.641      546.870      14.290        3.461      644.626
```

Table 9.11 Longley data regression. Best four-predictor model after eliminating two predictors using the manual stepwise approach.

```
> longley4.lm <- lm(Employed ~
+                   GNP + Unemployed + Armed.Forces + Year,
+                   data=longley)

> summary(longley4.lm)

Call:
lm(formula = Employed ~ GNP + Unemployed + Armed.Forces + Year,
    data = longley)

Residuals:
    Min      1Q   Median       3Q      Max
-0.4217  -0.1246  -0.0242   0.0837   0.4527

Coefficients:
               Estimate Std. Error t value Pr(>|t|)
(Intercept)   -3.60e+03   7.41e+02   -4.86  0.00050 ***
GNP           -4.02e-02   1.65e-02   -2.44  0.03283 *
Unemployed    -2.09e-02   2.90e-03   -7.20  1.7e-05 ***
Armed.Forces  -1.01e-02   1.84e-03   -5.52  0.00018 ***
Year           1.89e+00   3.83e-01    4.93  0.00045 ***
---
Signif. codes:  0 '***' 0.001 '**' 0.01 '*' 0.05 '.' 0.1 ' ' 1

Residual standard error: 0.279 on 11 degrees of freedom
Multiple R-squared:  0.995,Adjusted R-squared:  0.994
F-statistic:  590 on 4 and 11 DF,  p-value: 9.5e-13


> anova(longley4.lm)
Analysis of Variance Table

Response: Employed
             Df Sum Sq Mean Sq F value  Pr(>F)
GNP           1  179.0   179.0  2292.7    4e-14 ***
Unemployed    1    2.5     2.5    31.5  0.00016 ***
Armed.Forces  1    0.8     0.8    10.5  0.00779 **
Year          1    1.9     1.9    24.3  0.00045 ***
Residuals    11    0.9     0.1
---
Signif. codes:  0 '***' 0.001 '**' 0.01 '*' 0.05 '.' 0.1 ' ' 1

> vif(longley4.lm)
      GNP   Unemployed Armed.Forces         Year
  515.124       14.109        3.142      638.128
```

Table 9.12 Longley data regression. Best three-predictor model after eliminating three predictors using the manual stepwise approach.

```
> longley5.lm <- lm(Employed ~
+                   Unemployed + Armed.Forces + Year,
+                   data=longley)

> summary(longley5.lm)

Call:
lm(formula = Employed ~ Unemployed + Armed.Forces + Year,
   data = longley)

Residuals:
    Min      1Q   Median      3Q      Max
-0.5729 -0.1199  0.0409  0.1398  0.7530

Coefficients:
              Estimate Std. Error t value Pr(>|t|)
(Intercept)  -1.80e+03   6.86e+01  -26.18  5.9e-12 ***
Unemployed   -1.47e-02   1.67e-03   -8.79  1.4e-06 ***
Armed.Forces -7.72e-03   1.84e-03   -4.20   0.0012 **
Year          9.56e-01   3.55e-02   26.92  4.2e-12 ***
---
Signif. codes:  0 '***' 0.001 '**' 0.01 '*' 0.05 '.' 0.1 ' ' 1

Residual standard error: 0.332 on 12 degrees of freedom
Multiple R-squared:  0.993,Adjusted R-squared:  0.991
F-statistic:  555 on 3 and 12 DF,  p-value: 3.92e-13


> anova(longley5.lm)
Analysis of Variance Table

Response: Employed
             Df Sum Sq Mean Sq F value  Pr(>F)
Unemployed    1   46.7    46.7     424 1.0e-10 ***
Armed.Forces  1   57.0    57.0     517 3.1e-11 ***
Year          1   79.9    79.9     725 4.2e-12 ***
Residuals    12    1.3     0.1
---
Signif. codes:  0 '***' 0.001 '**' 0.01 '*' 0.05 '.' 0.1 ' ' 1

> vif(longley5.lm)
  Unemployed Armed.Forces         Year
       3.318        2.223        3.891
```

9.12.2 Automated Stepwise Regression

The second approach to model selection is stepwise regression. This automated approach is recommended when the number of predictors is so large that the manual approach becomes unacceptably laborious. We illustrate here how it is used to reach the same model that we found with the manual procedure. A stepwise approach that examines all subsets of predictors is viable if the number of predictors p is less than 10 to 12. If $p > 12$, then forward selection or backward elimination is preferred.

The three basic methods for automated stepwise regression are

forward selection: Predictors are added to the model one at a time until a stopping rule is satisfied.

backward elimination: All predictors are initially placed in the model. Predictors are removed from the model one at a time until a stopping rule is satisfied.

all subsets: All $2^p - 1$ possible models, where p is the number of predictors, are attempted and the best is identified. This method is viable only for "small" values of p. Efficient algorithms exist that avoid actually examining every such model.

The literature contains many hybrids and refinements of these basic methods.

Each of the automated stepwise methods uses a criterion for choosing the next step or stopping the algorithm.

Such criteria may relate to appreciable R^2_{adj} or F-statistic improvement or detriment, substantial mean square error decrease or increase, or size of change in Daniel–Mallows' C_p statistic discussed below. Another possibility is to look, at each step, at the p-value for the variables already in the model and for the potential next variable to be brought in to the model. If the largest p-value of the variables already in the model is larger than the threshold, then remove it. If the smallest p-value of the potential variables is larger than the threshold, then stop. Otherwise, bring in a new variable and repeat the process.

Computer algorithms allow the option of accepting or overriding default criterion values or thresholds for appreciable change.

Each of the automated stepwise methods uses one or more criteria for choosing among competing models. Here is a list of possible criteria.

p Models containing fewer predictors are easier to interpret and understand. It is desirable that the number of predictors p be as small as possible.

$\hat{\sigma}^2$ We also require that the predictors account for most of the variability in the response. Equivalently, we wish that the residual mean square, MSE $= \hat{\sigma}^2$, be as small as possible, preferably not much larger than for the model containing all candidate predictors. This criterion is easier to meet with more predictors rather than few; hence it asks that the number of predictors p be as large as possible and competes with the goal of minimizing p.

The above criteria address one of the two competing objectives at a time. Other criteria jointly address the two objectives.

R_{adj}^2 Unadjusted R^2 is not used as a model selection criterion because it necessarily increases as the number of predictors increases. A model can have R^2 close to 1 but be unacceptable due to severe collinearity. Instead we use R_{adj}^2, which is R^2 adjusted downward for the number of predictors,

$$R_{adj}^2 = 1 - \left(\frac{n-1}{n-p-1} \right)(1 - R^2) \tag{9.27}$$

which increases as R^2 increases but provides a penalty for an excessive number of predictors p. Models with higher R_{adj}^2 are preferred to ones with lower R_{adj}^2.

C_p Daniel–Mallows' C_p statistic is another criterion that addresses both the fit of the model and the number of predictors used. Consistent with customary notation, in the context of the C_p statistic but nowhere else in this chapter, p is the number of regression coefficient *parameters*, equal to the number of predictors *plus 1*. The original definition is

$$C_p = (\text{SS}_{\text{Res}}/\hat{\sigma}_{\text{full}}^2) + 2p - n \tag{9.28}$$

where SS_{Res} is the residual sum of squares for the reduced model under discussion (fewer X-variables than the full model) and the $\hat{\sigma}_{\text{full}}^2$ is the error mean square for the full model containing all candidate predictors. If the extra X-variables are noise, rather than useful, then the ratio $\text{SS}_{\text{Res}}/\hat{\sigma}_{\text{full}}^2 \approx ((n-p)\sigma^2)/\sigma_{\text{full}}^2 \approx n - p$. If the extra X-variables are useful, then the numerator $\sigma^2 \gg \sigma_{\text{full}}^2$ and the ratio will be much larger than $n - p$. The extra terms $2p - n$ make the entire C_p approximate p when the extra X-variables are not needed.

A desirable model has $C_p \approx p$ for a small number of parameters p. (If p_{max} denotes p for a model containing all candidate predictors, then necessarily $C_{p_{\text{max}}} = p_{\text{max}}$, but such a model is almost never acceptable.) C_p results are often conveyed with a C_p plot, that is, a plot of C_p vs p, with each point labeled with an identifier for its model and the diagonal line having equation $C_p = p$ added to the plot. Desirable models are those close to or under this diagonal line.

AIC The Akaike information criterion is proportional to the C_p statistic. The AIC is scaled in sum of squares units.

F At each step we can look at the p-value associated with the F-statistic for the variables already in the model and for the potential next variable to be brought in to the model. If the largest p-value of the variables already in the model is larger than the threshold, then remove it. If the smallest p-value of the potential variables is larger than the threshold, then stop. Otherwise, bring in a new variable and repeat the process.

9.12.3 Automated Stepwise Modeling of the Longley Data

Table 9.13 contains the results of an R stepwise regression analysis considering all subsets of the predictors, with printouts of the properties of two models of each size having smallest residual sum of squares among models having $C_p < 10$. Figure 9.9 is a plot of the C_p-values for all models with $C_p < 10$. The acronymic plot symbols in Figure 9.9 are decoded in Table 9.13. According to Table 9.13, the best parsimonious model is the one with the four predictors GNP, Unemployed, Armed.Forces, and Year displayed in Table 9.11. This model has C_p close to p, and a smaller AIC and larger adjusted R^2 than any of the other models in Table 9.13. Unlike the model we selected with our manual approach, this one includes the predictor GNP. The algorithm underlying Table 9.13 suggests inclusion of GNP despite its high correlation with Year and high VIF shown in Table 9.11.

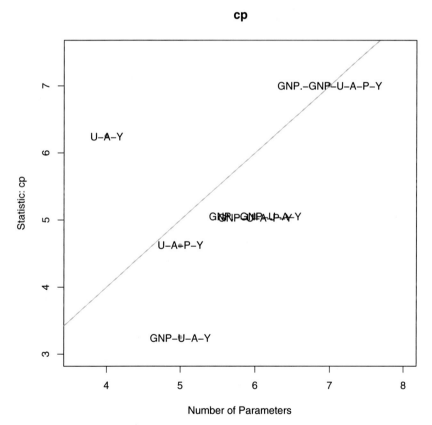

Fig. 9.9 C_p Plot for Longley data. See Table 9.13 for interpretations of the acronyms used to label points. The overplotting occurs because, as seen in Table 9.13, two models have almost identical values of C_p.

Table 9.13 Longley data regression. Model 7 with the four predictors GNP, Unemployed, Armed.Forces and Year is competitive with respect to C_p and other criteria. This model has the largest adjusted R^2 and the smallest C_p. This is the same model we found in Table 9.11.

```
> longley.subsets <-
+    leaps::regsubsets(Employed ~ GNP.deflator + GNP +
+                      Unemployed +
+                      Armed.Forces + Population + Year,
+                      data=longley, nbest=2)

> longley.subsets.Summary <- summaryHH(longley.subsets)

> ## longley.subsets.Summary
> tmp <- (longley.subsets.Summary$cp <= 10)

> longley.subsets.Summary[tmp,]
                   model p   rsq   rss adjr2   cp   bic stderr
5              U-A-Y 4 0.993 1.323 0.991 6.24 -68.0  0.332
7          GNP-U-A-Y 5 0.995 0.859 0.994 3.24 -72.1  0.279
8            U-A-P-Y 5 0.995 0.986 0.993 4.61 -69.9  0.299
9        GNP-U-A-P-Y 6 0.995 0.839 0.993 5.03 -69.7  0.290
10   GNP.-GNP-U-A-Y 6 0.995 0.841 0.993 5.05 -69.7  0.290
11 GNP.-GNP-U-A-P-Y 7 0.995 0.836 0.992 7.00 -67.0  0.305

Model variables with abbreviations

                                                           model
GNP                                                          GNP
Y                                                          Year
U-Y                                            Unemployed-Year
GNP-U                                          GNP-Unemployed
U-A-Y                             Unemployed-Armed.Forces-Year
GNP-U-A                           GNP-Unemployed-Armed.Forces
GNP-U-A-Y                    GNP-Unemployed-Armed.Forces-Year
U-A-P-Y              Unemployed-Armed.Forces-Population-Year
GNP-U-A-P-Y          GNP-Unemployed-Armed.Forces-Population-Year
GNP.-GNP-U-A-Y        GNP.deflator-GNP-Unemployed-Armed.Forces-Year
GNP.-GNP-U-A-P-Y GNP.deflator-GNP-Unemployed-Armed.Forces-Population-Year

model with largest adjr2
7

Number of observations
16
```

Which model is preferred, the one in Table 9.11 containing four predictors including GNP or the three predictor model in Table 9.12 that excludes GNP? Our answer to this question demonstrates our preference for the manual approach. The coefficient of GNP in Table 9.11 is negative. This model says that holding Unemployed, Armed.Forces and Year constant, GNP and Employed are *negatively* associated. This statement conflicts with our expectation that this association is positive, and is a strong argument against the four-predictor model in Table 9.11.

9.13 Residual Plots

Partial residual plots and *added variable plots* are visual aids for interpreting relationships between variables used in regression. They can serve as additional components of our manual approach for variable selection.

Figure 9.10 shows four different types of plots.

- Row 1 shows the response variable Y=Employed against each of the six predictors X_j.

- Row 2 shows the ordinary residuals $e = Y - \hat{Y}$ from the regression on all six variables against each of the six predictors.

- Row 3 shows the "partial residual plots", the partial residuals e^j for each predictor against that predictor. See Section 9.13.1 for construction of the partial residuals and Section 9.13.2 for construction of the partial residual plots.

- Row 4 shows the "added variable plots", the partial residuals e^j against the partial residuals $X_{j|1,2,...,j-1,j+1,...,p}$ of X_j regressed on the other five predictors. See Section 9.13.3 for the definition of partial correlation, and Section 9.13.4 for construction of the $X_{j|1,2,...,j-1,j+1,...,p}$ and the added variable plots.

We discuss the interpretation of the all four types of plots in Section 9.13.5. We recommend the discussions of partial residual plots and added variable plots in Weisberg (1985) and Hamilton (1992).

9.13.1 Partial Residuals

The partial residuals e^j for variable X_j in a model with p predictor variables X_j are defined

$$e^j = Y - \hat{Y}_{1,2,...,j-1,j+1,...,p} \tag{9.29}$$

and calculated with

$$e^j = X_j \hat{\beta}_j + e \tag{9.30}$$

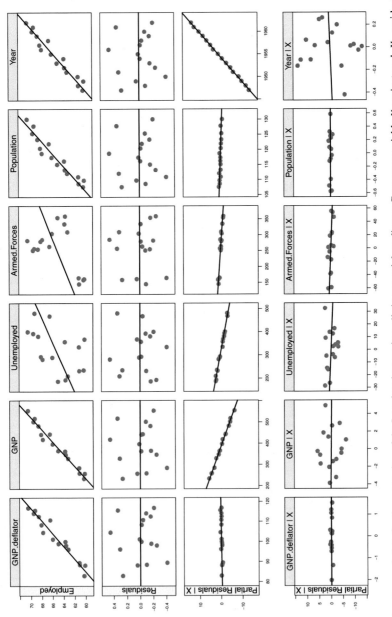

Fig. 9.10 Four types of plots for the regression of the Longley data against all six potential predictors: Response variable Y against each X_j, residuals e against each X_j, partial residuals plots of e^j against each X_j, added variable plots of e^j against the residuals of each X_j adjusted for the other X columns. The slopes shown in the panels of both bottom rows are equal to the regression coefficients from the regression shown in Table 9.9.

or equivalently

$$e_i^j = X_{ij}\hat{\beta}_j + e_i \quad \text{for } i = 1, \ldots, n \tag{9.31}$$

where $e = (e_i)$ are the ordinary residuals from the model with all p predictors

$$e = Y - \hat{Y}_{1,2,\ldots,p} \tag{9.32}$$

The partial residuals are interpreted as the additional information available for X_j to pick up after all X except X_j have been included in the model.

9.13.2 Partial Residual Plots

Partial residual plots are the set of plots of e^j against X_j for all j. Each panel's slope has exactly the numerical value of the corresponding regression coefficient.

We show the partial residual plots for the Longley data in Row 3 of Figure 9.10.

9.13.3 Partial Correlation

The partial correlation $r(X_1, X_2 | X_3, X_4, X_5)$ between X_1 and X_2, after correction for the effect of X_3, X_4, X_5, is the correlation coefficient between X_1 and X_2 after the (linear) effects of X_3, X_4, X_5 have been removed from both X_1 and X_2. When X_1 through X_5 are multivariate data, we can compute the sample partial correlation coefficient as follows:

- Regress X_1 on X_3, X_4, X_5. Get the residuals E_1.
- Regress X_2 on X_3, X_4, X_5. Get the residuals E_2.
- Find the (usual) correlation coefficient between E_1 and E_2. This turns out to be $r(X_1, X_2 | X_3, X_4, X_5)$.

In R, we use

```
partial.corr(cbind(X1,X2),
             cbind(X3,X4,X5))
```

using the function `partial.corr` defined in the **HH** package.

9.13.4 Added Variable Plots

The added variable plots are the set of plots of $E_1 = e^j$ against $E_2 = X_{j|1,2,...,j-1,j+1,...,p}$ for all j. We define $\hat{X}_{1,2,...,j-1,j+1,...,p}$ to be the predicted value of X_j after regressing X_j against all the other X-variables in the model. We define the residual

$$X_{j|1,2,...,j-1,j+1,...,p} = X_j - \hat{X}_{1,2,...,j-1,j+1,...,p} \tag{9.33}$$

to be the additional information in X_j after removing the information provided by all the other X in the model. Thus the added variable plots are the plots of the E_1 and E_2 defined by regressing Y and X_j against all the other X-variables. Each panel's slope has exactly the numerical value of the corresponding regression coefficient.

We show the added variable plots for the Longley data in Row 4 of Figure 9.10.

9.13.5 Interpretation of Residual Plots

9.13.5.1 Response Variable Against Each of the Predictors

Row 1 of Figure 9.10, the plots of the response variable Y=Employed against each of the six predictors X_j, is almost identical to the top row of the splom in Figure 9.7. The only difference is the explicit one-x regression line in Figure 9.10. If there is no visible slope in any of these panels, then we can effectively eliminate that x-variable from further consideration as a potential explanatory variable. This row is essentially the same as the first step of a stepwise-forward procedure. In this example, we cannot eliminate any of the potential predictors at this stage.

9.13.5.2 Residuals Against Each of the Predictors

Row 2 of Figure 9.10, the plots of the ordinary residuals $e = Y - \hat{Y}$ (from the complete regression of the response on all six potential predictors X_j), against each of the X_j shows horizontal slopes. This is by construction, as the least-squares residuals are orthogonal to all X-variables. In this example, we see no structure in the plots. The types of structure we look for are

Curvature.   Plot the residuals from the quadratic fit in the left side of Figure 9.5 against the predictor density and note that the residuals are predominantly above the $y = 0$ axis at the left and right ends of the range and predominantly below the axis in the middle of the range. Curvature in the residual plots often suggests that additional predictors, possibly powers of existing predictors, are needed in the model.

Nonuniformity of variance. The life.exp ˜ ppl.per.tv panel of Figure 4.14 shows high variability in life.exp for low values of ppl.per.tv and very low variability for high values of ppl.per.tv. Nonuniformity of variance in the residual plots often suggests power transformations of one or more of the variables. Transformations of both the response and predictor variables need to be considered.

Bunching or granularity. See the residuals ˜ lime panel of Figure 11.11 where we see that lime has only two levels and there are different variances for each.

9.13.5.3 Partial Residuals

Both Rows 3 and 4 use the partial residuals of the response as the y-variable of each plot. Since "partial" means "adjusted for all the other x-variables", each column of Rows 3 and 4 is different. Column 1 is adjusted for $X_2, X_3, \ldots, X_6$. Column 2 is adjusted for $X_1, X_3, \ldots, X_6$. Similarly through Column 6, which is adjusted for $X_1, \ldots, X_5$.

In Row 3, the *partial residual plots*, the x-variables are the observed x-variables X_j.

In Row 4, the *added residual plots*, the x-variables are the adjusted-x variables, that is, "adjusted for all the other x-variables". Thus the x-variable in Column 1 of Row 4 is $X_{1|2,\ldots,6}$, that is, X_1 adjusted for $X_2, \ldots, X_6$.

In both Rows 3 and 4 the slope of the two-dimensional least-squares line in panel j is exactly the value of the regression coefficient β_j for the complete regression of Y on all the X-variables in the model.

9.13.5.4 Partial Residual Plots

In Row 3, the partial residuals e^j are plotted against the observed x-variables X_j. Since the partial residuals e^j are specific to each X_j, the values for the y-range are unique to each panel. The x-range of the x-variables in Row 3 is the same as it is in Rows 1 and 2 of this display.

We look for the tightness of the points in each plot around their least-squares line. High variability around the two-dimensional least-squares line indicates low significance for the corresponding regression coefficient. Low variability around the least-squares line indicates a significant regression coefficient.

In Row 3 of Figure 9.10, we see that Columns 1 (GNP.deflator) and 5 (Population) have high variability around their least-squares lines. This is a reflection of the high p-value that we see for those regression coefficients in Table 9.9. The remaining four columns all look like their points are tightly placed against their least-squares lines, an indication of possible significance. Note that Column 2 (GNP) looks tight, even though its p-value is the nonsignificant 0.3127. We really do need the tabular results to completely understand what the graph is showing us.

9.13.5.5 Added Variable Plots

In Row 4, the partial residuals e^j are plotted against the adjusted x-variables $X_{j|1,2,\ldots,j-1,j+1,\ldots,p}$. In Row 4, both the x- and y-variables in each column have been adjusted for all the other X-variables. Therefore, both the x- and y-ranges are unique to each panel. The partial residuals, the y-variables in the added variable plots, are identical to the y-variables in the partial residual plots; hence the y-ranges are identical for corresponding columns of Rows 3 and 4.

We look at the slope of the two-dimensional least-squares line in each plot. A nearly horizontal line indicates low significance for the corresponding regression coefficient. A nonzero slope indicates a significant regression coefficient.

The three x-variables with significant regression coefficients in Table 9.9 have visible nonzero slopes to their least-squares lines in Row 4 of Figure 9.10. The three x-variables with nonsignificant regression coefficients have almost horizontal least-squares lines.

9.14 Example—U.S. Air Pollution Data

Exercise 4.2 introduces the data set data(usair) on causes of air pollution in U.S. cities. A scatterplot matrix of these data appears in Figure 9.11. Here we seek to develop a model to explain the response SO2, SO_2 content of air, using a subset of six available explanatory variables.

In Figure 9.11 we see that the three variables SO2, mfgfirms, and popn are all pushed against their minimum value with a long tail toward the maximum value. This pattern suggests a log transformation to bring these three distributions close to symmetry. Following these transformations, Figure 9.12 shows the new response variable lnSO2 and the revised list of six potential explanatory variables.

For pedagogical purposes we approach this problem in two different ways. We first use the automated stepwise regression approach and then consider the manual approach.

U.S. Air Pollution Data with SO$_2$ response variable

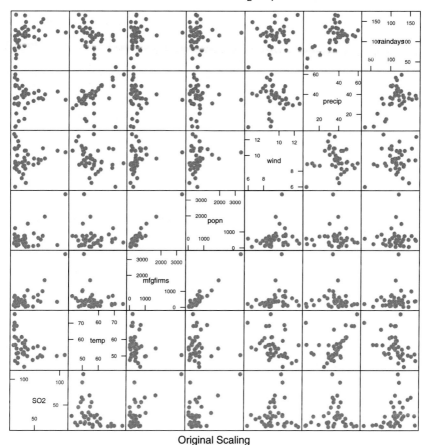

Original Scaling

Fig. 9.11 Scatterplot matrices for air pollution data with the original scaling.

We illustrate the automated approach with the leaps::regsubsets function in R, using the exhaustive method that considers all subsets. In this problem there are only a small number, $2^6 - 1 = 31$, of subsets to consider, so this method is viable. We request the best two subsets for each possible value of the number of included explanatory variables. The tabular and graphical results of the stepwise analysis are displayed in Table 9.14 and Figure 9.13. The model with the four predictors temp, lnmfg, wind, and precip seems best. It has $C_p \approx p$, the smallest AIC of contenders, the largest R^2_{adj}, and one of the smallest values of SS$_{\text{Res}}$.

In Table 9.15 we look at the detail for the selected model. We observe that all VIFs are small and the p-values are below 0.01 for all model coefficients. The signs of the estimated coefficients are reasonable or defensible. United States cities with high average annual temperature are located in the Sunbelt and tend to have less

U.S. Air Pollution Data with ln(SO₂) response variable

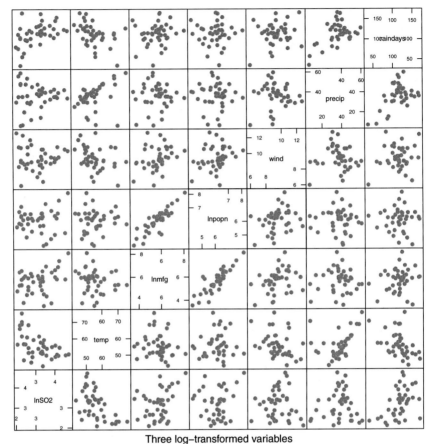

Three log–transformed variables

Fig. 9.12 Scatterplot matrices for air pollution data with improved symmetry after a log transformation of the three variables: SO2, mfgfirms, popn.

pollution-causing heavy industry than colder temperature cities well north of the Sunbelt. We are not surprised that greater amounts of manufacturing are associated with more pollution or that wind dissipates pollution.

We can arrive at the same model without a formal stepwise approach. We notice from Figure 9.12 that lnmfg and lnpopn are highly correlated, so it would be redundant to include both in the model. The variables precip and raindays seem quite similar, so again, it is unlikely that both are needed. Inspection of the C_p plot in Figure 9.13 indicates that the model with temp, lnmfg, wind, and precip has C_p close to p and only one member of each pair of similar predictors.

Table 9.14 Stepwise regression analysis of U.S. air pollution data. See also Figure 9.13.

```
> usair.regsubset <- leaps::regsubsets(
+       lnSO2 ~ lnmfg + lnpopn + precip + raindays + temp + wind,
+       data=usair, nbest=2)

> usair.subsets.Summary <- summaryHH(usair.regsubset)

> tmp <- (usair.subsets.Summary$cp <= 10)

> usair.subsets.Summary[tmp,]
                 model p   rsq   rss adjr2   cp    bic stderr
5              lnm-t-w 4 0.456 10.74 0.412 8.15 -10.09  0.539
6                p-t-w 4 0.446 10.94 0.401 8.93  -9.33  0.544
7            lnm-p-t-w 5 0.543  9.02 0.492 3.58 -13.51  0.501
8            lnm-r-t-w 5 0.513  9.61 0.459 5.82 -10.93  0.517
9        lnm-lnp-p-t-w 6 0.550  8.88 0.486 5.03 -10.46  0.504
10         lnm-p-r-t-w 6 0.543  9.02 0.477 5.58  -9.80  0.508
11   lnm-lnp-p-r-t-w 7 0.550  8.87 0.471 7.00  -6.78  0.511

Model variables with abbreviations
                                                          model
t                                                          temp
r                                                      raindays
p-t                                                 precip-temp
r-t                                               raindays-temp
lnm-t-w                                        lnmfg-temp-wind
p-t-w                                          precip-temp-wind
lnm-p-t-w                                lnmfg-precip-temp-wind
lnm-r-t-w                              lnmfg-raindays-temp-wind
lnm-lnp-p-t-w                      lnmfg-lnpopn-precip-temp-wind
lnm-p-r-t-w                     lnmfg-precip-raindays-temp-wind
lnm-lnp-p-r-t-w lnmfg-lnpopn-precip-raindays-temp-wind

model with largest adjr2
7


Number of observations
41
```

cp

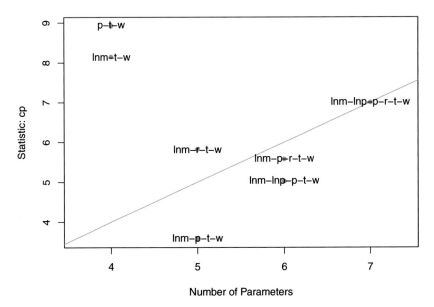

Fig. 9.13 C_p plot. Model "lnm-p-t-w" (lnmfg, precip, temp, wind) has the smallest C_p value and the largest R_{adj}^2. See also Table 9.14.

9.15 Exercises

We recommend that for all exercises involving a data set, you begin by examining a scatterplot matrix of the variables.

9.1. Use matrix algebra to prove the assertion in Equation (9.11) that the sum of the calculated residuals is also zero in multiple regression. We proved the assertion for simple linear regression in Exercise 8.9.

Hint: Write the vector of residuals as $e = (I - H)Y$, verify that $X = HX$, and use the fact that in a model with a nonzero intercept coefficient, as in Equation (9.1) and following, the first column of X is a column of ones.

9.2. Davies and Goldsmith (1972), reprinted in Hand et al. (1994), investigated the relationship between the abrasion loss of samples of rubber (in grams per hour) as a function of hardness and tensile strength (kg/cm²). Higher values of hardness indicate harder rubber. The data are accessed as data(abrasion).

Table 9.15 Fit of recommended model for U.S. air pollution data.

```
> usair.lm7 <- lm.regsubsets(usair.regsubset, 7)

> anova(usair.lm7)
Analysis of Variance Table

Response: lnSO2
          Df Sum Sq Mean Sq F value  Pr(>F)
lnmfg      1   2.26    2.26    9.00  0.0049 **
precip     1   0.03    0.03    0.11  0.7396
temp       1   6.21    6.21   24.77 1.6e-05 ***
wind       1   2.21    2.21    8.84  0.0052 **
Residuals 36   9.02    0.25
---
Signif. codes:  0 '***' 0.001 '**' 0.01 '*' 0.05 '.' 0.1 ' ' 1

> summary(usair.lm7)

Call:
lm(formula = lnSO2 ~ lnmfg + precip + temp + wind, data = usair)

Residuals:
    Min      1Q  Median      3Q     Max
-0.8965 -0.3405 -0.0854  0.2963  1.0321

Coefficients:
            Estimate Std. Error t value Pr(>|t|)
(Intercept)  6.89138    1.07009    6.44 1.8e-07 ***
lnmfg        0.23999    0.08677    2.77  0.0089 **
precip       0.01930    0.00738    2.62  0.0129 *
temp        -0.07304    0.01283   -5.69 1.8e-06 ***
wind        -0.18437    0.06203   -2.97  0.0052 **
---
Signif. codes:  0 '***' 0.001 '**' 0.01 '*' 0.05 '.' 0.1 ' ' 1

Residual standard error: 0.501 on 36 degrees of freedom
Multiple R-squared:  0.543,Adjusted R-squared:  0.492
F-statistic: 10.7 on 4 and 36 DF,  p-value: 8.23e-06


> vif(usair.lm7)
 lnmfg precip   temp   wind
 1.115  1.204  1.373  1.253
```

a. Produce a scatterplot matrix of these data. Based on this plot, does it appear that `strength` would be helpful in explaining `abrasion`?

b. Calculate the fitted regression equation.

c. Find a 95% prediction interval for the abrasion corresponding to a new rubber sample having hardness 60 and strength 200.

9.3. Narula and Wellington (1977) provide data on the sale price of 28 houses in Erie, Pennsylvania, in the early 1970s, along with 11 possible predictors of these prices. The data are accessed as `data(hpErie)`. The variables are:

`price:`   price in $100's

`taxes:`   taxes in dollars

`bathrm:`   number of bathrooms

`lotsize:`   lot size in square feet

`sqfeet:`   square footage of living space

`garage:`   number of cars for which there is garage space

`rooms:`   number of rooms

`bedrm:`   number of bedrooms

`age:`   age in years

`type:`   type of house
brick, brick and frame, aluminum and frame, frame

`style:`   2 story, 1.5 story, ranch

`fireplac:`   number of fireplaces

In parts a–d, exclude factors `type` and `style` from the analysis.

a. Produce a scatterplot matrix for these data. Notice that two houses had a sale price much higher than the others.

b. Use a stepwise regression technique to formulate a parsimonious model for sale price. Do the arithmetic signs of your model's regression coefficients make economic sense?

c. Redo part a with the two large-priced houses excluded. Compare your answer with that of part a.

d. Add a new variable `sqfeetsq` (defined as the square of `sqfeet`) to the list of variables. Perform the stepwise regression allowing for this new variable. Does its presence change the preferred model?

e. For the model you found in part d, provide plots of the residuals vs the fitted response for each of the 12 combinations of type and style. Use Figure 13.1 and its code included in HHscriptnames(13) as a template for constructing these plots. Based on these plots, does it appear that including either of the variables type or style would contribute to the model fit?

9.4. World Almanac and Book of Facts (2001) lists the winning times for the men's 1500-meter sprint event for the Olympics from years 1900 through 2000. The data are accessed as data(sprint).

a. Plot the data.

b. Use linear regression to fit the winning times to the year, producing a plot of the residuals vs the fitted values.

c. The residual plot suggests that an additional predictor should be added to the model. Refit this expanded model and compare it with the model you found in part b.

d. Interpret the sign of the coefficient of this additional predictor.

9.5. A company wished to model the number of minutes required to unload shipments of drums of chemicals at its warehouse as a function of the number of drums and the total shipment weight in hundreds of pounds. The data from 20 consecutive shipments, from Neter et al. (1996), are accessed as data(shipment).

a. Regress minutes on drums and weight, storing the residuals.

b. Interpret the regression coefficients of drum and weight.

c. Provide and discuss plots of the residuals against the fitted values and both predictors, and a normality plot.

d. Provide a 90% prediction interval for the time it would take to unload a new shipment of 10 drums weighing 1000 pounds.

9.6. The dataset data(uscrime) is introduced in Exercise 4.3. Use a stepwise regression approach to develop a model to explain R. Your solution should not have a collinearity problem, all predictor regression coefficients should be significantly different from zero and have an arithmetic sign consistent with common knowledge of the model variables, and no standard residual plots should display a problem.

9.7. It is desired to model the manhours needed to operate living quarters for U.S. Navy bachelor officers. Candidate explanatory variables are listed below. The data in data(manhours) are from Freund and Littell (1991) and Myers (1990), and originally from Navy (1979). Perform a thorough regression analysis, including relevant plots. Note that at least initially, there is a minor collinearity problem to be addressed. Show that, no matter how the collinearity is addressed, the predictions are similar. Only the interpretation of the effects of the x-variables is affected.

manhours: monthly manhours needed to operate the establishment

occupanc: average daily occupancy

checkins: average monthly number of check-ins

svcdesk: weekly hours of service desk operation

common: common use area, in square feet

wings: number of building wings

berthing: operational berthing capacity

rooms: number of rooms

9.A Appendix: Computation for Regression Analysis

regr2.plot

The regr2.plot function does the same type of plot for bivariate regression, one y-variable and two x-variables. The function is based on the persp perspective plotting function in R. We designed the regr2.plot function with options to display grids for the base plane and the two back planes in addition to the observed points and the regression plane and the fitted points. We turned off the default plot of the 3-dimensional box. The function regr2.plot uses the functions defined in our function persp.hh.s.

Chapter 10
Multiple Regression—Dummy Variables, Contrasts, and Analysis of Covariance

Any analysis of variance model (for example, anything in Chapters 6, 12, 13, or 14) can be expressed as a regression with dummy variables. The dummy variables are usually based on a set of contrasts. The algebra of individual contrast vectors is discussed in Section 6.9. Many software procedures and functions make explicit use of this form of expression. Here we explore this equivalence of different representations of the contrasts associated with a factor. The notation in Chapter 10 is that used in Sections I.4.2, 9.3, and 9.4.1.

Section 10.1 introduces dummy variables. Section 10.3 looks at the equivalence of different sets of dummy variable codings for factors. Section 13.5 shows how the R and SAS languages express the dummy variable coding schemes. Table 13.18 shows the notation for applying them to describe models with two or more factors.

10.1 Dummy (Indicator) Variables

Dummy variables, also called indicator variables, are a way to incorporate qualitative predictors into a regression model. If we have a qualitative predictor A with a distinct values, we will need $a - 1$ distinct dummy variables to code it. For example, suppose we believe that the gender of the subject may impact the response. We could define $X_{\text{female}} = 1$ if the subject is female and $X_{\text{female}} = 0$ if the subject is male. Then we interpret the estimated regression coefficient $\hat{\beta}_{\text{female}}$ as the estimated average amount by which responses for females exceed responses for males, assuming the values of all other predictors are unchanged. If $\hat{\beta}_{\text{female}} > 0$, then on average females will tend to have a higher response than males; if $\hat{\beta}_{\text{female}} < 0$, then the average male response will exceed the average female response. There are $g = 2$ levels to the classification variable gender, hence we defined $g - 1 = 1$ dummy variable to code that information. We pursue this example in Section 10.2.

© Springer Science+Business Media New York 2015
R.M. Heiberger, B. Holland, *Statistical Analysis and Data Display*,
Springer Texts in Statistics, DOI 10.1007/978-1-4939-2122-5_10

As another example, suppose one of the predictor variables in a model is the nominal variable ResidenceLocation, which can take one of $r = 3$ values: urban, suburban, or rural. If a qualitative predictor has r categories, we must assign $r - 1$ dummy variables to represent it adequately. Otherwise, we may be imposing an unwarranted implicit constraint. It would be incorrect to code this with a single numeric variable $X_{RL} = 0$ for urban, 1 for suburban, and 2 for rural, as that would imply that the difference between average urban and suburban responses must equal the difference between average suburban and rural responses, which is probably not justifiable.

One correct coding is to let $X_{RLu} = 1$ if urban and 0 otherwise and let $X_{RLs} = 1$ if suburban and 0 otherwise. Then the coefficient $\hat{\beta}_{RLu}$ of X_{RLu} is interpreted as the average difference between urban and rural response, and the coefficient $\hat{\beta}_{RLs}$ of X_{RLs} is interpreted as the average difference between suburban and rural response. The difference between the coefficients $\hat{\beta}_{RLu}$ and $\hat{\beta}_{RLs}$ is the average difference between the urban and suburban response. Here we used rural as the reference response. The results of the analysis would have been the same had we used either urban or suburban as the reference response. See Section 10.3 for the justification of this statement. See Exercise 10.3 to apply the justification to this example.

This type of coding is done automatically in R's linear modeling functions (lm and aov when variables have been defined as factors with the factor() function.

The PROC ANOVA and PROC GLM in SAS require use of the CLASSES command within the PROC specification. SAS's PROC REG requires explicit coding to construct the dummy variables in the DATA step.

Any pair of independent linear combinations of X_{RLu} and X_{RLs} would be equally as valid. R gives the user choice with the contrasts() and related functions. SAS gives the user choice with the estimate and test statements on the PROC ANOVA and PROC GLM commands.

10.2 Example—Height and Weight

10.2.1 Study Objectives

In the fall of 1998, one of us (RMH) collected the height, weight, and age of the 39 students in one of his classes. The data appear in file data(htwt). While this example does give information on the comparative height distributions of men and women, the primary intent then, and now, is to use this example to illustrate how the techniques of statistics give us terminology and notation for discussing ordinary observations.

10.2.2 Data Description

feet: height in feet rounded down to an integer

inches: inches to be added to the height in feet

lbs: weight in pounds

months: age in months

sex: m or f

meters: height in meters

10.2.3 Data Problems

From the stem-and-leaf in Table 10.1 we see that even in this small dataset, collected with some amount of care, there are data problems. There are 39 observations, yet only 38 made it to the stem-and-leaf and one of those has a missing value. Further investigation of the data file shows that one student reported her height in meters and another didn't indicate sex. For the remaining figures and tables in this chapter we converted meters to inches for the one. For the other we had the good fortune to have access to the sample population at the next class meeting and were able to fill in the missing value (m in this case) by checking the data forms directly with the students. We were lucky in this example that the data file was investigated soon enough after collection that the data anomalies could be resolved. That is not always possible. We describe techniques for dealing with missing data in Section 2.4.

We show a splom of the completed data in Figure 10.1. The age range in our class was 18–28 for women and 19–24 for men. There is no visible relation between age and either height or weight. There is a clear difference in height ranges between men and women and a visible, but less strong, difference in weight ranges. We investigate this further by expanding the lbs ˜ ht panel in Figure 10.2.

Table 10.1 Stem-and-leaf of Heights from class observation. We used this display to detect the two missing values. Note that this is an edited version of the output. We placed the two distributions adjacent to each other and added additional lines to the high end of the female distribution and to the low end of the male distribution to make the two stem-and-leaf displays align correctly.

```
> data(htwt)

> levels(factor(htwt$sex, exclude=NULL))
[1] "f" "m" NA

> any(is.na(htwt$ht))
[1] TRUE

> for (h in tapply(htwt$ht, factor(htwt$sex, exclude=NULL), c))
+    stem(h, scale=1.5)

  The decimal point is at the |

        Female                      Male
  58 | 0                    58 |
  60 |                      60 |
  62 | 00000                62 | 0
  64 | 000000000            64 | 0
  66 | 000008               66 | 000
  68 | 0                    68 | 00
  70 |                      70 |
  72 |                      72 | 000000
  74 |                      74 | 00
```

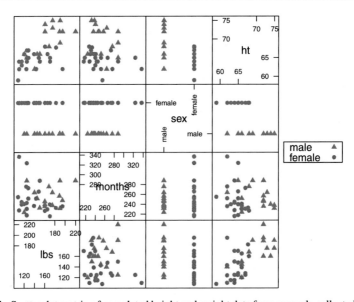

Fig. 10.1 Scatterplot matrix of completed height and weight data from example collected in class.

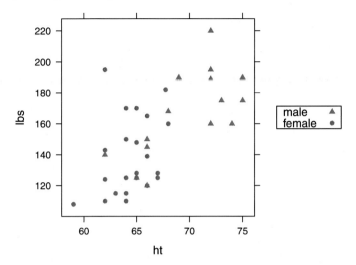

Fig. 10.2 Expansion of `lbs ~ ht` panel of Figure 10.1. There is visibly less overlap in the range for the heights of men and women than for their weights.

Table 10.2 One-way analysis of variance of heights from class observation.

```
> ## one-way analysis of variance
> htwt.aov <- aov(ht ~ sex, data=htwt)

> summary(htwt.aov)
            Df Sum Sq Mean Sq F value  Pr(>F)
sex          1    282   282.3    30.8 2.5e-06 ***
Residuals   37    339     9.2
---
Signif. codes:  0 '***' 0.001 '**' 0.01 '*' 0.05 '.' 0.1 ' ' 1

> model.tables(htwt.aov, type="means")
Tables of means
Grand mean

66.71

  sex
        f     m
    64.47 69.94
rep 23.00 16.00
```

10.2.4 Three Variants on the Analysis

Table 10.2 uses the techniques of Chapter 6 to compare the means of two distributions. The specific features that we will look at are the various values in the ANOVA table and the mean heights for each of the groups. We will follow by using regression on two different sets of dummy variables to duplicate those numbers.

We initially use the $g - 1 = 1$ dummy variable X_{female} with the $(1, 0)$ coding scheme suggested above, with value 1 for females and value 0 for males. We display the results of an ordinary linear regression of height on the dummy variable X_{female} in Table 10.3. The estimated intercept $\hat{\beta}_0 = 69.9375$ is the mean height for males. The

Table 10.3 Regression analysis of heights from class observation on the dummy variable coding sex as `female=1` for female and `female=0` for male.

```
> ## dummy variable
> htwt$female <- as.numeric(htwt$sex == "f")

> htwt.lm <- lm(ht ~ female, data=htwt)

> summary(htwt.lm, corr=FALSE)

Call:
lm(formula = ht ~ female, data = htwt)

Residuals:
    Min     1Q Median     3Q    Max
 -7.938 -2.202  0.533  2.062  5.062

Coefficients:
            Estimate Std. Error t value Pr(>|t|)
(Intercept)   69.938      0.757   92.42  < 2e-16 ***
female        -5.470      0.985   -5.55 2.5e-06 ***
---
Signif. codes:  0 '***' 0.001 '**' 0.01 '*' 0.05 '.' 0.1 ' ' 1

Residual standard error: 3.03 on 37 degrees of freedom
Multiple R-squared:  0.454,Adjusted R-squared:  0.44
F-statistic: 30.8 on 1 and 37 DF,  p-value: 2.54e-06


> anova(htwt.lm)
Analysis of Variance Table

Response: ht
          Df Sum Sq Mean Sq F value  Pr(>F)
female     1    282   282.3    30.8 2.5e-06 ***
Residuals 37    339     9.2
---
Signif. codes:  0 '***' 0.001 '**' 0.01 '*' 0.05 '.' 0.1 ' ' 1
```

estimated regression coefficient for the X_{female} predictor, $\hat{\beta}_{female} = -5.4701$, is the increment to the intercept that produces the mean height for females. The ANOVA table in Table 10.3 is identical to the ANOVA table in Table 10.2.

There are many other dummy variable coding schemes that we could use to get exactly the same ANOVA table and the same estimated mean heights for the two groups. We show another in Table 10.4. In this coding, the dummy variable X_{treat} has the value 1 for females and the value -1 for males. The estimated intercept $\hat{\beta}_0 = 67.2024$ is the average of the mean heights for females and males. The estimated regression coefficient for the X_{treat} predictor, $\hat{\beta}_{treat} = -2.7351$, is the amount that

Table 10.4 Regression analysis of heights from class observation on the dummy variable coding sex as `treat`=1 for female and `treat`=−1 for male.

```
> ## dummy variable
> htwt$treat <- (htwt$sex == "f") - (htwt$sex == "m")

> htwtb.lm <- lm(ht ~ treat, data=htwt)

> summary(htwtb.lm, corr=FALSE)

Call:
lm(formula = ht ~ treat, data = htwt)

Residuals:
   Min    1Q Median    3Q    Max
-7.938 -2.202  0.533  2.062  5.062

Coefficients:
            Estimate Std. Error t value Pr(>|t|)
(Intercept)   67.202      0.493  136.40  < 2e-16 ***
treat         -2.735      0.493   -5.55  2.5e-06 ***
---
Signif. codes:  0 '***' 0.001 '**' 0.01 '*' 0.05 '.' 0.1 ' ' 1

Residual standard error: 3.03 on 37 degrees of freedom
Multiple R-squared:  0.454,Adjusted R-squared:  0.44
F-statistic: 30.8 on 1 and 37 DF,  p-value: 2.54e-06


> anova(htwtb.lm)
Analysis of Variance Table

Response: ht
          Df Sum Sq Mean Sq F value  Pr(>F)
treat      1    282   282.3    30.8 2.5e-06 ***
Residuals 37    339     9.2
---
Signif. codes:  0 '***' 0.001 '**' 0.01 '*' 0.05 '.' 0.1 ' ' 1
```

added to the intercept produces the mean height for females and subtracted from the intercept produces the mean height for males. The ANOVA table in Table 10.4 is also identical to the ANOVA table in Table 10.2.

10.3 Equivalence of Linear Independent X-Variables (such as Contrasts) for Regression

It is not an accident that the ANOVA tables in Tables 10.2, 10.3, and 10.4 are identical. We explore here why that is the case.

Please review the definition of linear dependence in Section I.4.2.

The X matrix in the linear regression presentation of the one-way analysis of variance model with one factor with a categories must have a leading column of ones $X_0 = \mathbf{1}$ for the intercept and at least $a - 1$ additional columns, for a total of $c \geq a$ columns. The entire X matrix can be summarized by a *contrast matrix* $\underset{a \times c}{W}$ consisting of a unique rows, one for each level of the factor.

We explore the relationship between several different contrast matrices W in the case $a = 4$. The principles work for any value a. The matrix X of dummy variables itself consists of n_i copies of the ith row of W (where $n = \sum_{i=1}^{a} n_i$):

$$\underset{n \times c}{X} = \underset{n \times 4}{\begin{pmatrix} n_1\{(1\,0\,0\,0) \\ n_2\{(0\,1\,0\,0) \\ n_3\{(0\,0\,1\,0) \\ n_4\{(0\,0\,0\,1) \end{pmatrix}} \underset{4 \times c}{W} = \underset{n \times 4}{N}\ \underset{4 \times c}{W} \tag{10.1}$$

Any contrast matrix W with $a = 4$ rows and with rank 4 (which means it must have at least 4 columns) is equivalent for linear regression in the senses that

1. Any two such matrices W_1 and W_2 with dimensions $(4 \times c_1)$ and $(4 \times c_2)$ where $c_i \geq 4$ are related by postmultiplication of the first matrix by a full-rank matrix $\underset{c_1 \times c_2}{A}$, that is,

$$\underset{4 \times c_1}{W_1}\ \underset{c_1 \times c_2}{A} = \underset{4 \times c_2}{W_2}$$

Equivalently, any two such dummy variables matrices X_1 and X_2 with dimensions $(n \times c_1)$ and $(n \times c_2)$ are similarly related by

$$\underset{n \times c_1}{X_1}\ \underset{c_1 \times c_2}{A} = \underset{n \times c_2}{X_2}$$

Examples (R code for all the contrast types in these examples is included in file HHscriptnames(10)):

1a. A simple overparameterized matrix (5 columns with rank=4) (this is the SAS default):

$$W_{\substack{\text{simple} \\ 4\times(1+4)}} = \begin{pmatrix} 1\ 1\ 0\ 0\ 0 \\ 1\ 0\ 1\ 0\ 0 \\ 1\ 0\ 0\ 1\ 0 \\ 1\ 0\ 0\ 0\ 1 \end{pmatrix}$$

1b. Treatment contrasts (4 columns with rank=4) (R contr.treatment. This is the R default for factors. These are not 'contrasts' as defined in the standard theory for linear models as they are not orthogonal to the intercept.):

$$W_{\substack{\text{simple} \\ 4\times(1+4)}} \quad A_{\substack{ \\ (1+4)\times(1+3)}} \ = \ W_{\substack{\text{treatment} \\ 4\times(1+3)}}$$

$$\begin{pmatrix} 1\ 1\ 0\ 0\ 0 \\ 1\ 0\ 1\ 0\ 0 \\ 1\ 0\ 0\ 1\ 0 \\ 1\ 0\ 0\ 0\ 1 \end{pmatrix} \begin{pmatrix} 1\ 0\ 0\ 0 \\ 0\ 0\ 0\ 0 \\ 0\ 1\ 0\ 0 \\ 0\ 0\ 1\ 0 \\ 0\ 0\ 0\ 1 \end{pmatrix} = \begin{pmatrix} 1\ 0\ 0\ 0 \\ 1\ 1\ 0\ 0 \\ 1\ 0\ 1\ 0 \\ 1\ 0\ 0\ 1 \end{pmatrix}$$

1c. Helmert contrasts (4 columns with rank=4) (R contr.helmert):

$$W_{\substack{\text{simple} \\ 4\times(1+4)}} \quad A_{\substack{ \\ (1+4)\times(1+3)}} \quad = \quad W_{\substack{\text{helmert} \\ 4\times(1+3)}}$$

$$\begin{pmatrix} 1\ 1\ 0\ 0\ 0 \\ 1\ 0\ 1\ 0\ 0 \\ 1\ 0\ 0\ 1\ 0 \\ 1\ 0\ 0\ 0\ 1 \end{pmatrix} \begin{pmatrix} 1 & 1 & 1 & 1 \\ 0 & -2 & -2 & -2 \\ 0 & 0 & -2 & -2 \\ 0 & -1 & 1 & -2 \\ 0 & -1 & -1 & 2 \end{pmatrix} = \begin{pmatrix} 1 & -1 & -1 & -1 \\ 1 & 1 & -1 & -1 \\ 1 & 0 & 2 & -1 \\ 1 & 0 & 0 & 3 \end{pmatrix}$$

1d. Sum contrasts (4 columns with rank=4) (R contr.sum):

$$W_{\substack{\text{simple} \\ 4\times(1+4)}} \quad A_{\substack{ \\ (1+4)\times(1+3)}} \quad = \quad W_{\substack{\text{sum} \\ 4\times(1+3)}}$$

$$\begin{pmatrix} 1\ 1\ 0\ 0\ 0 \\ 1\ 0\ 1\ 0\ 0 \\ 1\ 0\ 0\ 1\ 0 \\ 1\ 0\ 0\ 0\ 1 \end{pmatrix} \begin{pmatrix} 1 & 0 & 0 & 0 \\ 0 & 1 & 0 & 0 \\ 0 & 0 & 1 & 0 \\ 0 & 0 & 0 & 1 \\ 0 & -1 & -1 & -1 \end{pmatrix} = \begin{pmatrix} 1 & 1 & 0 & 0 \\ 1 & 0 & 1 & 0 \\ 1 & 0 & 0 & 1 \\ 1 & -1 & -1 & -1 \end{pmatrix}$$

1e. Polynomial contrasts (4 columns with rank=4) (R `contr.poly`. This is the R default for ordered factors.):

$$
\begin{array}{ccc}
W_{\text{simple}} & A & = W_{\text{polynomial}} \\
4\times(1+4) & (1+4)\times(1+3) & 4\times(1+3)
\end{array}
$$

$$
\begin{pmatrix}
1 & 1 & 0 & 0 & 0 \\
1 & 0 & 1 & 0 & 0 \\
1 & 0 & 0 & 1 & 0 \\
1 & 0 & 0 & 0 & 1
\end{pmatrix}
\begin{pmatrix}
0.8 & 0.0000 & 0.0 & 0.0000 \\
0.2 & -0.6708 & 0.5 & -0.2236 \\
0.2 & -0.2236 & -0.5 & 0.6708 \\
0.2 & 0.2236 & -0.5 & -0.6708 \\
0.2 & 0.6708 & 0.5 & 0.2236
\end{pmatrix}
=
$$

$$
\begin{pmatrix}
1 & -0.6708 & 0.5 & -0.2236 \\
1 & -0.2236 & -0.5 & 0.6708 \\
1 & 0.2236 & -0.5 & -0.6708 \\
1 & 0.6708 & 0.5 & 0.2236
\end{pmatrix}
$$

2. The hat matrices are the same.

$$
H_1 = (X_1(X_1'X_1)^{-1}X_1') = (X_2(X_2'X_2)^{-1}X_2') = H_2
$$

An equivalent statement is that both X matrices span the same column space.

Proof. For the special case that $c = a$, hence the $X'X$ and A matrices are invertible:

$$
\begin{aligned}
H_2 &= \\
X_2(X_2'X_2)^{-1}X_2' &= \\
(X_1A)\big((X_1A)'(X_1A)\big)^{-1}(X_1A)' &= \\
(X_1A)(A'X_1'X_1A)^{-1}(A'X_1') &= \\
X_1(X_1'X_1)^{-1}X_1' &= \\
H_1
\end{aligned}
$$

When $c > a$, the step from line 4 to line 5 involves matrix algebra manipulations that we do not discuss here. Effectively, we are dropping any redundant columns.

3. The predicted values are the same.

$$
\hat{Y} = H_1 Y = H_2 Y
$$

4. The regression coefficients are related by premultiplication of the second set of coefficients by the same matrix A,

$$
\beta_1 = A\beta_2
$$

Proof.

$$E(Y) = X_2\beta_2 = (X_1A)\beta_2 = X_1(A\beta_2) = X_1\beta_1$$

5. The ANOVA (analysis of variance) table is the same:

Source	Sum of Squares		
Regression	$SS_{Reg} =$	$Y'H_1Y$	$= Y'H_2Y$
Residual	$SS_{Res} = Y'(I - H_1)Y$	$= Y'(I - H_2)Y$	

Exercise 10.1 gives you the opportunity to explore the equivalence of the two coding schemes in Section 10.2.

As a consequence of the equivalence up to multiplication by a matrix A, the regression coefficients in regression analyses with factors (which means most experiments) are uninterpretable unless the definitions of the dummy variables have been provided.

10.4 Polynomial Contrasts and Orthogonal Polynomials

Ott (1993) reports an experiment that uses an abrasives testing machine to test the wear of a new experimental fabric. The machine was run at six different speeds (measured in revolutions per minute). Forty-eight identical square pieces of fabric were prepared, 8 of which were randomly assigned to each of the 6 machine speeds. Each square was tested for a three-minute period at the appropriate machine setting. The order of testing was appropriately randomized. For each square, the amount of wear was measured and recorded. The data from file data(fabricwear) are displayed in Figure 10.3. The initial ANOVA is in Table 10.5.

From Figure 10.3 we see that the assumption in Equation (6.3) of approximately constant variance across groups is satisfied by this dataset, hence ANOVA is an appropriate technique for investigating the data. We also note one outlier at speed=200. We will return to that data point later.

The ANOVA table in Table 10.5 shows that speed is significant. From the table of means we see that the means increase with speed and the increase is also faster as speed increases. Figure 10.3 shows the same and suggests that the means are increasing as a quadratic polynomial in speed.

There are several essentially identical ways to check this supposition. We start with the easiest to do and then expand by illustrating the arithmetic behind it. When we defined speed as a factor in Table 10.5, we actually did something more specific,

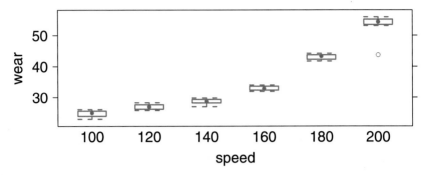

Fig. 10.3 Fabric wear as a function of speed. We see constant variance and a curved uphill trend. There is one outlier

Table 10.5 ANOVA and means for wear as a function of speed.

```
> fabricwear.aov <- aov(wear ~ speed, data=fabricwear)

> summary(fabricwear.aov)
            Df Sum Sq Mean Sq F value Pr(>F)
speed        5   4872     974     298 <2e-16 ***
Residuals   42    137       3
---
Signif. codes:  0 '***' 0.001 '**' 0.01 '*' 0.05 '.' 0.1 ' ' 1

> model.tables(fabricwear.aov, "mean")
Tables of means
Grand mean

34.93

 speed
speed
  100   120   140   160   180   200
24.78 26.96 28.68 32.93 43.05 53.19
```

we declared it to be an *ordered factor*. This means that the dummy variables are the orthogonal polynomials for six levels. We display the orthogonal polynomials in Figure 10.4 and Table 10.6. See the discussion in Section I.4 for an overview of orthogonal polynomials and their construction.

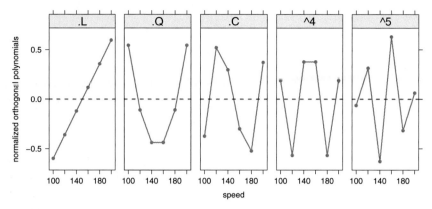

Fig. 10.4 Orthogonal polynomials for speed.

From the panels in Figure 10.4 we see that the linear polynomial plots as a straight line against the speed. The quadratic polynomial plots as a discretization of a parabola. The higher-order polynomials are rougher discretizations of their functions. In Table 10.6 we see that the orthogonal polynomials are scaled so their cross product is the identity matrix, that is, it is a diagonal matrix with 1s on the diagonal. Compare this (in Exercise 10.2) to a matrix of the simple powers of the integers $(1, 2, 3, 4, 5, 6)$. The columns of the simple powers span the same linear space as the orthogonal properties. Because they are not orthogonal (their cross product is not diagonal), their plots are harder to interpret and they may show numerical difficulties when used as predictor variables in a regression. See Appendix G for further discussion on the numerical issues.

In Table 10.7 we show two variants of an expanded display of the ANOVA from Table 10.5. The top of the table shows the regression coefficients for the regression against the orthogonal polynomials used as the dummy variables. Here we see that the linear and quadratic terms are highly significant. The cubic term is not significant. Based on our reading of the graph, and the comparison of the p-value for the quartic term to that of the quadratic term, we will interpret the quartic term as not significant and do all continuing work with the quadratic model.

In the bottom of Table 10.7 we show the partitioned ANOVA table with the linear, quadratic, and cubic terms isolated. By dint of the orthogonality the F-values are the square of the t-values for the coefficients ($36.3580^2 = 1321.903$) and the p-values are identical.

What happens when we redo the analysis without the outlier noted in Figure 10.3? The residual mean square goes down by a factor of 4; consequently, all the t-values go up. While the p-values for the cubic and quartic terms now show significance at .0001, we will continue to exclude them from our recommended model because the p-values for the linear and quadratic terms are orders of magnitude smaller ($< 10^{-16}$). See Exercise 10.8.

Table 10.6 Orthogonal polynomials for speed. The slightly complex algorithm shown here for scaling the orthogonal polynomials, with attention paid to computational precision by use of the zapsmall function, is necessary for factors with an odd number of levels. See Appendix G for further discussion on the numerical issues.

```
> tmp.c <- zapsmall(contrasts(fabricwear$speed), 14)

> dimnames(tmp.c)[[1]] <- levels(fabricwear$speed)

> tmp.c
          .L       .Q       .C       ^4       ^5
100 -0.5976   0.5455  -0.3727   0.1890  -0.06299
120 -0.3586  -0.1091   0.5217  -0.5669   0.31497
140 -0.1195  -0.4364   0.2981   0.3780  -0.62994
160  0.1195  -0.4364  -0.2981   0.3780   0.62994
180  0.3586  -0.1091  -0.5217  -0.5669  -0.31497
200  0.5976   0.5455   0.3727   0.1890   0.06299

> zapsmall(crossprod(tmp.c), 13)
   .L .Q .C ^4 ^5
.L  1  0  0  0  0
.Q  0  1  0  0  0
.C  0  0  1  0  0
^4  0  0  0  1  0
^5  0  0  0  0  1

> min.nonzero <- function(x, digits=13) {
+    xx <- zapsmall(x, digits)
+    min(xx[xx != 0])
+ }

> tmp.min <- apply(abs(tmp.c), 2, min.nonzero)

> sweep(tmp.c, 2, tmp.min, "/")
     .L .Q    .C ^4  ^5
100 -5   5 -1.25  1  -1
120 -3  -1  1.75 -3   5
140 -1  -4  1.00  2 -10
160  1  -4 -1.00  2  10
180  3  -1 -1.75 -3  -5
200  5   5  1.25  1   1
```

Table 10.7 Regression coefficients on dummy variables, and partitioned ANOVA table.

```
> summary(fabricwear.aov,
+           split=list(speed=list(speed.L=1, speed.Q=2,
+                      speed.C=3, rest=4:5)))
                  Df Sum Sq Mean Sq F value  Pr(>F)
speed              5   4872     974  297.70 < 2e-16 ***
  speed: speed.L   1   4327    4327 1321.90 < 2e-16 ***
  speed: speed.Q   1    513     513  156.76 9.1e-16 ***
  speed: speed.C   1      7       7    2.10   0.154
  speed: rest      2     25      13    3.88   0.028 *
Residuals         42    137       3
---
Signif. codes:  0 '***' 0.001 '**' 0.01 '*' 0.05 '.' 0.1 ' ' 1

> summary.lm(fabricwear.aov)

Call:
aov(formula = wear ~ speed, data = fabricwear)

Residuals:
   Min     1Q Median     3Q    Max
-9.487 -0.653  0.181  0.825  2.712

Coefficients:
            Estimate Std. Error t value Pr(>|t|)
(Intercept)   34.929      0.261  133.76  < 2e-16 ***
speed.L       23.256      0.640   36.36  < 2e-16 ***
speed.Q        8.009      0.640   12.52  9.1e-16 ***
speed.C        0.928      0.640    1.45    0.154
speed^4       -1.677      0.640   -2.62    0.012 *
speed^5       -0.600      0.640   -0.94    0.354
---
Signif. codes:  0 '***' 0.001 '**' 0.01 '*' 0.05 '.' 0.1 ' ' 1

Residual standard error: 1.81 on 42 degrees of freedom
Multiple R-squared:  0.973,Adjusted R-squared:  0.969
F-statistic:  298 on 5 and 42 DF,  p-value: <2e-16
```

10.4.1 Specification and Interpretation of Interaction Terms

Example—consider a model

$$E(Y) = \beta_0 + \beta_1 X_1 + \beta_2 X_2 + \beta_3 X_3 + \beta_4 X_4 + \beta_{34} X_3 X_4$$

to "explain" determinants of annual salary Y in dollars for workers in some population. Here X_1 is age in years, X_2 is gender (1 if female, 0 if male), X_3 is race (1 if white, 0 if nonwhite), and X_4 is number of years of schooling. (Discussion: What other variables might such a model include to explain salary?)

The existence of the interaction terms allows for the possibility that the degree of enhancement of education on schooling differs for whites and nonwhites.

Consider a white and a nonwhite of the same age and gender and having the same amount of schooling. Then:

- β_4 is the expected increase in annual salary for a nonwhite attributable to an additional year of schooling.

- $\beta_4 + \beta_{34}$ is the expected increase in annual salary for a white attributable to an additional year of schooling.

- β_{34} is the expected amount by which a white's salary increase as a result of an additional year of schooling exceeds a nonwhite's salary increase as a result of an additional year of schooling.

Also, still assuming the same age and gender,

- $\beta_3 + \beta_{34} X_4$ is the difference between white and nonwhite expected salary.
- β_3 is the component of this difference that does not depend on years of schooling and is attributable only to difference in race.

We examine this model further in Exercise 10.7.

10.5 Analysis Using a Concomitant Variable (Analysis of Covariance—ANCOVA)

In some situations where we seek to compare the differences in the means of a continuous response variable across levels of a factor A, we have available a second continuous variable that can be used to improve our ability to distinguish among the levels. Historically this extended model has been called the *analysis of covariance* model because the second variable varies along with the first. To avoid confusion with the concept of covariance introduced in Chapter 3, we prefer to call this approach *analysis using a concomitant variable*. Nevertheless we will retain use of the term covariate as a shorthand term for concomitant variable and the acronym ANCOVA as an abbreviation for this method.

If X_{ij} denotes the jth observation of the covariate at the ith level of factor A, our original ANOVA model in Equation (6.1) generalizes to

$$Y_{ij} = \mu + \alpha_i + \beta(X_{ij} - \bar{\bar{X}}) + \epsilon_{ij} \tag{10.2}$$
$$\text{for} \quad i = 1,\ldots,a \quad \text{and} \quad j = 1,\ldots,n_i$$

where $\bar{\bar{X}}$ is the grand mean of the X_{ij}'s and all other terms are as defined in Equation (6.1). The model in Equation (10.2) has separate intercepts α_i for each level of A but retains a common slope. The differences between the intercepts α_i are identical to the vertical differences between the parallel lines (to be illustrated in Figure 10.8). Equation (10.2) is the classical ANCOVA model.

The logic of this approach is that if X_{ij} is related to Y_{ij} then the ϵ's of the model in Equation (10.2) will be measured from a different regression line for each level of A rather than from a different horizontal line as in model (6.1). This will give the ϵ's less variability than those of Equation (6.1), thereby sharpening our inferences on the α_i's. The α_i's estimated from Equation (10.2) are said to be *adjusted* for the covariate. Quite frequently the range of observed X_{ij} differs for each level of A_i and therefore the $\bar{Y}_i$ means from Equation (6.1) reflect the difference in the X-values more than the differences attributable to the change in levels of A.

The next level of generalization allows the slopes to differ, i.e., replace the common β in Equation (10.2) with β_i:

$$Y_{ij} = \mu + \alpha_i + \beta_i(X_{ij} - \bar{\bar{X}}) + \epsilon_{ij} \tag{10.3}$$
$$\text{for} \quad i = 1,\ldots,a \quad \text{and} \quad j = 1,\ldots,n_i$$

We illustrate models Equations (10.2) and (10.3) in Section 10.6. In Section 10.6.5 we will use the model in Equation (10.3) to test the assumption that the lines are parallel. Formally, we will test whether the lines have the same slope

$$H_0: \beta_1 = \beta_2 = \beta_3 \tag{10.4}$$
$$H_1: \text{Not all } \beta_i \text{ are identical}$$

or the same intercept

$$H_0: \alpha_1 = \alpha_2 = \alpha_3 \tag{10.5}$$
$$H_1: \text{Not all } \alpha_i \text{ are identical}$$

or both (in which case the lines coincide). We illustrate each model by an appropriate graph. We construct a single meta-graph in Figure 10.12 to illustrate the comparison of all the models we consider.

These ideas can be extended to situations with more than one covariate variable and to more complicated experimental designs such as those discussed in Chapters 12 through 14.

10.6 Example—Hot Dog Data

10.6.1 Study Objectives

Hot dogs based on poultry are said to be healthier than ones made from either meat (beef and pork) or all beef. A basis for this claim may be the lower-calorie (fat) content of poultry hot dogs. Is this advantage of poultry hot dogs offset by a higher sodium content than meat hot dogs?

Researchers for *Consumer Reports* analyzed three types of hot dog: beef, poultry, and meat (mostly pork and beef, but up to 15% poultry meat). The data in file data(hotdog) come from Consumer Reports (1986) and were later used by Moore and McCabe (1989).

10.6.2 Data Description

Type: Type of hot dog (beef, meat, or poultry)

Calories: Calories per hot dog

Sodium: Milligrams of sodium per hot dog

10.6.3 One-Way ANOVA

We start by comparing the Sodium content of the three hot dog Types by the methods of Chapter 6 in Figure 10.5 and in Table 10.8. We see that the three Types have similar Sodium content.

Figure 10.6 shows the response Sodium plotted against the covariate Calories by Type. Within each panel we plot a horizontal line at the mean of the Sodium values for that `Type`. The analysis of variance in Table 10.8 compares the vertical distance between these horizontal lines. It ignores the most evident feature of this plot, that the three `Types` have very different fat contents with `Poultry` low, `Beef` intermediate, and `Meat` high. We wish to see if knowledge about `Calories` affects our understanding about `Sodium`.

10.6.4 Concomitant Explanatory Variable—ANCOVA

It is possible that our finding of similar `Sodium` content is attributable in part to a need to add sodium to enhance the flavor of higher-fat hot dogs. The `Calories` information can be incorporated into the analysis by adding `Calories` to the model as a concomitant explanatory variable. Then in this revised model, comparisons between the mean `Sodium` contents of the three `Types` will have been *adjusted for* differing `Calories` contents. In this way, comparisons between the three `Types` will be made on the basis that each `Type` has the mean `Calories` content of all `Types`.

We illustrate this revised analysis in two steps. Initially, in Figure 10.7 and Table 10.9, we show the regression (Chapter 8) of `Sodium` on `Calories` ignoring the `Types`. The common regression line makes some sense in the **Superpose** panel but very clearly has the wrong slope and wrong intercept in all three of the individual panels.

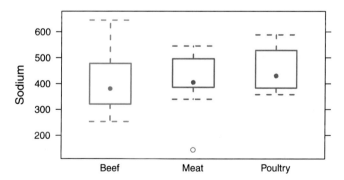

Fig. 10.5 Boxplots comparing the `Sodium` content of three `Types` of hot dogs. See Table 10.8.

Sodium ~ Type, x=Calories

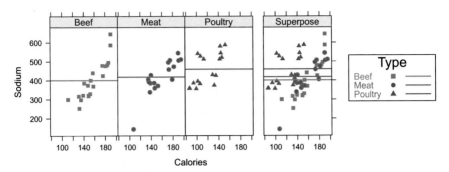

Fig. 10.6 Sodium ~ Type, x=Calories. Horizontal lines at Sodium means for each Type. $Y_{ij} = \mu + \alpha_i + \epsilon_{ij}$. See Table 10.8. The intent of the notation is twofold: The arithmetic of the analysis is based on the one-way ANOVA of Sodium ~ Type. The graph is more complex. The points in the graph show y=Sodium plotted against x=Calories separately for each level of Type. The horizontal line in each panel is the mean of the levels of Sodium at each level of Type.

Table 10.8 Hot dog ANOVA and means. This is the one-way ANOVA of Chapter 6. See Figures 10.5 and 10.6.

```
> aovStatementAndAnova(TxC)
> anova(aov(Sodium ~ Type, data = hotdog))
Analysis of Variance Table

Response: Sodium
          Df Sum Sq Mean Sq F value Pr(>F)
Type       2  31739   15869    1.78   0.18
Residuals 51 455249    8926

> model.tables(TxC, type="means")
Tables of means
Grand mean

424.8

 Type
     Beef  Meat Poultry
    401.1 418.5     459
rep  20.0  17.0      17
```

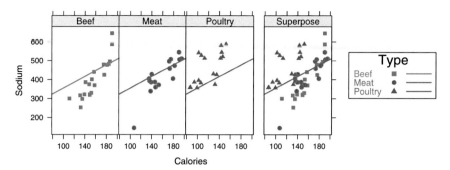

Fig. 10.7 Sodium ~ Calories, groups=Type. Common regression line that ignores Type. $Y_{ij} = \mu + \beta(X_{ij} - \bar{\bar{X}}) + \epsilon_{ij}$. See Table 10.9. The intent of the notation is twofold: The arithmetic of the analysis is based on the simple linear regression of Sodium ~ Calories. The graph is more complex. The points in the graph show y=Sodium plotted against x=Calories separately for each level of Type. The common regression line in all panels ignores Type.

Table 10.9 Hot dog ANCOVA with a common regression line that ignores Type. See Figure 10.7.

```
> aovStatementAndAnova(CgT, warn=FALSE)
> anova(aov(Sodium ~ Calories, data = hotdog))
Analysis of Variance Table

Response: Sodium
          Df Sum Sq Mean Sq F value  Pr(>F)
Calories   1 106270  106270    14.5 0.00037 ***
Residuals 52 380718    7321
---
Signif. codes:  0 '***' 0.001 '**' 0.01 '*' 0.05 '.' 0.1 ' ' 1
```

Figure 10.8 and Table 10.10 show parallel regression lines for each type. They have separate intercepts and a common slope. This model is the standard *analysis of covariance* model. We are interested in the vertical distance between the parallel lines. Equivalently, we are interested in the distance between the intercepts. We see from the $F = 37.07433$ with $p = 1.3_{10}^{-10}$ in the first part of Table 10.10 that the vertical distance is significant.

The original preliminary conclusion based on Table 10.8 was misleading because it left out the critical dependence of y=Sodium on the x=Calories variable.

It is possible (see Exercise 10.5 for an example) for the covariate to be significant and not the grouping factor. In this example both are significant.

Sodium ~ Calories + Type

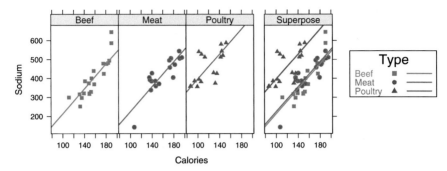

Fig. 10.8 Sodium ~ Calories + Type. Parallel lines. $Y_{ij} = \mu + \alpha_i + \beta(X_{ij} - \bar{\bar{X}}) + \epsilon_{ij}$. See Table 10.10. This illustrates the standard ANCOVA model.

Table 10.10 Hot dog ANCOVA with parallel lines and separate intercepts. See Figure 10.8.

```
> aovStatementAndAnova(CpT)
> anova(aov(Sodium ~ Calories + Type, data = hotdog))
Analysis of Variance Table

Response: Sodium
          Df Sum Sq Mean Sq F value  Pr(>F)
Calories   1 106270  106270    34.6 3.3e-07 ***
Type       2 227386  113693    37.1 1.3e-10 ***
Residuals 50 153331    3067
---
Signif. codes:  0 '***' 0.001 '**' 0.01 '*' 0.05 '.' 0.1 ' ' 1
```

We construct Figure 10.9 and Table 10.11 to show the means for the response Sodium adjusted for the covariate Calories. The adjustment maintains the same vertical distance between the fitted lines that we observe in Figure 10.8. From the ANOVA table in Table 10.11 we see that the adjusted means have the same residual sum of squares as the unadjusted means. The residual degrees of freedom are wrong because the analysis doesn't know that the effect of the Calories variable has already been removed. The Type sum of squares is not what we anticipated because we did not adjust the Type dummy variables for the covariate; we only adjusted the response variable.

Now that we have shown the factor Type to be important, we show in Table 10.12 and Figure 10.10 the results of multiple comparisons analysis using the Tukey procedure. These show that Meat and Beef are indistinguishable and that Poultry differs from both Meat and Beef.

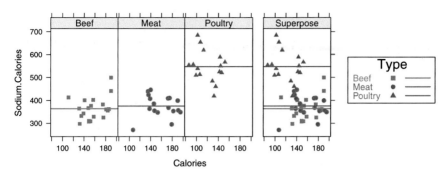

Fig. 10.9 Sodium.Calories ~ Type, x=Calories. Horizontal lines after adjustment for the covariate. $\left(Y_{ij}|X_{ij}\right) = \mu + \alpha_i + \epsilon_{ij}$. See Table 10.11. The vertical distance from each point to its line is identical in this figure to the vertical distances shown in Figure 10.8.

Table 10.11 Horizontal lines after adjustment for the covariate. See Figure 10.9.

```
> aovStatementAndAnova(T.C)
> anova(aov(Sodium.Calories ~ Type, data = hotdog))
Analysis of Variance Table

Response: Sodium.Calories
          Df Sum Sq Mean Sq F value  Pr(>F)
Type       2 368463  184232    61.3 2.7e-14 ***
Residuals 51 153331    3006
---
Signif. codes:  0 '***' 0.001 '**' 0.01 '*' 0.05 '.' 0.1 ' ' 1
```

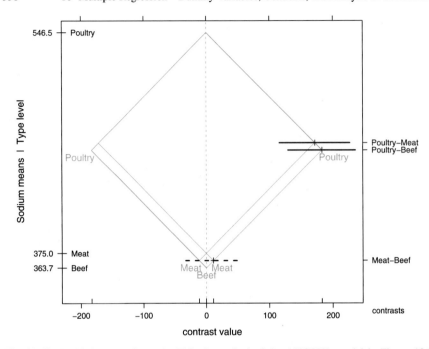

Fig. 10.10 Multiple comparisons by Tukey's method of the ANCOVA model in Figure 10.8 and Table 10.10 comparing the mean Sodium content of three Types of hot dogs adjusted for Calories. See also Figures 10.8 and 10.9 and Table 10.12.

Table 10.12 Multiple comparisons by Tukey's method of the ANCOVA model in Figure 10.8 and Table 10.10 comparing the mean Sodium content of three Types of hot dogs adjusted for Calories. See also Figure 10.10.

```
> CpT.mmc <- mmc(aov.trellis(CpT))

> CpT.mmc
Tukey contrasts
Fit: aov(formula = Sodium ~ Calories + Type, data = hotdog)
Estimated Quantile = 2.41
95% family-wise confidence level
$mca
              estimate stderr  lower   upper height
Poultry-Meat   171.47  23.13 115.73 227.21  460.8
Poultry-Beef   182.76  22.19 129.30 236.22  455.1
Meat-Beef       11.29  18.28 -32.75  55.34  369.4
$none
         estimate stderr lower upper height
Poultry    546.5  16.07 507.8 585.2  546.5
Meat       375.0  14.13 341.0 409.1  375.0
Beef       363.7  12.94 332.6 394.9  363.7
```

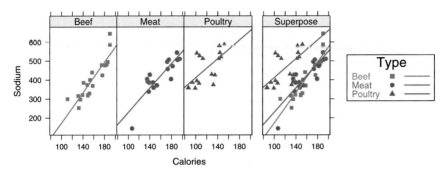

Fig. 10.11 Sodium ~ Calories * Type. Separate regression lines. $Y_{ij} = \mu + \alpha_i + \beta_i(X_{ij} - \bar{\bar{X}}) + \epsilon_{ij}$. See Table 10.13.

Table 10.13 Hot dog ANCOVA with separate regression lines (slopes and intercepts). See Figure 10.11.

```
> aovStatementAndAnova(CsT)
> anova(aov(Sodium ~ Calories * Type, data = hotdog))
Analysis of Variance Table

Response: Sodium
              Df Sum Sq Mean Sq F value  Pr(>F)
Calories       1 106270  106270   35.69 2.7e-07 ***
Type           2 227386  113693   38.18 1.2e-10 ***
Calories:Type  2  10402    5201    1.75    0.19
Residuals     48 142930    2978
---
Signif. codes:  0 '***' 0.001 '**' 0.01 '*' 0.05 '.' 0.1 ' ' 1
```

10.6.5 Tests of Equality of Regression Lines

In Section 10.6.4 we assume the constant slope model (10.2) and test whether the intercepts differed by testing (10.5) about α_i. We can also work with the separate slope model (10.3) and test (10.4) about β_i.

Figure 10.11 and Table 10.13 show separate regression lines for each group. These have separate intercepts and slopes. The F-test of Calories:Type in Table 10.13 having p-value = .185 addresses the null hypothesis that the regression lines for predicting Sodium from Calories are parallel.

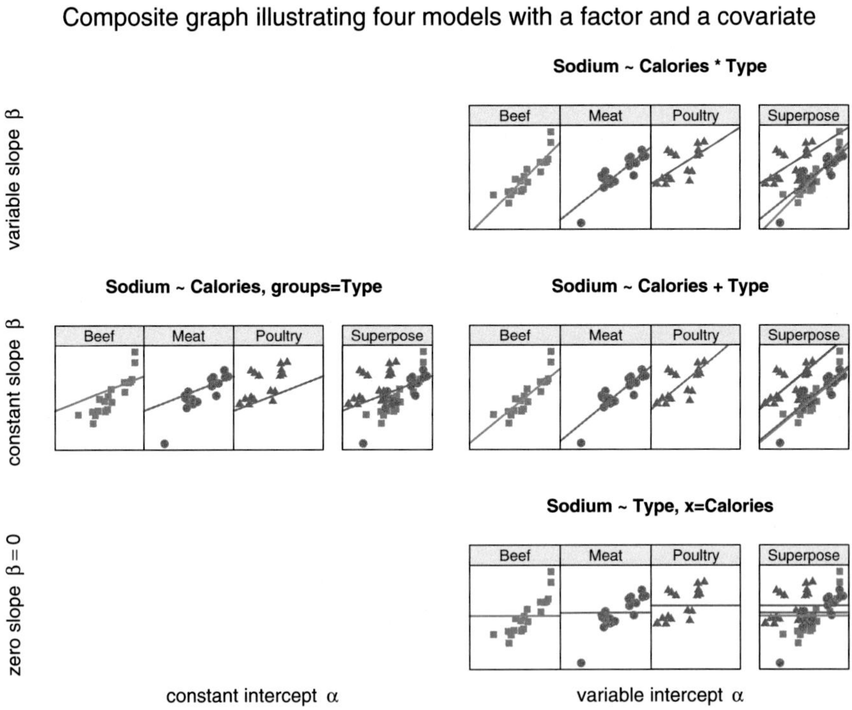

Composite graph illustrating four models with a factor and a covariate

Fig. 10.12 Four models for the hot dog data, arranged in two columns corresponding to the two possibilities for the intercept in the model and three rows corresponding to the three possibilities for the slope in the model. The models are often described as

	Constant intercept α	Variable intercept α
Variable slope β		Analysis of covariance with interaction of the factor and the covariate.
Constant slope β	Linear regression, ignoring the factor.	Standard analysis of covariance with constant slope and variable intercept.
Zero slope $\beta = 0$		Analysis of variance, ignoring the covariate.

Observe in Figure 10.11 that the slopes of the lines for the regressions of Sodium on Calories appear to differ for the three Types of hot dog. This null hypothesis is expressed as two equalities in Equation (10.4) and is tested in Table 10.13 using the two degree-of-freedom sum of squares for the interaction Calories:Type. The p-value for this test, 0.185, implies that the null hypothesis cannot be rejected and therefore that the three slopes are homogeneous. Any difference among them is too small to detect with the sample sizes in this data set.

Conditional on the homogeneity of the three slopes, the two degree-of-freedom sum of squares for Type in Table 10.10 tests the hypothesis that the three regression lines have a common intercept, a null hypothesis expressed in Equation (10.5). The zero p-value for this test implies that the intercepts are not identical.

10.7 ancovaplot **Function**

The ANCOVA plot has been calculated with the ancovaplot function, one of the functions that we provide in the **HH** package. The ancovaplot function constructs the appropriate trellis graphics commands for the plot. The specific feature that requires a separate function is its handling of the x= and groups= arguments respectively for the one-way ANOVA and the simple regression models. The result of the function is an ancovaplot object, which is essentially an ordinary trellis object with a different class. We have provided methods for ancova and related functions that will operate directly on the ancovaplot object.

The four basic options are shown in Table 10.14. Output from each is shown in Figures 10.7, 10.6, 10.8, and 10.11 and Tables 10.9, 10.8, 10.10, and 10.13. Figure 10.12 shows the graphs from all four in a single coordinated display.

Table 10.14 Four ways to use the ancovaplot function. See Figure 10.12 for a coordinated placement of all four of these plots on the same page.

```
data(hotdog, package="HH")
data(col3x2, package="HH")

## constant line across all groups
## y ~ x
ancovaplot(Sodium ~ Calories, groups=Type, data=hotdog, col=col3x2)

## different horizontal line in each group
## y ~ a
ancovaplot(Sodium ~ Type, x=Calories, data=hotdog, col=col3x2)

## constant slope, different intercepts
## y ~ x + a   or   y ~ a + x
ancovaplot(Sodium ~ Calories + Type, data=hotdog, col=col3x2)

## different slopes, and different intercepts
## y ~ x * a   or   y ~ a * x
ancovaplot(Sodium ~ Calories * Type, data=hotdog, col=col3x2)
```

10.8 Exercises

We recommend that for all exercises involving a data set, you begin by examining a scatterplot matrix of the variables.

10.1. Demonstrate that the two coding schemes

$$W_{\text{female}} = \begin{pmatrix} 1 & 1 \\ 1 & 0 \end{pmatrix} \quad \text{and} \quad W_{\text{treat}} = \begin{pmatrix} 1 & 1 \\ 1 & -1 \end{pmatrix}$$

in Section 10.2 are equivalent for regression in the sense of Section 10.3 by finding the A matrix that relates them.

10.2. Demonstrate that the orthogonal polynomials in Table 10.6 span the same column space as the matrix whose columns are the simple polynomials $x = (1, 2, 3, 4, 5, 6)$, x^2, x^3, x^4, x^5. Plot the columns of the matrix and compare the plot to Figure 10.4.

10.3. Demonstrate that the two coding schemes for the ResidenceLocation example in Section 10.1 are equivalent by defining the corresponding W variables and finding the A matrix that relates them.

10.4. We first investigated the dataset data(water) in Exercise 4.4.

a. Plot mortality vs calcium, using separate plot symbols for each value of derbynor. Does it appear from this plot that derbynor would contribute to explaining the variation in mortality?

b. Perform separate regressions of mortality on calcium for each value of derbynor. Compare these to the estimated coefficients in a multiple regression of mortality on both calcium and derbynor.

c. Interpret the regression coefficients in the multiple regression in terms of the model variables.

d. Suggest the public health conclusions of your analysis.

10.5. Do an analysis of covariance with model (10.2) of the simple dataset

y	x	a
1	1	1
2	2	1
3	3	2
4	4	2
5	5	3
6	6	3

Show that covariate x is significant and the grouping factor a is not.

10.6. The Erie house-price data data(hpErie) is introduced in Exercise 9.3. That exercise invites examination of the impact of two high-priced houses by comparing analyses with these houses included or omitted. Revisit these data, adding a dummy variable highprice defined as 1 if one of the two high-priced houses and 0 otherwise. Perform a stepwise regression analysis including this new variable and compare your results with those in Exercise 9.3.

10.7. Reconsider the salary model in Section 10.4.1.

a. Interpret, in terms of the model variables salary, age, gender, etc., the finding that β_2 is significantly less than zero.

b. Write the null hypothesis in terms of the β_j's:

> $E(Y)$ for whites with 12 years of schooling is the same as $E(Y)$ for nonwhites with 16 years of schooling.

c. Write the null hypothesis in terms of the β_j's:

> $E(Y)$ increases at the rate of $2,000 per year of schooling for whites and at the rate of $2,500 per year of schooling for nonwhites.

d. If the gender and race are interpreted as factors, rather than as arbitrarily coded dummy variables, then the generated dummy variables differ from the 0 and 1 coding used in Section 10.4.1. Therefore, the estimated $\hat{\beta}_j$ will differ. Explain why the t-tests and the F-test will remain the same.

10.8. Rerun the polynomial contrasts for the data(fabricwear) example in Table 10.7 without the outlier noted in Figure 10.3.

Chapter 11

Multiple Regression—Regression Diagnostics

In Chapter 9 we show how to set up and produce an initial analysis of a regression model with several predictors. In this chapter we discuss ways to investigate whether the model assumptions are met and, when the assumptions are not met, ways to revise the model to better conform with the assumptions. We also examine ways to assess the effect on model performance of individual predictors or individual cases (observations).

11.1 Example—Rent Data

11.1.1 Study Objectives

Alfalfa is a high-protein crop that is suitable as food for dairy cows. There are two research questions to ask the data in file data(rent) (from file (alr162) in Weisberg (1985)). It is thought that rent for land planted to alfalfa relative to rent for other agricultural purposes would be higher in areas with a high density of dairy cows and rents would be lower in counties where liming is required, since that would mean additional expense.

11.1.2 Data Description

The data displayed in the scatterplot matrices (sploms) in Figure 11.1 were collected to study the variation in rent paid in 1977 for agricultural land planted to alfalfa. The unit of analysis is a county in Minnesota; the 67 counties with appreciable rented

© Springer Science+Business Media New York 2015
R.M. Heiberger, B. Holland, *Statistical Analysis and Data Display*,
Springer Texts in Statistics, DOI 10.1007/978-1-4939-2122-5_11

farmland are included. Note that we automatically conditioned the splom on the factor lime. The original data include:

rnt.alf: average rent per acre planted to alfalfa

rnt.till: average rent paid for all tillable land

cow.dens: density of dairy cows (number per square mile)

prop.past: proportion of farmland used as pasture

lime: "lime" if liming is required to grow alfalfa; "no.lime" otherwise
            (Lime is a calcium oxide compound that is spread on a field as a fertilizer.)

We added one more variable

alf.till: the ratio of rnt.alf to rnt.till

to investigate the relative rent question.

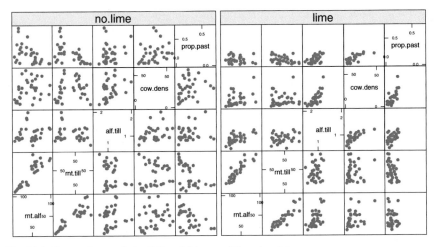

Fig. 11.1 Scatterplot matrices of all variables conditioned on lime.

11.1.3 Rent Levels

It is immediately clear from the sploms in Figure 11.1 that lime is very important in the distribution of cow.dens and prop.past as neither has any large values in the lime splom. The ratio alf.till is slightly higher in the no.lime splom.

lime does not seem to have an effect on either of the rent variables rent.alf or rent.till, as their panels have similar distributions in both sploms. The regression analysis of rent.alf in Table 11.1 supports that impression as lime has a very low *t*-value. prop.past also has a very low *t*-value.

Table 11.1 rent.alf regressed against all other observed variables.

```
> rent.lm31 <-
+     lm(rnt.alf ~ rnt.till + cow.dens + prop.past + lime,
+         data=rent)

> summary(rent.lm31)

Call:
lm(formula = rnt.alf ~ rnt.till + cow.dens + prop.past + lime,
    data = rent)

Residuals:
    Min      1Q  Median      3Q     Max
-21.229  -4.869  -0.029   4.755  27.767

Coefficients:
             Estimate Std. Error t value Pr(>|t|)
(Intercept)    -3.334      4.101   -0.81  0.41931
rnt.till        0.883      0.069   12.80  < 2e-16 ***
cow.dens        0.432      0.108    4.00  0.00017 ***
prop.past     -11.380     11.894   -0.96  0.34236
lime1          -0.506      1.425   -0.36  0.72371
---
Signif. codes:  0 '***' 0.001 '**' 0.01 '*' 0.05 '.' 0.1 ' ' 1

Residual standard error: 9.31 on 62 degrees of freedom
Multiple R-squared:  0.84,Adjusted R-squared:  0.83
F-statistic: 81.6 on 4 and 62 DF,  p-value: <2e-16


> anova(rent.lm31)
Analysis of Variance Table

Response: rnt.alf
          Df Sum Sq Mean Sq F value  Pr(>F)
rnt.till   1  25824   25824  297.89  <2e-16 ***
cow.dens   1   2386    2386   27.53  2e-06 ***
prop.past  1     74      74    0.85   0.36
lime       1     11      11    0.13   0.72
Residuals 62   5375      87
---
Signif. codes:  0 '***' 0.001 '**' 0.01 '*' 0.05 '.' 0.1 ' ' 1
```

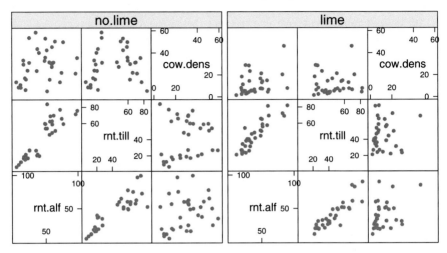

Fig. 11.2 Scatterplot matrices of `rnt.alf` with 2 *X*-variables, conditioned on `lime`.

We therefore look at a simpler model, without the `prop.past` predictor but with the `cow.dens:lime` interaction, in Figure 11.2 and Table 11.2. Although the regression analysis shows the `lime` coefficient as not significant, it shows the interaction of lime with cow density to be on the edge of significance ($p = .055$). We left both in the model because there appears to be much higher variability in the residuals for high values of `rnt.till` and lower variability in the residuals for low values of `cow.dens` in the `no.lime` counties as indicated in Figure 11.3.

Our conclusion from this portion of the analysis is that rent for alfalfa is related to rent for tillage and to cow density. The relationship with cow density may depend on the need for lime. We need to investigate the variability of the residuals.

Table 11.2 rent.alf regressed against all variables except prop.past, and including the interaction of cow density with lime.

```
> rent.lm4ln <- lm(rnt.alf ~ rnt.till + cow.dens +
+                   lime + cow.dens:lime, data=rent)

> summary(rent.lm4ln)

Call:
lm(formula = rnt.alf ~ rnt.till + cow.dens + lime + cow.dens:lime,
    data = rent)

Residuals:
    Min      1Q  Median      3Q     Max
-24.346  -4.251  -0.194   4.151  27.193

Coefficients:
               Estimate Std. Error t value Pr(>|t|)
(Intercept)     -5.9584     3.0117   -1.98    0.052 .
rnt.till         0.9269     0.0536   17.28  < 2e-16 ***
cow.dens         0.4567     0.0991    4.61 2.1e-05 ***
lime1           -3.6034     2.1642   -1.66    0.101
cow.dens:lime1   0.1926     0.0986    1.95    0.055 .
---
Signif. codes:  0 '***' 0.001 '**' 0.01 '*' 0.05 '.' 0.1 ' ' 1

Residual standard error: 9.1 on 62 degrees of freedom
Multiple R-squared:  0.847, Adjusted R-squared:  0.838
F-statistic: 86.1 on 4 and 62 DF,  p-value: <2e-16


> anova(rent.lm4ln)
Analysis of Variance Table

Response: rnt.alf
              Df Sum Sq Mean Sq F value  Pr(>F)
rnt.till       1  25824   25824  311.61 < 2e-16 ***
cow.dens       1   2386    2386   28.80 1.3e-06 ***
lime           1      5       5    0.07   0.799
cow.dens:lime  1    316     316    3.81   0.055 .
Residuals     62   5138      83
---
Signif. codes:  0 '***' 0.001 '**' 0.01 '*' 0.05 '.' 0.1 ' ' 1
```

11.1.4 Alfalfa Rent Relative to Other Rent

Returning to the sploms in Figure 11.1, we see that that lime puts an upper bound
on the alf.till ratio. The ratio does seem to go up with cow density and seems
to have a variance relation with proportion in pasture. In Table 11.3, a regression
of the alf.till ratio against the non-rent variables, we see that we can drop the
prop.past variable.

We continue with Table 11.4 and Figure 11.4, which show an ordinary analysis
of covariance with model

$$alf.till \sim cow.dens * lime \qquad (11.1)$$

The ANOVA table in Table 11.4 shows the interaction is not quite significant.

We choose to investigate individual points by looking at plots of the residuals in Figure 11.5 (with the QQ-plot expanded in Figure 11.9) and the regression
diagnostics in Figure 11.6. These show the three points (19, 33, 60) in the no.lime
group and the single point (49) in the lime group as being potentially influential.
Figure 11.6, produced with our functions lm.case.s and plot.case.s, includes
boundaries for the standard recommended thresholds for the various diagnostic measures discussed in Section 11.3.

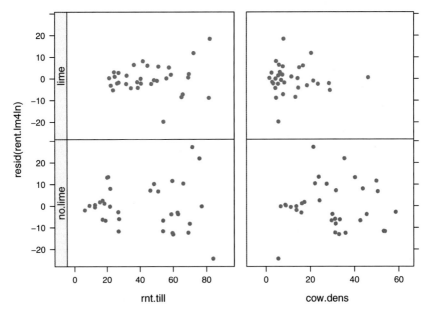

Fig. 11.3 Residuals from rnt.alf ~ rnt.till + cow.dens*lime (in Table 11.2 and
Figure 11.2) plotted against the X-variables conditioned on lime.

Table 11.3 `alf.till` ratio regressed against cow density | `lime` and proportion in pasture.

```
> rent.lm12p <- lm(alf.till ~ lime * cow.dens + prop.past, data=rent)

> summary(rent.lm12p, corr=FALSE)

Call:
lm(formula = alf.till ~ lime * cow.dens + prop.past, data = rent)

Residuals:
    Min      1Q  Median      3Q     Max
-0.3342 -0.1247 -0.0203  0.1045  0.7853

Coefficients:
                Estimate Std. Error t value Pr(>|t|)
(Intercept)      0.78957    0.05637   14.01  < 2e-16 ***
lime1           -0.09686    0.05333   -1.82  0.07419 .
cow.dens         0.00944    0.00259    3.64  0.00056 ***
prop.past        0.18989    0.22670    0.84  0.40546
lime1:cow.dens   0.00391    0.00242    1.62  0.11063
---
Signif. codes:  0 '***' 0.001 '**' 0.01 '*' 0.05 '.' 0.1 ' ' 1

Residual standard error: 0.223 on 62 degrees of freedom
Multiple R-squared:  0.366,Adjusted R-squared:  0.325
F-statistic: 8.94 on 4 and 62 DF,  p-value: 9.17e-06


> anova(rent.lm12p)
Analysis of Variance Table

Response: alf.till
              Df Sum Sq Mean Sq F value  Pr(>F)
lime           1  0.846   0.846   17.03 0.00011 ***
cow.dens       1  0.754   0.754   15.19 0.00024 ***
prop.past      1  0.045   0.045    0.91 0.34503
lime:cow.dens  1  0.130   0.130    2.62 0.11063
Residuals     62  3.078   0.050
---
Signif. codes:  0 '***' 0.001 '**' 0.01 '*' 0.05 '.' 0.1 ' ' 1
```

We locate the potentially influential points in Figure 11.7 and see them as the three counties with the highest ratios and the one `lime` county with an unusually high cow density. In Section 11.3 we will discuss the statistics displayed in Figures 11.5 and 11.6 as well as their interpretation.

We redo the analysis without these four points in Table 11.6 and Figure 11.8. After isolating these four counties we see significantly different slopes in the `no.lime` and `lime` counties.

Table 11.4 alf.till ratio regressed against cow density | lime. See Figure 11.4.

```
> rent.lm12m <- aov(alf.till ~ lime * cow.dens, data=rent)

> anova(rent.lm12m)
Analysis of Variance Table

Response: alf.till
               Df Sum Sq Mean Sq F value  Pr(>F)
lime            1  0.846   0.846   17.11 0.00011 ***
cow.dens        1  0.754   0.754   15.26 0.00023 ***
lime:cow.dens   1  0.140   0.140    2.84 0.09708 .
Residuals      63  3.113   0.049
---
Signif. codes:  0 '***' 0.001 '**' 0.01 '*' 0.05 '.' 0.1 ' ' 1

> summary.lm(rent.lm12m)

Call:
aov(formula = alf.till ~ lime * cow.dens, data = rent)

Residuals:
    Min      1Q  Median      3Q     Max
-0.3296 -0.1362 -0.0139  0.0877  0.8408

Coefficients:
                Estimate Std. Error t value Pr(>|t|)
(Intercept)      0.80653    0.05248   15.37  < 2e-16 ***
lime1           -0.10424    0.05248   -1.99    0.051 .
cow.dens         0.01024    0.00241    4.25  7.1e-05 ***
lime1:cow.dens   0.00405    0.00241    1.68    0.097 .
---
Signif. codes:  0 '***' 0.001 '**' 0.01 '*' 0.05 '.' 0.1 ' ' 1

Residual standard error: 0.222 on 63 degrees of freedom
Multiple R-squared:  0.359,Adjusted R-squared:  0.328
F-statistic: 11.7 on 3 and 63 DF,  p-value: 3.32e-06
```

Our conclusion at this step is that for most counties, there is a linear relationship of the rent ratio to the cow density, with the slope depending on the need for lime. The three no.lime counties and the one lime county need additional investigation.

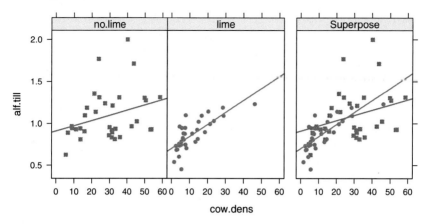

Fig. 11.4 ANCOVA `rnt.alf/rnt.till ~ cow.dens | lime`. See Table 11.4.

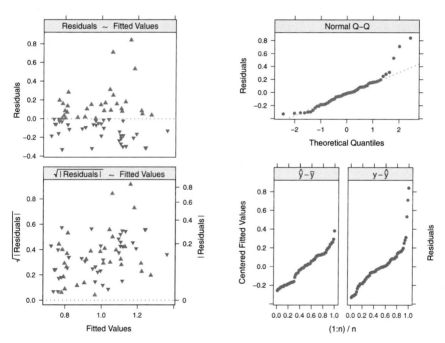

Fig. 11.5 Residuals from ANCOVA (`rnt.alf/rnt.till`) `~ cow.dens | lime`. See Table 11.4 and Figure 11.4. The structure of the panels in this figure is discussed in Section 8.4. The figure itself is similar to Figure 8.6.

Table 11.5 Case diagnostics for model in Table 11.4. The diagnostics are plotted in Figure 11.6. The case numbers for the noteworthy cases are listed here.

```
> rent.case12m <- case(rent.lm12m)

> rent.case12m.trellis <-
+     plot(rent.case12m, rent.lm12m, par.strip.text=list(cex=1.2),
+          layout=c(3,3), main.cex=1.6, col=likertColor(2)[2], lwd=4)

> rent.case12m.trellis ## display both graph and list of noteworthy cases
                        Noteworthy Observations
Student del resid       19 33
deleted std dev         19 33
h                       13 40 49 56
Cook's distance
dffits                   5 19 32 33 49 60 66
DFBETAS (Intercept)      5 19
DFBETAS lime1            5 19
DFBETAS cow.dens         5 32 33 49 60
DFBETAS lime1:cow.dens   5 32 33 49 60
```

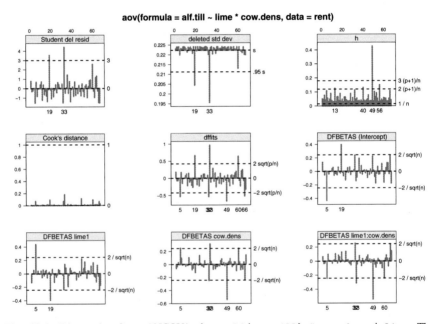

Fig. 11.6 Diagnostics from ANCOVA (`rnt.alf/rnt.till`) ~ `cow.dens` | `lime`. The model is displayed in Table 11.4 and Figure 11.4. Each of the statistics in these panels is discussed in Section 11.3 and shown enlarged in Figures 11.12–11.17. To work around the problem that identification in the graph's x-axis of noteworthy cases often suffers from overprinting, the `plot.case` function returns and prints a list of noteworthy cases. We show the list in Table 11.5.

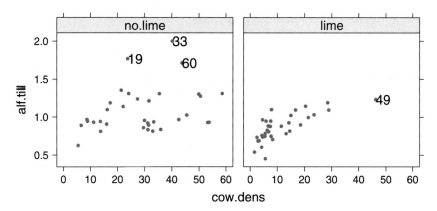

Fig. 11.7 Identified points in ANCOVA (rnt.alf/rnt.till) ~ cow.dens | lime.

Table 11.6 ANCOVA of `alf.till` ratio regressed against cow density and lime with four removed observations. See Figure 11.8. Compare to Table 11.4.

```
> rent.lm12ms.aov <- aov(alf.till ~ lime * cow.dens,
+                        data=rent[-c(19, 33, 60, 49),])

> anova(rent.lm12ms.aov)
Analysis of Variance Table

Response: alf.till
               Df Sum Sq Mean Sq F value  Pr(>F)
lime            1  0.428   0.428   17.81 8.5e-05 ***
cow.dens        1  0.395   0.395   16.43 0.00015 ***
lime:cow.dens   1  0.233   0.233    9.67 0.00288 **
Residuals      59  1.419   0.024
---
Signif. codes:  0 '***' 0.001 '**' 0.01 '*' 0.05 '.' 0.1 ' ' 1
```

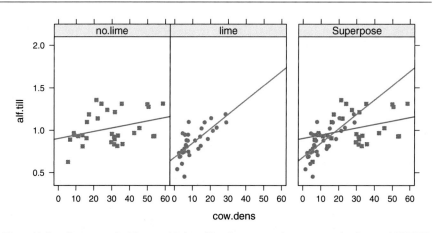

Fig. 11.8 Repeat of Figure 11.4 with four counties removed from ANCOVA ancovaplot(alf.till ~ lime * cow.dens, data=rent[-c(19, 33, 60, 49),]). See Table 11.6. Compare to Figure 11.4.

11.2 Checks on Model Assumptions

We assume in Section 9.3 that the model error terms $\epsilon_i \sim \text{NID}(0, \sigma^2)$ (Normal Independently Distributed), that is that they have the same variance σ^2 for all cases, are mutually uncorrelated or independent, and are normally distributed. In order for the conclusions from our analyses to be valid, these assumptions must be true. Therefore, we discuss ways to verify the assumptions and then suggest some remedies when assumptions are not met.

11.2.1 Scatterplot Matrix

We previously mentioned the importance of routinely producing scatterplot matrices as part of analyses involving several variables. We produced many such plots in our discussion in Section 11.1. Here we focus on the rows of the scatterplot matrix that correspond to the response variables. The panels in these rows, the plots of the response y vs each of the explanatory variables x_j, should each be approximately linear. In Section 11.1.3 the response is shown in the `rnt.alf` row in Figure 11.1 and in Figure 11.2. In Section 11.1.4 the response is the `alf.till` row in Figure 11.1 and in Figure 11.4. If the plot of y against any explanatory variable suggests curvature in the relationship, the analyst should consider transforming either the response variable or that explanatory variable so that following transformation the plot of y vs the transformed x_j is close to linear. A successful transformation suggests the use of this transformed predictor rather than the original in the regression model. Exercise 11.5 explores this idea.

11.2.2 Residual Plots

Before a model can be accepted for use in explanation or prediction, the analyst should produce and examine plots involving the residuals calculated from the fit of the model to the data. The residuals e_i should be plotted vs each of the following, one plot point per case:

- the fitted values of the response $\hat{y}_i$
- each of the model's explanatory variables x_j
- possibly other variables under consideration for the model but not yet a part of it
- time, if the data are time-ordered

In addition, the partial residuals (see Section 9.13.1) should be plotted against the corresponding predictors and against the residuals from regressing each predictor against the other predictors (added variable plots; see Section 9.13.4). Ideally, each of these plots should exhibit no systematic character and have random scatter about the horizontal line at 0, the mean of the e_i.

In order to check for normality, the analyst should produce a normal probability plot of the residuals. If there is doubt that this plot confirms normality, the analyst can request the p-value from an all-purpose test of normality having good power against a variety of alternatives, such as the Shapiro–Wilk test mentioned in Section 5.7.

If a residual plot suggests that an assumption is not met, the analyst must seek a remedy following which the assumption is met.

We show in Figure 11.9 the normal probability plot for the rent ratio alf.till analysis in Table 11.4 and Figure 11.4. It does not look normal. Compare this plot to Figure 11.10, which shows probability plots of six normal and six non-normal variables.

From the cow.dens column, we again see similar behavior in Rows 1 and 3. We also note the higher variability in Y for the higher densities. We get a sense of why we see that difference in variability from the interaction lime:cow.dens column. Here we see, most clearly in the partial residuals plot in Row 3, that the high variability is observed when the interaction variable is negative, corresponding to the no.lime counties.

Figure 11.11 shows several plots of the residuals and partial residuals from the model in Table 11.4 and Figure 11.4. From the lime column, we see that the ratio alf.till is higher for lime=−1 (no lime) than for lime=1 (lime). The pattern is similar in the observed variable plots in Row 1 and the partial residuals plots in Row 3, suggesting that the lime effect is independent of the other variables.

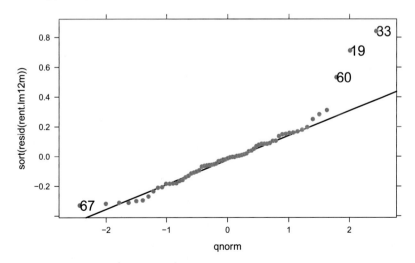

Fig. 11.9 Normal plot of residuals from ANCOVA rnt.alf/rnt.till ~ cow.dens | lime. See Table 11.4 and Figure 11.4. The results do not look normal. We ran the Shapiro-Wilk normality test with statistic W=0.8969 and $p = 4 \cdot 10^{-5}$. We identified the four most extreme points. Three of them are the three no.lime counties that we had previously identified.

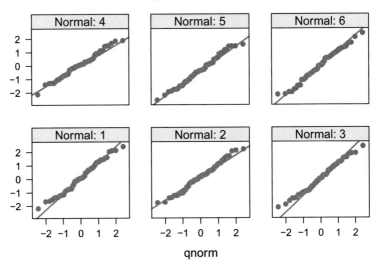

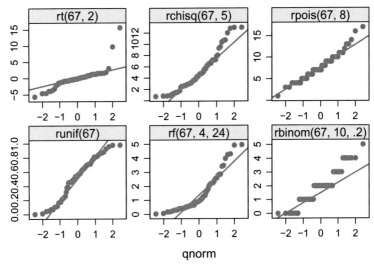

Fig. 11.10 Normal plot of six randomly generated normal variables and six randomly generated nonnormal variables. These plots are placed here to help you calibrate your eye to what normal and nonnormal distributions look like when plotted against the normal quantiles. t: long left and right tails as indicated by points below the diagonal on the left and above the diagonal on the right. Chi-square: short left and long right tails. Poisson: discrete appearance and long tail on the right. Uniform: short tails on left and right as indicated by points above the diagonal line on the left and below the diagonal line on the right. F: short tail on the left and long tail on the right. Binomial: discrete positions on the y-axis, with short tail on the left; this example with $p = .2$ is not symmetric and we see more points on the left.

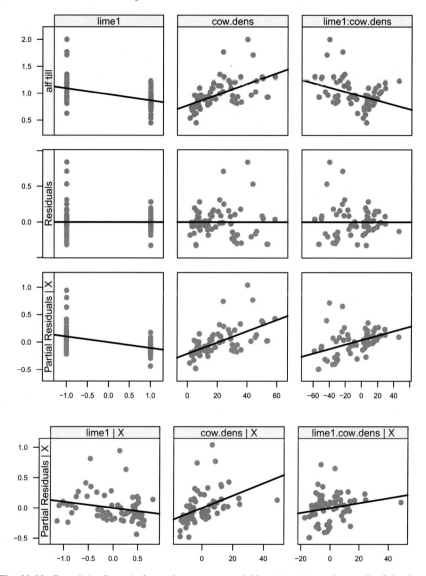

Fig. 11.11 Row 1 (at the top) shows the response variable alf.till against each of the three predictors. Row 2 shows the ordinary residuals $e = Y - \hat{Y}$ from the regression on all three variables against each of the three predictors. Row 3 shows the "partial residuals plots", the partial residuals for each predictor against that predictor. Row 4 shows the "added variable plots", the partial residuals against the residuals of X_j regressed on the other two predictors. The slope for both rows 3 and 4, the partial residuals and the added variables, is exactly the regression coefficient for that term.

Table 11.7 Regression Diagnostics Formulas

Name	Notation and definition	Sequenced calculation formulas	Description
Observed response variable	$\underset{n\times 1}{Y}$		
Observed predictor variables	$\underset{n\times(1+p)}{X}$	$X = [\mathbf{1}\,X_1 X_2 \ldots X_p]$	All n observations
Fitted value	$\hat{Y} = (\hat{Y}_i)$	$X\hat{\beta}$	
Residual	e_i	$Y_i - \hat{Y}_i$	
Standard deviation	$s = \sqrt{\text{MSE}} = \sqrt{\text{var}(Y_i|X)}$	$\sqrt{\sum e_i^2/(n-p-1)}$	
Leverage	$h_i = h_{ii} = \frac{\partial \hat{y}_i}{\partial y_i}$	$\text{diag}\left(X(X'X)^{-1}X'\right)$	
Variance of e_i	$\text{var}(e_i)$	$s^2(1-h_{ii})$	$Y_i = \hat{Y}_i + e_i$
Variance of $\hat{Y}_i$	$\text{var}(\hat{Y}_i)$	$s^2 h_i$	$\text{var}(Y_i) = \text{var}(\hat{Y}_i) + \text{var}(e_i)$
Standardized residual	e_i^*	$e_i/(s\sqrt{1-h_i})$	$e_i/(\sigma\sqrt{1-h_i}) \sim N(0,1)$ when H_0 is true
Data with i^{th} row deleted	$\underset{(n-1)\times(1+p)}{X_{(i)}}$	$X_{[(1,2,\ldots,i-1,i+1,\ldots,n),(0,1,\ldots,p)]}$	
Deleted regression coefficients	$\hat{\beta}_{(i)} = (X'_{(i)}X_{(i)})^{-1}X'_{(i)}Y_{(i)}$	See description in Section 11.3.6.	Estimation of β based on $n-1$ observations, all except i. The definition isn't efficient. Use the algorithm in Section 11.3.6.
Deleted standard deviation	$s_{(i)} = \sqrt{\text{MSE}_{(i)}}$	$\sqrt{\left((n-p)s^2 - e_i^2/(1-h_i)\right)/(n-p-1)}$	$n-1$ observations, all except i.
Deleted predicted value	$\hat{Y}_{(i)}$	$X_{[i,(0,\ldots,p)]}\hat{\beta}_{(i)}$	Prediction Y_i based on the remaining $n-1$ observations

Studentized deleted residual	t_i	$e_i/\left(s_{(i)}\sqrt{1-h_i}\right)$	$t_i \sim t_{n-p-2}$ when H_0 is true
Cook's distance	$D_i = \left(\hat{Y} - \hat{Y}_{(i)}\right)'\left(\hat{Y} - \hat{Y}_{(i)}\right)/(p\,s^2)$ $= \left(\hat{\beta} - \hat{\beta}_{(i)}\right)' X'X \left(\hat{\beta} - \hat{\beta}_{(i)}\right)/(p\,s^2)$	$\dfrac{e_i^2}{p\,\mathsf{MSE}}\left(\dfrac{h_i}{(1-h_i)^2}\right)$	
Inverse of crossproduct of X	$C = (c_{ij})$	$(X'X)^{-1}$	
Covariance of coefficients	$s^2 C$	$s^2(X'X)^{-1}$	Estimated covariance matrix of regression coefficients
DFBETAS	$\mathsf{DFBETAS}_{ik} = (\hat{\beta}_k - \hat{\beta}_{k(i)})/(s_{(i)}\sqrt{c_{kk}})$		Standardized $\Delta\hat{\beta}_k$ when observation i is deleted
DFFITS	$\mathsf{DFFITS}_i = (\hat{Y}_i - \hat{Y}_{i(i)})/(s_{(i)}\sqrt{h_i})$	$\left(\dfrac{n-p-1}{\mathsf{SSE}\,(1-h_i)-e_i^2}\right)^{-\frac{1}{2}}\left(\dfrac{h_i}{1-h_i}\right)^{-\frac{1}{2}}$	Standardized $\Delta\hat{Y}_i$ when observation i is deleted

11.3 Case Statistics

Many of the diagnostics discussed in this chapter fall under the heading *case statistics*, i.e., they have a value for each of the n cases in the data set. If a case statistic has a value that is unusual, based on thresholds we discuss, the analyst should *scrutinize* the case. One action the analyst might take is to delete the case. This is justified if the analyst determines the case is not a member of the same population as the other cases in the data set. But deletion is just one possibility. Another is to determine that the flagged case is unusual in ways apart from those available in its information in the present data set, and this may suggest a need to add one or more additional predictors to the model.

There are many case statistics used in regression diagnostics. The concepts are complex and the notation more so. We summarize the notation in Table 11.7. We discuss each of the formulas and illustrate them with the diagnostic plots for the rent data that we originally showed in Figure 11.6. We reproduce each of the panels in that figure as a standalone plot here as part of the discussion.

We focus on five distinct case statistics, each having a different function and interpretation. (One of these, DFBETAS, is a vector with a distinct value for each regression coefficient including the intercept coefficient.) For small data sets the analyst may choose to display each of these case statistics for all cases. For larger data sets we suggest that the analyst display only those values of the case statistics that exceed a threshold, or flag, indicating that the case is unusual in some way. Recommended thresholds are mentioned in the following sections.

Leverage   measures how unusual a case is with respect to the values of its predictors, i.e., whether the values of a case's predictors are an outlying point in the p-dimensional space of predictors. Unlike the other case statistics, leverage does not involve the response variable.

Studentized deleted residuals   suggest how unusual cases are with respect to the case's value of the response variable.

Cook's distance   is a combined measure of the unusualness of a case's predictors and response. It sometimes happens that a case is flagged by Cook's distance but not quite flagged by leverage or Studentized deleted residuals.

DFFITS   indicates the extent to which deletion of the case impacts predictions made by the model.

DFBETAS   (one for each regression coefficient) show the extent to which deletion of a case would perturb that regression coefficient.

In the following sections we discuss these statistics in turn, presenting two formulas for each of them. The first, the definitional formula, is intended to be intuitive. It is used to explain to the reader what the formula measures and why it is helpful to view it in an analysis. It is also inefficient and should not be used as a computational formula. The second formula, the computational formula, is an order of magnitude more efficient for computation. It is not intuitive. We leave for Exercise 11.8 the proofs that the two sets of formulas are equivalent.

11.3.1 Leverage

The calculation of leverages is briefly addressed in 9.3.1. Leverages measure how unusual a case is with respect to its set of predictors. Unlike other measures in this chapter, leverages do not involve the response variable. The leverage h_{ii} of case i, usually abbreviated to h_i, is the ith diagonal entry of the *hat matrix* $H = X(X'X)^{-1}X'$. This matrix has come to be called the hat matrix because in matrix notation the predicted response is $\hat{Y} = X(X'X)^{-1}X'Y = HY$, i.e., H transforms Y to $\hat{Y}$ by placing a "hat" on the Y. It can be shown (see Exercise 11.9) that all leverages satisfy $\frac{1}{n} \leq h_i \leq 1$. If a model contains p predictors, an excessively large leverage is one for which

$$h_i > \frac{2(p+1)}{n} \quad \text{or} \quad h_i > \frac{3(p+1)}{n} \tag{11.2}$$

These suggested rules derive from the fact that the average of all n leverages is $\frac{p+1}{n}$, so they are based on exceeding 2 or 3 times this average. A case that is flagged because its leverage exceeds one or both of these thresholds has a value for at least one predictor that is unusual compared to values of such predictors for other cases. We can show that

$$h_{ii} = \frac{\partial \hat{y}_i}{\partial y_i} \quad \text{and} \quad h_{ij} = \frac{\partial \hat{y}_i}{\partial y_j}$$

The leverage h_i of case i is geometrically interpreted as the generalized (Mahalanobis) distance of $X_{i.}$ (the ith row of X) from the $(p + 1)$-dimensional centroid of all n rows of X.

More complicated forms of leverage have been devised to diagnose a group of cases that when considered together are unusual but when considered individually are not unusual.

Figure 11.12 displays the leverages for each case of the fit of the rent data using Model (11.1). This figure includes horizontal dotted lines demarking the two leverage thresholds given above. We observe that county 49 exceeds both thresholds, telling us that this county (requiring `lime`) has an unusually large `cow.dens`.

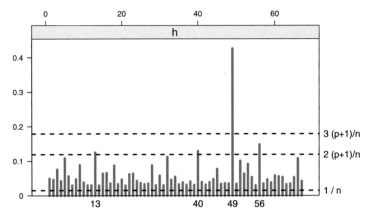

Fig. 11.12 Leverage for Model (11.1) for rent data.

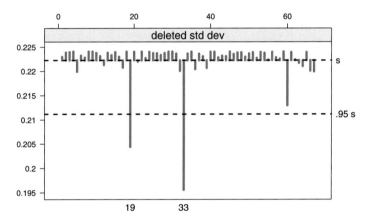

Fig. 11.13 Deleted standard deviations for Model (11.1) for rent data.

11.3.2 Deleted Standard Deviation

The deleted standard deviation $s_{(i)}$ is defined to be the value of s calculated from the same regression model using all cases *except* case i. Because the primary use of the $s_{(i)}$ is in the definition of the Studentized deleted residuals, there are no standard rules for interpreting these values themselves.

We compare the $s_{(i)}$ values to two thresholds, $.95s$ and $1.05s$. If deletion of an observation shifts the estimated standard deviation by 5% in either direction, we note it on the graph and choose to investigate the observation.

Figure 11.13 shows the deleted standard deviations for the rent data. We see two observations, 19 and 33, that are below our lower threshold.

11.3.3 Standardized and Studentized Deleted Residuals

The standardized and Studentized residuals help to assess the effect of each individual case on the calculated regression relationship. For case i the standardized residual

$$e_i^* = e_i / \sqrt{\widehat{\mathrm{var}}(e_i)} \tag{11.3}$$

is the calculated residual, e_i, standardized by dividing by its estimated standard error

$$\sqrt{\widehat{\mathrm{var}}(e_i)} = s \sqrt{1 - h_i} \tag{11.4}$$

Note that because this standard error depends on i, it differs slightly from case to case. The standardized residual is also called the *internally standardized residual* because the calculation of s includes case i.

The *Studentized deleted residual,* also called the *externally standardized residual,* for case i is calculated from the regular residuals, the deleted standard deviations, and the hat diagonals.

$$t_i = \frac{e_i}{s_{(i)} \sqrt{1 - h_i}} \tag{11.5}$$

As implied by this notation, t_i has a Student's t distribution with $n - p - 1$ degrees of freedom. Considering the t distribution with moderate degrees of freedom, we say that case i's response value is "unusual" (the actual response differs "appreciably" from the predicted response) if its absolute Studentized deleted residual exceeds 2 or 3. Such a case may be termed an *outlier.* We recommend a threshold of 2 for small data sets and 3 for large data. The reason for this recommendation is that for a large data set, 2 is the approximate 97.5[th] percentile of the t distribution so that when the model assumptions are satisfied for all cases, approximately 5% of these residuals will exceed 2 by chance alone.

We prefer the use of Studentized deleted residuals rather than standardized residuals because the former are interpretable as t statistics but the latter are not. A reason is that the numerator and denominator of t_i are statistically independent, but the numerator and denominator of the standardized residuals e_i^* are not independent.

It can be shown (see Exercise 11.8c) that the Studentized deleted residual defined intuitively in Equation (11.5) can be calculated more efficiently by the computational formula

$$t_i = e_i \left(\frac{n - p - 1}{\mathsf{SSE}\,(1 - h_i) - e_i^2} \right)^{\frac{1}{2}} \tag{11.6}$$

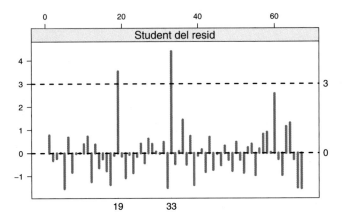

Fig. 11.14 Studentized deleted residuals for Model (11.1) for rent data.

where SSE is the error sum of squares under the full model having n cases. All terms in this expression are available from a single fitting with the n cases. Therefore, in calculating the n t_i's it is not necessary to refit the model n times corresponding to deleting each case in turn.

For our modeling of the rent data in Table 11.4, Figure 11.14 displays the Studentized (deleted) residuals for each case. We see that counties 19 and 33 both exceed the threshold 3, indicating that these counties have unusually large values of `alf.till`.

11.3.4 Cook's Distance

While leverage addresses the unusualness of a case's predictor variables, and Studentized deleted residuals address (primarily) the unusualness of a case's response variable, the Cook's distance D_i of a case assesses the unusualness of both its response and predictors. The Cook's distance D_i for case i can be interpreted in two ways.

Let $\hat{Y}$ be the n-vector of fitted values using all n cases and $\hat{Y}_{(i)}$ be the n-vector of fitted values when case i is not used in fitting. Then

$$D_i = \frac{\left(\hat{Y} - \hat{Y}_{(i)}\right)'\left(\hat{Y} - \hat{Y}_{(i)}\right)}{p\,\text{MSE}} \tag{11.7}$$

This illustrates the interpretation that Cook's distance for case i measures the change in the vector of predicted values when case i is omitted.

Let $\hat{\beta}_{(i)}$ be the vector of estimated regression coefficients estimated without case i. Then

$$D_i = \frac{\left(\hat{\beta} - \hat{\beta}_{(i)}\right)' X'X \left(\hat{\beta} - \hat{\beta}_{(i)}\right)}{p\,\mathsf{MSE}} \tag{11.8}$$

This representation shows that D_i measures the change in the vector of estimated regression coefficients when case i is omitted.

As with the Studentized deleted residual, the n Cook's distances can be calculated without running n regressions omitting each case in turn. It can be shown that

$$D_i = \frac{e_i^2}{p\,\mathsf{MSE}} \left(\frac{h_i}{(1-h_i)^2}\right) \tag{11.9}$$

From this formula it is apparent that a case with a large Cook's distance has either a large residual, a large leverage, or some combination of these two.

We recommend that a case be regarded as unusual if its Cook's distance exceeds 1. This threshold for what constitutes an unusually large value of Cook's distance D_i follows the recommendation of Weisberg (1985) (page 120).

> Since for most F distributions the 50% point is near 1, a value of $D_i = 1$ will move the estimate to the edge of about a 50% confidence region, a potentially important change. If the largest D_i is substantially less than 1, deletion of a case will not change the estimate of β by much. To investigate the influence of a case more closely, the analyst should delete the large D_i case and recompute the analysis to see exactly what aspects of it have changed.

There are also arguments, for example in Fox (1991), for a much smaller threshold $4/(n - p - 1)$ or $4/n$ that decreases with increasing sample size. We are unconvinced by these arguments.

Figure 11.15 displays the Cook's distances for the rent data. Counties 5, 19, 32, 33, 49, 60, and 66 have much larger Cook's distances than the other counties, but none of these 7 counties approaches the threshold of 1 that would flag a county as unusual. Therefore, Cook's distance flags no data points fitted by `alf.till ~ lime*cow.dens`.

11.3.5 DFFITS

DFFITS, shown in Figure 11.16, is an abbreviation for "difference in fits". DFFITS_i is a standardized measure of the amount by which predicted value $\hat{Y}_i$ for case i changes when the data on this case is deleted from the data set. A flag for a case with large DFFITS is one having absolute value greater than $2\sqrt{p/n}$.

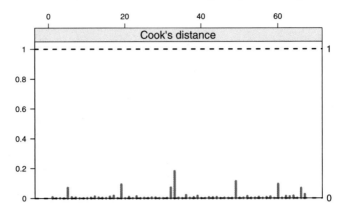

Fig. 11.15 Cook's distances for Model (11.1) for rent data.

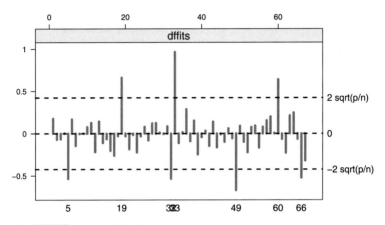

Fig. 11.16 DFFITS for Model (11.1) for rent data.

The interpretation of DFFITS$_i$ is apparent from the formula

$$\text{DFFITS}_i = \frac{\hat{Y}_i - \hat{Y}_{i(i)}}{\sqrt{\text{MSE}_{(i)}\, h_i}} \qquad (11.10)$$

where, as before, an (i) in a subscript means that the quantity is calculated with case i omitted from the data. As is seen from

$$\text{DFFITS}_i = \left(\frac{n - p - 1}{\text{SSE}\,(1 - h_i) - e_i^2} \right)^{\frac{1}{2}} \left(\frac{h_i}{1 - h_i} \right)^{\frac{1}{2}} \qquad (11.11)$$

DFFITS$_i$ can be calculated from the output of the regression using all n cases.

11.3.6 DFBETAS

$DFBETAS_{ik}$ is a standardized measure of the amount by which the kth regression coefficient changes if the ith observation is omitted from the data set. A case is considered to have a large such measure if its absolute DFBETAS is greater than $2/\sqrt{n}$. Since a regression analysis has np DFBETAS in all, a request for DFBETAS in a large complicated regression analysis will generate a lot of output.

$DFBETAS_{ik}$ is defined by

$$DFBETAS_{ik} = \frac{\hat{\beta}_k - \hat{\beta}_{k(i)}}{\sqrt{MSE_{(i)}\, c_{kk}}}$$

for $k = 0, 1, \ldots, p$, where c_{kk} is the kth diagonal entry in $(X'X)^{-1}$. The terms $\hat{\beta}_{k(i)}$ are called the deleted regression coefficients.

An efficient calculation algorithm is

1. Let $\hat{\beta}$ be the regression coefficients from regressing y on x.

2. Let X be the matrix of predictors including the column $\mathbf{1}$.

3. Factor $X = QR$. See Section I.4.7 for details.

4. Multiply the ith row of Q by $z_i = e_i/(1 - h_i)$. Call the result Q_z.

5. Solve $R\,\Delta b = Q'_z$ for Δb.

6. Then $\hat{\beta}_{(i)} = \hat{\beta} - \Delta b_i$, where Δb_i is the ith column of Δb.

This algorithm is efficient because it does the hard work of solving a linear system only once, when it factors $X = QR$ to construct the orthogonal matrix Q and the triangular matrix R. The backsolve in step 5 is not hard work because it is working with a triangular system. All the remaining steps are simple linear adjustments to the original solution.

Another efficient algorithm, shown in Table 11.8, is essentially the same although with the steps in a different order. This is the algorithm used by R in function `stats:::dfbetas.lm`.

Figure 11.17 gives one DFBETAS plot for each predictor in the model in Table 11.4. We do not ordinarily interpret DFBETAS for the intercept term. Figure 11.6 shows that cases 5 and 19 impact the regression coefficient of `lime`, cases 33 and 49 impact the regression coefficient of `cow.dens`, and that these four counties plus county 32 are primarily responsible for the difference in slopes of the two regression lines in Figure 11.4.

Table 11.8 R's algorithm for dfbetas. The function `chol2inv` inverts a symmetric, positive definite square matrix from its Choleski decomposition. Equivalently, it computes $(X'X)^{-1}$ from the (R part) of the QR decomposition of X. The value `infl$sigma` is a vector whose i^{th} element contains the estimate of the residual standard deviation obtained when the ith case is dropped from the regression. The value returned by the `stats:::dfbeta` function is the changes in the coefficients which result from dropping each case. Function `stats:::dfbeta` does the scaling.

```
> stats:::dfbetas.lm
function (model, infl = lm.influence(model, do.coef = TRUE),
    ...)
{
    qrm <- qr(model)
    xxi <- chol2inv(qrm$qr, qrm$rank)
    dfbeta(model, infl)/outer(infl$sigma, sqrt(diag(xxi)))
}
<bytecode: 0x10a0fa708>
<environment: namespace:stats>
```

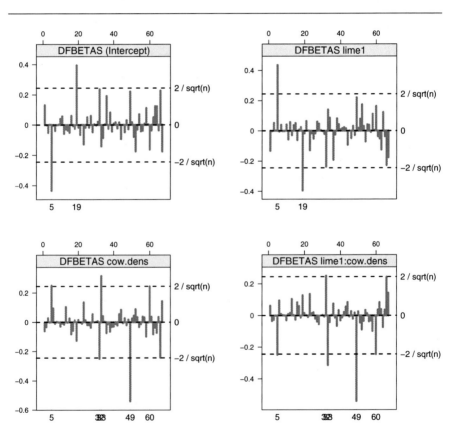

Fig. 11.17 DFBETAS for all four predictors in Model (11.1) for the rent data: the column of 1s for the intercept, the factor `lime`, the covariate `cow.dens`, and the interaction `lime:cow.dens`.

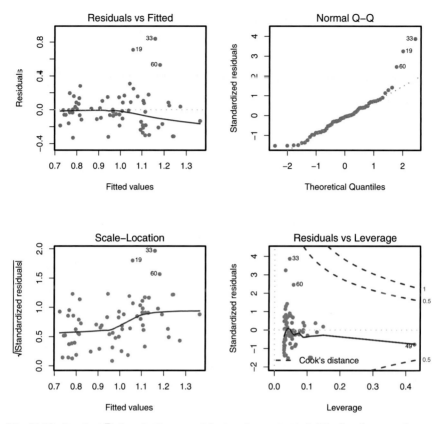

Fig. 11.18 Standard R plot of a linear model: `plot(rent.lm12m)`. The first three panels were discussed in Figure 11.5. The fourth is discussed in Figure 11.19. R by default fits a smooth curve to the points in these plots. The three largest residuals are indicated. "Largest" means larger than the others. There is no statistical significance associated with an identified point.

11.3.7 Residuals vs Leverage

We show R's standard set of regression diagnostic plots in Figure 11.18. The first three are essentially the same as the first three included in our Figure 11.5 constructed with `lmplot` from the **HH** package. R by default fits a smooth curve through both the plots of Residuals vs Fitted and $\sqrt{|Residuals|}$ vs Fitted. The fourth standard R plot, shown enlarged in Figure 11.19, shows the Residuals plotted against the leverage and includes contours of Cook's distance.

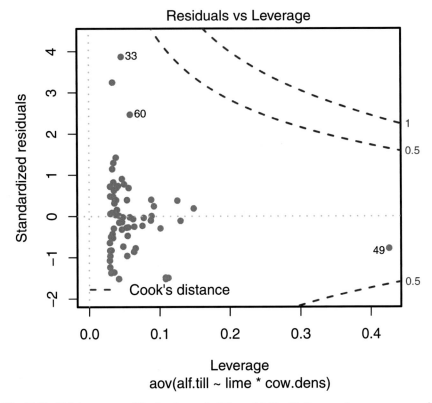

Fig. 11.19 This is a repeat of the fourth panel of Figure 11.18 with the smooth curve suppressed. The "Residuals vs Leverage" plot shows the standardized residuals e^* against the leverage h_i along with contours of Cook's distance. Cook's distance, a combined measure of the "unusualness" of a case's predictors and response, is discussed in Section 11.3.4. The contours of constant Cook's distance c are calculated as $\sqrt{c\,p\,(1-h_i)/h_i}$, where p is the number of estimated regression coefficients ($p = 2$ for simple linear regression). By default, contours are plotted for the two c-values 0.5 and 1. Note on the graph that the contour lines are closer to the 0-residual horizontal line for higher leverage values (corresponding to points farther away from $\bar{x}$) than for lower leverage values.

11.3.8 Calculation of Regression Diagnostics

Regression diagnostics are calculated from the matrix formulation of the equations in the "Sequenced calculation formulas" column of Table 11.7.

In R see the documentation for the functions dfbetas, lm.influence, and plot.lm. See also our functions lm.case and plot.case in the **HH** package.

Regression diagnostics in **SAS** are computed by adding the option INFLUENCE to the MODEL statement in PROC REG.

11.4 Exercises

We recommend that for all exercises involving a data set, you begin by examining a scatterplot matrix of the variables.

11.1. Data from Brooks et al. (1988), reprinted in Hand et al. (1994), relate the number of monthly man-hours associated with the anesthesiology service for 12 U.S. Naval hospitals to the number of surgical cases, the eligible population per thousand, and the number of operating rooms. The data appear in the file data(hospital).

a. Construct and examine a scatterplot matrix of these data. Does it appear that multicollinearity will be a problem?

b. Fit the response to all three predictors, calculating the VIFs. Based on the analysis thus far, which predictor is the best candidate for removal? Why?

c. Fit the response with the predictor in part (b) removed.

d. Calculate the Studentized residuals, leverages, and Cook's distances for the model in part (c). Based on these calculations, what action would you recommend?

11.2. We previously encountered the dataset data(hardness) in Section 9.7 and Exercise 4.5. Since density is easily measured but hardness is not, it is desired to model hardness as a function of density.

a. Construct a histogram of hardness and confirm that a transformation is required in order to use this chapter's regression modeling procedures.

b. Regress the transformation of hardness you chose based on either part (a) or Exercise 4.5. For this regression, produce a scatterplot of the residuals vs the fitted values and of the residuals vs density. Conclude from these plots that a quadratic regression is appropriate.

c. We illustrate a linear and a quadratic fit of the hardness data in Figure 9.5 and Table 9.4. Produce residual plots and regression diagnostics for both models.

11.3. The dataset data(concord) is described in Exercise 4.6. Use multiple regression analysis to model water81 as a function of a subset of the five candidate predictors. Consider transforming variables to assure that the assumption of regression analysis are well satisfied. Carefully interpret, in terms of the original model variables, all regression coefficients in your final model.

11.4. Creatine clearance is an important but difficult to measure indicator of kidney function. It is desired to estimate clearance from more readily measured variables. Neter et al. (1996) discuss data, originally from Shih and Weisberg (1986), relating clearance to serum clearance concentration, age, and weight. The datafile is data(kidney).

a. Regress clearance on each of the three individual predictors. Investigate the adequacy of this model.

b. Improve on the model in part (a) by adding to the set of candidate predictors the squares and pairwise products of the three original predictors. Conclude that the addition of one of these six new candidates improves the original model.

c. Investigate the adequacy of this model.

d. Carefully interpret each of the four estimated regression coefficients in terms of the model variables.

11.5. Heavenrich et al. (1991) provide data on the gasoline mileage (MPG) of 82 makes and models of automobiles as well as 4 potential predictors of MPG. The data appear in data(mileage). The potential predictors are

WT: vehicle weight in 100 lbs

HP: engine horsepower

SP: top speed in mph

VOL: cubic feet of cab space

We wish to use them to model MPG.

a. Produce a scatterplot matrix and comment on the plots of MPG vs HP and of HP vs SP.

b. Regress MPG on WT, HP, and SP. Are the signs of the estimated regression coefficients as expected? Explain what is causing the anomaly.

c. First regress MPG on WT and SP and then regress MPG on WT and HP. Which of these two regressions is preferred?

d. For the model you prefer in part (c), produce a normal plot of the residuals and a plot of the residuals vs the fitted values. What do you conclude?

e. Regress the log of MPG on WT and SP and also the log of MPG on the log of WT and SP. Produce residual plots and normal probability plots from both of these runs. Based on the numerical output and plots, explain which model is preferred.

f. For the preferred model, produce case diagnostics. For each flagged case, indicate what is unusual about it.

11.6. Neter et al. (1996) discuss a dataset relating the amount of life insurance carried in thousands of dollars (lifeins) to average annual income in thousands of dollars (anninc) and risk aversion score (riskaver), for 18 managers, where higher scores connote greater risk aversion. The data are contained in the file data(lifeins).

a. Produce a scatterplot matrix. Which of `anninc` and `riskaver` appears to be more closely related to `lifeins`?

b. Regress `lifeins` on `anninc` and `riskaver`, storing the residuals.

c. From a scatterplot of these residuals vs `anninc`, conclude that the relationship between `lifeins` and `anninc` is nonlinear. Define the square of average annual income, `annincsq` = `anninc`^{2}. Regress `lifeins` on the three predictors `anninc`, `annincsq`, and `riskaver`. Plot the residuals from this run against `anninc`. Based on this plot, discuss whether addition of the curvature term seems worthwhile.

d. Identify cases (managers) whose values indicate either high influence or high leverage. Also note whether these cases have high values of any of the measures Cook's distance, DFFITS, or DFBETAS. If so, interpret such high values in terms of the model variables.

11.7. Refer to `data(hpErie)`, previously considered in Exercise 9.3.

a. Rerun the regression for the final model you found in Exercise 9.3b, this time requesting a complete set of regression diagnostics.

b. Closely examine the values of the diagnostics for the two high-priced houses that are the focus of Exercise 9.3c. Would you recommend both of these houses or just one of them for special scrutiny?

11.8. Prove the equivalence of the intuitive and computational formulas for the following case statistics:

a. DFFITS in Equations (11.10) and (11.11)

b. Cook's distance in either intuitive Equation (11.7) or (11.8), and computational Equation (11.9)

c. Studentized deleted residual in Equations (11.5) and (11.6)

11.9. Explore the diagonals of the hat matrix $H = X(X'X)^{-1}X'$.

a. Prove that all leverages satisfy $\frac{1}{n} \leq h_i \leq 1$. Since H is a projection matrix, show that the upper bound on the diagonals is 1. Since the column $X_0 = 1$ is included in the X matrix, show that the lower bound on the diagonals is $\frac{1}{n}$.

b. Show that the average leverage

$$\frac{\sum_i h_i}{n} \equiv (p+1)/n$$

Chapter 12
Two-Way Analysis of Variance

In Chapter 6 we consider situations where a response variable is measured on groups of observations classified by a single factor and look at ways to compare the changes in the mean of the response variable attributable to the various levels of this factor. Here we extend this to situations where there are two factors. In Chapters 13 and 14 we will discuss instances where there are more than two factors.

12.1 Example—Display Panel Data

12.1.1 Study Objectives

An air traffic controller must be able to respond quickly to an emergency condition indicated on her display panel. It was desired to compare three types of display panel. Each panel was tested under four simulated emergency situations. Two well-trained controllers were assigned to each of the 12 combinations of emergency condition and display panel type; 24 controllers in all. The data in `data(display)` are from Bowerman and O'Connell (1990). It is clear that the type of display panel is a fixed factor, but unclear from this reference whether emergency situation is a fixed or random factor (review these concepts in Sections 6.2 and 6.4). That is, do these four situations represent the totality of incidents to which air traffic controllers might be exposed, or are they four of far more situations? In the former case, emergency situation is a fixed factor; in the latter case, emergency situation is a random factor.

© Springer Science+Business Media New York 2015
R.M. Heiberger, B. Holland, *Statistical Analysis and Data Display*,
Springer Texts in Statistics, DOI 10.1007/978-1-4939-2122-5_12

12.1.2 Data Description

The data in data(display) is structured as 24 rows with four variables.

time: the response variable, time in seconds

panel: factor with three levels indicating the panel being tested

emergenc: factor describing four simulated emergencies

panel.ordered: repeat of the panel factor with the levels reordered to match the order of the response means.

12.1.3 Analysis Goals

We seek to determine whether the three panels afford significantly different display times and whether such conclusions are consistent across different types of emergency.

Exhibited here are graphs and tables that will aid in answering these questions. Discussion of this output is deferred until Section 12.11.

Figure 12.1 shows plots for assessing interaction between panel and emergenc as well as boxplots for examining the main effects of these factors. The concept of *interaction* is introduced in Section 12.2. The structure of the interaction plot in Figure 12.1 is discussed in Section 12.4.

Table 12.1 shows the aov and anova statements assuming that emergenc is a fixed factor. Table 12.2 and Figures 12.2 and 12.3 show the panel means and the results of the multiple comparisons by the Tukey method. As will be explained in Section 12.11, the conclusion derived from this table is that there is a significant difference in response times for the three panels. Panel 3 affords a significantly longer response time than panels 1 or 2; response times for panels 1 and 2 do not differ significantly.

Table 12.3 shows the aov and summary statements assuming that emergenc is a random factor.

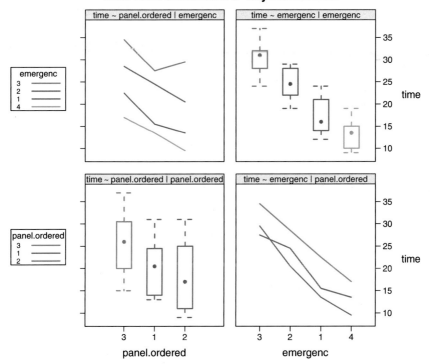

Fig. 12.1 Interaction plot for display panel experiment. The nearly parallel traces suggest the absence of interaction between `panel` and `emergenc`. Note that we reordered the emergencies and the panels by average time in order to simplify the appearance of the plot. The structure of the interaction plot is discussed in Section 12.4.

Table 12.1 Display panel data: ANOVA table with test of `panel` appropriate if `emergenc` is fixed. The test of `panel` is from the "both factors fixed" column of Table 12.8. That is, all sums of squares are compared to the `Residuals` line of the ANOVA table. The listing is continued in Table 12.2.

```
> displayf.aov <- aov(time ~ emergenc * panel, data=display)

> anova(displayf.aov)
Analysis of Variance Table

Response: time
               Df  Sum Sq Mean Sq F value     Pr(>F)
emergenc        3 1052.46  350.82 60.5731 1.612e-07 ***
panel           2  232.75  116.38 20.0935 0.0001478 ***
emergenc:panel  6   28.92    4.82  0.8321 0.5675015
Residuals      12   69.50    5.79
---
Signif. codes:  0 '***' 0.001 '**' 0.01 '*' 0.05 '.' 0.1 ' ' 1
```

Table 12.2 Display panel data: ANOVA table with test of `panel` appropriate if `emergenc` is fixed. Multiple comparisons of `panel` by Tukey method. The standard deviation for the comparison is based on the `Residuals` line of the ANOVA table in Table 12.1. We show plots of the multiple comparisons in Figures 12.3 and 12.2.

```
> displayf.mmc <- mmc(displayf.aov, focus="panel")

> displayf.mmc
Tukey contrasts
Fit: aov(formula = time ~ emergenc * panel, data = display)
Estimated Quantile = 2.668615
95% family-wise confidence level
$mca
      estimate   stderr      lower       upper  height
3-1      5.375 1.203294   2.163871   8.586129 22.9375
3-2      7.375 1.203294   4.163871  10.586129 21.9375
1-2      2.000 1.203294  -1.211129   5.211129 19.2500
$none
    estimate    stderr     lower     upper height
3     25.625 0.8508574  23.35439  27.89561 25.625
1     20.250 0.8508574  17.97939  22.52061 20.250
2     18.250 0.8508574  15.97939  20.52061 18.250
```

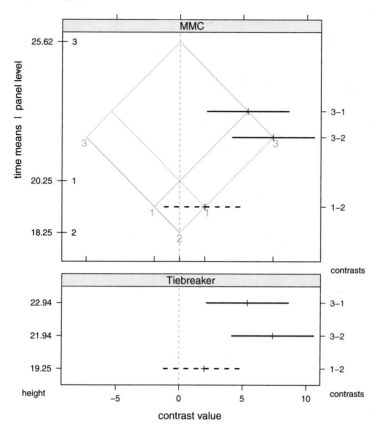

Fig. 12.2 MMC plot of pairwise comparisons of `panel` means by the Tukey method. The top panel shows the `panel` means along the y-axis and the confidence intervals for the differences along the x axis. The Tiebreaker plot in the bottom panel shows the contrasts equally spaced along the y-axis and in the same sequence as the top panel. The heights displayed as the y-axis tick labels in the Tiebreaker panel are the actual heights along the y-axis for the contrasts in the MMC panel. These heights are the weighted averages of the means being compared by the contrasts. The Tiebreaker panel is not needed in this example.

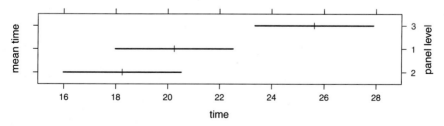

Fig. 12.3 Confidence intervals on each of the `panel` means.

Table 12.3 Display panel data: ANOVA table with test of panel appropriate if emergenc is random. In this example, the test is from the "A fixed, B random" column of Table 12.8 with panel taking the role of A. That is, the sum of squares for panel is compared to the panel:emergenc interaction line of the ANOVA table.

```
> displayr.aov <- aov(time ~ Error(emergenc/panel) + panel,
+                     data=display)

> summary(displayr.aov)

Error: emergenc
          Df Sum Sq Mean Sq F value Pr(>F)
Residuals  3   1052   350.8

Error: emergenc:panel
          Df Sum Sq Mean Sq F value  Pr(>F)
panel      2 232.75  116.38   24.15 0.00135 **
Residuals  6  28.92    4.82
---
Signif. codes:  0 '***' 0.001 '**' 0.01 '*' 0.05 '.' 0.1 ' ' 1

Error: Within
          Df Sum Sq Mean Sq F value Pr(>F)
Residuals 12   69.5   5.792
```

12.2 Statistical Model

To model an experiment with two factors, we begin by calling the factors A and B, where A has a levels and B has b levels. We use n_{ij} to denote the number of observations taken from cell (i, j), i.e., the treatment combination corresponding to level i of A and level j of B, $i = 1, \ldots, a$ and $j = 1, \ldots, b$. Our discussion in this chapter is confined to the case where the n_{ij} are equal for all i, j, and sometimes $n_{ij} = 1$. We extend the notation of Equation (6.1) by replacing the singly indexed symbol α_i with a doubly indexed set of symbols $\alpha_i + \beta_j + (\alpha\beta)_{ij}$ and model the kth observation at the ith level of A, j^{th} level of B, as

$$Y_{ijk} = \mu + \alpha_i + \beta_j + (\alpha\beta)_{ij} + \epsilon_{ijk} = \mu_{ij} + \epsilon_{ijk} \tag{12.1}$$

for $1 \leq i \leq a$, $1 \leq j \leq b$, and $1 \leq k \leq n_{ij}$. The expectations for the cell means are denoted

$$E(Y_{ijk}) = \mu_{ij} = \mu + \alpha_i + \beta_j + (\alpha\beta)_{ij} \tag{12.2}$$

We assume the errors $\epsilon_{ijk} \sim \text{NID}(0, \sigma^2)$, that is they are assumed to be normally independently distributed with a common variance σ^2. The parameter μ represents the grand mean of all ab populations.

Each of the factors A and B can be either fixed or random. If A is fixed, then we assume that $\sum_i \alpha_i = 0$. If A is random, we assume that each $\alpha_i \sim N(0, \sigma_A^2)$. Similarly, if B is fixed, then we assume that $\sum_j \beta_j = 0$ and if B is random, we assume that each $\beta_j \sim N(0, \sigma_B^2)$.

The term $(\alpha\beta)_{ij}$ models the possibility of interaction between the two factors. If A and B are both fixed factors, then the sum of $(\alpha\beta)_{ij}$ over either i or j is zero. If both factors are random, then $(\alpha\beta)_{ij} \sim N(0, \sigma_{AB}^2)$. In the case of a mixed model, where for concreteness we have A fixed and B random, $(\alpha\beta)_{ij} \sim N(0, \frac{a-1}{a}\sigma_{AB}^2)$ subject to $\sum_i (\alpha\beta)_{ij} = 0$ for each $j = 1, \ldots, b$.

Factors A and B are said to *interact* if the difference in response between two levels of A differs according to the level of B. Equivalently, there is *interaction* between factors A and B if the difference in response between two levels of B differs according to the level of A. Graphically, the traces for each level of factor A across levels of B are parallel if there is no interaction, and are not parallel when there is interaction. Equivalently, the traces for each level of B across levels of A are parallel if there is no interaction. In Figure 12.1 we see essentially parallel traces, consistent with the non-significance of the test of the interaction in Table 12.1. In Figure 12.12 we will see nonparallel, actually crossing, traces consistent with the significance of the interaction in Table 12.12.

12.3 Main Effects and Interactions

As in one-way ANOVA, we are interested in comparing the means of observations in each *cell*, that is for each *treatment combination* (combination of factor levels), in the design, and for combinations of cells. We work with the *cell means*

$$\bar{Y}_{ij} = \sum_k Y_{ijk}/n_{ij} \tag{12.3}$$

and the *marginal means*. The marginal means for the rows are calculated by averaging the cell means in each row over the columns. The marginal means for the columns are calculated by averaging the cell means in each column over the rows:

$$\bar{Y}_{i.} = \sum_j \bar{Y}_{ij}/b \tag{12.4}$$

$$\bar{Y}_{.j} = \sum_i \bar{Y}_{ij}/a \tag{12.5}$$

Table 12.4 Table of means for the rhizobium clover experiment of Section 12.14. Means from Table 12.12 have been arranged in a two-way table to display the cell means in the body of the table, the marginal means on the margins of the table, and the grand mean as the margin of the marginal means. clover and clover+alfalfa are the two levels of the factor comb. The left side of the table shows the means symbolically using $\bar{Y}_{ij}$ notation. The right side show the numerical values from Table 12.12.

Strain	Clover	Clover+alfalfa	Mean		Strain	Clover	Clover+alfalfa	Mean
3DOk1	$\bar{Y}_{11}$	$\bar{Y}_{12}$	$\bar{Y}_{1.}$		3DOk1	29.04	28.41	28.72
3DOk5	$\bar{Y}_{21}$	$\bar{Y}_{22}$	$\bar{Y}_{2.}$		3DOk5	36.29	27.44	31.86
3DOk4	$\bar{Y}_{31}$	$\bar{Y}_{32}$	$\bar{Y}_{3.}$	$=$	3DOk4	21.35	23.98	22.66
3DOk7	$\bar{Y}_{41}$	$\bar{Y}_{42}$	$\bar{Y}_{4.}$		3DOk7	22.93	24.96	23.95
3DOk13	$\bar{Y}_{51}$	$\bar{Y}_{52}$	$\bar{Y}_{5.}$		3DOk13	22.49	24.30	23.39
k.composite	$\bar{Y}_{61}$	$\bar{Y}_{62}$	$\bar{Y}_{6.}$		k.composite	25.97	24.92	25.45
Mean	$\bar{Y}_{.1}$	$\bar{Y}_{.2}$	$\bar{Y}_{..}$		Mean	26.35	25.67	26.01

where $n_{i.} = \sum_j n_{ij}$, $n_{.j} = \sum_i n_{ij}$, and $n_{..} = \sum_{ij} n_{ij}$. Marginal means get their name because they are often displayed on the margins of a two-way table of cell means, as in Table 12.4. We also use the *grand mean*:

$$\bar{Y}_{..} = \sum_i n_{i.} \bar{Y}_{i.} \Big/ n_{..} = \sum_j n_{.j} \bar{Y}_{.j} \Big/ n_{..} = \sum_{ijk} Y_{ijk} \Big/ n_{..} \tag{12.6}$$

When more than one factor is present, there are three principal types of comparisons that we will investigate.

Main effects are comparisons of the marginal means for one of the factors, for example, $\bar{Y}_{1.} - \bar{Y}_{2.}$. It is usually valid to compare main effects only when there is no interaction.

Interactions (or interaction effects) are comparisons of the cell means across levels of both factors, for example, $(\bar{Y}_{13} - \bar{Y}_{23}) - (\bar{Y}_{14} - \bar{Y}_{24})$. When interaction is present, that is when differences in the cell means across rows depend on the column or equivalently, when comparisons of the form indicated here are significantly different from 0, we usually must use simple effects, not main effects, to discuss the factors.

Simple effects are separate comparisons of the cell means across levels of one factor for some or all levels of the other factor, for example, $\bar{Y}_{13} - \bar{Y}_{23}$. See Section 13.3.

The analyst should be alert to the possibility that interaction is present. The nature of the analysis when interaction exists is different from that when interaction is absent.

Without interaction, the analysis proceeds similarly to the procedures for one-way analysis. The marginal means are calculated and compared, perhaps by using one of the multiple comparisons techniques discussed in Sections 6.3, 7.1.3, or 7.1.4.1. The advantage of the two-way analysis in this case is in the efficiency, hence increased power, of the comparisons. Because we use the same residual sum of squares for the denominator of both F-tests (for the rows and for the columns), we can run the combined experiment to test the effect of both factors for less expense than if we were to run two separate experiments.

When interaction between two factors is present, it is not appropriate to compare the main effects, the levels of one of these factors averaged over the levels of the other factor. It is possible, for example, that the mean of Y increases over factor B for level 1 of factor A and decreases over factor B for level 2 of factor A. Averaging over the levels of factor A would mask that behavior of the response.

We explore main effects, interactions, and simple effects with the rhizobium data in Section 12.14.

12.4 Two-Way Interaction Plot

The two-way interaction plot, first shown in Figure 12.1 and used throughout the remainder of this book, shows all main effects and two-way interactions for designs with two or more factors. We construct it using the `interaction2wt` function in the **HH** package by analogy with the `splom` (scatterplot matrix) function in the **lattice** package. The rows and columns of the two-way interaction plot are defined by the Cartesian product of the factors.

1. Each main diagonal panel shows a boxplot for the marginal effect of a factor.

2. Each off-diagonal panel is a standard interaction plot of the factors defining its position in the array. Each point in the panel is the mean of the response variable conditional on the values of the two factors. Each line in the panel connects the cell means for a constant level of the *trace* factor. Each vertically aligned set of points in the panel shows the cell means for a constant value of the *x*-factor.

3. Panels in mirror-image positions interchange the trace- and *x*-factors. This duplication is helpful rather than redundant because one of the orientations is frequently much easier to interpret than the other.

4. The rows are labeled with a key that shows the line type and color for the trace factor by which the row is defined.

5. Each box in the boxplot panels has the same color, and optionally the same line type, as the corresponding traces in its row.

6. The columns are labeled by the *x*-factor.

12.5 Sums of Squares in the Two-Way ANOVA Table

Table 12.5 presents the structure of the analysis of variance table for a balanced two-way ANOVA with a levels of the A factor, b levels of the B factor, and n observations at each of the ab AB-treatment combinations, analogous to Table 6.2 for one-way ANOVA. If the test F_{AB} shows that AB interaction is present, the F-tests on A and B are not interpretable.

If the AB interaction is not significant, then the form of the tests for the main effects A and B depends on whether the factors A and B are fixed or random factors. See the discussion in Section 12.10 where Table 12.8 lists the expected mean squares and F-tests under various assumptions.

Table 12.5 Two-way ANOVA structure with both factors representing fixed effects.

Analysis of Variance of Dependent Variable y					
Source	Degrees of Freedom	Sum of Squares	Mean Square	F	p-value
Treatment A	df_A	SS_A	MS_A	F_A	p_A
Treatment B	df_B	SS_B	MS_B	F_B	p_B
AB Interaction	df_{AB}	SS_{AB}	MS_{AB}	F_{AB}	p_{AB}
Residual	df_{Res}	SS_{Res}	MS_{Res}		
Total	df_{Total}	SS_{Total}			

Terms of the table are defined by:

Treatment A	
df_A	$a - 1$
SS_A	$bn \sum (\bar{Y}_{i.} - \bar{Y}_{..})^2$
MS_A	SS_A / df_A
F_A	MS_A / MS_{Res}
p_A	$1 - \mathcal{F}_F(F_A \mid df_A, df_{Res})$

Treatment AB	
df_{AB}	$(a - 1)(b - 1)$
SS_{AB}	$n \sum (\bar{Y}_{ij} - \bar{Y}_{..})^2 - SS_A - SS_B$
MS_{AB}	SS_{AB} / df_{AB}
F_{AB}	MS_{AB} / MS_{Res}
p_{AB}	$1 - \mathcal{F}_F(F_{AB} \mid df_{AB}, df_{Res})$

Treatment B	
df_B	$b - 1$
SS_B	$an \sum (\bar{Y}_{.j} - \bar{Y}_{..})^2$
MS_B	SS_B / df_B
F_B	MS_B / MS_{Res}
p_B	$1 - \mathcal{F}_F(F_B \mid df_B, df_{Res})$

Residual	
df_{Res}	$ab(n - 1)$
SS_{Res}	$\sum_i \sum_j (Y_{ijk} - \bar{Y}_{ij})^2$
MS_{Res}	SS_{Res} / df_{Res}

Total	
df_{Total}	$abn - 1$
SS_{Total}	$\sum_i \sum_j \sum_k (Y_{ijk} - \bar{Y}_{..})^2$

Table 12.5 shows the F-statistics and their p-values for tests on the main effects A and B under the assumption that both factors represent fixed effects. Most ANOVA programs calculate these values by default whether or not they are appropriate.

12.6 Treatment and Blocking Factors

Treatment factors are those for which we wish to determine if there is an effect. *Blocking* factors are those for which we believe there is an effect. We wish to prevent a presumed blocking effect from interfering with our measurement of the treatment effect.

An experiment with two factors may have either two treatment factors or one treatment factor and one blocking factor. The primary objective of a factorial experiment is comparisons of the levels of treatment factors. By contrast, a blocking factor is set up in order to enhance one's ability to distinguish between the levels of treatment factors. The term *block* was chosen by analogy to two of the dictionary definitions: a rectangular section of land bounded on each side by consecutive streets; or a set of similar items sold or handled as a unit, such as shares of stock.

We are not interested in comparing the *blocks*, i.e., the levels of a blocking factor. In a well-designed experiment, we anticipate that the response differs across the levels of a blocking factor because if the levels of this factor cover a variety of experimental conditions, this broadens the scope of our inferences about treatment differences. Multiple comparisons across blocks are not meaningful because we know in advance that the blocks are different. In general, blocking is advisable and successful as an experimental and analytical technique if the experimental units can reasonably be grouped into blocks such that the units within every block are homogeneous, while the units in any given block are different from those in any other block. By homogeneous units, we mean that they will tend to respond alike if treated alike. Usually, there is no interaction between blocking and treatment factors; otherwise blocking will not have accomplished its objective and the analysis will be much less able to detect significant differences than if blocking were properly done.

Blocking is the natural extension to three or more treatments of the matched pairs design introduced in Section 5.5. The F-test of the treatment effect against the residual is the generalization of the paired t-test. It is exactly true that a blocked design with two levels of the treatment factor and with many blocks of size two is identical to the matched pairs design.

For example, in an experiment on tire wear, the location of the tire on the car (say, Right Front) is a treatment effect and the specific car (of the many used in the experiment) is a blocking effect.

12.7 Fixed and Random Effects

As mentioned in Sections 6.2 and 6.4, treatment factors may be regarded as either fixed or random. The levels of a fixed factor are the only levels of interest in the experiment, and we wish to see if the response is homogeneous across these levels. The levels of a random factor are a random sample from some large population of levels, and we are interested in assessing whether the variance of responses over this population of levels is essentially zero. Block factors are almost always regarded as random.

The levels of a treatment factor can be either categorical or quantitative. For example, in an experiment where the `fertilizer` treatment has four levels, the experimental levels of fertilizer could be four different fertilizer compounds, or four different applications per acre of one fertilizer compound. When the levels are quantitative, it is usually preferable to regard the factor as a single degree-of-freedom predictor variable.

12.8 Randomized Complete Block Designs

A randomized complete block design (RCBD) has one treatment factor involving t treatment levels and one blocking factor having b levels. The b blocks each contain experimental units arranged according to the principles discussed in Section 12.6. That is, experimental units in the same block are expected to respond alike if treated alike, while the blocks should reflect a variety of experimental conditions to broaden the scope of conclusions to be drawn from inferences about the treatments. It is assumed that blocks and treatments do not interact. This assumption permits us to compare the treatment levels when each block contains exactly t experimental units, i.e., there is no replication of treatments within any block. If there are $n > 1$ observations on each treatment within each block, then additional degrees of freedom are available for comparing treatments. We outline the effect of larger sample size, which usually means more degrees of freedom in the denominator of statistical tests, in Section 3.10. In summary, more degrees of freedom move us up the t-table or F-table or χ^2-table and the critical value gets smaller.

The model for the RCBD with one observation on each treatment in each block is

$$y_{ij} = \mu + \tau_i + \rho_j + \epsilon_{ij} \tag{12.7}$$

where μ represents the overall mean, τ_i is the differential effect of treatment level i, ρ_j is the differential effect of block j, and the ϵ's are random $N(0, \sigma^2)$ residuals. We further define

$$\bar{y}_i = \sum_i y_{ij}/b, \ \bar{y}_j = \sum_j y_{ij}/t, \text{ and } \bar{\bar{y}}_{..} = \sum_i \sum_j y_{ij}/bt \tag{12.8}$$

Table 12.6 ANOVA table structure for a randomized complete block design with no replication.

Analysis of Variance of Dependent Variable y					
Source	Degrees of Freedom	Sum of Squares	Mean Square	F	p-value
Blocks	df_{Blk}	SS_{Blk}	MS_{Blk}		
Treatments	df_{Tr}	SS_{Tr}	MS_{Tr}	F_{Tr}	p_{Tr}
Residuals	df_{Res}	SS_{Res}	MS_{Res}		
Total	df_{Total}	SS_{Total}			

The terms of the table are defined by:

Blocks		Residual	
df_{Blk}	$b-1$	df_{Res}	$(b-1)(t-1)$
SS_{Blk}	$\sum_i \sum_j (\bar{y}_j - \bar{\bar{y}}_{..})^2$	SS_{Res}	$\sum_i \sum_j (y_{ij} - \bar{y}_i - \bar{y}_j + \bar{\bar{y}}_{..})^2$
MS_{Blk}	SS_{Blk}/df_{Blk}	MS_{Res}	SS_{Res}/df_{Res}

Treatments		Total	
df_{Tr}	$t-1$	df_{Total}	$bt-1$
SS_{Tr}	$\sum_i \sum_j (\bar{y}_i - \bar{\bar{y}}_{..})^2$	SS_{Total}	$\sum_i \sum_j (y_{ij} - \bar{\bar{y}}_{..})^2$
MS_{Tr}	SS_{Tr}/df_{Tr}		
F_{Tr}	MS_{Tr}/MS_{Res}		
p_{Tr}	$1 - \mathcal{F}_F(F_{Tr} \mid df_{Tr}, df_{Res})$		

The setup of the ANOVA table for an RCBD with $n = 1$ is shown in Table 12.6. Some ANOVA programs also display an F-statistic and p-value for blocks, but it is inappropriate to interpret these since the experiment is designed in such a way that responses will differ across blocks and artificially force high F values for blocks. We could do efficiency of blocking calculations. See, for example, Cochran and Cox (1957) (Section 4.37).

12.9 Example—The Blood Plasma Data

12.9.1 Study Objectives

The dataset data(plasma) comes from Anderson et al. (1981) and is reproduced in Hand et al. (1994). The data are measurements on plasma citrate concentrations in micromols/liter obtained from 10 subjects at 8 am, 11 am, 2 pm, 5 pm, and 8 pm.

To what extent is there a normal profile for the level in the human body during the day?

This experiment is viewed as an RCBD with treatment factor time and blocking factor id. It is desirable here that the subjects (blocks) be as unlike as possible in order to broaden the scope of the conclusion about normal profiles as much as possible. The no-interaction assumption amounts to assuming that the daily response profile is constant across subjects.

12.9.2 Data Description

The data in data(plasma) is structured as 50 rows with three variables.

plasma: the response variable, plasma citrate concentrations in micromols/liter

time: factor with five values: 8 am, 11 am, 2 pm, 5 pm, and 8 pm

id: factor with 10 levels, one per subject

12.9.3 Analysis

We begin our analysis with the interaction plots in Figure 12.4. There seem to be anomalies for id=3 at 8 pm and for id=6 at 11 am, but otherwise both sets of traces look reasonably parallel.

We proceed with an additive model in Table 12.7 and discover that the ratio of the id stratum Residual Mean Square to the Within stratum Residual Mean Square ($1177/147.5 = 7.98$) is large (had this been a valid test, which it is not because id is a blocking factor, it would have been $F = 7.98$), confirming our decision to block on patients. This is not a hypothesis test, because we know at the beginning of our analysis that patients are different from each other.

The test of differences due to time rejects the null hypothesis that the response at all times is the same. Since there appears to be no interaction, we can act as if there is a single pattern that applies to everyone. We investigate the time pattern with the MMC plot in Figure 12.5. The only significant single contrast is between the low at 5PM and the high at 11AM. The low at 5PM is clearly visible in the plasma ~ time | id panel of Figure 12.4. The high at 11AM is hinted at in Figure 12.4.

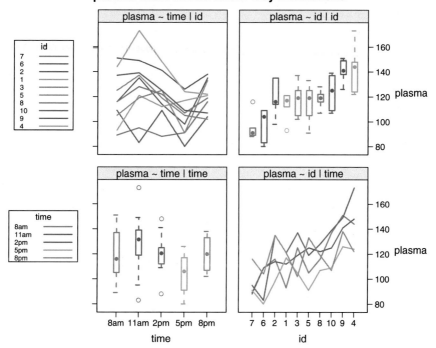

Fig. 12.4 Interaction Plot for Plasma Citrate. The `id` factor has been sorted by median `plasma` value. The `time` factor must be displayed in chronological order.

Table 12.7 ANOVA Table for Plasma Citrate Experiment

```
> plasma.aov <- aov(plasma ~ Error(id) + time, data=plasma)

> summary(plasma.aov)

Error: id
          Df Sum Sq Mean Sq F value Pr(>F)
Residuals  9  10593    1177

Error: Within
          Df Sum Sq Mean Sq F value  Pr(>F)
time       4   2804   701.0   4.754 0.00349 **
Residuals 36   5308   147.5
---
Signif. codes:  0 '***' 0.001 '**' 0.01 '*' 0.05 '.' 0.1 ' ' 1
```

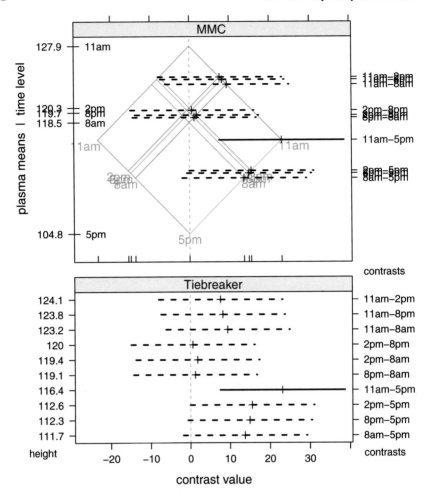

The MMC panel shows informative overprinting. Please see Tiebreaker panel and caption.

Fig. 12.5 MMC plot and Tiebreaker plot of `time` in the plasma data. The low at 5PM is clearly visible in the `plasma ~ time | id` panel of Figure 12.4. The high at 11AM is hinted at in Figure 12.4. The Tiebreaker plot in the bottom panel is imperative for this example. The means at many of the levels of `time` are very close and therefore their labels are overprinted. As a consequence, the heights of the contrasts are similar and their labels are also overprinted. The Tiebreaker plot shows the contrasts equally spaced along the *y*-axis and in the same sequence as the top panel. The heights displayed as the *y*-axis tick labels in the Tiebreaker panel are the actual heights along the *y*-axis for the contrasts in the MMC panel.

12.10 Random Effects Models and Mixed Models

In Section 6.4, we compare two analyses of the same data assuming the single factor is fixed or random. There we indicate that a table of expected mean squares may be used to formulate the correct mean square ratio to test the hypothesis of interest. We also show that in the single factor case, while the same ratio is used in both the fixed and random cases, the hypothesis tested about the factor differs in the two cases.

When we have two or more factors and interactions, the test statistics as well as the hypotheses depend on whether the factors are fixed or random. The formulas for standard errors for comparing the levels of fixed factors also depend on whether the other factor(s) are fixed or random.

Table 12.8 is an algebraically derived table of expected mean squares for an experiment with two possibly interacting factors A and B and equal sample sizes $n_{ij} = n \geq 2$ at each of the ab treatment combinations under each of three assumptions: the fixed model where both factors are fixed, the mixed model where one factor is fixed and the other factor is random, and the random model where both factors are random. Each entry in the table is derived by evaluating, for example (using the notation of Table 12.5), the statement

$$E(\text{MS}_\text{A}) + E\left(bn \sum (\bar{Y}_{i.} - \bar{Y}_{..})^2\right)/(a-1)$$

where we model Y_{ijk} and $E(Y_{ijk})$ by Equations 12.1 and 12.2.

From the lineups of the expected mean squares, we see that for testing the A main effect, the appropriate denominator mean square is the Residual mean square when factor B is fixed (from the "Both factors fixed" column, $\text{EMS(A)} = \sigma^2 + nbk_A^2$ and $\text{EMS(Residual)} = \sigma^2$).

Table 12.8 Expected mean squares in two-way analysis of variance. Compare to Tables 6.4, 12.5, and 13.11. See Section 12.10 for the discussion on when to use each of the columns.

Source	df	Both factors fixed		A fixed, B random		Both factors random	
Treatment A	$a-1$	σ^2	$+ nbk_A^2$	$\sigma^2 + n\sigma_{AB}^2$	$+ nbk_A^2$	$\sigma^2 + n\sigma_{AB}^2$	$+ nb\sigma_A^2$
Treatment B	$b-1$	σ^2	$+ nak_B^2$	σ^2	$+ na\sigma_B^2$	$\sigma^2 + n\sigma_{AB}^2 + na\sigma_B^2$	
AB Interaction	$(a-1)(b-1)$	$\sigma^2 + nk_{AB}^2$		$\sigma^2 + n\sigma_{AB}^2$		$\sigma^2 + n\sigma_{AB}^2$	
Residual	$ab(n-1)$	σ^2		σ^2		σ^2	
Total	$abn-1$						

where $\quad k_A^2 = \dfrac{\sum_i \alpha_i^2}{a-1} \quad k_B^2 = \dfrac{\sum_j \beta_j^2}{b-1} \quad k_{AB}^2 = \dfrac{\sum_i \sum_j (\alpha\beta)_{ij}^2}{(a-1)(b-1)}$

The appropriate denominator mean square for testing the A main effect is the AB-interaction mean square when B is random (from the other two columns EMS(A) $= \sigma^2 + n\sigma_{AB}^2 + nb\, f(A)$ and EMS(AB) $= \sigma^2 + n\sigma_{AB}^2$, where $f(A) = \kappa_A^2$ when A is fixed and $f(A) = \sigma_A^2$ when A is random). The ratio of these mean squares is appropriate for testing equality of the levels of factor A because the corresponding ratio of these expected mean squares exceeds one if and only if $\sigma_A^2 > 0$ or $\kappa_A^2 > 0$. Use of the Residual mean square as the denominator of the F-test would be inappropriate because such a ratio would exceed one if there is an AB interaction effect.

The conclusions for testing the B main effect follow from interchanging "A" and "B" in the previous sentence.

12.11 Example—Display Panel Data—Continued

In Section 12.1 we introduced the display panel example illustrating a two-way analysis of variance. We continue here with the analysis by discussing Figures 12.1–12.3 and Tables 12.1–12.3.

In Figure 12.1 we display two-way interaction plots and boxplots for the factors panel and emergenc. The two interaction plots in the off-diagonal panels contain equivalent information, but in general, one of them is more readily interpretable than the other. In this instance, the close-to-parallel traces suggest the absence of interaction between panel and emergenc. This is anticipated because emergenc is a block factor and confirmed by the large p-value for the interaction test in Table 12.1. One set of boxplots in Figure 12.1 evinces a greater response time with panel 3 than with either panel 1 or panel 2. The other set of boxplots shows substantial differences in the response times of the four emergencies; this is anticipated since emergenc is regarded as a blocking factor and differences in response across blocks are expected by design.

The simplest ANOVA specification in Table 12.1 assumes all factors are fixed. We see that when emergenc is a fixed factor, the F-statistic for panel is 20.09 on 2 and 12 degrees of freedom. The small corresponding p-value suggests that response time varies with the type of panel.

If emergenc is a random factor, as in Table 12.3, the pattern of expected mean squares in Table 12.8 indicates that the appropriate denominator mean square for testing panel is the interaction mean square. This test is specified by placing emergenc/panel inside the Error() function in the model formula. We see that panel is tested with $F = 24.15$ on 2 and 6 degrees of freedom.

The F-statistic for panel corresponds to a small p-value under either assumption on emergenc. Therefore, in this example, we reach the same conclusion under both assumptions: that response time differs across panels. However, since in general the F-statistic differs in the two cases, the ultimate conclusion concerning a fixed factor may depend crucially on our assumption concerning the other factor. If emergenc is a fixed factor, the conclusions regarding panels applies to these four emergencies only. If emergenc is a random factor, the panel conclusions apply to the entire population of emergencies from which these four emergencies are assumed to be a random sample.

The F-test for interaction between panel and emergenc when emergenc is a random factor is the same test as when emergenc is a fixed factor.

Since panel is a fixed factor, an appropriate follow-up is a Tukey test to compare the response time for each display panel. This is shown in Table 12.2 for the case where emergenc is fixed. The means are in the estimate column of the $none section. We find that both display panel 1 and display panel 2 have significantly shorter response times than display panel 3, but panels 1 and 2 are not significantly different. Therefore, we conclude that display panel 3 can safely be eliminated from further consideration. The absence of interaction tells us that these conclusions are consistent over emergencies. If interaction had existed in this experiment, one would have concluded that the optimal panel differs according to the type of emergency. Then one would need to make separate panel recommendations for each emergency type. Since we will normally select just one panel type for the entire facility, and since we have no control over emergencies, the decision process would become more difficult.

The confidence intervals in the $mca section of Table 12.2 and in both panels Figure 12.2 display the differences between all pairs of panel means using the two-sided Tukey multiple comparisons procedure introduced in Section 6.3. The $mca (the term *mca* stands for *multiple comparisons analysis*) section of Table 12.2 shows the results of the $\binom{3}{2} = 3$ pairwise tests. The negative lower bound and positive upper bound for the 1-2 comparison indicates that the confidence interval for the difference between the corresponding population means ($\bar{y}_1 = 20.250$ and $\bar{y}_2 = 18.250$) includes zero, hence the difference is not significant. The comparisons between $\bar{y}_3 = 25.625$ and the other two panel means have positive lower and upper bounds, hence these confidence intervals exclude zero. This indicates that the population mean of panel 3 is significantly different from both other population means.

Figure 12.2 provides two confidence interval displays for pairwise comparisons of the population means of the three panels. Both contain the confidence intervals on each pairwise difference taken directly from the $mca section of Table 12.2. A pairwise difference of means is significantly different from zero; equivalently, the two means differ significantly if the confidence interval for the pairwise difference excludes zero. If this confidence interval includes zero, then conclude that the two population means do not significantly differ. Thus the "1–2" interval says that these two panel means are indistinguishable. The "1–3" and "2–3" intervals says that the mean of panel 3 differs from the means of the other two panels. The top panel is an MMC plot (see Chapter 7) with the contrasts displayed on the isomeans grid as a background that shows the individual panel means. The bottom panel uses equal vertical spacing between contrasts.

The $none (the term *none* indicating no contrasts) section of Table 12.2 shows the results for the individual group means. Figure 12.3 contains simultaneous confidence intervals for the three population means, where the confidence coefficient, here 95%, is the probability that each interval contains its respective population mean. If two of these confidence intervals overlap, then the corresponding population means are not significantly different. Since panels 1 and 2 have overlapping intervals, these two panel means are not distinguishable. If a pair of these confidence intervals does not overlap, then the corresponding population means are declared to differ significantly. Since the panel 3 interval does not overlap the other two, we conclude that the mean of panel 3 differs from the means of the other two panels.

12.12 Studentized Range Distribution

The tabled values of the Studentized Range Distribution (see Section J.1.10) of a set of a means are scaled for the random variable $Q = (\bar{y}_{(a)} - \bar{y}_{(1)})/s_{\bar{y}}$. The denominator is the standard error of a single $\bar{y}$. The estimated quantile (critical point) shown in Table 12.2 and used in Figure 12.2 is 2.668. This is not the Studentized range tabular value $q_{.05}$ but instead $q_{.05}/\sqrt{2}$. Details of the R calculation can be followed in file HHscriptnames(12). The equivalent SAS code reports the Studentized range $q_{.05}$.

We use the tabled values in two places in the MMC display. In Table 12.2, $q_{.05} = 3.77278$, the *Estimated Quantile* is $\frac{q_{.05}}{\sqrt{2}} = 2.668$, $MS_{Res} = 5.791667$ (from Table 12.1), and $m = 8$.

In the $none section of Table 12.2 we show the sample means $\bar{y}_i$ in the estimate column and the standard error $s_{\bar{y}}$ of an individual $\bar{y}$ in the stderr column. We must adjust the Q value by dividing by $\sqrt{2}$. The formula for the simultaneous 95% confidence intervals on individual means is

$$\mu_i: \qquad \bar{Y}_i \pm \frac{q_{.05}}{\sqrt{2}} \sqrt{\frac{MS_{Res}}{m}} \tag{12.9}$$

where $q_{.05}$ is the 95th percentile of the Studentized range distribution and m is the common sample size used in calculating each sample mean. The "minimum significant difference" in this table is the "±" part of formula (12.9),

$$\frac{q_{.05}}{\sqrt{2}} \sqrt{\frac{MS_{Res}}{m}} = 2.27 \tag{12.10}$$

In the $mca section of Table 12.2 we show the differences $\bar{y}_i - \bar{y}_j$ in the estimate column and the standard error $\sqrt{2}s_{\bar{y}}$ of the difference in the stderr column. Again we must adjust the Q value by dividing by $\sqrt{2}$. The formula for the simultaneous 95% confidence intervals on pairwise mean differences shown in Figure 12.2 is

$$\mu_i - \mu_j: \quad \bar{Y}_i - \bar{Y}_{i'} \pm \frac{q_{.05}}{\sqrt{2}} \sqrt{2\frac{\mathsf{MS_{Res}}}{m}} = \bar{Y}_i - \bar{Y}_{i'} \pm q_{.05} \sqrt{\frac{\mathsf{MS_{Res}}}{m}} \quad (12.11)$$

The "minimum significant difference" in this table is the "±" part of formula (12.11),

$$q_{.05} \sqrt{\frac{\mathsf{MS_{Res}}}{m}} = 3.21 \quad (12.12)$$

12.13 Introduction to Nesting

In the previous examples the two factors have a *crossed* relationship. Saying that factors *A* and *B* are crossed indicates that each level of *A* may be observed in a treatment combination with any level of *B*. Alternatively, two factors may have a *nested* or hierarchical relationship. When *B* is nested within *A*, the levels of *B* are similar but not identical for different levels of *A*.

12.13.1 Example—Workstation Data

A small electronics firm wishes to compare three methods for assembling an electronic device. For this purpose, the plant has available six different workstations. The study is conducted by randomly assigning $s = 2$ workstations to each of the $m = 3$ assembly methods. At each workstation–method combination $w = 5$ randomly selected production workers will assemble the device for one hour using the appropriate assembly method. The response is the number of devices produced in one hour. The data from Bowerman and O'Connell (1990) (p. 890) are accessible as data(workstation) and are displayed in Figure 12.6.

12.13.2 Data Description

The data in data(plasma) is structured as 30 rows with three variables.

method: factor with three levels describing the assembly methods

station: factor with two levels describing the workstations

devices: response variable, number of devices produced in one hour.

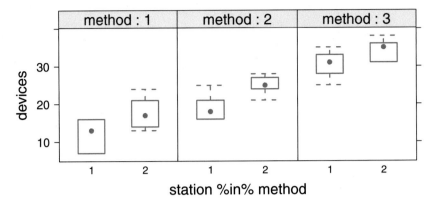

Fig. 12.6 Boxplot of workstation data. The significance of method and station within method are confirmed in Table 12.10.

12.13.3 Analysis Goals

Note that the workstations assigned to any assembly method are different from those assigned to any other method. As a consequence, the factors (which we call station and method) are not crossed with one another, and an analysis using a model we have previously studied would be incorrect. The factor station is said to be *nested* within the factor method because each workstation is associated with exactly one of the methods.

Our analysis assumes that station is a fixed factor. If instead station were assumed to be a random factor, the code would have to be modified to force station to be tested against the station within method mean square instead of against the Residuals mean square. The procedures for doing so are demonstrated in the data analysis in Section 13.4.

The basic structure of the ANOVA table is in Table 12.9. In R, we use the formula

```
devices ~ method / station
```

to indicate that station is nested within method. The analysis is in Table 12.10.

We conclude that when using at least one of the three methods, the two workstations for that method produced a significantly different number of devices. We also conclude that the three methods produced significantly different numbers of devices.

In this example there is balanced sampling. That is, each method has the same number of workstations and each workstation has the same number of workers. Without much additional difficulty, the above nested factorial analysis can be extended to situations with unbalanced sampling. (In contrast, when one has unbalanced sampling and crossed factors, the analysis is considerably more difficult than with balanced sampling.)

Table 12.9 Basic structure of the ANOVA table for a nested design with $m = 3$, $s = 2$, and $w = 5$.

Source	df		MS	F
	Algebra	Example		
Method	$m - 1$ = 3−1	= 2	MS_m	$\dfrac{MS_m}{MS_w}$
Station within Method	$m(s - 1)$ = 3×(2−1)	= 3	MS_s	$\dfrac{MS_s}{MS_w}$
Worker within Station (Residual)	$ms(w - 1)$ = 3×2×(5−1) = 24		MS_w	
Total	$msw - 1$ = 3×2×5 − 1 = 29			

Table 12.10 Workstation data. ANOVA table and means.

```
> workstation.aov <- aov(devices ~ method / station,
+                         data=workstation)

> summary(workstation.aov)
               Df Sum Sq Mean Sq F value  Pr(>F)
method          2 1545.3   772.6  51.452 2.09e-09 ***
method:station  3  210.2    70.1   4.666   0.0105 *
Residuals      24  360.4    15.0
---
Signif. codes:  0 '***' 0.001 '**' 0.01 '*' 0.05 '.' 0.1 ' ' 1

> model.tables(workstation.aov, "means", se=TRUE)
Tables of means
Grand mean

23.06667

 method
method
    1    2    3
14.8 22.1 32.3

 method:station
       station
method 1    2
     1 11.8 17.8
     2 19.2 25.0
     3 30.4 34.2

Standard errors for differences of means
         method method:station
          1.733           2.451
replic.      10               5
```

12.14 Example—The *Rhizobium* Data

12.14.1 Study Objectives

Erdman (1946) discusses experiments to determine if antibiosis occurs between *Rhizobium Meliloti* and *Rhizobium Trifolii*. Rhizobium is a bacteria, growing on the roots of clover and alfalfa, that fixes nitrogen from the atmosphere into a chemical form the plants can use. The research question for Erdman was whether there was an interaction between the two types of bacteria, one specialized for alfalfa plants and the other for clover plants. If there was an interaction, it would indicate that clover bacteria mixed with alfalfa bacteria changed the nitrogen fixing response of alfalfa to alfalfa bacteria or of clover to clover bacteria. The biology of the experiment says that interaction indicates antibiosis or antagonism of the two types of rhizobium. That is, the goal was to test whether the two types of rhizobium kill each other off. If they do, then there will be less functioning bacteria in the root nodules and consequently nitrogen fixation will be slower.

Erdman ran two sets of experiments in parallel. In one the response variable was the nitrogen content in clover plants, in the other the nitrogen content in alfalfa plants. The treatments were combinations of bacterial cultures in which the plants were grown. As a historical note, beginning with Steel and Torrie (1960), the one-way analysis of the clover plus alfalfa combination of the Clover experiment has been frequently used as an example to illustrate multiple comparisons procedures. Here we examine the complete data from two related two-way experiments.

12.14.2 Data Description

Both experiments are two-way factorial experiments with two treatment factors:

strain: one of six rhizobium cultures, five pure strains and one a mixture of all five strains. Five strains of alfalfa rhizobium were used for the alfalfa plants and five strains of clover rhizobium were used for the clover plants.

comb: a factor at two levels. At one level the rhizobium cultures consisted of only strains specialized for the host plant. At the other level each of the six cultures was combined with a mixture of rhizobium strains specialized for the other plant.

12.14.3 First Rhizobium Experiment: Alfalfa Plants

Five observations on the response variable, nitrogen content, were taken at each of the 12 strain*comb treatment combinations. Primary interest was in the differences in responses to the six rhizobium treatments. Erdman originally analyzed the response variable "milligrams of nitrogen per 20 plants". After studying his analysis and his discussion we choose to analyze a related response variable, "milligrams of nitrogen per gram of dry plant weight". We give the original analysis as Exercise 12.1.

12.14.4 Second Rhizobium Experiment: Clover Plants

Five observations on the response variable, nitrogen content, were taken at each of the 12 strain*comb treatment combinations. Primary interest was in the differences in responses to the six rhizobium treatments. Erdman originally analyzed the response variable "milligrams of nitrogen per 10 plants". After studying his analysis and his discussion, we choose to analyze a related response variable, "milligrams of nitrogen per gram of dry plant weight". We give the original analysis as Exercise 12.2.

12.14.5 Initial Plots

Datasets data(rhiz.alfalfa) and data(rhiz.clover) contain the complete data for both experiments. The alfalfa data is plotted in Figure 12.7. The clover data is plotted in Figure 12.8. Erdman's response variable is shown as nitro in both figures. Our response variable is shown as Npg. The single most evident feature from the clover boxplots is the large response to the pure culture 3DOk5. This observation is the one that caused us to consider the alternate response variable. There were fewer plants, hence larger plants, for this strain. We posit that the reported values were scaled up, that is reported as grams per 10 plants. We hope that analyzing the ratio, milligrams of nitrogen per gram of plants, rather than the reported rate, milligrams per 10 plants, will adjust for the outliers. Nothing in the alfalfa plots is as clear.

As a graphical aside, we looked at four different layouts for these plots. In Figures 12.7 and 12.8 we show vertical boxplots by strain conditioned on comb. We also looked at vertical boxplots by comb conditioned on strain and horizontal boxplots with both conditionings. We chose this one because we have a preference for the response variable on the vertical axis and because we believe the patterns

Alfalfa Experiment

Combination

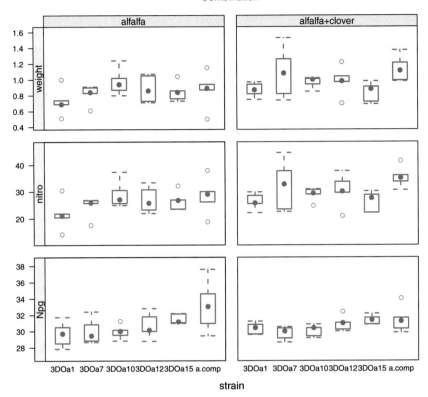

Fig. 12.7 Boxplot of alfalfa data. The Npg response variable has the least variability. We shall continue our analysis with Npg.

are easier to see when this example is conditioned on the factor comb. The other three layouts for the data in Figure 12.8 can be viewed by running the code in file HHscriptnames(12). Also see the discussion in Section 13.A.

Clover Experiment

Combination

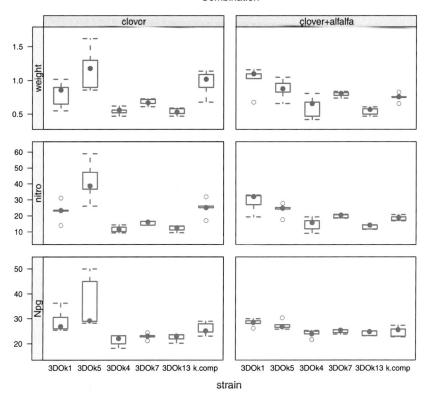

Fig. 12.8 Boxplot of clover data. The Npg response variable has the least variability. We shall continue our analysis with Npg. The large variability in the 3DOk5 strain is visible in all three response variables.

12.14.6 Alfalfa Analysis

The ANOVA table and table of means for the alfalfa experiment are in Table 12.11. Since there was no interaction with the combination of clover strains of bacteria (strain:comb interaction *p*-value = .53 in Table 12.11), there is no evidence of antibiosis or antagonism.

Since only the strain main effect is significant, we confine our investigation to differences among the means for strain. Figures 12.9, 12.10, and 12.11 display the results of the Tukey multiple comparison procedure for comparing strain mean differences. Since strain has 6 levels, we simultaneously examine all $\binom{6}{2} = 15$ pairwise mean differences. Any mean difference having a confidence interval in Figure 12.10 that doesn't include 0 is declared a significantly differing pair. There are three such confidence intervals, therefore we conclude that a.composite has a

Table 12.11 ANOVA table and table of means for alfalfa experiment. See Figure 12.7.

```
> ## unset position(rhiz.alfalfa$comb) for glht
> data(rhiz.alfalfa) ## fresh copy of the data.

> rhiz.alfalfa.aov <- aov(Npg ~ strain * comb, data=rhiz.alfalfa)

> summary(rhiz.alfalfa.aov)
            Df Sum Sq Mean Sq F value  Pr(>F)
strain       5  46.22   9.244   4.565 0.00174 **
comb         1   0.57   0.573   0.283 0.59714
strain:comb  5   8.44   1.687   0.833 0.53275
Residuals   48  97.21   2.025
---
Signif. codes:  0 '***' 0.001 '**' 0.01 '*' 0.05 '.' 0.1 ' ' 1

> alf.means <- model.tables(rhiz.alfalfa.aov, type="means",
+                           se=TRUE, cterms="strain")

> alf.means
Tables of means
Grand mean

30.73547

 strain
strain
 3DOa1  3DOa7 3DOa10 3DOa12 3DOa15 a.comp
 30.00  29.89  30.00  30.82  31.43  32.27

Standard errors for differences of means
         strain
         0.6364
replic.     10
```

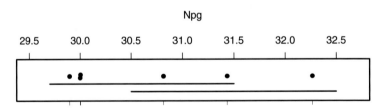

3DOa7: 29.8932   3DOa12: 30.8171   a.comp: 32.2650
  3DOa10: 29.9997   3DOa15: 31.4347
  3DOa1: 30.0031

Fig. 12.9 Means for alfalfa experiment. Dots that appear over a common horizontal line correspond to population means that do not differ significantly according to the Tukey multiple comparisons procedure with simultaneous 95% confidence intervals. Compare this figure to Figure 12.10.

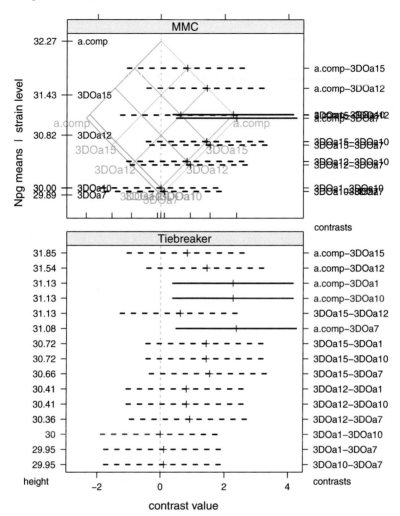

The MMC panel shows informative overprinting. Please see Tiebreaker panel and caption.

Fig. 12.10 Mean–mean multiple comparisons plot and Tiebreaker plot showing Tukey contrasts for alfalfa data. The MMC panel shows informative overprinting. Note that 3DOa1, 3DOa10, and 3DOa7 have almost identical means. Consequently (i) their means overprint on the left axis, (ii) their differences overprint on the right axis, and (iii) their contrasts are displayed at the same vertical position in the MMC panel. Most of the overprinting contrasts cross zero and are not significant. Their details are displayed in the Tiebreaker panel. The only significant contrasts (the red solid lines) are on the right corner of the isomeans grid. All three are contrasts of a.comp with the three almost identical means of the lower three strains. Again the details are clear in the Tiebreaker panel. The MMC panel displays the contrasts at heights constructed as the average of the two means being compared. The Tiebreaker panel shows the contrasts in the same data-dependent vertical order as the MMC panel. The Tiebreaker panel breaks the ties in the MMC panel by placing the confidence intervals at equally spaced vertical positions. See also Figure 12.11 where we have constructed a set of orthogonal contrasts to capture and illustrate the relationships among the levels.

significantly higher mean response than each of 3D0a1, 3D0a7, and 3D0a10; these were the only significant differences detected. The inference is that any of the three treatments with high response (3D0a12, 3D0a15, or a.composite) should be used.

Equivalent information is contained in Figure 12.9, where two population means are declared significantly different if their corresponding sample means are *not* underlined by the same line. (Such an underlining display may be used only when all samples have the same size, as is the case here.)

Figure 12.10 is very busy because it shows 15 pairwise contrasts for only 5 degrees of freedom. In Figure 12.11 we provide a graphical summary of our conclusions by constructing an orthogonal basis for the contrasts. We believe the orthogonal contrasts in Figure 12.11 are easier to use in expository settings. The detail of Figure 12.10 is needed to help us construct a useful and meaningful set of orthogonal contrasts. We see that the single comparison between a.composite and the average of the three strains with low means (3D0a1, 3D0a7, and 3D0a10) is the only significant effect.

Figure 12.10 and 12.11 each have two panels. The MMC (mean–mean multiple comparisons) panel shows the MMC plot discussed in Chapter 7. There is severe overprinting of the confidence intervals and their labeling because so many of the means and estimates of their differences have almost identical values. The overprinting is itself information on similarity of level means. Nonetheless we need a tiebreaker that will return legibility to the plot. We provide the tiebreaker in the Tiebreaker panel, an ordinary multiple comparisons plot of the individual contrasts placed at equally spaced vertical positions, sorted to be in the same data-dependent order that is used in the MMC panel. This sort order is based on the values of the level means. The standard sort order used by both R (see for example Table 6.3) and SAS is based on the names of the levels.

12.14.7 Clover Analysis

The ANOVA table and table of means for the clover experiment are shown in Table 12.12. In this experiment the strain:comb interaction effect is significant and the comb main effect is not significant.

The significance of the strain:comb in Table 12.12 (p-value $< .01$) implies that we can't immediately, if at all, interpret the main effect of strain. Main effect comparisons of the levels of comb and strain are inappropriate because the difference in response to two levels of strain will differ according to the level of comb. From the table of means in Table 12.12 and the interaction plots in Figure 12.12 we discover, again, that strain 3D0k5 is the anomaly. The interaction is made evident by the lack of parallel profiles in both interaction plot panels. The three points marked

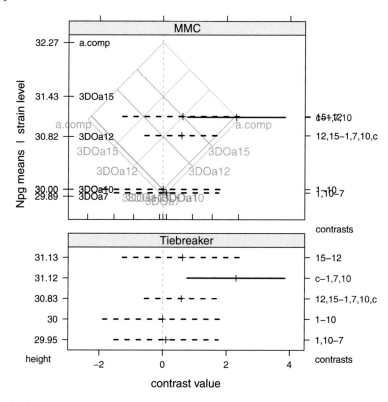

The MMC panel shows informative overprinting. Please see Tiebreaker panel and caption.

Fig. 12.11 MMC plot and Tiebreaker plot of an orthogonal set of Tukey contrasts for alfalfa data. There are six strains, hence five independent comparisons. This orthogonal set has been chosen to summarize the information in Figure 12.10 and show that only one comparison is significantly different from 0. We see now that the three strains with low means are indistinguishable, that the two intermediate strains are indistinguishable, and that these two clusters are not significantly different from each other. The only significant comparison is from the mean of the composite to the mean of the cluster of three strains with low means. The MMC panel shows the same overprinting discussed in the caption to Figure 12.10. The Tiebreaker panel shows clearly the single significant comparison of the composite mean to the mean of the cluster of three strains with low means.

as outliers in both boxplot panels are the points that drive much of the remaining analysis. We show the simple effects in Figure 12.13.

Once we decide that main effects are not meaningful in the presence of strong interaction, we must look at the behavior separately for each level of the comb factor. We continue to do so in the context of a single analysis because we are still able to use the residual term constructed from all levels of comb. This residual term has 48 degrees of freedom. Had we been forced to run separate analyses each would have had a residual with much fewer degrees of freedom. Recall from Section 3.10 that tests are more powerful when the denominator has higher degrees of freedom.

Table 12.12 ANOVA table and table of means for clover experiment. See Figure 12.8.

```
> rhiz.clover.aov <- aov(Npg ~ strain * comb, data=rhiz.clover)

> summary(rhiz.clover.aov)
            Df Sum Sq Mean Sq F value   Pr(>F)
strain       5  642.3  128.45   9.916 1.47e-06 ***
comb         1    6.9    6.88   0.531  0.46955
strain:comb  5  228.2   45.65   3.524  0.00857 **
Residuals   48  621.8   12.95
---
Signif. codes:  0 '***' 0.001 '**' 0.01 '*' 0.05 '.' 0.1 ' ' 1

> model.tables(rhiz.clover.aov, type="means", se=TRUE)
Tables of means
Grand mean

26.00674

 strain
strain
 3DOk1  3DOk5  3DOk4  3DOk7 3DOk13 k.comp
 28.72  31.86  22.66  23.95  23.39  25.45

 comb
comb
        clover clover+alfalfa
        26.345         25.668

 strain:comb
        comb
strain   clover clover+alfalfa
  3DOk1  29.04  28.41
  3DOk5  36.29  27.44
  3DOk4  21.35  23.98
  3DOk7  22.93  24.96
  3DOk13 22.49  24.30
  k.comp 25.97  24.92

Standard errors for differences of means
        strain   comb strain:comb
        1.6096 0.9293      2.2763
replic.     10     30           5
```

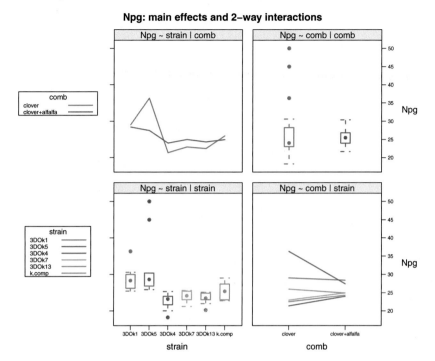

Fig. 12.12 Interaction plot for clover experiment. The three points marked as outliers in clover 3DOk5 are the points that drive much of the remaining analysis. We show the simple effects for this interaction in Figure 12.13.

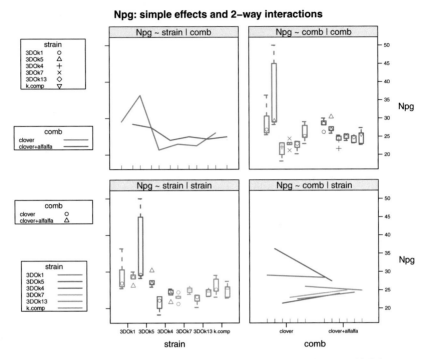

Fig. 12.13 Simple effects plot for the clover experiment interaction in Figure 12.12. It is now even more clear that the clover 3DOk5 points differ from the others.

We therefore repartition the sums of squares in Table 12.13 and look separately at the simple effect of strain within each of the levels of comb. The notation in Table 12.13 is the mechanics by which the 10 degrees of freedom are separated into two meaningful groups of 5 degrees of freedom. The differences in the clover strains of rhizobium alone are significant. The differences with the combination clover and alfalfa strains of rhizobium are not. Therefore, we examine only the *simple effects* within the clover strains. These simple effects are the differences between pairs of means of strain within the clover level of the factor comb. We examine and report on those such differences that are statistically significant. Since the simple effect for strain within the clover+alfalfa level of comb is not significant, we do not look further at those means.

Erdman's interpretation of the analysis shows that bacteria strain 3DOk5 showed antibiosis with the alfalfa bacteria strains. With 3DOk5 the response was strong alone and suppressed when combined with the alfalfa bacteria culture.

Table 12.13 ANOVA table showing simple effects for strain in clover experiment. We partitioned the sums of squares for the nesting with the `split` argument to the `summary` function. We needed to display the names of the individual regression coefficients in order to determine which belonged to each of the levels of `comb`. In this example the `comb` and `strain` effects are orthogonal, hence the partitioning is valid. The individual degrees of freedom are usually not interpretable.

```
> rhiz.clover.nest.aov <-
+     aov(Npg ~ comb/strain, data=rhiz.clover)

> summary(rhiz.clover.nest.aov)
            Df Sum Sq Mean Sq F value Pr(>F)
comb         1    6.9    6.88   0.531   0.47
comb:strain 10  870.5   87.05   6.720  2e-06 ***
Residuals   48  621.8   12.95
---
Signif. codes:  0 '***' 0.001 '**' 0.01 '*' 0.05 '.' 0.1 ' ' 1

> old.width <- options(width=35)

> names(coef(rhiz.clover.nest.aov))
 [1] "(Intercept)"
 [2] "combclover+alfalfa"
 [3] "combclover:strain3DOk5"
 [4] "combclover+alfalfa:strain3DOk5"
 [5] "combclover:strain3DOk4"
 [6] "combclover+alfalfa:strain3DOk4"
 [7] "combclover:strain3DOk7"
 [8] "combclover+alfalfa:strain3DOk7"
 [9] "combclover:strain3DOk13"
[10] "combclover+alfalfa:strain3DOk13"
[11] "combclover:straink.comp"
[12] "combclover+alfalfa:straink.comp"

> options(old.width)

> summary(rhiz.clover.nest.aov,
+          split=list("comb:strain"=
+           list(clover=c(1,3,5,7,9),
+             "clover+alf"=c(2,4,6,8,10))))
                        Df Sum Sq Mean Sq F value    Pr(>F)
comb                     1    6.9    6.88   0.531     0.470
comb:strain             10  870.5   87.05   6.720  2.00e-06 ***
  comb:strain: clover    5  788.4  157.68  12.172  1.22e-07 ***
  comb:strain: clover+alf 5  82.1   16.42   1.268     0.293
Residuals               48  621.8   12.95
---
Signif. codes:  0 '***' 0.001 '**' 0.01 '*' 0.05 '.' 0.1 ' ' 1
```

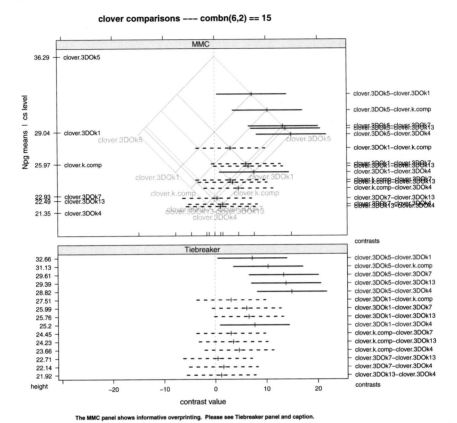

Fig. 12.14 MMC plot and Tiebreaker plot of Tukey simple effect contrasts for comb="clover" data. It is visually quite clear that the strain 3DOk5 differs from the rest (very strongly for the bottom three strains and less so for the middle two strains.) There is also one marginally significant contrast between the second largest mean and the smallest mean. We illustrate this observation in Figure 12.15 with an appropriately chosen set of orthogonal contrasts. The Tiebreaker plot in the bottom panel is imperative for this example. The means at many of the levels of cs are very close and therefore their labels are overprinted. The Tiebreaker plot shows the contrasts equally spaced along the *y*-axis and in the same sequence as the top panel.

Table 12.14 shows the dummy variables and Table 12.15 shows the regression coefficients for the simple effects of strain in the clover experiment displayed in Table 12.13. The names for the columns of the dummy variables generated by the program are excessively long and would force the matrix of dummy variables to occupy many pages just to accommodate the column names. Therefore, we abbreviated them. We see the nesting structure in the dummy variables as the cmb*n* columns for pure strains and the cm+*n* columns for combination strains are identical in structure. Only the cmb*n* regression coefficients are significant. The dummy variables are constructed from the default treatment contrasts.

Since there is interaction in the clover experiment, we must look at the multiple comparisons for the simple effects of strain at each value of comb.

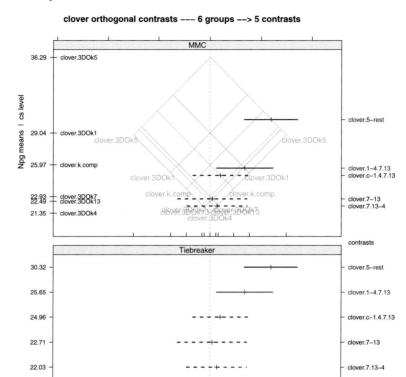

Fig. 12.15 MMC plot and Tiebreaker plot of an orthogonal basis set of Tukey simple effect contrasts for comb="clover" data. We summarize the conclusions from Figure 12.14. The strongest contrast compares 3DOk5 to the rest. There is one other marginally significant contrast. Two contrasts show that the three strains with the lowest means are indistinguishable. The Tiebreaker plot in the bottom panel is imperative for this example. The means at many of the levels of cs are very close and therefore their labels are overprinted. The Tiebreaker plot shows the contrasts equally spaced along the *y*-axis and in the same sequence as the top panel.

Figure 12.14 shows the simple effects for comb="clover". The only strongly significant contrasts are the ones centered on the upper right isomeans grid line (clover.3DOk5) comparing 3DOk5 to the rest of the strains. There is one other borderline significant contrast. The Tiebreaker panel makes it slightly easier to identify the names of the contrasts. The set of orthogonal contrasts in Figure 12.15 shows that the single contrast comparing 3DOk5 to the others carries almost all the significance in Figure 12.15.

Figure 12.16 shows that there are no significant contrasts in the simple effects for comb="clover+alfalfa". We forced Figure 12.16 to be on the same scale as Figure 12.14.

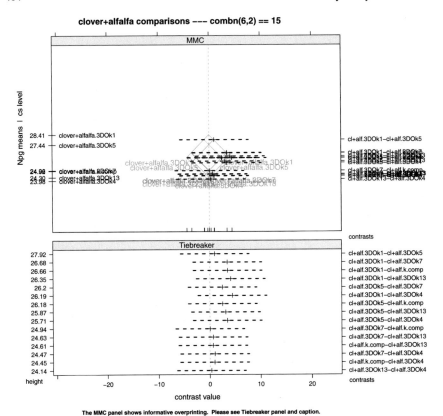

Fig. 12.16 MMC plot and Tiebreaker plot of Tukey simple effect contrasts for comb="clover+alfalfa" data. This plot is on the same scale as Figure 12.14. This common scale emphasizes the disparity between 3DOk5 in comb="clover" and any values of strain in comb="clover+alfalfa". None of the simple effects for strain within the clover+alfalfa level of comb are significant. The Tiebreaker plot in the bottom panel is imperative for this example as all the means are almost identical and therefore their labels are overprinted.

Table 12.14 Dummy variables for simple effects of `strain` in clover experiment. These dummy variables are based on the treatment contrasts. The sums of squares from these dummy variables are displayed in Table 12.13. The regression coefficients are in Table 12.15. The dummy variables and regression coefficients have been reordered to place the within-`clover` values together and the within-`clover+alfalfa` values together.

```
> ## Look at the contrasts, their generated dummy variables,
> ## and their regression coefficients.
> ## Abbreviate their names for presentation.
> tmp <- abbreviate(names(coef(rhiz.clover.nest.aov)))

> ## tmp
>
> ## contrasts(rhiz.clover$comb)
> ## contrasts(rhiz.clover$strain)
>
> cnx <- aov(Npg ~ comb/strain, data=rhiz.clover, x=TRUE)$x

> dimnames(cnx)[[2]] <- tmp

> ## cnx
> cnx[seq(1,60,5), c(1,2,  3,5,7,9,11)]
    (In) cmb+ c:3D05 c:3D04 c:3D07 c:3D01 cm:.
1     1    0    0      0      0      0     0
6     1    0    1      0      0      0     0
11    1    0    0      1      0      0     0
16    1    0    0      0      1      0     0
21    1    0    0      0      0      1     0
26    1    0    0      0      0      0     1
31    1    1    0      0      0      0     0
36    1    1    0      0      0      0     0
41    1    1    0      0      0      0     0
46    1    1    0      0      0      0     0
51    1    1    0      0      0      0     0
56    1    1    0      0      0      0     0

> cnx[seq(1,60,5), c(4,6,8,10,12)]
    c+:3D05 c+:3D04 c+:3D07 c+:3D01 c+:.
1      0       0       0       0     0
6      0       0       0       0     0
11     0       0       0       0     0
16     0       0       0       0     0
21     0       0       0       0     0
26     0       0       0       0     0
31     0       0       0       0     0
36     1       0       0       0     0
41     0       1       0       0     0
46     0       0       1       0     0
51     0       0       0       1     0
56     0       0       0       0     1
```

Table 12.15 Regression coefficients for simple effects of `strain` in clover experiment. The contrasts and dummy variables are displayed in Table 12.14. The dummy variables and regression coefficients have been reordered to place the within-`clover` values together and the within-`clover+alfalfa` values together.

```
> cnxb <- round(coef(summary.lm(rhiz.clover.nest.aov)), 3)

> dimnames(cnxb)[[1]] <- tmp

> ## cnxb
> cnxb[c(1,2,  3,5,7,9,11, 4,6,8,10,12),]
          Estimate Std. Error t value Pr(>|t|)
(In)        29.042      1.610  18.043    0.000
cmb+        -0.637      2.276  -0.280    0.781
c:3D05       7.243      2.276   3.182    0.003
c:3D04      -7.688      2.276  -3.378    0.001
c:3D07      -6.110      2.276  -2.684    0.010
c:3D01      -6.556      2.276  -2.880    0.006
cm:.        -3.070      2.276  -1.349    0.184
c+:3D05     -0.966      2.276  -0.424    0.673
c+:3D04     -4.430      2.276  -1.946    0.057
c+:3D07     -3.441      2.276  -1.512    0.137
c+:3D01     -4.106      2.276  -1.804    0.078
c+:.        -3.482      2.276  -1.530    0.133
```

12.15 Models Without Interaction

Experiments with two factors are normally designed with a large enough sample size to investigate the possibility that the factors interact. When the analyst has previous experience with these factors, or subject area knowledge that the factors are unlikely to interact, it is possible to set up the model without an interaction term:

Algebra	$Y_{ijk} = \mu + \alpha_i + \beta_j + \epsilon_{ijk}$	
R	$Y \sim$	$A + B$
SAS	$Y =$	$A \quad B$

The residual portion of this no-interaction model includes the $(a - 1)(b - 1)$ degrees of freedom that would otherwise have been attributable to the AB interaction. If the no-interaction assumption is correct, the no-interaction model provides a more precise estimate of the residual than a model incorporating interaction and this in turn implies more power for tests involving the individual main effects or the means of their levels. With this model, comparisons among the levels of factor A or among the levels of factor B are undertaken in much the same way as in a one-way experiment, but using this model's residual sum of squares and degrees of freedom.

When we initially posit a model containing the two-factor interaction, it may happen that the analysis of variance test for interaction leads to non-rejection of the no-interaction hypothesis. If the evidence for no interaction is sufficiently strong (a large p-value for this test and/or no strong subject area feeling about the existence of interaction), the analyst may feel comfortable about reverting to the no-interaction model and proceeding with the analysis as above. This amounts to pooling a nonsignificant interaction sum of squares with the previous residual sum of squares (calculated under the now rejected assumption of an interaction) to produce a revised residual mean square (under the assumption of no interaction). This combined or pooled estimate is justified because in the absence of interaction, the interaction mean square estimates the same quantity, the residual variance, as does the residual mean square. The pooled estimate of the residual variance is an improvement over the individual estimates because it is constructed with additional degrees of freedom. Therefore, the pooled estimate provides more powerful inferences on the level means of the two factors than would a residual mean square in a model including interaction. See Section 5.4.2 for further discussion of pooling.

12.16 Example—Animal Feed Data

12.16.1 Study Objectives

A manufacturer of animal feed investigated the influence on the amount of vitamin A retained in feed. The manufacturer considered 15 treatment combinations formed from 5 amounts of feed supplement and 3 levels of temperature at which the supplements were added to the feed. Two samples were selected at each treatment combination. The data from Anderson and McLean (1974), accessible as data(feed), are said to be on transformed scales that this reference does not specify. The response variable is retained and the two factors are temp and supp.

12.16.2 Analysis

The data is displayed in the interaction plot in Figure 12.17. The profiles in the interaction plot are sufficiently close to parallel to suggest that there is no interaction between temp and supp.

Table 12.16 Feed data: ANOVA table for model with interaction. The interaction is not significant.

```
> feed.int.aov <- aov(retained ~ temp * supp, data=feed)

> anova(feed.int.aov)
Analysis of Variance Table

Response: retained
          Df Sum Sq Mean Sq F value    Pr(>F)
temp       2 1479.2  739.60 26.0423 1.321e-05 ***
supp       4 3862.1  965.53 33.9977 2.334e-07 ***
temp:supp  8  243.5   30.43  1.0716    0.4313
Residuals 15  426.0   28.40
---
Signif. codes:  0 '***' 0.001 '**' 0.01 '*' 0.05 '.' 0.1 ' ' 1
```

Initially, in Table 12.16, we fit an interaction model leading to an interaction *p*-value of 0.43, confirming our impression from the interaction plot that temp and supp do not interact. It is not unreasonable to conclude that temperature affects each concentration of feed supplement in roughly the same way. Therefore, we abandon the assumption of interaction and move to a no-interaction model.

Table 12.17 Feed data: ANOVA with main effects and their polynomial contrasts.

```
> feed.aov <- aov(retained ~ temp + supp, data=feed)

> anova(feed.aov)
Analysis of Variance Table

Response: retained
          Df Sum Sq Mean Sq F value    Pr(>F)
temp       2 1479.2  739.60  25.410 1.499e-06 ***
supp       4 3862.1  965.53  33.172 3.037e-09 ***
Residuals 23  669.5   29.11
---
Signif. codes:  0 '***' 0.001 '**' 0.01 '*' 0.05 '.' 0.1 ' ' 1

> summary(feed.aov, split=
+         list(temp=list(linear=1, quadratic=2),
+              supp=list(linear=1, quadratic=2, rest=3:4)))
                Df Sum Sq Mean Sq F value    Pr(>F)
temp             2   1479   739.6  25.409 1.50e-06 ***
  temp: linear   1    370   369.8  12.705  0.00165 **
  temp: quadratic 1  1109  1109.4  38.114 2.68e-06 ***
supp             4   3862   965.5  33.172 3.04e-09 ***
  supp: linear   1   2912  2912.1 100.046 7.61e-10 ***
  supp: quadratic 1   947   946.7  32.525 8.30e-06 ***
  supp: rest     2      3     1.7   0.058  0.94418
Residuals       23    669    29.1
---
Signif. codes:  0 '***' 0.001 '**' 0.01 '*' 0.05 '.' 0.1 ' ' 1

> model.tables(feed.aov, type="means", se=TRUE)
Tables of means
Grand mean

68.8

 temp
temp
  40    80   120
60.2  77.4  68.8

 supp
supp
    2     4     6     8    10
48.33 64.67 76.00 79.00 76.00

Standard errors for differences of means
        temp  supp
        2.413 3.115
replic.    10     6
```

retained: main effects and 2–way interactions

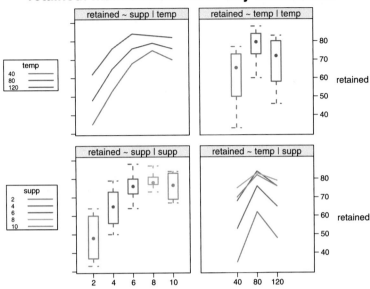

Fig. 12.17 Feed data interaction plots.

The fit of the no-interaction model in Table 12.17 suggests that both temp and supp impact significantly on retained. Since both of these factors have quantitative levels, our analysis of the nature of the mean differences involves modeling the response retained as polynomial functions of both temperature and the amount of supplement. The method for accomplishing such modeling was introduced in Section 10.4.

The interaction plot in Figure 12.17 suggests that the response to changes in the level of supp is quadratic in nature and that possibly the response to changes in the level of temp is quadratic as well. Therefore, for both of these factors we calculated the one degree-of-freedom tests on the linear and quadratic contrasts among the factor levels, and show the results in Table 12.17. Since the p-values for both quadratic contrasts are close to 0, there is strong evidence that the response of vitamin A retention is a quadratic function of both temperature and amount of feed supplement.

We show the MMC plot for supplement in Figure 12.18 for all pairwise contrasts and in Figure 12.19 for the orthogonal contrasts. The Tiebreaker panel is needed because two of the supplement means are identical. The MMC plot of the orthogonal polynomial contrasts shows the linear and quadratic effects are significant.

Our findings implies that for maximum vitamin A retention we should recommend intermediate amounts of temp and supp, perhaps in the vicinity of temp=80 and supp=6. An enlargement of this experiment could more accurately determine the optimal values.

If the analyst had been told, prior to the design of the experiment, that the primary goal was to determine the optimizing combination of the inputs temp and supp, the analyst would have considered using a *response surface design*, the most efficient design for this purpose. A brief introduction to such designs is contained in Montgomery (2001).

12.17 Exercises

12.1. Do the original Erdman alfalfa analysis of Section 12.14.3 with nitro as the response variable. Use the data accessible as data(rhiz.alfalfa).

12.2. Do the original Erdman clover analysis of Section 12.14.4 with nitro as the response variable. Use the data accessible as data(rhiz.clover).

12.3. Analyze the two factor experiment with data accessible as data(testing). This is a 3×3 design with 4 observations per treatment combination. The factors are breaker at levels 1 to 3 and Gauger at levels 1 to 3. The observations are strengths of cement. The cement is "gauged" or mixed with water and worked by three different gaugers before casting it into cubes. Three testers or "breakers" later tested the cubes for compressive strength, measured in pounds per square inch. Each gauger gauged 12 cubes, which were divided into 3 sets of 4, and each breaker tested one set of 4 cubes from each gauger. Breakers and gaugers are fixed in this experiment. Breakers and gaugers are people, not machines. Are there differences in the strength of the cement that depend on the handling by the breakers and gaugers?

We got the data from Hand et al. (1994). The data originally appeared in Davies and Goldsmith (1972). There the data were coded by $.1(X - 1000)$ before analysis.

The term *coded data* means that they have been centered and scaled to make the numerical values easier to work with by hand. The F-tests in the ANOVA table and the t-tests for regression coefficients with coded data are identical to the tests for the original data.

12.4. An agronomist compared five different sources of nitrogen fertilizer and a control (no fertilization) on the yield of barley. A randomized block design was used with four types of soil as the blocks. Each soil type was randomly divided into six plots, and the six treatments were randomly assigned to plots within each type. The treatments were, respectively, $(NH_4)SO_4$, NH_4NO_3, $CO(NH_2)_2$, $Ca(NO_3)_2$, $NaNO_3$, and control. The data, taken from Peterson (1985), are accessible as data(barleyp).

a. Plot the data. Does it appear from the plot that yield is related to treatment? Does it appear from the plot that blocking was successful?

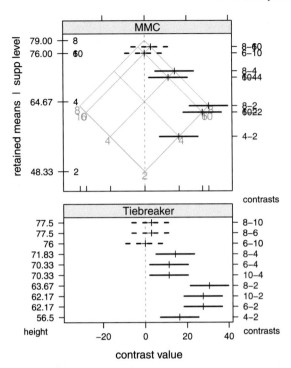

The MMC panel shows informative overprinting. Please see Tiebreaker panel and caption.

Fig. 12.18 Feed data MMC plot for pairwise contrasts of supplement. The means of `retained` at levels of feed 6 and 10 are identical. We need the Tiebreaker panel to distinguish them. Since Table 12.17 shows that the polynomial contrasts are very significant, we show the MMC plot for orthogonal polynomial contrasts in Figure 12.19.

b. Set up the two-way analysis of variance table for these data and explain what you conclude from it.

c. Use the Dunnett procedure, introduced in Section 7.1.3, to compare the five fertilizers with the control. Report your findings to the agronomist.

12.5. A chemist compared the abilities of three chemicals used on plywood panels to retard the spread of fire. Twelve panels were obtained, and each chemical was randomly sprayed on four of these twelve panels. Two pieces were cut from each panel and the time was measured for each piece to be completely consumed by a standard flame. (Thus Panel is nested within Chemical and Sample is nested within Panel.) The data, from Peterson (1985), are accessible as `data(retard)`. Carefully noting the relationship between the factors `chemical` and `panel`, and considering whether these factors are fixed or random, set up an analysis of variance and followup analysis of `chemical` means in order to make a recommendation of the best chemical for retarding the spread of plywood fires.

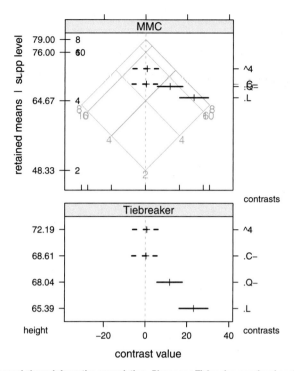

The MMC panel shows informative overprinting. Please see Tiebreaker panel and caption.

Fig. 12.19 Feed data MMC plot for orthogonal polynomial contrasts of supplement. As indicated in the ANOVA table, the linear and quadratic contrasts are significant and the cubic and quartic are not. We show the Tiebreaker panel even though it is not really needed in this example.

12.6. The judging of the ice skating events at the 2002 Winter Olympics in Salt Lake City was very controversial. The data, accessible as data(skateslc), are taken from Olympic Committee (2001). The dataset contains the scores on both technique and presentation of the five leading skaters, assigned by each of nine judges. We have recoded the data with $10(X-5)$. Perform a two-way analysis of variance where the response is the total of both scores. Do further analysis and comment on the consistency of the nine judges across skaters.

12.7. Box and Cox (1964), reprinted in Hand et al. (1994), present the results of a 3× 4 factorial experiment with four replications per treatment combination to illustrate the importance of investigating the normality assumption underlying analyses of variance. The original response variable is the survival time, survtime of each of four antidotes, treatment to each of three poisons. The data are accessible as data(animal).

a. Perform a two-way analysis of variance using survtime as the response, taking care to save the calculated cell residuals.

b. Produce a normal probability plot (described in Chapter 5) of the cell residuals and use it to conclude that the residuals are not normally distributed.

c. Redo the two-way analysis of variance with a reciprocal transformation of the response variable survtime, and again save the cell residuals. From a normal probability plot of these cell residuals, conclude that these new residuals are normally distributed and hence the transformation was successful.

d. Report your findings to the antidote researchers.

12.8. An experiment was constructed to compare the effects on etchings of wafers of four etching compounds and heat treatment by each of three furnaces. The experiment was reported in Johnson and Leone (1967) and the data are accessible as data(furnace). Viewing furnace as a random factor and allowing for the possibility of interaction, provide a thorough analyses of these data.

12.9. Anemia, caused by iron deficiency in the blood, is common in some countries. It was hypothesized that food cooked in iron pots would gain iron content from the pot and, hence when eaten, contribute to alleviation of iron deficiency. Research performed by Adish et al. (1999) compares the iron content (mg/100g) of three types of (traditional Ethiopian) food when cooked in pots of aluminum, clay, or iron. The data, accessible as data(ironpot), give the Iron content in mg/100g of food, the type of Pot, and the type of Food. Perform a two-way analysis of variance and provide interaction plots. Based on your analysis, is the hypothesis supported? Does your conclusion apply to all Foods studied?

12.10. To check the consistency of new car fuel efficiency, the miles per gallon of gasoline consumed was recorded for each of 5 cars of the same year and brand, on each of 10 randomly selected days. The investigation was reported in Johnson and Leone (1967) and the data are accessible as data(mpg). Viewing car as a random treatment factor and day as a random blocking factor, analyze the data and carefully state your conclusions. Suggest ways to elaborate on and improve this experiment.

12.11. Williams (1959), originally in Sulzberger (1953), examined the effects of temperature on the strength of wood, plywood, and glued joints. The data are accessible as data(hooppine). The studied wood came from hoop pine trees. The response is compressive strength parallel to the grain, and the treatment factor is temperature in degrees C. An available covariate is the moisture content of the wood, and tree is a blocking factor.

a. Fit a full model where both strength and moisture are adjusted for the blocking factor tree, allowing for the possibility that temp interacts with moisture.

b. Conclude that the interaction term can be deleted from this model. Reanalyze without this term. Carefully state your conclusions.

c. Investigate the nature of the relationship between strength and temp. Conclude that a linear fit will suffice.

d. Provide plots illustrating the conclusion from part a and the final model in parts b and c.

12.A Appendix: Computation for the Analysis of Variance

When there is more than a single factor, the discussion in this chapter is usually limited to the case where the sample size n_{ij} is the same for each cell. The programs we use for the computation do not usually have this limitation. We will discuss more general cases in Chapters 13 and 14.

With R we will be using aov for the calculations and anova and related commands for the display of the results. aov can be used with equal or unequal cell sizes. Model (12.1)

$$Y_{ijk} = \mu + \alpha_i + \beta_j + (\alpha\beta)_{ij} + \epsilon_{ijk} = \mu_{ij} + \epsilon_{ijk} \tag{12.1}$$

is denoted in R either by the formula
    Y ~ A + B + A:B
which uses the operator + to indicate the sum of two terms and the operator : to indicate the interaction of two factors, or by the formula
    Y ~ A * B
which uses the operator * to denote the crossing of two factors. The operator ~ is read as "is modeled by". The second formula is syntactically expanded by the program to the first formula before it is evaluated. We usually prefer the more compact notation Y ~ A * B because it more closely captures the English statement, "Y is modeled by the crossing of factors A and B."

With SAS we use PROC ANOVA and PROC GLM. PROC ANOVA is limited to the equal sample size cases (actually, to balanced designs; see the SAS documentation for details). Where there are at least two factors and unequal cell sizes [that is, the n_{ij} are not constrained to be equal and some cells may be empty (with $n_{ij} = 0$)] PROC GLM should be used. PROC ANOVA may not give sensible answers in such cases. Model (12.1) is denoted in SAS either by the expression
    Y = A   B   A*B
which uses a space to indicate the sum of two terms and the operator * to indicate the interaction term, or by the expression
    Y = A | B
which uses the operator | to denote the crossing of two factors. The operator = is read as "is modeled by". The second expression is syntactically expanded by the program to the first expression before it is evaluated. We usually prefer the more compact notation Y = A | B because it more closely captures the English statement, "Y is modeled by the crossing of factors A and B."

The intercept term μ and the error term ϵ_{ijk} are assumed in both statistical languages. The existence of the subscripts is implied and the actual values are specified by the data values.

The formula language also includes a notation for nesting of factors. We introduce nesting in Section 12.13 and say more in Section 13.5, especially in Tables 13.18 and 13.21. In R use the formula

    Y ~ A + A:B

or the formula (which will be expanded to the first formula)

    Y ~ A / B

which uses the / to indicate that A is an outer factor and B is nested within A.

SAS doesn't have the equivalent of the second formula. In SAS, use either the equivalent of the first formula

    Y = A   A*B

or an alternative notation

    Y = A   B(A)

which uses the parenthesis notation to indicate that B is nested within A.

Note that the A:B notation (or A*B in SAS) tells about the relation of the levels of the factors, not the degrees of freedom. Degrees of freedom depend on the linear dependencies with earlier terms in the model formula.

Chapter 13

Design of Experiments—Factorial Designs

Designs are often described by the number of factors. Chapter 6, "One-Way Analysis of Variance", discusses designs with one factor. Chapter 12, "Two-Way Analysis of Variance", discusses designs with two factors. More generally, we speak of "three-way" or "higher-way" designs and talk about main effects (one factor), two-way interactions (two factors), three-way interactions, four-way interactions, and so forth. Factors can have crossed or nested relationships. A factor can be fixed or random. When interaction is significant, its nature must be carefully investigated. If higher-order interactions, meaning those involving more than two factors, can be assumed to be negligible, it is often possible to design experiments that require observations on only a fraction of all possible treatment combinations.

Section 13.1 discusses a three-way ANOVA design with a covariate and polynomial contrasts. Section 13.2 introduces Latin squares. Section 13.3 introduces simple effects for interaction analyses. Section 13.4 discusses a nested factorial experiment with both crossed and nesting relationships among the factors. Section 13.6.1 discusses the SAS terminology for types of sums of squares used in sequential and conditional ANOVA tables. Related topics are discussed in Chapter 14.

13.1 A Three-Way ANOVA—Muscle Data

Cochran and Cox (1957) report on an experiment to assess the effect of electrical stimulation to prevent the atrophy of muscle tissue in rats. The dataset is available as data(cc176). The response wt.d is the weight of the treated muscle. There were three fixed factors: the number of treatments daily, n.treat, 1, 3, or 6; the duration of treatments in minutes, 1, 2, 3, or 5; and the four types of current used. A concomitant variable, the weight of corresponding muscle on the opposite untreated side of the rat, wt.n, was also made available. There were two replications of the entire experiment.

© Springer Science+Business Media New York 2015
R.M. Heiberger, B. Holland, *Statistical Analysis and Data Display*,
Springer Texts in Statistics, DOI 10.1007/978-1-4939-2122-5_13

The analysis is constructed with code in HHscriptnames(13). The data are plotted in Figure 13.1. The ANCOVA and adjusted means are in Table 13.1. Also included in Table 13.1 is a partitioning of the 2 degrees of freedom for n.treats into linear and quadratic components, taking account of the unequal spacing of the quantitative levels of n.treats.

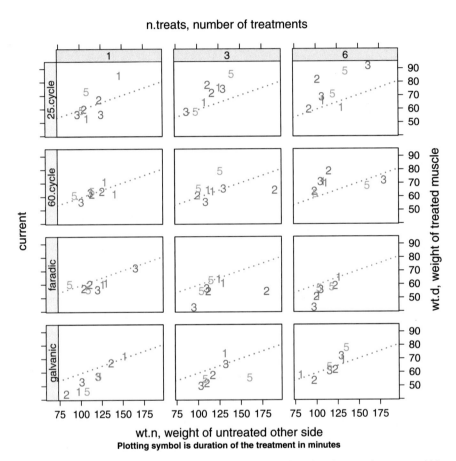

Fig. 13.1 Muscle data. The response variable wt.d is plotted against the covariate wt.n within each current×n.treats experimental condition. The plotting symbol is the duration of the treatment in minutes. The ANCOVA and adjusted means are in Table 13.1. We also plotted the common regression line (ignoring experimental conditions) of the response against the covariate. The presence of a covariate wt.n effect is evident from the graph by noting that the points in all panels approximate the uphill slope of the regression slope. The absence of a minutes effect is evident since there is no systematic pattern among the plotting symbols.

Table 13.1 suggests that after adjusting for the concomitant variable wt.n, there are no significant interactions and the effect of minutes is not significant. This table shows that n.treat contributes significantly to explaining wt.d,

Table 13.1 Muscle data. ANCOVA and adjusted means. The covariate wt.n, the linear effect of n.treats, and the current are the significant treatment effects. We show the calculation of y.adj, the response variable adjusted for the covariate, and the adjusted means.

```
> ## y=wt.d with x=wt.n as covariate
> ## (get essentially the same ANOVA as the approximate (y-bx)^2
> ## ANOVA table in Cochran and Cox)
> cc176.aov <- aov(wt.d ~ rep + wt.n + n.treats*minutes*current,
+                  data=cc176)

> ## summary(cc176.aov)
> summary(cc176.aov,
+          split=list(n.treats=list(n.treats.lin=1,
+                                    n.treats.quad=2)),
+          expand.split=FALSE)
                          Df Sum Sq Mean Sq F value  Pr(>F)
rep                        1    605     605   12.58 0.00091 ***
wt.n                       1   1334    1334   27.74 3.6e-06 ***
n.treats                   2    439     219    4.56 0.01557 *
  n.treats: n.treats.lin   1    438     438    9.11 0.00413 **
  n.treats: n.treats.quad  1      1       1    0.01 0.91048
minutes                    3    184      61    1.28 0.29409
current                    3   2114     705   14.66 7.8e-07 ***
n.treats:minutes           6    198      33    0.69 0.66051
n.treats:current           6    492      82    1.70 0.14163
minutes:current            9    383      43    0.88 0.54627
n.treats:minutes:current  18   1022      57    1.18 0.31542
Residuals                 46   2212      48
---
Signif. codes:  0 '***' 0.001 '**' 0.01 '*' 0.05 '.' 0.1 ' ' 1

> ##
> ## adjust y for x
> cc176$y.adj <- cc176$wt.d -
+   (cc176$wt.n - mean(cc176$wt.n))*coef(cc176.aov)["wt.n"]

> ## duplicate CC Table 5.17
> cc176.means <- tapply(cc176$y.adj,
+                       cc176[,c("current","n.treats")], mean)

> cc176.means
          n.treats
current        1     3     6
  galvanic 56.03 59.08 65.29
  faradic  59.95 55.79 57.27
  60.cycle 63.26 63.92 68.58
  25.cycle 64.47 71.78 73.20

> apply(cc176.means, 1, mean)
galvanic  faradic 60.cycle 25.cycle
   60.13    57.67    65.25    69.82
```

p–value = .0000036. The visible upward trend in all panels of Figure 13.1 suggests that response wt.d increases linearly with n.treats and differs according to the type of current used. The response variable y.adj in Figure 13.2 is constructed by adjusting the response wt.d for the covariable wt.n. We see in both Figures 13.1 and 13.2 a larger response when current is 25 cycle than when current is at one of its other three levels. Inclusion of wt.n reinforces these conclusions. The parallel traces in Figure 13.2 correspond to the absence of interaction between n.treat and current, a finding also suggested by the large p-value for this interaction in Table 13.1.

y.adj: main effects and 2–way interactions

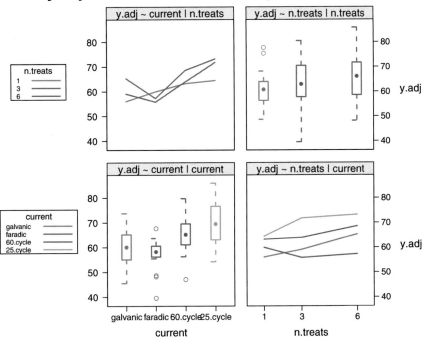

Fig. 13.2 Muscle data. Two-way interactions of significant main effects from the ANCOVA in Table 13.1. The adjusted response y.adj increases linearly with n.treats and differs according to the type of current used.

In Table 13.2 we display a microplot of horizontal boxplots that compares the distributions of responses for each level of the factor current. This is the same set of boxplots that appears in the y.adj ~ current | current panel of Figure 13.2.

Boxplots capture comparative information better than numbers. They don't have to take much space, therefore they can fit into tables of numbers and satisfy both the convention (perhaps mandated) of displaying numbers and the legibility of displaying graphs. We call the plots *microplots* when they are deliberately sized to fit into a table of numbers without interfering with the overall layout. When small plots are placed into a table of numbers, they can carry the same or more information per cm^2 as the numbers themselves.

Table 13.2 Muscle data: Distribution statistics and boxplots for adjusted weights. The statistics show only a few values. The boxplot shows the entire distribution.

Treatment	Min	m−sd	Mean	m+sd	Max	boxplot
25.cycle	54.21	61.04	69.82	78.59	85.92	
60.cycle	47.07	57.93	65.25	72.58	79.69	
faradic	39.55	51.98	57.67	63.37	67.64	
galvanic	45.52	52.82	60.13	67.44	73.72	

A display comparable to Figure 13.3 could be used to determine the nature of a 3-way interaction. Such an interaction does not exist in this example.

Figure 13.4 shows four different models for the relationship between the response wt.d and the covariate wt.n. The figure is similar to Figure 10.12 which showed four sets of panels for the simpler dataset with only one factor. The overall conclusion is that the relation between wt.d and wt.n differs according to the levels of current and n.treat. More detail appears in the caption of this figure.

Figure 13.5 is a Tukey procedure MMC plot examining the six pairwise differences among the four levels of current. As summarized in the caption of this figure, four of these six differences are declared statistically significant. Inspection of Figure 13.5 and the means of the levels of current in Table 13.1 reveals that 25.cycle and 60.cycle current, indistinguishable from each other, correspond to significantly greater treated muscle weight wt.d than either galvanic or faradic current.

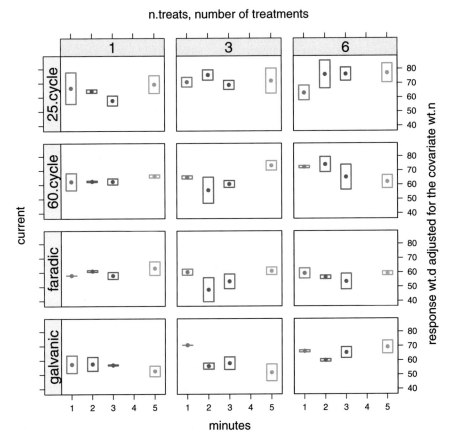

Fig. 13.3 Muscle data. Three-way interactions of all effects. One of the (3! = 6) possible orderings. The three-way interaction is not significant in this example. If there were a significant three-way interaction, the patterns in boxplots in adjacent rows and columns would not be the same. For example, we note a hint of a difference in the y.adj ~ minutes behavior across panels. It has a negative slope in the galvanic ~ 3 panel and a positive slope in the faradic ~ 3 panel, but a positive slope in the galvanic ~ 6 panel and a negative slope in the faradic ~ 6 panel. The ANOVA table tells us these differences in slope are not significant. These boxplots are all based on samples of size 2. Such boxplots are a well-defined but uncustomary way to display such a sample.

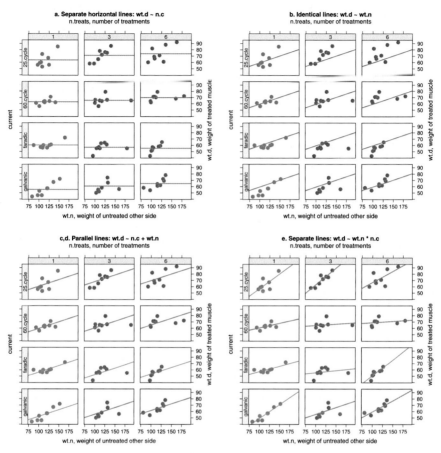

Fig. 13.4 Muscle data. ANCOVA plots with four different models. Panel a ignores `wt.n` and shows the average value of `wt.d`. Panel b fits a common regression of `wt.d` on `wt.n` on all combinations of `n.treat` and `current` and differs from Figure 13.1 only in its choice of plot symbol. Panel c,d allows for different intercepts but forces common slopes. The difference in intercepts corresponds to the small *p*-value for `wt.n` in Table 13.1. Panel e shows distinct regressions of `wt.d` on `wt.n` for each combination of `current` and `n.treat`. It suggests that the relationship between `wt.d` and `wt.n` differs according to the levels of `n.treat` and `current`. The term `n.c` in the title for the graphs is an abbreviation for the interaction (`n.treats*current`).

Table 13.3 Muscle data. ANCOVA with the simpler model using only the significant terms from Table 13.1 plus an additional interaction. We show the MMC plot for the current effect in Figure 13.5.

```
>   cc176t <- cc176

>   for (i in names(cc176t))
+     if (is.factor(cc176t[[i]]))
+       contrasts(cc176t[[i]]) <-
+         contr.treatment(length(levels(cc176t[[i]])))

>   sapply(cc176t, class)
$wt.d
[1] "numeric"

$wt.n
[1] "numeric"

$n.treats
[1] "positioned" "ordered"      "factor"

$current
[1] "positioned" "ordered"      "factor"

$minutes
[1] "positioned" "ordered"      "factor"

$rep
[1] "factor"


>   cc176t.aov <- aov(wt.d ~ rep + wt.n + n.treats + wt.n*current,
+                     data=cc176t)

>   summary(cc176t.aov)
              Df Sum Sq Mean Sq F value  Pr(>F)
rep            1    605     605   14.19 0.00030 ***
wt.n           1   1334    1334   31.29 2.6e-07 ***
n.treats       2    439     219    5.15 0.00776 **
current        3   2114     705   16.53 1.5e-08 ***
wt.n:current   3    867     289    6.78 0.00038 ***
Residuals     85   3624      43
---
Signif. codes:  0 '***' 0.001 '**' 0.01 '*' 0.05 '.' 0.1 ' ' 1
```

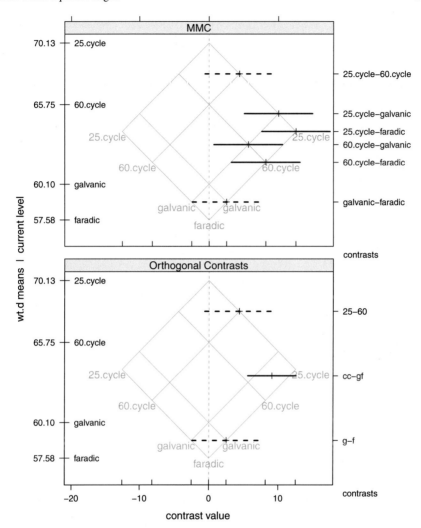

Fig. 13.5 Muscle data. MMC plot for the analysis in Table 13.3. The top panel shows the Tukey 95% intervals for all pairwise contrasts of treatment means wt.d adjusted for wt.n at the four levels of current. The Tukey procedure leads to the conclusions illustrated by the orthogonal contrasts in the bottom panel. The adjusted treatment means of both of 25.cycle and 60.cycle exceed both of galvanic and faradic; 25.cycle is indistinguishable from 60.cycle; and galvanic is indistinguishable from faradic. The Tiebreaker plot is not needed in this example.

13.2 Latin Square Designs

This design is useful when we have three factors having the same number, say r, of levels and the factors do not interact. Although there are r^3 treatment combinations, the Latin Square design permits us to run the experiment with a carefully chosen

Table 13.4 Sample 4×4 Latin square design. The rows represent tire positions: LF is Left-Front, RF is Right-Front, LR is Left-Rear, and RR is Right-Rear.

		Car		
Position	1	2	3	4
LF	C	D	A	B
RF	B	C	D	A
LR	A	B	C	D
RR	D	A	B	C

Table 13.5 Sample ANOVA for 4×4 Latin square design.

Source	df	Sum of Sq	Mean Sq	F Value	Pr(> F)
Row	$r-1$	SS_{Row}	MS_{Row}		
Column	$r-1$	SS_{Col}	MS_{Col}		
Treatment	$r-1$	SS_{Trt}	MS_{Trt}	MS_{Trt}/MS_{Res}	$1 - \mathcal{F}_{df_{Trt}, df_{Res}}(F)$
Residual	$(r-1)(r-2)$	SS_{Res}	MS_{Res}		
Total	r^2-1	SS_{Total}			

subset of r^2 of them while retaining the ability to conduct tests on all main effects. A Latin square is a square array of r Latin letters $A, B, C, \ldots$ such that each letter appears exactly once in each row and once in each column. Typically, the treatment factor is associated with these letters, and both *row* and *column* are blocking type factors. For example, if an experiment is run at r selected times of day on each of r days, then each row could represent one of the days, and each column one of the selected times. As another example, displayed in Table 13.4, if we have four cars available to compare the wear of four brands of tire, the rows of the square could represent the wheel position on the car, the columns represent the selected car, and the letters the tire brands.

The basic structure of the ANOVA table is in Table 13.5. Since we are using the Row and Columns factors as blocks, there is no test for those terms in the table. The only test we are justified in making is the test of the Treatment. The purpose of including the Row and Column factors is to pick up some of the Total Sum of Squares and thereby reduce the size of the residual mean square.

The arithmetic of the Latin square design depends on the assumption of no interaction between Row, Column, and Treatment. The arithmetic of the interaction of Row and Column gives $(r-1)^2$ df to the interaction and 0 df for an error term. By assuming no interaction, we gain the ability to split the $(r-1)^2$ df into two components: Treatment with $r-1$ df and Error with $(r-1)^2 - (r-1) = (r-1)(r-2)$ df. If the no-interaction assumptions hold, this is a very efficient design.

Almost always, $5 \leq r \leq 8$, for if $r < 5$ there are too few df for error, and one is unlikely to encounter situations where one has three factors each having $r > 8$ levels, two of which are blocking factors. However, it is possible to run an experiment containing several 3×3 squares or several 4×4 squares, each of which is considered a block, in order to achieve sufficient error df.

Catalogs of Latin squares appear in Cochran and Cox (1957) and elsewhere. In practice, one selects a square from a catalog and randomizes it by randomly assigning levels of one of the blocking factors to the rows of the square, randomly assigning levels of the other blocking factor to the columns of the square, and then randomly assigning treatment levels to the letters.

13.2.1 Example—Latin Square

The dataset data(tires), from Hicks and Turner (1999, page 115), is displayed in Table 13.6 alongside the original Latin square. A boxplot of the data is in Figure 13.6.

An initial ANOVA run in Table 13.7 revealed significant differences among cars and brands, but not among positions. Here $r = 4$, allowing just 6 df for estimating error. Hence the denominator df of the F-tests is also 6, which as discussed in Section 5.4.4 implies that these tests have little power. Nevertheless, the differences in this example among cars and brands are large enough for the F-tests to detect them.

Table 13.6 Latin square of tire wear experiment. The Latin square from Table 13.4 is repeated here. On the left with letters and on the right with the observed response values.

	Car					Car			
Position	1	2	3	4	Position	1	2	3	4
LF	C	D	A	B	LF	12	11	13	8
RF	B	C	D	A	RF	14	12	11	13
LR	A	B	C	D	LR	17	14	10	9
RR	D	A	B	C	RR	13	14	13	9

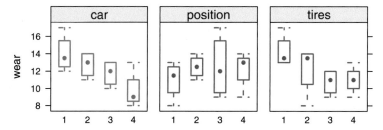

Fig. 13.6 Tires data. Boxplots of the response variables wear against the three factors.

To learn about the nature of the brand differences, we reran with a request for Tukey multiple comparisons tests on the brand means (Tables 13.7 and 13.8 and Figure 13.7). We find that brand 1 had significantly greater wear than brands 3 and 4, but the improvement in wear of brand 1 over brand 2 was not significant. We also see that cars 1, 2, and 3 all had significantly greater wear than car 4; no significant difference in tire wear was detected among cars 1, 2, and 3.

In this example the primary interest is studying the differences between brands of tires. Both car and position are blocking factors. We assume different cars will have different effects on tires because each person who owns a car drives different routes and puts the car through different wear patterns. We know there are differences in position on the car. Front tires are used for steering, rear tires just follow. In some cars only the front tires get power directly from the engine. In other cars only the rear tires, and in 4-wheel drive vehicles both front and rear tires get power. The goal of the Latin square experiment is to reduce the residual sum of squares by absorbing some of the variation into known blocking factors. This makes the comparisons of interest, those on brand, more precise because they can be made with a smaller standard deviation (based on the residual mean square).

Table 13.7 Latin square design. `tires` data. Differences in the blocks `car` are large, justifying blocking. Differences in `brand` are significant. We investigate further in Table 13.8.

```
>   data(tires)

>   tires.aov <- aov(wear ~ car + position + brand, data=tires)

>   summary(tires.aov)
            Df Sum Sq Mean Sq F value  Pr(>F)
car          3  38.69  12.896  14.395 0.00378 **
position     3   6.19   2.062   2.302 0.17695
brand        3  30.69  10.229  11.419 0.00683 **
Residuals    6   5.38   0.896
---
Signif. codes:  0 '***' 0.001 '**' 0.01 '*' 0.05 '.' 0.1 ' ' 1

>   tapply(tires$wear, tires$car, "mean")
    1     2     3     4
14.00 12.75 11.75  9.75

>   tapply(tires$wear, tires$position, "mean")
    1     2     3     4
11.00 12.50 12.50 12.25

>   tapply(tires$wear, tires$brand, "mean")
    1     2     3     4
14.25 12.25 10.75 11.00
```

Table 13.8 Continuation of analysis in Table 13.7. Latin square design. `tires` data. Brand 1 shows significantly greater mean `wear` than brand 3 or brand 4.

```
>    tires.mmc.brand <- mmc(tires.aov, linfct=mcp(brand="Tukey"))

>    ## print(tires.mmc.brand)
>    brand.lmat <- cbind("1-43" =c( 2, 0,-1,-1),
+                        "4-3"  =c( 0, 0,-1, 1),
+                        "143-2"=c( 1,-3, 1, 1))

>    dimnames(brand.lmat)[[1]] <- levels(tires$brand)

>    tires.mmc.brand <- mmc(tires.aov, linfct=mcp(brand="Tukey"),
+                           focus.lmat=brand.lmat)

>    print(tires.mmc.brand)
Tukey contrasts
Fit: aov(formula = wear ~ car + position + brand, data = tires)
Estimated Quantile = 3.462
95% family-wise confidence level
$mca
      estimate stderr   lower upper height
1-2      2.00 0.6693 -0.3173 4.317  13.25
1-4      3.25 0.6693  0.9327 5.567  12.62
1-3      3.50 0.6693  1.1827 5.817  12.50
2-4      1.25 0.6693 -1.0673 3.567  11.63
2-3      1.50 0.6693 -0.8173 3.817  11.50
4-3      0.25 0.6693 -2.0673 2.567  10.88
$none
   estimate stderr  lower upper height
1    14.25 0.4732 12.611 15.89  14.25
2    12.25 0.4732 10.611 13.89  12.25
4    11.00 0.4732  9.361 12.64  11.00
3    10.75 0.4732  9.111 12.39  10.75
$lmat
        estimate stderr  lower upper height
1-43     3.375 0.5796  1.368 5.382  12.56
2-143    0.250 0.5465 -1.642 2.142  12.13
4-3      0.250 0.6693 -2.067 2.567  10.88

>    contrasts(tires$brand) <- brand.lmat

>    tires.aov <- aov(wear ~ car + position + brand, data=tires)

>    summary(tires.aov, split=list(brand=list("1-43"=1, rest=2:3)))
              Df Sum Sq Mean Sq F value Pr(>F)
car            3   38.7   12.90   14.40 0.0038
position       3    6.2    2.06    2.30 0.1769
brand          3   30.7   10.23   11.42 0.0068
  brand: 1-43  1   30.4   30.37   33.91 0.0011
  brand: rest  2    0.3    0.16    0.17 0.8441
Residuals      6    5.4    0.90
```

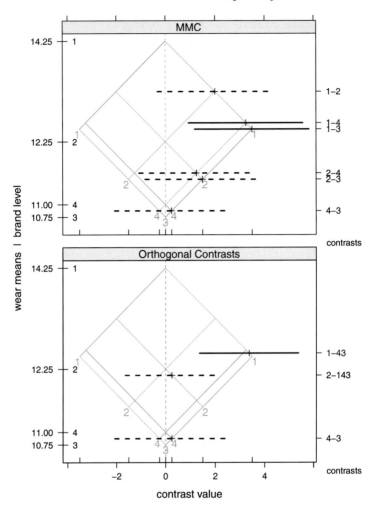

Fig. 13.7 Tires data. The top panel shows the Tukey 95% intervals for all pairwise contrasts of wear means at the four levels of brand. The Tukey procedure leads to the conclusions illustrated by the orthogonal contrasts in the bottom panel. Brand 1 shows significantly greater mean wear than brand 3 or brand 4. The Tiebreaker plot is not needed in this example.

The results of the F-test on a blocking factor are not ordinarily presented in the discussion because block differences are expected, and multiple comparisons on block means are not usually performed. Nevertheless, when blocks are significant, it is an indication that the blocking was worthwhile. Most experimental design texts contain formulas for the efficiency attributable to blocking in Latin square, randomized complete block, and other experimental designs; see, for example, Cochran and Cox (1957) (Section 4.37).

We continue with this example in Exercise 13.6 where we use dummy variables to illustrate the linear dependence of the treatment (brand) sum of squares on the interaction of the two blocking factors.

13.3 Simple Effects for Interaction Analyses

When a low p-value for an interaction term in an ANOVA table leads the analyst to believes that interaction exists, it is necessary to study the nature of the interaction.

We re-emphasize that in this situation, tests of the main effects of the factors comprising the interaction are inappropriate. Instead we seek to analyze the *simple effects*, which we now define. An analysis of simple effects, along with interaction plots of cell means which we've previously discussed, are the correct tools for investigating interaction. We note which simple effects are appreciable, either by examining confidence intervals on the simple effects or by testing whether the simple effects are zero. Since such activities involve simultaneous inferences, it is desirable to give attention to use either simultaneous confidence levels or a familywise error rate for simultaneous tests. The importance of studying individual simple effects rather than the overall interaction effect is comparable to the ecological fallacy introduced in Section 4.2 and the cautions resulting from Simpson's paradox, discussed in Section 15.3.

We confine attention here to the case of two factors, say A and B, at a and b levels, respectively. Continuing with the notation of Equation (12.1), let μ_{ij} denote the mean response of the treatment combination where A is at its ith level and B is at its jth level. Then a simple effect for B is a pairwise comparison of levels of B at a particular level of A, for example, $\mu_{12} - \mu_{13}$. Similarly, a simple effect for A is a pairwise comparison of levels of A at a particular level of B, such as $\mu_{32} - \mu_{12}$.

This assumes that the levels of a factor for which we are calculating simple effects are qualitative in nature. If instead the levels of factor B are quantitative, then a different analysis is called for, namely, a comparison of comparable polynomial contrasts of the cell means at each level of factor A. This analysis is superior to performing a separate one-way analyses comparing the levels of B because we are pooling the information from all levels of A to estimate the common error variance. This enables us to compare the levels of B with maximum available power.

In experiments with three or more factors, there is a potential for 3-factor interaction. If there are three factors (A, B, C) that interact, this may be interpreted as saying that the nature of the AB interaction varies according to the particular level of factor C. An analysis of such an interaction is more complicated than is the case for two interacting factors.

13.3.1 Example—The filmcoat Data

We illustrate the use of simple effects in analyzing interaction with the dataset data(filmcoat) from Iman (1994, pp. 768–778).

13.3.2 Study Objectives

Chemical vapor deposition is a process used in the semiconductor industry to deposit thin films of silicon dioxide and photoresist on substrates of wafers as they are manufactured. The films must be as thin as possible and have a uniform thickness, which is measured by a process called infrared interference.

A process engineer evaluated a low-pressure chemical vapor deposition (LPCVD) process that reduces costs and increases productivity. The engineer set up an experiment to study the effect of chamber temperature and pressure on film thickness. Three temperatures and three pressures were selected to represent the low, medium, and high levels of operating conditions for both factors. The experiment was conducted by randomly selecting one of the temperature–pressure combinations and determining the thickness of the film coating after processing is completed. This experiment was repeated three times with each temperature–pressure combination. The engineer wanted to determine the joint effect of temperature and pressure on the mean film thickness. The response was thickness (in Ångström units) of film coatings applied to wafers.

13.3.3 Data Description

temprt: temperature: low, medium, and high levels

pressure: pressure: low, medium, and high levels

coat: thickness of film coat

The data are displayed in Table 13.9 and Figure 13.8. The table of means and ANOVA table are in Table 13.10. The plots of the means and interactions are in Figure 13.9.

We observe that the temprt × pressure interaction is moderately significant. Therefore, conclusions about which level of temprt tends to minimize coat depend on the level of pressure. This statement is supported by Figure 13.9, which suggests that for low and high pressure, the response coat is minimized at medium temprt while for medium pressure, coat is minimized at high temprt. In addition, it is suggested that for low and medium temprt, coat is minimized at low pressure while for high temprt, coat is minimized at medium pressure.

Table 13.9 `filmcoat` data. Thickness of film coat at various settings of temperature and pressure.

	Pressure		
Temperature	Low	Medium	High
Low	42, 43, 39	45, 43, 45	45, 44, 47
Medium	36, 34, 37	39, 39, 37	40, 42, 38
High	38, 37, 37	35, 36, 33	40, 41, 42

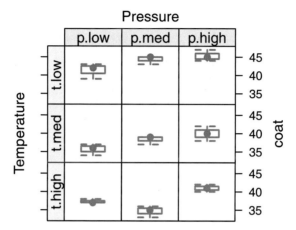

Fig. 13.8 a Filmcoat data. Each box is based on three data points. The interaction between temperature and pressure is very visible in the `t.high ~ p.med` panel. Along the top two rows of the plot, the boxes move up from left to right. In the bottom row, the second box is below the other two.

13.3.4 Data Analysis

These informal visual impressions are formally investigated by examining the simultaneous confidence intervals on the simple effects displayed in Figures 13.10 and 13.11. Control of the simultaneous confidence level at 95% within each of the two sets of nine intervals is maintained by using simulation-generated critical points for this procedure as recommended by Edwards and Berry (1987). These simultaneous confidence intervals are produced in R with the `glht` function in the **multcomp** package. The simultaneous confidence of the collection of all 18 confidence intervals is closer to 90%.

Table 13.10 Means and ANOVA table for filmcoat data. The moderately significant interaction between pressure and temperature requires that we examine simple effects rather than main effects.

```
>    reshape2::acast(filmcoat, temprt ~ pressure, mean,
+                    value.var="coat", margins=TRUE)
           p.low    p.med    p.high    (all)
t.low   41.33333 44.33333 45.33333 43.66667
t.med   35.66667 38.33333 40.00000 38.00000
t.high  37.33333 34.66667 41.00000 37.66667
(all)   38.11111 39.11111 42.11111 39.77778

>    film.aov1 <- aov(coat ~ temprt*pressure, data=filmcoat)

>    summary(film.aov1)
                Df Sum Sq Mean Sq F value   Pr(>F)
temprt           2 204.67  102.33  47.638 6.46e-08 ***
pressure         2  78.00   39.00  18.155 4.83e-05 ***
temprt:pressure  4  37.33    9.33   4.345   0.0124 *
Residuals       18  38.67    2.15
---
Signif. codes:  0 '***' 0.001 '**' 0.01 '*' 0.05 '.' 0.1 ' ' 1
```

We examine which of these intervals excludes zero and for those that do, whether the interval lies above or below zero. In Figure 13.10, we see that at medium pressure,

- the confidence interval on mean coat at low temprt minus mean coat at high temprt lies entirely above zero,
- the confidence interval on mean coat at medium temprt minus mean coat at high temprt is closer to zero, but is also entirely above zero.

From these two statements we can conclude that at medium pressure, mean coat is minimized at high temprt. This is the only firm conclusion we can draw about coat minimization because many of the other intervals in these two figures overlap zero, indicating nonsignificant differences of means. We are unable to formally confirm our other graphical impressions that for low and high pressure, the response coat is minimized at medium temprt. Nor are we able to confirm our initial graphical impressions about levels of pressure that minimize coat at each level of temprt.

In summary, while we can make some confident assertions about differences in coating between some of the combinations of temperature and pressure, it is not possible to infer from these data an overall recommendation of the optimal combination of temperature and pressure. It is possible that a larger experiment would have led to such a conclusion.

coat: simple effects and 2-way interactions

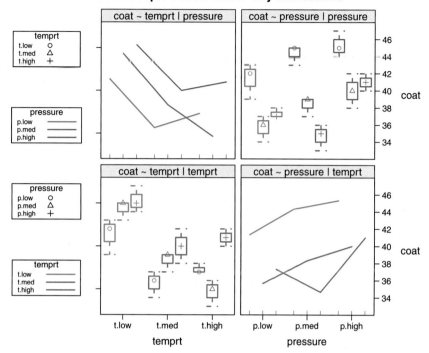

Fig. 13.9 Simple effects plot for `filmcoat` data. At medium `pressure`, mean `coat` is minimized at high `temprt`. No other firm conclusions can be drawn because many of the simultaneous confidence intervals in Figures 13.10 and 13.11 overlap zero. We saw in Section 13.3 that main effects are not defined in the presence of interaction. The simple effects plot here shows individual boxplots for each level of pressure conditioned on the level of temperature in the lower left panel. The plot shows individual boxplots for each level of temperature conditioned on the level of pressure in the upper right panel.

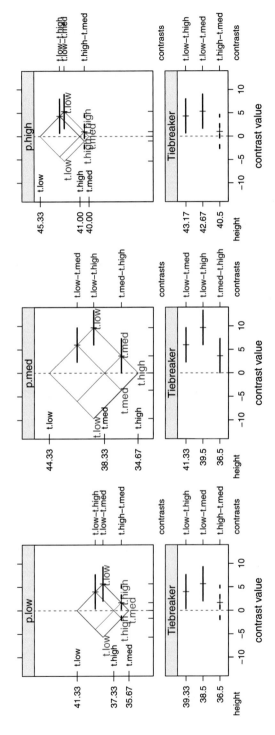

Fig. 13.10 Simultaneous 95% simulation-based confidence intervals for the filmcoat data: Simple effects of temperature at each level of pressure. The Tiebreaker plots in the bottom panels are imperative for this example.

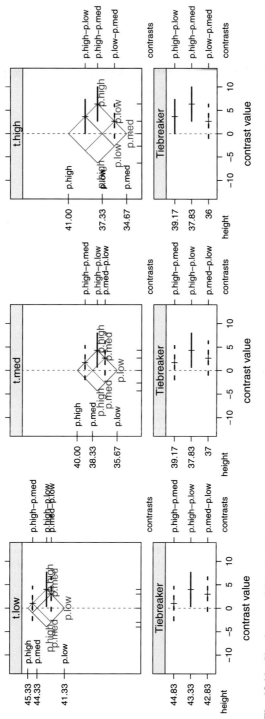

Fig. 13.11 Simultaneous 95% simulation-based confidence intervals for the filmcoat data: Simple effects of pressure at each level of temperature. The Tiebreaker plots in the bottom panels are imperative for this example.

13.4 Nested Factorial Experiment

Thus far we have considered situations where the relationships among the factors are either completely crossed or completely nested. It is also possible to have an experiment with three or more factors having both crossed and nesting relationships. Such an arrangement is called a nested factorial experiment.

13.4.1 Example—Gunload Data

We illustrate one possible arrangement with an example taken from Hicks and Turner (1999). It was desired to improve the number of rounds per minute that could be fired from a large naval gun. There are two levels of loading method, 1=new and 2=old, and three groups defining the physiques of the loaders, 1=slight, 2=average, 3=heavy. From each of these groups the experimenter selected three equal sized teams of men. Thus there are three teams of men having slight build, three teams of men having average build, and three teams of men having heavy build. Using both of the two methods, each of the nine teams fired the gun on two separate occasions. It is seen that team is nested within group, and that method is crossed with both group and team within group. The factors method and group are fixed, while team is a random factor. The data are contained in the dataset data(gunload). We display the data in Figure 13.12.

If all three factors were fixed factors, then the residual mean square would serve as the denominator for all analysis of variance table F-tests on main effects and interactions. When at least one factor is random, the F-test denominators are sometimes another mean square in the ANOVA table. As with our use of Table 12.8 to determine the correct denominator for an analysis with two crossed factors where one or both could be random factors, we construct Table 13.11 to aid in our analysis of the gunload data. The table is constructed by writing the sums of squares as quadratic forms in the Y_{ijkl} defined in Equation (13.1), and using Equation (I.6) in Appendix I for finding the expected values of these quadratic forms. Then as with Table 12.8, the ANOVA F-test of any effect uses as the denominator the mean square having expectation identical to the expected mean square of the effect, apart from the term for the effect itself.

In Table 13.11, we have three factors, which we call M, G, and T for easy association with method, group, and team in the gunload example. We use the corresponding Greek letters θ (for meTHod), γ, and τ for the population effects. Here T is nested within G, and M is crossed with both G and T. In this table, the factors M, G, and T have m, g, and t levels, respectively, and the common number of replications of each treatment combination is n. In the gunload example, there are $n = 2$ occasions.

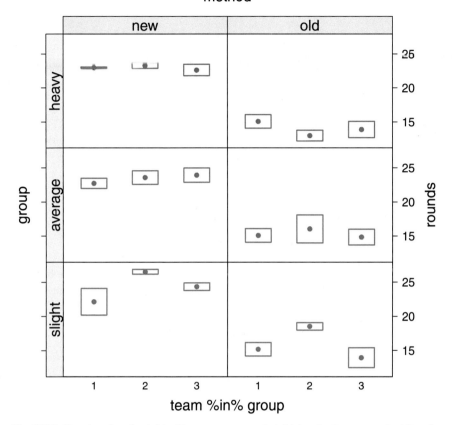

Fig. 13.12 Boxplot of gunload data. The response rounds is higher for the new method than the old method and does not appear to differ across the three physique groups. These findings are consistent with the small p-value for method and the large p-value for group in Table 13.13.

The statistical model associated with this analysis may be written as

$$Y_{ijkl} = \mu + \gamma_i + \tau_{j(i)} + \theta_k + (\gamma\theta)_{ik} + (\tau\theta)_{jk(i)} + \epsilon_{ijkl} \tag{13.1}$$

Here γ_i is the effect of group level i, $\tau_{j(i)}$ is the effect of level j of team nested within level i of group, θ_k represents level k of method, $(\gamma\theta)_{ik}$ represents the interaction of group and method, $(\tau\theta)_{jk(i)}$ represents the interaction of team and method within group level i, and ϵ_{ijkl} is the residual error.

For ease of presentation, we use the convention that

$$\sigma_A^2 = \frac{\sum_i \alpha_i^2}{a-1} \quad \text{if A is a fixed factor}$$

and

$$\sigma^2_{AB} = \frac{\sum_{ij}(\alpha\beta)^2_{ij}}{(a-1)(b-1)} \quad \text{if A and B are both fixed factors.}$$

We use the indicator function I_A defined as 1 if A is a random factor or 0 if A is a fixed factor. (This is not the same convention we used in Table 6.4. There we used the notation σ^2_A for random effects and κ^2_A for fixed effects.)

We illustrate the use of Table 13.11 by considering the test of the main effect for group, factor G. Since in this example method is fixed and team is random, we have $I_M = 0$ and $I_T = 1$. Therefore, the expected value of the mean square for G is

$$\sigma^2 + nm\sigma^2_T + nmt\sigma^2_G \tag{13.2}$$

and the expected value of the mean square for T nested in G is

$$\sigma^2 + nm\sigma^2_T \tag{13.3}$$

The ratio of these mean squares is appropriate for testing equality of the levels of group, factor G, because the corresponding ratio of these expected mean squares exceeds one if and only if $\sigma^2_G > 0$. Use of the residual mean square as the denominator of the F-test would be inappropriate because such a ratio would exceed one if there is a G effect, a T (team) effect, or both effects.

Note that if instead method were a random factor and team were a fixed factor, the pattern of expected mean squares would be quite different from those in Table 13.11, with different denominators appropriate for some of the ANOVA F-tests.

The model specifications for the sums of squares in the gunload example are shown for both R and SAS in Table 13.12. Discussion of the operators in this table appears in Section 13.5 and Table 13.18.

We overrode the default choices of the denominator mean squares for the F-tests for method, group, and method*group. These new choices are necessitated by the facts that one of the factors is random and there are both mixing and crossing of factors. Our conclusions here are that after correcting for both loaders' physiques and other person-to-person differences, the two methods have significantly different loading speeds. The new method averaged 23.59 rounds per minute compared with 15.08 rounds per minute for the old method. The analysis also shows a secondary finding that loading times do not differ significantly across physique groups.

Table 13.11 Expected mean squares for a three-factor nested factorial ANOVA. See also Tables 6.4 and 12.8.

Source	df	Expected Mean Square
G	$g - 1$	$\sigma^2 + nI_M\sigma^2_{TM} + ntI_M\sigma^2_{GM} \qquad\qquad + nmI_T\sigma^2_T + nmt\sigma^2_G$
T within G	$g(t - 1)$	$\sigma^2 + nI_M\sigma^2_{TM} \qquad\qquad\qquad\qquad + nmI_T\sigma^2_T$
M	$m - 1$	$\sigma^2 + nI_T\sigma^2_{TM} + ntI_G\sigma^2_{GM} + ngt\sigma^2_M$
GM	$(m - 1)(g - 1)$	$\sigma^2 + nI_T\sigma^2_{TM} + nt\;\sigma^2_{GM}$
TM within G	$g(m - 1)(t - 1)$	$\sigma^2 + n\;\sigma^2_{TM}$
Residual	$mgt(n - 1)$	σ^2
Total	$mgtn - 1$	

Table 13.12 Nested factorial model specifications in R and SAS. Specifications for the sums of squares for the gunload example are shown here. The tests are specified separately. In Table 13.13 we show use of the `Error` function in R.

Algebra	$Y_{ijkl} = \mu + \gamma_i +$	$\tau_{j(i)}$	$+ \theta_k +$	$(\gamma\theta)_{ik} +$	$(\tau\theta)_{jk(i)}$	$+ \epsilon_{ijkl}$	
R	`Y`	`~`	`G + T%in%G + M + G:M + T:M%in%G`				
SAS	`Y`	`=`	`G`	`T(G)`	`M`	`G*M`	`T*M(G)`

13.4.2 Example—Turkey Data (Continued)

We continue the discussion of the turkey data `data(turkey)` from Section 6.8.

Contrasts of the form used in Table 6.8 are so important in the design of experiments and in their analysis that we have a simple terminology and notation to describe them. In this experiment there are three distinct factors, one with two levels and two with three levels:

`trt.vs.control`:   with levels `control` and `treatment`

`additive`:   with levels `control`, A, and B

`amount`:   with levels 0, 1, and 2

occurring in the pattern shown in Table 13.14. The algebraic formula describing the model is

$$Y_{mijk} = \mu + \tau_m + \alpha_i + \beta_j + (\alpha\beta)_{ij} + \epsilon_{ijk} = \mu_{ij} + \epsilon_{mijk} \tag{13.4}$$

Table 13.13 Gunload data. The *F*-tests that appear without the `Error` function are incorrect. Here we produce correct tests by using the `Error` function to override the default choice of denominators of *F*-tests.

```
> gunload.aov <-
+    aov(rounds ~ method*group + Error((team %in% group)/method),
+        data=gunload)

> summary(gunload.aov)

Error: team:group
          Df Sum Sq Mean Sq F value Pr(>F)
group      2  16.05   8.026   1.227  0.358
Residuals  6  39.26   6.543

Error: team:group:method
             Df Sum Sq Mean Sq F value    Pr(>F)
method        1  652.0   652.0 364.841 1.33e-06 ***
method:group  2    1.2     0.6   0.332     0.73
Residuals     6   10.7     1.8
---
Signif. codes:  0 '***' 0.001 '**' 0.01 '*' 0.05 '.' 0.1 ' ' 1

Error: Within
          Df Sum Sq Mean Sq F value Pr(>F)
Residuals 18  41.59   2.311

> model.tables(gunload.aov, type="means")
Tables of means
Grand mean

19.33333

 method
method
   new    old
23.589 15.078

 group
group
 slight average   heavy
 20.125  19.383  18.492

 method:group
      group
method slight average heavy
   new 24.350  23.433 22.983
   old 15.900  15.333 14.000
```

Several issues are raised here to be discussed. How do these factors relate to each other? How does describing a design in terms of the factors specify the analysis?

Factors can be related in several ways (see Table 13.18). In the turkey example, we illustrate two relations: crossing and nesting.

crossing: Every level of additive appears at every level of amount. In this example, Additive A appears at Amounts 1 and 2, as does Additive B.

nesting: Some Additive–Amount combinations (A1, A2, B1, B2) appear in only the treatment level of trt.vs.control. Other Additive–Amount combinations (control-0) appear in only the control level of trt.vs.control. The factors additive and amount are then said to be *nested* within the factor trt.vs.control.

When we add these factors to the dataset, for example with commands in Table 13.15, we can write a much simpler model formula that automatically produces the easily readable ANOVA table in Table 13.16. Notice that the four 1-degree-of-freedom sums of squares in Table 13.16 are a decomposition of the 4-degree-of-freedom sum of squares in Table 6.7. The significance of the corresponding F-test in Table 6.7 is a rationale for producing and interpreting Table 13.16. We illustrate the structure with the table of means in Table 13.17 and the boxplots in Figure 13.13. See the discussion on orientation of boxplots in Section 13.A.

Table 13.14 Factor structure for turkey data.

Treatment	Level			Trt.vs.Cont	Treatment	Level		
	0	1	2			0	1	2
control	×			C	control	×		
A		×	×	T	A		×	×
B		×	×	T	B		×	×

In the turkey example there does not seem to be serious interaction ($p \approx .06$). In other situations the interaction dominates the analysis. An example with prominent interaction is the analysis of the *Rhizobium* clover data in Section 12.14.7.

Table 13.15 R commands to create factors for the turkey data introduced in Section 6.8.

```
> data(turkey)

> turkey[c(1,7,13,19,25),]
      diet wt.gain
1  control     4.1
7       A1     5.2
13      A2     6.3
19      B1     6.5
25      B2     9.5

> turkey$trt.vs.control <-
+     factor(rep(c("control","treatment"), c(6,24)))

> contrasts(turkey$trt.vs.control) <- c(4,-1)

> turkey$additive <- factor(rep(c("control","A","B"), c(6,12,12)),
+                           levels=c("control","A","B"))

> contrasts(turkey$additive) <- c(0,1,-1)

> turkey$amount <- factor(rep(c(0,1,2,1,2), c(6,6,6,6,6)))

> contrasts(turkey$amount) <- c(0,1,-1)

> turkey[c(1,7,13,19,25),]
      diet wt.gain trt.vs.control additive amount
1  control     4.1        control  control      0
7       A1     5.2      treatment        A      1
13      A2     6.3      treatment        A      2
19      B1     6.5      treatment        B      1
25      B2     9.5      treatment        B      2
```

Table 13.16 ANOVA for turkey data with the crossing of the `additive` and `amount` factors nested within the `trt.vs.control` factor. Interaction is borderline nonsignificant. The main effects of `additive` and `amount` nested within `trt.vs.control` are significant.

```
> turkey3.aov <- aov(wt.gain ~ trt.vs.control / (additive*amount),
+                        data=turkey, x=TRUE)

> summary(turkey3.aov)
                           Df Sum Sq Mean Sq F value    Pr(>F)
trt.vs.control              1  56.58   56.58 179.395 6.58e-13 ***
trt.vs.control:additive     1  22.81   22.81  72.337 7.56e-09 ***
trt.vs.control:amount       1  22.43   22.43  71.105 8.88e-09 ***
trt.vs.control:additive:amount  1   1.21    1.21   3.852   0.0609 .
Residuals                  25   7.88    0.32
---
Signif. codes:  0 '***' 0.001 '**' 0.01 '*' 0.05 '.' 0.1 ' ' 1
```

Table 13.17 Means for turkey data.

```
> print(na.print="",
+ tapply(turkey$wt.gain,
+        turkey[,c("additive","amount")],
+        mean)
+ )
         amount
additive        0   1        2
  control 3.783333
  A               5.5 6.983333
  B               7.0 9.383333
```

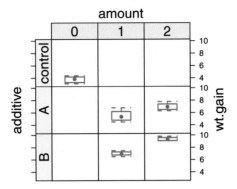

Fig. 13.13 Turkey data with factor structure. The main effect for `amount` is visible as the consistent increase in `wt.gain` as `amount` goes from 1 to 2. The main effect for `additive` is visible as the consistent increase in `wt.gain` as `additive` goes from A to B. The hint of borderline nonsignificant Interaction is seen as the different slope connecting the median dots in the A row and the B row of the display. This figure shows the boxplots in vertical orientation. Figure 13.17 shows the same boxplots in horizontal orientation. We discuss orientation in Section 13.A.

13.5 Specification of Model Formulas

Dummy variables (discussed in Section 10.1), and the contrasts they code for, are so important that all statistical languages have constructs for describing them and the relations between them. The model specification operators in R and SAS are detailed in Table 13.18.

Let us explore the meaning of the concepts of crossed and nested factors with a set of simple examples using the data in Table 13.19. The dataset `data(abc)` has two factors, A with three levels and B with four levels. Tables 13.20 and 13.21 show visual interpretations of the structure of the dataset with several different assumptions about the relation of the factors.

We show the model formula specifications, the generated dummy variables, the ANOVA tables, and the estimated a_i, b_j, $(ab)_{ij}$, and $b_{j(i)}$ values for several differently structured models. Table 13.22 contains the complete R input for these models. We recommend that you read these examples closely and experiment with them on your computer.

The simplest set of dummy variables (that is, easiest to understand) is the set of treatment contrasts. The most frequently used is the set of sum contrasts. We show both in Table 13.23.

Table 13.18 Model specification operators. In `R`, the two notations `a:b` and `b%in%a` are equivalent.

	R		SAS		Algebra
	abbrev	expanded	abbrev	expanded	
Double index	`a:b`		`a*b`		$(ab)_{ij}$
Sum	`a+b`		`a b`		$a_i + b_j$
Cross	`a*b`	`a+b+a:b`	`a|b`	`a b a*b`	$a_i + b_j + (ab)_{ij}$
Nested	`b%in%a`		`b(a)`	`a*b`	$b_{j(i)}$
Nest	`a/b`	`a + b%in%a`	`a b(a)`	`a` `a*b`	$a_i + b_{j(i)}$

Table 13.19 Sample data used to explore concepts of crossed and nested factors. Factor A has three levels and factor B has four levels. The interaction AB has 12 levels named by crossing the level names of A and B. The nested factor BwA (B within A) has 12 levels named without reference to factor A.

obs	A	B	AB	BwA	y
r.w	r	w	r.w	c	0.17
r.x	r	x	r.x	d	2.25
r.y	r	y	r.y	e	−1.57
r.z	r	z	r.z	f	−1.55
s.w	s	w	s.w	g	−0.24
s.x	s	x	s.x	h	1.71
s.y	s	y	s.y	i	0.38
s.z	s	z	s.z	j	−1.26
t.w	t	w	t.w	k	0.34
t.x	t	x	t.x	l	−0.15
t.y	t	y	t.y	m	−1.70
t.z	t	z	t.z	n	−1.93

Table 13.20 Rearrangements of abc data to show different assumptions about the relation of the factors: data, one factor, two factors crossed.

```
> data(abc)

> abc
      A B  AB BwA      y
r.w  r w r.w   c -0.02
r.x  r x r.x   d  1.19
r.y  r y r.y   e -0.02
r.z  r z r.z   f  0.23
s.w  s w s.w   g  0.67
s.x  s x s.x   h  1.95
s.y  s y s.y   i -0.71
s.z  s z s.z   j -0.40
t.w  t w t.w   k -0.56
t.x  t x t.x   l  0.01
t.y  t y t.y   m  0.13
t.z  t z t.z   n  1.19

> abc.oneway <- ## one-way
+ with(abc,
+      matrix(y, 4, 3, dimnames=list(1:4, A=unique(A)))
+      )

> abc.oneway
   A
       r     s     t
 1 -0.02  0.67 -0.56
 2  1.19  1.95  0.01
 3 -0.02 -0.71  0.13
 4  0.23 -0.40  1.19

> abc.crossed <- ## crossed
+ with(abc,
+      matrix(y, 3, 4, byrow=TRUE,
+             dimnames=list(A=unique(A), B=unique(B)))
+      )

> abc.crossed
   B
A      w    x     y     z
  r -0.02 1.19 -0.02  0.23
  s  0.67 1.95 -0.71 -0.40
  t -0.56 0.01  0.13  1.19
```

Table 13.21 Rearrangements of abc data to show different assumptions about the relation of the factors: two factors nested, two factors doubly indexed.

```
> abc.nested <- ## nested
+ with(abc,
+        matrix(c(y[1:4],      rep(NA,8),
+                 rep(NA,4), y[5:8],      rep(NA,4),
+                 rep(NA,8),              y[9:12]),
+                 3, 12, byrow=TRUE,
+                 dimnames=list(A=unique(A), BwA=BwA))
+                 )

> print(abc.nested, na.print="")
  BwA
A      c    d     e    f    g    h     i    j     k    l    m    n
  r -0.02 1.19 -0.02 0.23
  s                        0.67 1.95 -0.71 -0.4
  t                                              -0.56 0.01 0.13 1.19

> abc.double.indexed <- ## doubly-indexed
+    abc[,"y",drop=FALSE]

> abc.double.indexed
        y
r.w -0.02
r.x  1.19
r.y -0.02
r.z  0.23
s.w  0.67
s.x  1.95
s.y -0.71
s.z -0.40
t.w -0.56
t.x  0.01
t.y  0.13
t.z  1.19
```

Table 13.22 Specification of several models using one or both variables in the abc dataset.

```
## one-way
abc.A.aov <- aov(y ~ A, data=abc)
anova(abc.A.aov)
coef(abc.A.aov)
contrasts(abc$A)
model.matrix(abc.A.aov)


## crossed: no interaction
abc.ApB.aov <- aov(y ~ A+B, data=abc)
anova(abc.ApB.aov)
coef(abc.ApB.aov)
contrasts(abc$A)
contrasts(abc$B)
model.matrix(abc.ApB.aov)


## crossed: with interaction
abc.AsB.aov <- aov(y ~ A*B, data=abc)
anova(abc.AsB.aov)
coef(abc.AsB.aov)
contrasts(abc$A)
contrasts(abc$B)
contrasts(abc$AB)
model.matrix(abc.AsB.aov)


## nested
abc.BwA.aov <- aov(y ~ A/B, data=abc)
anova(abc.BwA.aov)
coef(abc.BwA.aov)
contrasts(abc$A)
contrasts(interaction(abc$A, abc$B))
model.matrix(abc.BwA.aov)


## doubly-indexed
abc.AB.aov <- aov(y ~ AB, data=abc)
anova(abc.AB.aov)
coef(abc.AB.aov)
contrasts(abc$AB)
model.matrix(abc.AB.aov)
```

Table 13.23 These dummy variables are constructed for a fit of the form $\hat{y}_i = m + a_i$ to a model of the form $y_{ij} = \mu + \alpha_i + \epsilon_{ij}$. With treatment contrasts, m is an estimate of $\mu + \alpha_r$, a_s is an estimate of $\alpha_s - \alpha_r$, and a_t is an estimate of $\alpha_t - \alpha_r$. With sum contrasts, m is an estimate of μ, a_1 is an estimate of α_r, and a_2 is an estimate of α_s. There is no need for an a_3 because of the constraint on the parameters $\alpha_t = -(\alpha_r + \alpha_s)$

```
> model.matrix(~A, data=abc,              > model.matrix(~A, data=abc,
+     contrasts=                          +     contrasts=
+         list(A=contr.treatment))        +         list(A=contr.sum))
      (Intercept) A2 A3                         (Intercept) A1 A2
r.w             1  0  0                   r.w             1  1  0
r.x             1  0  0                   r.x             1  1  0
r.y             1  0  0                   r.y             1  1  0
r.z             1  0  0                   r.z             1  1  0
s.w             1  1  0                   s.w             1  0  1
s.x             1  1  0                   s.x             1  0  1
s.y             1  1  0                   s.y             1  0  1
s.z             1  1  0                   s.z             1  0  1
t.w             1  0  1                   t.w             1 -1 -1
t.x             1  0  1                   t.x             1 -1 -1
t.y             1  0  1                   t.y             1 -1 -1
t.z             1  0  1                   t.z             1 -1 -1
attr(,"assign")                          attr(,"assign")
[1] 0 1 1                                [1] 0 1 1
attr(,"contrasts")                       attr(,"contrasts")
attr(,"contrasts")$A                     attr(,"contrasts")$A
   2 3                                        [,1] [,2]
r  0 0                                    r     1    0
s  1 0                                    s     0    1
t  0 1                                    t    -1   -1
```

13.5.1 Crossing of Two Factors

We provide a detailed discussion here of just one model, the crossing of two factors. The dummy variables for interaction in the crossing model

$$\text{Algebra:} \quad y_i = \mu + \alpha_i + \beta_j + (\alpha\beta)_{ij} + \epsilon_{ij}$$
$$\text{R:} \quad Y \sim \quad A + B + A{:}B$$
$$\text{SAS:} \quad Y = \quad A \quad B \quad A{*}B$$

are constructed as the outer product of the rows of the dummy variables for each of the main effects. We illustrate in Table 13.24 with the sum contrasts.

Table 13.24 Dummy variables for the interaction $(\alpha\beta)_{ij}$ constructed as the outer product of the rows of the dummy variables for the two main effects A and B. We continue with the data of Table 13.19 and the sum contrasts defined in the right side of Table 13.23. For example, in row r.z the value in column A1:B1 is the product of the 1 in column A1 and the -1 in column B1. There are two degrees of freedom, hence two dummy variables for the A effect. There are three dummy variables for the B effect. Therefore, there are 2×3 dummy variables for the A:B interaction.

```
> old.width <- options(width=70)

> mm <- model.matrix(~A*B, data=abc,
+               contrasts=list(A=contr.helmert, B=contr.helmert))

> mm[,]
    (Intercept) A1 A2 B1 B2 B3 A1:B1 A2:B1 A1:B2 A2:B2 A1:B3 A2:B3
r.w           1 -1 -1 -1 -1 -1     1     1     1     1     1     1
r.x           1 -1 -1  1 -1 -1    -1    -1     1     1     1     1
r.y           1 -1 -1  0  2 -1     0     0    -2    -2     1     1
r.z           1 -1 -1  0  0  3     0     0     0     0    -3    -3
s.w           1  1 -1 -1 -1 -1    -1     1    -1     1    -1     1
s.x           1  1 -1  1 -1 -1     1    -1    -1     1    -1     1
s.y           1  1 -1  0  2 -1     0     0     2    -2    -1     1
s.z           1  1 -1  0  0  3     0     0     0     0     3    -3
t.w           1  0  2 -1 -1 -1     0    -2     0    -2     0    -2
t.x           1  0  2  1 -1 -1     0     2     0    -2     0    -2
t.y           1  0  2  0  2 -1     0     0     0     4     0    -2
t.z           1  0  2  0  0  3     0     0     0     0     0     6

> print(AA <- mm["s.y", c("A1","A2")])
A1 A2
 1 -1

> print(BBB <- mm["s.y", c("B1","B2","B3")])
B1 B2 B3
 0  2 -1

> outer(AA, BBB)
   B1 B2 B3
A1  0  2 -1
A2  0 -2  1

> as.vector(outer(AA, BBB))
[1]  0  0  2 -2 -1  1

> mm["s.y", c("A1:B1","A2:B1","A1:B2","A2:B2","A1:B3","A2:B3")]
A1:B1 A2:B1 A1:B2 A2:B2 A1:B3 A2:B3
    0     0     2    -2    -1     1

> options(old.width)
```

13.5.2 Example—Dummy Variables for Crossed Factors Nested Within Another Factor—Turkey Data (Continued Again)

A model formula specifies a set of dummy variables. Just as in one-way analysis of variance, we control the structure of the dummy variables with the contrast matrix assigned to each factor. Let us look at the dummy variables generated for us by the model formula

```
wt.gain ~ trt.vs.control / (additive*amount)
```

We do so in R by adding the argument x=TRUE to the aov statement in Table 13.16 and then displaying the x component of the resulting aov object in Table 13.25.

There are some complications in the display in Table 13.25. The generation of the x matrix of dummy variables doesn't know about the actual degrees of freedom for each effect. It assumes the maximum possible if all implied cells were observed (in this example there are $2 \times 3 \times 3 = 18$ cells implied by the complete crossing of trt.vs.control, additive, and amount). Only five of those implied cells actually have observations. The match and the relabeling are used to find just the ones that matter in this example and to give them more reasonable names. See Exercise 13.10 for guidance on discovering how the match function is used.

The predicted value for an observation i is calculated, as with any linear model, as the inner product of the regression coefficients with the dummy variables in row i. In this example, we predict the weight gain for observation 7 as

$$\hat{y}_7 = (1 \ -1 \ 1 \ 1 \ 1) \begin{pmatrix} 6.5300 \\ -0.6867 \\ -0.9750 \\ -0.9667 \\ 0.2250 \end{pmatrix} = 5.5$$

13.6 Sequential and Conditional Tests

When there are two or more predictors in a model, they are usually not orthogonal to each other. Therefore, the interpretation given to the relative importance of each predictor depends on the order in which they enter the model. One of the important goals of designed experiments is the choice of combinations of levels for factors that will make the dummy variables for each factor or interaction orthogonal to the others. Most of the examples in this book in Chapters 12, 13, and 14 have orthogonal effects.

When the data for an example have continuous predictors or covariates, or are classified by factors with unequal numbers of observations per cell, the effects are

Table 13.25 Regression coefficients and dummy variables for turkey data. This table is a continuation of Tables 13.15 and 13.16.

```
> match(dimnames(coef(summary.lm(turkey3.aov)))[[1]],
+        dimnames(turkey3.aov$x)[[2]])
[1]  1  2  4  8 12

> turkey[c(1,7,13,19,25),]
      diet wt.gain trt.vs.control additive amount
1  control    4.1        control  control      0
7       A1    5.2      treatment        A      1
13      A2    6.3      treatment        A      2
19      B1    6.5      treatment        B      1
25      B2    9.5      treatment        B      2

> turkey3.coef <- summary.lm(turkey3.aov)$coef

> turkey3.x <- turkey3.aov$x

> term.names <-
+     c("(Intercept)","trt.vs.control","additive","amount",
+       "additive:amount")

> dimnames(turkey3.coef)[[1]] <- term.names

> dimnames(turkey3.x)[[2]][c(1,2,4,8,12)] <- term.names

> zapsmall(turkey3.coef)
                 Estimate Std. Error   t value Pr(>|t|)
(Intercept)       6.53000    0.10253  63.68585   0.0000
trt.vs.control   -0.68667    0.05127 -13.39386   0.0000
additive         -0.97500    0.11464  -8.50510   0.0000
amount           -0.96667    0.11464  -8.43241   0.0000
additive:amount   0.22500    0.11464   1.96272   0.0609

> turkey3.x[c(1,7,13,19,25), c(1,2,4,8,12)]
   (Intercept) trt.vs.control additive amount additive:amount
1            1              4        0      0               0
7            1             -1        1      1               1
13           1             -1        1     -1              -1
19           1             -1       -1      1              -1
25           1             -1       -1     -1               1
```

usually not orthogonal. Most of the examples in Chapters 9, 10, and 11 have continuous predictor variables and therefore do not have orthogonal effects.

When effects are not orthogonal, the sequence in which they are entered into the model affects the interpretation of the effects. See Sections 9.6 (Partial F-Tests) and 9.13 (Residual Plots) for techniques used to investigate the relative importance of the predictors.

The sequential ANOVA table depends on the order in which the effects are entered into the model. Each row of the table is calculated under the assumption that all effects in higher rows have already been included and that all effects in lower rows have not. R normally prints the sequential ANOVA table. SAS calls the sequential ANOVA table the table of Type I sums of squares.

There are several types of conditional ANOVA tables. One of the most frequently used is Yates' weighted squares of mean, what SAS calls Type III sums of squares, in which each row of the table is calculated under the assumption that all other rows—both higher and lower in their placement in the ANOVA table—have already been included. We, and many others, have difficulty with Type III sums of squares when used with designed experiments because they violate the principle of marginality that says it is usually not meaningful to test, estimate, or interpret main effects of explanatory variables when the variables interact. The principle was stated by Nelder (1977) and strongly supported by Venables (1998).

Another method, Yates' Method of Fitting Constants, what SAS calls Type II sums of squares, makes different assumptions for each class of effect. ANOVA table rows for main effects assume all other main effects are already included in the model. ANOVA table rows for two-way interactions assume all main effects and other two-way interactions are in the model. Higher-order interactions assume all lower-order effects and interactions are already in the model, hence are consistent with the marginality principle.

13.6.1 *SAS Terminology for Conditional Sums of Squares*

The terminologies Type I, Type II, and Type III sum of squares originated by SAS have become so widely known that they are used nowadays even outside the context of interpreting SAS listing files. In order that readers be able to request and interpret SAS analysis of variance presentations, we provide here more details on these types in two contexts, the context of designed experiments having two factors and the context of regression analysis with continuous predictor variables.

Suppose the response is Y, the two factors are A and B, and the SAS model statement reads

```
Model Y = A B A*B;
```

Since no particular types of sums of squares were requested, SAS provides by default Types I and III. If the user wishes to override the default, particular types can be requested as illustrated here:

```
Model Y = A B A*B /ss1 ss2;
```

The sequential (Type I) sum of squares for each effect is the portion of model sum of squares attributable to that effect above and beyond what is attributable to all effects listed prior to it in the expanded model statement. It is conditional on all the previous terms already being in the model. Thus in the illustration, the Type I sum of squares for B is the marginal contribution of factor B conditional on factor A already being in the model. Use of this sum of squares is appropriate if a model containing factor A without factor B makes sense, but a model containing factor B makes no sense unless the model already includes factor A.

For each main effect in the model statement, the Type II sum of squares is the marginal contribution of that effect beyond the sum of squares attributable to all other main effects in the model statement. The Type II sum of squares for A*B is the portion of model sum of squares attributable to this interaction after the main effects A and B are already in the model. Yates (1934) gave the name *method of fitting constants* to what is now called Type II sums of squares. In the absence of interaction, this method produces the maximum power tests for the main effects.

Note that while the Type I sums of squares for A, B and A*B add up to the model sum of squares, the Type II sums of squares for these three effects are not in general an orthogonal partitioning of the model sum of squares and hence do not in general sum to the model sum of squares. An exception occurs when the data are balanced (for example, each of the *ab* cells contain the same number of observations); in this case the Type I and Type II sums of squares coincide.

Type III sums of squares can be used with the above model statement provided that each of the *ab* cells contains at least one observation. The Type III sum of squares for any effect, including A*B, is adjusted for *all* other effects in the model. If the sampling is balanced, the Type III sums of squares for main effects coincide with the sums of squares for Type I and Type II. The Type III partitioning provides what is known as the Yates' weighted squares of means analysis of unbalanced data; see Searle (1971).

Type IV sums of squares coincide with Type III sums of squares when all cells contain observations. This partitioning is used when some cells are empty, a situation we do not pursue in this text. The Type IV sum of squares partitioning is not unique, a feature that makes many analysts uncomfortable with their use.

The nomenclature Type I and Type III (as well as Type II and Type IV) was originated by SAS in Goodnight (1978) and summarized in SAS Institute, Inc. (1999).

13.6.2 Example—Application to Clover Data

In Table 12.12 we showed the two-way ANOVA table for the clover data. The data in
data(rhiz.clover) is balanced. As a consequence all three forms of conditional
sums of squares give exactly the same results (as long as proper contrasts (with
columns orthogonal to the constant column) are used for all factors. The treatment
contrasts (the default) are not proper contrasts and may show strange results).

In Table 13.26 we illustrate the three conditional sets of sums of squares using
an unbalanced dataset, constructed by deleting two observations from the balanced
clover data. The data are displayed in Figure 13.14.

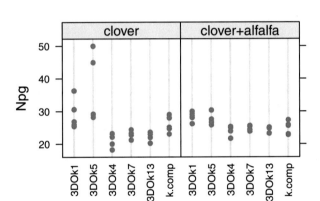

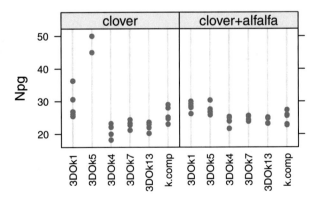

Fig. 13.14 Full clover dataset on top. Two observations removed on the bottom.

Table 13.26 Display of conditional sums of squares (Sequential/Type I, Method of Fitting Constants/Type II, and Weighted Squares of Means/Type III) for an unbalanced dataset. The car::Anova function requires contrasts orthogonal to the constant column. Compare the Sum Sq values for strain and comb. The left column shows the model formula Npg ~ strain * comb and the right column shows Npg ~ comb * strain. The main effects in the sequential tables (using anova) depend on the order. The main effects in the method of fitting constants tables car::Anova(type=2) do not depend on order and are the same as the second position of the sequential tables. The main effects in the weighted squares of means tables car::Anova(type=3) do not depend on order; they violate the principal of marginality because they are conditional on taking out the interaction effect first. The Residuals and interaction comb:strain sums of squares are identical in all six tables.

```
> data(rhiz.clover)

> ## drop two observation to illustrate Type II and III sums of squares
> ## I am dropping the non-outlier observations in 3D0k5
> cloverD <- rhiz.clover[-c(7,9,10),]

> old.opt <- options(show.signif.stars=FALSE, digits=3)
```

```
> cloverDsc.aov <-                              > cloverDcs.aov <-
+   aov(Npg ~ strain * comb,                    +   aov(Npg ~ comb * strain,
+       data=cloverD,                           +       data=cloverD,
+       contrasts=                              +       contrasts=
+           list(strain=contr.sum,              +           list(strain=contr.sum,
+                comb=contr.sum))               +                comb=contr.sum))
```

```
> anova(cloverDsc.aov)[,c(2,1,4,5)]             > anova(cloverDcs.aov)[,c(2,1,4,5)]
            Sum Sq Df F value  Pr(>F)                       Sum Sq Df F value  Pr(>F)
strain         657  5   29.28 4.2e-13           comb            2  1    0.52    0.48
comb            20  1    4.46    0.04           strain        675  5   30.07 2.7e-13
strain:comb    594  5   26.47 2.3e-12          comb:strain   594  5   26.47 2.3e-12
Residuals      202 45                          Residuals      202 45
```

```
> Anova(cloverDsc.aov, type=2)                  > Anova(cloverDcs.aov, type=2)
Anova Table (Type II tests)                    Anova Table (Type II tests)

Response: Npg                                   Response: Npg
            Sum Sq Df F value  Pr(>F)                       Sum Sq Df F value  Pr(>F)
strain         675  5   30.07 2.7e-13           comb           20  1    4.46    0.04
comb            20  1    4.46    0.04           strain        675  5   30.07 2.7e-13
strain:comb    594  5   26.47 2.3e-12          comb:strain   594  5   26.47 2.3e-12
Residuals      202 45                          Residuals      202 45
```

```
> Anova(cloverDsc.aov, type=3)                  > Anova(cloverDcs.aov, type=3)
Anova Table (Type III tests)                   Anova Table (Type III tests)

Response: Npg                                   Response: Npg
            Sum Sq Df F value  Pr(>F)                       Sum Sq Df F value  Pr(>F)
(Intercept) 38711  1  8621.1 < 2e-16           (Intercept) 38711  1  8621.1 < 2e-16
strain        1049  5    46.7 < 2e-16           comb           86  1    19.3 6.8e-05
comb            86  1    19.3 6.8e-05           strain       1049  5    46.7 < 2e-16
strain:comb    594  5    26.5 2.3e-12          comb:strain   594  5    26.5 2.3e-12
Residuals      202 45                          Residuals      202 45

                                               > options(old.opt)
```

13.6.3 Example—Application to Body Fat Data

We revisit in Table 13.27 the analysis begun in Section 9.2 of a portion of the body fat data data(fat) using the two predictors abdomin and biceps of the response bodyfat.

The F-value 54.92 applies to the composite hypothesis $H_0 : \beta_1 = \beta_2 = 0$ against the alternative that at least one β_i is nonzero, where the β_i are the coefficients of the two predictors. The corresponding small p-value indicates that either abdomin or biceps or both are linearly related to bodyfat. The $R^2 = 0.714$ tells us that 71.4% of the variability in these subjects' bodyfat is accounted for by their abdomin and biceps measurements. The remaining 28.6% of bodyfat variability is explained by other measurable variables not presently in the model as well as the random error component of the model in Equation (9.1).

The Type I sums of squares are sequential in that the sum of squares 2440.5 for the first listed predictor, abdomin, is calculated assuming that this is the only predictor in the model, while the sum of squares 209.32 for the second listed predictor, biceps, is calculated assuming that the first listed predictor is already in the model. In general, the top to bottom ordering of sources of variation in the Type I sum of squares table is the same as the ordering of these sources in the Model statement.

Each predictor's Type III sum of squares is calculated assuming that all other predictors are already in the model. Thus the Type III sum of squares for abdomin, 1823.02, is *conditional* in the sense that it is calculated under the assumption that the model already contains the other predictor biceps. In general, any entry in a Type III sum of squares table is conditioned on the existence in the model of all sources above it in this table.

The parameter estimates in both Table 9.1 and Table 13.27 are based on this same "last-in" rule, corresponding to the Type III sums of squares. The Type III F-value for abdomin, 75.57, is the square of the t-value for abdomin, 8.69, in the Parameter section of Table 13.27 and in Table 9.1. This corresponds to the interpretation of the t-tests for the regression coefficients, each of which measures the marginal contribution of its predictor variable conditional on all the other predictor variables already being in the model.

In this context, it is preferable to work with the Type I analysis if the investigator believes that a model containing biceps makes no sense unless abdomin is already in the model. Otherwise, with continuous predictor variables, the Type III approach is preferred. In this example, each predictor has a statistically significant impact on bodyfat after the other predictor has already been included in the model. In general, it is possible for one predictor to have an insignificant additional impact on the response when other more prominent predictors are already in the model. See Section 9.11 on collinearity for a discussion of this issue.

Table 13.27 car::Anova from the **car** package display for two-X regression of bodyfat. The sequential sums of squares (Type I sums of squares) correspond to the display in Table 9.1. The sequential sums of squares are an orthogonal partitioning of the model sum of squares. The weighted squares of means (Type III sums of squares) is not an orthogonal partitioning. The sum of the two values for abdomin and biceps is not equal to the model sum of squares.

```
> library(car)

> data(fat)

> fat.lm <- lm(bodyfat ~ abdomin + biceps, data=fat)

> ## regression coefficients
> coef(summary(fat.lm))
              Estimate Std. Error   t value     Pr(>|t|)
(Intercept) -14.5937363 6.69222199 -2.180701 3.459548e-02
abdomin       0.6829379 0.07855885  8.693329 4.168613e-11
biceps       -0.9221544 0.31304822 -2.945726 5.133920e-03

> ## sequential sums of squares (Type I)
> anova(fat.lm)
Analysis of Variance Table

Response: bodyfat
          Df  Sum Sq Mean Sq  F value    Pr(>F)
abdomin    1 2440.50 2440.50 101.1718 5.581e-13 ***
biceps     1  209.32  209.32   8.6773  0.005134 **
Residuals 44 1061.38   24.12
---
Signif. codes:  0 '***' 0.001 '**' 0.01 '*' 0.05 '.' 0.1 ' ' 1

> ## weighted squares of means (Type III)
> Anova(fat.lm, type="III")
Anova Table (Type III tests)

Response: bodyfat
             Sum Sq Df F value    Pr(>F)
(Intercept)  114.71  1  4.7555  0.034595 *
abdomin     1823.02  1 75.5740 4.169e-11 ***
biceps       209.32  1  8.6773  0.005134 **
Residuals   1061.38 44
---
Signif. codes:  0 '***' 0.001 '**' 0.01 '*' 0.05 '.' 0.1 ' ' 1

> ## model sum of squares
> var(fat$bodyfat) * (nrow(fat)-1) - sum(fat.lm$residuals^2)
[1] 2649.817
```

13.7 Exercises

13.1. Consider an experiment to determine which of the four types of `valve` used in an artificial heart maximizes blood pressure control as measured by maximum `flow` gradient (mm Hg). `Flow` was maintained at each of the same six `pulse` rates for each valve type. Two `runs` were made for each valve type. The order of the eight runs at the four valve types was randomized. Note that `run` is a random factor, nested within `valve`. The dataset `data(heartvalve)` comes from Anderson and McLean (1974). Perform a thorough analysis including plots of the data.

13.2. An experiment reported in Lewin and Shakun (1976) investigated whether an Octel filter (`type=2`) or a standard filter (`type=1`) provided superior suppression of `noise` produced by automobile exhaust systems. The experiment considered three vehicle `sizes` coded 1 small, 2 medium, 3 large; and both the right 1 and left 2 `side` of cars. The dataset is available as `data(filter)`. Perform a thorough analysis leading to a recommendation of which filter to use under the various experimental conditions.

13.3. An experiment explored the abilities of six commercial laboratories to accurately measure the percentage `fat` content in samples of powdered eggs. A pair of samples from a single can was sent to each lab. The labs were told that the samples were of two `types`, but in fact they were from the same can. Each lab assigned two `technicians` to analyze each type. The dataset from Bliss (1967) is `data(eggs)`. Analyze the data in order to recommend which lab(s) have superior or inferior abilities to ascertain the fat content of powdered eggs.

13.4. Box and Cox (1964), reprinted in Hand et al. (1994), reported the results of a 3^3 factorial experiment. The dataset is `data(wool)`. The response is the `cycles` under tension to failure of worsted yarn. The three factors are `length` of test specimen (250, 300, 350 mm), `amplitude` of loading cycle (8, 9, 10 mm), and `load` (40, 45, 50 g). The levels of all three factors are coded as -1, 0, 1 in the data file. The authors recommend a preliminary log transformation of the response. Perform an analysis to determine the influences of the factors on the response. This dataset is from the paper defining the Box–Cox transformation.

13.5. A 5×3×4 factorial experiment is designed to compare the wear resistance of vulcanized rubber Davies (1954, p. 192). The three factors are

   filler: 5 qualities

   pretreatment: 3 methods

   raw: 4 qualities

wear: main effects and 2–way interactions

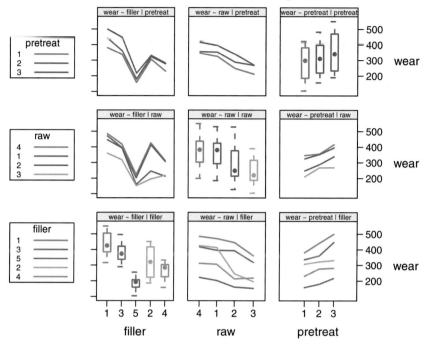

Fig. 13.15 Main effects and two-way interactions for wear resistance of vulcanized rubber.

There is only one replicate; thus the assumption must be made that the three-factor interaction is negligible and the three-factor sum of squares can be used for the error term.

The data are available as data(vulcan). The graph of the main effects and two-way interactions is in Figure 13.15. The simple effects of filler and raw are in Figure 13.16.

a. Determine from the ANOVA table whether any of the main effects or two-way interactions are significant.

b. Why can't we test the three-way interaction?

c. From the figures and tables of means, determine if any levels of any factors can be eliminated from further consideration. Assume that we are looking for big numbers for the best wear resistance.

d. Of the treatment combinations that are left for consideration, are any clearly dominant? Would we need to make a conditional recommendation to the client?

wear: simple effects and 2–way interactions

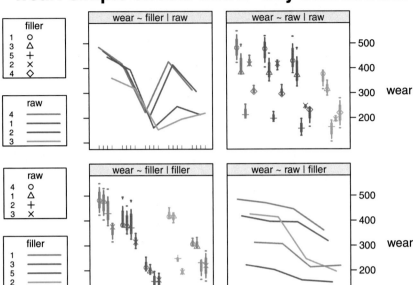

Fig. 13.16 Main effects and simple effects of `filler` and `raw` for wear resistance of vulcanized rubber.

13.6. Continue the Latin square example using the dataset `data(tires)` in Section 13.2. The treatment sum of squares with 3 degrees of freedom is linearly dependent on the Row×Column interaction with $(r - 1) \times (c - 1)$ degrees of freedom. Demonstrate the dependency by showing that each of the dummy variables for `brand` has a zero residual when regressed on the dummy variables for `car*position`.

You may use chunk 39 of file `HHscriptnames(13)`, reproduced in Table 13.28, as a starting point. Explain why the residual sum of squares in `tr1.lm` (and the analogous `tr2.lm` and `tr3.lm`) is 0.

13.7. Peterson (1985) discusses an experiment to assess the effects on strengths of spot welds (psi $\times 10^{-3}$) created by robots on automobile assembly lines. On each of two assembly `lines` (blocks) there were three fixed treatment factors: `maker` at two levels, rod `diameters` at three levels (30 mm, 60 mm, 90 mm), and `chromium` content at three levels (1%, 1.5%, 2%). The 18 treatment combinations were randomly assigned to 18 robots on each assembly line. The dataset is `data(weld)`. Analyze the data, including

Table 13.28 Chunk 39 of file HHscriptnames (13) to be used as starting point for Exercise 13.6.

```
## R defaults to treatment contrasts for factors.
## We need an orthogonal set of factors for this exercise.
##
data(tires)
contrasts(tires$car)        <- contr.helmert(4)
contrasts(tires$position) <- contr.helmert(4)
contrasts(tires$brand)      <- contr.helmert(4)

tires.aov <- aov(wear ~ car + position + brand, data=tires, x=TRUE)
anova(tires.aov)
tires.rc.aov <- aov(wear ~ car * position, data=tires, x=TRUE)
anova(tires.rc.aov)

t(tires.aov$x[,8:10])
t(tires.rc.aov$x[,8:16])
tr1.lm <- lm(tires.aov$x[,8] ~ tires.rc.aov$x[,8:16])
anova(tr1.lm)
```

a. a discussion of interaction among the treatment factors.

b. a recommendation of the combination of the treatment factors for maximizing strength. Explain how you know that your recommendation for diameter is distinctly better than the next-best choice of diameter.

13.8. Peterson (1985) describes an investigation to compare the abilities of seven washday products to remove dirt in cloth:

- liquid detergent
- granular detergent
- detergent flakes
- liquid detergent plus phosphate
- granular detergent plus phosphate
- detergent flakes with phosphate
- soap

Each of these seven products was assigned to three bedsheets soiled in a standard way, and the amount of dirt removed (mg) from each bedsheet was recorded. The data is in data(washday). Analyze these data.

a. Perform a one-way analysis of variance to assess whether the mean amount of dirt removed is the same for all seven products.

b. Partition the 6 degrees of freedom sum of squares for product into 6 mutually orthogonal 1 degree-of-freedom sums of squares, each of which has an interpretation based on the similarities and differences among the products.

c. Estimate each of the six corresponding contrasts.

d. The six levels (hence five contrasts) within detergent can be specified as the crossing of three levels of form (liquid, granular, flakes) and two levels of ingredient (none, phosphate). Rewrite the model as a crossing of form and phosphate nested within soap.vs.detergent.

e. Assuming that the costs per wash are roughly the same for all seven products, provide recommendations for consumers.

13.9. Neter et al. (1996) describe an experiment to compare the work of market research firms. The dataset is data(market). It was desired to evaluate the effects on quality of work performed by 48 firms of the factors of the three crossed factors fee level (feelevel), scope, and supervision. Fee level has three levels (1 = high, 2 = average, 3 = low), scope has two levels (1 = all performed in-house; 2 = some contracted out), and supervision has two levels (1 = local supervisors, 2 = traveling supervisors). Construct an analysis of variance table. Produce and interpret interaction plots for any interaction found significant in the table. Compare the means of the levels of any factors not involved in a significant interaction.

13.10. In Table 13.25 we use the R match function to identify which of the implied dummy variables in a nested design are actually used. The complete command using the match function is

```
match(dimnames(coef(summary.lm(turkey3.aov)))[[1]],
      dimnames(turkey3.aov$x)[[2]])
```

Study the command by picking up pieces of it and dropping them into the Console window. For example, assuming you have already defined all the variables by running the R statements leading up to Table 13.25, open file HHscriptnames(13) in your editor and highlight and run the pieces of code corresponding to the lines in Table 13.29.

13.11. It is desired to compare a response variable dimvar, dimensional variability, of a component produced by each of three machines. Each machine is comprised of two spindles, and four components are selected at random from each spindle. This example is attributable to Montgomery (2001), and the dataset is data(spindle). Perform an analysis to determine the effects of spindle and machine on dimvar, assuming that both factors are fixed.

Table 13.29 Isolated code fragments to be run one line at a time to help learn what the complete statement is doing. See Exercise 13.10 for more detail.

```
                  summary.lm(turkey3.aov)
             coef(summary.lm(turkey3.aov))
       dimnames(coef(summary.lm(turkey3.aov)))
       dimnames(coef(summary.lm(turkey3.aov)))[[1]]

               turkey3.aov$x
       dimnames(turkey3.aov$x)
       dimnames(turkey3.aov$x)[[2]]

match(dimnames(coef(summary.lm(turkey3.aov)))[[1]],
       dimnames(turkey3.aov$x)[[2]])
```

13.12. In an experiment reported by Montgomery (2001), the response variable is a measure of surface finish of a metal part. Each part is produced by one of four machines, a fixed factor. Three operators are assigned to produce parts on each machine. The operators are selected at random and a total of 12 different operators are chosen for the 4 machines. Analyze the data in data(surface) to determine the effects on surface of machine and operator.

13.A Appendix: Orientation for Boxplots

We display the boxplots for the turkey data in two orientations in Figures 13.13 and 13.17. We prefer the vertical orientation for the values of the response variable because it accords with how we have been trained to think of functions—levels of the independent variable along the abscissa and the response variable along the ordinate. Most of the graphs in this book are oriented with the response variable in the vertical direction.

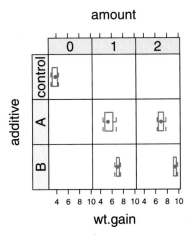

Fig. 13.17 Turkey data with factor structure. This figure is essentially the same as Figure 13.13, but with horizontal boxplots instead of vertical boxplots.

We chose to display Figures 12.7, and 12.8 in vertical orientation. In Section 12.14.5 we discuss three other options for varying the horizontal and vertical orientation and varying the conditioning variable. One of our concerns was legibility of the labels when there are too many long labels on the abscissa. Code chunks for viewing the options are included in file HHscriptnames(12).

Chapter 14

Design of Experiments—Complex Designs

In this chapter we introduce some additional topics in experimental design beyond those discussed in Chapters 6, 12, and 13. The principle of confounding is used to design efficient experiments having many factors but using only a small subset of all possible treatment combinations. Split plot designs involve placing a restriction on the randomization of treatments to experimental units in order to achieve more precision for comparisons involving levels of one factor in exchange for reduced precision for comparisons involving levels of another factor. We illustrate crossover designs that allow for the estimation of treatment effects that can linger across time periods. We show how to test for interaction in two-way designs having exactly one observation at each treatment combination. We show how to extend ANCOVA to designs with blocking factors.

14.1 Confounding

In order to understand the following sections on fractional factorial designs and split plot designs, one must become familiar with the concept of *confounding* of factors.

Two effects (main effects or interactions) are said to be *confounded* if they cannot be independently estimated. (English language equivalents of the statistics term *confounded* include intermixed, intertwined, and confused.) Two completely confounded effects are said to be *aliases* of one another. Each such effect is referred to as an alias of the other effect or is said to be aliased with the other effect. The whole plot effect (indexing on the physical location of the plot) and the treatment effect (indexing on the level of the treatment assigned to the whole plot) in a split plot design (see Sections 14.2 and 14.3) are completely confounded. Effects can also be partially confounded. See a design of experiments text (Cochran and Cox, 1957, for example) for a more complete discussion of confounding.

© Springer Science+Business Media New York 2015
R.M. Heiberger, B. Holland, *Statistical Analysis and Data Display*,
Springer Texts in Statistics, DOI 10.1007/978-1-4939-2122-5_14

If the analyst must be able to estimate separately the effects of both of the two factors or interactions, it is essential that these factors not be confounded. On the other hand, if the effects of some interactions can be *assumed to be negligible,* an effective design strategy may be to confound such negligible interactions with non-negligible factors or interactions. By doing so, the analyst strives to be able to estimate all effects of interest with a much smaller experiment than would be required without using confounding. The fractional factorial designs in Section 14.4 illustrate confounding of interactions with blocks.

To illustrate the importance of avoiding the confounding of nonnegligible factors, let's return to the turkey data data(turkey) analysis in Section 13.4.2. In that experiment there were five groups of six turkeys per group. The turkeys in each group were fed one of five diets, a control diet and four experimental diets A1 A2 B1 B2. The naming convention for the experimental diets refers to four combinations of the two factors additive, with levels A or B, and amount, with levels 1 or 2. Both of these factors were a priori believed likely to impact on the response, wt.gain. Suppose instead that a novice investigator without training in statistics fed 12 turkeys diet A1 and 12 other turkeys diet B2. If the novice then subtracts the mean weight gain on A1 from the mean weight gain on B2, there is no way to tell whether the result is attributable to the difference between amounts 2 and 1, the difference between diets B and A, or some combination of these two factors. In this poorly designed experiment, additive and amount are confounded. With the correctly designed experiment, it is possible to estimate separately the effects of additive and amount, as well as the interaction between these factors.

As another illustration of confounding, consider an experiment involving three factors A, B, C each having two levels, where blocks of homogeneous experimental units are of size at most 4, so that we can examine just four treatment combinations (t.c.'s) in any block. Also suppose that we are able to assume that all interactions are negligible, i.e., the response is additive with respect to the three factors.

A word on notation. For each factor we arbitrarily designate one of the levels as the upper, or 1 level, and the other level as the lower, or 0 level. A t.c. may be written by listing the lowercase letters of all factors observed at their 1 level. Thus if there are four factors A, B, C, D, the t.c. bd is that with factors A and C at their lower levels and factors B and D at their upper levels. The t.c. where all factors are at their lower level is denoted (1).

Returning to our three-factor experiment, suppose we run just the four t.c.'s a, b, c, and abc in any block. Then we can estimate the A main effect as $\frac{1}{2}(abc + a - b - c)$, and the other two main effects similarly. In this setup, we say that factor A is confounded with the BC interaction because BC would be estimated with this same estimate. Since we are assuming that this interaction is negligible, we are estimating only the A main effect. Similarly, the estimate of the B main effect would be confounded with the negligible AC interaction.

In this way we can estimate all main effects with just four observations. However, no degrees of freedom are available to estimate the residual sum of squares needed to produce confidence intervals and conduct tests. This can be handled by replicating runs of the above four t.c.'s or their mirror image *ab*, *bc*, *ac*, and (1), in additional blocks. The set of four t.c.'s run in each block is called a *fractional replicate* of the set of all possible treatment combinations. In Section 14.4 we study fractional factorial designs, where an entire experiment consists of one large fractional replicate, usually arranged in blocks consisting of smaller fractional replicates.

In more complicated designs it is common for main effects and interactions to have several aliases. A design may be described by providing an equation that specifies its aliasing structure.

14.2 Split Plot Designs

This design involves placing a *restriction* on the randomization of treatments to experimental units. Sometimes it is easiest to administer an experiment by applying one treatment factor to groups of experimental units, called *plots*, and another treatment factor to the individual experimental units, referred to as *subplots*. Designs with such a restriction on the randomization are called *split plot designs*. This design strategy is especially useful if the experimenter wants to gain greater precision for inferences involving the treatment applied to subplots, the subplot treatment, at the expense of lower precision for inferences on the treatment applied to whole plots, the whole plot treatment. We confine our attention to the simple case of one blocking factor and two fixed treatment factors. However, the principles behind this restricted randomization approach can be applied to other design types such as ones without blocks, or Latin squares, factors that are random rather than fixed, and situations involving more than two treatment factors. The terminology *split-split plot* refers to the possibility of further splitting the plot into sub-subplots within subplots to accommodate three or more treatment factors.

We model this experiment as follows, where Y_{ijk} is the yield of the observation in block i receiving level j of fixed treatment A, and level k of fixed treatment B. Let the number of blocks be r, the number of levels of A be a, and the number of levels of B be b.

$$Y_{ijk} = \mu + \rho_i + \alpha_j + \eta_{ij} + \beta_k + (\alpha\beta)_{jk} + \epsilon_{ijk} = \mu_{ijk} + \epsilon_{ijk} \qquad (14.1)$$

where $1 \leq i \leq r$, $1 \leq j \leq a$, and $1 \leq k \leq b$. Note that this model contains *two* random error components, $\eta_{ij} \sim N(0, \sigma_\eta^2)$ and $\epsilon_{ijk} \sim N(0, \sigma_\epsilon^2)$, the first associated with the plots and the second associated with the subplots. If (as expected) $\sigma_\epsilon^2 < \sigma_\eta^2$, then comparisons of the levels of B will be performed with greater precision at the cost of less precision for comparisons of the levels of A. Also note that within each block, treatment A is completely confounded with whole plots.

Although the example below is an agricultural experiment, split plot designs are widely used in other application areas including industry, clinical trials, and the social sciences. The agricultural terminologies plot and subplot are usually retained when working in other application areas. The terminology "repeated measures" is used in the social sciences for similar designs.

14.3 Example—Yates Oat Data

This example comes from Yates (1937). We are interested in examining the effects of nitrogen fertilizer and seed variety on the yield of an oat crop. A total of 72 experimental units are arranged in 6 blocks of size 12. Each block is randomly subdivided into three plots, and each plot is further randomly subdivided into four subplots. In each block, the three varieties of seed are randomly assigned to the three plots. Then within each plot, the four levels of nitrogen are randomly assigned to the four subplots. Thus the randomization proceeds in two stages. By assigning varieties to plots and nitrogen to subplots it is implicit that there is more interest in the comparison of nitrogen levels than the comparison of varieties. The nitrogen levels are equally spaced amounts of a single fertilizer, 0, .2, .4, and .6.

The data are available as data(oats). Note that this dataset contains six variables: those for yield, blocks, variety, plots, nitrogen, and subplots.

The design layout for this example is in Table 14.1. The physical locations of the blocks, plots, and subplots are indicated positionally. The random assignment of variety to whole plots is visible since each column within a block contains only one level of variety. The random assignment of nitrogen to subplots is made visible since each column within a block contains all four levels of nitrogen.

Here we have two fixed factors bearing a crossed relationship. In this situation the restricted randomization requires that the plot factor variety must be tested with denominator mean square for plots(blocks) or "whole plot error" as in Tables 14.2 and 14.3. This specification usually requires a statement to override of the default choice of the denominator of the F-test.

Testing variety against the residual mean square is incorrect in this example because that test assumes an unrestricted randomization. Table 14.4 is therefore not correct.

The interaction plot of the two treatment variables is in Figure 14.1. The correct tabular analyses support the visual impressions from this figure:

- The factors variety and nitrogen do not interact.

- The mean Yield increases linearly as the amount of nitrogen increases. nitrogen linear has a small p-value; the p-value of nitrogen quadratic is large.

- The mean Yield does not differ significantly across the three levels of variety.

Table 14.1 Experimental layout for oat yield data with display of randomization scheme. Within each block, the `variety` factor is randomly assigned to an entire plot. Within each `block/plot`, the `nitrogen` factor is randomly assigned to the subplots.

```
, , B1                              , , B4

    P1      P2      P3                  P1      P2      P3
S1 V3:N.6 V1:N.0 V2:N.0           S1 V3:N.4 V2:N.0 V1:N.2
S2 V3:N.4 V1:N.2 V2:N.2           S2 V3:N.6 V2:N.4 V1:N.4
S3 V3:N.2 V1:N.6 V2:N.4           S3 V3:N.0 V2:N.6 V1:N.6
S4 V3:N.0 V1:N.4 V2:N.6           S4 V3:N.2 V2:N.2 V1:N.0


, , B2                              , , B5

    P1      P2      P3                  P1      P2      P3
S1 V3:N.4 V1:N.6 V2:N.2           S1 V2:N.6 V1:N.4 V3:N.4
S2 V3:N.0 V1:N.0 V2:N.0           S2 V2:N.0 V1:N.6 V3:N.6
S3 V3:N.2 V1:N.2 V2:N.4           S3 V2:N.4 V1:N.0 V3:N.2
S4 V3:N.6 V1:N.4 V2:N.6           S4 V2:N.2 V1:N.2 V3:N.0


, , B3                              , , B6

    P1      P2      P3                  P1      P2      P3
S1 V2:N.2 V3:N.6 V1:N.0           S1 V1:N.4 V2:N.6 V3:N.0
S2 V2:N.4 V3:N.2 V1:N.6           S2 V1:N.0 V2:N.4 V3:N.2
S3 V2:N.6 V3:N.4 V1:N.2           S3 V1:N.6 V2:N.0 V3:N.4
S4 V2:N.0 V3:N.0 V1:N.4           S4 V1:N.2 V2:N.2 V3:N.6
```

Table 14.2 Correct analysis for oats split plot design. The ANOVA table was constructed by specifying the denominators for the appropriate F-tests in the `Error` formula.

```
> yatesppl.aov <-
+     aov(y ~ variety*nitrogen + Error(blocks/plots/subplots),
+         data=yatesppl)

> summary(yatesppl.aov)

Error: blocks
          Df Sum Sq Mean Sq F value Pr(>F)
Residuals  5  15875    3175

Error: blocks:plots
          Df Sum Sq Mean Sq F value Pr(>F)
variety    2   1786     893    1.49   0.27
Residuals 10   6013     601

Error: blocks:plots:subplots
                 Df Sum Sq Mean Sq F value  Pr(>F)
nitrogen          3  20021    6674    37.7 2.5e-12 ***
variety:nitrogen  6    322      54     0.3    0.93
Residuals        45   7969     177
---
Signif. codes:  0 '***' 0.001 '**' 0.01 '*' 0.05 '.' 0.1 ' ' 1
```

Table 14.3 Table of means and effects for oats split plot. The **stats** package in R (as of version 3.2.2) provides standard errors for effects, but not for means, in multi-strata designs.

```
> model.tables(yatesppl.aov, type="means")
Tables of means
Grand mean

104

 variety
variety
     1      2      3
 97.63 104.50 109.79

 nitrogen
nitrogen
     1      2      3      4
 79.39  98.89 114.22 123.39

 variety:nitrogen
        nitrogen
variety 1      2      3      4
      1  71.50  89.67 110.83 118.50
      2  80.00  98.50 114.67 124.83
      3  86.67 108.50 117.17 126.83

> model.tables(yatesppl.aov, type="effects", se=TRUE)
Tables of effects

 variety
variety
     1      2      3
-6.347  0.528  5.819

 nitrogen
nitrogen
     1      2      3      4
-24.583 -5.083 10.250 19.417

 variety:nitrogen
        nitrogen
variety 1      2      3      4
      1 -1.542 -2.875  2.958  1.458
      2  0.083 -0.917 -0.083  0.917
      3  1.458  3.792 -2.875 -2.375

Standard errors of effects
          variety nitrogen variety:nitrogen
            5.006    3.137            5.433
replic.        24       18                6
```

Table 14.4 Incorrect specification for oats split plot design that ignores the split plot. The test for `variety` is incorrectly tested against the 45-df Residual and incorrectly shows as significant. The `nitrogen` and interaction tests are correct. We placed the model formula inside the `terms` function, and used the `keep.order=TRUE` argument, to force the ANOVA table to display the terms in the specified order. We are using the same order as appears in the split plot analysis. We do need this specification in order to calculate the multiple comparisons of the `nitrogen` (more generally, the subplot) effects using `glht` and `mmc`. See Table 14.7 and Figure 14.2 for the details.

```
> yatesppl.wrong.aov <-
+    aov(terms(y ~ (blocks*variety) + (nitrogen*variety),
+             keep.order=TRUE),
+       data=yatesppl)

> summary(yatesppl.wrong.aov)
                 Df Sum Sq Mean Sq F value  Pr(>F)
blocks            5  15875    3175   17.93 9.5e-10 ***
variety           2   1786     893    5.04  0.0106 *
blocks:variety   10   6013     601    3.40  0.0023 **
nitrogen          3  20021    6674   37.69 2.5e-12 ***
variety:nitrogen  6    322      54    0.30  0.9322
Residuals        45   7969     177
---
Signif. codes:  0 '***' 0.001 '**' 0.01 '*' 0.05 '.' 0.1 ' ' 1
```

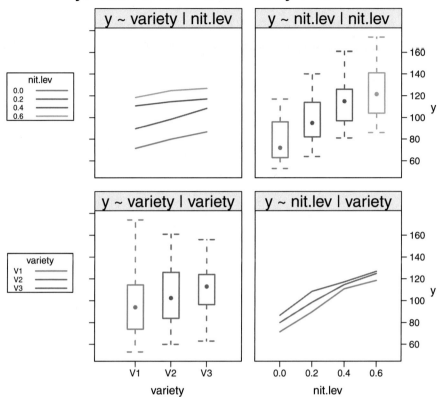

Fig. 14.1 Interaction plot for Yates split plot on oats. The linear effect of nitrogen level is clearly visible in both right-hand panels. The lack of effect of variety, confirmed in Table 14.2, is also visible.

14.3.1 Alternate Specification

Our presentation of the Yates oat data in Table 14.1 and our emphasis in the analysis in Table 14.2 show five distinct factors in the split plot design. We believe this is the best way to illustrate the concepts of restricted randomization, of different precisions for different comparisons, and the logistics and practical details of running an experiment. Many texts and examples show only three factors by suppressing the explicit identification of the plots and subplots.

The arithmetic of the analysis and the interpretation of the results are identical whether or not the structure is made explicit with the extra factors. The two ANOVA tables generated by the statements in Tables 14.2 and 14.5 are identical. The specification in Table 14.5 leads the reader to ask nonsense questions like "how can nitrogen be crossed with variety and nested within variety at the same time?" The specification in Table 14.2, by explicitly naming the random plots and subplots factors and explicitly showing the random assignment of the fixed variety and nitrogen factors to the random factors, makes the distinction clear. The plot structure shows the subplots nested in the plots. The treatment structure shows the variety crossed with the nitrogen. They are different factors. We are therefore not surprised that they have different relationships.

The *ANOVA table* and *interpretation* of the analysis are identical with either specification. The *logic* behind the EMS (expected mean squares) calculations is displayed when the plots and subplots factors are visible. The statistical justification for the appropriate *F*-tests is cryptic at best when the plots and subplots factors are suppressed. We further explore the equivalence of these two formulations in Exercise 14.7.

The common practice of suppressing the structure is a legitimate response to the computing technology at the time (1930s) when the split plot design was in-

Table 14.5 Alternate specifications of design. The hypothesis tests in the ANOVA table here are identical to the tests in Table 14.2. The use of the same factor names in the specification of both the treatment model and the Error model can lead to confusion for the person reading the model specification and the ANOVA table.

```
> yatesppl2.anova <-
+     aov(y ~ variety*nitrogen + Error(blocks/variety/nitrogen),
+         data=yatesppl)

> summary(yatesppl2.anova)

Error: blocks
          Df Sum Sq Mean Sq F value Pr(>F)
Residuals  5  15875    3175

Error: blocks:variety
          Df Sum Sq Mean Sq F value Pr(>F)
variety    2   1786     893    1.49   0.27
Residuals 10   6013     601

Error: blocks:variety:nitrogen
                  Df Sum Sq Mean Sq F value  Pr(>F)
nitrogen           3  20020    6674    37.7 2.5e-12 ***
variety:nitrogen   6    322      54     0.3    0.93
Residuals         45   7969     177
---
Signif. codes:  0 '***' 0.001 '**' 0.01 '*' 0.05 '.' 0.1 ' ' 1
```

vented. The calculation of the analysis with 3 explicit factors costs $O(26^3)$ multiplications. It costs $O(72^3)$ multiplications [that is, $(72/26)^3 \approx 21$ times as many] with 5 explicit factors. (The "big O" notation is defined in Appendix Section I.4.1 in the "operation count" discussion.) When ANOVA analyses were routinely performed with handcrank-operated calculating equipment, the time savings was well worth the ambiguity in notation.

14.3.2 Polynomial Effects for Nitrogen

Note that, as anticipated, the mean square for comparing the levels of the whole plot factor variety, 601, is greater than the mean square for comparing the levels of the levels of the subplot factor nitrogen, 177. There is no evidence of interaction between variety and nitrogen. The large p-value for variety suggests that the three varieties do not differ significantly, but the small p-value for the subplot factor, nitrogen, tells us yield is significantly affected by the amount of nitrogen used.

We further investigate the nature of the relationship between nitrogen and yield by decomposing the 3-df sum of squares for nitrogen into orthogonal contrasts for linear, quadratic, and cubic effects. In R we use the polynomial contrast function cont.poly to assign the contrasts to the nitrogen factor. See Table 14.6 for details. (In SAS we explicitly define the contrasts in contrast statements.) Since only the linear contrast is significant (p-value < .01) we conclude that yield increases linearly with nitrogen. This finding suggests a need for further experimentation to determine the amount of nitrogen that should be used to maximize yield.

Response surface methodology is the experimental design technique used to determine the combination of inputs that maximizes or minimizes output. We do not pursue this further but recommend the interested reader consult either Montgomery (2001) or Box et al. (1978).

We calculate the MMC for the nitrogen effect in Table 14.7 using the ANOVA from Table 14.4. The subplot tests in Table 14.4 are correct. The multiple comparisons plots for nitrogen are in Figure 14.2. The linear contrast appears to be carrying all the significance.

Table 14.6 Continuation of the split plot analysis from Table 14.2. The 3-df `nitrogen` effect is partitioned into polynomial contrasts. The linear effect carries almost the entire sum of squares and is the only significant contrast.

```
> ## polynomial contrasts in nitrogen
> contrasts(yatesppl$nitrogen)
  2 3 4
1 0 0 0
2 1 0 0
3 0 1 0
4 0 0 1

> contrasts(yatesppl$nitrogen) <- contr.poly(4)

> contrasts(yatesppl$nitrogen)
        .L    .Q         .C
1 -0.6708204  0.5 -0.2236068
2 -0.2236068 -0.5  0.6708204
3  0.2236068 -0.5 -0.6708204
4  0.6708204  0.5  0.2236068

> ## split plot analysis with polynomial contrasts
> yatespplp.aov <-
+   aov(y ~ variety*nitrogen + Error(blocks/plots/subplots),
+       data=yatesppl)

> summary(yatespplp.aov,
+         split=list(nitrogen=list(linear=1, quad=2, cub=3)),
+         expand.split=FALSE)

Error: blocks
          Df Sum Sq Mean Sq F value Pr(>F)
Residuals  5  15875    3175

Error: blocks:plots
          Df Sum Sq Mean Sq F value Pr(>F)
variety    2   1786   893.2   1.485  0.272
Residuals 10   6013   601.3

Error: blocks:plots:subplots
                 Df Sum Sq Mean Sq F value   Pr(>F)
nitrogen          3  20021    6674  37.686 2.46e-12 ***
  nitrogen: linear  1  19536   19536 110.323 1.09e-13 ***
  nitrogen: quad    1    480     480   2.713    0.106
  nitrogen: cub     1      4       4   0.020    0.887
variety:nitrogen  6    322      54   0.303    0.932
Residuals        45   7969     177
---
Signif. codes:  0 '***' 0.001 '**' 0.01 '*' 0.05 '.' 0.1 ' ' 1
```

Table 14.7 MMC of nitrogen effect in oats split plot design calculated using the ANOVA from Table 14.4. The subplot tests in Table 14.4 are correct. Pairwise contrasts, observed means, and polynomial contrasts are shown in Figure 14.2.

```
> yatesppl.mmc <- mmc(yatesppl.wrong.aov, focus="nitrogen")

> nitrogen.lmat <- contr.poly(4)

> rownames(nitrogen.lmat) <- levels(yatesppl$nitrogen)

> yatesppl.mmc <- mmc(yatesppl.wrong.aov, focus="nitrogen",
+                     focus.lmat=nitrogen.lmat)
mmc: At least one reversed contrast name did not have a '-' sign.
     We appended a '-' sign.

> yatesppl.mmc
Tukey contrasts
Fit: aov(formula = terms(y ~ (blocks * variety) + (nitrogen *
          variety), keep.order = TRUE), data = yatesppl)
Estimated Quantile = 2.667
95% family-wise confidence level
$mca
      estimate stderr  lower upper height
4-3    9.167   4.436 -2.662 21.00 118.81
4-2   24.500   4.436 12.671 36.33 111.14
3-2   15.333   4.436  3.504 27.16 106.56
4-1   44.000   4.436 32.171 55.83 101.39
3-1   34.833   4.436 23.004 46.66  96.81
2-1   19.500   4.436  7.671 31.33  89.14
$none
   estimate stderr  lower  upper height
4   123.39  3.137 115.02 131.75 123.39
3   114.22  3.137 105.86 122.59 114.22
2    98.89  3.137  90.52 107.25  98.89
1    79.39  3.137  71.02  87.75  79.39
$lmat
    estimate stderr  lower  upper height
.C-    0.500  3.507 -8.852  9.852  105.3
.Q-    5.167  3.137 -3.198 13.531  104.0
.L    36.833  3.507 27.482 46.185  102.7
```

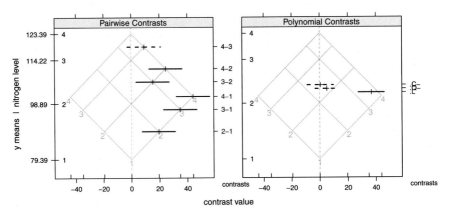

Fig. 14.2 MMC plot of nitrogen effect in oats split plot design. Calculations are shown in Table 14.7. The 45° isomeans grid lines (and the `nitrogen` means they reflect) are approximately equally spaced. This suggests that the linear contrast may be carrying all the significance. The polynomial contrasts displayed in the right panel agree. The contrasts in the right panel are all in the vertical center of the plot. This frequently happens with contrasts that have approximately equal weights for both high and low means. The Tiebreaker plot isn't needed in this example.

14.4 Introduction to Fractional Factorial Designs

The idea behind fractional factorial designs is to substantially reduce the number of *experimental units* (e.u.'s) required for the experiment by purposely confounding (see Section 14.1) all effects of interest with only effects that are not of interest and that can be assumed negligible. One is almost always able to assume that high-order interactions are negligible. This strategy permits estimation of main effects and low-order interactions while experimenting on only a small proportion of all possible *treatment combinations* (t.c.'s). The resulting experimental plan is called a *fractional replicate*. Implementation of this class of designs involves carefully selecting a fractional replicate subset of the possible t.c.'s so as to purposely confound high-order factor interactions with one another (and with blocks, if any), while maintaining the unconfoundedness of main effects and of lower-order interactions of interest.

Since it is frequently the case that there are more t.c.'s to be run than there are homogeneous experimental units (e.u.'s), a blocking scheme is usually part of this type of design. A fractional replicate of the complete experiment is run within each homogeneous block. The assignment of t.c.'s to e.u.'s purposely confounds higher-order interactions of the treatment factors with the block effects.

Note that the $r \times r$ Latin square design in Section 13.2 is a special case of fractional replication, a $\frac{1}{r}$ replicate of an r^3 experiment.

Our discussion will be limited to the situation where there are n factors each at 2 levels and only 2^k e.u.'s are available, $k < n$; this is referred to as a $1/(2^{n-k})$ fractional replication. The design is called a $2^{n-(n-k)}$ design. Fractional factorial designs exist when all factors have a common number of levels greater than 2, or the factors have varying numbers of levels, for example three factors with 2 levels each and four factors each having 3 levels. The need to consider a situation with many 2-level factors is not uncommon, for the 2 levels can be the presence or absence of a particular condition.

14.4.1 Example—2^{8-2} Design

Suppose we have 8 factors (denoted by the letters A through H), each with 2 levels, and we have enough experimental units to run $2^6 = 32$ of the $2^8 = 128$ possible t.c.'s. Further assume that the maximum sized set of homogeneous experimental units is $2^4 = 16$, so that the 64 selected t.c.'s will be arranged in 4 blocks, each containing 16 e.u.'s. Table 14.8 is an experimental layout for the $2^{8-2} = (2^8)/4 = 64$ design, prior to randomization.

Table 14.8 Experimental layout for the $2^{8-2} = (2^8)/4 = 64$ design, prior to randomization. This design follows from a permutation of the factor labels in Plan 6A.16 of Cochran and Cox (1957). This design specification is available as data(Design_2.8_2).

Block 1	Block 2	Block 3	Block 4
(1)	ab	ce	de
ach	bch	aeh	acdeh
aef	bef	acf	adf
cefh	abcefh	fh	cdfh
bdh	adh	bcdeh	beh
abcd	cd	abde	abce
abdefh	defh	abcdfh	abfh
bcdef	acdef	bdf	bcf
beg	aeg	bcg	bdg
abcegh	cegh	abgh	abcdgh
abfg	fg	abcefg	abdefg
bcfgh	acfgh	befgh	bcdefgh
degh	abdegh	cdgh	gh
acdeg	bcdeg	adg	acg
adfgh	bdfgh	acdefgh	aefgh
cdfg	abcdfg	defg	cefg

The treatment combinations have been very carefully chosen so that, provided one can assume that all 3-factor and higher-order interactions are negligible, one can "cleanly" estimate all main effects and 2-factor interactions with a sufficient number of error df to assure tests of reasonable power. Blocks, all main effects, and all two-factor interactions are confounded only with three-factor and higher interactions. This property makes the 2^{8-2} a Resolution V design, and it is often denoted as 2_V^{8-2}.

The basic form of the ANOVA table for the 2^{8-2} design in Table 14.8 is in Table 14.9. Estimates are not available for interaction effects that are confounded with blocks.

Table 14.9 ANOVA table for the $2^{8-2} = (2^8)/4$ design from Table 14.8. The main effects and 2-factor interactions are unconfounded with blocks.

Source	df	Comments
Blocks	3	Blocks are aliased with acf, bdg, and cdh
Main effects	8	Unconfounded
2-factor interactions	28	$\binom{8}{2}$ terms, unconfounded
Residuals	24	Aliased with 3-factor and higher interactions
Total	63	

If instead we had been required to maintain a maximum block size of 8, an additional 4 degrees of freedom would go to the blocks. It would have been necessary to completely confound 2 of the 28 2-factor interactions with blocks, and 2 Residuals df would have moved to blocks leaving only 22 for the Residuals.

The analysis of such data is very straightforward as shown in Table 14.10. We need to ensure that the model statement declares only blocks, the 8 main effects, and the 28 interactions.

Table 14.10 Specification and analysis of a $2^{8-2} = (2^8)/4$ fractional factorial design. The code shows the specification of the dummy variables for the design in R282 and specification of the analysis using the model formula shown in the aov statement. The displayed formula generates the dummy variables for the complete set of main effects and all 2-factor interactions.

```
R282 <- t(sapply(strsplit(Design_2.8_2$trt,""),
               function(trtcomb)
                  as.numeric(letters[1:8] %in% trtcomb)))
dimnames(R282) <- list(Design_2.8_2$trt, letters[1:8])
R282 <- data.frame(blocks=Design_2.8_2$blocks, R282)
R282
data(R282.y) ## R282.y was randomly generated
R282.aov <- aov(R282.y ~ blocks + (a+b+c+d+e+f+g+h)^2, data=R282)
anova(R282.aov)
model.matrix(R282.aov)
## confirm aliasing
R282E.aov <- aov(R282.y ~ Error(blocks) + (a+b+c+d+e+f+g+h)^2,
                data=R282)
summary(R282E.aov)
```

14.4.2 Example—2^{5-1} Design

Five factors involved in a manufacturing process for an integrated circuit were investigated. For brevity we refer to the five factors as A, B, C, D, E. Resources were available to examine only 16 of the 2^5 treatment combinations. A particular half-replicate of the complete experiment was used such that all main effects are confounded with four-factor interactions and all two-factor interactions are confounded with three-factor interactions. Based on experience with these factors, the investigator was very confident that only factors A, B, C, and the AB interaction were likely to have an appreciable effect on the process yield. This is confirmed by examining the interaction plot in Figure 14.3 or the means in Table 14.11. The output in Table 14.11 also contains two tables of means demonstrating that the D main effect and AC interaction are not significant. The data from Montgomery (2001) is in data(circuit). We find that there is a significantly higher yield at the higher level of each of A, B, C than at their respective lower levels, and that the simple effect of B at the higher level of A is significantly greater than the simple effect of B at the lower level of A.

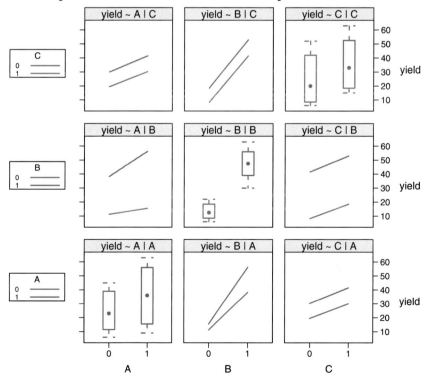

Fig. 14.3 Interaction plot for circuit 2^{5-1} design. All three main effects are visibly significant, as is the $A \times B$ interaction. The traces in the $A \times C$ and $B \times C$ panels are parallel, hence not significant.

Table 14.11   Analysis of $2^{5-1} = (2^5)/2$ fractional factorial design.

```
> circuit.aov <- aov( yield ~ A + B + C + A:B, data=circuit)

> summary(circuit.aov)
          Df Sum Sq Mean Sq F value  Pr(>F)
A          1    495     495   193.2 2.5e-08 ***
B          1   4590    4590  1791.2 1.6e-13 ***
C          1    473     473   184.6 3.2e-08 ***
A:B        1    189     189    73.8 3.3e-06 ***
Residuals 11     28       3
---
Signif. codes:  0 '***' 0.001 '**' 0.01 '*' 0.05 '.' 0.1 ' ' 1

> model.tables(circuit.aov, type="means")
Tables of means
Grand mean

30.31

 A
A
   0     1
24.75 35.88

 B
B
    0     1
13.38 47.25

 C
C
    0     1
24.88 35.75

 A:B
   B
A   0     1
  0 11.25 38.25
  1 15.50 56.25

> tapply(circuit[,"yield"], circuit[,"D"], mean)
    0     1
30.75 29.88

> tapply(circuit[,"yield"], circuit[,c("A","C")], mean)
   C
A      0    1
  0 19.50 30.0
  1 30.25 41.5
```

14.5 Introduction to Crossover Designs

This is a subclass of *repeated measures designs*, used when one applies two or more treatments to each of several subjects over the course of two or more periods, and needs to account for the possibility that a *carryover* or *residual* effect of a treatment lingers into the following period (and possibly beyond it). Thus a subject's response may be attributable to both the treatment given in the period and the treatment administered in the preceding period. One seeks to be able to provide unconfounded estimates of both the direct and residual effects. These designs are also referred to as changeover or residual effects designs.

The possible existence of residual effects is easy to imagine in medical or agricultural experiments. It also must be accounted for in meteorological experiments involving cloud seeding intended to induce precipitation.

Intuitively, a "good" design is one in which

1. Each treatment occurs equally often on each subject.

2. Each treatment occurs equally often in each period.

3. Each treatment follows each other treatment the same number of times.

But the available numbers of treatments, subjects, and periods often make it impossible to satisfy all three criteria. As a simple example, suppose we have two treatments, say A and B, to compare in three periods on two experimental animals. Consider the following two designs:

	Design 1		Design 2	
Period	Animal 1	Animal 2	Animal 1	Animal 2
1	A	B	A	B
2	B	A	B	A
3	A	B	B	A

Which design is preferred?

It turns out that both the direct and residual treatment effects can be estimated much more precisely in Design 2 than in Design 1. Design 1 has the deficiency that each treatment is always preceded by the other treatment, never by itself. Only Design 2 satisfies the third of the above intuitive criteria. This design is a member of a class of crossover designs constructed as a Latin square with the last row repeated once. This class has the property that the estimation of direct and residual treatment effects are orthogonal to one another.

14.5.1 Example—Two Latin Squares

This design, from Cochran and Cox (1957) and Heiberger (1989) uses two 3×3 Latin squares to estimate the residual effects as well as the direct effects of milk yields resulting from three treatments to dairy cows. The design and data from data(cc135) are in Table 14.12 and Figure 14.4.

Table 14.12 Two 3 × 3 Latin squares for crossover design with display of the residual effect. The factor nores is an indicator for observations that do not have a residual effect because there is no preceding treatment. The residual treatment factor restreat has the value of the treatment treat for the preceding period with the same square and sequence.

a. Design arranged to show the Latin square structure.

square	1	2
cow	1 2 3	4 5 6
period 1	A B C	A B C
period 2	B C A	C A B
period 3	C A B	B C A

b. Design and data arranged by observation.

period	square	sequence	cow	treat	yield	nores	restreat
1	1	1	1	A	38	0	0
1	1	2	2	B	109	0	0
1	1	3	3	C	124	0	0
1	2	1	4	A	86	0	0
1	2	2	5	B	75	0	0
1	2	3	6	C	101	0	0
2	1	1	1	B	25	1	A
2	1	2	2	C	86	1	B
2	1	3	3	A	72	1	C
2	2	1	4	C	76	1	A
2	2	2	5	A	35	1	B
2	2	3	6	B	63	1	C
3	1	1	1	C	15	1	B
3	1	2	2	A	39	1	C
3	1	3	3	B	27	1	A
3	2	1	4	B	46	1	C
3	2	2	5	C	34	1	A
3	2	3	6	A	1	1	B

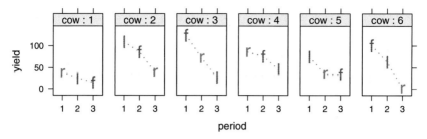

Fig. 14.4 Each panel shows the observations per period for one cow. In all treatment plans yield goes down over time. Note that cows 3 and 6 who start with full.grain have the highest initial yield. The drop in yield from anything else to full.grain is smaller than the drop to either of the other treatments.

We specify the design as if the residual effects are attributable to another factor. The residual effects `restreat` are explicitly set to the value of the direct effects `treat` in the preceding `period`. A new dummy variable `nores` is set to 0 for the first period (in which there are no residual treatments), and to 1 for the remaining periods (in which it is meaningful to speak about residual effects from the previous period). The new variable `nores` is confounded with one degree of freedom of `period`. `restreat` is nested in `nores`. The arithmetic does not need the identification of the `nores` dummy variable. The analysis is easier to follow if `nores` is explicitly identified.

The analysis specification statements includes two sets of `aov` statements, with different orderings for the direct effects `treat` and residual effects `res.treat`. The ANOVA table edited from the output is in Table 14.13.

The direct and residual effects are not orthogonal to each other. Their sum is partitioned in two different ways in Table 14.13. The additional sum of squares for the residual effects after accounting for the direct effects, for which we use the first model statement and the results of which are reflected in the first set of braced lines in the ANOVA table, tells us whether there are longer-term differences that need to be accounted for. The additional sum of squares for the direct effects after accounting for the residual effects, for which we use the second model statement and the results of which are reflected in the second set of braced lines in the ANOVA table, tells us whether the single period effects are themselves important. Since the residual effects after the direct effects are borderline significant with $p = .06$, we conclude that isolating the residual effect is important. Had we not done so the power for detecting the direct effects would have been reduced because the `res.treat` sum of squares would have been left in the residual. It would have inflated the residual, hence decreased the F-value and increased the p-value for the direct effects. We can approximate this effect by calculating $MS_{resid\,(approx)} = (616+199)/(2+4) = 136$ and $F_{approx} = 1138/136 = 8.36$ on 2 and 6 df with $p = .018$. This is an approximation because that isn't how the experiment was done.

Table 14.13 Latin Square for Residual Effects design cc135. The braced expressions are two different partitionings of the same 4 df for the combined `treat + res.treat` effects. This ANOVA table is manually constructed from the output of the programs.

```
data(cc135)
a1c <-  aov(terms(yield ~ cow + square:period + treat + res.treat,
                 keep.order=TRUE), data=cc135)
summary(a1c)

a1cr <- aov(terms(yield ~ cow + square:period + res.treat + treat,
                 keep.order=TRUE), data=cc135)
summary(a1cr)
```

Source	Df	Sum of Sq	Mean Sq	F Value	Pr(F)
cow	5	5781.111	1156.222	23.211	0.005
period in square	4	11489.111	2872.278	57.662	0.001
treat+res.treat	4	2892.972			
⎰ treat	2	2276.778	1138.389	22.853	0.006
⎱ res.treat after treat	2	616.194	308.097	6.185	0.060
⎰ res.treat	2	38.422	19.211	0.386	0.703
⎱ treat after res.treat	2	2854.550	1427.275	28.653	0.004
Residuals	4	199.250	49.812		
Total	17	20362.444			

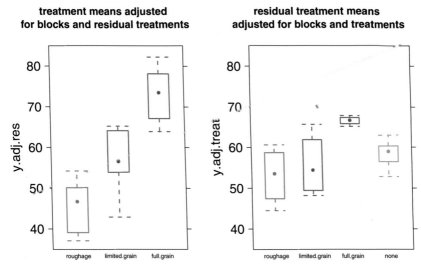

Fig. 14.5 Boxplots for residual effects design, treatments and residual treatments, each adjusted for the blocking factors and for each other. From the left panel, we see that the full.grain treatment is much better than the other two. From the right panel, we see that the residual effect of the full.grain treatment is much stronger than the others. This is the explanation for the reduced drop in yield we commented on in Figure 14.4.

The boxplots of the adjusted treatments and adjusted residual treatments are in Figure 14.5.

14.6 ANCOVA with Blocks: Example—Apple Tree Data

In Section 10.5 we studied the analysis using a concomitant variable (covariate) in an experiment having only one factor. In this example we demonstrate the use of this technique when there are two factors.

In Section 10.6 we showed a composite figure with four distinct models for an ANCOVA design with one covariate and one treatment factor. All four models were displayed together in Figure 10.12. In this section we have a more complex design, with a block factor in addition to the covariate and treatment factor. We therefore have a more complex composite figure with six models in Figure 14.6.

14.6.1 Study Objectives

Pearce (1983), later reprinted in Hand et al. (1994), describes a randomized block experiment to determine the effects of six ground cover treatments on the yield of apple trees. A concomitant variable, the *volume of crop* during the four years prior to treatment, is available. Perform the analysis both using and ignoring the concomitant variable. Provide recommendations as to treatment.

14.6.2 Data Description

The data are accessed with data(apple). The variables are

treat: ground cover treatments. Treatments 1–5 are experimental treatments; Treatment 6 is a control

block: four randomized blocks

yield: pounds over a four-year period following treatment

pre: volume of crop over a four-year period prior to treatment

14.6.3 Data Analysis

The strategy is to begin with the most complex model and then progress toward simpler ones. This systematic approach assures that the ultimately selected model will include all significant effects but be no more complex than necessary. The ancova plots for the full set of models we will use are in Figure 14.6. The index to the ANOVA tables for each of these models is in Table 14.14. We will discuss each of the models.

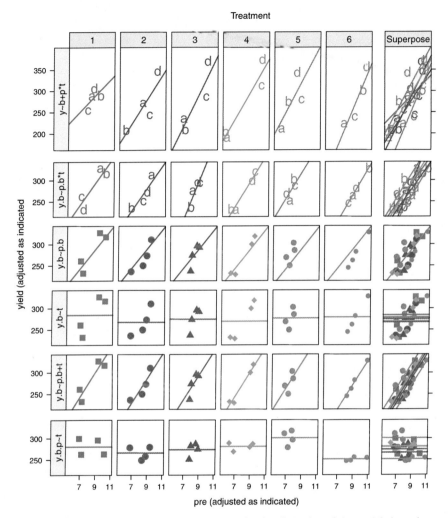

Fig. 14.6 Several models for `yield`. See the text for the discussion of the models in each row. The concluding model is illustrated in the bottom row. Treatment 6 is significantly lower on the vertical axis than the other treatments. The color in each panel represents the treatment. The plotting symbol in the top two rows indicates the block. The plotting symbol in the bottom four rows indicates the treatment (and is intentionally redundant with the column number and color). The left strip shows the row number and an abbreviated model name. The full model formulas are shown in Table 14.14. All six rows are on the same y-scale. Row 1 has a wider y-range in the y-scale because its response variable is `yield` which reflects the block effect. Rows 2–6 show `yield.block`, the yield adjusted for blocks. The bottom two rows have the same vertical distances between the fitted line and the points. The bottom two rows are magnified in Figure 14.8.

Table 14.14 Index to analysis of covariance models for the apple data in Section 14.6.

Model Number	Model Name Abbreviated	Model					ANOVA Table
1	y~b+p*t	yield	~ block +	pre	*	treat	14.15
2	y.b~p.b*t	yield.block	~	pre.block	*	treat	14.16
2q		yield.block	~	pre.block	+	treat	14.17
				+ (pre:treat).block			
3	y.b~p.b	yield.block	~	pre.block			14.18
4	y.b~t	yield.block	~			treat	14.19
5	y.b~p.b+t	yield.block	~	pre.block	+	treat	14.20
5b	y~b+p+t	yield	~ block +	pre	+	treat	14.21
5c	y.b~b+p.b+t	yield.block	~ block +	pre.block	+	treat	14.22
6	y.b.p~t	yield.block.pre	~			treat	14.23

14.6.4 Model 1: `yield ~ block + pre * treat`

Model 1 allows for the possibility that the covariate `pre` and the treatment factor `treat` interact. This model allows for differing slopes in the simple regressions of `yield` on `pre` across the levels of `treat`. We draw two conclusions from the ANOVA of Model 1.

1. The ANOVA table for Model 1 in Table 14.15 shows that most of the sum of squares for the data is attributable to the blocking factor ($SS_{block}/SS_{Total} = 47852/72034 = 66\%$). Note that we can discuss the proportion of total sum of squares for the blocking factor `block`, but not the F- and p-values—even though the program calculated and printed them in the ANOVA table (most ANOVA programs print these values). We assumed that blocks are important for this study and designed the study by stratifying the sample within blocks. Because the study was designed under the assumption that blocks are important, there is no testable hypothesis about blocks. We do have the right to calculate an estimate of the efficiency of the blocking. Most experimental design texts contain formulas for the efficiency attributable to blocking in Latin Square, Randomized Complete Block, and other experimental designs. See, for example, Cochran and Cox (1957) (Section 4.37).

   We investigate the block factor in Figure 14.7. The left column of panels indicates that the means of both `yield` and `pre` are heterogeneous across blocks. Therefore, in subsequent analysis we adjust both `yield` and `pre` for blocks. For example, the quantity `yield.block` in Table 14.14 represents `yield` adjusted for `blocks`. This is the original data vector of `yields` minus the vector of least-squares estimates of `block` effects.

$$\texttt{yield.block}_{ij} = Y_{ij} - \hat{\beta}_i,$$

Table 14.15 Analysis of variance for apple data. Model 1 'y~b+p*t', full model with blocks and interaction of covariate and treatment. The interaction between the treatment and the covariate (pre:treat) is not significant. The model specifies separate lines in each panel of row 1 in Figure 14.6. The lines in row 1 are dominated by the block differences (block "d" is always at the top).

```
> anova(aov(yield ~ block + pre * treat, data = apple))
Analysis of Variance Table

Response: yield
          Df Sum Sq Mean Sq F value  Pr(>F)
block      3  47853   15951   80.82 7.9e-07 ***
pre        1  15944   15944   80.78 8.6e-06 ***
treat      5   4353     871    4.41   0.026 *
pre:treat  5   2109     422    2.14   0.152
Residuals  9   1776     197
---
Signif. codes:  0 '***' 0.001 '**' 0.01 '*' 0.05 '.' 0.1 ' ' 1
```

where $\hat{\beta}_i$ is the least-squares estimate of the effect of block i in the simple one-way ANOVA model

$$Y_{ij} = \mu + \beta_i + \epsilon_{ij}$$

Comparable definitions apply to pre.block and yield.block.pre. These adjustments are in the same sense used in several places in Section 9.13, and allow us to focus our attention on the remaining variables in the model. From the right column of panels of Figure 14.7 we see that the adjusted values are more homogeneous. The adjusted response variable (in Model 2, with ANOVA table in Table 14.16) has the same sums of squares for the treatment and covariate effects as the unadjusted response variable in Table 14.15.

The heterogeneity of responses across blocks is also visible in Row 1 of Figure 14.6 where we display separate regressions of yield on pre for each of the 6 levels of treat. The plotting characters a,b,c,d represent blocks 1–4. We see that block 4 ("d") has a consistently higher yield than the other 3 blocks.

2. Since the ANOVA table shows $p = .152$ for the interaction pre:treat we conclude that the slopes do not significantly differ. We will show the model without interaction in Section 14.6.8.

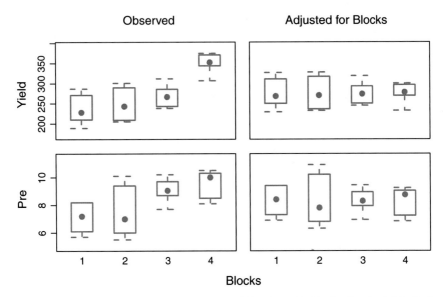

Fig. 14.7 Apple `yield` and `pre` by `block`. In the left column the block effect is visible. In the right column, in which we have adjusted the data for the block effect, we see that the data appear more homogeneous.

14.6.5 Model 2: `yield.block ~ pre.block * treat`

In Model 2 (row 2 in Figure 14.6), we have removed the block effect. The range of the observations in the y-direction is reduced and the highly visible block effect (with "d" at the top) is gone. The regression lines in each panel, repeated in the Superpose panel, are similar. We will eventually drop the interaction term in Model 5 (row 5 of in Figure 14.6) and note that the fits are similar. First we must look at the ANOVA table for Model 2. We do so twice, naively and incorrectly in Table 14.16 and more sophisticatedly and correctly in Table 14.17. Table 14.16 naively constructs the interaction dummy variables by crossing the covariate adjusted for blocks with the dummy variables for treatment. This does not span the same subspace as the construction in Table 14.17, where we take the dummy variables for the interaction from Table 14.15 and adjust them for blocks. See Exercise 14.8 for further details.

Table 14.16 Analysis of variance for apple data. Model 2 'y.b~p.b*t', model with response yield and covariate pre adjusted for blocks, and with interaction of adjusted covariate and treatment. There are separate lines in each panel. The block effect shows as zero because the block variation has been removed. The covariate and treatment sums of squares are identical to those terms in Table 14.15. The interaction dummy variables are not correctly adjusted by this procedure. The linear space spanned by the crossing of pre.block and treat is not the same as the linear space specified by adjusting the dummy variables of the crossing of block and treat for block. Hence the interaction sums of squares is incorrect.

```
> apple$yield.block <-
+    apple$yield - proj(lm(yield ~ block, data=apple))[,"block"]

> apple$pre.block   <-
+    apple$pre   - proj(lm( pre ~ block, data=apple))[,"block"]

> ## wrong interaction sum of squares
> anova(aov(yield.block ~ block + pre.block * treat,
+        data = apple))
Analysis of Variance Table

Response: yield.block
                Df Sum Sq Mean Sq F value  Pr(>F)
block            3      0       0    0.00 1.00000
pre.block        1  15944   15944   42.56 0.00011 ***
treat            5   4353     871    2.32 0.12840
pre.block:treat  5    514     103    0.27 0.91600
Residuals        9   3372     375
---
Signif. codes:  0 '***' 0.001 '**' 0.01 '*' 0.05 '.' 0.1 ' ' 1
```

Table 14.17 Analysis of variance for apple data. Model 2q 'y.b˜p.b*t' has adjusted the response yield, the covariate pre, and the dummy variables for interaction of adjusted covariate and treatment for blocks. There are separate lines in each panel. The interaction dummy variables are correctly adjusted by this procedure. Hence the interaction sums of squares is correct and matches the interaction sum of squares for Model 1 in Table 14.15. The degrees of freedom for the Residuals is incorrect. It includes the degrees of freedom that belong to the block term.

```
> applebpst.aov <- aov(yield ~ block + pre * treat, data=apple,
+                          x=TRUE)

> appleQ <- qr.Q(qr(applebpst.aov$x))

> '(pre.block:treat).block' <- appleQ[,11:15]

> ## correct anova for 'y.b~p.b*t'
> anova(aov(yield.block ~ pre.block + treat +
+                          '(pre.block:treat).block', data=apple))
Analysis of Variance Table

Response: yield.block
                         Df Sum Sq Mean Sq F value  Pr(>F)
pre.block                 1  15944   15944  107.71 2.4e-07 ***
treat                     5   4353     871    5.88  0.0057 **
'(pre.block:treat).block' 5   2109     422    2.85  0.0637 .
Residuals                12   1776     148
---
Signif. codes:  0 '***' 0.001 '**' 0.01 '*' 0.05 '.' 0.1 ' ' 1
```

14.6.6 Model 3: yield.block ~ pre.block

Model 3 is linear regression of the adjusted yield.block by the adjusted covariate pre.block and ignoring the treat factor. The points are plotted in panels by treatment in row 3 of Figure 14.6 even though treatment is not included in the construction of the common line nor in the ANOVA table in Table 14.18. This line is not a good fit to the points in most of these panels, suggesting that treat cannot be ignored as an explanatory variable.

Table 14.18 Analysis of variance for apple data. Model 3 'y.b˜p.b', simple linear regression model with yield and pre adjusted for blocks, and ignoring treatment. There is one common regression line in all panels. The degrees of freedom for the Residuals is incorrect. It includes the degrees of freedom that belong to the block term.

```
> anova(aov(yield.block ~ pre.block, data = apple))
Analysis of Variance Table

Response: yield.block
           Df Sum Sq Mean Sq F value  Pr(>F)
pre.block   1  15944   15944    42.6 1.5e-06 ***
Residuals  22   8238     374
---
Signif. codes:  0 '***' 0.001 '**' 0.01 '*' 0.05 '.' 0.1 ' ' 1
```

Table 14.19 Analysis of variance for apple data. Model 4 'y.b˜t', one-way ANOVA of yield adjusted for blocks by treatment, and ignoring covariate. Separate horizontal lines in each panel.

```
> anova(aov(yield.block ~ treat, data = apple))
Analysis of Variance Table

Response: yield.block
           Df Sum Sq Mean Sq F value Pr(>F)
treat       5    750     150    0.12   0.99
Residuals  18  23432    1302
```

14.6.7 Model 4: `yield.block ~ treat`

Model 4 is one-way analysis of variance of the adjusted `yield.block` by the treatment. The covariate is ignored. The ANOVA table in Table 14.19. We can see in the figure that most of the points are not close to the horizontal lines in any of the panels.

14.6.8 Model 5: `yield.block ~ pre.block + treat`

Model 5, `yield.block ~ pre.block + treat`, in row 5 of Figure 14.6, with ANOVA in Tables 14.20, 14.21, and 14.22, forces parallel regression lines with possibly differing intercepts for the adjusted response against the adjusted covariate and the treatment. The value $p = 0.0417$ for `treat` tells us that there is a significant difference at $\alpha = .05$ in the `treat` adjusted means. Stepping forward for a moment,

Table 14.20 Analysis of variance for apple data. Model 5 'y.b~p.b+t', model with response yield and covariate pre adjusted for blocks, and with no interaction of adjusted covariate and treatment. Separate parallel lines in each panel. There is no block term in this model, hence the Residuals line includes the degrees of freedom that should be in the block effect.

```
> anova(aov(yield.block ~ pre.block + treat, data = apple))
Analysis of Variance Table

Response: yield.block
           Df Sum Sq Mean Sq F value Pr(>F)
pre.block   1  15944   15944   69.76 2e-07 ***
treat       5   4353     871    3.81 0.017 *
Residuals  17   3885     229
---
Signif. codes:  0 '***' 0.001 '**' 0.01 '*' 0.05 '.' 0.1 ' ' 1
```

to be justified by the time we get there, the intercepts of the parallel regression lines in rows 5 and 6 of Figure 14.6 are significantly different.

Model 5b, yield ~ block + pre + treat, gives the same regression coefficients, fitted values, and residuals as model 5. The ANOVA tables for Models 5 and 5b show different degrees of freedom for the residuals, as Model 5 does not account for the prior adjustment for the blocking factor.

Model 5c, yield.block ~ block + pre.block + treat, gives the same regression coefficients, fitted values, and residuals as Models 5 and 5b. The ANOVA tables for Model 5c shows the correct degrees of freedom for the residuals. Model 5c does account for the prior adjustment for the blocking factor by including the block factor in the model equation solely to absorb the degrees of freedom.

In Model 5, the sum of squares for treat after adjustment for the covariate pre is deemed significant. In Model 4, the effect of treat is assessed prior to consideration of pre and found to be not significant. Therefore, taking account of the available covariate pre has enabled us to detect differences in the adjusted mean yield at the levels of treat. Without the presence of this covariate, we would not have detected treat differences.

If the covariate pre had not been available for this analysis, the sums of squares for block and for treat would be identical to those in the anova table for Model 4. But without pre in the model, the corresponding terms in the anova table would have lower F statistics and higher p-values than those in Table 14.16 because the sum of squares for pre, that should have been assigned to a term for pre, would instead remain as a large component of the sum of squares for Residuals.

The finding of a significant difference for the treatment in Model 5 (row 5 of Figure 14.6) means that we can interpret the difference in intercepts among the panels of row 5, or equivalently the vertical difference between the regression lines at any specified value of the adjusted covariate.

Table 14.21 Analysis of variance for apple data. Model 5b 'y~b+p+t', model with original unadjusted response yield and covariate pre, and with no interaction of covariate and treatment. Separate parallel lines in each panel. This table gives the full analysis of this model and shows the correct degrees of freedom for the Residuals. The pre.block and treat sums of squares in Table 14.20 match the sums of squares here.

```
> anova(aov(yield ~ block + pre + treat, data=apple))
Analysis of Variance Table

Response: yield
          Df Sum Sq Mean Sq F value  Pr(>F)
block      3  47853   15951   57.48 4.1e-08 ***
pre        1  15944   15944   57.45 2.6e-06 ***
treat      5   4353     871    3.14   0.042 *
Residuals 14   3885     278
---
Signif. codes:  0 '***' 0.001 '**' 0.01 '*' 0.05 '.' 0.1 ' ' 1
```

Table 14.22 Analysis of variance for apple data. Model 5c 'y.b~b+p.b+t', model with response yield and covariate pre adjusted for blocks, and with no interaction of adjusted covariate and treatment. Separate parallel lines in each panel. The block effect has been removed from both response and covariate and therefore shows with a 0 sum of squares in the ANOVA table. Placing the block term in the model gets the correct degrees of freedom for the Residuals.

```
> anova(aov(yield.block ~ block + pre.block + treat, data=apple))
Analysis of Variance Table

Response: yield.block
          Df Sum Sq Mean Sq F value  Pr(>F)
block      3      0       0    0.00   1.000
pre.block  1  15944   15944   57.45 2.6e-06 ***
treat      5   4353     871    3.14   0.042 *
Residuals 14   3885     278
---
Signif. codes:  0 '***' 0.001 '**' 0.01 '*' 0.05 '.' 0.1 ' ' 1
```

14.6.9 Model 6: yield.block.pre ~ treat

We have one more step. Most people have trouble seeing vertical distances between nonhorizontal parallel lines. Therefore, we use Row 6 of Figure 14.6 to make the lines horizontal and retain the same vertical differences. We do so by making an additional adjustment to the response variable, subtracting out the effect of the covariate pre.block from the response variable yield.block. Details are in Table 14.23.

Row 6 (Model 6, `yield.block.pre ~ treat`) of Figure 14.6 adjusts `yield` for both `treat` and `pre`. The differences in the y-coordinate of the fitted horizontal lines drawn at adjusted the `yield` means for each `treat` level displays the extent of the effect of `treat` on `yield` after accounting for both `block` and `pre`. The difference in intercepts of the horizontal regression lines in Row 6 is identical to the difference in intercepts in Row 5. The ANOVA table for Model 6 is in Table 14.23.

The plot of Model 6 is constructed from the plot of Model 5 by vertically shifting the points and the regression lines about a fixed point at the mean of the covariate. We redraw the treatment panels of rows 5 and 6 from Figure 14.6 as Figure 14.8 to illustrate the shift. The lengths of the residual lines from the points to the fitted line are identical in rows 5 and 6.

Table 14.23 Analysis of variance for apple data. Model 6 '`y.b.p~t`', model with response `yield` adjusted for both blocks and the covariate `pre`, Separate horizontal lines in each panel. The vertical distance between these horizontal lines is identical to the vertical distance between the parallel lines in Model 5. The degrees of freedom for the `Residuals` is incorrect. It includes the degrees of freedom that belong to the `block` and `pre` terms.

```
> apple.aov.4 <-  aov(yield.block ~ pre.block + treat, data=apple)

> apple$yield.block.pre <- apple$yield.block -
+            predict.lm(apple.aov.4, type="terms", terms="pre.block")

> anova(aov(yield.block.pre ~ treat, data = apple))
Analysis of Variance Table

Response: yield.block.pre
          Df Sum Sq Mean Sq F value Pr(>F)
treat      5   5472    1094    5.07 0.0045 **
Residuals 18   3885     216
---
Signif. codes:  0 '***' 0.001 '**' 0.01 '*' 0.05 '.' 0.1 ' ' 1
```

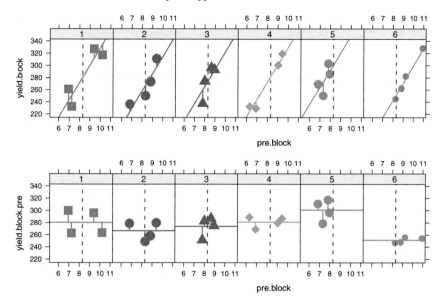

Fig. 14.8 Vertical difference between points and the parallel lines from rows 5 and 6 of Figure 14.6. It is easily seen in the bottom row that the control treatment 6 has a lower adjusted mean than the others. The light gray vertical reference line is at the mean of the adjusted `pre` values. The vertical reference line indicates the fulcrum of the vertical shifting of the points and parallel regression lines. The lengths of the residual lines from the points to the fitted line are identical in rows 5 and 6.

14.6.10 Multiple Comparisons

Now that we have detected `treat` differences, we wish to investigate their nature. Since level 6 of `treat` has been designated as a control, we follow up the finding of significant `treat` with Dunnett's multiple comparison procedure to simultaneously compare the mean adjusted `yield` of `treat` levels 1 through 5 with the mean adjusted `yield` of `treat` level 6. We illustrate the multiple comparisons in Figure 14.9 with the mean–mean display including the Tiebreaker panel and in Table 14.24. These displays tell us that the adjusted mean `yield` of `treat` level 5 is significantly greater than the adjusted mean `yield` of the control (`treat` level 6). No other significant differences with `treat` level 6 are uncovered.

Table 14.24 Multiple comparisons of Treatments for Model 5 (yield ~ block + pre + treat) by Dunnett's method. The comparison of treatment 5 with the control treatment 6 is the only significant comparison. The output section labeled $mca is the set of contrasts comparing the means of the experimental treatments 1–5 to the mean of the control treatment 6. The section labeled none (meaning no contrasts) are the means for each individual treatment.

```
> apple5.aov <- aov(yield ~ block + pre + treat, data=apple)

> anova(apple5.aov)
Analysis of Variance Table

Response: yield
          Df Sum Sq Mean Sq F value  Pr(>F)
block      3  47853   15951   57.48 4.1e-08 ***
pre        1  15944   15944   57.45 2.6e-06 ***
treat      5   4353     871    3.14   0.042 *
Residuals 14   3885     278
---
Signif. codes:  0 '***' 0.001 '**' 0.01 '*' 0.05 '.' 0.1 ' ' 1

> apple5d.mmc <- mmc(apple5.aov,
+     linfct=mcp(treat=contrMat(table(apple$treat),
+                               type="Dunnett", base=6)))

> apple5d.mmc
Dunnett contrasts
Fit: aov(formula = yield ~ block + pre + treat, data = apple)
Estimated Quantile = 2.821
95% family-wise confidence level
$mca
     estimate stderr   lower upper height
5-6     49.58  13.30  12.065 87.10  276.1
4-6     29.80  12.67  -5.924 65.53  266.2
1-6     29.14  12.13  -5.062 63.34  265.9
3-6     22.73  12.21 -11.711 57.17  262.7
2-6     15.23  12.21 -19.211 49.67  259.0
$none
   estimate stderr lower upper height
5     300.9  8.794 276.1 325.7  300.9
4     281.1  8.430 257.4 304.9  281.1
1     280.5  8.343 256.9 304.0  280.5
3     274.1  8.331 250.6 297.6  274.1
2     266.6  8.331 243.1 290.1  266.6
6     251.3  8.980 226.0 276.7  251.3
```

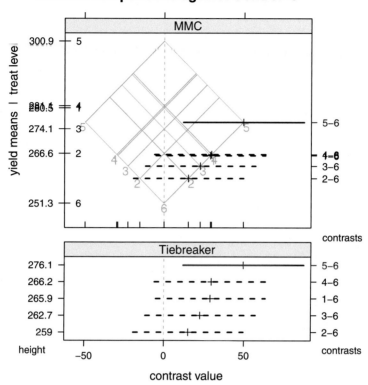

Dunnett comparisons against Control=6

The MMC panel shows informative overprinting. Please see Tiebreaker panel and caption.

Fig. 14.9 Multiple comparisons of Treatments for Model 5 (yield ~ block + pre + treat) by Dunnett's method. The comparison of treatment 5 with the control treatment 6 is the only significant comparison. The means for treatments 1 and 4 are very close; therefore the contrasts 4-6 and 1-6 are almost collinear in the MMC panel. We need the Tiebreaker panel to distinguish them.

14.7 Example—`testscore`

The example in Section 14.6 was modeled with one covariate. This example requires
two covariates.

14.7.1 Study Objectives

The dataset `data(testscore)` from Johnson and Tsao (1945), also reprinted in
Anderson and Bancroft (1952), discuss the modeling of a final test score using four
factors and two concomitant variables. Their goal was to explain the response variable `final` score by using the information in the factors and continuous variables.
As a secondary goal they used this paper to illustrate the advantages of using concomitant variables (covariates) to increase the sensitivity of their experiment and the
precision with which estimates and predictions can be made.

14.7.2 Data Description

`final:`   response variable: final test score

`sex:`   male=1, female=0

`scholastic standing:`   good=1, average=2, poor=3

`order:`   order on a battery of standardized tests: high=1, medium=2, low=3

`grade:`   10, 11, 12

`initial:`   initial test score

`mental.age:`   mental age

Note that this was not a longitudinal study: Different students were assessed at each
grade level.

The study has no replication. The original authors used the four-way interaction
as their initial error term. They found that all of the three-way interactions and some
of the two-way interactions were not significant, so they continued by pooling all
the nonsignificant interactions. Our analysis begins by assuming that the four-way
and all three-way interactions are negligible. We proceed to investigate the two-way
interactions and main effects, both with and without adjustment for the continuous
variables.

14.7.3 Analysis—Plots

The goal of the analysis is to explain the variability in the final scores based on the initial score, the covariates, and the factors. We start the analysis by looking at the scatterplot matrix of the data in Figure 14.10. Two things are immediately visible. First, the plots of the factor levels against each other form simple lattices. This reflects the design of the study in which the factors were chosen to be balanced. Second, we see that the response variable final is positively correlated with the two continuous variables initial and mental.age and negatively correlated with two of the factors standing and order. When we look closer at the factors, we discover that they are coded with high scores first. Therefore, the visual negative correlation is an artifact of the coding. We recode both standing and order to place the low scores first and thereby have all variables coded in a consistent direction.

We redraw the plot with the recoded variables in Figure 14.11. While here, we also changed the order of the variables, placing all the continuous variables together and all the factors together. There is now a clear display of the positive correlations among all continuous variables and ordered factors. We would display Figure 14.11 rather than Figure 14.10 in our final report.

14.7.4 Analysis—ANOVA

In an analysis with both factors and continuous concomitant variables, it is necessary to determine the contribution of each. We therefore fit four different models,

1. factors only,

2. continuous first: fitting factors after continuous predictors,

3. continuous second: fitting continuous predictors after factors,

4. continuous only.

These four models are summarized in Table 14.25. The full analyses can be reproduced with the code in file HHscriptnames(14). The first thing to notice is the residual mean square s^2 for the models. The reduction in sum of squares from "factors only" to both factors and continuous (observed in either column "continuous first" or "continuous second") has an $F_{2,26} = ((97.704 - 28.057)/2)/1.0791 = 32.27$ with $p = 10^{-7}$. The reduction from continuous only to both has $F_{25,26} = ((131.291 - 28.057)/25)/1.0791 = 3.827$ with $p = 0.0006$.

We look further at a follow-up set of listings for the continuous-variables-first analysis in Table 14.26. We see that the initial score all by itself has $R^2 = 1468.423/1607.333 \approx 0.9$. We determined from the results of testscore2.aov in Table 14.25 that none of the two-way interactions (with the possible exception

Original ordering of factor values

Fig. 14.10 testscore data, original ordering of variables and factor levels. The plots of the factor levels against each other form simple lattices. This reflects the design of the study in which the factors were chosen to be balanced. The response variable final is positively correlated with the two continuous variables initial and mental.age and negatively correlated with two of the factors standing and order.

of sex:order) is significant; therefore, we dropped them. We also see the main effect sex is not significant as the first factor; therefore, we moved it last. In the ANOVA testscore5.aov we see that grade is the least important of the main effects, so we moved it to next-to-last. In the ANOVA testscore6.aov we see that we don't need grade at all. This brings us to the ANOVA testscore7.aov with three significant main effects after adjusting for the covariates.

Let us plot the information in ANOVA table testscore7.aov. As is usual with analysis of covariance, we are interested in the difference between the intercepts for each of the groups assuming parallel planes defined by the coefficients of the covariates. In the testscore7.aov setting we have 2 covariates and 18 groups

Revised ordering of factor values

Fig. 14.11 testscore data, revised ordering of variables and factor levels. The continuous variables are collected together in the lower left. We recoded both standing and order to place the low scores first. Now the response variable final is positively correlated with all covariates and factors. This is the only splom we would show the client.

with model

$$Y_{ijk\ell} = \hat{\mu}_{ijk} + \hat{\beta}_1 X_{1,ijk\ell} + \hat{\beta}_2 X_{2,ijk\ell} + \hat{\epsilon}_{ijk\ell}$$

We move the X terms to the other side to get final.adj, the final scores adjusted for the covariates X_1=initial and X_2=mental.age.

$$\hat{Y}_{adj} = \hat{Y}_{ijk\ell} - \hat{\beta}_1 X_{1,ijk\ell} - \hat{\beta}_2 X_{2,ijk\ell} = \hat{\mu}_{ijk} + \hat{\epsilon}_{ijk\ell}$$

We plot the initial variable (X_1), the observed final variable (Y), and the adjusted final.adj ($\hat{Y}_{adj}$) in Figure 14.12. The standard error of the difference of two observations in the plot is approximately $\sqrt{(2)(1.324)/3} \approx 0.94$. Therefore, the

Table 14.25 Comparison of sequential p-values for four different models of the `testscore` data. In each of the models, the response variable is the `final` score.

	Factors only testscore1		Continuous first testscore2		Continuous second testscore3s		Continuous only testscore4	
	Df	Pr(F)	Df	Pr(F)	Df	Pr(F)	Df	Pr(F)
initial			1	0.0000			1	0.0000
mental.age			1	0.0132			1	0.0914
sex	1	0.0069	1	0.2522	1	0.0000		
grade	2	0.0010	2	0.0029	2	0.0000		
standing	2	0.0000	2	0.0005	2	0.0000		
order	2	0.0000	2	0.0000	2	0.0000		
sex:grade	2	0.0184	2	0.6567	2	0.0000		
sex:standing	2	0.7728	2	0.7506	2	0.4427		
sex:order	2	0.6719	2	0.0792	2	0.2885		
grade:standing	4	0.0519	4	0.1519	4	0.0001		
grade:order	4	0.8250	4	0.1319	4	0.3304		
standing:order	4	0.6085	4	0.2426	4	0.0952		
initial					1	0.0000		
mental.age					1	0.1066		
Residuals Df	28		26		26		51	
Residuals Mean Sq		3.4894		1.0791		1.0791		2.5743
Residuals Sum Sq		97.704		28.057		28.057		131.291
Total Sum Sq	53	1607.333	53	1607.333	53	1607.333	53	1607.333

visible differences in the plot are significant. We repeat the Adjusted panel of Figure 14.12 with greater magnification as the top panel of Figure 14.13. We can easily see all three main effects of the factors `standing`, `order`, and `sex` in the top panel of Figure 14.13. We can also see the hint of interaction between `sex` and `order` in the reversal of direction in the `medium` panel: the male scores go up from poor to good and the female scores go down from poor to good.

14.7.5 Summary of ANOVA

We conclude the search for a model with our final model, `testscore7.aov`. We find three main effects: `standing`, `order`, and `sex` after accounting for the two covariates: `initial` and `mental.age`. These are the same variables the original authors found.

The original investigation of this data used the analysis to predict `final` scores given the variables in the data description.

Table 14.26 Continuous variables first, and further refinements.

```
> ## after looking at all of above
> ## Total Sum of Squares
> var(testscore$final) * (length(testscore$final)-1)
[1] 1607

> testscore5.aov <- aov(final ~ initial + mental.age +
+                       grade + standing + order + sex
+                       + sex:order,
+                       data=testscore)

> summary(testscore5.aov)
            Df Sum Sq Mean Sq F value  Pr(>F)
initial      1   1468    1468 1176.02 < 2e-16
mental.age   1      8       8    6.10 0.01765
grade        2     16       8    6.34 0.00393
standing     2     21      10    8.23 0.00096
order        2     30      15   11.82 8.5e-05
sex          1      7       7    5.69 0.02169
order:sex    2      6       3    2.35 0.10811
Residuals   42     52       1

> testscore6.aov <- aov(final ~ initial + mental.age +
+                       standing + order + grade + sex,
+                       data=testscore)

> summary(testscore6.aov)
            Df Sum Sq Mean Sq F value  Pr(>F)
initial      1   1468    1468 1108.18 < 2e-16
mental.age   1      8       8    5.75 0.02080
standing     2     29      15   10.96 0.00014
order        2     34      17   12.77 4.2e-05
grade        2      3       2    1.13 0.33094
sex          1      7       7    5.36 0.02535
Residuals   44     58       1

> testscore7.aov <- aov(final ~ initial + mental.age +
+                       standing + order + sex,
+                       data=testscore)

> summary(testscore7.aov)
            Df Sum Sq Mean Sq F value  Pr(>F)
initial      1   1468    1468 1108.96 < 2e-16
mental.age   1      8       8    5.75 0.02056
standing     2     29      15   10.97 0.00013
order        2     34      17   12.77 3.9e-05
sex          1      7       7    5.66 0.02153
Residuals   46     61       1
```

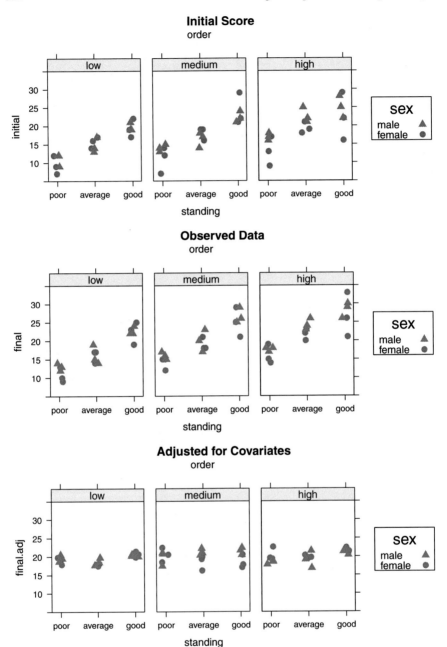

Fig. 14.12 `testscore` data, `initial` scores, `final` scores, and `final` scores adjusted for the covariates using model `testscore7.aov`. All panels are on the same scale. The plotting position of the `standing` variable have been jittered to avoid overstriking of the points. The `initial` scores are responsible for most of the variability. The adjusted scores still show the effects of the three factors `standing`, `order`, `sex`. Figure 14.13 shows the adjusted scores at greater magnification.

Table 14.27 and Figure 14.13 contain predictions for the final test score for the average values of the two covariates and for all possible values of the three factors we retained.

Table 14.27 Predictions of final score for each of the categories using model testscore7.aov.

```
> newdata <- cbind(initial=mean(testscore$initial),
+                  mental.age=mean(testscore$mental.age),
+                  testscore[c(1:9,28:36),
+                            c("standing","order","sex")])

> newdata[c(1,2,3,4,18),]
   initial mental.age standing   order    sex
1    17.48      44.06     good    high   male
2    17.48      44.06     good  medium   male
3    17.48      44.06     good     low   male
4    17.48      44.06  average    high   male
36   17.48      44.06     poor     low female

> final.pred <- predict(testscore7.aov, newdata=newdata)

> final.pred.table <- tapply(final.pred, newdata[,3:5], c)

> final.pred.table
, , sex = female

          order
standing    low medium  high
   poor   16.05  17.69 18.69
   average 17.48  19.12 20.12
   good    20.37  22.02 23.01

, , sex = male

          order
standing    low medium  high
   poor   16.81  18.46 19.45
   average 18.24  19.89 20.88
   good    21.14  22.79 23.78


> ## now summarize this over each factor to get predicted values
> apply(final.pred.table, 1, mean) ## each scholastic standing
  poor average   good
 17.86   19.29  22.19

> apply(final.pred.table, 2, mean) ## each individual order
   low medium   high
 18.35  20.00  20.99
```

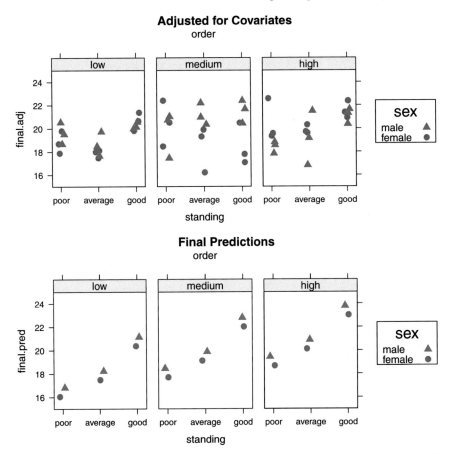

Fig. 14.13 `testscore` data, The `final.adj` panel has identical content to the third panel in Figure 14.12, but on a larger vertical scale. The Prediction panel here contains the same information as Table 14.27 displayed on the same scale as the `final` scores adjusted for the covariates.

14.8 The Tukey One Degree of Freedom for Nonadditivity

What can we do when linearity cannot be assumed and there are not enough observations to study interaction? We construct a single degree of freedom variable that represents the linear by linear component of interaction.

14.8.1 Example—Crash Data—Study Objectives

Does the tendency of teenage drivers to be involved in automobile accidents increase dramatically with the number of passengers in the car? Believing this to be so, legislators in several states require that junior license holders be prohibited from driving with more than one passenger. Williams (2001) presents the data `data(crash)`.

14.8.2 Data Description

crashrate: crashes per 10,000 trips

agerange: driver age: 16–17, 18–19, 30–59

passengers: number of passengers: 0, 1, 2, 3+

14.8.3 Data Analysis

The original author displayed the data in the barplot format of Figure 14.14. We see several difficulties with barplots used in this situation:

1. There are two factors, driver age and passenger presence. They are not treated symmetrically.

2. The marginal main effects of the factors are not displayed.

3. The response variable is a rate. Barplots were designed for counts and don't make much sense for rates.

4. The majority of the plotting surface is used to display the region where the data doesn't appear.

5. The symmetry in uncertainty is hidden by the asymmetry between the heavy bar below each observed value and the empty region above.

6. Superimposed error bars (not shown here) on a barplot add to the asymmetry.

We prefer the interaction plot format of Figure 14.15 to display both the main effects of each factor and the interactions between the two factors. Figure 14.15 addresses all the difficulties noted above for barplots. As usual, when we present an interaction plot we show two versions, interchanging the trace- and x-factors

Crash Rates by Driver Age and Passenger Presence per 10,000 Trips

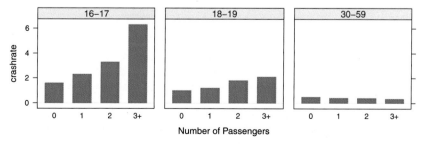

Fig. 14.14 Barplot of crash data.

Crash Rates by Driver Age and Passenger Presence per 10,000 Trips

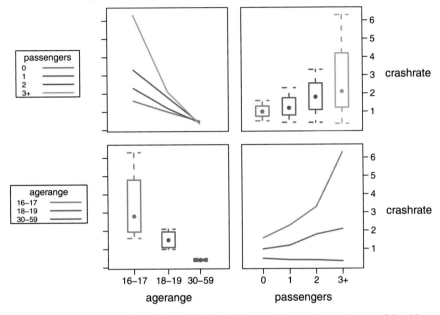

Fig. 14.15 Interaction plot of crash data. The lines in lower right panel trace the age of the driver. These lines show the same information as the heights of the bars in Figure 14.14. The lines in the upper left panel trace the number of passengers. The boxplots on the main diagonal show wide disparity in variance.

Table 14.28 ANOVA for crash data.

```
> crash.aov <- aov(crashrate ~ agerange + passengers, data=crash)

> summary(crash.aov)
            Df Sum Sq Mean Sq F value Pr(>F)
agerange     2 17.943   8.971   7.238 0.0252 *
passengers   3  6.230   2.077   1.675 0.2701
Residuals    6  7.437   1.239
---
Signif. codes:  0 '***' 0.001 '**' 0.01 '*' 0.05 '.' 0.1 ' ' 1
```

between them. In this example the traces in the lower right panel match the heights of the bars in each panel of Figure 14.14. We find the lack of parallelism in the traces in the same panel to be more compelling than the same information spread across several panels.

Table 14.28 contains an initial analysis of variance. It suggests that the number of passengers does not impact significantly on crashrate. The figures show completely different means and variances for different levels of passengers, implying that some type of interaction is present. The usual way of investigating interaction can't be used for this example because there is only one observation for each combination of agerange and passengers. The entire residual sum of squares has only 6 degrees of freedom. If all of them are used to estimate the interaction, then none are left for the residual, and testing becomes impossible.

The Tukey *One Degree of Freedom for Nonadditivity* (ODOFFNA) Tukey (1949) and Hoaglin et al. (1983) is the proper response to this impasse. Tukey constructs the *comparison values*, the linear-by-linear component of the interaction. He identifies it as one of the degrees of freedom in the residual and tests it for significance against the remaining degrees of freedom in the residual. There are several possible options for dealing with a significant ODOFFNA: Find an outlier, transform the data, regress on the comparison value.

We construct the comparison value by partitioning the data into four components. We augment the $r \times c$ table with a column of row margins $r_i^0 = 0$ and a row of column margins $c_j^0 = 0$ and a summary number in the last row–last column $\hat{\mu}^0 = 0$. The table at the right symbolically indicates the four sections of the augmented table. Note that each value of the table has been partitioned into four pieces, one in each section of the augmented table. We can trivially reconstruct the original table with $Y_{ij} = \hat{\mu}^0 + r_i^0 + c_j^0 + Y_{ij}$.

	0	1	2	3+	row
16–17	1.60	2.30	3.30	6.30	0.00
18–19	1.00	1.20	1.80	2.10	0.00
30–59	0.49	0.41	0.40	0.34	0.00
col	0.00	0.00	0.00	0.00	0.00

$=$

	0 1 2 3+	row
16–17		
18–19	Y_{ij}	r_i^0
30–59		
col	c_j^0	$\hat{\mu}^0$

In the second display of the table, we find the row means r_i^1 for each of the rows of the first table and place them in the column of row margins. We also subtract the row means from each value of the original table. Now the reconstruction of the original table with $Y_{ij} = \hat{\mu}^0 + r_i^1 + c_j^0 + e_{ij}^1$ is still possible, but is no longer trivial.

	0	1	2	3+	row
16–17	−1.775	−1.075	−0.075	2.925	3.375
18–19	−0.525	−0.325	0.275	0.575	1.525
30–59	0.080	0.000	−0.010	−0.070	0.410
col	0.0000	0.0000	0.0000	0.000	0.000

$=$

	0 1 2 3+	row
16–17		
18–19	$e_{ij}^1 = Y_{ij} - r_i^1$	$r_i^1 = \bar{Y}_{i.}$
30–59		
col	c_j^0	$\hat{\mu}^0$

In the third display of the table, we find the column means c_j for each of the columns of the second table and place them in the row of column margins. We also subtract these column means from each value of the body of the second table. Now the reconstruction of the original table with $Y_{ij} = \hat{\mu} + r_i + c_j + e_{ij}$ is meaningful. For example, the entry 1.60 in the upper-left corner of the original table is calculated with components in the final table: $1.770 + 1.605 - .740 - 1.035$.

	0	1	2	3+	row
16–17	−1.035	−0.6083	−0.1383	1.7817	1.605
18–19	0.215	0.1417	0.2117	−0.5683	−0.245
30–59	0.820	0.4667	−0.0733	−1.2133	−1.360
col	−0.740	−0.4667	0.06333	1.1433	1.770

$=$

	0 1 2 3+	row
16–17		
18–19	$e_{ij} = e_{ij}^1 - c_j$	$r_i = r_i^1 - \hat{\mu}$
30–59		
col	$c_j = \bar{e}_{\cdot j}^1$	$\hat{\mu} = \bar{r}_{\cdot}^1 = \bar{Y}_{\cdot \cdot}$

This process is called row and column polishing by means. The notation was introduced by Tukey with medians and is described in the "Median Polish" chapter of Hoaglin et al. (1983). Median polish is an iterative technique that usually takes at least two cycles. The analogous mean polish illustrated here uniquely converges on just the one cycle illustrated here. Medians are resistant to outliers whereas means are not. The advantage of means is that they correspond to least-squares fits.

We construct the comparison value by recombining the components

$$\text{comparison value} = \frac{r_i \times c_j}{\hat{\mu}} = x_{\text{cv}}$$

	0	1	2	3+
16–17	−0.6710	−0.4232	0.05743	1.0368
18–19	0.1024	0.0646	−0.00877	−0.1583
30–59	0.5685	0.3586	−0.04866	−0.8785

We then compute the ANOVA table (in Table 14.29) with the comparison value as a covariate.

The regression coefficient of the comparison value, in this example $\hat{\beta}_{\text{cv}} = 1.5516$, is identical to the regression of the residuals from the first ANOVA against the comparison value as illustrated in Figure 14.16. Tukey calls this the *diagnostic plot*.

There are several things to do with this plot. If most of the points lie in a horizontal band, then the ones that don't are potential outliers and need to be carefully investigated. If there is a clear nonhorizontal line, then the slope of the line gives a hint as to an appropriate transformation. When the slope is $\hat{\beta}_{\text{cv}}$, then an appropriate power

Table 14.29 ANOVA with the comparison value as a covariate. See also Figure 14.16. The regression coefficient for the comparison value is 1.5516.

```
> crash2.aov <- aov(crashrate ~ agerange + passengers +
+                    as.vector(cv), data=crash)

> summary(crash2.aov)
              Df Sum Sq Mean Sq F value  Pr(>F)
agerange       2  17.94    8.97   161.2 2.9e-05 ***
passengers     3   6.23    2.08    37.3 0.00076 ***
as.vector(cv)  1   7.16    7.16   128.7 9.3e-05 ***
Residuals      5   0.28    0.06
---
Signif. codes:  0 '***' 0.001 '**' 0.01 '*' 0.05 '.' 0.1 ' ' 1

> coef(crash2.aov)
  (Intercept)      agerange.L      agerange.Q  passengers.L
      1.77000        -2.09657         0.30006       1.38189
 passengers.Q    passengers.C   as.vector(cv)
      0.40333         0.06559         1.55165
```

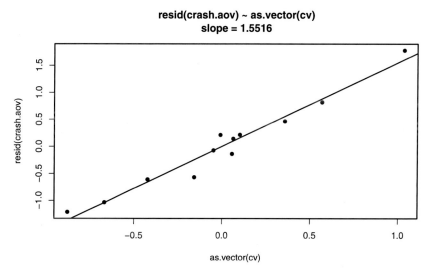

Fig. 14.16 Diagnostic plot for crash data. See also Table 14.29. The data follow the regression line. Therefore, the slope of 1.5516 suggests a power transformation of $1 - 1.5516 = -.5516$.

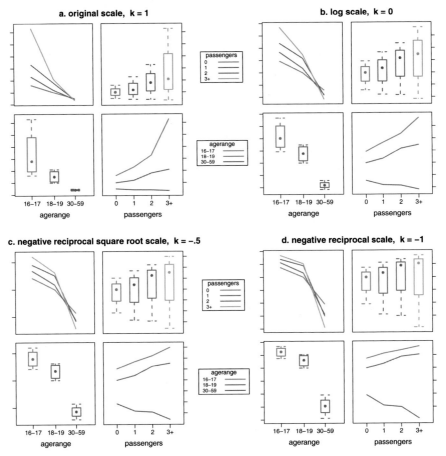

Fig. 14.17 Main effect and interaction plots of the crash data with several different power transformations. The $k = -.5$ and $k = -1$ plots look similar. Since it is easy to explain the reciprocal—it is measured in units of trips per crash—we select the reciprocal for further analysis and display in Table 14.30. Panel a is identical to Figure 14.15. Figure 14.18 is an enlarged version of panel d. The cover illustration of the first edition is based on panel d.

transformation that will stabilize the variance of the groups is given by $k = 1 - \hat{\beta}_{cv}$. In this example, the suggested power is $1 - 1.5516 = -.5516$. As with the Box–Cox transformations in Section 4.8, we usually look at graphs of the transformed values to help make the decision. We show in Figure 14.17 the data distributions and interaction plots for four power transformations: the original scale ($k = 1$), and powers $k = (0, -.5, -1)$.

We see in Figure 14.17 that our transformation has accomplished several of its purposes. Moving from panel a to b to c to d, we see that the variance has been stabilized. In panel a, the boxplots show strong changes in height; in panels c and d, the boxes are almost the same height. In panel a, the lines in the subpanel

Trips per Crash by Driver Age and Passenger Presence

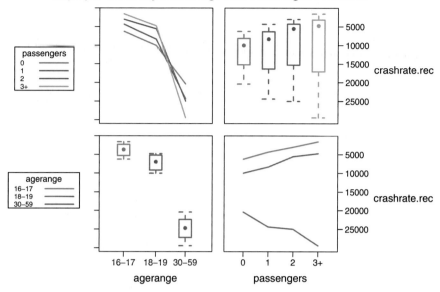

Fig. 14.18 Interaction plot of reciprocal crash data. The message is best seen in the bottom row. The boxplots in the lower-left panel show a difference in the small number of trips per crash for teens compared to the larger number of trips per crash for the adults. The lines in the lower-right panel show that adults increase the number of trips per crash with more passengers, meaning that they drive more safely with more passengers. But teens decrease the number of trips per crash with more passengers, meaning that they drive less safely with more passengers. This figure is identical to panel d in Figure 14.17.

Table 14.30 ANOVA of the reciprocal of crash rate. The passengers effect is not shown as significant and the single degree of freedom for the linear by linear component of the interaction isn't shown at all.

```
> crashi.aov <-
+    aov(10000/crashrate ~ agerange + passengers, data=crash)

> summary(crashi.aov)
            Df   Sum Sq  Mean Sq F value  Pr(>F)
agerange     2 1.02e+09 5.09e+08   44.97 0.00024 ***
passengers   3 2.42e+06 8.06e+05    0.07 0.97325
Residuals    6 6.79e+07 1.13e+07
---
Signif. codes:  0 '***' 0.001 '**' 0.01 '*' 0.05 '.' 0.1 ' ' 1
```

tracing passengers are not parallel and the lines in the subpanel tracing agerange are uphill or nearly horizontal. In panels c and d, the lines in the subpanel tracing passengers are parallel and downward sloping. The lines tracing agerange are uphill and parallel for the two teenage classifications and downhill for the adults.

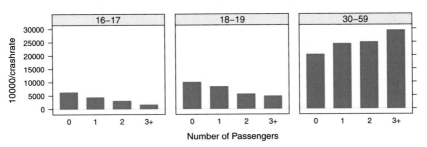

Fig. 14.19 Barplot of reciprocal crash data. The values displayed here are identical to the traces in the lower-right panel of Figure 14.18. Again we see that adults increase the number of trips per crash with more passengers, meaning that they drive more safely with more passengers. But teens decrease the number of trips per crash with more passengers, meaning that they drive less safely with more passengers.

The simplistic ANOVA table for the reciprocal displayed in Table 14.30 correctly shows `agerange` as significant, but it also shows `passengers` as not significant. We still need more work to make an ANOVA table of the reciprocal `crashrate` show the interaction between `agerange` and `passengers`, an interaction that we can see in Figures 14.15 and 14.18.

We choose the reciprocal over the reciprocal square root for the presentation of the analysis to the client even though both produce similar graphs. The reciprocal is measured in units of Trips per Crash, units that are easily interpreted. The reciprocal square root, in units of Square Root of Trips per Crash, is not easily interpreted. We use the reciprocal, not the negative reciprocal, when we present the results to the client. We do not need to maintain for the client the monotonicity of the series of transformations that we need when we compare several powers chosen from the ladder of powers.

The transformation of the crash rate to the reciprocal scale of trips per crash also works in the barplot format of Figure 14.19. The downhill and parallel pattern for the teens is evident. This contrasts sharply with the larger values and uphill pattern of the adults.

Now that we can see the different behavior for the `passengers` conditional on the `agerange`, let us make the ANOVA table show it. We do so by isolating the linear contrast for `passengers` and nesting it in the `agerange`. We combine the two teenage groups, which show a parallel upward trend over `passengers`. The resulting ANOVA table is in Table 14.31. The *p*-values in Table 14.31 are comparable to those in Table 14.29. Instead of an overall passenger effect, we are now looking at the linear contrast of passenger within the ageranges.

Table 14.31 ANOVA of the reciprocal of crash rate with linear effect of number of passengers nested within age range. We are partitioning the previous passenger + residual effects into the agerange/passenger effect plus a redefined residual. We see that the passengers within age range is highly significant for both age ranges, teens and adults.

```
> pass <- as.numeric(crash$passengers)

> crashinlin.aov <- aov(10000/crashrate ~ agerange/pass,
+                         data=crash)

> ## summary(crashinlin.aov)
> ## coef(summary.lm(crashinlin.aov))
> print(coef(summary.lm(crashinlin.aov))[4:6,], digits=4)
                   Estimate Std. Error t value  Pr(>|t|)
agerange16-17:pass    -1531        335  -4.569 0.0038164
agerange18-19:pass    -1849        335  -5.520 0.0014874
agerange30-59:pass     2762        335   8.244 0.0001721

> print(digits=3,
+ summary(crashinlin.aov,
+      split=list("agerange:pass"=list(teens=1:2, adults=3))))
                       Df   Sum Sq  Mean Sq F value   Pr(>F)
agerange                2 1.02e+09 5.09e+08   906.9  3.6e-08 ***
agerange:pass           3 6.70e+07 2.23e+07    39.8  0.00024 ***
  agerange:pass: teens  2 2.88e+07 1.44e+07    25.7  0.00115 **
  agerange:pass: adults 1 3.81e+07 3.81e+07    68.0  0.00017 ***
Residuals               6 3.37e+06 5.61e+05
---
Signif. codes:  0 '***' 0.001 '**' 0.01 '*' 0.05 '.' 0.1 ' ' 1
```

What actually did we do in Table 14.31 that we were unable to do in Table 14.30? The basic idea of the ODOFFNA is to look at the linear by linear component of the interaction. There aren't enough degrees of freedom with only one observation per cell to do more than that. From the graph in Figure 14.17d we see that the two teenage groups show a clear uphill trend and the adults a clear downhill trend. When we combine those two effects by averaging over the ages, we lose the distinction between age groups. We therefore isolated just one degree of freedom of the passengers effect, the linear component. We nested the linear effect within the ageranges and easily see in Table 14.31 that the teens and adults have strong, and opposite, linear components of the passengers effect.

In summary, adults have many more trips between crashes than teenagers. Adults have fewer crashes as the number of passengers increases. Teens have more crashes as the number of passengers increases.

14.8.4 Theory

The heart of the analysis in Section 14.8.1 involves partitioning the residual sum of squares in Table 14.28 into two components in Table 14.29. The sum of squares for one of these components involves the vector of comparison values, the linear-by-linear effect constructed in Section 14.8.1. By construction, the vector of comparison values is orthogonal to the dummy variables for the row and column effects. Hence, it must be in the space of the interaction of the row and column effects. In this section we outline the theory justifying that the statistic for testing nonadditivity in Table 14.29, the ratio of nonadditivity mean square to remaining residual mean square, has an F-distribution under the null hypothesis of nonadditivity.

If in the model in Equation (12.1) we have $n_{ij} = 1$ for all i and j, that is, there is exactly one observation at each treatment combination, then no degrees of freedom are available to estimate the interactions $(\alpha\beta)_{ij}$ as we would have if there were at least two observations at each treatment combination. However, for this situation Tukey (1949) developed a test for the special case of multiplicative interaction, that is, a test of

$$H_0: \theta = 0 \quad \text{vs} \quad H_1: \theta \neq 0 \tag{14.2}$$

in the model

$$Y_{ij} = \mu + \alpha_i + \beta_j + \theta\alpha_i\beta_j + \epsilon_{ij} \quad \text{for} \quad i = 1, \ldots, a \quad \text{and} \quad j = 1, \ldots, n_i$$

where $\sum_i \alpha_i = 0$, and $\sum_j \beta_j = 0$. In this situation, unbiased estimators are

$$\hat{\alpha}_i = \bar{Y}_{i.} - \bar{\bar{Y}}_{..}$$
$$\hat{\beta}_j = \bar{Y}_{.j} - \bar{\bar{Y}}_{..}$$
$$\hat{\theta}\alpha_i\beta_j = Y_{ij} - \bar{Y}_{i.} - \bar{Y}_{.j} + \bar{\bar{Y}}_{..}$$

Therefore a reasonable estimator of θ is

$$\hat{\theta} = \frac{\sum_{i,j}(\bar{Y}_{i.} - \bar{\bar{Y}}_{..})(\bar{Y}_{.j} - \bar{\bar{Y}}_{..})(Y_{ij} - \bar{Y}_{i.} - \bar{Y}_{.j} + \bar{\bar{Y}}_{..})}{\sum_{i,j}(\bar{Y}_{.j} - \bar{\bar{Y}}_{..})^2(\bar{Y}_{.j} - \bar{\bar{Y}}_{..})^2}$$

Since the numerator estimates

$$\sum_{i,j} \alpha_i\beta_j(\theta\alpha_i\beta_j) = \theta \sum_{i,j} \alpha_i^2\beta_j^2$$

and the denominator estimates

$$\sum_{i,j} \alpha_i^2\beta_j^2$$

it follows that

$$\hat{\theta} = \frac{\Sigma_{i,j}(\bar{Y}_{i.} - \bar{\bar{Y}}_{..})(\bar{Y}_{.j} - \bar{\bar{Y}}_{..})(Y_{ij})}{\Sigma_{i,j}(\bar{Y}_{.j} - \bar{\bar{Y}}_{..})^2(\bar{Y}_{.j} - \bar{\bar{Y}}_{..})^2}$$

From this it can be shown that if $\theta = 0$,

$$S_N = \frac{\Sigma_{i,j}(\bar{Y}_{i.} - \bar{\bar{Y}}_{..})(\bar{Y}_{.j} - \bar{\bar{Y}}_{..})(Y_{ij})^2}{\Sigma_{i,j}(\bar{Y}_{.j} - \bar{\bar{Y}}_{..})^2(\bar{Y}_{.j} - \bar{\bar{Y}}_{..})^2} \qquad (14.3)$$

has a χ^2 distribution with 1 df, and if $\theta \neq 0$, (14.3) has a noncentral χ^2 distribution. Tukey (1949) showed that if SS_{Res} denotes the usual residual Sum of Squares with $(a-1)(b-1)$ df, then

$$\frac{S_N}{\left(\dfrac{SS_{Res} - S_N}{ab - a - b}\right)} \qquad (14.4)$$

has an F distribution with 1 and $ab - a - b$ degrees of freedom if the null hypothesis in Equation (14.2) is true and a noncentral F distribution otherwise. Therefore, the test of the no-multiplicative-interaction hypothesis is given by Equation (14.4).

Experience with this test has shown that it is sensitive to the presence of other forms of interaction.

14.9 Exercises

14.1. Peterson (1985) reports an experiment to determine the effect of the annealing (heating to be followed by slow cooling) `temperature` on the `strength` of three metal `alloys`. Four temperatures were used: 675, 700, 725, and 750 degrees Fahrenheit. It was only possible to make one run per day in each of the four ovens available for the experiment, but all three alloys were accommodated in each oven run. The experimenter regarded this as a split plot design with days as blocks and ovens as plots. The data are in `data(anneal)`. Perform a complete analysis including data plots and an investigation of the borderline significant interaction.

14.2. Peterson (1985) discusses an experiment to compare the `yield` in kg/plot of four varieties of beans (New Era, Big Green, Little Gem, and Red Lake), also taking into account 3 `spacings` between rows, 20, 40, and 60 cm. A randomized block design was used, with four blocks. The data are in `data(bean)`. Analyze the data including an investigation of the simple effects of `variety` at each level of spacing.

14.3. Barnett and Mead (1956), also in Johnson and Leone (1967), discuss an experiment to determine the efficiency of radioactivity decontamination. The response is a measure (alpha activity) of remaining contamination, modeled by these four 2-level factors:

B: added barium chloride

A: added aluminum sulfate

C: added carbon

P: final pH

The maximum possible block size consisted of eight plots, but four blocks were used. The four-factor interaction was confounded with blocks, so that an even number of the four factors were at their upper levels in blocks 1 and 3, and an odd number of the four factors were at their upper levels in blocks 2 and 4. The experimental layout is implicit in data(radioact). Assuming that all three-factor interactions are negligible and that treatments do not interact with blocks, perform an analysis of variance and produce estimated contrasts to determine which factors and two-factor interactions contribute most to the remaining contamination. Produce a written recommendation for the client.

14.4. Neter et al. (1996) describe an experiment to examine the effect of two irrigation methods and two fertilizers on the yield of wheat. The data are in data(wheat). Five fields were available to conduct this experiment. Each field was divided into two Plots to which the two irrigation methods were randomly assigned. Each Plot was subdivided into two Subplots to which the fertilizers were randomly assigned. Perform a complete analysis. Produce a written recommendation for the farmer. (The Plot and Subplot factors were not in our source. We added them to the dataset to be consistent with our emphasis in Section 14.3 on the importance of distinguishing between the plot structure of the experimental material and the assigned treatments.)

14.5. In Section 14.3 it is stated that use of the split plot design resulted in increased precision for inferences involving the subplot factor nitrogen at the cost of reduced precision for the whole plot factor variety. Demonstrate this by reanalyzing the data under the assumption that the experimental design was a completely random one, with no randomization restrictions. That is, compare the F-tests for nitrogen and for variety under both design assumptions.

14.6. Fouts (1973) reports on an experiment in which times in minutes were recorded for each of 4 chimpanzees to learn each of 10 signs of American Sign Language. The data are in data(chimp).

a. Use the Tukey one-degree-of-freedom test to check whether there is interaction between sign and chimpanzee.

b. Is there evidence that the signs differ in the time required for the chimps to securely learn them? Discuss.

14.7. In Section 14.3.1 we note that the three-factor representation of the split plot design gives the same ANOVA table as the more complete five-factor representation. In this exercise we ask you to verify that statement by looking at the column space spanned by the various dummy variables implicitly defined by the model formula. Review the definitions of column space and orthogonal bases in Appendix Section I.4. We continue with the example in Section 14.3. The R code for this exercise is in chunk 53 of file HHscriptnames(14).

a. The whole plot column space is defined by the
        plots %in% blocks
   dummy variables generated by the alternate residuals formula: orthogonal contrasts are critical.

```
data(yatesppl)
yatesppl.resida.aov <-
    aov(y ~ blocks/plots,
        data=yatesppl, x=TRUE,
        contrasts=list(blocks=contr.helmert,
                       plots=contr.helmert))
summary(yatesppl.resida.aov)
t(yatesppl.resida.aov$x)
```

b. This is the same column space defined by the
        variety + blocks:variety
   dummy variables generated by the computational shortcut

```
yatesppl.short.aov <-
   aov(terms(y ~ blocks + variety + blocks*variety +
             nitrogen + variety*nitrogen,
             keep.order=TRUE),
             ## try it without keep.order=TRUE
       data=yatesppl, x=TRUE)
summary(yatesppl.short.aov)
t(yatesppl.short.aov$x)
```

c. We illustrate this by regressing the response variable y on the
        variety + blocks:variety

dummy variables:

```
## project y onto blocks/plots dummy variables
plots.aov <-
  lm(y ~ yatesppl.resida.aov$x[,7:18], data=yatesppl)
summary.aov(plots.aov)
y.bp <- predict(plots.aov)
variety.aov <-
  aov(y.bp ~ blocks*variety, data=yatesppl)
summary(variety.aov)
```

and seeing that we reproduce the
        plots %in% blocks
stratum of the ANOVA table

```
Error: plots %in% blocks
            Df Sum of Sq Mean Sq F Value      Pr(F)
  variety   2  1786.361 893.1806 1.48534 0.2723869
Residuals 10  6013.306 601.3306
```

obtained from the complete five-factor specification.

```
## split plot analysis
yatesppl.anova <-
  aov(y ~ variety*nitrogen +
          Error(blocks/plots/subplots),
      data=yatesppl)
summary(yatesppl.anova)
```

14.8. Tables 14.16 and 14.17 show two different analyses for Model 2 yield.block ~ pre.block * treat. Confirm that the difference between them is in the choice of dummy variables for the pre.block:treat interaction term. Do so by constructing both sets of dummy variables and regressing one against the other. The residuals are nonzero. File HHscriptnames(14) contains code for both sets of dummy variables.

Chapter 15

Bivariate Statistics—Discrete Data

In this chapter we discuss bivariate discrete distributions. Bivariate means that there are two factors (categorical variables) defining cells. The response values are frequencies, that is, counts or instances of observations, at each cell.

It is convenient to arrange such data in a contingency table, that is, a table with r rows representing the possible values of one categorical variable and c columns representing the possible values of the other categorical variable. Each of the rc cells of the table contains an integer, the number of observations having the levels of the two variables specified by the cell location. We give extra attention to the special case where $r = c = 2$, that is, a 2×2 contingency table.

This data type is different from situations with two categorical variables (factors) described in Chapters 12 through 14. In those chapters the response variables are one or more continuous measurements at each cell.

We introduce a relatively new form of graph, the mosaic plot, which along with related plots we access through the R package **vcd** "Visualizing Categorical Data". In simplest terms, a mosaic plot consists of a tiling of rectangles (hence the name) where the height, width, and area (product of height and width) are each interpretable.

15.1 Two-Dimensional Contingency Tables—Chi-Square Analysis

15.1.1 Example—Drunkenness Data

Table 15.1 shows the number of persons convicted of drunkenness in two London courts during the first six months of 1970. The data come from Cook (1971), later reprinted in Hand et al. (1994). The dataset is accessed as data(drunk). There are

© Springer Science+Business Media New York 2015
R.M. Heiberger, B. Holland, *Statistical Analysis and Data Display*,
Springer Texts in Statistics, DOI 10.1007/978-1-4939-2122-5_15

two rows, males and females. The five columns are five age categories. The question of interest is whether the age distribution of convicted offenders is the same for both genders, or equivalently, as illustrated in the bar chart of Figure 15.1, if the proportion of female offenders is the same for all age categories.

We show the mosaic plot of the same data in Figure 15.2.

Table 15.1 Persons convicted of drunkenness in two London courts during the first six months of 1970.

Age group	0–29	30–39	40–49	50–59	≥60
Number of males	185	207	260	180	71
Number of females	4	13	10	7	10

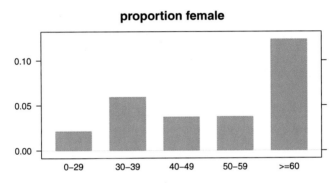

Fig. 15.1 Proportion female for drunkenness data.

The tabular display from the chi-square analysis (to be described in Section 15.1.2) is in Table 15.2.

The p-value .0042 for the chi-square test strongly suggests an association between age and gender of convicted offenders, that is, the proportion of females in each age group is not identical. From the "Cell Chi-square" values in Table 15.2 and their square roots displayed in Figure 15.3, it seems that only 1 of the 10 cells contributes appreciably to total chi-square. The cell for female offenders aged at least 60 contributes 10.34, or 68%, of the total chi-square value 15.25. We observe 10 female offenders aged at least 60, but under the null hypothesis of independence we expect only 3.8 offenders. No other cell is suggestive of dependence.

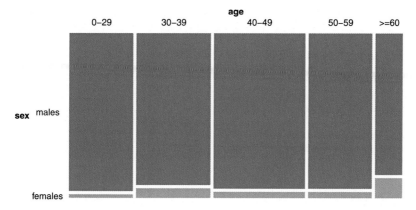

Fig. 15.2 Mosaic plot of the drunkenness data. The widths of the sets of tiles for each age group are proportional to the counts of all persons in that age groups. The heights of the bottom set of tiles (females) are the same as the heights of the bars in Figure 15.1 and are the proportions female within each age group. The area of each tile is proportional to the count of individuals of that sex and age group.

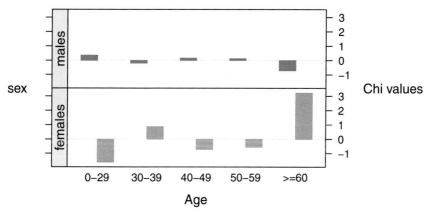

Fig. 15.3 Cell chi-deviations (residuals from Table 15.2, also the signed square root of the cell chi-square values $\chi_{ij} = (n_{ij} - e_{ij})/\sqrt{e_{ij}}$) for drunkenness data. The signed heights of these bars are the same as the signed heights of the bars in the association plot in Figure 15.4.

We must be careful not to overinterpret this finding as meaning that older females have a greater tendency toward this crime than older males. We believe that the finding may be an artifact of the demographic distribution. The population proportion of females under the age of 60 is roughly 50%. This proportion tends to increase after age 60 because of higher male mortality beginning at approximately that age. Therefore, it is possible that this study could have been improved by adjusting the responses to a per capita basis.

Table 15.2 Chi-square analysis of Drunkenness Data.

```
> drunk.chisq <- chisq.test(drunk)

> drunk.chisq

Pearson's Chi-squared test

data:  drunk
X-squared = 15.25, df = 4, p-value = 0.004217


> drunk.chisq$observed
         age
sex       0-29 30-39 40-49 50-59 >=60
  males    185   207   260   180   71
  females    4    13    10     7   10

> drunk.chisq$expected
         age
sex          0-29   30-39  40-49   50-59   >=60
  males   180.219 209.78 257.46 178.312 77.237
  females   8.781  10.22  12.54   8.688  3.763

> drunk.chisq$residuals    ## cell chi values
         age
sex          0-29    30-39   40-49   50-59    >=60
  males     0.3562 -0.1918  0.1586  0.1264 -0.7096
  females -1.6135  0.8690 -0.7185 -0.5728  3.2148

> drunk.chisq$residuals^2 ## cell chi-square values
         age
sex          0-29    30-39    40-49    50-59     >=60
  males    0.12686 0.03679  0.02516  0.01599  0.50357
  females  2.60344 0.75512  0.51626  0.32814 10.33473
```

15.1.2 Chi-Square Analysis

When we work with two-dimensional tables, such as this example, we often want to test whether the row and column classifications are independent, that is, whether the probability of an entry's being in a particular row is independent of the entry's column.

When $r = 2$, the test is essentially asking whether the proportion of data in Row 1 is homogeneous across the c columns, i.e., whether c binomial populations have the same (unspecified) proportion parameter. Therefore, this is a generalization of the inferences comparing $c = 2$ population proportions discussed in Chapter 5 to $c \geq 2$ population proportions.

If the total number of observations n is sufficiently large and if none of the rc expected cell frequencies is less than somewhere between 3 and 5, the chi-square distribution may be used to test the hypothesis of independence. The logical idea behind this test is to compare, in each of the cells,

n_{ij} the observed frequency in the cell in row i and column j

with

e_{ij} the expected frequency calculated under the assumption
    that the independence null hypothesis is true.

The test statistic is a function of the aggregate discrepancy between the n_{ij}'s and e_{ij}'s.

Define

$$\begin{cases} n_{i.} & = \sum_j n_{ij} \text{ the row totals} \\\\ n_{.j} & = \sum_i n_{ij} \text{ the column totals} \\\\ n = n_{..} & = \sum_{i,j} n_{ij} \text{ the grand total} \end{cases} \tag{15.1}$$

Under the null hypothesis of independence between rows and columns, the expected frequency for the cell in row i and column j is

$$e_{ij} = \frac{n_{i.}\, n_{.j}}{n} \tag{15.2}$$

The cell residuals, also called chi-deviations or scaled deviations,

$$\chi_{ij} = \frac{n_{ij} - e_{ij}}{\sqrt{e_{ij}}} \tag{15.3}$$

are displayed in Figure 15.3. The squares, the cell chi-square values χ_{ij}^2, are displayed in Figure 15.3 and also displayed in Table 15.2.

The assoc function in the **vcd** package provides an alternative plot of the chi-deviations. In Figure 15.4 we see the association plot for the drunk dataset. The heights of the rectangles are proportional to the cell residual values χ_{ij}, the widths of the rectangles are proportional to the square root of the cell expected values $\sqrt{e_{ij}}$, and therefore the areas are proportional to the difference in observed and expected frequencies $n_{ij} - e_{ij}$.

The test statistic is the sum of the scaled deviations squared

$$\hat{\chi}^2 = \sum_{ij} \frac{(n_{ij} - e_{ij})^2}{e_{ij}} \tag{15.4}$$

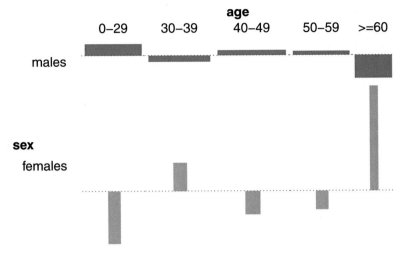

Fig. 15.4 Association plot for the drunk dataset. The heights of the rectangles are proportional to the cell residual values $\chi_{ij} = (n_{ij} - e_{ij})/\sqrt{e_{ij}}$ (the same heights as in Figure 15.3), the widths of the rectangles are proportional to the square root of the cell expected values $\sqrt{e_{ij}}$, and therefore the areas are proportional to the difference in observed and expected frequencies $n_{ij} - e_{ij}$.

where the sum is taken over all rows and columns. If the null hypothesis is true, $\hat{\chi}^2$ has, approximately, a chi-square distribution with $(r-1)(c-1)$ degrees of freedom, and the p-value is $1 - \mathcal{F}_{\chi^2}(\hat{\chi}^2 \mid (r-1)(c-1))$, the chi-square tail probability associated with $\hat{\chi}^2$. Most authorities agree that the chi-square approximation is good if almost all e_{ij} are at least 5 and none is less than 3. The degrees-of-freedom formula derives from the fact that if all marginal totals are given, knowledge of $(r-1)(c-1)$ interior values uniquely determines the remaining $r + c - 1$ interior values.

Apart from the degrees-of-freedom calculation, the form of this test, comparing observed and expected frequencies, is identical to the goodness-of-fit test described in Section 5.7.1. This methodology can be extended to contingency tables having more than two dimensions.

Further analysis is required to assess the nature of any lack of independence that is suggested by the chi-square test. One approach to this is discussed in the next paragraph. Another is a multivariate display technique called *correspondence analysis*. See Greenacre (1984) for an introduction to this topic.

The chi-square test of independence, shown in Table 15.2, is the default display from chisq.test. The result of the test also contains each cell's observed value n_{ij}, expected value e_{ij}, residual (square root of its contribution to the chi-square statistic) χ_{ij}.

Cells with a sizeable cell chi-square value have an appreciable discrepancy between their observed and expected frequency. Scrutiny of such cells leads to interpretation of the nature of the dependence between rows and columns. A cell chi-square is calculated as $(n - e)^2/e$, where n and e are, respectively, the cell's observed frequency and expected frequency under the null hypothesis. Under the model that observations are randomly assigned to cells with a Poisson distribution, the variance of n is also e. Hence $(n - e)^2/e$ has the form

$$\frac{(n - E(n))^2}{\text{var}(n)}$$

and, using the normal approximation, we interpret it as approximately a one-df chi-square statistic. Since the 95th and 99th percentiles of this chi-square distribution are 3.84 and 6.63, we recommend reporting the discrepancy between the observed and expected frequency for all cells with cell chi-square exceeding the higher value. Also, consideration should be given to reporting cells having chi-square between 3.84 and 6.63 when the discrepancy between the cell's observed and expected frequency can be meaningfully interpreted.

15.2 Two-Dimensional Contingency Tables—Fisher's Exact Test

An alternative to the approximate chi-square statistic discussed above is Fisher's exact test, which uses the exact hypergeometric distribution probabilities calculated for all tables at least as extreme in the alternative hypothesis direction as the existing table. Since it is exact, this procedure, when available, is preferable to the chi-square test, but it is extremely computer intensive for all but the smallest tables, even by contemporary standards.

Fisher's exact test is available in R as `fisher.test(x)`, where x is a two-dimensional contingency table in matrix form. For tables larger than 2×2, only the two-sided test is available.

15.2.1 Example—Do Juvenile Delinquents Eschew Wearing Eyeglasses?

Weindling et al. (1986) discuss a small study of juvenile delinquent boys and a control group of nondelinquents. The data in Table 15.3 also appear in Hand et al. (1994). All of these subjects failed a vision test. The boys were also classified according to whether or not they wore glasses.

Table 15.3 Wearing prescribed glasses and juvenile delinquency.

	Juvenile delinquents	Nondelinquents
Wears glasses	1	5
Doesn't wear glasses	8	2

Clearly, the data set is much too small to use the chi-square analysis of the previous section. Therefore, we request an analysis using Fisher's exact test, shown in Table 15.4. We are interested in the two-sided p-value, .0350. Since this falls between the two thresholds .01 and .05, we can say there is suggestive but inconclusive evidence that a smaller proportion of delinquents than nondelinquents wear glasses.

Table 15.4 Fisher's exact test of glasses data. Is the proportion of delinquents who wear glasses the same as the proportion of nondelinquents who wear glasses Fisher's exact test for glasses data. The p-value for the two-sided exact test is .035. Compare this to $p = .0134$ from the (uncorrected) chi-square approximation that R warns might be incorrect.

```
> fisher.test(glasses)

Fisher's Exact Test for Count Data

data:  glasses
p-value = 0.03497
alternative hypothesis: true odds ratio is not equal to 1
95 percent confidence interval:
 0.0009526 0.9912282
sample estimates:
odds ratio
   0.06464
```

The calculation of the p-values for Fisher's exact test utilizes the hypergeometric probability distribution discussed in Appendix J to calculate the probability of obtaining the observed table and "more extreme" tables assuming that the table's marginal totals are fixed. We illustrate the calculations in Table 15.5. The observed table is shown in Column **1**. The remaining columns, indexed by the [1,1] cell count, show all possible tables with the same row and column margins. The probability of observing the counts in Table 15.3 (identical to Column **1**, marked "*", in Table 15.5) given this table's marginal totals is

$$\frac{\binom{9}{1}\binom{7}{5}}{\binom{16}{6}} = 0.0236$$

The one-sided p-value is the sum of this probability and the probability, 0.0009, of the more extreme table **0** on the same tail of the distribution.

Table 15.5 All possible 2×2 tables with the same margins as the observed glasses table. In these tables `glasses` denotes "wears glasses" and `del` denotes "delinquent." We show the probabilities of each under the assumption of a hypergeometric distribution for the `[1,1]` position and the common row and column cell margins. The observed table **1** is marked with "∗" and the more extreme tables in the same tail **0** and opposite tail **6** are marked "<". The mosaic plots of each of the tables are shown in Figure 15.5. The probabilities of each of the tables are graphed in Figure 15.6.

glasses	[1,1] cell count												
	0		**1**		**2**		**3**		**4**		**5**		**6**	
	del	no.del	del	no.del	del	no.del	del	no.del	del	no.del	del	no.del	del	no.del
glasses	0	6	1	5	2	4	3	3	4	2	5	1	6	0
no.glasses	9	1	8	2	7	3	6	4	5	5	4	6	3	7
probability	0.0009		0.0236		0.1573		0.3671		0.3304		0.1101		0.0105	
which	<		*										<	

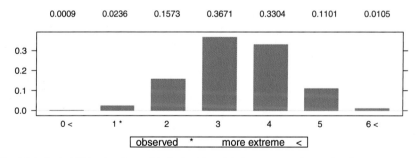

Fig. 15.5 All possible 2×2 tables with the same margins as the observed glasses table. The numerical values corresponding to this figure are in Table 15.5. The observed table is Table **1**. In each table we are comparing the proportion of `glasses` within `delinq` to the proportion of `glasses` within `non.del`. The widths of each bar are proportional to the count of people in that category of `delinquent`. The heights are percents of `wearer` within that category of `delinquent` and therefore add up to 100% in each column.

Fig. 15.6 Probabilities for all possible 2×2 tables with the same margins as the observed `glasses` table.

Table **6**, the more extreme on the opposite tail of the distribution, has probability

$$\frac{\binom{9}{6}\binom{7}{0}}{\binom{16}{6}} = 0.0105$$

The two-sided p-value is the sum of the probabilities of the observed table and both more extreme tables, $0.0236 + 0.0009 + 0.0105 = 0.0350$. The R code for the probability calculations is included in file `HHscriptnames(15)`.

15.3 Simpson's Paradox

Simpson's paradox, a counterintuitive situation, occurs when the presence of a third variable is unexpectedly responsible for a change, or even reversal, of the relationship between two categorical variables. The following example taken from Blyth (1972), with dataset `data(blyth)`, illustrates this phenomenon.

The data, including the margin summed over location, are shown in Table 15.6 and Figure 15.7. A medical researcher selected 11,000 human subjects at location A and 10,100 subjects at location B. At A, 1,000 of the subjects were randomly assigned to the standard treatment (**standard**) and the remaining 10,000 subjects were assigned to a new treatment (**new**). At B, 10,000 of the subjects were randomly assigned to **standard** and the remaining 100 subjects were assigned to **new**. Eventually, each subject was classified as not-survived (**not**) or survived (**survive**).

Table 15.6. The intent is to show for each Location that the percentage surviving with the **new** Treatment is larger than with the **standard** treatment, but that

Table 15.6 Blyth's data illustrating Simpson's Paradox. Within each location (A and B), the new treatment has a higher survival rate than standard treatment. Summed over locations (A&B combined), new has a lower survival rate than standard. We show several different style graphs of the Counts and Proportions for this dataset in Figure 15.7.

		Location					
		A		B		A&B combined	
Summary	Survival	standard	new	standard	new	standard	new
Count	not	950	9000	5000	5	5950	9005
	survive	50	1000	5000	95	5050	1095
Percent	not	95	90	50	5	54	89
	survive	5	10	50	95	46	11

the reverse is true when the two Locations are combined. This portrayal fails for these data because it is not possible to distinguish the very small counts 5, 50, and 95 in three of the eight cells in Table 15.6 when they are displayed on a common numerical scale with counts ranging from 950 to 9,005 in the table's other cells.

The paradox is better communicated by Figure 15.7 panels b,c,d,f which graph the percentages themselves. We observe that in Location A, the percent surviving following the new Treatment was 10% as compared to 5% with the **standard** treatment. In Location B, the **new** Treatment improved the survival percentage to 95% from 50% for the **standard** Treatment. It seems that the **new** Treatment was very successful. Now look at the summary results for both Locations combined. The survival rate for the **standard** Treatment is 46%, but it is only 11% for the **new** Treatment. The combined results suggest that the **new** Treatment is a disaster!

The substantive reason for this finding is that the subjects in Location A were much less healthy than those in B and the **new** Treatment was given mostly to subjects in A, where it could not be expected to fare as well as with subjects in B. That is, the factors Treatment and Location are not independent. When this is so, it can happen as here that

$$P(\texttt{survive} \mid \texttt{new}) < P(\texttt{survive} \mid \texttt{standard})$$
$$\text{while both}$$
$$P(\texttt{survive} \mid \texttt{new} \cap A) \geq P(\texttt{survive} \mid \texttt{standard} \cap A)$$
$$\text{and}$$
$$P(\texttt{survive} \mid \texttt{new} \cap B) \geq P(\texttt{survive} \mid \texttt{standard} \cap B)$$

corresponding to .11 < .46 but .10 > .05 and .95 > .50 in this example.

The Percent panels b,c,d,f of Figure 15.7 display the disparity better than the Count panels a,e because they transform the observed data from the scale reported by the client to the proportion scale, a scale in which the reversal is visible. In the proportion scale, most strongly in the mosaic plot in Panel d, we can easily see that the combined location information is almost the same as the B–standard and A–new information. In retrospect we can also see the same information in the Count panels. We can explain it by noting that there is almost no data in the A–standard and B–new cells; hence the combined information really is just the B–standard and A–new information plus a little noise.

When analyzing contingency table data, we should be alert to the possibility illustrated in this example that results for tables individually can differ from those when these tables are combined.

What is the resolution in situations such as this where individual results contradict combined results? Almost always, the individual results have more credence because combining such individuals cannot be adequately justified. In Blyth's example, the disparity between individual and combined results could have been attributable to different baseline health status of the patients at the two locations. Or it could be an artifact of the radically different treatment allocation patterns at the two locations.

a. Stacked Bar Chart of Counts

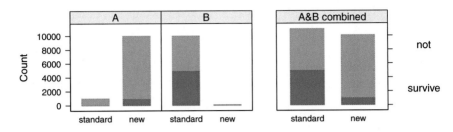

b. Stacked Bar Chart of Percents

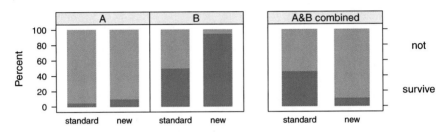

c. Bar Chart of Percent Surviving

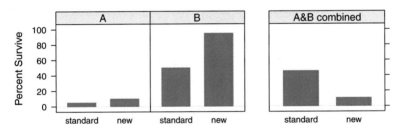

Fig. 15.7 Blyth's data illustrating Simpson's paradox. Within each location (A and B), the new treatment has a higher survival rate than standard treatment. Summed over locations, new has a lower survival rate than standard. We note that the combined location information is almost the same as the B–standard and A–new information. The great disparity in counts among the four Location–Treatment groups makes it difficult to see the survival rates in the stacked bar chart in Panel a. The rates are easier to see in the stacked bar chart of percents in Panel b. They are less easy to see in the bar chart of only survive rates in Panel c—these bars are identical to the bottom bars in Panel b, but the visual sense of proportion is missing.

d. Mosaic plot of Counts and Percents

e. Likert Plot of Counts

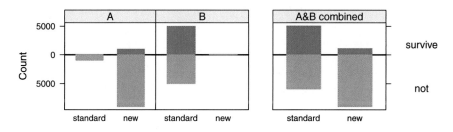

f. Likert Plot of Percents

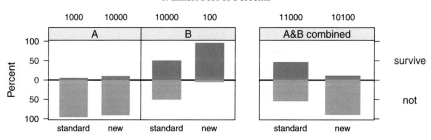

Fig. 15.7 continued The mosaic plot in Panel d shows three summaries: Counts of treatment assignments—as widths of each set of bars, Percents surviving conditional on Location–Treatment—as the relative heights of the bars in each Location–Treatment combination, and Counts of each Location–Treatment combination—as areas of each rectangle. In Panel e, the Likert plot of Counts, we again can barely see the **B–new** bars. In Panel f, the Likert plot of Percents, we print the observation counts for the Location–Treatment combinations above the bars. In the Likert plots (diverging stacked barcharts), the not survive values are shown diverging down from zero. In the other plots, the vertical range is from 0% to 100% and the not survive values are stacked above the survive values.

Simpson's paradox is the discrete analogue of the ecological fallacy discussed in Section 4.2. It is also related to the need to examine simple effects in the presence of interaction of qualitative factors, discussed in Chapters 12 to 14, since both problems refer to the importance of distinguishing overall conclusions from conclusions for subgroups.

15.4 Relative Risk and Odds Ratios

Analysis of data arranged in a 2×2 table is equivalent to comparing two proportions. However, analyzing the difference $p_1 - p_2$ via a CI or test as outlined in Chapter 5 is often not an appropriate way to compare them. Instead a measure of *relative* difference is appropriate. Consider two cases, the first with $p_1 = .02$ and $p_2 = .07$ and the second with $p_1 = .50$ and $p_2 = .55$. In both cases, $p_2 - p_1 = .05$. However, in the first case, p_2 is 250% more than p_1, but in the second case p_2 is only 10% more than p_1, and from this point of view it is inadequate to merely consider differences of proportions, particularly proportions close to either 0 or 1.

We discuss two additional measures for comparing two proportions. The first, the *relative risk*, is simply the ratio of the two proportions, $\hat{p}_1 / \hat{p}_2$.

The *odds ratio* is a widely used measure of relative difference. It is more informative than a chi-square test for a 2×2 table because it measures the magnitude of difference between two proportions. Unlike the chi-square test, the odds ratio is minimally affected by the size of the sample.

Based on the definition of odds in Equation (3.2), if $\hat{p}$ is an estimated probability of success, the estimated *odds in favor* of success are $\hat{\omega} = \hat{p}/(1 - \hat{p})$. For comparing two estimated proportions, $\hat{p}_1$ and $\hat{p}_2$ in a 2×2 contingency table, the estimated ratio of two odds, the odds ratio, is

$$\hat{\Psi} = \hat{\omega}_2 / \hat{\omega}_1 \tag{15.5}$$

A quick way to hand-calculate the estimated odds ratio is $\hat{\Psi} = (n_{11}n_{22})/(n_{21}n_{12})$, and for this reason the odds ratio is also known as the *cross-product ratio*.

If the odds ratio exceeds 1 so does the relative risk, and conversely.

15.4.1 Glasses (Again)

For example, reconsider the data of Section 15.2.1. The relative risk is

$$\frac{\left(\dfrac{8}{10}\right)}{\left(\dfrac{1}{6}\right)} = 4.8$$

This says that based on these data, nonwearers of glasses are four times more likely to become delinquent than wearers of glasses.

The odds ratio is

$$\hat{\Psi} = \frac{\left(\frac{\frac{5}{7}}{1-\frac{5}{7}}\right)}{\left(\frac{\frac{1}{9}}{1-\frac{1}{9}}\right)} = 20$$

This means that the odds that a nondelinquent wears glasses are estimated to be 20 times the odds that a delinquent wears glasses. Alternatively, the odds that a delinquent wears glasses are estimated as 1/20 times the odds that a nondelinquent wears glasses. If this ratio had been maintained for a larger sample, an implication might have been that police needn't pay much attention to boys wearing glasses.

Often investigators report the log of the odds ratio since the change in the reference group simply reverses the sign of the log odds: $\ln(20) = 2.996$ and $\ln\left(\frac{1}{20}\right) = -2.996$.

15.4.2 Large Sample Approximations

A useful property of both the odds ratio and the relative risk is that for large sample sizes, the log of the estimated odds ratio and the log of the estimated relative risk are approximately normally distributed.

15.4.2.1 Odds Ratio

From Agresti (1990) Equation (3.15) we find the log of the estimated odds ratio is approximately normally distributed

$$\ln(\hat{\Psi}) \sim N\left(\ln(\Psi), \sigma^2_{\ln(\hat{\Psi})}\right) \tag{15.6}$$

with mean equal to the log of the population odds ratio and estimated variance

$$\hat{\sigma}^2_{\ln(\hat{\Psi})} = \frac{1}{n_{11}} + \frac{1}{n_{12}} + \frac{1}{n_{21}} + \frac{1}{n_{22}} \tag{15.7}$$

These facts lead to large sample confidence intervals and hypothesis tests for odds ratios. A test of $H_0: \ln(\Psi) = 0$, or equivalently, $\Psi = 1$, is based on

$$z_{calc} = \frac{\ln(\hat{\Psi})}{\sqrt{\frac{1}{n_{11}} + \frac{1}{n_{12}} + \frac{1}{n_{21}} + \frac{1}{n_{22}}}}$$

An approximate $100(1 - \alpha)\%$ confidence interval on $\ln(\Psi)$ is

$$\text{CI}\big(\ln(\Psi)\big) = \ln(\hat{\Psi}) \pm z_{\frac{\alpha}{2}} \sqrt{\frac{1}{n_{11}} + \frac{1}{n_{12}} + \frac{1}{n_{21}} + \frac{1}{n_{22}}} = (L, U) \qquad (15.8)$$

If we denote this interval by (L, U), the approximate confidence interval on the odds ratio Ψ_{21} is (e^L, e^U).

15.4.2.2 Relative Risk

From Agresti (1990) Equation (3.18) we find the log of the estimated relative risk is approximately normally distributed

$$\ln(\hat{p}_1/\hat{p}_2) \sim N\left(\ln(p_1/p_2), \sigma^2_{\ln(\hat{p}_1/\hat{p}_2)}\right) \qquad (15.9)$$

with mean equal to the log of the population relative risk and estimated variance

$$\hat{\sigma}^2_{\ln(\hat{p}_1/\hat{p}_2)} = \sqrt{\frac{1 - p_1}{p_1 \, n_1} + \frac{1 - p_2}{p_2 \, n_2}} \qquad (15.10)$$

These facts lead to large sample confidence intervals and hypothesis tests for relative risks. A test of $H_0: \ln(\hat{p}_1/\hat{p}_2) = 0$, or equivalently, $\hat{p}_1/\hat{p}_2 = 1$, is based on

$$z_{calc} = \frac{\ln(\hat{p}_1/\hat{p}_2)}{\sqrt{\frac{1-p_1}{p_1 n_1} + \frac{1-p_2}{p_2 n_2}}}$$

An approximate $100(1 - \alpha)\%$ confidence interval on $\ln(\hat{p}_1/\hat{p}_2)$ is

$$\text{CI}\big(\ln(\hat{p}_1/\hat{p}_2)\big) = \ln(\hat{p}_1/\hat{p}_2) \pm z_{\frac{\alpha}{2}} \sqrt{\frac{1 - p_1}{p_1 n_1} + \frac{1 - p_2}{p_2 n_2}} = (L, U) \qquad (15.11)$$

If we denote this interval by (L, U), the approximate confidence interval on the relative risk Ψ_{21} is (e^L, e^U).

15.4.3 Example—Treating Cardiac Arrest with Therapeutic Hypothermia

Holzer (2002) supervised a multicenter trial of patients who were randomly assigned to receive or not receive therapeutic hypothermia (lowered body temperature) to assist in recovery following resuscitation from cardiac arrest. Six months after cardiac arrest, patients were classified as having a favorable neurologic outcome or not. Of the 136 patients treated with hypothermia, 75 had a favorable neurological outcome. Of the 137 patients not treated with hypothermia, 54 had a favorable neurological outcome (see Table 15.7 and Figure 15.8 for the counts and Figure 15.9 for the odds and log of the odds). All patients received standard treatment for cardiac arrest apart from hypothermia and the treatment was blinded from the assessors of the outcome.

Table 15.7 Results of therapeutic hypothermia investigation Holzer (2002).

	Favorable neurological outcome		Total
	Yes	No	Total
Treated	75	61	136
Control	54	83	137

For a patient who receives the therapeutic hypothermia treatment, the estimated odds in favor of a favorable neurologic outcome are

$$\hat{\omega}_2 = \left(\frac{\frac{75}{136}}{1 - \frac{75}{136}} \right) \approx 1.23$$

For a patient who does not receive this treatment, the estimated odds in favor of a favorable neurological outcome are

$$\hat{\omega}_1 = \left(\frac{\frac{54}{137}}{1 - \frac{54}{137}} \right) \approx 0.65$$

The odds and log odds are plotted in Figure 15.9.

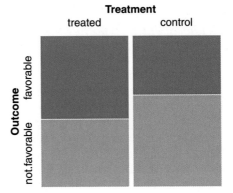

Fig. 15.8 `mosaic(Outcome ~ Treatment)`

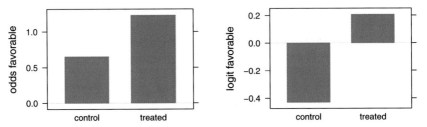

Fig. 15.9 Barchart of odds and barchart of logit.

The estimated odds ratio is

$$\hat{\Psi} = \frac{\hat{\omega}_2}{\hat{\omega}_1} = \frac{(75)(83)}{(54)(61)} \approx 1.8898$$

The reader can verify that the estimated standard deviation of the log of the odds ratio is 0.246, and that this leads to an approximate 95% confidence interval $(1.168, 3.058)$ for the population odds ratio.

For the hypothermia example with $\alpha = .05$, we have

$$\text{CI}\big(\ln(\Psi)\big) = \ln(1.8898) \pm 1.96 \sqrt{\frac{1}{75} + \frac{1}{54} + \frac{1}{61} + \frac{1}{83}} \approx (0.1552, 1.1177)$$

We therefore have an approximate confidence interval on the odds ratio of

$$\text{CI}(\Psi) \approx (1.168, 3.058)$$

This means that the odds of a favorable neurological outcome for a patient receiving the therapeutic hypothermia treatment are estimated to be between 1.17 and 3.06 times the odds of a favorable neurological outcome for a patient not receiving this particular treatment. Further, the calculated z-statistic for a test of the one-sided alternative hypothesis that the population odds ratio exceeds 1 is 2.592, with corresponding p-value less than 0.01. Therefore, there is strong evidence that the therapeutic hypothermia treatment improves probability of a successful neurological outcome.

For fixed probability of favorable outcome for control patients of $54/137 = .3942$, corresponding to fixed odds of favorable outcome of $.3942/.6058 = .6506$, the point estimate for the odds of favorable outcome for treated patients is given by $1.8898 \times .6506 = 1.2295$, and an estimated confidence interval for the odds by $(1.168 \times .6506, 3.058 \times .6506) = (0.7599, 1.9895)$. The corresponding point and interval estimates for the probabilities of outcome for treated are

estimate	favorable		
point	1.2295/2.2295	=	0.5515
interval	(0.7599/1.7599, 1.9895/2.9895)	=	(0.4318, 0.6655)

The point estimate of the probability of favorable outcome for treated is exactly the observed proportion $75/136=0.5515$, and the confidence interval of the proportion of favorable outcomes excludes the observed proportion for the control group.

We can extend this discussion by assuming any fixed probability of favorable outcome for treatment and then calculating the confidence interval for the probability of favorable outcome for control. We do so in Figure 15.10 for the set of fixed probabilities $p_1 = (0, .05, \ldots, 1)$.

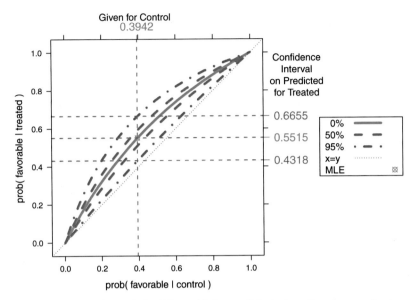

Fig. 15.10 Confidence intervals for *P*(favorable | treated) in the hypothermia example given the odds ratio. The confidence intervals are calculated from an assumed *P*(favorable | control) and the given odds ratio using the odds ratio formula in Equation 15.8. Details for *P*(favorable | control) = 54/137 = .3942 and *P*(not.favorable | control) = .6058, corresponding to given odds of favorable outcome of .3942/.6058 = .6506, are shown in Section 15.4.3. Symmetrically, we can assume any fixed probability of favorable outcome for control and then calculate the confidence interval for the probability of favorable outcome for treatment.

15.5 Retrospective and Prospective Studies

Consider two possible experiments to assess whether vitamin C supplementation prevents occurrences of the common cold.

In the first experiment we select 100 people who have had a cold during the past two months and 100 people who have not had a cold during the past two months. We then ask these people whether or not they have taken a daily vitamin C supplement during this period. In the second experiment we select 200 volunteers, assigning them to take no vitamin C supplementation apart from that offered by the study. We randomly assign 100 of these subjects to receive the study's vitamin C supplement and the other 100 subjects to receive a placebo, indistinguishable by these subjects from vitamin C. Then, two months later, we ask the subjects whether or not they have had a cold since the experiment began.

The first experiment is an example of a retrospective study, also called a case–control study. Subjects having a condition are called cases, subjects not having a condition are termed controls, and subjects are cross-classified with a risk factor (present or absent). In the above example, the risk factor is presence or absence of vitamin C supplementation. In retrospective studies, the subjects are selected after the events in question have already occurred, in this case having contracted one or more colds. Such studies are common in medical research because they generally assure a larger number of subjects than prospective studies.

The second experiment is an example of a prospective study, also known as a cohort study. Samples are taken from a population of subjects classified according to two risk factors (events) defined prior to initiating sampling, in this case assignment to vitamin C or placebo. Such studies often require that subjects be followed for a period of time until the subjects are determined to have a condition or not.

In the cold example, the analysis of the retrospective study can be done immediately, but analysis of the prospective study must wait two months to see if colds develop. Prospective studies often run over a long time period; 5 to 10 years is not unusual. It is not uncommon for subjects to withdraw or be lost to the study. For this reason, it is more difficult to obtain sizeable samples from prospective studies than from retrospective ones. Prospective studies are more informative than retrospective studies. Investigators have more control over the risk factor in prospective studies than in retrospective studies. In prospective studies investigators are often able to obtain information on important confounding variables that bear on the response. Such information is usually unavailable in retrospective studies. The experiment discussed in Section 15.4.3 is an example of a prospective study.

Odds ratios are particularly important in the analysis of experiments involving retrospective studies. In a retrospective study it is unlikely that the cases can be considered a random sample of all persons afflicted with the condition. In the context of our example, we cannot be sure that the 100 selected people with colds are representative of all people with colds. Therefore, in such a study we cannot estimate the proportion of people having the risk factor who have the condition, or the proportion of people without the risk factor who have the condition. Nevertheless, in a retrospective study we are able to measure the odds ratio and we can claim that the sample odds ratio estimates the population odds ratio.

15.6 Mantel–Haenszel Test

Analysts are often called on to interpret k 2×2 contingency tables, related to one another by the fact that each table has the same row and column categories. The k tables usually represent the k levels of a third (categorical) factor in addition to the two-level factors specified by rows and columns. For example, we look in Table 15.8 and Figure 15.11 at data studying the effectiveness of the Salk vaccine for polio protection for $k = 6$ different age groups. Each of the $k = 6$ 2×2 tables in the "**Observed**" column shows the response (paralysis or no.paralysis) for subjects who were or were not vaccinated (vac or no.vac). The complete discussion of the dataset and the table are in Section 15.7.

Table 15.8 Detail for calculation of the Cochran–Mantel–Haenszel test of the polio example. See the discussion in Section 15.7, where we find that the Mantel–Haenszel chi-square test without the continuity correction is 16.54.

Age Group	Vaccination	Observed		Expected		prob	Chi-Square		[1,1] position for Mantel–Haenszel test						
		no.par	par	no.par	par	no.par	chisq	p.chisq	O	E	O–E	var	n	dev	mh
0–4	no.vac	10	24	15.00	19.00	0.294	5.965	0.015	10	15.00	–5.00	4.25	68	–2.42	5.88
	vac	20	14	15.00	19.00	0.588									
5–9	no.vac	3	15	7.20	10.80	0.167	6.806	0.009	3	7.20	–4.20	2.65	45	–2.58	6.65
	vac	15	12	10.80	16.20	0.556									
10–14	no.vac	3	2	3.00	2.00	0.600	0.000	1.000	3	3.00	0.00	0.67	10	0.00	0.00
	vac	3	2	3.00	2.00	0.600									
15–19	no.vac	1	6	3.11	3.89	0.143	4.219	0.040	1	3.11	–2.11	1.12	18	–2.00	3.99
	vac	7	4	4.89	6.11	0.636									
20–39	no.vac	7	5	8.44	3.56	0.583	1.501	0.221	7	8.44	–1.44	1.44	27	–1.20	1.45
	vac	12	3	10.56	4.44	0.800									
40+	no.vac	3	2	3.33	1.67	0.600	0.600	0.439	3	3.33	–0.33	0.22	6	–0.71	0.50
	vac	1	0	0.67	0.73	1.000									

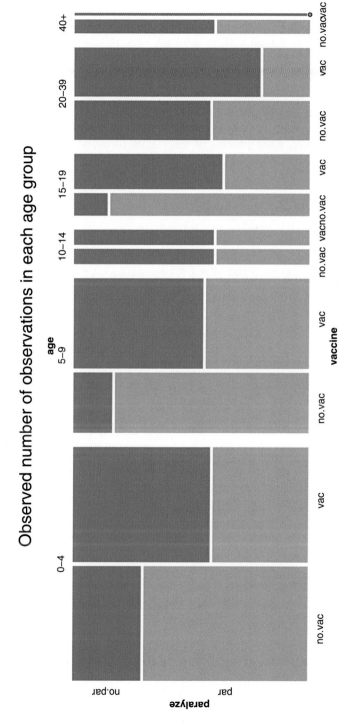

Fig. 15.11 Mosaic plot of the salk dataset. Within each age group we see that the observed proportion of non-paralysis (blue height as a fraction of total height for the bars) for the Vaccinated group is greater than or equal to the proportion of non-paralysis for the Not.Vaccinated group. The bar widths are proportional to the number of people in that age–vaccination grouping. See discussion in Section 15.7.

In earlier sections of this chapter we consider procedures for testing independence of the row and column categories for individual tables. We are now interested in testing the hypothesis that all k tables show the same pattern of relation of the rows to the columns: Either all tables show independence of rows and columns or all show the same dependency structure.

The Mantel–Haenszel test, also referred to as the Cochran–Mantel–Haenszel test, tests the hypothesis that row vs column independence holds *simultaneously in each table*. It is designed to be sensitive to an overall consistent pattern. It has low power for detecting association when patterns of association for some strata are in the opposite direction of other strata.

Let us now look at the algebra of the test statistic. Since we now have k 2×2 tables, we require a third subscript on the n's. Let the kth table be

n_{11k}	n_{12k}	$n_{1.k}$
n_{21k}	n_{22k}	$n_{2.k}$
$n_{.1k}$	$n_{.2k}$	$n_{..k}$

Also define

$$e_{ijk} = \frac{(n_{i.k})\,(n_{.jk})}{n_{..k}}$$

to be the expected (i, j) cell count under independence in table k, and

$$V(n_{11k}) = \frac{n_{1.k}\,n_{2.k}\,n_{.1k}\,n_{.2k}}{n_{..k}^2\,(n_{..k} - 1)} = \frac{e_{11k}\,e_{22k}}{(n_{..k} - 1)}$$

to be the estimated variance of n_{11k} under the assumption of a hypergeometric distribution of a 2×2 table with fixed margins. Then we make a normal approximation and work with

$$n_{11k} \sim N\big(e_{11k}, V(n_{11k})\big)$$

The sum $\sum_k n_{11k}$ is also approximately normal with mean $\sum_k e_{11k}$ and variance $\sum_k V(n_{11k})$. We therefore use as the test statistic the quantity

$$M^2 = \frac{\left[\sum_k n_{11k} - \sum_k e_{11k}\right]^2}{\sum_k V(n_{11k})} \tag{15.12}$$

and the p-value of the test is $1 - \mathcal{F}_{\chi^2}(M^2 \mid 1)$, the corresponding tail percentage of the chi-square distribution with 1 df.

Sometimes we will wish to use a variant of M^2 with a continuity correction and then we use

$$M^2 = \frac{\left[\left|\sum_k n_{11k} - \sum_k e_{11k}\right| - .5\right]^2}{\sum_k V(n_{11k})} \tag{15.13}$$

Inspecting the form of M^2 tells us that significance can occur under either of two conditions:

1. Most or all tables must have the observed $(1, 1)$ cell count at least as large as expected under the null hypothesis.

2. Most or all tables must have the observed $(1, 1)$ cell count at most as small as expected under the null hypothesis.

Equivalently, most or all of the tables must have an odds ratio either

1. at least 1, or

2. at most 1.

Note that M^2 is **not** the same as the chi-square statistic one gets from the 2×2 table formed as the sum of the k tables.

15.7 Example—Salk Polio Vaccine

Chin et al. (1961), also in Agresti (1990), discuss 174 polio cases classified by age of subject, whether or not the subject received the Salk polio vaccine, and whether the subject was ultimately paralyzed by polio. The dataset is in data(salk). We wish to learn if symptom status (paralysis or not) is independent of vaccination status after controlling for age.

Each of the $k=6$ "Observed" subtables in Table 15.8, one for each of $k=6$ age ranges, shows two estimated probabilities of no paralysis, for subjects without vaccine and subjects with vaccine. In the "0–4" subtable, for example, we see $p_{\text{no.vac}}(\text{no.par})=.294$ and $p_{\text{vac}}(\text{no.par})=.588$. In all cases the observed proportion with vaccine is higher. The "chi-square" column shows the ordinary contingency table chi-square for each subtable. The four subtables with older subjects do not have many observations and do not strongly support the conclusion that vaccine is better. The Cochran–Mantel–Haenszel test provides a way of combining the information, properly weighted, from all six subtables to get a stronger conclusion. The "O", "E", and "O−E" columns show the [1,1] or [no.vac,no.par] position from the "Observed" and "Expected" tables. "O–E" is the weighted difference of the row probabilities $O_i - E_i = w_i(p_{no.vac}(no.par) - p_{vac}(no.par))$ [with weights $w_i = 1/(1/n_1 + 1/n_2)$ for the ith subtable, where n_j is the total count on the jth row]. While we choose to focus on the counts of the [1,1] cells, an identical conclusion would be reached if the focus were on any of the three other cells of the 2×2 table.

The "var" column shows the variance of "O–E" under the assumption of the hypergeometric distribution for O_i assuming both row and column margins of the ith table are fixed. The "dev" column is a standardized deviation, $(O-E)/\sqrt{var}$, and the "mh" column is the squared standardized deviation. We plot the standardized deviations in Figure 15.12. The squared standardized deviation is the Mantel–Haenszel statistic for the subtable. The MH statistic for a subtable is very close to the chi-square statistic.

The Cochran–Mantel–Haenszel (CMH) test for the set of all $k=6$ subtables is constructed as a weighted combination of the same components used for the subtable statistics. Since each "O–E" is a random variable with mean and variance, we use Equations (3.8) and (3.9) to combine them. The CMH statistic is constructed from the sum of the "O–E" for the subtables, divided by the standard deviation of the sum, which is the square root of the sum of the variances: $\sum(O-E)/\sqrt{\sum(var)}$. Then the whole is squared. Thus the CMH statistic for this example is

$$\frac{\left(\sum(O-E)\right)^2}{\sum(var)} = \frac{(-5 - 4.20 - 0 - 2.11 - 1.44 - .33)^2}{(4.25 + 2.65 + .67 + 1.12 + 1.44 + .22)} = 16.54$$

For each of the age ranges with a sufficiently large sample, Fisher's exact test performed on the 2×2 tables, shown in Table 15.9, detects a positive association between symptom and vaccination status: Persons vaccinated had a significantly

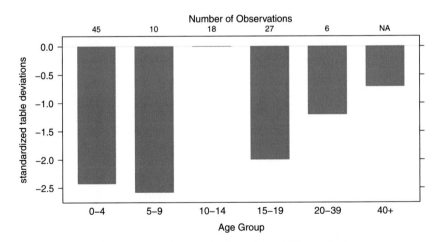

Fig. 15.12 Standardized deviations for individual table Mantel–Haenszel values.

lower incidence of paralysis than persons not vaccinated. In addition, each of the six 2×2 tables has an odds ratio $\big((\text{no.par/par})_{\text{no.vac}}\big)/\big((\text{no.par/par})_{\text{vac}}\big)$ of at most 1. Therefore, the Mantel–Haenszel test can be used. The Mantel–Haenszel test statistic has value 16.54, which is highly significant. This means that the relationship between symptom and vaccination status is consistent over all age ranges.

15.8 Example—Adverse Experiences

Evaluation of adverse experience data is a critical aspect of all clinical trials. This dotplot of incidence and relative risk, and the specific example, are taken from Amit et al. (2008). We proposed graphics for exploratory data analysis or signal identification, and for adverse experiences (AEs, also read as adverse events) that may result from a compound's mechanism of action or events that are of special interest to regulators.

Figure 15.13 is a two-panel display of the AEs most frequently occurring in the active arm of the study. The first panel displays their incidence by treatment group, with different symbols for each group. The second panel displays the relative risk of an event on the active arm relative to the placebo arm, with 95% confidence intervals (as defined in Equation 15.11) for a 2×2 table. The panels have the same vertical coordinates and different horizontal coordinates. R code for the construction of this plot is available as the `AEdotplot` function in the **HH** package.

Table 15.9 Run Fisher's exact test on each of the 2×2 tables in the "Observed" column of Table 15.8. Each of the six 2×2 tables has an odds ratio of at most 1. Therefore, the Mantel–Haenszel test can be used.

```
> data(salk)

> salk2 <- tapply(salk$Freq, salk[c(2,3,1)], c)

> class(salk2) <- "table"

> ## salk2  ## salk2 is structured as a set of 2x2 tables
> lt <- apply(salk2, 3, fisher.test, alternative="less")

> ## odds ratio and p-value
> sapply(lt, '[', c("estimate","p.value"))
          0-4      5-9       10-14  15-19   20-39   40+
estimate 0.2973   0.1669    1      0.1098  0.3645  0
p.value  0.01359  0.009521  0.7381 0.05656 0.2116  0.6667
```

Most Frequent On–Therapy Adverse Events Sorted by Relative Risk

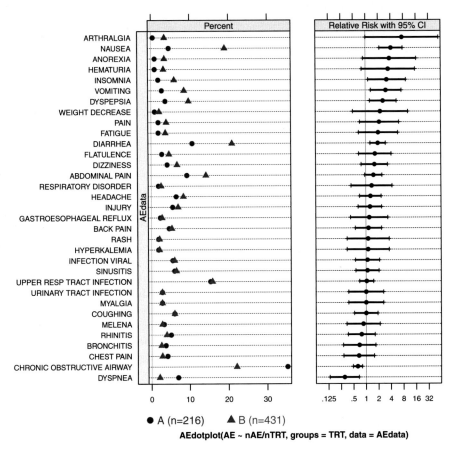

Fig. 15.13 Adverse Events dotplot for the clinical trial discussed in Amit et al. (2008). The left panel shows the observed percents of patients who reported each adverse event. The center panel shows the relative risk of Drug B relative to Drug A with 95% confidence intervals for a 2×2 table. By default, the AEs are ordered by relative risk so that events with the largest increases in risk for the active treatment are prominent at the top of the display.

If the display is not for regulatory purposes, intervals showing ±1 s.e. may be preferred. If confidence intervals are presented, multiple comparison issues should be given consideration, particularly if there is interest in assessing the statistical significance of differences of the relative risk for so many types of events. However, the primary goal of this display is to highlight potential signals by providing an estimate of treatment effect and the precision of that estimate. The criteria for including specific AEs in the display should be carefully considered. In Figure 15.13, the criterion used was that AEs have at least 2% incidence in the active arm.

The graphical presentation of this form of data has a very strong impact. The AEs are ordered by relative risk so that events with the largest increases in risk for the active treatment are prominent at the top of the display. Unlike with a table, it is immediately obvious to the reader which are the most serious AEs. This could be reversed to put the largest increases at the bottom, or the order could be defined by the actual risk in one of the treatment arms rather than the relative risk. Depending on the desired message, other sorting options include: magnitude of relative risk, incidence of event in a given treatment arm, or total incidence of events across all arms. We do not recommend ordering alphabetically by preferred term, which is the likely default with routine programming, because that makes it more difficult to see the crucial information of relative importance of the AEs.

15.9 Ordered Categorical Scales, Including Rating Scales

Ordered Categorical Scales, including Rating scales such as Likert scales, are very common in marketing research, customer satisfaction studies, psychometrics, opinion surveys, population studies, and numerous other fields. We recommend diverging stacked bar charts as the primary graphical display technique for Likert and related scales. We also show other applications where diverging stacked bar charts are useful. Many examples of plots of Likert scales are given. We discuss the perceptual issues in constructing these graphs.

An ordered categorical scale is an ordered list of mutually exclusive terms. Ordered categorical scales are used, for example, in questionnaires where each respondent is asked to choose one of the terms as a response to each of a series of questions. The usual summary is a table that shows the number of respondents who chose each term for each question. The summary table is a special case of a contingency table where the rows are individual questions and the columns are the ordered set of potential responses.

A rating scale is a form of psychometric scale commonly used in questionnaires. The most familiar rating scale is the Likert scale (Likert, 1932), which consists of a discrete number of choices per question among the sequence: "strongly disagree", "disagree", "no opinion", "agree", "strongly agree". Likert-type scales may use other sequences of bipolar adjectives: "not important" to "very important"; "evil" to "good". These scales sometimes have an odd number of levels, permitting a neutral choice. Sometimes they have an even number of levels, forcing the respondent to make a directional choice. Some ordered categorical scales are uni-directional— age ranges or population quantiles, for example—for which negative and neutral interpretations are not meaningful.

For concreteness we present in Section 15.9.1 a dataset from a survey for which a natural display is a coordinated set of diverging stacked bar charts. We introduce the dataset by showing and discussing a multi-panel plot of the entire dataset. Then we move to the construction and interpretation of individual panels.

15.9.1 Display of Professional Challenges Dataset

Our primary data example is from an *Amstat News* article (Luo and Keyes, 2005) reporting on survey responses to a question on job satisfaction. A total of 565 respondents replied to the survey. Each person answered one of five levels of agreement or disagreement with the question "Is your job professionally challenging?" The respondents were partitioned into nonoverlapping subsets by several different criteria. For each of the criteria, the original authors were interested in comparing the percent agreement by that criterion's groups.

In Figure 15.14, we show the complete results of the survey as a coordinated set of diverging stacked bar charts. In this section we concentrate on the appearance of the plot for its function of representing the meaning of the dataset.

There are six panels in the plot. The top panel shows "All Survey Respondents". The remaining panels show different partitions of the 565 respondents. In the second panel from the top, for example, the criterion name "Employment sector" is in the left strip label. The respondents self-identify to one of the five employment groups named in the left tick labels. The number of people in each group is indicated as the right-tick label. Each stacked bar is 100% wide. Each is partitioned by the percent of that employment group who have selected the agreement level indicated in the legend below the body of the plot. The legend is ordered by the values of the labels. Darker colors indicate stronger agreement. Gray indicates the neutral position, in this example, "No Opinion". The bar for the neutral position is split, half to the left side of the vertical zero reference line and half to the right side. The reference line is placed behind the bars to prevent it from artificially splitting the neutral bar into two pieces. The default color palette has red on the left for disagreement and blue on the right for agreement. See Section 15.9.2 for a discussion of color palettes.

The intent of this plot is to compare percents within subgroups of the survey population; consequently we made all bars have equal vertical thickness. The panel heights are proportional to the number of bars in the panel. The *x*-axis labels are displayed with positive numbers on both sides. The bars within each panel have been sorted by the percent agreeing (totaled over all levels of agreement). We usually prefer horizontal bars, as shown here, because the group labels and the names of the groups are easier to read when they are displayed horizontally on the *y*-axis.

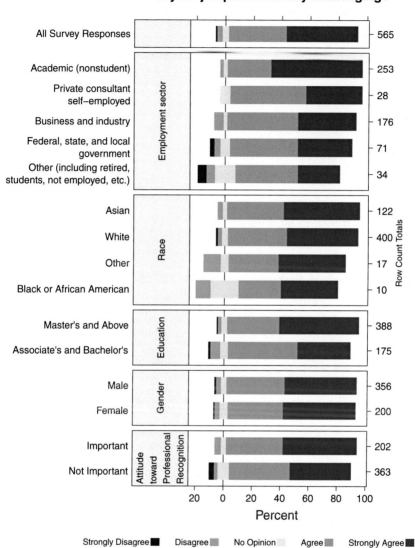

Fig. 15.14 Survey responses to a question on job satisfaction (Luo and Keyes, 2005). A total of 565 respondents replied to the survey. Each person answered one of five levels of agreement or disagreement with the question "Is your job professionally challenging?" Each panel of the plot shows a breakdown of the respondents into categories defined by the criterion listed in its left strip label.

Table 15.10 The respondents have been divided into five employment categories. The rows (employment categories) are displayed in the original order: alphabetical plus other. Columns are displayed sequentially, with disagreement to the left and agreement to the right.

	Strongly Disagree	Disagree	No Opinion	Agree	Strongly Agree
Employment Sector					
Academic (nonstudent)	0	5	8	78	162
Business and industry	0	11	5	88	72
Federal, state, an local government	2	3	5	34	27
Private consultant/self-employed	0	0	2	15	11
Other (including retired, students, not employed, etc.)	2	2	5	15	10

15.9.2 Single-Panel Displays

In this section, we look at the data for just the Employment panel of the full plot in Figure 15.14. Table 15.10 shows the respondents divided into five employment categories and the counts for each agreement level within each employment category.

Diverging stacked bar charts are easily constructed from Likert scale data. Each row of the table is mapped to a stacked bar in a bar chart. Usually the bars are horizontal. The counts (or percentages) of respondents on each row who agree with the statement are shown to the right of the zero line in one color; the counts (or percentages) who disagree are shown to the left in a different color. Agreement levels are coded from light (for closer to neutral) to dark (for more distant from neutral). The counts (or percentages) for respondents who neither agree nor disagree are split down the middle and are shown in a neutral color. The neutral category is omitted when the scale has an even number of choices. Our default color palette is the (Red–Blue) palette constructed by the `diverge_hcl` function in the **colorspace** package in R. The colors in the diverging palettes have equal intensity for equal distances from the center. The base colors Red and Blue have been chosen to avoid ambiguity for those with the most prevalent forms of color vision deficiencies.

It is difficult to compare lengths without a common baseline; see pages 54–57 of Robbins (2013) and the reference therein to Cleveland and McGill (1984). We are primarily interested in the total count (or percent) to the right or left of the zero line; the breakdown into strongly or not is of lesser interest so that the primary comparisons do have a common baseline of zero.

Figure 15.15 shows a direct translation of the counts in Table 15.10 to a plot. The strongest message in this presentation is that the sample has a very large percentage of academics. It is harder to compare relative proportions in the employment categories because the total counts in each row are quite disparate.

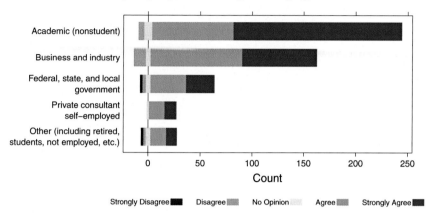

Fig. 15.15 This plot is a direct translation of the numerical values in Table 15.10 to graphical form. Blue is agree, red is disagree, gray is no opinion. The strongest message in this presentation is that the sample has a very large percentage of academics.

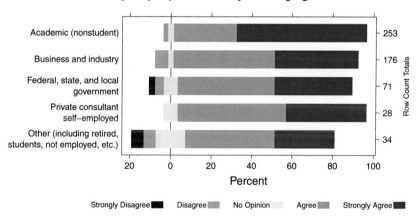

Fig. 15.16 In this variant of the plot, we display the row percents. We don't want to lose information about the uneven selection of respondents from the employment sectors so we display the counts as the right axis labels. Now we see that "Academic (nonstudent)" stands out as the largest percentage of dark blue on the graph.

Figure 15.16 displays the percents within each row. Now it is easy to see that a large majority of the people in all employment categories have a positive answer to the survey question. We don't want to lose the disparity in row totals, so we use the row count totals as the right-axis tick labels.

For plots such as Figure 15.16 with a single panel, and also for multiple-panel plots where the rows are distinct in each panel, we can still do better. Figure 15.17 shows the same scaling as Figure 15.16, but this time the row order is data-

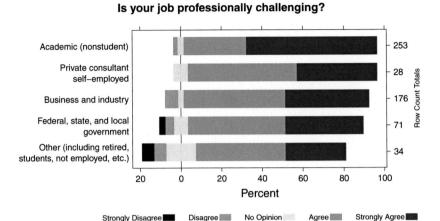

Fig. 15.17 In our third presentation of the same table, we now sort the rows of the table by the total percent agree (dark blue + light blue + $\frac{1}{2}$ gray). Data-dependent ordering is usually more meaningful than alphabetical ordering for unordered categories.

dependent. Rows are now ordered by the total percent of positive responses. This allows the reader to recognize groupings among rows (in this example, groupings of employment categories) that show similar responses.

15.9.3 Multiple-Panel Displays

Our illustration in Figures 15.15–15.17 is a single panel. It displays information on a single question for a partition of the respondents into several groups based on employment. Figure 15.14 is a multiple-panel display containing Figure 15.17 combined with other partitions of the same set of respondents.

15.9.3.1 One Question with Multiple Subsets of the Sample

Figure 15.14 shows responses to the same question for the same population of respondents partitioned into several series of groups based on additional characteristics.

The partitions have different numbers of groups. In order to retain the same vertical spacing between parallel bars, the vertical space allocated for the panels must differ. The different panels have been labeled by the name of the partitioning characteristic. In this example the panels are identified by a left strip label. Within each panel the bars have been sorted by the total percent of positive responses.

15.9.3.2 One or More Subpopulations with Multiple Questions

Figure 15.18 is differently structured. The data are the responses to a survey sponsored by the New Zealand Ministry of Research Science and Technology (New Zealand Ministry of Research Science and Technology, 2006). Here we have two different sets of questions that have been asked of the same set of respondents. Sometimes the respondents can be subdivided. If we had the data separately for the Male and Female subsets, we could have a plot similar to Figure 15.18 but with two columns of panels, left and right, one for each subset.

15.9.3.3 Common Structure at Multiple Times—Population Pyramids

Population pyramids are used in demographic studies and in epidemiological studies. The pyramid is a pair of back-to-back bar charts, one for males and one for females. We display the population pyramid as a Likert-type scale with two levels, male and female, for each age range. Figure 15.19 shows five pyramids at ten-year intervals years 1939–1979, with the y-labels on both the left and right axes. A **shiny** app that cycles through all years 1900–1979 is available by entering

```
shiny::runApp(system.file("shiny/PopulationPyramid",
                          package="HH"))
```

The data is from the USAage dataset in the **latticeExtra** package.

15.10 Exercises

15.1. Hand et al. (1994) revisit a dataset attributed to Karl Pearson, `data(crime)`, that examines the relationship between type of `crime` committed and whether the perpetrator was a `drinker` or abstainer. Investigate whether these two classifications are independent, and if they are not, discuss the nature of the dependence.

15.2. Senie et al. (1981), also in Hand et al. (1994), investigate whether the frequency of breast self-examination is related to age group. The data appear accessed as `data(selfexam)`. Do you agree that there is a relationship? If so, describe it.

15.3. Sokal and Rohlf (1981), later in Hand et al. (1994), concern an experiment to determine the preference of invading ant colonies on two species of acacia tree. A total of 28 trees were made available for the study, 15 of species A and 13 of species B. Initially each tree was treated with insecticide to remove all existing colonies. Then 16 ant colonies were invited to invade any of the trees they chose. By construction, the 2×2 data, in `data(acacia)`, have both margins fixed. Use

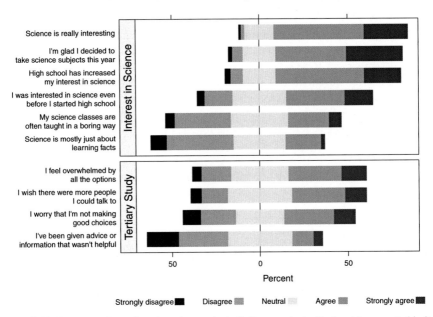

Fig. 15.18 Two sets of questions have been asked of all respondents. Each set is presented in its own panel.

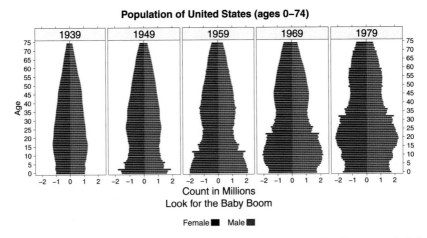

Fig. 15.19 Five population pyramids at ten-year intervals years 1939–1979. We can see the baby boom start at the bottom of the population graph for 1949 and work its way up over time. We have placed the age tick labels on both the left and right axes of the set of panels.

Fisher's exact test to determine if the ants have a significant preference for one species over the other.

15.4. Fleiss (1981) presents the dataset in `data(mortality)` concerning mortality following 37,840 live births to nonwhite mothers in New York City in 1974. In the file, rows are birth weights, ≤2500 grams or >2500 grams, and columns are outcomes one year after birth, dead or alive. 2500 grams is 5.5 pounds. Construct and carefully interpret the sample odds ratio for these data and construct a 95% confidence interval for the population odds ratio.

15.5. Wynder et al. (1958), later in Fleiss (1981), report a retrospective study of factors associated with cancer of the oral cavity. In this study there were 34 women with this cancer and 214 women, matched by age, without it. It was found that 24% of the cases but 66% of the controls were nonsmokers. The dataset is available as `data(oral)`. Construct and carefully interpret the sample odds ratio for these data and then construct a 95% confidence interval on the population odds ratio.

15.6. Braungart (1971) refers to `data(political)`, also found in Bishop et al. (1975), in which 271 college students of the 1960s who admitted to extreme political leanings were cross classified according to the style of parental decision making they received, authoritarian or democratic, and their political leaning, left or right. Construct and carefully interpret the sample odds ratio for these data and also construct a 95% confidence interval on the population odds ratio.

Table 15.11 Jury pool composition data for the Rotorua and Nelson districts.

	Rotorua		Nelson		Combined	
	Maori	Non-Maori	Maori	Non-Maori	Maori	Non-Maori
Jury pool	79	258	1	56	80	314
Others	8,810	23,751	1,328	32,602	10,138	56,353

15.7. Westbrooke (1998) discusses claims that Maori are underrepresented on juries in districts in New Zealand. Jury pool composition data for the Rotorua and Nelson districts are shown in Table 15.11, and accessed as `data(jury)`, along with totals for these two districts combined.

From this table it is easy to verify the following:

- The population of Rotorua is 27.0% Maori, but this district's jury pool is only 23.4% Maori.

- The population of Nelson is 3.9% Maori, but this district's jury pool is only 1.7% Maori.

- However, the combined population of these two districts is 15.3% Maori, but the combined jury pools of these districts is 20.3% Maori.

Discuss whether Maori are indeed underrepresented on the juries of these two districts.

15.8. Brochard et al. (1995) report a prospective study of patients with chronic obstructive pulmonary disease, assessing the effect of noninvasive ventilation therapy on reducing the need for subsequent invasive intubation. A total of 85 patients were recruited at five centers. The data are in Table 15.12 and dataset data(intubate). The data in the file are arranged differently than in the table: The first column is the center number; the second column entries are yes if received ventilation therapy and no if didn't receive this therapy; the third column entries are yes if required invasive intubation and no if didn't require invasive intubation; and the entries in column 4 are the number of patients in the categories specified in columns 1–3.

Compute the odds ratio at each center. Use the Mantel–Haenszel test to produce a carefully stated conclusion of the combined data.

Table 15.12 Chronic Obstructive Pulmonary Disease.

	Center									
	1		2		3		4		5	
Ventilation therapy	invasive intubation	not inv	invasive intubation	not inv	invasive intubation	not inv	invasive intubation	not inv	invasive intubation	not inv
ventilation	3	6	2	3	1	7	0	5	5	11
not vent	9	0	5	1	4	5	3	1	10	4

Chapter 16

Nonparametrics

16.1 Introduction

Most of the statistical procedures we've introduced in the previous 15 chapters require an assumption about the form of a probability distribution, often the Normal distribution. When such assumptions are unjustified, the consequences of the procedure are dubious at best. In situations where distributional assumptions cannot be justified, even after a well-chosen data transformation, the analyst should consider another approach.

Nonparametric statistical procedures are ones that do not require the assumption of a specific probability distribution. (In contrast, procedures that do make distributional assumptions are referred to as parametric procedures.) In exchange for not requiring a detailed distributional assumption, nonparametric testing procedures have less power, and nonparametric confidence intervals are wider than their parametric analogues in the same problem situation using the same error control. In addition, the parameter(s) of interest in a nonparametric procedure may not be identical to those of corresponding parametric procedures. For example, we may use a nonparametric procedure for comparing two population medians as a substitute for a parametric procedure for comparing two means. For these reasons, parametric procedures are preferred if their assumptions are reasonably well met.

In the previous chapters we have discussed a number of procedures for inferring about one or more population means. Those procedures required assumptions that underlying populations are normally distributed, at least approximately, and that when inferring about the means of two or more populations, these populations have a common variance. Often transformations such as those discussed in Chapter 4 can be used to make such assumptions tenable once the transformations have been applied. Sometimes this is not possible, for example, when the data contain outliers whose elimination cannot be justified. In such instances, a nonparametric approach

© Springer Science+Business Media New York 2015
R.M. Heiberger, B. Holland, *Statistical Analysis and Data Display*,
Springer Texts in Statistics, DOI 10.1007/978-1-4939-2122-5_16

may be considered. Our discussion here is limited to hypothesis tests. Analogous confidence interval estimates cannot be described succinctly and are discussed in Lehmann (1998) and Desu and Raghavarao (2003).

Many nonparametric procedures use statistics based on *ranks* of the data rather than the data themselves. Such procedures require only that the data be on at least an ordered scale. In contrast, parametric procedures generally carry the more stringent requirement that the data be measured on an interval or ratio scale. The various scale types are defined in Section 2.1.

We've actually encountered some nonparametric procedures in earlier chapters. For example, the two-sample Kolmogorov–Smirnov goodness-of-fit test discussed in Section 5.9 does not require that we specify the natures of the two populations being sampled. In this chapter we introduce some additional commonly used non-parametric procedures. Our examples include checks of the assumptions of competing parametric procedures and comparisons of the nonparametric and parametric results.

16.2 Sign Test for the Location of a Single Population

The example in Section 5.2.1 discusses a parametric approach to a problem involving `data(vocab)`, concerning whether $\mu = 10$ is consistent with a random sample of 54 test scores. Since Figure 5.3 shows that the sample had one high outlier, there was at least some doubt that the assumptions underlying the parametric analysis were correct.

The nonparametric approach here is the *sign test* to assess the hypothesis that the population median equals 10. If the population is not symmetric so that its mean and median are not identical, then the analogy between the nonparametric and parametric inferences is imperfect. We look again at the data in Table 16.1.

The logic of this nonparametric test stems from the insight that if the population median η is indeed 10, we would expect half of the sample values not exactly equal to 10 to fall on either side of 10. If appreciably more than half exceed 10, this suggests that the median exceeds 10. Let n be the number of sample items different from 10 and m be the number of these exceeding 10. The formal test of

$$H_0: \eta = 10$$

vs

$$H_1: \eta > 10$$

is based on the distribution of a binomial random variable X with n trials and success probability .5. The test has p-value $= P(X \geq m)$. The two-sided test, with the same null hypothesis but having alternative hypothesis

Table 16.1 $p = P(X \geq 49) = P(X \geq 48.5) = 1 - \mathcal{F}_{Bi}(48.5 \mid 50, .5) \approx 4{10}^{-14}$

```
> data(vocab)

> table(vocab)
vocab
 9 10 11 12 13 14 15 16 17 19
 1  4 13  7  9  9  4  5  1  1

> table(vocab$score - 10)

-1  0  1  2  3  4  5  6  7  9
 1  4 13  7  9  9  4  5  1  1

> table( sign(vocab$score-10) )

-1  0  1
 1  4 49

> pbinom(48.5, 50, .5, lower=FALSE)
[1] 4.53e-14

> 1 - pbinom(48.5, 50, .5)
[1] 4.53e-14
```

$$H_1 : \eta \neq 10$$

has p-value $2P\big(X \geq \max(n - m, m)\big)$.

This test is called the sign test because it is based on the arithmetic signs of the differences between the data and the null hypothesized median. Some authors refer to it as the binomial test.

In data(vocab), 4 of the 54 scores equalled the null value $\eta=10$, and $n=50$ scores were not exactly equal to the null value of $\eta=10$. We observed $m=49$ scores that exceeded 10 and $n - m=1$ score less than 10. If the null were true, then X comes from the distribution Bin($n = 50, p = .5$) with discrete density shown in Figure 16.1. The one-sided p-value is the probability of observing 49 or more (that is, 49 or 50) larger scores from this distribution. This probability is the sum of the probabilities for the bars at 49 and 50. Therefore, the one-sided p-value is

$$P(X \geq 49) = 1 - \mathcal{F}_{Bi}(48 \mid 50, .5)$$

using the binomial distribution with $n = 50$ and $p = .5$. This number $p \approx 4{10}^{-14}$ is calculated in R by

```
pbinom(48.5, 50, .5, lower.tail=FALSE)
```

a. Binomial(n=50, p=.5)

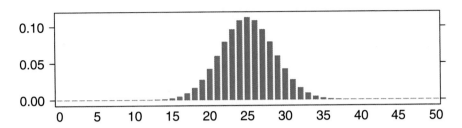

b. Binomial(n=50, p=.5) on log scale

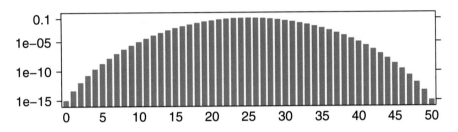

c. magnified Binomial(n=50, p=.5)
to emphasize x=49 and x=50

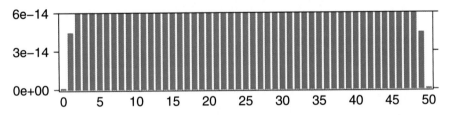

Fig. 16.1 Panel a shows the Bin($n = 50, p = .5$) discrete density. Panel b shows the same discrete density on a log scale. Panel c magnifies (and truncates) the probability scale to emphasize that there are nonzero values at $x = 49$ and $x = 50$. The p-value for the sign test is the probability that a random selection from this distribution would yield the observed $x = 49$ or the larger $x = 50$. From the graph we see that this is the very small number $p \approx 4_{10}^{-14}$.

$p \approx 4_{10}^{-14}$ is overwhelming evidence that the population median η exceeds 10, consistent with the strong parametrically based conclusion in Section 5.2.1 that the population mean μ exceeds 10.

The procedure we just developed is an example of an *exact test* or of a *randomization test*. In this form of testing procedure, a discrete model for the distribution of the observed statistic under the null hypothesis is postulated. Then the probability of obtaining the observed value (or larger) from that model is calculated and used as the p-value of the test.

16.3 Comparing the Locations of Paired Populations

Just as the parametric paired t-test is equivalent to a one-sample t-test on the pair differences, the locations of paired populations can be compared by inferring about the median of the population consisting of the differences (in a consistent direction) between the two individual populations.

16.3.1 Sign Test

Exercise 5.13 requests a comparison of the population means of pre- and post-treatment measurements based on a random sample of 48 patients. However, the density and histogram plots of the post-treatment measurements in Figure 16.2 show a number of both low and high outliers, suggesting that the post population may not be close to normally distributed, and calling into question the validity of an analysis based on a two-sample t-test. Therefore, a nonparametric approach should be considered.

We show the sign test for paired observations in Table 16.2. Analogous to Section 16.2, let n be the number of pairs of observations, excluding tied pairs. For testing against the one-sided alternative hypothesis that the post-treatment median is less than the pretreatment median, let m be the number of sample pairs where the post-treatment measurement is less than the pretreatment measurement. Then for binomially distributed X with n trials and success probability 0.5, the p-value is $P(X \leq m)$. For these data, $n = 46, m = 15$ and we use R to calculate p =pbinom(15,46,.5) $\approx$ .0129. This p-value indicates moderate evidence that in the population of patients from which this sample was selected, the median post-treatment angle is less than the median pretreatment angle. For the analogous parametric paired t-test requested in Exercise 5.13, the p-value is considerably less than 0.01. Because we question the validity of the t-test, we believe the p-value from the t-test is way too small.

Table 16.2 Sign test applied to relative rotation angle data. The conclusion of the test is that the median post-treatment angle is less than the median pretreatment angle.

```
> table(sign(har1$Post - har1$Pre))

-1  0  1
31  2 15

> pbinom(15, 46, .5, lower=TRUE)
[1] 0.01295
```

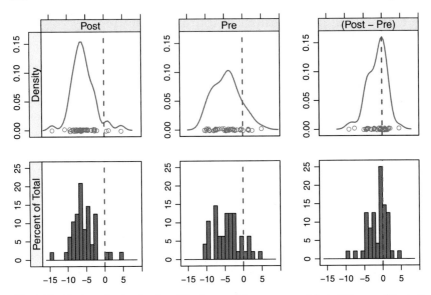

Fig. 16.2 We show the density plot of the Post and Pre scores, and their difference (Post-Pre), in the top set of panels. We show the histogram of the same variables in the bottom panels. The individual jittered points in the density plots may be thought of as a vertical view of the histogram. Both graphical displays show that the data does not appear to be normally distributed. The Post scores have outliers are well outside the vicinity of the bulk of the data. The Pre scores have a relatively flat distribution. Post scores are generally smaller than Pre scores. The (Post-Pre) differences are predominantly negative.

16.3.2 Wilcoxon Signed-Ranks Test

The sign test in Section 16.3.1 uses the arithmetic sign of differences between members of each pair but ignores the magnitude of these differences. Such information can be crucial, particularly if the samples contain outlying values. The Wilcoxon signed-ranks test uses information on the magnitude of the differences and therefore is more sensitive than the sign test. The Wilcoxon test assumes an interval or ratio measurement scale involving continuous variables. If the data are ordinal but not interval, the Wilcoxon test cannot be used.

The construction of the signed-rank test for data(har1) is depicted in Table 16.3. For testing hypotheses involving the difference between the medians of two populations, say the X = har1$Pre population and Y = har1$Post population, let $D_i = X_i - Y_i$ = har$diff be the observed difference for the ith pair. Let $n = 46$ be the number of such differences that are not zero; hereafter ignore any zero differences. Rank the n nonzero absolute differences ($|D_i|, i = 1, \ldots, n$) = har$abs in ascending order from 1 to n and store the ranks in har$rank. (If there are ties, assign the average rank. For example, if the first and second largest absolute differences are both equal to .3, assign rank 1.5 to both absolute differences.) Copy the ranks for the positive differences into har$prnk. Then the test statistic is the sum of the ranks in har$prnk. The idea here is that if the population medians are equal, we expect half of the differences to be positive and so the test statistic should be approximately one half the sum of the ranks of the nonzero differences, or $n(n + 1)/4$, where n is now the number of nonzero differences. A test statistic much different from this value suggests that the difference between the population medians is nonzero. The signed rank distribution is discussed in Section J.3.7. We usually show the normal approximation. The variance is related to the variance of the discrete uniform distribution (see Appendix J.3.1). We outline the calculation of the variance in Exercise 16.6.

Figure 16.3 shows the differences and the ranks of the absolute values of these differences used in a Wilcoxon signed-ranks test for the data in data(har1). The positive differences (between the pre- and post-treatment angles) in this example generally have larger absolute values than the negative differences. The magnitudes of these differences is relevant to the test result—a large positive difference is greater evidence that pre exceeds post than a small positive difference. These magnitudes are ignored by the sign test but accounted for by the Wilcoxon signed-ranks test.

Assuming one-sided alternative hypothesis as in Section 16.2, this test is run in R with

```
wilcox.test(X, Y, alternative="greater",
            paired=TRUE, exact=TRUE)
```

If restrictive conditions are met, R calculates the p-value based on the exact distribution of the test statistic. These conditions are $n < 50$ and no tied ranks. Otherwise, as in the analysis of the data(har1) dataset, the p-value is based on a normal approximation. The results of the wilcox.test command applied to data(har1) are shown in Table 16.5.

The fact that the p-value for the Wilcoxon test is much less than that of the sign test for the same data demonstrates that the signed-ranks test is more powerful than the sign test.

Table 16.3 Detail for construction of the Wilcoxon signed-ranks test applied to relative rotation angle data. The diff column contains the differences Pre - Post. The abs column contains the absolute value of the differences. NA values were assigned to the 0-valued differences. This has the effect of placing them last in the sort order. Only nonzero differences are used in the test. In this example with 48 observations, there are $n = 46$ nonzero differences. The rank column contains the ranks of the absolute differences. The prnk column contains the same ranks, but only for rows that have positive differences. The test statistic in Tables 16.5 and 16.4 is the sum of the prnk column.

```
> har <- data.frame(diff=har1$Pre - har1$Post)

> har$abs <- abs(har$diff)

> har$abs[har$abs==0] <- NA

> har$rank <- rank(har$abs)

> har$rank[har$diff == 0] <- 0

> har$prnk <- har$rank     ## rank for positive differences

> har$prnk[har$diff < 0] <- 0

> har[order(har$abs),]                    ## manually edit into columns
     diff abs rank prnk                      diff abs rank prnk
4    0.3 0.3  1.5  1.5            11 -1.9 1.9 25.0  0.0
22   0.3 0.3  1.5  1.5            33  2.1 2.1 26.0 26.0
13  -0.4 0.4  3.5  0.0            18  2.2 2.2 27.0 27.0
35   0.4 0.4  3.5  3.5            36  2.2 2.2 28.5 28.5
1    0.4 0.4  5.0  5.0            44 -2.2 2.2 28.5  0.0
17  -0.4 0.4  7.0  0.0            46  2.3 2.3 30.0 30.0
21  -0.4 0.4  7.0  0.0            25  2.3 2.3 31.0 31.0
39   0.4 0.4  7.0  7.0            5   2.9 2.9 32.0 32.0
19   0.5 0.5  9.5  9.5            43  3.2 3.2 33.0 33.0
45   0.5 0.5  9.5  9.5            6   3.2 3.2 34.0 34.0
16   0.7 0.7 11.5 11.5            23  3.8 3.8 35.0 35.0
26  -0.7 0.7 11.5  0.0            31  3.9 3.9 36.0 36.0
12  -0.7 0.7 13.0  0.0            47  4.1 4.1 37.0 37.0
15  -0.8 0.8 14.0  0.0            40  4.2 4.2 38.0 38.0
42   0.9 0.9 15.0 15.0            38  4.3 4.3 39.0 39.0
37   0.9 0.9 16.0 16.0            14  4.5 4.5 41.0 41.0
20  -1.0 1.0 17.0  0.0            24  4.5 4.5 41.0 41.0
28  -1.1 1.1 18.0  0.0            30  4.5 4.5 41.0 41.0
27  -1.1 1.1 19.0  0.0            34 -4.6 4.6 43.0  0.0
48  -1.1 1.1 20.0  0.0            29  5.0 5.0 44.0 44.0
10   1.4 1.4 21.0 21.0            8   7.5 7.5 45.0 45.0
32  -1.5 1.5 22.0  0.0            41  9.0 9.0 46.0 46.0
2    1.7 1.7 23.0 23.0            3   0.0  NA  0.0  0.0
7   -1.7 1.7 24.0  0.0            9   0.0  NA  0.0  0.0
```

Table 16.4 The sum of the positive ranks from Table 16.3 is first calculated. Then we construct the normal approximation. These values match the values in Table 16.5.

```
> ## calculate the statistic
> sum.prnk <- sum(har$prnk)

> n <- sum(har$diff != 0)

> sum.prnk
[1] 808.5

> n
[1] 46

> ## normal approximation
> mean.prnk <- n*(n+1)/4

> numerator <- sum.prnk - mean.prnk

> var.prnk1 <- n * (n + 1) * (2 * n + 1)/24

> NTIES <- table(har$abs[1:46]) ## non-zero differences

> TIEadjustment <- sum(NTIES^3 - NTIES)/48

> var.afterTIES <- var.prnk1 - TIEadjustment

> z <- (numerator - .5) / sqrt(var.afterTIES)

> z
[1] 2.923

> p.val <- pnorm(z, lower.tail = FALSE)

> p.val
[1] 0.001733
```

Table 16.5 Wilcoxon signed-ranks test applied to relative rotation angle data. The conclusion of the test is that the median post-treatment angle is less than the median pretreatment angle.

```
> wilcox.test(har1$Pre, har1$Post, alternative="greater",
+             paired=TRUE, exact=FALSE)

    Wilcoxon signed rank test with continuity correction

data:  har1$Pre and har1$Post
V = 808.5, p-value = 0.001734
alternative hypothesis: true location shift is greater than 0
```

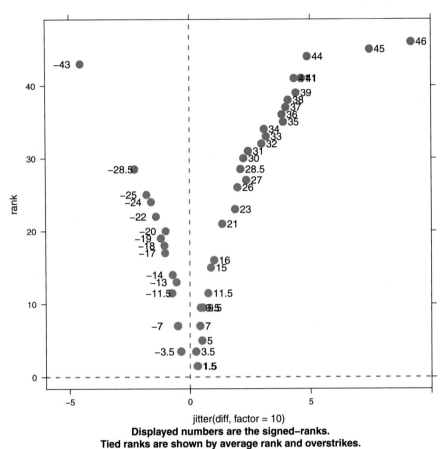

jitter(diff, factor = 10)
Displayed numbers are the signed–ranks.
Tied ranks are shown by average rank and overstrikes.

Fig. 16.3 The differences and ranks from Table 16.3 used in the Wilcoxon test of data(har1). Observe that the positive differences between pre- and post-treatment angles tend to have larger magnitudes than the negative differences. The Wilcoxon test, which takes account of these magnitudes, has more power than the sign test, which ignores the magnitudes. The test statistic is the sum of the ranks for positive differences (the sum of the displayed numbers in the upper-right quadrant).

16.4 Mann–Whitney Test for Two Independent Samples

Some authors refer to this test as the Wilcoxon–Mann–Whitney test because Frank Wilcoxon initiated its development. It is analogous to the parametric two-sample t-test but compares medians rather than means. We wish to compare the medians of two populations based on independent random samples from each. It is assumed that the measurement scale is at least ordinal. Combine the two samples and then rank the resulting observations in ascending order. As in Section 16.3.2, if there are tied observations, each should be assigned the average rank. The test statistic

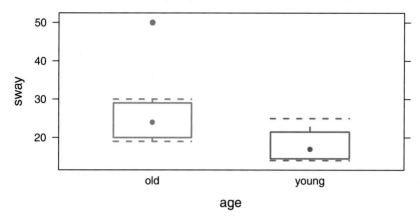

Fig. 16.4 Balance data. The extreme outlier in the `age=="old"` group makes the two-sample *t*-test invalid. See Table 16.6.

T is the sum of the ranks of the smaller sample. Assume the smaller sample is assigned index 1 and the larger sample index 2. Denoting the observed value of T as t_{calc}, the *p*-value depends on the form of the alternative hypothesis. If the alternative hypothesis is $\eta_1 > \eta_2$, then *p*-value = $P(T > t_{\text{calc}})$. For the alternative $\eta_1 < \eta_2$, the *p*-value is = $P(T < t_{\text{calc}})$. For the two-sided alternative, *p*-value = $2 \min\big(P(T > t_{\text{calc}}), P(T < t_{\text{calc}})\big)$.

The data in `data(balance)`, taken from Teasdale et al. (1993), illustrate this test. These authors sought to compare the forward/backward sway in mm of 9 elderly and 8 young subjects who took part in an investigation of their reaction times. We see in Figure 16.4 that one of the `elderly` measurements is an extreme outlier, and if it is retained in the analysis the two-sample *t*-test is invalid. The alternative hypothesis is that the median sway of elderly subjects exceeds the median sway of young subjects. The R analysis uses `wilcox.test` as in Section 16.3.2.

In Table 16.6 we list the two samples and their ranks after they are combined. The analysis with the R function `wilcox.test` is in Table 16.7 and "by hand" with R in Table 16.8. The Wilcoxon distribution is described in Section J.3.8. We conclude that there is moderate evidence that, on average, elderly subjects have greater sway than young subjects.

Table 16.6 Mann–Whitney test applied to balance data. See Figure 16.4.

old		young	
sway	ranks	sway	ranks
19	6.5	25	13.5
30	16.0	21	9.5
20	8.0	17	4.5
19	6.5	15	3.0
29	15.0	14	1.5
25	13.5	14	1.5
21	9.5	22	11.0
24	12.0	17	4.5
50	17.0		

49.0 = rank sum, smaller sample

Table 16.7 Mann–Whitney test applied to balance data by R. R does not compute the exact test when some of the ranks are tied. The ranks are plotted in Figure 16.5.

```
> wilcox.test(balance$sway[balance$age=="young"],
+             balance$sway[balance$age=="old"],
+             alternative="less", exact=FALSE)

Wilcoxon rank sum test with continuity correction

data:  balance$sway[balance$age == "young"] and
       balance$sway[balance$age == "old"]
W = 13, p-value = 0.01494
alternative hypothesis: true location shift is less than 0
```

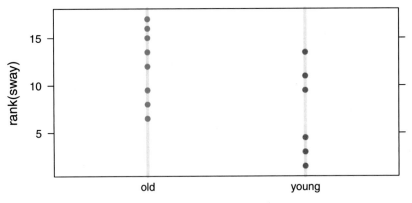

Fig. 16.5 Balance data. The plot of the ranks shows a distinct difference in the medians ranks of the two age groups.

Table 16.8 Mann–Whitney test applied "by hand" to balance data.

```
> data(balance)

> unname(n.old <- table(balance$age)["old"])
[1] 9

> unname(n.young <-  table(balance$age)["young"])
[1] 8

> (r <- rank(balance$sway))
 [1]  6.5 16.0  8.0  6.5 15.0 13.5  9.5 12.0 17.0 13.5  9.5
[12]  4.5  3.0  1.5  1.5 11.0  4.5

> (NTIES <- table(r))
r
 1.5    3  4.5  6.5    8  9.5   11   12 13.5   15   16   17
   2    1    2    2    1    2    1    1    2    1    1    1

> unname(tie.adjustment <-
+           sum(NTIES^3 - NTIES) /
+           ((n.old + n.young) * (n.old + n.young - 1)))
[1] 0.1103

> unname(SIGMA <-
+           sqrt((n.old * n.young/12) *
+                ((n.old + n.young + 1) - tie.adjustment)))
[1] 10.36

> unname(STATISTIC.old <-
+           c(W = sum(r[1:9]) - n.old * (n.old + 1)/2))
[1] 59

> unname(z.old <- STATISTIC.old - n.old * n.young/2)
[1] 23

> unname(z.old <- (z.old - .5)/SIGMA)
[1] 2.172

> unname(pnorm(z.old, lower.tail = FALSE))
[1] 0.01494

> unname(STATISTIC.young <-
+           c(W = sum(r[10:17]) - n.young * (n.young + 1)/2))
[1] 13

> unname(z.young <- STATISTIC.young - n.old * n.young/2)
[1] -23

> unname(z.young <- (z.young + .5)/SIGMA)
[1] -2.172

> unname(pnorm(z.young))
[1] 0.01494
```

16.5 Kruskal–Wallis Test for Comparing the Locations of at Least Three Populations

This test is the nonparametric analogue of the F-test for equality of the means of several populations. A natural generalization of the Mann–Whitney test, it tests the null hypothesis that all k populations have a common median. It is assumed that the measurement scale is at least ordered and that the k random samples are mutually independent.

Suppose n_i is the size of the sample from population i and $n. = \sum_i n_i$. Rank all $n.$ observations from 1 to $n.$ and let R_i be the sum of the ranks from sample i. As usual, in case of ties, assign average ranks to the tied values. The test statistic is

$$T = \frac{12}{n.(n.+1)} \sum_{i=1}^{k} \frac{[R_i - n_i(n.+1)/2]^2}{n_i} \qquad (16.1)$$

The idea behind this formula is that if the null hypothesis is exactly true, the expected sum of the ranks of sample i is $E(R_i) = n_i(n.+1)/2$.

This test is conducted in R with the command
    kruskal.test(y ~ g, dataframe)
where y is the numeric vector of sample observations and g is the same sized factor indicating the population number from which the observation came. As usual dataframe is the data.frame containing the observations.

Exercise 6.2 refers to an experiment comparing the pulse rates of workers while performing six different tasks data(pulse). From Figure 16.6 we see that the normality assumption required for a standard one-way analysis of variance is somewhat questionable as most tasks seem to show a uniform distribution of pulses. We therefore investigate the data analysis via the Kruskal–Wallis test.

Table 16.9 Kruskal–Wallis rank sum test applied to pulse rate data by R.

```
> kruskal.test(pulse ~ task, data=pulse)

Kruskal-Wallis rank sum test

data:  pulse by task
Kruskal-Wallis chi-squared = 16, df = 5, p-value = 0.007
```

Table 16.9 shows the chi-square approximation to the distribution of the Kruskal–Wallis statistic yields a p-value of .007. This is not too dissimilar from the one-way

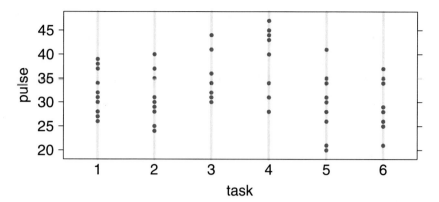

Fig. 16.6 Pulse data. The uniform distributions within the groups make the one-way ANOVA inappropriate. See Table 16.9.

ANOVA p-value of .0015. We conclude, unsurprisingly, that the pulse rate of workers differs according to the task performed immediately prior to the pulse reading.

16.6 Exercises

16.1. A study of Darwin (1876), discussed in Hand et al. (1994) with the data in data(darwin), compared the ultimate heights of plants grown from otherwise comparable seedlings that were either cross-fertilized or self-fertilized. Compare the heights using the Wilcoxon signed-ranks procedure. Would a paired t-test have been appropriate?

16.2. High levels of carbon monoxide transfer are a risk factor for contracting pneumonia. Ellis et al. (1987), also in Hand et al. (1994), studied the levels of carbon monoxide transfer in 7 chicken pox patients who were smokers. They were measured upon hospital admission and one week later. The data are in the file data(pox).

a. Verify the inappropriateness of a paired t-test for these data. Discuss your reasoning.

b. Analyze using the Wilcoxon signed-ranks procedure.

16.3. Simpson et al. (1975), also in Chambers et al. (1983), studied the amount of rainfall (measured in acre-feet) following cloudseeding. They seeded 26 clouds with silver nitrate and 26 clouds with a placebo seeding (that is, the airplane went up but didn't release the silver nitrate). The data are contained in the file data(seeding). Assume these data represent independent random samples.

a. Use the Mann–Whitney test to assess whether cloudseeding impacted rainfall.

b. Construct the ladder of powers graph (see Figure 4.19) to find a power transformation that makes the histograms of post-transformed samples symmetric and bell-shaped. Redo part a using the two-sample t-test instead of the Mann–Whitney test.

c. Discuss which of these procedures is preferred.

16.4. VanVliet and Gupta (1973), later cited by Hand et al. (1994), compared the birthweights in kg of 50 infants having severe idiopathic respiratory distress syndrome. Twenty-seven (27) of these infants subsequently died and the remainder lived. The data appear in the file data(distress). Perform a nonparametric test to address whether deaths from this cause are associated with low birthweight.

16.5. Exercise 6.8 requested a comparison of the mean disintegration times of four types of pharmaceutical tablets. The data are in file data(tablet1). Compare the results of a Kruskal–Wallis test with the conclusion of the F-test for that exercise.

16.6. Calculate the variance of the null distribution for the Wilcoxon signed-ranks statistic, assuming no ties. The set of all signed ranks, not just the positive values, in the null distribution consists of the integers $\{-n$ to $-1, 1$ to $n\}$. This has mean 0 and variance proportional to the $\sum i^2$. The variance of just the positive values is 1/4 times the variance of all of them. The mean of just the positive signed ranks is proportional to $\sum i$.

Chapter 17

Logistic Regression

Logistic regression is a technique similar to multiple regression with the new feature that the predicted response is a probability. Logistic regression is appropriate in the often-encountered situation where we wish to model a dependent variable which is either

dichotomous: The dependent variable can assume only the two possible values 0 and 1 (often as a coding of a two-valued categorical variable such as Male/Female or Treatment/Control).

sample proportion: The dependent variable is a probability and hence confined to the interval $(0, 1)$.

The methodology of ordinary multiple regression analysis cannot cope with these situations because ordinary regression assumes that the dependent variable is continuous on the infinite interval $(-\infty, \infty)$. When this assumption on the dependent variable is not met, we must employ a suitable transformation—a *link* function—to change its range. One such transformation (shown in Equation (17.1) and Figure 17.1) from the closed interval $[0, 1]$ to the set of all real numbers is the logarithm of the odds, known as the logit transformation

$$y = \text{logit}(p) = \ln\left(\frac{p}{1-p}\right) \tag{17.1}$$

The logit transformation is the key to logistic regression. R functions for the logit and its inverse

$$p = \text{antilogit}(y) = \frac{e^y}{1 + e^y} \tag{17.2}$$

are defined in the **HH** package.

© Springer Science+Business Media New York 2015
R.M. Heiberger, B. Holland, *Statistical Analysis and Data Display*,
Springer Texts in Statistics, DOI 10.1007/978-1-4939-2122-5_17

The model for logistic regression is

$$\text{logit}(p) = X\beta + \epsilon \tag{17.3}$$

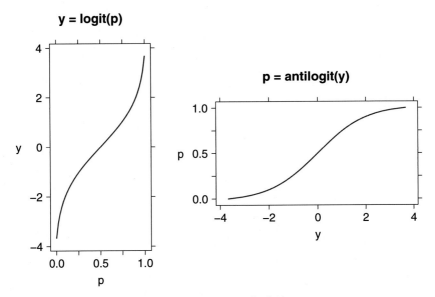

Fig. 17.1 Left panel: $y = \text{logit}(p)$. Right panel: $p = \text{antilogit}(y)$.

where

    p The response is either a binary 0/1 variable indicating failure/success or a number in the range [0, 1] indicating an observed proportion of successes.

    X matrix of predictor variables.

    β vector of regression coefficients.

    ϵ vector of discrepancies assumed to have a binomial distribution.

In the special case of a single continuous predictor, model (17.3) specializes to

$$\text{logit}(p) = \beta_0 + \beta_1 x + \epsilon \tag{17.4}$$

Logistic regression is a special case of a *generalized linear model*. There are two components to the generalization.

1. In ordinary linear models, the response variable is assumed to be linearly related to the predictor models. In generalized linear models a function (the link function) of the response variable is assumed to be linearly related to the predictor models. In logistic models we usually use the logit function.

2. In ordinary linear models, the variance of the residuals is assumed to be normal. In *glm* (generalized linear models), the variance function is usually something else. With logistic models the variance function is assumed to be binomial.

Probit regression is another type of generalized linear model. Instead of using the logit link function, probit regression uses the inverse of the normal c.d.f., Φ^{-1}, defined in Section 3.9, to map from $(0, 1)$ to the set of all real numbers. As with logistic regression, the variance function for probit models is usually assumed to be binomial. Probit regression and logistic regression give very similar results unless many of the estimated probabilities are close to either 0 or 1.

The predictor variables in logistic regression are the same types of continuous variables and factors that we use in ordinary multiple regression. All the familiar operations on the predictor variables (nesting, crossing, polynomial powers) are also appropriate with generalized linear models.

17.1 Example—The Space Shuttle Challenger Disaster

17.1.1 Study Objectives

The NASA space shuttle has two booster rockets, each of which has three joints sealed with O-rings. A warm O-ring quickly recovers its shape after a compression is removed, but a cold one will not. An inability of an O-ring to recover its shape can allow a gas leak, which may lead to disaster. On January 28, 1986, the Space Shuttle Challenger exploded during the launch.

The coldest previous launch temperature was 53 degrees Fahrenheit. The temperature forecast for time of launch of Challenger on the morning of January 28, 1986, was 31 degrees Fahrenheit. On the evening of January 27, a teleconference was held among people at Morton Thiokol, Marshall Space Flight Center, and Kennedy Space Center. There was a substantial discussion among engineers over whether the flight should be cancelled. No statistician was present for any of these discussions.

17.1.2 Data Description

The input dataset in data(spacshu) from Dalal et al. (1989) contains two columns.

tempF: temperature in degrees Fahrenheit at the time of launch

damage: 1 if an O-ring was damaged and 0 otherwise

Each launch has six cases, one for each O-ring. There are a total of $23 \times 6 = 138$ cases (the O-rings for one flight were lost at sea).

17.1.3 Graphical Display

The five panels in Figures 17.2 and 17.3 show the relationship between number of damaged O-rings and launch temperature for space shuttle flights prior to the Challenger disaster. They clearly suggest a temperature effect. Logit regression can be used to model the probability of O-ring damage as a function of launch temperature, and hence estimate the probability that any one particular O-ring is damaged at launch temperature 31°F.

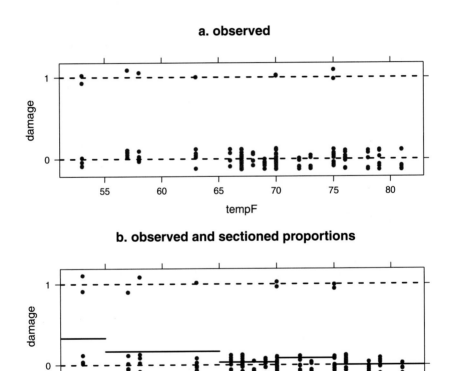

Fig. 17.2 Panel a. Original data. Panel b. Sectioned fit. (The panel naming continues through Figures 17.3, 17.4, and 17.6.)

c. observed and sectioned proportions
appropriate temperature scale

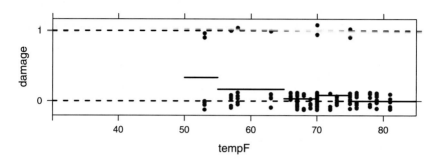

d. glm logit fit with pi, estimating p(damage in one ring)

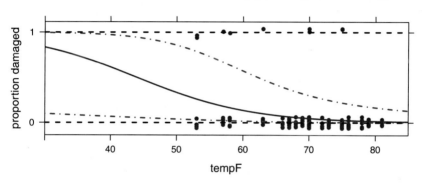

e. glm logit fit with pi, estimating number of damaged rings

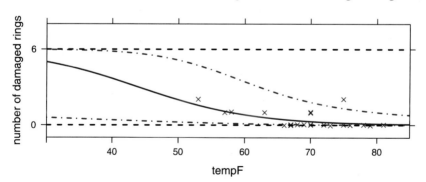

Fig. 17.3 Panel c. Sectioned fit with appropriate temperature scale. Panel d. Logit fit with 95% prediction band focusing on estimating probability of damage in one ring on the next launch. Panel e. Logit fit with 95% prediction band focusing on estimating number of damaged rings in the next launch.

The logistic curves in Figures 17.3d and 17.3e decrease as the predictor variable tempF increases. This behavior differs from that displayed in Figure 17.1b because the logistic regression coefficient of tempF is negative. Models with a single predictor and a positive logistic regression coefficient have logit fits resembling Figure 17.1b.

The dataset data(spacshu) from Dalal et al. (1989) includes one observation for each O-ring. Figure 17.2a shows the observed data, jittered (by adding random noise to break ties) so multiple O-rings at the same launch (hence same temperature) are visible. Note that the tempF scale includes only the observed temperatures. Figure 17.2b shows a simplistic model fit to the data. Within each 5-degree interval we have calculated the proportion of O-rings that were damaged. We see a strong indication that the proportion of damaged rings goes up as the temperature goes down. The graph suggests that the probability of damage will be high when the temperature reaches 31°F. There are several limitations to this inference. First, the graph doesn't extend to 31°F. Second, the model doesn't extend to 31°F.

Figure 17.3 extends the axes to include the temperature of the launch day in question. It is easier to see the suggested inference of high probability of damage even in panel c. We need a different model than simply averaging over a 5-degree range to clarify the impression. Panels d and e show the prediction bands for estimating probabilities of damage to individual O-rings (• in panel d) and for estimating the number of O-rings damaged per flight (× in panel e). Note this is an extrapolation. The fitted response values and their standard errors are calculated in Table 17.1

Figure 17.3d shows confidence bands for the number of damaged O-rings. The shape of the prediction bands in panels d and e are the same because the expected number of damaged rings is the constant 6 (the number of O-rings on a shuttle) times the probability of damage to an individual ring.

At the January 28, 1986, teleconference, they displayed not Figure 17.3, but Figure 17.4, a figure showing only those launches with at least one damaged O-ring.

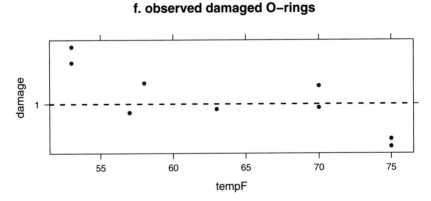

Fig. 17.4 Panel f. Observed damaged O-rings, without the information about total number of rings.

Figure 17.4 is essentially just the top of our Figure 17.2a, the portion with damage=1. One cannot tell from such a chart that there is a temperature/damage relationship. Information about temperatures for launches without damage is highly relevant. Tragically, the assembled engineers did not realize the vital importance of seeing the number of launches without damage. They okayed the launch on January 28.

17.1.4 Numerical Display

We show in Table 17.1 a logistic regression analysis of these data to the model

$$\text{logit}(p) = \beta_0 + \beta_{\text{tempF}} \, \texttt{tempF} + \epsilon \qquad (17.5)$$

where $p = P(\texttt{damage})$. Discussion of the terminology used in this table appears in Section 17.2.

The fitted equation is $\text{logit}(\hat{p}) = 5.085 - .1156\ \texttt{tempF}$. A test of hypothesis that the coefficient of $\texttt{tempF}$ is zero is the 1 d.f. χ^2 statistic 6.144 with corresponding p-value .0132. This p-value suggests that there is a moderately significant relationship between $\texttt{temp}$ and $\texttt{damage}$. Inserting $\texttt{tempF}=31$ gives $\text{logit}(\hat{p}) = 1.5014 \Rightarrow \hat{p} = .8178$. If the six O-rings on a Space Shuttle fail independently of one another (a roughly true assumption), we could have expected $6 \times .8178 = 4.9$ failures of the six O-rings for the launch. This analysis could have been performed prior to launch!

The interpretation of logistic regression coefficients is less straightforward than interpretations of linear regression coefficients. In this problem, an increase in launch temperature of 1°F multiplies the expected odds in favor of O-ring failure by $e^{-0.1156} = 0.8908$. Equivalently, each 1°F decrease in launch temperature corresponds to multiplication of the expected odds of O-ring failure by $e^{0.1156} = 1.1225$. In this problem the intercept coefficient 5.085 is not readily interpretable because $\texttt{tempF} = 0$ is not a feasible launch temperature.

We must study Table 17.1 together with the $\texttt{logit.p}$ ~ $\texttt{tempF}$ plot in Figure 17.5a. The model says that $\texttt{logit.p}=\text{logit}(p)$ is linearly related to temperature $\texttt{tempF}$. The slope $\beta_{\text{tempF}} = -0.1156$ and intercept $\beta_0 = 5.0848$ describe the straight line in Figure 17.5a.

Table 17.2 compares the three scales used in logistic regression analysis. Figure 17.5 shows the predicted probability of failure on each scale. Logits are hard to interpret as they are not in a scale that we are comfortable thinking about. Two alternate transformations are the odds ratio in the $\texttt{odds scale}$ panel, and the probability in the $\texttt{probability scale}$ panel.

Table 17.1 Logistic regression of Challenger data.

```
> spacshu.bin.glm <- glm(damage ~ tempF, data=spacshu, family=binomial)

> spacshu.bin.glm

Call:  glm(formula = damage ~ tempF, family = binomial, data = spacshu)

Coefficients:
(Intercept)         tempF
      5.085        -0.116

Degrees of Freedom: 137 Total (i.e. Null);   136 Residual
Null Deviance:      66.5
Residual Deviance: 60.4   AIC: 64.4

> anova(spacshu.bin.glm, test="Chi")
Analysis of Deviance Table

Model: binomial, link: logit

Response: damage

Terms added sequentially (first to last)


        Df Deviance Resid. Df Resid. Dev Pr(>Chi)
NULL                      137       66.5
tempF  1      6.14        136       60.4     0.013 *
---
Signif. codes:
0 '***' 0.001 '**' 0.01 '*' 0.05 '.' 0.1 ' ' 1

> coef(summary(spacshu.bin.glm))
            Estimate Std. Error z value Pr(>|z|)
(Intercept)   5.0850    3.05248   1.666  0.09574
tempF        -0.1156    0.04702  -2.458  0.01396

> ## prediction on response scale, in this case (0,1).
> ## leading to Figure spaceshuttle-d.pdf, Panel d
> spacshu.pred <-
+    interval(spacshu.bin.glm, newdata=data.frame(tempF=30:85),
+             type="response")
```

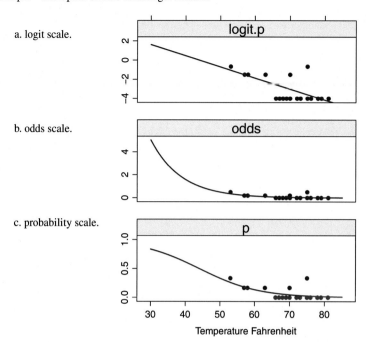

Fig. 17.5 Observed damage and estimated proportion of damaged O-rings in three scales plotted against temperature. Panel a shows the logit scale, where we see the linear relationship that logit(p) is proportional to temperature. Panel b shows the odds scale, where we see the tendency for the odds in favor of damage to go up as temperature goes down. Panel c shows the probability scale, where we see the tendency for the probability of damage to go up as temperature goes down.

Table 17.2 Three scales used in logistic regression.

probabilities	$\hat{p}$	`p.hat <- predict(spacshu.bin.glm` ` type="response")`
odds	$\dfrac{\hat{p}}{1-\hat{p}}$	`odds.hat <- p.hat/(1-p.hat)`
logit	$\text{logit } \hat{p} = \log\left(\dfrac{\hat{p}}{1-\hat{p}}\right)$	`logit.p.hat <- log(odds.hat)`

The calculations for the 95% prediction bands based on the linear model for the logit(p) are shown in the proportion scale and the logit scale in Table 17.3. The bands are displayed in the proportion scale in Figure 17.3d and in the logit scale in Figure 17.6g.

Table 17.3 Details of logistic regression of Challenger data. The top table is in the *link* scale where the model is linear. The bottom table is in the *response* scale where the units are more easily interpreted; they are probabilities in this example.

```
> ## prediction on link scale, in this case (-Inf, Inf)
> ## leading to Figure spaceshuttle-g.pdf Panel g
> spacshu.pred.link <-
+    interval(spacshu.bin.glm, newdata=data.frame(tempF=30:85),
+              type="link")

> cbind(tempF=30:85, round(spacshu.pred.link, digits=2))[c(1:3,54:56),]
     tempF    fit ci.low ci.hi pi.low pi.hi
1       30   1.62  -1.66  4.90  -2.21  5.45
2       31   1.50  -1.69  4.69  -2.25  5.26
3       32   1.39  -1.71  4.49  -2.29  5.06
54      83  -4.51  -6.37 -2.65  -7.23 -1.79
55      84  -4.63  -6.57 -2.68  -7.40 -1.85
56      85  -4.74  -6.77 -2.71  -7.58 -1.90

> cbind(tempF=30:85, round(spacshu.pred, digits=2))[c(1:3,54:56),]
     tempF  fit ci.low ci.hi pi.low pi.hi
1       30 0.83   0.16  0.99   0.10  1.00
2       31 0.82   0.16  0.99   0.10  0.99
3       32 0.80   0.15  0.99   0.09  0.99
54      83 0.01   0.00  0.07   0.00  0.14
55      84 0.01   0.00  0.06   0.00  0.14
56      85 0.01   0.00  0.06   0.00  0.13
```

g. glm logit fit with logit(pi), estimating logit(p(damage in one ring))

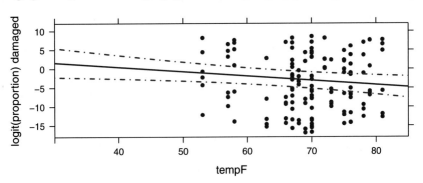

Fig. 17.6 Panel g. Logit fit on logit scale with 95% prediction band.

Lavine (1991) disagreed with the analysis in Dalal et al. (1989), stating that the 31°F temperature at Challenger's launch is too far below the lowest previously observed launch temperature, 53°F, to assume that the logistic relationship persisted for the Challenger launch. He claimed that without additional engineering input, one can only say that the probability of O-ring failure at 31°F is greater than or equal to the probability of O-ring failure at 53°F. We agree that the extrapolation from 53°F to 31°F is too great for anyone to be highly confident in our $\hat{p} = .8178$ estimate of the failure probability at 31°F.

17.2 Estimation

The computations for calculating the parameter estimates $\hat{\beta}$ are usually done by the method of maximum likelihood with an iterative computer program. The likelihood is defined algebraically as the joint probability of the observations viewed as a function of the parameters conditional on the data.

Let us use the notation that the ith case, for $i: 1, \ldots, n$, consists of a single response value y_i and a single predictor variable x_i. The likelihood with a single predictor variable is written as

$$L(\beta|y; x) = L\left((\beta_0, \beta_1) \,\middle|\, (y_1, \ldots, y_n) ; (x_1, \ldots, x_n)\right)$$

$$= \prod_{i=1}^{n} f\left(y_i \,\middle|\, x_i ; \beta_0, \beta_1\right) \tag{17.6}$$

$$= \prod_{i=1}^{n} p_i^{y_i}(1 - p_i)^{1-y_i}$$

where

$$p_i = \frac{\exp(\beta_0 + \beta_1 x_i)}{1 + \exp(\beta_0 + \beta_1 x_i)} = \text{antilogit}(\beta_0 + \beta_1 x_i) \tag{17.7}$$

or equivalently,

$$\text{logit}(p_i) = \log\left(\frac{p_i}{1 - p_i}\right) = \beta_0 + \beta_1 x_i \tag{17.8}$$

The logistic link function appears in the likelihood equation with the expression of p as the antilogit of the linear function of the predictor x-variables.

In general, the terminology *link* function refers to the transformation of the response variable such that the transformed response achieves an approximate linear relationship with explanatory variables. After selecting a link function, it is also necessary to specify a variance function, usually by accepting the software's default choice. A binomial variance is the usual choice with logistic or probit links. A Poisson variance is associated with a logarithmic link function. See McCullagh and Nelder (1983) and the software help files for more details.

The binomial error function appears in the likelihood expression (17.6) as the product of Bernoulli (binomial with $n = 1$) terms, one for each of the n observations. The binomial distribution is thus involved in the likelihood equations for estimating the β coefficients.

The log of the likelihood, called the *loglikelihood*, is written as

$$\ell(\beta|y; x) = \log\big(L(\beta|y; x)\big) \tag{17.9}$$

$$= \sum_{i=1}^{n}\Big(y_i \log(p_i) + (1 - y_i)\log(1 - p_i)\Big)$$

We estimate the parameters β by differentiating the log of the likelihood with respect to the vector of parameters β, setting the derivatives to 0, and solving for β. Substitute Equation 17.7 into Equation 17.9 and differentiate with respect to β_0 and β_1 to get:

$$\frac{\partial}{\partial \beta_0}\, \ell(\beta|y; X) = \sum_{i=1}^{n}(y_i - p_i) \quad = 0$$

$$\frac{\partial}{\partial \beta_1}\, \ell(\beta|y; X) = \sum_{i=1}^{n}x_i(y_i - p_i) = 0 \tag{17.10}$$

Exercise 17.1 gives you an opportunity to follow the steps in detail.

Logistic regression is a special case of a *generalized linear model,* hence the model specification is easily interpreted.

glm: The statement in Table 17.1

```
spacshu.bin.glm <- glm(damage ~ tempF, data=spacshu,
                family=binomial)
```

says that the logit of the expected value of the response variable damage is to be modeled by a linear function of the predictor variable tempF, that is,

$$\mathrm{logit}(E(\hat{p}_j)) = \log\left(\frac{E(\hat{p}_j)}{1 - E(\hat{p}_j)}\right) = \beta_0 + \beta_x x$$

The family=binomial argument says that the error term has a binomial distribution and that the link function is the logit. The logit is the default link for the binomial family; we could have specified the link explicitly with

```
spacshu.bin.glm <- glm(damage ~ tempF, data=spacshu,
                family=binomial(link=logit))
```

R places the results of the fitting process into a "glm" object (for this example called spacshu.bin.glm) that we can look at.

Two principal functions are used to display textual information about the object.

anova: The anova function

```
anova(spacshu.bin.glm, test="Chi")
```

displays the *analysis of deviance table*, an analogue to the *analysis of variance table* in ordinary linear models. Table 17.1 is consistent with what we see in Figure 17.5. There is a clearly visible effect in the figure. The ANOVA table shows the linear dependence on tempF has p-value 0.013, highly significant at $\alpha = .05$. The table is interpreted similarly to an ordinary analysis of variance table. The *deviance* for tempF is twice the drop in the loglikelihood from the model without the tempF term to the model with the tempF term, $6.144035 = 66.54037 - 60.39634$, with degrees of freedom the difference in the number of parameters in the two models, $1 = 137 - 136$. We use the χ^2 table, hence $1 - \mathcal{F}_{\chi^2}(6.144035 \mid 1) = 0.01318561$.

Deviance is very general concept. The formula for deviance specializes to variance for the normal error function with the identity link function. In this, and in many other ways, least squares is a special case of maximum likelihood.

summary: The summary function

```
summary(spacshu.bin.glm)
```

shows the table of regression coefficients. It is less helpful than the deviance table in situations with only a single predictor variable. In this example it says that the t-value for tempF is -0.1155985. This number does not have an exact t-distribution, which is why no p-value is associated with it. Since $|-2.46| > 2$, it appears to be significant, but we must look at the chi-square value in the analysis of deviance table to make a valid inference.

17.3 Example—Budworm Data

An experiment discussed in Collett (1991) and Venables and Ripley (1997) was performed to model the dose–response of a particular insecticide required to kill or incapacitate budworms, insects that attack tobacco plants. A potential additional factor was the budworm's sex. In particular, it was desired to estimate the lethal dose proportions LD25, LD50, LD75. LD50, lethal dose 50, is an abbreviation for the dose that is lethal for 50% of the budworms. Twenty moths of each sex were exposed to each of seven experimental doses. The data, accessible as data(budworm), present ldose, the $\log_2$ of dose, along with the number of moths of each sex that

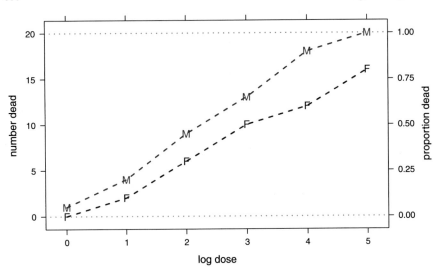

Fig. 17.7 Observed number (out of 20) of dead Male and Female budworms as a function of log(dose). The goal of the study is the prediction of the proportion dead, so we included a right axis with the data rescaled values to proportion. The horizontal reference lines are placed at the limits `proportion=0` and `proportion=1`. The dashed lines connecting the observed values are a visual aid to the reader of the graph.

were killed after three days of exposure. We show the observed data in Figure 17.7. For each dose of insecticide, male budworms have higher mortality than females.

We model the data with a logistic regression calculated as a generalized linear model with a binomial error term and a logit link function. Table 17.4 shows two models, with and without an interaction term. The *p*-value of the interaction is high, so we will continue with the no-interaction model.

We read the predicted LD50 (on $\log_2$ scale) by placing a horizontal line at `prob=.5` on the `prob ~ log.dose` panel of Figure 17.8. For each dose of insecticide, male budworms have higher mortality than females. We look at the `log.dose` coordinate of the intersection of the horizontal line with the fitted lines and find that the LD50 is slightly over 2 for males and slightly over 3 for females. Calculations for the LD50, as well as ones for LD25 and LD75, using the **MASS** function `dose.p` are illustrated in Table 17.5.

Figure 17.9 shows the relationship of predicted probability of kill, odds, and logit to the observed data. The `logit ~ log.dose` panel shows the assumed linear relationship of $\text{logit}(p) = \log(p/(1 - p))$.

Table 17.4 Logisitic regression of the budworm data with and without an interaction term. The *p*-value shows that we not need the interaction term, and we do need both the other terms.

```
> SF <- cbind(numdead=budworm$numdead,
+              numalive = 20 - budworm$numdead)

> ## model with interaction term for sex and logdose, from VR
> budworm.lg <-
+   glm(SF ~ sex*ldose,
+       data=budworm,
+       family = binomial)

> anova(budworm.lg, test="Chisq")
Analysis of Deviance Table

Model: binomial, link: logit

Response: SF

Terms added sequentially (first to last)


          Df Deviance Resid. Df Resid. Dev Pr(>Chi)
NULL                        11      124.9
sex        1      6.1        10      118.8    0.014 *
ldose      1    112.0         9        6.8   <2e-16 ***
sex:ldose  1      1.8         8        5.0    0.184
---
Signif. codes:
0 '***' 0.001 '**' 0.01 '*' 0.05 '.' 0.1 ' ' 1

> ## model with no interaction term
> budworm.lg0 <- glm(SF ~ sex + ldose - 1,
+                    data=budworm,
+                    family = binomial)

> anova(budworm.lg0, test="Chisq")
Analysis of Deviance Table

Model: binomial, link: logit

Response: SF

Terms added sequentially (first to last)


      Df Deviance Resid. Df Resid. Dev Pr(>Chi)
NULL                    12      126.2
sex    2      7.4        10      118.8    0.024 *
ldose  1    112.0         9        6.8   <2e-16 ***
---
Signif. codes:
0 '***' 0.001 '**' 0.01 '*' 0.05 '.' 0.1 ' ' 1
```

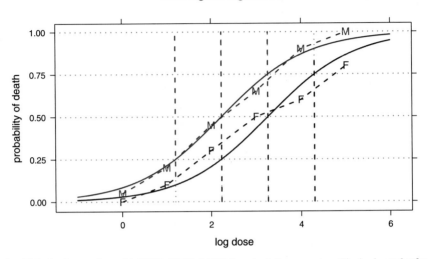

Fig. 17.8 Budworm data with LD25, LD50, LD75 from logistic regression. The horizontal reference lines at probabilities .25, .50, and .75 are used to find the LD values. For example, the horizontal line at p=.50 intersects the fitted Male curve at log dose=2.229. Thus the LD50 for Male is 2.229. Similarly the horizontal line at p=.50 intersects the fitted Female curve at log dose=3.264. Thus the LD50 for Female is 3.264.

Table 17.5 Budworm data with LD25, LD50, LD75 from logistic regression. See also Figure 17.8.

```
> ## LD25 LD50 LD75
> xp.M <- MASS::dose.p(budworm.lg0, cf = c(2,3), p = 1:3/4)

> xp.M
            Dose      SE
p = 0.25:  1.197  0.2635
p = 0.50:  2.229  0.2260
p = 0.75:  3.262  0.2550

> xp.F <- MASS::dose.p(budworm.lg0, cf = c(1,3), p = 1:3/4)

> xp.F
            Dose      SE
p = 0.25:  2.231  0.2499
p = 0.50:  3.264  0.2298
p = 0.75:  4.296  0.2747
```

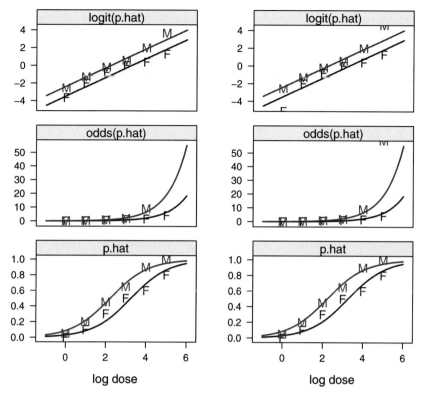

Fig. 17.9 Predicted probabilities of kill, and the odds and logit=log(odds), as a function of log(dose). A constant difference in the logit scale, seen as parallel lines in the `logit ~ log.dose` panel, corresponds to a constant ratio in the odds. The logit scale plot of this example, with one continuous covariate and one factor, is analogous to the analysis of covariance for continuous response and normally distributed error term as seen in Figure 10.8. The panels on the left for odds and logit have missing symbols where the transformation created infinite values $\left(\text{odds}(1) = \infty, \text{logit}(0) = -\infty, \text{logit}(1) = \infty\right)$. In the panels on the right we placed partial symbols on the edge of the panel to indicate that there is an observation and that it is off the scale. The `p.hat ~ log.dose` is the same on both sides and is the same as Figure 17.8.

17.4 Example—Lymph Nodes

These data come from Brown (1980). The problem is outlined in this reference as follows:

> When a patient is diagnosed as having cancer of the prostate, an important question in deciding on treatment strategy for the patient is whether or not the cancer has spread to the neighboring lymph nodes. The question is so critical in prognosis and treatment that it is customary to operate on the patient for the sole purpose of examining the nodes and removing tissue samples to examine under the microscope for evidence of cancer. However, certain variables that can be measured without surgery are predictive of the nodal involvement; and the purpose of the study presented here was to examine the data for 53

prostate cancer patients receiving surgery, to determine which of five preoperative variables are predictive of nodal involvement, and how accurately the prediction can be made.

In particular, the medical investigator was interested in whether or not an elevated level of acid phosphatase in the blood serum would be of added value in the prediction of whether or not the lymph nodes were affected, given the other four more generally used variables.

17.4.1 Data

The variables accessed as data(lymph) are described in Table 17.6. The data are displayed in Figure 17.10, constructed as a plot of the two continuous x-variables conditioned on the three 2-level factors and with the plotting symbol chosen to reflect the 2-level response variable. The investigator wants to know if the predicted proportion of nodes (symbol "1") in each panel (that is, conditional on grade, stage, X-ray, and age) is affected by the additional information on acid phosphatase.

Our first impression from Figure 17.10 is that age seems not to make a difference. Therefore, our subsequent analysis considers only the three remaining factors and one continuous predictor.

17.4.2 Data Analysis

In Table 17.7 we model the logit transformation of nodes as a function of X.ray, stage, grade, and acid.ph. Figure 17.11 contains plots of jittered nodes vs acid.ph along with the model's predicted probability of nodal involvement partitioned according to the $8 = 2^3$ combinations of the factors X.ray, stage, and grade. The predicted relationship is plausible in 5 of these 8 panels, where there

Table 17.6 Lymph Data from Brown (1980)

Response Variable	
nodes	1 indicates nodal involvement found at surgery,
	0 no nodal involvement
Predictor Variables	
X.ray	1 is serious, 0 is less serious
stage	measure of size and location of tumor,
	1 is serious, 0 is less serious
grade	pathology reading of a biopsy, 1 is serious, 0 is less serious
age	at diagnosis
acid.ph	level of serum acid phosphatase

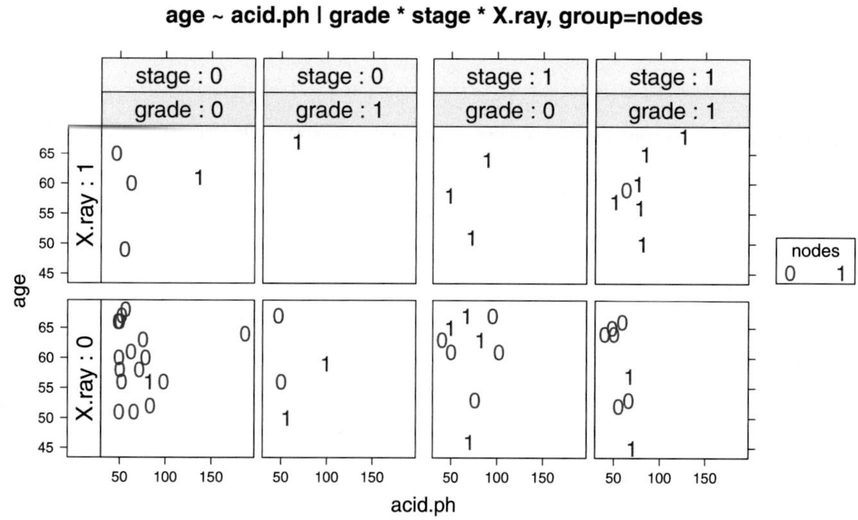

age ~ acid.ph | grade * stage * X.ray, group=nodes

Fig. 17.10 Much of the relation between variables is visible in this set of panels conditioned on three factors, with the plotting symbol representing a fourth factor. The top row with X-ray=1 has a higher proportion of nodes=1. The right two columns with stage=1 have a higher proportion of nodes=1.

is a visible tendency for the nodes=1 points to be farther to the right, that is to have higher acid.ph scores, than the nodes=0 cases. Two of the panels have only nodes=1 cases and can't be used for this comparison. In only one panel (X.ray=0, grade=0, stage=1) is there a balanced overlap of ranges. This suggests that acid.ph is positively associated with nodes and therefore may be a useful additional predictor.

The p-value for acid.ph in Table 17.7, 0.075, indicates that acid.ph is a borderline predictor of the presence of nodes. A tentative interpretation is that acid.ph is a potential predictor for some but not all combinations of X.ray, stage, and grade. In this same table, the p-value for grade is 0.45, suggesting that this factor can be dropped from the model. However, we choose to retain gradea because a rerun of the model without grade actually increases the p-value of acid.ph. Investigation of this issue is requested in Exercise 17.10.

Figure 17.12 shows the predicted probability of nodal involvement as a function of acid.ph, and the transformation of these predicted probabilities to predicted odds and predicted logits for the additive model in Table 17.7, conditioned on the levels of X.ray, stage, and grade. Figure 17.12 uses different plot symbols for the two levels of stage. The left column shows all $(2 \times 2 \times 2)$ = eight groups. A separate line is shown for predictions for each of the $8 = 2^3$ combinations of the three factors X.ray, stage, and grade. In the logit.p.hat ~ acid.ph panels there

Table 17.7 Logistic regression with three factors and one continuous predictor. See also Figures 17.11 and 17.12.

```
> lymph3.glm <- glm(nodes ~ X.ray + stage + grade + acid.ph,
+                    data=lymph, family=binomial)

> anova(lymph3.glm, test="Chisq")
Analysis of Deviance Table

Model: binomial, link: logit

Response: nodes

Terms added sequentially (first to last)


         Df Deviance Resid. Df Resid. Dev Pr(>Chi)
NULL                        52       70.3
X.ray     1    11.25        51       59.0   0.0008 ***
stage     1     5.65        50       53.4   0.0175 *
grade     1     0.57        49       52.8   0.4489
acid.ph   1     3.16        48       49.6   0.0753 .
---
Signif. codes:  0 '***' 0.001 '**' 0.01 '*' 0.05 '.' 0.1 ' ' 1
```

are 8 parallel lines, one for each value of acid.ph*X.ray*nodes. The right two columns partition the lines into separate sets of panels for each level of X.ray. The logit.p.hat ~ acid.ph plots in the top panels in this figure suggest that there is positive association between nodal involvement and acid.ph. The intercepts in the right two columns and the top row show the clear distinction attributable to the level of X-ray. Within the X-ray=0 and X-ray=1 columns, the stage=1 lines are both higher than the stage=2 lines. Within each of the panels the grade=1 line is higher than the grade=0 line. The distinctions easily seen in the top row with the logit scale are reflected in the odds.hat and p.hat rows.

17.4.3 Additional Techniques

In the remainder of this section we consider a simpler logistic regression model for nodes primarily to further illustrate the variety of modeling and plotting techniques available to analysts. We model the probability of nodal involvement as a function of just acid.ph and X.ray, initially requiring constant slope and allowing for the possibility that the intercept in logistic relationship between nodes and acid.ph differs according to whether X.ray = 0 or 1.

jittered observed and predicted probability(nodes)

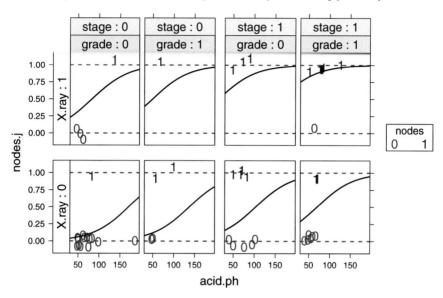

Fig. 17.11 Jittered and predicted nodes vs acid.ph for each combination of values of stage, grade, and X.ray. The asymptotes for fit at nodes=0 and nodes=1 are indicated with a gray dotted line. The significant terms in Table 17.7 are visible here. The fitted line is higher and to the right in the X-ray=1 than in the X-ray=0 panels. The fitted line is higher and to the right in the stage=1 panels than in the stage=0 panels. All panels show higher fitted values for higher acid.ph values. We can even see the small effect of grade in each pair of panels at the same X-ray and stage. All eight fitted lines shown here are displayed together in the first column and p.hat row of Figure 17.12.

Figure 17.13 shows nodes as a function of acid.ph separately for each value of X.ray. Except for the single point with acid.ph=187, to be investigated later, the illustration shows clearly that high acid.ph predicts nodes=1 for X.ray=1, and that high acid.ph doesn't help very much for X.ray=0.

We can get a sense of this interpretation in Figure 17.14 by partitioning the *x*-axis (acid.ph) into sections 20 units wide and plotting the proportion of 1s in each section. The line segments plotted in each *x*-section mostly represent increasing proportions of 1s as acid.ph goes up. Compare this to the logistic regression technique in Table 17.8 and Figure 17.15 which uses a continuous *x*-axis, thus no segmentation, and forces the fit to be monotone increasing.

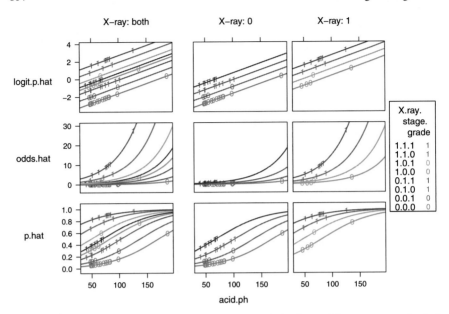

Fig. 17.12 Plot of `p.hat`, `odds(p.hat)`, `logit(p.hat)` for the model `nodes ~ acid.ph |`
`X.ray + stage + grade`, with plotting symbol indicating `stage` and placed at the predicted
values for the observed `acid.ph`. The left column shows the fit for both values of `X.ray`. The
right columns are the same fitted values with the illustration conditioned on the value of `X.ray`. The
panel in the `p.hat` row and the `X-ray:both` column shows the eight fitted lines from Figure 17.11.

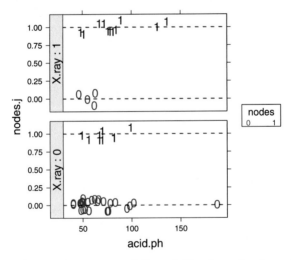

Fig. 17.13 `xyplot(nodes.j ~ acid.ph | X.ray)`. The observed nodes are jittered to break
the ties. The dashed lines indicate the location of the actual 0 and 1 values.

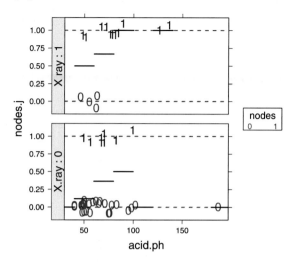

Fig. 17.14 Figure 17.13 plus proportion of 1s in sections 20 units wide.

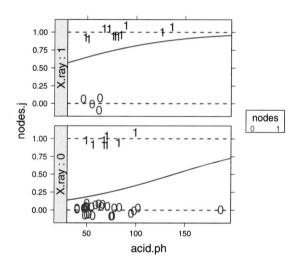

Fig. 17.15 Figure 17.13 plus fitted logistic curve for the simple model `glm(nodes ~ X.ray + acid.ph, data=lymph, family=binomial)`.

In Figure 17.15 with logistic regression we model the data in Figure 17.13 by fitting a linear model with x=`acid.ph` to the logit of the response variable

$$\text{logit}(E(\hat{p}_j)) = \log\left(\frac{E(\hat{p}_j)}{1 - E(\hat{p}_j)}\right) = \beta_0 + \beta_j + \beta_x x$$

with the error distribution assumed to be binomial with

$$p_j = P(\text{nodes} = 1 \mid \text{X.ray} = j) \tag{17.11}$$

Table 17.8 Logistic regression with one factor and one continuous predictor. See also Figure 17.15 .

```
> lymph1.glm <- glm(nodes ~ X.ray + acid.ph, data=lymph, family=binomial)

> anova(lymph1.glm, test="Chisq")
Analysis of Deviance Table

Model: binomial, link: logit

Response: nodes

Terms added sequentially (first to last)


          Df Deviance Resid. Df Resid. Dev Pr(>Chi)
NULL                        52       70.3
X.ray      1    11.25       51       59.0   0.0008 ***
acid.ph    1     1.94       50       57.1   0.1634
---
Signif. codes:  0 '***' 0.001 '**' 0.01 '*' 0.05 '.' 0.1 ' ' 1

> summary(lymph1.glm)$coef
            Estimate Std. Error z value Pr(>|z|)
(Intercept) -2.33567    0.96831  -2.412 0.015861
X.ray1       2.11596    0.70911   2.984 0.002845
acid.ph      0.01685    0.01262   1.336 0.181701
```

The model in Equation (17.11) is structurally similar to an ANCOVA model, with common slope β_x and with different intercepts. β_0 is the reference intercept and β_j is the offset of the intercept for group j. The ANOVA table and plot for this model are in Table 17.8 and Figure 17.15. We see the appropriateness of this model in the logit.p.hat ~ acid.ph panel of Figure 17.16 and, even more strikingly, in the logit.p.hat ~ acid.ph panels of Figure 17.12.

The results of fitting the model to the data of Figure 17.13 are displayed in Figures 17.15 (just the probability scale) and 17.16 (with the probability, odds, and logit scales). Note that the farther right we move in each panel, the higher the proportion of dots that appear in nodes=1 and therefore the higher the predicted probability that a case will have nodes=1.

Figure 17.14 displays partitioned fits by value of X.ray, suggesting that forcing a common slope (in the logit scale) might not be the best two-variable model for these data. Let us try adding the interaction of X.ray and acid.ph. We see in Table 17.9 that the X.ray:acid.ph interaction has a p-value of .0605. The fitted curves for this model are shown in the probability scale in Figure 17.17. The transformation to odds and logit scales is shown in Figure 17.18. Note the completely different appearance of the curve for the X.ray=1 group when it is not constrained to be parallel to the X.ray=0 curve in the logit scale.

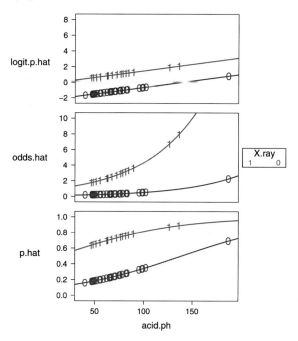

Fig. 17.16 Three scalings of fit to nodes ~ acid.ph | X.ray with the responses at both values of X.ray superposed on the same plot. The parallel lines in the logit.p.hat panel show the analogy with ANCOVA.

Table 17.9 Logistic regression with interaction of X.ray and acid.ph. See also Figure 17.17.

```
> lymph1Xa.glm <- glm(nodes ~ X.ray * acid.ph, data=lymph, family=binomial)

> anova(lymph1Xa.glm, test="Chisq")
Analysis of Deviance Table

Model: binomial, link: logit

Response: nodes

Terms added sequentially (first to last)


              Df Deviance Resid. Df Resid. Dev Pr(>Chi)
NULL                            52        70.3
X.ray          1    11.25       51        59.0  0.0008 ***
acid.ph        1     1.94       50        57.1  0.1634
X.ray:acid.ph  1     3.52       49        53.5  0.0605 .
---
Signif. codes:  0 '***' 0.001 '**' 0.01 '*' 0.05 '.' 0.1 ' ' 1
```

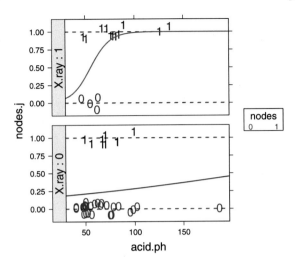

Fig. 17.17 Figure 17.13 plus fitted logistic curves for the interaction model `glm(nodes ~ X.ray * acid.ph, data=lymph, family=binomial)`.

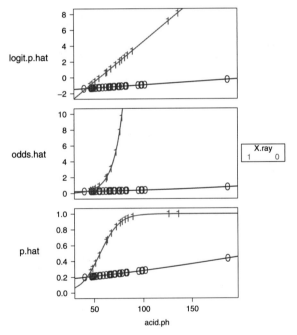

Fig. 17.18 Three scalings of fit to `nodes ~ X.ray * acid.ph` with the responses at both values of `X.ray` superposed on the same plot. The non-parallel lines in the `logit.p.hat` panel show the analogy with interaction in the ANCOVA setting. Two of the points in the `odds.hat` panel are beyond the limits we chose for the display. Their values in the `odds.hat` scale are 1256 and 3472.

Figure 17.17 displays the overlay of the (nodes ~ acid.ph | X.ray) from Figure 17.13 and the fitted values from the model lymph1Xa.glm from Table 17.9. Figure 17.17 actually displays the entire fitted line for the p.hat panels, not just the specific points corresponding to the observed data. Compare this to Figure 17.15.

Each panel of Figure 17.17 is constructed as the superposition of the subpanels for observed data (nodes ~ acid.ph | X.ray) and for predicted probability (p.hat ~ acid.ph | X.ray) from Table 17.9. We see that the probability of nodal involvement as a function of acid.ph depends on the level of X.ray.

When we look at Figure 17.16 we see that the logit.p.hat ~ acid.ph panel shows two distinct parallel lines. These parallel lines correspond to the predicted probabilities for the two levels of X.ray.

Remembering that logit($\hat{p}$) = log(odds($\hat{p}$)), we see that the parallel lines have common slope (0.01685 logit.p.hat units per acid.ph) and different intercepts 2.116 units apart. Translating the difference in intercepts back to odds ratios, we find the odds ratio attributable to X.ray is $e^{2.116}$ = 8.298, and we can see from the (odds.hat ~ acid.ph) panel of Figure 17.16 that the X.ray=1 line is about 8 times higher than the X.ray=0 line.

17.4.4 Diagnostics

All the usual diagnostic calculations, statistics, and graphs developed for linear regression are used with logit regression to help determine goodness of fit. Details are discussed in Hosmer and Lemeshow (2000).

One of the most important diagnostics for this dataset is the comparison of the graphs in Figures 17.15 and 17.17 where we note that the outlier for acid.ph=187 is a critical point. In Figure 17.14, with the sectioned fit, we see that that single point in the X.ray=0 panel flattens the logistic curve for both levels of X.ray. Compare the similar curves for both levels of X.ray in Figures 17.15 and 17.16 without the interaction term to the very different curves in Figures 17.17 and 17.18 with the interaction term.

17.5 Numerical Printout

We calculate the fitted values from a logistic regression model in the three scales (probability, odds, and logit) defined in Table 17.2 and plot them against the continuous predictor variables. The fit in lymph1.glm is calculated in Table 17.8 and illustrated in Figures 17.15 and 17.16. The logit.p.hat ~ acid.ph panel in Figure 17.16 shows the straight-line fit of logit $\hat{p}$ to acid.ph. There are two parallel lines in the panel, one for each value of the factor X.ray. The slope is given by the coefficient $\beta_{\text{acid.ph}}$. The difference in intercepts is given by the coefficient $\beta_{\text{X.ray1}}$ (note the dependence on the dummy variable coding scheme). The fitted curves in the p.hat ˜ acid.ph panel are the same, now transformed to the probability scale. The p.hat ˜ acid.ph panel of Figure 17.16 consists of the curves in both panels of Figure 17.15. Figure 17.15 overlays the logit.p.hat ˜ acid.ph panels on the nodes ˜ acid.ph panels of Figure 17.13.

The more complete model with all three factors in Figure 17.11 is calculated in Table 17.7.

17.6 Graphics

The datasets for this chapter are displayed with several types of graphs. Most of the graphs are constructed with multiple panels to display several variables in a coordinated fashion.

17.6.1 Conditioned Scatterplots

The initial display of the lymph dataset in Figure 17.10 is a scatterplot of the two continuous variables age on the y-axis and acid.ph on the x-axis, conditioned on the three binary-valued x-factors and using the binary-valued *nodes*-variable as the plotting character. All eight panels of the display are scaled alike in order to help the eye distinguish the important features. Common scaling is critical for making comparisons.

Figure 17.11 is similarly constructed, although in this case with age suppressed and with nodes on the y-axis. This graph presents the same view of the data, with important differences. We have jittered the nodes variable to counter the overprinting. Also, in Figure 17.11, we have overlaid the observed data with the predicted probabilities from the logistic model. By doing so we get an immediate sense of how the predictions are affected by the values of the conditioning factors and can see directly in each panel that higher acid.ph values correspond to higher predicted probabilities and to a higher proportion of points with observed response nodes=1.

The displays of the simplified models using only the single factor X.ray in Figures 17.13–17.18 are constructed similar to Figure 17.11. In these simpler models there are only two panels rather than eight.

Figure 17.12 shows two sets of conditioned scatterplots. The right two columns are conditioned on the level of X.ray. The left column is marginal to the right two columns. Each of the left column's panels shows the sum of the corresponding panels in the right two columns.

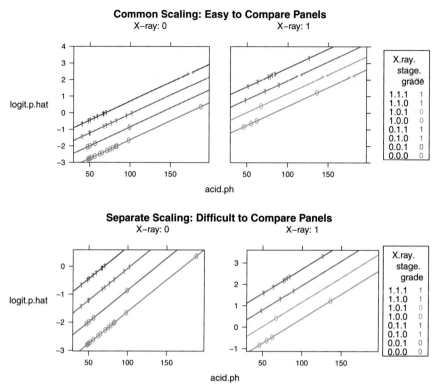

Fig. 17.19 Two views of the `logit.p.hat` row and the `X-ray: 0` and `X-ray: 1` columns of Figure 17.12. The top two panels with common scaling make it easy to see the effect of the `X.ray` level on the logit. With a common scale we see that the `X.ray=0` predictions are mostly to the left of and below the `X.ray=0` observations. The bottom two panels, with idiosyncratic scaling for each panel, hinder the ability of the reader to understand the effect of `X.ray` level. With separate scales, each set of points is individually centered in its own plotting window, and the ability to compare them visually is lost.

17.6.2 Common Scaling in Comparable Plots

Let us illustrate the importance of common scaling by looking in Figure 17.19 at the predictions of Figure 17.11 in the logit scale both with common scaling and with idiosyncratic scaling. In the logit scale all eight lines have identical slope. With common scaling we see the parallelism. We also see the outlier at `X.ray=0` and `acid.ph=187`. With idiosyncratic scaling we see two unrelated parallel structures, and worse, we think the distant points at `X.ray=1` and `acid.ph≈130` are as far out as the real outlier at `X.ray=0` and `acid.ph=187`. Only when the viewer looks very closely at the printed scale, a requirement that indicates a poorly designed graphic, is it possible to see the major message in this idiosyncratically scaled pair of graphs.

17.6.3 Functions of Predicted Values

Figures 17.15 and 17.16 show the predicted values for the simplest of the logistic regression models for the lymph dataset. The observed x=acid.ph are the same in both graphs. The panels in Figure 17.15 are conditioned on the level of X.ray. Each panel in Figure 17.16 shows both levels of X.ray and a different transformation of the predicted probability. Figure 17.15 superposes the predicted p.hat values in the probability scale onto the jittered nodes.j data first shown in Figure 17.13. The variable $\hat{p}$=p.hat is in the same scale (0 to 1) as the observed data. In this scale, the S-shaped logistic curve is apparent. The p.hat panel of Figure 17.16 shows only the predicted probability of success and does not show the observed node values.

The odds transformation $\frac{\hat{p}}{1-\hat{p}}$=odds.hat in the odds.hat panel of Figure 17.16 shows that the predicted odds for the two groups have a constant ratio. Constant ratios are difficult to see in a graph. The third transformation, the logit transformation $logit(\hat{p}) = log(\frac{\hat{p}}{1-\hat{p}})$=logit.p.hat, most clearly shows the model assumptions: The two lines for the two groups are parallel. A constant difference in the log(odds) corresponds to a constant ratio in the odds.

17.7 Model Specification

Logistic regression is a special case of generalized linear modeling; hence the programming constructs developed for ordinary linear models can be used.

The ANOVA tables and predicted values will agree across programs. The regression coefficients and their standard deviations depend on the choice of coding schemes for the factors. R and S-Plus use the treatment contrasts (defined in Section 10.3) with argument base=1 as the default (In R notation, contr.treatment(k, base=1)). SAS uses the treatment contrasts with base=k for a k-level factor. In R notation, SAS uses contr.treatment(k, base=k).

17.7.1 Fitting Models When the Response Is Dichotomous

The response variable nodes in the lymph dataset is dichotomous, that is, it can take only the two values 0 or 1.

17.7.1.1 R and S-Plus

The basic R and S-Plus statements for estimating logistic models with dichotomous response variables as in the example in Table 17.7 and Figures 17.11 and 17.12, are a special case of generalized linear models,

```
lymph3.glm <- glm(nodes ~ X.ray + stage + grade + acid.ph,
                  data=lymph, family=binomial)
anova(lymph3.glm, test="Chisq")
summary(lymph3.glm)
```

The user needs to use ordinary data management statements to arrange for the appropriate subjects and variables to appear in the data.frames. Graphical display of the results also requires data management statements to select columns of the data and to construct the transformations of the predicted values.

17.7.1.2 SAS

The basic SAS statements for estimating logistic models with dichotomous response variables as in the example in Table 17.7 and Figure 17.11, use PROC LOGISTIC

```
proc logistic data=lymph descending;
    title "lymph 3" ;
    model Y = X_ray  stage  grade  acid_ph ;
    output out=probs3 predicted=prob xbeta=logit;
run;
```

The user needs to use ordinary DATA step data management statements to arrange for the appropriate subjects and variables to appear in the datasets. Graphical display of the results also requires data management statements to select columns of the data and to construct the transformations of the predicted values. SAS PROC LOGISTIC doesn't accept the nesting and crossing statements of PROC GLM. For more complex models we will need to construct interaction variables acidXry0 and acidXry1 in an extra DATA step. The descending option is needed to make PROC LOGISTIC model $P(\text{nodes} = 1)$. The default is to model $P(\text{nodes} = 0)$.

17.7.2 Fitting Models When the Response Is a Sample Proportion

The examples in Sections 17.1 and 17.4 both have dichotomous (0 or 1) responses. As illustrated in Section 17.3, responses for logistic regression models can also be proportions. Models with a proportion-valued response variable are requested in Exercises 17.8 and 17.9. Specifying such models differs slightly from specifications of models with dichotomous responses.

Suppose the total number of observations is in a column named n.total for R or S-Plus (or n_total for SAS), and the number of observations having the attribute under study is in a column named n.attribute (or n_attribute for SAS). Further assume there are two explanatory variables X1 and X2; extension to situations with more than two explanatory variables will be obvious.

R and S-Plus use a syntax based on the odds ratio and SAS uses a syntax based on the proportion for this type of problem.

17.7.2.1 R and S-Plus

glm() uses the symbolic representation of the *odds ratio*
        cbind(n.attribute, n.total-n.attribute)
for the response in the model formula. Here is the entire model formula

```
glm(cbind(n.attribute, n.total-n.attribute) ~ X1 + X2,
    family=binomial)
```

17.7.2.2 SAS

PROC GENMOD uses the symbolic representation of the *proportion*
        n_attribute/n_total
for the response in the model statement. Here is the entire model statement

```
MODEL n_attribute/n_total = X1  X2 / dist=binomial;
```

17.8 LogXact

The model estimates shown in Section 17.2 and confidence limits and p-values in computer listings throughout this chapter are based on maximum likelihood estimation, and asymptotic (large-sample) standard errors and normality. LogXact[©] Cytel Software Corporation (2004) is a software package that uses algorithms from Mehta et al. (2000) to perform exact, rather than asymptotic, estimation and inference for logistic regression models. Documentation for LogXact states that its exact results sometimes differ appreciably from the asymptotic results provided by R, S-Plus, or SAS. We recommend that readers who work extensively with logistic regression models become familiar with this package.

17.9 Exercises

We recommend that for all exercises involving a data set, you begin by examining a scatterplot matrix of the variables.

17.1. In Equations (17.6) and (17.9) we show the likelihood and loglikelihood equations for the logistic regression model (17.4). For the dataset

x	y
1	1
2	0
3	1
4	0

estimate the parameters β_0 and β_1. This problem is almost doable by hand. The derivatives in Equation (17.10) are tedious (remember to substitute the data values for x and y and to take the derivatives with respect to β_j). Once you have simplified the derivatives (an easy task), you have two nasty fractions in β_0 and β_1 to set equal to zero and solve simultaneously (a hard task). Verify that you have done the algebra and arithmetic correctly by comparing the results to a computer program. You can also substitute these answers into your simultaneous equations and verify that they are both satisfied.

Computer programs usually use iterative techniques rather than the brute-force technique suggested here.

17.2. Complete the analysis of the data in Exercise 17.1. Plot the data and the analysis and interpret the numerical values and the graph.

17.3. Mendenhall et al. (1989), subsequently discussed in several other places, studied the effect of radiotherapy on the absence (response = 1) or presence (response = 0) of tongue carcinoma three years after the completion of radiotherapy. The explanatory variable days is the number of days across which the fixed total dose of radiotherapy was administered. The data is accessed as data(tongue).

a. Use logistic regression to model response as a function of days. Is the coefficient of days significantly different from 0? Estimate the change in the odds of response resulting from one additional day of radiotherapy.

b. Produce a plot of the fitted equation with 95% prediction bands analogous to Figure 17.3d.

c. The negative arithmetic sign of the coefficient of days may at first glance seem counterintuitive. Offer a possible explanation for this result.

17.4. Dataset data(icu), from Hosmer and Lemeshow (2000), presents the ICU data, a selection of cases from a larger study of survival in an intensive care unit.

Table 17.10 Code Sheet for the ICU Study. This is Table 1.5 from Hosmer and Lemeshow (2000).

Variable	Description	Codes/Values	Name
1	Identification Code	ID Number	ID
2	Vital Status	0 = Lived, 1 = Died	STA
3	Age	Years	AGE
4	Sex	0 = Male, 1 = Female	SEX
5	Race	1 = White, 2 = Black, 3 = Other	RACE
6	Service at ICU Admission	0 = Medical, 1 = Surgical	SER
7	Cancer Part of Present Problem	0 = No, 1 = Yes	CAN
8	History of Chronic Renal Failure	0 = No, 1 = Yes	CRN
9	Infection Probable at ICU Admission	0 = No, 1 = Yes	INF
10	CPR Prior to ICU Admission	0 = No, 1 = Yes	CPR
11	Systolic Blood Pressure at ICU Admission	mm Hg	SYS
12	Heart Rate at ICU Admission	Beats/min	HRA
13	Previous Admission to an ICU Within 6 Months	0 = No, 1 = Yes	PRE
14	Type of Admission	0 = Elective, 1 = Emergency	TYP
15	Long Bone, Multiple, Neck, Single Area, or Hip Fracture	0 = No, 1 = Yes	FRA
16	PO2 from Initial Blood Gases	$0 = > 60, 1 = \leq 60$	PO2
17	PH from Initial Blood Gases	$0 = \geq 7.25, 1 = < 7.25$	PH
18	PCO2 from Initial Blood Gases	$0 = \leq 45, 1 > 45$	PCO
19	Bicarbonate from Initial Blood Gases	$0 = \geq 18, 1 < 18$	BIC
20	Creatinine from Initial Blood Gases	$0 = \leq 2.0, 1 = > 2.0$	CRE
21	Level of Consciousness at ICU Admission	0 = No Coma or Deep Stupor, 1 = Deep Stupor, 2 = Coma	LOC

The major goal of the study was to develop a logistic regression model to predict the probability of survival to hospital discharge of these patients. The code sheet for the data is in Table 17.10. We will initially look at just two variables, AGE and SEX. There are a few young males and no young females who did not survive, suggesting that SEX might be an important indicator.

a. Plot the response variable STA (survival status) against the continuous variable AGE conditioned on SEX.

b. Fit a logistic regression model to STA with predictor variables AGE and SEX.

c. Plot the logistic fit of STA against AGE conditioned on SEX, comparable in style to Figure 17.17.

d. Interpret your results.

17.5. For the same dataset data(icu) used in Exercise 17.4, we will also look at CPR status. Did the patient receive CPR prior to being admitted?

a. Plot the response variable STA (survival status) against the continuous variable AGE conditioned on CPR.

b. Fit a logistic regression model to STA with predictor variables AGE and CPR.

c. Plot the logistic fit of STA against AGE conditioned on CPR, comparable in style to Figure 17.17.

d. Interpret your results.

17.6. The dataset data(esr) from Collett (1991) deals with the relationship between erythrocyte sedimentation rate (ESR) and two other blood chemistry measures, fibrin, the level of plasma fibrinogen (g/liter), and gamma globulin level (g/liter). ESR is the settlement rate of red blood cells from suspension in blood. Healthy individuals have ESR below 20 mm/hr. In this analysis the variable ESR is dichotomized to be 1 if ESR < 20 and 0 if ESR ≥ 20. Use logistic regression to determine if fibrin or globulin impact on ESR.

17.7. Lee (1980) describes an investigation to determine if any of six prognostic variables can be used to predict whether a patient will respond to a treatment for acute myeloblastic leukemia. The dataset data(leukemia) contains the prognostic variables in columns 1–6 and response = 1 if responds to treatment and 0 if doesn't respond to treatment in column 7. (This data file also contains variables in columns 8–9 that are to be ignored in this exercise.) The prognostic variables are

age: in years

smear: smear differential (%)

infiltrate: absolute infiltrate (%)

labeling: labeling index (%)

blasts: absolute blasts ($\times 10^3$)

temp: temperature ($\times 10$ degrees F)

a. Construct scatterplot matrices of the prognostic variables conditioned on the two values of response. Which prognostic variables seem most closely associated with response?

b. Justify a log transformation of `blasts`.

c. Find a good-fitting model to explain `response`. Interpret all estimated logistic regression coefficients.

17.8. Higgens and Koch (1977), later reprinted in Hand et al. (1994), discuss a study of workers in the U.S. cotton industry to discover factors that relate to contraction of the lung disease byssinosis. In the dataset `data(byss)` the columns are

`yes:`   number suffering from byssinosis

`no:`   number not suffering from byssinosis

`dust:`   dustiness of workplace, 1 = high, 2 = medium, 3 = low

`race:`   1 = white, 2 = other

`sex:`   1 = male, 2 = female

`smoker:`   1 = yes, 2 = no

`emp.length:`   length of employment:
   1 = "<10 years", 2 = "10–20 years", 3 = ">20 years"

a. Fit a logistic model to explain what factors affect the probability of contracting byssinosis. Since the response variable involves the counts in the two response categories rather than a dichotomous indicator, you will need to use model syntax similar to that described in Section 17.7.2. Look at the main effects and the two-way interactions.

b. Produce a plot showing all the significant main effects and interactions. One such plot is a multipanel display of the observed proportion of byssinosis sufferers against each of the significant effects.

c. Carefully state all conclusions.

17.9. Murray et al. (1981), also in Hand et al. (1994), report on a survey that explored factors affecting the pattern of psychotropic drug consumption. The columns of the dataset `data(psycho)` are

`sex:`   0 = male, 1 = female

`age.group`

`mean.age`

`GHQ:`   General Health Questionnaire, 0 = low, 1 = high

`taking:`   number taking psychotropic drugs

`total:`   total number

a. Use logistic regression to model the proportion taking psychotropic drugs as a function of those explanatory variables among `sex`, `agegroup`, and `GHQ` and their interactions that are significant. Since the response variable is a sample proportion, not a dichotomous indicator, you will have to fit models using the syntax described in Section 17.7.2.

b. Produce a plot showing all the significant main effects and interactions. One such plot is a multipanel display of the observed proportion of byssinosis sufferers against the significant effects.

c. Interpret the logistic regression coefficients of all main effects.

17.10. Table 17.7 shows a *p*-value of 0.45 for the test that the coefficient of `grade` is zero.

a. Fit the model `logit.p.hat ~ X.ray + stage + acid.ph`.

b. Account for the difference between the *p*-values in Table 17.7 and in part 17.10a by fitting and interpreting a model containing the term `acid.ph %in% grade`.

Chapter 18

Time Series Analysis

18.1 Introduction

Time series analysis is the technique used to study observations that are measured over time. Examples include natural phenomena (temperature, humidity, wind speed) and business variables (price of commodities, stock market indices) that are measured at regular intervals (hourly, daily).

Like regression analysis, time series analysis seeks to model a response variable as a function of one or more explanatory variables. Time series analysis differs from other forms of regression analysis in one fundamental way. Previously we have assumed the observations are uncorrelated with each other, except perhaps through their dependency on the explanatory variables. Thus, in regression, we would model

$$Y = X\beta + \epsilon$$

and make the assumption that corr(ϵ_i, ϵ_j) = 0. In time series analysis we do not make the assumption of independence. Instead we make an explicit assumption of dependence and the task of the analysis is to model the dependence.

Conventionally, the notation used to denote a time series is $X_t, t = 1, 2, \ldots, n$. In this chapter we assume that the successive times at which observations are taken are equally spaced apart, for example, monthly, quarterly, or annual observations. (Extensions to non-equally spaced observations are in the **zoo** package in R.) Initially, we also assume that X_t is a *stationary* zero-mean time series. (A time series is said to be stationary if the distribution of $\{X_t, \ldots, X_{t+n}\}$ is the same as that of $\{X_{t+k}, \ldots, X_{t+n+k}\}$ for any choice of t, n, and k.) This means that as time passes, the series does not drift away from its mean value. Often when stationarity is absent it can be achieved by analyzing differences between successive terms of the time series rather than the original time series itself.

© Springer Science+Business Media New York 2015   631
R.M. Heiberger, B. Holland, *Statistical Analysis and Data Display*,
Springer Texts in Statistics, DOI 10.1007/978-1-4939-2122-5_18

The term *lag* is used to describe earlier observations in a sequence. We indicate lagged observations with the *backshift* operator B, defined to mean

$$BX_t = B^1 X_t = X_{t-1}$$

and by extension

$$B^2 X_t = B(BX_t) = BX_{t-1} = X_{t-2}$$

Often B is used as the argument of a polynomial function. For example, if

$$\psi(B) = 3B^2 - 4B + 2$$

then

$$\psi(B)X_t = 3X_{t-2} - 4X_{t-1} + 2X_t$$

There is a distinction between the residual error ϵ in regression analysis and ε components of time series models. A time series $X_t, t = 1, 2, \ldots, n$, is a special case of a large class of models known as *stochastic processes*. Random variables $\varepsilon_t, \varepsilon_{t-1}, \ldots$ are *random shocks* to the process. This shock concept is distinct from regression residuals ϵ that represent the inability of model predictors to completely explain the response. In some time series models a linear combination of lagged ε's, $\sum \theta_k \varepsilon_{t-k}$, can be viewed as an error concept analogous to ϵ in regression.

The ARIMA class of models discussed in this chapter can be fit to most regular time series that exhibit systematic behavior with random perturbations that are small compared to the systematic components. They are not appropriate for modeling time series having irregular cyclical behavior (such as the business cycle when modeling Gross National Product) or irregular sizeable shocks (such as federal spending for relief from natural disasters such as major hurricanes, floods, earthquakes, etc.).

Standard time-related data manipulations are easier when the time parameter is built into the data object. Fundamental operations like comparisons of two series or merging two series (`ts.union` or `ts.intersect`) are easily specified and the program automatically aligns the time parameter. Table 18.1 illustrates the two alignment options.

18.2 The ARIMA Approach to Time Series Modeling

In this chapter we introduce the Box–Jenkins ARIMA approach to time series modeling Box and Jenkins (1976). This methodology involves two primary types of dependence structures, autoregression and moving averages, as well as the concept of differencing. We assume throughout that the independent random shocks ε_t are distributed with mean 0 and a common variance σ^2.

Table 18.1 Alignment of time series with the (`ts.union` or `ts.intersect`) functions.

```
> x <- ts(sample(10), start=1978)

> y <- ts(sample(6), start=1980)

> x
Time Series:
Start = 1978
End = 1987
Frequency = 1
 [1] 10  6  9  1  4  3  2  7  8  5

> y
Time Series:
Start = 1980
End = 1985
Frequency = 1
[1] 3 4 6 1 2 5
```

```
> ts.union(x,y)                    > ts.intersect(x,y)
Time Series:                       Time Series:
Start = 1978                       Start = 1980
End = 1987                         End = 1985
Frequency = 1                      Frequency = 1
        x  y                              x y
1978   10 NA                       1980   9 3
1979    6 NA                       1981   1 4
1980    9  3                       1982   4 6
1981    1  4                       1983   3 1
1982    4  6                       1984   2 2
1983    3  1                       1985   7 5
1984    2  2
1985    7  5
1986    8 NA
1987    5 NA
```

18.2.1 AutoRegression (AR)

The equation describing the first-order autoregression model AR(1) is

$$X_t = \phi X_{t-1} + \varepsilon_t \tag{18.1}$$

Each observation X_t is correlated with the preceding observation (at lag=1) X_{t-1} and, to a lesser extent, with all earlier observations. In the AR(1) model, each observation X_t has correlation ϕ with the preceding (at lag=1) observation X_{t-1}. The correlation of X_t with X_{t-k} is

$$\text{corr}(X_t, X_{t-k}) = \phi^k, \quad k = 1, 2, \ldots \tag{18.2}$$

That is, the correlation decreases exponentially with the length of lag. For example,

$$\text{corr}(X_t, X_{t-1}) = \text{corr}(\phi X_{t-1} + \varepsilon_t, X_{t-1})$$
$$= \phi$$

and

$$\text{corr}(X_t, X_{t-2}) = \text{corr}(\phi X_{t-1} + \varepsilon_t, X_{t-2})$$
$$= \text{corr}(\phi(\phi X_{t-2} + \varepsilon_{t-1}) + \varepsilon_t, X_{t-2})$$
$$= \phi^2 \text{corr}(X_{t-2}, X_{t-2}) + \text{corr}(\phi \varepsilon_{t-1} + \varepsilon_t, X_{t-2})$$
$$= \phi^2 + 0$$

The AR(1) model is further discussed in Section 18.5.2.

With p-order lags, the autoregression equation is written as

$$\Phi_p(B)X_t = \varepsilon_t \tag{18.3}$$

where

$$\Phi_p(B) = \phi(B) = 1 - \phi_1 B - \ldots - \phi_p B^p$$

is a pth-degree polynomial. This model is referred to as AR(p). The AR(1) model in Equation (18.1) is the special case where $\Phi_p(B) = \Phi_1(B) = 1 - \phi B$.

18.2.2 Moving Average (MA)

The equation describing the first-order moving average model MA(1) is

$$X_t = \varepsilon_t - \theta \varepsilon_{t-1} \tag{18.4}$$

This model is called "moving average" because the right-hand side is a weighted moving average of the independent random shock ε_t at two adjacent time periods.

Each observation X_t in the MA(1) model is correlated with the preceding observation X_{t-1} and is uncorrelated with earlier observations. For example,

$$\text{corr}(X_t, X_{t-1}) = -\theta/(1 + \theta^2)$$

and

$$\text{corr}(X_t, X_{t-2}) = \text{corr}(\varepsilon_t - \theta \varepsilon_{t-1}, \varepsilon_{t-2} - \theta \varepsilon_{t-3})$$
$$= 0$$

With q-order lags, the equation is written as

$$X_t = \Theta_q(B)\varepsilon_t \tag{18.5}$$

where

$$\Theta_q(B) = \theta(B) = 1 - \theta_1 B - \ldots - \theta_q B^q$$

is a qth-degree polynomial. This model is denoted MA(q). The MA(1) model in Equation (18.4) is the special case where $\Theta_q(B) = \Theta_1(B) = 1 - \theta B$. The MA(1) model is further discussed in Section 18.5.3.

18.2.3 Differencing

Differencing of order 1 is defined by

$$\nabla X_t = (1 - B)X_t = X_t - X_{t-1} \tag{18.6}$$

Simple models are written for the differenced data, for example,

$$\nabla X_t = \varepsilon_t - \theta \varepsilon_{t-1} \tag{18.7}$$

or, equivalently

$$X_t - X_{t-1} = \varepsilon_t - \theta \varepsilon_{t-1}$$

Model (18.7) is structurally the same as Model (18.4) in that it has the same right-hand side. The left-hand sides differ. Model (18.7) uses the differenced time series ∇X_t as its response variable where Model (18.4) used the observed variable X_t. Differencing removes nonstationarity in the mean. More complicated models involving higher-order differencing are denoted by a polynomial

$$\nabla^d(B) = (1 - B)^d$$

The interpretation is

$$\nabla^1(B)X_t = X_t - X_{t-1}$$

18.2.4 Autoregressive Integrated Moving Average (ARIMA)

We work with both AR(p) and MA(q) with lags greater than or equal to 1, and with a combined situation called ARIMA(p, d, q) (autoregressive integrated moving average). The term *integrated* means that we use the AR and MA techniques on *differenced* data. The general form of the ARIMA(p, d, q) model is

$$\Phi_p(B)\, \nabla^d X_t = \Theta_q(B)\, \varepsilon_t \tag{18.8}$$

where ε_t is a random shock with mean zero and var$(\varepsilon_t) = \sigma_\varepsilon^2$.

There are many important special cases.

ARIMA(1,0,0) = AR(1)   model is in Equation (18.1).

ARIMA(0,0,1) = MA(1)   model is in Equation (18.4).

ARIMA(0,1,0)   is the first difference model in Equation (18.6).

ARIMA(0,1,1)   model is shown in Equation (18.7).

ARIMA(1,1,1)   model looks like

$$\Phi_1(B)\nabla X_t = \Theta_1(B)\varepsilon_t$$
$$(1 - \phi B)(1 - B)X_t = (1 - \theta B)\varepsilon_t$$
$$(1 - (1 + \phi)B + \phi B^2)X_t = (1 - \theta B)\varepsilon_t$$
$$X_t - (1 + \phi)X_{t-1} + \phi X_{t-2} = \varepsilon_t - \theta\varepsilon_{t-1}$$

ARIMA($p, 0, q$)   with $d = 0$, hence no differencing, is also called an ARMA(p, q) model (autoregressive moving average)).

18.3 Autocorrelation

Two principal tools for studying time series are the autocorrelation function (ACF) and the partial autocorrelation function (PACF). The ACF assists in the diagnosis of MA models. The PACF is used in the diagnosis of AR models.

18.3.1 Autocorrelation Function (ACF)

The defining equation for the lag-k autocorrelation coefficient ρ_k is

$$\rho_k = \text{acf}(k) = \text{corr}(X_t, X_{t-k})$$

The discrete function $\{\rho_k\}$ indexed by the lag k is called the autocorrelation function of the series Z. The sample estimators $\{r_k\}$ are defined by

$$\bar{X} = \frac{1}{n} \sum_{t=1}^{n} X_t$$

$$c_k = \frac{1}{n} \sum_{t=k+1}^{n} (X_t - \bar{X})(X_{t-k} - \bar{X}) \quad \text{autocovariance}$$

$$r_k = c_k/c_0 \qquad\qquad\qquad \text{autocorrelation}$$

Note that the division is always by n.

The ACF for a time series $z = (x_1, \ldots, x_n)$ is calculated in R by

$$\text{acf}(z)$$

18.3.2 Partial Autocorrelation Function (PACF)

The defining equation for the PACF is

$$\phi_{kk} = \text{pacf}(k) = \text{corr}(X_t, X_{t-k} \mid X_{t-1}, X_{t-2}, \ldots, X_{t-(k-1)})$$

The sample estimators are defined by solving the Yule–Walker equations that hold for an AR(p) process (see Box and Jenkins (1976) for details). An illustrative (but not practical) estimator is shown in the R function in Table 18.2. The PACF is calculated in R by

$$\text{acf}(z, \text{type="partial"})$$

Note that $\rho_1 = \phi_{11}$, that is pacf(1) = acf(1), in mathematics notation. In R notation, the same statement is

$$\text{acf}(x)["1"]\$acf \ == \ \text{acf}(x, \text{type="partial"})["1"]\$acf$$

for a numeric vector x.

18.4 Analysis Steps

There are three main steps in time series analysis using the ARIMA models of the Box–Jenkins approach.

Identification: choice of the proper transformations to apply to the time series, consisting of variance-stabilizing transformations and of differencing. Determining the number of model parameters: d, the order of differencing; p, the number of autoregressive parameters; and q, the number of moving average parameters.

Table 18.2 Illustrative definition of the PACF. Do **NOT** use in actual calculations!

```
> ## This function illustrates the definition of the pacf.
> ## Do NOT use in actual calculations!
> ##
> ## my.pacf requires a detrended series, otherwise the answer is
> ## nonsense, as it starts losing precision after the first few lags.
>
> my.pacf <- function(z, k=2) {
+   z <- z - mean(z)
+   x <- ts.intersect(z, lag(z,-1))
+   if (k==1) return(cor(x[,1], x[,2]))
+   for (kk in 2:k) x <- ts.intersect(x, lag(z,-kk))
+   nr <- nrow(x)
+   nc <- ncol(x)
+   r1 <- lm(x[,1]   ~ -1 + x[,-c(1,nc)])$resid
+   r2 <- lm(x[,nc]  ~ -1 + x[,-c(1,nc)])$resid
+   cor(r1,r2)
+ }

> my.pacf(ozone.subset, 2)
[1] -0.2665

> acf(ozone.subset, type="partial", plot=FALSE)$acf[2]
[1] -0.2583
```

Estimation: estimation of the parameters of the identified model, usually by maximum likelihood.

Diagnostics: verification that the estimated model and parameters do indeed capture the essence of the behavior of the data.

We offer these recommendations for interpreting sequence plots, ACF plots, and PACF plots. They are based on the Box–Jenkins methodology described in the texts by Box and Jenkins (1976) and Wei (1990).

1. Trends in the sequence plot must be removed by differencing. This is required before attempting to interpret the ACF and PACF plots. The interpretation below of ACF and PACF plots depends on stationarity.

2. No correlation—white noise
   The ACF and PACF are negligible at all lags.

3. AR(p)
   The ACF decays slowly.
   The PACF cuts off at lag p.

4. MA(q)
   The ACF cuts off at lag q.
   The PACF decays slowly.

5. ARMA(p, q)
   The ACF decays slowly from lag $\max(q - p, 0)$ on.
   The PACF decays slowly from lag $\max(p - q, 0)$ on.

   The orders p and q usually can't be read directly from these plots. Looking at the ACF and PACF plots for models with larger values of p and q can be helpful. The ESACF (extended sample autocorrelation function) (see, for example, Wei (1990), p. 128) can also be helpful.

   Several additional tools are used to identify well-fitting models.

- The *Akaike information criterion* (AIC) for a particular model is defined as $-2(\ln L) + 2m$, where L is the model's loglikelihood and m is the number of parameters needed to estimate the model. Like the C_p statistic used to decide among multiple regression models, introduced in Equation (9.28), the AIC is the sum of a goodness-of-fit component and a penalty for lack of simplicity. Low values of AIC are preferred to large values.

- The *portmanteau goodness-of-fit test* for a particular model at lag ℓ is actually a collection of tests, one for each $k = 1, 2, \ldots, \ell$. Each of these individual tests is a test of the negligibility of the autocorrelations of the model residuals up to and including lag ℓ. In a well-fitting model, these hypotheses should be retained because such a model should have negligible autocorrelations. Therefore, for well-fitting models the p-values of these tests should not be small.

- The highest-order AR and MA parameters of well-fitting models are significantly different from zero, indicating that the corresponding model terms and terms of lower order are needed. Such significance is suggested by a corresponding t statistic that exceeds 2 in absolute value. For example, a model with $p = 2$ must have ϕ_2 significantly different from zero, but it is not essential that ϕ_1 be significantly nonzero.

- A well-fitting model has an estimated residual variance $\hat{\sigma}^2$ at least as small as those of competing models. The estimated residual variance and the AIC carry similar information. The AIC is usually preferred to the residual variance because the AIC includes a penalty for lack of simplicity and the residual variance does not.

In most situations these diagnostics will point to the same uniquely best model or subset of equivalently well-fitting models.

18.5 Some Algebraic Development, Including Forecasting

Usually, the ultimate purpose of finding a well-fitting time series model is the production of forecasts and forecast intervals h periods beyond the final observation of the existing series. While we have shown how to produce forecasts and intervals in R, here we provide a brief introduction to the algebra behind such forecasts. The algebra is intractable by hand for all but a few special cases.

18.5.1 The General ARIMA Model

The time series model for a 0-mean time series X_t, with $E(X_t) = 0$ and $\text{var}(X_t) = \sigma^2$, is

$$\phi(B)\nabla^d X_t = \theta(B)\,\varepsilon_t \tag{18.9}$$

One way to rewrite (18.9) is

$$\varepsilon_t = \theta^{-1}(B)\phi(B)\nabla^d X_t$$

Once the coefficients of ϕ and θ have been estimated by maximum likelihood, the fitted model is expressed in terms of the calculated residuals as

$$\hat{\varepsilon}_t = \hat{\theta}^{-1}(B)\hat{\phi}(B)\nabla^d X_t$$

where $\hat{\theta}(\cdot)$ and $\hat{\phi}(\cdot)$ are the polynomials in B after the coefficient estimates have been substituted into $\theta(\cdot)$ and $\phi(\cdot)$.

The calculated residuals $\hat{\varepsilon}_t$ will be used in many subsequent calculations.

The model (18.9) can also be rewritten as

$$
\begin{aligned}
X_t &= \phi^{-1}(B)\nabla^{-d}\,\theta(B)\,\varepsilon_t \\
&\overset{\text{def}}{=} \psi(B)\,\varepsilon_t \\
&= (1 + \psi_1 B + \ldots)\,\varepsilon_t \\
&= \varepsilon_t + \psi_1\varepsilon_{t-1} + \ldots + \psi_k\varepsilon_{t-k} + \ldots
\end{aligned}
$$

where $\psi(B)$ may have an infinite number of terms. The number of terms is finite with purely MA models (where $\psi = \theta$) and infinite when there are AR or differencing factors. In order that the model be stationary and invertible (that is, explicitly solvable for X_t), it is required that the roots of both polynomials $\phi(B)$ and $\psi(B)$ lie outside the unit circle. In addition $\phi(B)$ and $\theta(B)$ must have no roots in common. If the polynomials have common roots, these roots can be factored out.

The nonzero-mean case is essentially the same. Let the nonzero-mean time series be $Y_t = X_t + \mu$. We can subtract the mean from the observed Y_t-values to construct a 0-mean times series $X_t = Y_t - \mu$ and then proceed.

When we use the model for forecasting h steps ahead, we use the equation

$$\hat{X}_{t+h} = E(X_{t+h}|X_t, X_{t-1}, \ldots) = (\psi_h + \psi_{h+1}B + \ldots)\varepsilon_t$$

with forecast error

$$e_{t+h} = X_{t+h} - \hat{X}_{t+h} = \varepsilon_{t+h} + \psi_1\varepsilon_{t+h-1} + \ldots + \psi_{h-1}\varepsilon_{t+1}$$

and with variance of the forecast error

$$\text{var}(e_{t+h}) = \sigma^2(1 + \psi_1^2 + \psi_2^2 + \ldots + \psi_{h-1}^2)$$

The forecast error for h-step ahead forecasts, and its variance, have exactly h terms. The ε_t are uncorrelated. The forecast errors are correlated.

Probability limits for the forecasts are calculated as

$$\hat{X}_{t+h} \pm z_{\alpha/2}\,\hat{\sigma}\,\sqrt{1 + \psi_1^2 + \psi_2^2 + \ldots + \psi_{h-1}^2}$$

18.5.2 Special Case—The AR(1) Model

Starting from $X_2 = \phi X_1 + \varepsilon_2$, incrementing the subscripts on X_t, and then back-substituting [for example, $X_3 = \phi(\phi X_1 + \varepsilon_2) + \varepsilon_3$], we eventually get

$$X_{t+h} = \phi^h X_t + \phi^{h-1}\varepsilon_{t+1} + \ldots + \phi\varepsilon_{t+h-1} + \varepsilon_{t+h}$$

As a consequence, we take $\hat{X}_{t+h} = \hat{\phi}^h X_t$. Further,

$$\text{var}(X_{t+h}) = \sigma^2(1 + \phi^2 + \phi^4 + \ldots + \phi^{2(h-1)})$$

$$= \sigma^2\left(\frac{1 - \phi^{2h}}{1 - \phi^2}\right)$$

A $100(1 - \alpha)\%$ prediction interval for X_{t+h} is

$$\hat{X}_{t+h} \pm z_{\alpha/2}\,\hat{\sigma}\,\sqrt{\frac{1 - \hat{\phi}^{2h}}{1 - \hat{\phi}^2}}$$

18.5.3 Special Case—The MA(1) Model

Here we have $X_{t+1} = \varepsilon_{t+1} - \theta_1 \varepsilon_t$, and the general formulas simplify to

$$
\begin{aligned}
\hat{X}_{t+1} &= -\theta_1 \hat{\varepsilon}_t && \text{for } h = 1 \\
\hat{X}_{t+h} &= 0 && \text{for } h > 1 \\
\text{var}(\hat{X}_{t+1}) &= \sigma^2 && \text{for } h = 1 \\
\text{var}(\hat{X}_{t+h}) &= \sigma^2(1 + \theta_1^2) && \text{for } h > 1
\end{aligned}
$$

In the MA(q) models the $\hat{\varepsilon}_{t+j}$-values are known for past observations (those for which $j \le 0$), hence they appear in the prediction equations. A $100(1 - \alpha)\%$ prediction interval for X_{t+h} is

$$
\begin{aligned}
\hat{X}_{t+1} &\pm z_{\alpha/2}\hat{\sigma} && \text{for } h = 1 \\
\hat{X}_{t+h} &\pm z_{\alpha/2}\hat{\sigma}\sqrt{1 + \theta_1^2} && \text{for } h > 1
\end{aligned}
$$

18.6 Graphical Displays for Time Series Analysis

We present a number of graphical displays to facilitate the identification and model checking steps of ARIMA(p, d, q) modeling. Much of this material previously appeared in Heiberger and Teles (2002). We discuss an extension of these displays to model time series with seasonal components in Section 18.8. A general discussion of the features of these graphs appears in Section 18.A of this chapter's appendix.

Table 18.3 summarizes the nine achievable models formed by possible combinations of the number of AR parameters ($p = 0, 1, 2$) and MA parameters ($q = 0, 1, 2$). The appearance of the left-hand and right-hand side of the model equations is shown for each value of ($p, 0, q$).

Figures 18.1 and 18.3 are examples of coordinated plots useful for identifying an ARIMA time series model.

Figure 18.1 contains a plot of the original time series along with its autocorrelation function and partial autocorrelation function. (Figure 18.2 is comparable to Figure 18.1 but for a differenced time series.)

The set of plots in Figure 18.3 consists of the residual ACF and PACF, the portmanteau goodness-of-fit test statistic (GOF), the standardized residuals, and the Akaike information criterion (AIC). The panels in the first four sets of plots are indexed by the number of ARMA parameters p and q. The AIC plot uses p and q as plotting variables. The orders of differencing and the orders of the autoregressive and moving average operators have been limited to $0 \le p, d, q, \le 2$. While this limitation is usually reasonable in practice, it is not inherent in the software.

Table 18.3 3×3 layout for the ARIMA$(p, 0, q)$ models. All the time series diagnostic plots and summary tabular data are constructed on this pattern. The rows give the number of AR parameters ($p = 0, 1, 2$) and the corresponding left-hand side of the model equation. The columns give the number of MA parameters ($q = 0, 1, 2$) and the corresponding right-hand side of the model equation. For example, the $(1, 1)$ cell of the array shows the information for the ARIMA$(1, 0, 1)$ model:
$$X_t - \phi_1 X_{t-1} = \varepsilon_t - \theta_1 \varepsilon_{t-1}$$
In all the displays we show, the differencing parameter d ($d = 0$ in this example) and the seasonal parameters $(P, D, Q)_s$ (if any) are held constant.

		Moving average model — Right-hand side		
Autoregression model		$q = 0$	$q = 1$	$q = 2$
p	Left-hand side	ε_t	$\varepsilon_t - \theta_1 \varepsilon_{t-1}$	$\varepsilon_t - \theta_1 \varepsilon_{t-1} - \theta_2 \varepsilon_{t-2}$
$p = 0$	X_t	$(0,0,0)$	$(0,0,1)$	$(0,0,2)$
$p = 1$	$X_t - \phi_1 X_{t-1}$	$(1,0,0)$	$(1,0,1)$	$(1,0,2)$
$p = 2$	$X_t - \phi_1 X_{t-1} - \phi_2 X_{t-2}$	$(2,0,0)$	$(2,0,1)$	$(2,0,2)$

Each set of nine panels is systematically structured in a 3×3 array indexed by the number of AR parameters p and MA parameters q. All nine panels in a set are scaled identically. Thus the reader can scan a row or column of the array of panels and see the effect of adding one more parameter to either the AR or MA side of the model.

The graphics are used to analyze the monthly El Nino data in file data(elnino) from NIST (2005). The *El Nino* effect is thought to be a driver of world-wide weather. The *southern oscillation* is a predictor of El Nino. It is defined as the sea level barometric pressure at Tahiti minus the sea level barometric pressure at the Darwin Islands. Repeated southern oscillation values below -1 essentially defines an El Nino. Figures 18.1 and 18.2 show the reported data y_t=elnino and the first differences $\nabla y_t \overset{\text{def}}{=} y_t - y_{t-1}$. The horizontal dashed lines on the ACF and PACF plots are the critical values for $\alpha = .05$ tests of the hypothesis, at each individual lag k, that the correlation coefficient is zero. Spikes on these plots that fall outside these horizontal boundaries suggest the possibility of a nonzero correlation.

Figure 18.1 suggests that successive months' southern oscillations are positively associated: corr$(y_t, y_{t-1}) \approx 0.65$. To address the positive association between successive months we analyze first differences in Figure 18.2; this Figure does not suggest a need for additional differencing and its ACF and PACF for the first differences shows systematic behavior: the ACF cuts off at lag 1 and the PACF decays slowly. This suggests an ARIMA$(0,1,1)$ model for the original series.

Since the ACF and PACF show systematic behavior, we proceed to Figure 18.3, a collection of five sets of coordinated plots on a single page designed to facilitate identifying the best ARIMA$(p, 1, q)$ model based on fits of the nine models with

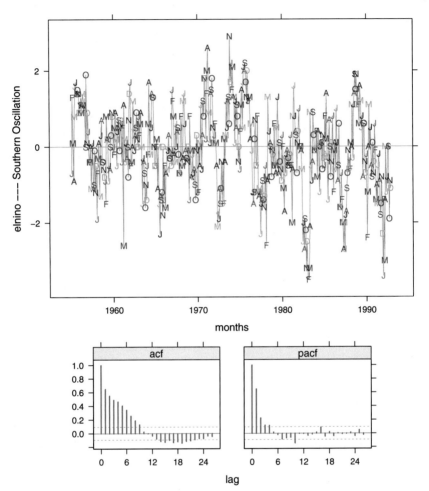

Fig. 18.1 Coordinated time series plot and ACF/PACF plots for the `elnino` time series: y_t. The response variable on the time series plot is the monthly value of the southern oscillation.

$0 \leq p, q \leq 2$. For each of these nine models, Figure 18.3 shows the ACF and PACF plots of the standardized residuals, a plot of the p-values of portmanteau goodness-of-fit tests at various lags, the Akaike information criterion arranged in the form of a pair of interaction plots, and the standardized residuals themselves. While in theory it is possible for p or q to exceed 2 (and the software permits larger values of p and/or q), this is unlikely to occur in practice provided that the data have been properly differenced and that any seasonal effects have been addressed.

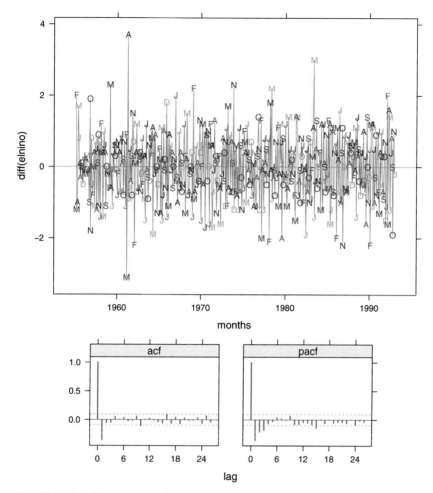

Fig. 18.2 Coordinated time series and ACF/PACF plots for the differenced `elnino` time series: ∇y_t.

Figure 18.3 confirms the choice $p = 0$, $q = 1$. For this parsimonious model,

- The ACF and PACF plots stay within the thresholds of significance.
- All p-values for the goodness-of-fit test exceed 0.05.
- The Akaike criterion is only slightly above that for all less parsimonious models.

Table 18.4 provides additional support for the ARIMA(0,1,1) model. The maximum likelihood estimate of the ARIMA parameters `coef` shows the MA(1) parameter to be 0.53. As this number is not close to 1, no further differencing is needed. The standardized value (denoted `t.coef`) of the MA(1) parameter is 13.2. This number greatly exceeds the usual critical value $|2|$.

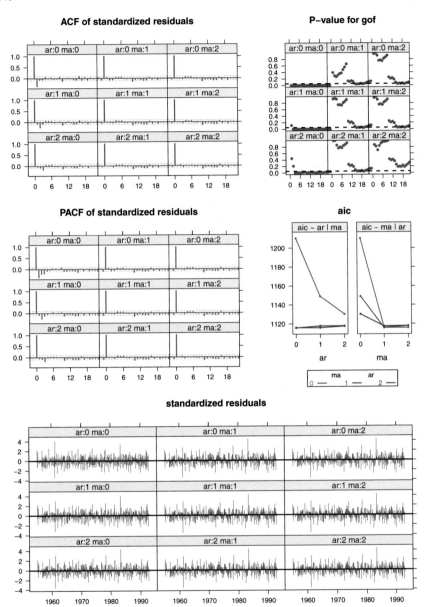

Fig. 18.3 Diagnostic plots for the set of models ARIMA(p,1,q) fit to the `elnino` data by maximum likelihood. Each set of nine panels is systematically structured in a 3×3 array with rows indexed by the number of AR parameters p and columns by the number of MA parameters q. All nine panels in a set are scaled identically. The AIC has been plotted as a pair of interaction plots: AIC plotted against q using line types defined by p; and AIC plotted against p, using line types defined by q.

Table 18.4 Estimation results for ARIMA($p, 1, q$) models fit to the elnino data.

```
> elnino.loop <- arma.loop(elnino, order=c(2,1,2))

> elnino.loop
$series
[1] "elnino"

$model
[1] "(p,1,q)"

$sigma2
        0      1      2
0 0.8333 0.6740 0.6707
1 0.7250 0.6708 0.6707
2 0.6925 0.6704 0.6678


$aic
     0    1    2
0 1210 1116 1116
1 1149 1116 1118
2 1130 1118 1118


$coef
              ar1       ar2      ma1       ma2
(0,1,0)        NA        NA       NA        NA
(1,1,0) -0.36153        NA       NA        NA
(2,1,0) -0.43720 -0.21245       NA        NA
(0,1,1)        NA        NA  -0.5258       NA
(1,1,1)  0.12397        NA  -0.6127       NA
(2,1,1)  0.07603 -0.04306  -0.5625       NA
(0,1,2)        NA        NA  -0.4868  -0.06870
(1,1,2) -0.02671        NA  -0.4604  -0.08237
(2,1,2)  0.73112 -0.23604  -1.2121   0.48199


$t.coef
              ar1       ar2       ma1      ma2
(0,1,0)        NA        NA        NA       NA
(1,1,0)  -8.2319        NA        NA       NA
(2,1,0)  -9.5122  -4.6147        NA       NA
(0,1,1)        NA        NA   -12.492       NA
(1,1,1)   1.4328        NA    -8.944       NA
(2,1,1)   0.6434  -0.5906    -5.099       NA
(0,1,2)        NA        NA   -10.391  -1.4911
(1,1,2)  -0.0642        NA    -1.115  -0.3805
(2,1,2)   3.0449  -2.0250    -5.266   2.6781
```

18.7 Models with Seasonal Components

One of the strengths of the Box–Jenkins method is its handling of seasonal parameters. Time series frequently show seasonal patterns in their correlation structure. Most economic series have annual, quarterly, monthly, or weekly patterns as well as daily patterns. For example, retail sales figures often shows a surge in activity in December; power consumption figures show seasonal patterns as heating is used in the winter months and air conditioning in the summer months, and a weekly pattern where weekday consumption differs systematically from weekend consumption.

18.7.1 Multiplicative Seasonal ARIMA Models

Seasonal parameters are handled similarly to the nonseasonal parameters, with the subscripts varying by increments of the season. With monthly data, an annual season $s = 12$ is denoted by using 12-month lags, that is, X_t and X_{t-12}. We use uppercase Greek letters Φ and Θ to denote autoregressive and moving average polynomials, respectively, in the seasonal backshift operator B^s. The polynomials in the backshift B^s are denoted $\Phi(B^s)$ and $\Theta(B^s)$, and the differences are $\nabla_s = 1 - B^s$.

The seasonal portion of a seasonal model is denoted ARIMA$(P, D, Q)_s$ (with, for example, the seasonal $s = 12$ used for annual seasons when the underlying data is monthly, and $s = 7$ for weekly seasons when the underlying data is daily), where

P is the number of lags in the seasonal AR portion of the model, equivalently the order of the polynomial

$$\Phi(B^s) = 1 - \Phi_1 B^s - \ldots - \Phi_P B^{sP}$$

Q is the number of lags in the seasonal MA portion of the model, equivalently the order of the polynomial

$$\Theta(B^s) = 1 - \Theta_1 B^s - \ldots - \Theta_Q B^{sQ}$$

D is the number of seasonal differences prior to the AR and MA modeling, equivalently the power of the differencing binomial

$$\nabla_s^D = (1 - B^s)^D$$

The general multiplicative seasonal model, denoted

$$\mathrm{ARIMA}(p, d, q) \times (P, D, Q)_s$$

is given by

$$\Phi(B^s)\phi(B)\nabla_s^D\nabla^d X_t = \Theta(B^s)\theta(B)\varepsilon_t \tag{18.10}$$

For various technical reasons, the roots of the seasonal polynomials $\Phi(B)$ and $\Psi(B)$ must satisfy certain conditions that parallel the restrictions on $\phi(B)$ and $\psi(B)$ mentioned in Section 18.5.1. The roots of $\Phi(B)$ and $\Psi(B)$ must lie outside the unit circle to assure that the model is stationary and invertible (solvable for $X(t)$). In addition, $\Phi(B)$ and $\Psi(B)$ must have no common roots. If these polynomials have common roots, these roots can be factored out.

18.7.2 Example—CO2 ARIMA$(0, 1, 1) \times (0, 1, 1)_{12}$ Model

The final model for the CO2 example discussed in Section 18.8 is written as ARIMA$(0, 1, 1) \times (0, 1, 1)_{12}$ model for X_t:

$$\nabla_{12}\nabla X_t = (1 - \theta_1 B)\left(1 - \Theta_1 B^{12}\right)\varepsilon_t \tag{18.11}$$

which expands to

$$X_t - X_{t-1} - X_{t-12} + X_{t-13} = \varepsilon_t - \theta_1\varepsilon_{t-1} - \Theta_1\varepsilon_{t-12} + \theta_1\Theta_1\varepsilon_{t-13}$$

18.7.3 Determining the Seasonal AR and MA Parameters

The procedure for determining the order P and Q of the seasonal parameters is comparable to the recommendations given in Section 18.4 for determining the order p and q of the nonseasonal parameters. As before, we work with the ACF and PACF for an appropriately differenced model. The distinction is that in examining the behavior of the ACF and PACF for seasonality, we examine only the values at seasonal intervals. For example, for monthly data with annual season ($s = 12$), these plots are examined at $t = 12, 24, 36, \ldots = 12 \times (1, 2, 3, \ldots)$, ignoring values at other times. We then visualize the cutoff or decay behavior where *lag* now refers to seasonal intervals. If the ACF decays slowly at $t = 12, 24, 36, \ldots = 12 \times (1, 2, 3, \ldots)$ and the PACF cuts off at $t = 24 = 12 \times 2$, then $P = 2$ and $Q = 0$. If the PACF decays slowly at $t = 12, 24, 36, \ldots = 12 \times (1, 2, 3, \ldots)$ and the ACF cuts off at $t = 12 \times 1$, then $P = 0$ and $Q = 1$.

18.8 Example of a Seasonal Model—The Monthly co2 Data

We extend the graphical displays discussed in Section 18.6 to the identification and model checking steps of ARIMA$(p, d, q) \times (P, D, Q)_s$ modeling. These graphs also first appeared in Heiberger and Teles (2002). A general discussion of the features of these graphs is deferred to Section 18.A.

The graphics are illustrated with one of the time series datasets distributed in the R package **datasets**, the Mauna Loa Carbon Dioxide Concentration series collected by the Scripps Institute of Oceanography, in La Jolla, California. The source is the climatology database maintained by the Oak Ridge National Laboratory Peterson (1990). These data represent monthly CO_2 concentrations in parts per million (ppm) from January 1959 to December 1990. Missing values have been filled in by linear interpolation.

Figures 18.4 through 18.7 are structured presentations of the plots of the series itself, of the ACF, and of the PACF. We show a magnified section of the plot for a five-year interval in Figure 18.5 Figure 18.4 displays the raw data series while Figure 18.6 displays the differenced series (monthly) and Figure 18.7 displays the twice-differenced series (monthly and annually).

18.8.1 Identification of the Model

Figure 18.4 is the plot of the observed data. The plot of the series itself shows a strong upward trend and a systematic labeling, with peaks occurring in the spring months and troughs in the autumn months. It is clear that the mean of this series is not constant over time. Both the ACF and PACF show systematic behavior. The ACF exhibits large values and a very slow decay with an annual periodicity. The PACF has large values and an annual periodicity. The conclusion is that the series is nonstationary, that is, it does not have a constant mean, and its autocorrelation function is time-dependent, implying that it shows nonrandom time-dependent behavior. Monthly differencing is required to model the nonstationarity, and annual differencing is necessary to remove the periodicity.

The time series and ACF and PACF plots for the differenced series ∇X_t in Figure 18.6 also show systematic annual behavior. The time series plot shows August/September troughs. The ACF exhibits a very slow decay at the seasonal lags, lags that are multiples of the seasonal period $s = 12$ months. This confirms that seasonal differencing with period 12 is required.

Figure 18.7 shows the time series (and the ACF and PACF) after non-seasonal and seasonal differencing $\nabla_{12} \nabla X_t$. There are no longer systematic components visible in the plot of the differences. The differenced series is stationary and it becomes

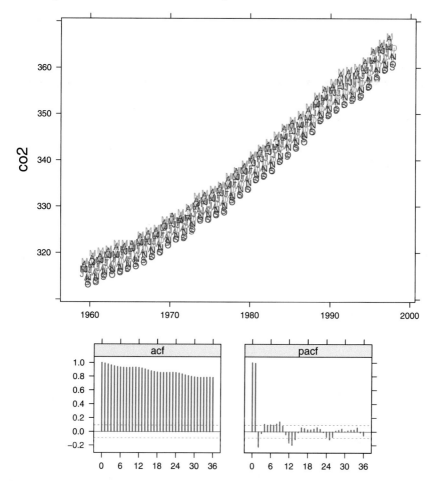

Fig. 18.4 Coordinated time series plot and ACF/PACF plots for the Mauna Loa CO_2 time series: X_t. The response variable on the time series plot is concentration in parts per million. We show a magnified section of the time series plot in Figure 18.5.

possible to identify a model for the series, that is, to look for the AR and MA parameters that best fit the twice-differenced data.

The nonseasonal component of the model of X_t is identified by looking at the first few monthly lags of the sample ACF and PACF of $\nabla_{12}\nabla X_t$ in Figure 18.7. The ACF seems to cut off after lag 1 and the PACF shows an exponential decay. The same type of behavior is seen at the seasonal lags, i.e., the ACF cuts off after lag 12 and the PACF shows an exponential decay at lags $12, 24, 36, \ldots$. These characteristics of the ACF and PACF suggest the ARIMA$(0, 1, 1) \times (0, 1, 1)_{12}$ model for X_t:

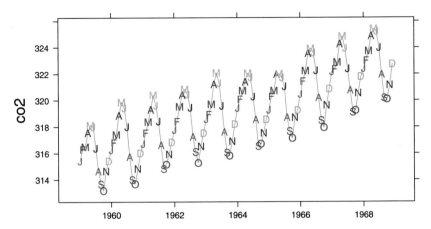

Fig. 18.5 Time series plot for 10 years of the Mauna Loa CO_2 time series: X_t. The data from 1959 to 1997 is shown in Figure 18.4.

$$\nabla_{12}\nabla X_t = (1 - \theta_1 B)\left(1 - \Theta_1 B^{12}\right)\varepsilon_t \qquad (18.12)$$

A closer look at the ACF in Figure 18.7 indicates that it too may show an exponential decay in the first lags, suggesting that the ARIMA$(1, 1, 1) \times (0, 1, 1)_{12}$ model

$$(1 - \phi_1 B)\nabla_{12}\nabla X_t = (1 - \theta_1 B)\left(1 - \Theta_1 B^{12}\right)\varepsilon_t \qquad (18.13)$$

might also be appropriate.

18.8.2 Parameter Estimation and Diagnostic Checking

In general, when analyzing seasonal time series data, initial guesses of at least some of the parameters p, q, P, Q may be provided from inspections of coordinated plots of original and differenced data such as Figures 18.4–18.7. Figures 18.8 and 18.9 each simultaneously consider nine models produced with the user function arma.loop described in Section 18.A.4. Figures in this class can be used to suggest seasonal parameters P and Q for a given set of nonseasonal and differencing parameters p, q, d, D, or to suggest nonseasonal parameters p and q for a given set of seasonal and differencing parameters P, Q, d, D. Alternating consideration of figures of both of these types can be used to settle on a final model.

Continuing with the co2 data, Figure 18.8 displays a set of diagnostic plots for several models without a seasonal component, the ARIMA$(p, 1, q) \times (0, 1, 0)_{12}$ models with $0 \le p, q \le 2$, that have been fit to the series $\nabla_{12}\nabla X_t$.

Since the co2 data exhibit a seasonal behavior, Figure 18.8 is expected to confirm that seasonal parameters are required in the model of $\nabla_{12}\nabla X_t$. All the residual ACF

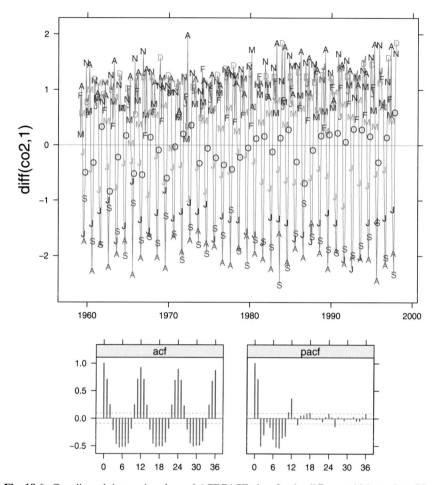

Fig. 18.6 Coordinated time series plot and ACF/PACF plots for the differenced Mauna Loa CO_2 time series: ∇X_t.

plots show a significant spike at lag=12 months, and all the GOF plots show a break at the same lag=12 months.

The residual ACF, PACF, and GOF plots in Figure 18.8 clearly confirm that seasonal parameters are necessary. The cutoff after the spike at lag=12 of the residual ACF, and the exponential decay of the residual PACF at the seasonal lags (those that are multiples of 12 months), show that a seasonal MA parameter is necessary. This agrees with the identification of candidate models (18.11) and (18.13).

Next consider the ARIMA$(p, 1, q) \times (0, 1, 1)_{12}$ models with $0 \leq p, q \leq 2$. The diagnostic plots for models including the seasonal MA parameter are in Figure 18.9.

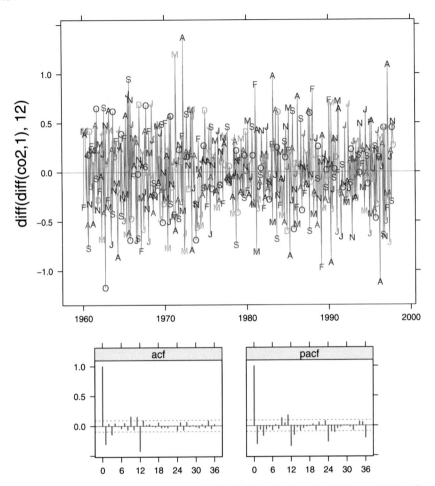

Fig. 18.7 Coordinated time series plot and ACF/PACF plots for the twice-differenced Mauna Loa CO_2 time series: $\nabla_{12}\nabla X_t$.

The $q = 0$ column of the residual ACF and GOF plots shows poor fits. The $q = 1$ column appears better than the $q = 2$ column. All three GOF plots for $q = 1$ are similar. The AIC plots show almost identical values when $q = 1$. This is seen as three almost coincident points at $q = 1$ in the "aic $\sim$ ma | ar" plot and as a horizontal line for $q = 1$ over all three values of p in the "aic $\sim$ ar | ma" plot. The conclusion is that one nonseasonal MA parameter is necessary.

Table 18.5 shows the AIC, the estimates of σ_ε^2, and the estimates of the ARMA parameters with their t-statistics for the set of ARIMA$(p, 1, q) \times (0, 1, 1)_{12}$ models. From the t.coef section of Table 18.5, the t statistics for both AR parameters in the

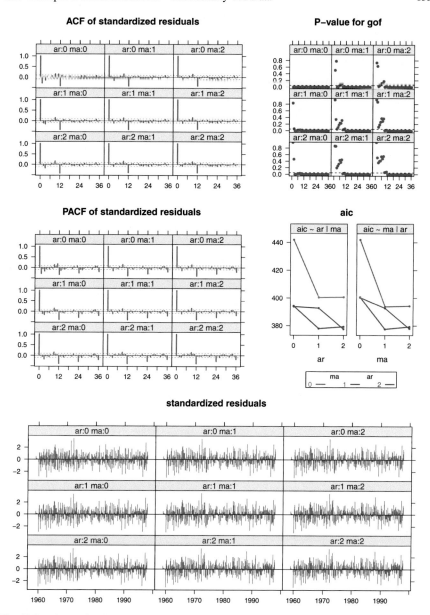

Fig. 18.8 Diagnostic plots for the set of models $ARIMA(p, 1, q) \times (0, 1, 0)_{12}$ fit to the CO_2 data by maximum likelihood. Each set of nine panels is systematically structured in a 3×3 array with rows indexed by the number of AR parameters p and columns by the number of MA parameters q. All nine panels in a set are scaled identically. The AIC is plotted as a pair of interaction plots: AIC plotted against q using line types defined by p; and AIC plotted against p, using line types defined by q.

ARIMA(2, 1, 1)×(0, 1, 1)$_{12}$ model are not significant and this model can be rejected. The t-statistic for the AR(1) parameter in the ARIMA(1, 1, 1) × (0, 1, 1)$_{12}$ model is marginally significant, leading us to consider both the models ARIMA(0, 1, 1) × (0, 1, 1)$_{12}$ and ARIMA(1, 1, 1) × (0, 1, 1)$_{12}$ for X_t. A different criterion is needed to distinguish between them. Both models are consistent with the analysis at the identification stage.

The detailed display of the estimation results for the ARIMA(0, 1, 1) × (0, 1, 1)$_{12}$ model is shown in Table 18.6 (as displayed by the new print method for arima objects). A similar display for the ARIMA(1, 1, 1) × (0, 1, 1)$_{12}$ model in Table 18.7 shows that the AR(1) and MA(1) parameters are highly correlated ($r = -.0167184/\sqrt{.02040605 \times .0152966} = -0.9463$). The ARIMA(1, 1, 1) × (0, 1, 1)$_{12}$ models can be discarded from further consideration.

The final step is the verification of the adequacy of the ARIMA(0, 1, 1)×(0, 1, 1)$_{12}$ model. The residual ACF and PACF plots exhibit no significant spikes and all the GOF p-values are also not significant, showing that the residuals are approximately white noise. The AIC values have dropped from 310 in Figure 18.8 to 136 in Figure 18.9. The standardized residuals in Figure 18.9 are not inconsistent with the normal distribution. The ARIMA(0, 1, 1)×(0, 1, 1)$_{12}$ model seems to be appropriate for X_t and the estimated model is

$$\nabla_{12}\nabla X_t = \left(1 - \hat{\theta}_1 B\right)\left(1 - \hat{\Theta}_1 B^{12}\right)\hat{\varepsilon}_t \tag{18.14}$$
$$= (1 - 0.36338\, B)(1 - 0.85806\, B^{12})\,\hat{\varepsilon}_t$$
$$\hat{\sigma}_\varepsilon^2 = 0.080299$$

Table 18.5 Estimation results for ARIMA$(p, 1, q) \times (0, 1, 1)_{12}$ models fit to the CO_2 data.

```
> ddco2.loopPQ <-
+       arma.loop(co2,
+                order=c(2,1,2),
+                seasonal=list(order=c(0,1,1), period=12))

> ddco2.loopPQ
$series
[1] "co2"

$model
[1] "(p,1,q)x(0,1,1)12"

$sigma2
        0       1       2
0 0.09063 0.08260 0.08242
1 0.08358 0.08221 0.08214
2 0.08315 0.08176 0.08162

$aic
      0     1     2
0 221.5 178.2 179.1
1 184.2 178.1 180.1
2 183.5 177.8 179.1

$coef
                        ar1       ar2     ma1      ma2    sma1
(0,1,0)x(0,1,1)12        NA        NA      NA       NA -0.8887
(1,1,0)x(0,1,1)12 -0.292654        NA      NA       NA -0.8603
(2,1,0)x(0,1,1)12 -0.317119 -0.07825      NA       NA -0.8551
(0,1,1)x(0,1,1)12        NA        NA -0.3501       NA -0.8506
(1,1,1)x(0,1,1)12  0.239889        NA -0.5710       NA -0.8516
(2,1,1)x(0,1,1)12  0.390518  0.10540 -0.7329       NA -0.8544
(0,1,2)x(0,1,1)12        NA        NA -0.3436 -0.0492 -0.8499
(1,1,2)x(0,1,1)12 -0.962631        NA  0.6204 -0.3571 -0.8440
(2,1,2)x(0,1,1)12  0.007095  0.23191 -0.3477 -0.2473 -0.8548

$t.coef
                        ar1    ar2     ma1    ma2   sma1
(0,1,0)x(0,1,1)12        NA     NA      NA     NA -36.66
(1,1,0)x(0,1,1)12  -6.43561     NA      NA     NA -34.04
(2,1,0)x(0,1,1)12  -6.64989 -1.650      NA     NA -33.66
(0,1,1)x(0,1,1)12        NA     NA -7.0529     NA -33.15
(1,1,1)x(0,1,1)12   1.67707     NA -4.6167     NA -33.29
(2,1,1)x(0,1,1)12   3.01877  1.503 -6.2632     NA -33.52
(0,1,2)x(0,1,1)12        NA     NA -7.2649 -1.036 -33.26
(1,1,2)x(0,1,1)12 -37.32018     NA 11.4769 -7.224 -31.34
(2,1,2)x(0,1,1)12   0.01896  1.883 -0.9297 -1.150 -33.30
```

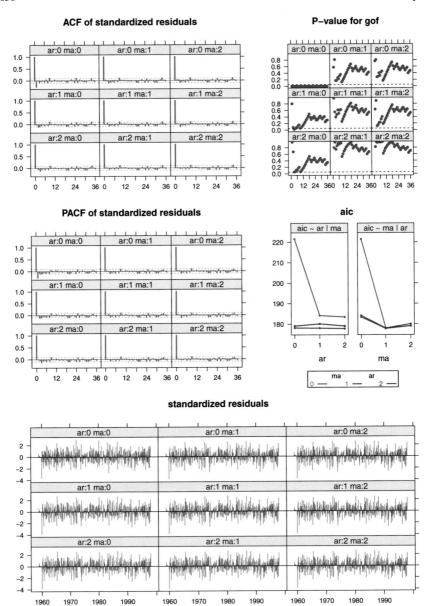

Fig. 18.9 Diagnostic plots for the set of models ARIMA$(p, 1, q) \times (0, 1, 1)_{12}$ fit to the CO_2 data by maximum likelihood.

Table 18.6 Estimation results for ARIMA$(0, 1, 1) \times (0, 1, 1)_{12}$ models fit to the CO_2 data.

```
> co2.arima <- ddco2.loopPQ[["0","1"]]

> co2.coef.t <- co2.arima$coef / sqrt(diag(co2.arima$var.coef))

> co2.arima

Call:
arima(x = x, order = c(0, 1, 1), seasonal = list(order = c(0, 1, 1),
      period = 12))

Coefficients:
        ma1    sma1
      -0.35  -0.851
s.e.   0.05   0.026

sigma^2 estimated as 0.0826:  log likelihood = -86.08,  aic = 178.2

> co2.coef.t
    ma1     sma1
 -7.053 -33.151

> vcov(co2.arima)
            ma1         sma1
ma1   0.0024638 -0.0002662
sma1 -0.0002662  0.0006583
```

Table 18.7 Estimation results for ARIMA$(1, 1, 1) \times (0, 1, 1)_{12}$ models fit to the CO_2 data.

```
> co2.arima11 <- ddco2.loopPQ[["1","1"]]

> co2.coef11.t <- co2.arima11$coef / sqrt(diag(co2.arima11$var.coef))

> co2.arima11

Call:
arima(x = x, order = c(1, 1, 1), seasonal = list(order = c(0, 1, 1),
      period = 12))

Coefficients:
         ar1      ma1     sma1
       0.240   -0.571   -0.852
s.e.   0.143    0.124    0.026

sigma^2 estimated as 0.0822:  log likelihood = -85.03,  aic = 178.1

> co2.coef11.t
    ar1      ma1     sma1
  1.677   -4.617  -33.287

> vcov(co2.arima11)
             ar1         ma1        sma1
ar1    0.0204605  -0.0167184  -0.0004893
ma1   -0.0167184   0.0152966   0.0002301
sma1  -0.0004893   0.0002301   0.0006544
```

18.8.3 Forecasting

The final plot in Figure 18.10 shows the last year of observed data and the forecasts, with their 95% forecast limits, obtained from the fitted model for the following year, i.e., for the months January through December 1998.

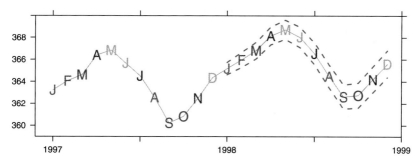

Fig. 18.10 CO_2—1997 observed, 1998 forecast + 95% CI.

18.9 Exercises

Many of the time series exercises ask you to construct and/or interpret plots of the time series itself, of the ACF and PACF, and of the diagnostics from a 3×3 set of ARIMA models. For Exercises 18.1, 18.2, and 18.3, go through this set of steps:

a. Describe the plot of the data and the ACF and PACF plots. Comment on whether you see anything systematic in the plot of the data. Are there spikes in the ACF and PACF plots. At which lags do they appear and what do they suggest? Do the ACF and PACF plots show any indication of a seasonal effect?

b. We chose to investigate a family of ARIMA$(p, 0, q)$ models, with $0 \leq p, q \leq 2$. Study the figures showing the diagnostic plots and the tables listing the parameter estimates. Describe each of the four sections of the diagnostic plot. What characteristics of each suggest a final model? Does the σ^2 (`sigma2`) section in the tables also suggest the same final model? Note for these three exercises, all of which are stationary and have zero mean, that the $(0, 0)$ panels of the ACF, PACF, and standardized residuals plots are essentially the same as the ACF, PACF, and time series plot of the data.

c. We printed the detail for the ARIMA$(1, 0, 1)$ model. Compare the ARIMA$(1, 0, 1)$ model to a simpler model with the closest σ^2. How do the AIC and the σ^2 compare? Would you recommend the simpler model? Why or why not?

18.1. Figure 18.11 shows the sequence, ACF, and PACF plots for a mystery time series X data(tser.mystery.X). Figure 18.12 and Table 18.8 show the diagnostics and estimated coefficients obtained by fitting the 3×3 set of ARIMA($p, 0, q$) models to the series. Table 18.9 shows the detail for the ARIMA($1, 0, 1$) model. Study the graphs and tables and explain why and how they indicate that one of these models seems better suited to explain the data than the others.

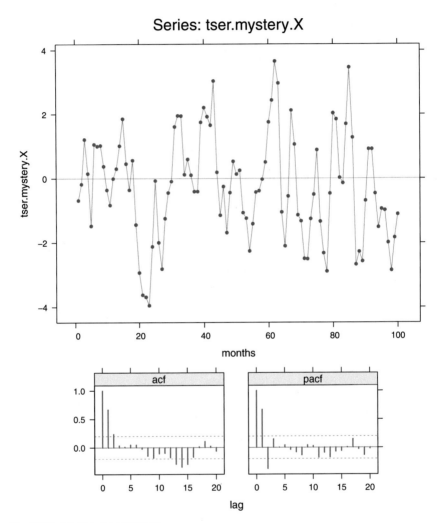

Fig. 18.11 Mystery time series X.

series: tser.mystery.X   model: (p,0,q)   by CSS–ML

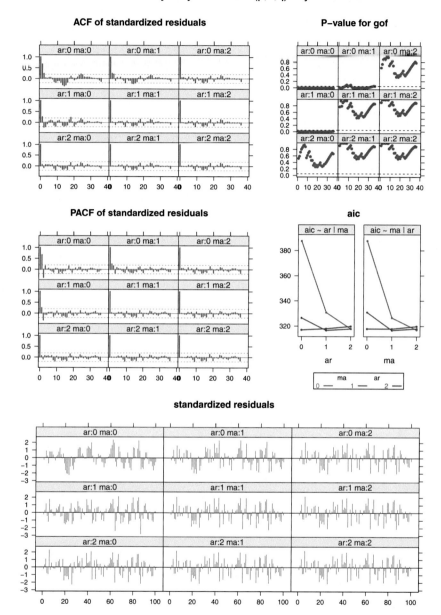

Fig. 18.12 Mystery time series X.

Table 18.8 Mystery time series X.

```
> X.loop
$series
[1] "tser.mystery.X"

$model
[1] "(p,0,q)"

$sigma2
       0     1     2
0 2.715 1.435 1.277
1 1.500 1.271 1.262
2 1.286 1.260 1.260

$aic
      0     1     2
0 387.7 326.6 317.1
1 330.9 316.7 318.0
2 317.9 317.9 319.8

$coef
            ar1      ar2     ma1      ma2 intercept
(0,0,0)      NA       NA      NA       NA        NA
(1,0,0) 0.6636       NA      NA       NA   -0.2744
(2,0,0) 0.9135  -0.3721      NA       NA   -0.2528
(0,0,1)      NA       NA 0.7264       NA   -0.2568
(1,0,1) 0.4299       NA 0.5221       NA   -0.2629
(2,0,1) 0.6213  -0.1781 0.3483       NA   -0.2571
(0,0,2)      NA       NA 0.9272  0.32958   -0.2548
(1,0,2) 0.2614       NA 0.7057  0.17341   -0.2588
(2,0,2) 0.5417  -0.1479 0.4280  0.04826   -0.2572

$t.coef
            ar1      ar2      ma1     ma2 intercept
(0,0,0)      NA       NA       NA      NA        NA
(1,0,0) 9.0295       NA       NA      NA   -0.7684
(2,0,0) 9.9262  -4.0464       NA      NA   -1.0257
(0,0,1)      NA       NA 13.2037      NA   -1.2471
(1,0,1) 3.8606       NA  5.1540      NA   -0.8828
(2,0,1) 2.6785  -0.9701  1.5262      NA   -0.9525
(0,0,2)      NA       NA 10.6280  3.5060   -1.0063
(1,0,2) 1.0949       NA  3.0019  0.8897   -0.9135
(2,0,2) 0.7891  -0.4658  0.6246  0.1267   -0.9478
```

Table 18.9 Mystery time series X.

```
> X.loop[["1","1"]]

Call:
arima(x = x, order = c(1, 0, 1))

Coefficients:
        ar1    ma1  intercept
      0.430  0.522     -0.263
s.e.  0.111  0.101      0.298

sigma^2 estimated as 1.27:  log likelihood = -154.3,  aic = 316.7
```

18.2. Figure 18.13 shows the sequence, ACF, and PACF plots for a mystery time series Y data(tser.mystery.Y). Figure 18.14 and Table 18.10 show the diagnostics and estimated coefficients obtained by fitting the 3×3 set of ARIMA$(p, 0, q)$ models to the series. Table 18.11 shows the detail for the ARIMA$(1, 0, 1)$ model. Study the graphs and tables and explain why and how they indicate that one of these models seems better suited to explain the data than the others.

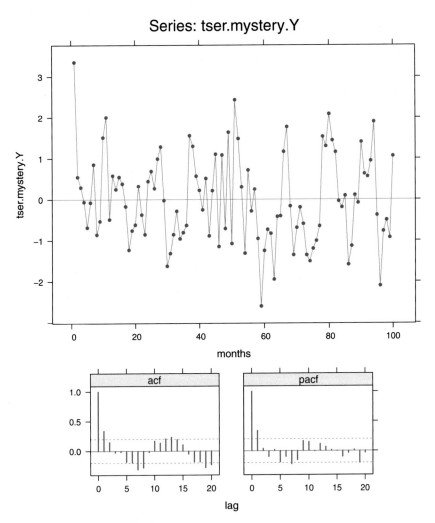

Fig. 18.13 Mystery time series Y.

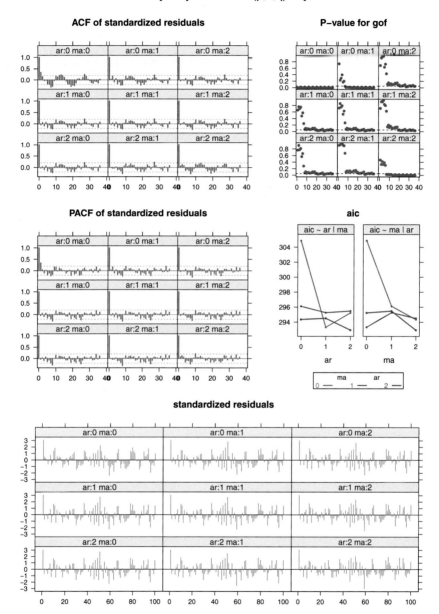

series: tser.mystery.Y   model: (p,0,q)   by CSS–ML

Fig. 18.14 Mystery time series Y.

Table 18.10 Mystery time series Y.

```
> Y.loop
$series
[1] "tser.mystery.Y"

$model
[1] "(p,0,q)"

$sigma2
      0     1      2
0 1.186 1.064 1.0243
1 1.035 1.034 1.0042
2 1.034 1.015 0.9502

$aic
      0     1     2
0 304.9 296.1 294.4
1 293.3 295.3 294.5
2 295.2 295.5 293.0

$coef
            ar1      ar2      ma1     ma2 intercept
(0,0,0)      NA       NA       NA      NA        NA
(1,0,0)  0.3740       NA       NA      NA   0.03050
(2,0,0)  0.3623  0.03129       NA      NA   0.03248
(0,0,1)      NA       NA  0.29684      NA   0.01777
(1,0,1)  0.4189       NA -0.05195      NA   0.03180
(2,0,1) -0.4676  0.38428  0.82758      NA   0.03363
(0,0,2)      NA       NA  0.37855  0.1854   0.02912
(1,0,2) -0.6314       NA  1.03122  0.4139   0.02815
(2,0,2) -0.9648 -0.54982  1.32535  0.9453   0.01509

$t.coef
            ar1      ar2      ma1     ma2 intercept
(0,0,0)      NA       NA       NA      NA        NA
(1,0,0)   3.817       NA       NA      NA    0.1884
(2,0,0)   3.433   0.2962       NA      NA    0.1944
(0,0,1)      NA       NA   3.6035      NA    0.1331
(1,0,1)   1.994       NA  -0.2342      NA    0.1924
(2,0,1)  -1.913   3.6606   3.3770      NA    0.1988
(0,0,2)      NA       NA   3.5048   2.022    0.1846
(1,0,2)  -2.958       NA   5.0593   3.723    0.1879
(2,0,2)  -8.179  -5.0155  21.2693  14.542    0.1191
```

Table 18.11 Mystery time series Y.

```
> Y.loop[["1","1"]]

Call.
arima(x = x, order = c(1, 0, 1))

Coefficients:
         ar1      ma1  intercept
       0.419   -0.052      0.032
s.e.   0.210    0.222      0.165

sigma^2 estimated as 1.03:  log likelihood = -143.6,  aic = 295.3
```

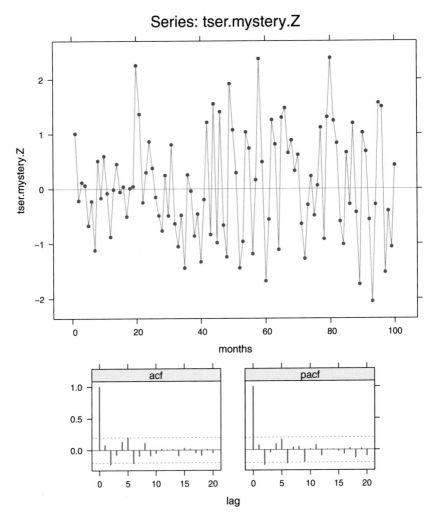

Fig. 18.15 Mystery time series Z.

18.3. Figure 18.15 shows the sequence, ACF, and PACF plots for a mystery time series Z data(tser.mystery.Z). Figure 18.16 and Table 18.12 show the diagnostics and estimated coefficients obtained by fitting the 3×3 set of ARIMA$(p, 0, q)$ models to the series. Table 18.13 shows the detail for the ARIMA$(1, 0, 1)$ model. Study the graphs and tables and explain why and how they indicate that one of these models seems better suited to explain the data than the others.

series: tser.mystery.Z model: (p,0,q) by CSS–ML

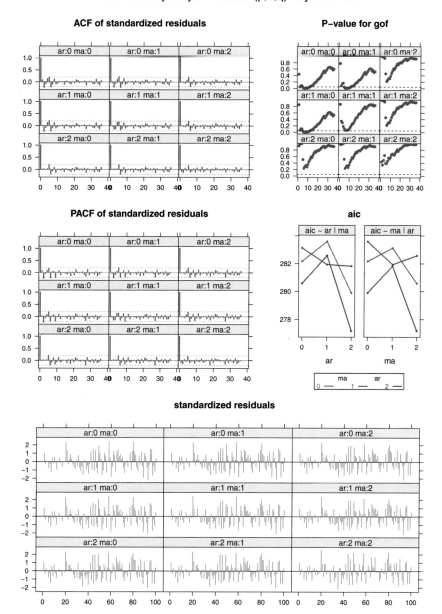

Fig. 18.16 Mystery time series Z.

Table 18.12 Mystery time series Z.

```
> Z.loop
$series
[1] "tser.mystery.Z"

$model
[1] "(p,0,q)"

$sigma2
        0      1      2
0 0.9454 0.9354 0.8930
1 0.9399 0.9039 0.8930
2 0.8868 0.8862 0.8088

$aic
      0     1     2
0 282.2 283.1 280.6
1 283.6 281.9 282.6
2 279.9 281.8 277.2

$coef
               ar1      ar2      ma1      ma2 intercept
(0,0,0)         NA       NA       NA       NA        NA
(1,0,0)    0.07625       NA       NA       NA   0.06379
(2,0,0)    0.09630  -0.2360       NA       NA   0.06402
(0,0,1)         NA       NA  0.13415       NA   0.06471
(1,0,1)   -0.78022       NA  0.91668       NA   0.06438
(2,0,1)    0.17084  -0.2424 -0.07901       NA   0.06405
(0,0,2)         NA       NA  0.09445  -0.2177   0.06464
(1,0,2)    0.01253       NA  0.08246  -0.2183   0.06466
(2,0,2)   -0.38096  -0.8567  0.57783   0.9732   0.06617

$t.coef
               ar1      ar2      ma1      ma2 intercept
(0,0,0)         NA       NA       NA       NA        NA
(1,0,0)    0.76442       NA       NA       NA    0.6083
(2,0,0)    0.98844   -2.424       NA       NA    0.7721
(0,0,1)         NA       NA   1.0571       NA    0.5905
(1,0,1)   -5.17277       NA   8.8387       NA    0.6291
(2,0,1)    0.57788   -2.458  -0.2664       NA    0.7885
(0,0,2)         NA       NA   0.9615   -2.171    0.7771
(1,0,2)    0.02659       NA   0.1788   -2.139    0.7786
(2,0,2)   -4.25626  -13.016   6.2170    9.171    0.6453
```

Table 18.13 Mystery time series *Z*.

```
> Z.loop[["1","1"]]

Call:
arima(x = x, order = c(1, 0, 1))

Coefficients:
          ar1     ma1  intercept
       -0.780   0.917      0.064
s.e.    0.151   0.104      0.102

sigma^2 estimated as 0.904:  log likelihood = -137,  aic = 281.9
```

18.4. The product data data(product) (originally from Nicholls (1979) and reproduced in Hand et al. (1994)) graphed in Figure 18.17 are the weekly sales of a plastic container used for the packaging of drugs in the United States. First differences were taken to produce the y_t, shown in Figure 18.18. The AR(1) model converged with AIC=2919, a larger number than for nonconverging models with more terms. The nonconverging models shown in Figure 18.19 and Table 18.14 showed high correlation between the estimates of the AR and MA coefficients.

a. Discuss why the above-mentioned findings and other results in Table 18.14 imply that an ARIMA($p, 1, q$) nonseasonal model is inappropriate for these data.

b. The peaks in the ACF and PACF plots of Figure 18.18 at 4 and 8 weeks suggest that there might be a monthly effect in this data. Examine and discuss the set of ARIMA($p, 1, q$) × $(1, 0, 0)_4$ models for these data.

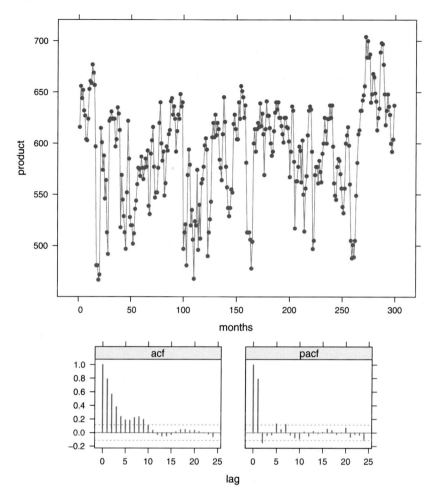

Fig. 18.17 Coordinated time series plot and ACF/PACF plots for the product time series: y_t. The response variable on the time series plot is weekly sales of the product.

Table 18.14 Estimation results for ARIMA($p, 1, q$) models fit to the product data.

```
> product.loop <- arma.loop(product, order=c(2,1,2))

> product.diags <- diag.arma.loop(product.loop, x=product, lag.max=60)

> product.loop
$series
[1] "product"

$model
[1] "(p,1,q)"

$sigma2
       0      1      2
0 1043 1041.8 1025.3
1 1042 1029.3  919.9
2 1033  917.8  916.3

$aic
     0    1    2
0 2919 2920 2918
1 2920 2919 2889
2 2920 2889 2890

$coef
               ar1       ar2      ma1      ma2
(0,1,0)         NA        NA       NA       NA
(1,1,0)  0.02806        NA       NA       NA
(2,1,0)  0.03066 -0.09105       NA       NA
(0,1,1)         NA        NA  0.03390       NA
(1,1,1) -0.82544        NA  0.88997       NA
(2,1,1)  0.90160 -0.15413 -0.98624       NA
(0,1,2)         NA        NA -0.02332 -0.1995
(1,1,2)  0.71744        NA -0.81250 -0.1736
(2,1,2)  1.11728 -0.32627 -1.20533  0.2183

$t.coef
               ar1      ar2       ma1      ma2
(0,1,0)         NA       NA        NA       NA
(1,1,0)  0.4833        NA        NA       NA
(2,1,0)  0.5299   -1.576        NA       NA
(0,1,1)         NA       NA    0.5341       NA
(1,1,1) -6.5556        NA    8.7970       NA
(2,1,1) 15.4832   -2.648  -61.7180       NA
(0,1,2)         NA       NA   -0.3401  -2.0824
(1,1,2) 12.9601        NA  -10.9925  -2.5043
(2,1,2)  4.6121   -1.697   -4.8510   0.8844
```

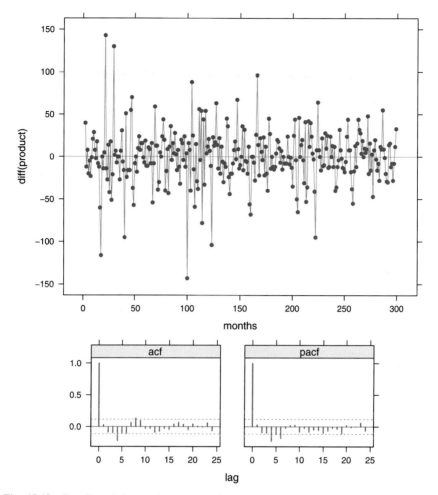

Fig. 18.18 Coordinated time series and ACF/PACF plots for the differenced product time series: ∇y_t.

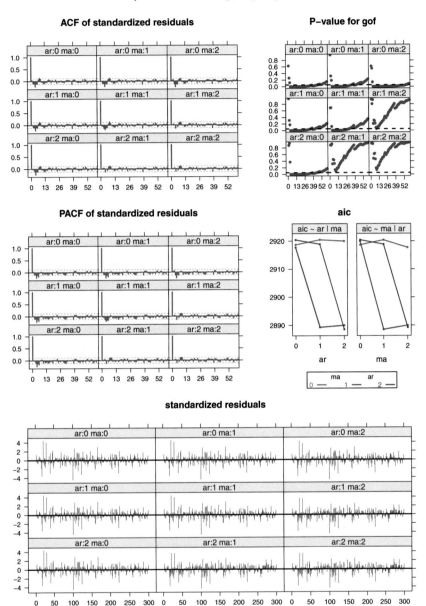

Fig. 18.19 Diagnostic plots for the set of models ARIMA(p,1,q) fit to the Product data by maximum likelihood. Each set of nine panels is systematically structured in a 3×3 array with rows indexed by the number of AR parameters p and columns by the number of MA parameters q. All nine panels in a set are scaled identically. The AIC has been plotted as a pair of interaction plots: AIC plotted against q using line types defined by p; and AIC plotted against p, using line types defined by q.

18.5. Figures 18.20, 18.21, and 18.22 and Tables 18.15 and 18.16 show the mean monthly air temperature in degrees Fahrenheit from January 1920 to December 1939 at Nottingham. R users can use the `nottem` data in the pkgdatasets package. We first got the data from Venables and Ripley (1997). The original source is "Meteorology of Nottingham" in *City Engineer and Surveyor*. We show the original series, the seasonally differenced series, the diagnostic display from the series of models ARIMA$(p, 0, q) \times (2, 1, 0)_{12}$, and numerical results from the set of all nine models table and detail on the recommended model ARIMA$(1, 0, 0) \times (2, 1, 0)_{12}$.

a. What are the most evident features of the plot of the original data?

b. Compare the plot of the seasonally differenced data to the original plot. What structure was captured by the differencing? What remains?

c. Compare the recommended model ARIMA$(1, 0, 0) \times (2, 1, 0)_{12}$ to the next most likely model ARIMA$(2, 0, 0) \times (2, 1, 0)_{12}$. Do you agree that the `ar(2)` term is not needed? Why?

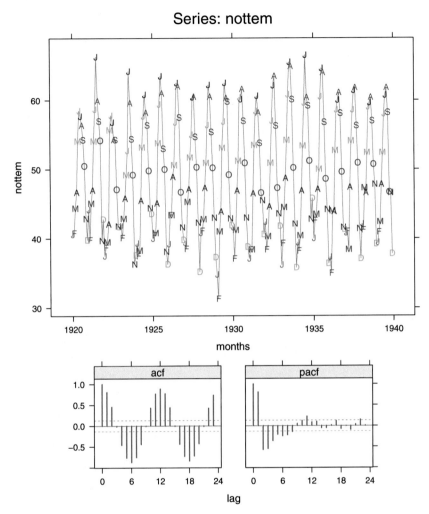

Fig. 18.20 Mean monthly air temperature in degrees Fahrenheit from January 1920 to December 1939 at Nottingham.

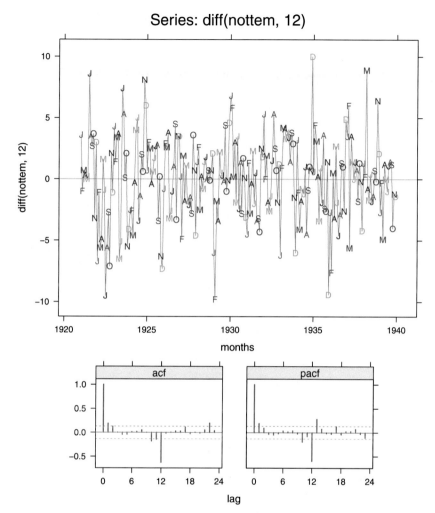

Fig. 18.21 Seasonal differences of mean monthly air temperature in degrees Fahrenheit from January 1920 to December 1939 at Nottingham.

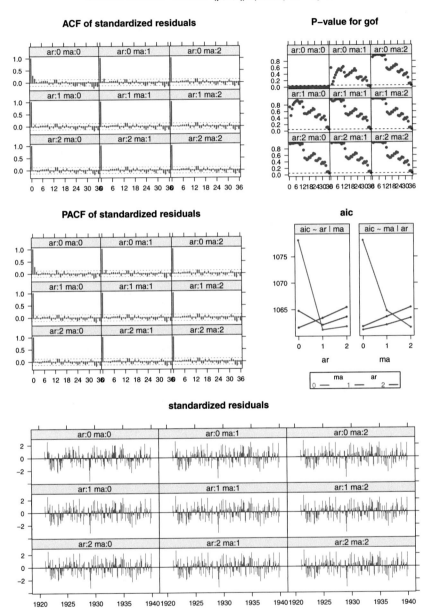

Fig. 18.22 Mean monthly air temperature in degrees Fahrenheit from January 1920 to December 1939 at Nottingham Castle.

Table 18.15 Nottingham temperature—models ARIMA$(p, 0, q) \times (2, 1, 0)_{12}$.

```
> nottem.loop <- arma.loop(nottem,  order=c(2,0,2),
+                          seasonal=list(order=c(2,1,0), period=12),
+                          method="ML")

> print(nottem.loop, digits=4)
$series
[1] "nottem"

$model
[1] "(p,0,q)x(2,1,0)12"

$sigma2
      0     1     2
0 6.219 5.799 5.661
1 5.702 5.674 5.656
2 5.666 5.661 5.656

$aic
     0    1    2
0 1078 1065 1062
1 1061 1062 1063
2 1062 1064 1065

$coef
                         ar1      ar2      ma1     ma2    sar1     sar2
(0,0,0)x(2,1,0)12         NA       NA       NA      NA -0.8220 -0.2931
(1,0,0)x(2,1,0)12   0.285599       NA       NA      NA -0.8598 -0.2963
(2,0,0)x(2,1,0)12   0.261443  0.07937       NA      NA -0.8602 -0.3074
(0,0,1)x(2,1,0)12         NA       NA  0.23606      NA -0.8505 -0.2866
(1,0,1)x(2,1,0)12   0.478197       NA -0.20979      NA -0.8603 -0.3033
(2,0,1)x(2,1,0)12   0.001805  0.15577  0.26037      NA -0.8608 -0.3100
(0,0,2)x(2,1,0)12         NA       NA  0.25718  0.1575 -0.8607 -0.3131
(1,0,2)x(2,1,0)12   0.169231       NA  0.09343  0.1200 -0.8611 -0.3127
(2,0,2)x(2,1,0)12   0.118763  0.01077  0.14252  0.1212 -0.8614 -0.3153

$t.coef
                         ar1      ar2     ma1    ma2    sar1     sar2
(0,0,0)x(2,1,0)12         NA       NA      NA     NA -13.07   -4.418
(1,0,0)x(2,1,0)12   4.451829       NA      NA     NA -13.46   -4.443
(2,0,0)x(2,1,0)12   3.899219   1.1772      NA     NA -13.54   -4.596
(0,0,1)x(2,1,0)12         NA       NA  4.0370     NA -13.32   -4.275
(1,0,1)x(2,1,0)12   2.668022       NA -1.0576     NA -13.52   -4.556
(2,0,1)x(2,1,0)12   0.003237   0.9644  0.4644     NA -13.56   -4.620
(0,0,2)x(2,1,0)12         NA       NA  3.9101  2.302 -13.58   -4.661
(1,0,2)x(2,1,0)12   0.448103       NA  0.2501  1.008 -13.58   -4.650
(2,0,2)x(2,1,0)12   2.887888      NaN  1.0990  1.015 -26.55  -29.916


> nottem.diag <-
+    rearrange.diag.arma.loop(diag.arma.loop(nottem.loop, nottem))
```

Table 18.16 Nottingham temperature—recommended model ARIMA$(1, 0, 0) \times (2, 1, 0)_{12}$.

```
> nottem.loop[["1","0"]]

Call:
arima(x = x, order = c(1, 0, 0), seasonal = list(order = c(2, 1, 0),
      period = 12),  method = "ML")

Coefficients:
         ar1      sar1     sar2
        0.286  -0.860  -0.296
s.e.  0.064   0.064    0.067

sigma^2 estimated as 5.7:  log likelihood = -526.6,  aic = 1061
```

18.6. We have a time series of size $n = 100$ for which we have determined that we have an ARIMA(1,0,0) model and have estimated $\hat{\mu} = 15$, $\hat{\phi} = .2$, and $\hat{\sigma}^2 = 3$. The last few observations in the series are

t	97	98	99	100
X_t	13	15	18	17

Forecast, with 95% forecast intervals, the values $\hat{X}_{101}$ and $\hat{X}_{102}$.

18.7. We have a nonseasonal time series in the dataset `data(tsq)` covering 100 periods. The time series and its ACF and PACF plots are displayed in Figure 18.23. Table 18.17 contains the R output from a 3×3 set of ARIMA models fit to the data. The `tsdiagplot` for these data is in Figure 18.24. Use this information to answer the following questions:

a. Recommend the $(p, 0, q)$ order for an ARIMA modeling of these data.

b. Write out the equation for the best-fitting model following your recommendation in part (a).

c. Use your model and the coefficient information in the R output to produce forecasts and 95% forecast intervals for the value of this series in periods 101 and 102.

Table 18.17 Three by three set of ARIMA models for Exercise 18.7.

```
> tsq.loop <- arma.loop(tsq, order=c(2,0,2))

> tsq.loop
$series
[1] "tsq"

$model
[1] "(p,0,q)"

$sigma2
      0      1      2
0 1.330 0.9469 0.9267
1 1.173 0.9328 0.8730
2 1.003 0.9175 0.8715

$aic
      0     1     2
0 316.3 285.2 285.0
1 305.8 285.6 283.2
2 292.6 286.0 284.9

$coef
            ar1      ar2      ma1      ma2 intercept
(0,0,0)      NA       NA       NA       NA        NA
(1,0,0)  0.3444       NA       NA       NA   0.09045
(2,0,0)  0.4894 -0.38808       NA       NA   0.08309
(0,0,1)      NA       NA   0.7524       NA   0.08843
(1,0,1) -0.1523       NA   0.8031       NA   0.09029
(2,0,1) -0.1087 -0.15017   0.7447       NA   0.08999
(0,0,2)      NA       NA   0.6043  -0.1653   0.09055
(1,0,2)  0.8468       NA  -0.2290  -0.7710   0.12449
(2,0,2)  0.7879  0.06726  -0.2076  -0.7924   0.12323

$t.coef
            ar1      ar2      ma1      ma2 intercept
(0,0,0)      NA       NA       NA       NA        NA
(1,0,0)  3.6551       NA       NA       NA    0.5505
(2,0,0)  5.1886  -4.0813       NA       NA    0.7431
(0,0,1)      NA       NA   11.139       NA    0.5208
(1,0,1) -1.2440       NA   12.664       NA    0.5992
(2,0,1) -0.8042  -1.2726    7.520       NA    0.6785
(0,0,2)      NA       NA    5.272   -1.493    0.6549
(1,0,2) 14.4918       NA   -3.308  -11.612    3.9454
(2,0,2)  6.4214   0.5445   -2.841  -11.185    3.7167
```

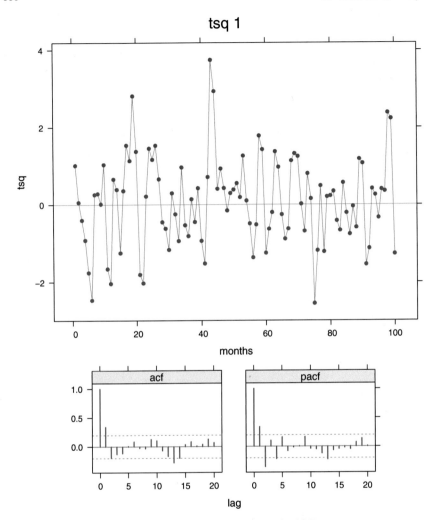

Fig. 18.23 Time series and its ACF and PACF plots for Exercise 18.7.

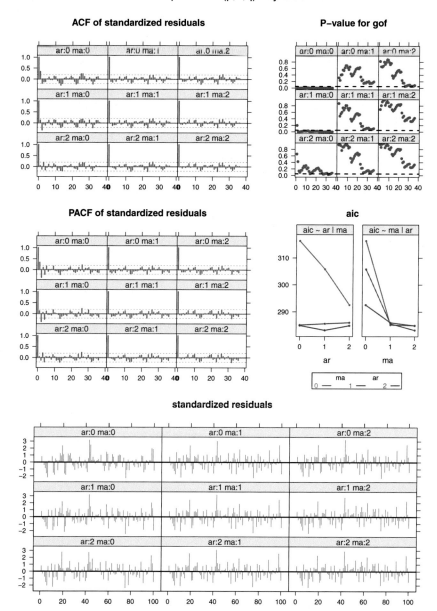

Fig. 18.24 Diagnostic plots for the 3×3 set of ARIMA models in Exercise 18.7.

18.8. Figure 18.25 contains time series, ACF, and PACF plots for monthly data on the thickness of the ozone layer (measured in Dobson units) at Arosa, Switzerland, from September 1931 through November 1953. Note the labeling of the months of the year (J = January, June, or July, F = February, etc.) at the plot points. The data in the dataset data(ozone) are from Andrews and Herzberg (1985), Table 12.1.

a. Comment on the seasonal nature of the time series plot and discuss how this is consistent with what you see in the ACF plot.

b. Notice that the variability of the series appears to increase, at least temporarily, in the early 1940s and around 1952 to 1953. For each of these periods, *identify historical events* that potentially impacted on the atmosphere to produce this increased variability.

18.9. $n = 100$ and $\bar{Z} = 25$. See Figure 18.26.

a. Identify a tentative underlying model in explicit form and justify your model.

b. Propose possible preliminary parameter estimates for your model.

c. Assume the residual sum of squares from the fitting of your model is 256, and $Z_{98} = 24, Z_{99} = 26, Z_{100} = 25$. Compute your forecasts for Z_{101} and Z_{102} and their 95% forecast intervals.

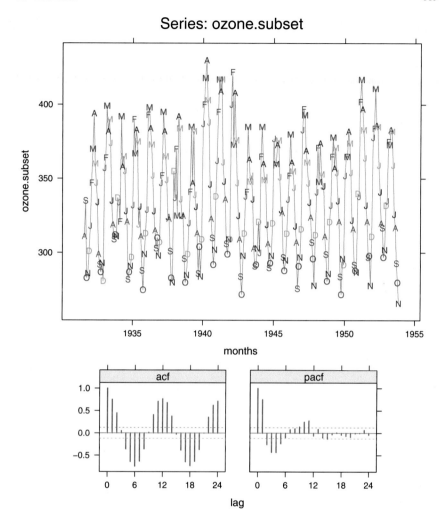

Fig. 18.25 Thickness of the ozone layer (measured in Dobson units) at Arosa, Switzerland, from September 1931 through November 1953.

	lag									
	1	2	3	4	5	6	7	8	9	10
acf(Z_t)	0.80	0.61	0.47	0.40	0.31	0.21	0.18	0.11	0.06	0.01
pacf(Z_t)	0.80	0.08	0.00	−0.11	0.00	−0.12	0.07	0.05	0.01	0.02

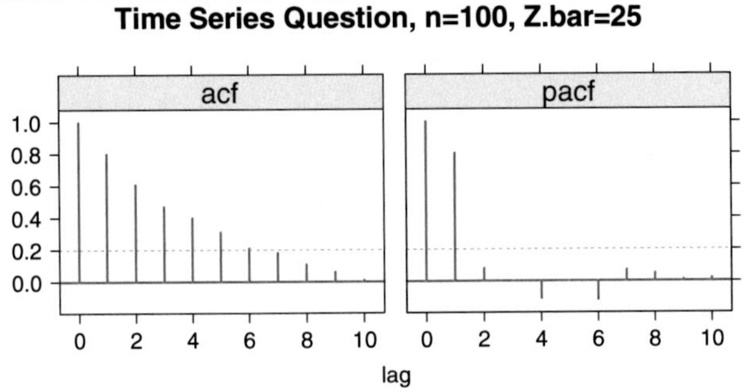

Time Series Question, n=100, Z.bar=25

Fig. 18.26 ACF and PACF for Exercise 18.9, $n = 100$ and $\bar{Z} = 25$. The same information is presented in both tabular and graphical form.

					lag					
	1	2	3	4	5	6	7	8	9	10
acf(Z_t)	0.93	0.92	0.90	0.90	0.87	0.86	0.85	0.84	0.82	0.80
acf(∇Z_t)	−0.57	−0.10	0.12	0.06	−0.12	0.09	0.05	−0.01	0.02	0.03

Time Series Question, n=100, Z.bar=60

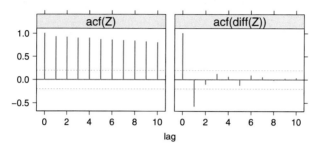

Fig. 18.27 ACF and PACF for Exercise 18.10, $n = 100$ and $\bar{Z} = 60$. The same information is presented in both tabular and graphical form.

18.10. $n = 100$ and $\bar{Z} = 60$. Identify a tentative underlying model in explicit form and justify your model. See Figure 18.27.

18.11. $n = 100$ and $\bar{Z} = 55$. Identify a tentative underlying model in explicit form and justify your model. See Figure 18.28.

18.12. Time series data differs from any other data type we have discussed in one important characteristic: The observations are not independent. What are the implications of that difference for modeling time series data? Be sure to discuss implications for each of

a. Modeling

b. Estimation

c. Prediction

	lag									
	1	2	3	4	5	6	7	8	9	10
$acf(Z_t)$	0.99	0.94	0.87	0.81	0.75	0.65	0.55	0.53	0.43	0.40
$acf(\nabla Z_t)$	0.43	0.28	0.51	0.80	0.65	0.44	0.31	0.77	0.30	0.20
$acf(\nabla_4 Z_t)$	0.72	0.67	0.55	0.32	0.38	0.23	0.24	0.23	0.18	0.13
$acf(\nabla\nabla_4 Z_t)$	0.30	0.07	0.32	0.50	0.20	0.01	−0.05	−0.01	0.02	0.03

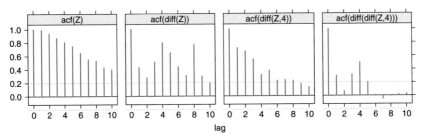

Fig. 18.28 ACF and PACF for Exercise 18.11, $n = 100$ and $\bar{Z} = 55$. The same information is presented in both tabular and graphical form.

18.13. Figure 18.29 shows the "United States of America Monthly Employment Figures for Males Aged 16–19 Years from 1948 to 1981". Dataset `data(employM16)` is Table T.65.1 from Andrews and Herzberg (1985). What are the features of this plot that you would try to capture in a time series model? Comment on

a. Seasonality

b. Trend

c. Aberrations

18.A Appendix: Construction of Time Series Graphs

This section discusses the technical aspects of the construction of the set of plots used to check the validity of the proposed model. The interpretation of the plots, and the discussion of how to use them to help identify the model that best fits the data, appear in Sections 18.6 and 18.8.

The graphical display techniques demonstrated in Sections 18.6 and 18.8 were developed by Heiberger and Teles (2002). The R functions from the **HH** package used to produce these displays are described in help files `?tsacfplots` and `?tsdiagplot`.

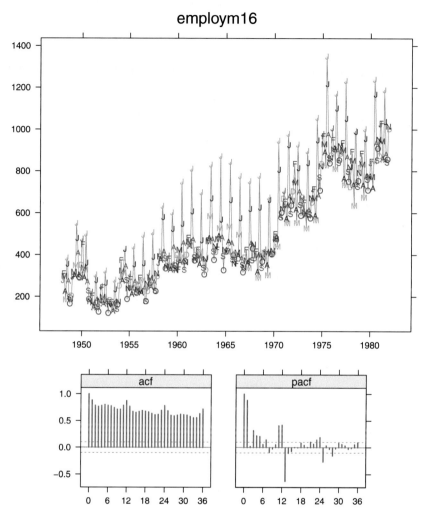

Fig. 18.29 United States of America Monthly Employment Figures for Males Aged 16–19 Years from 1948 to 1981, for use with Exercise 18.13.

The set of plots in Figure 18.8 consists of the residual ACF and PACF, the portmanteau goodness-of-fit test statistic (GOF), the standardized residuals, and the Akaike information criterion (AIC). The panels in the first four sets of plots are indexed by the number of nonseasonal ARMA parameters p and q for fixed values of the seasonal parameters P and Q. The AIC plot uses p and q as plotting variables. The orders of differencing and the orders of the autoregressive and moving average operators (both seasonal and nonseasonal) have been limited to $0 \leq p, d, q, P, D, Q \leq 2$. While this limitation is usually reasonable in practice, it is not inherent in the software.

Each set of nine panels is systematically structured in a 3×3 array indexed by the number of AR parameters and MA parameters. All nine panels in a set are scaled identically. Thus the reader can scan a row or column of the array of panels and see the effect of adding one more parameter to either the AR or MA side of the model.

Traditionally (that is, as constructed by the standard R tsdiag), the plots coordinated in Figure 18.8 are shown on nine separate pages, one page for each model. The standard display shows the standardized residuals, the residual ACF and PACF plots, and the portmanteau goodness-of-fit test. The nine sets of plots, each associated with a different model, will not necessarily be scaled alike. Even the GOF and ACF/PACF plots for the same model may have different lag scales.

Labeling the axis in months and putting the residual ACF and PACF plots and the GOF plot on the same set of lags make it easy to compare the plots for different models. In this example it is easy to see that something is happening at lag=12 months. The AIC plots for all the models in Figure 18.8 are similar, with AIC≈315. The AIC has been plotted as a pair of interaction plots: AIC plotted against q, the number of nonseasonal MA parameters, using line types defined by p, the number of nonseasonal AR parameters; and AIC plotted against p, using line types defined by q. These plots enable us to study the magnitudes of the differences in AIC of competing models.

18.A.1 Characteristics of This Presentation of the Time Series Plot

- Individual points are identified with a letter indicating the position of each observation according to the frequency of collection of the data. The user can control the choice of plotting characters. The default characters used are dependent on the frequency of collection of the data. For example, when the frequency is 12, the default plotting characters are the month abbreviations J,F,M,A,M,J,J,A,S,O,N,D. Otherwise they are chosen from the beginning of the lower case alphabet letters().

- The plotting characters are an explicit argument and can be chosen by the user (with pch.seq), or suppressed entirely with type="l".

- Color is often very helpful with the time series plots. Color plots show the seasonal pattern more strongly than the black and white. Figure 18.4 has a clear pattern of gold-May and red-April along its top and green-October along the bottom. The first differences in Figure 18.6 show a clear pattern of blue-August along the bottom and color-coded June–July–August–September below the axis. Figure 18.7 shows random behavior in colors.

18.A.2 Characteristics of This Presentation of the Sample ACF and PACF Plots

- The axes are coordinated and have the same scale.

- Lags are indicated in appropriate units (for example, months for monthly series).

- The ACF and PACF plots consistently both show, or do not show (at the user's option), the spike for correlation=1 at lag=0.

- The default tick marks are related to the frequency of collection of the data. The user has control over tick mark location.

- Most of the plotting surface is occupied by the body of the plot, and the amount of surface used for labeling is minimized.

We point out that the individual plots are accessible to the user. They can be placed on their own pages or displayed with other relative spacings. For details, see the function HH:::print.tsacfplots.

18.A.3 Construction of Graphical Displays

This section shows how to construct the two display types presented in this chapter. For brevity, only Figures 18.7 and 18.9 are described.

Figure 18.7, a single display with subgraphs, is constructed with the single command:

```
tsacfplots(diff(diff(co2,1), 12))
```

The figure uses the majority of the plotting surface to display the time series itself and a minority of the plotting surface to display the ACF and PACF plots drawn to the same scale.

All models in the family of ARIMA models under investigation are fit with a single command specified in standard R time series model notation:

```
ddco2.loopPQ <-
    arma.loop(co2,
              order=c(2,1,2),
              seasonal=list(order=c(0,1,1), period=12))
```

Figure 18.9 plots the family of models, again as a single display with coordinated sets of subgraphs, with another single command:

$$\texttt{tsdiagplot(armas=co2.loop)}$$

The series of plots in each set of subgraphs is displayed in the same systematic order. All plots of the same form are displayed to the same scale.

Fine control of plotting options and labeling is possible with optional arguments to the `tsacfplots` and `tsdiagplot` functions. Each of the individual subgraphs is also directly accessible to the user.

Formal and systematic display of a series of models makes it easy to recognize the structural differences in the series of models and to compare them.

18.A.4 Functions in the HH package for R

Several functions are provided and described here in terms of their role in the modeling. In addition to these functions, there are unexported functions that the primary functions call to do much of the work.

Primary Functions in the **HH** Package

> `tsacfplots:` Provides a single display (of the form of Figure 18.7) with the times series plot central and both the ACF and PACF plots on the same scale. It does so by calling `seqplot` (equivalent to `ts.plot` but with much finer control of labeling options) for the time series plot and then `acf.pacf.plot` for the coordinated ACF and PACF plots. These in turn are constructed by the R routine `acf`.

> `arma.loop:` Takes a time series and a model statement of the form

$$(p_{max}, d, q_{max}) \times (P, D, Q)_{\text{period}}$$

> It then loops through the family of models indexed by the model parameters $1{:}p_{max}$ and $1{:}q_{max}$, with d, P, D, Q held constant. Results are stored in a list indexed by the values of p and q.

> `arma.loop` also permits the model statement (note that order matters)

$$(P_{max}, D, Q_{max})_{\text{period}} \times (p, d, q)$$

> and then loops through the family of models indexed by the model parameters $1{:}P_{max}$ and $1{:}Q_{max}$, with p, d, q, D held constant. Results are stored in a list indexed by the values of P and Q.

diag.arma.loop: Produces an indexed list of the arima.diag results for each model in the result of the arma.loop. Diagnostics are calculated on the boundary values of the parameters p and q, and in particular for those functions defined in the special case $(p, d, q) = (0, 0, 0)$.

tsdiagplot: Takes a time series and a model statement and calls all the diagnostic plot routines. It makes sensible default choices for all the arguments and produces a graph similar to Figure 18.9. For printouts of any of the numerical tables, or finer control over the layout and labeling of the plots, the user should study the more detailed illustrations of function use in the demo("tsamstat").

Print Methods

print.arma.loop and summary.arma.loop: Produce tables similar to Table 18.5 from the result of the arma.loop function.

Individual Plot Functions

Each of the subgraphs in tsacfplots and tsdiagplot is directly accessible to the users. Each is fully parameterized.

tsacfplots: Figures 18.4 and 18.7.
     seqplot   Time series
     acf.pacf.plot Coordinated ACF and PACF

tsdiagplot: Figures 18.8 and 18.9.
     acfplot   ACF and PACF of residuals
     residplot   Standardized residuals
     gofplot   Portmanteau goodness-of-fit statistic (GOF)
     aicsigplot Interaction plot of AIC or σ^2

seqplotForecast: Figure 18.10. Data, forecasts, and confidence bands.

Additional functions

Not for direct use by users.

rearrange.diag.arma.loop: Rearranges the list of diagnostics indexed by model into a list of matrices of diagnostics, each matrix indexed by the models. The sole purpose of this rearrangement is for plotting.

Appendix A

R

R (R Core Team, 2015) is the *lingua franca* of data analysis and graphics. In this book we will learn the fundamentals of the language and immediately use it for the statistical analysis of data. We will graph both the data and the results of the analyses. We will work with the basic statistical tools (regression, analysis of variance, and contingency tables) and some more specialized tools. Many of our examples use the additional functions we have provided in the **HH** package (Appendix B) (Heiberger, 2015). The R code for all tables and graphs in the book is included in the **HH** package.

In later appendices we will also look at the **Rcmdr** menu system (Appendix C), RExcel integration of R with Excel (Appendix D), and the **shiny** package for integration of R with interactive html web pages (Appendix E).

A.1 Installing R—Initial Installation

R is an open-source publicly licensed software system. R is free under the GPL (Gnu Public License).

R is available for download on Windows, Macintosh OSX, and Linux computer systems. Start at http://www.R-project.org. Click on "download R" in the "Getting Started" box, pick a mirror near you, and download and install the most recent release of R for your operating system.

Once R is running, you will download several necessary contributed packages. You get them from a running R session while connected to the internet. This install.packages statement will install all the packages listed and additional packages that these specified packages need.

© Springer Science+Business Media New York 2015
R.M. Heiberger, B. Holland, *Statistical Analysis and Data Display*,
Springer Texts in Statistics, DOI 10.1007/978-1-4939-2122-5

A.1.1 Packages Needed for This Book—*Macintosh* and *Linux*

Start R, then enter

```
install.packages(c("HH","RcmdrPlugin.HH","RcmdrPlugin.mosaic",
                   "fortunes","ggplot2","shiny","gridExtra",
                   "gridBase","Rmpfr","png","XLConnect",
                   "matrixcalc", "sem", "relimp", "lmtest",
                   "markdown", "knitr", "effects", "aplpack",
                   "RODBC", "TeachingDemos",
                   "gridGraphics", "gridSVG"),
                dependencies=TRUE)

## This is the sufficient list (as of 16 August 2015) of packages
## needed in order to install the HH package.  Should
## additional dependencies be declared by any of these packages
## after that date, the first use of "library(HH)" after the
## installation might ask for permission to install some more
## packages.
```

The install.packages command might tell you that it can't write in a system directory and will then ask for permission to create a personal library. The question might be in a message box that is behind other windows. Should the installation seem to freeze, find the message box and respond "yes" and accept its recommend directory. It might ask you for a CRAN mirror. Take the mirror from which you downloaded R.

A.1.2 Packages and Other Software Needed for This Book—*Windows*

A.1.2.1 RExcel

If you are running on a Windows machine and have access to Excel, then we recommend that you also install RExcel. The RExcel software provides a seamless integration of R and Excel. See Appendix D for further information on RExcel, including the download statements and licensing information. See Section A.1.2.2 for information about using **Rcmdr** with RExcel.

A.1.2.2 **RExcel Users Need to Install Rcmdr as Administrator**

Should you choose to install **RExcel** then you need to install **Rcmdr** as Administrator. Otherwise you can install **Rcmdr** as an ordinary user.

Start R as Administrator (on Windows 7 and 8 you need to right-click the R icon and click the "Run as administrator" item). In R, run the following commands (again, you must have started R as Administrator to do this)

```
## Tell Windows that R should have the same access to the
## outside internet that is granted to Internet Explorer.
setInternet2()

install.packages("Rcmdr",
                 dependencies=TRUE)
```

Close R with q("no"). Answer with n if it asks
    Save workspace image? [y/n/c]:

A.1.2.3 **Packages Needed for This Book—Windows**

The remaining packages can be installed as an ordinary user. **Rcmdr** is one of the dependencies of **RcmdrPlugin.HH**, so it will be installed by the following statement (unless it was previously installed). Start R, then enter

```
## Tell Windows that R should have the same access to the
## outside internet that is granted to Internet Explorer.
setInternet2()

install.packages(c("HH","RcmdrPlugin.HH","RcmdrPlugin.mosaic",
                   "fortunes","ggplot2","shiny","gridExtra",
                   "gridBase","Rmpfr","png","XLConnect",
                   "matrixcalc", "sem", "relimp", "lmtest",
                   "markdown", "knitr", "effects", "aplpack",
                   "RODBC", "TeachingDemos",
                   "gridGraphics", "gridSVG"),
                 dependencies=TRUE)

## This is the sufficient list (as of 16 August 2015) of packages
## needed in order to install the HH package.  Should
## additional dependencies be declared by any of these packages
## after that date, the first use of "library(HH)" after the
## installation might ask for permission to install some more
## packages.
```

A.1.2.4 `Rtools`

`Rtools` provides all the standard Unix utilities (C and Fortran compilers, command-line editing tools such as `grep`, `awk`, `diff`, and many others) that are not included with the Windows operating system. These utilities are needed in two circumstances.

1. Should you decide to collect your R functions and datasets into a package, you will need `Rtools` to build the package. See Appendix F for more information.

2. Should you need to use the `ediff` command in Emacs for visual comparison of two different versions of a file (yesterday's version and today's after some editing, for example), you will need `Rtools`. See Section M.1.2 for an example of visual comparison of files.

You may download `Rtools` from the Windows download page at CRAN. Please see the references from that page for more details.

A.1.2.5 Windows Potential Complications: Internet, Firewall, and Proxy

When `install.packages` on a Windows machine gives an *Error* message that includes the phrase "`unable to connect`", then you are probably working behind a company firewall. You will need the R statement

    `setInternet2()`

before the `install.packages` statement. This statement tells the firewall to give R the same access to the outside internet that is granted to **Internet Explorer**.

When the `install.packages` gives a *Warning* message that says you don't have write access to one directory, but it will install the packages in a different directory, that is normal and the installation is successful.

When the `install.packages` gives an *Error* message that says you don't have write access and doesn't offer an alternative, then you will have to try the package installation as Administrator. Close R, then reopen R by right-clicking the R icon, and selecting "Run as administrator".

If this still doesn't allow the installation, then

1. Run the R line

    `sessionInfo()`

2. Run the `install.packages` lines.

3. Highlight and pick up the entire *contents* of the R console and save it in a text file. Screenshots are usually not helpful.

4. Show your text listing to an R expert.

A.1.3 Installation Problems—Any Operating System

The most likely source of installation problems is settings (no write access to restricted directories on your machine, or system wide firewalls to protect against offsite internet problems) that your computer administrator has placed on your machine.

Check the FAQ (Frequently Asked Questions) files linked to at
http://cran.r-project.org/faqs.html.

For Windows, see also Section A.1.2.5.

If outside help is needed, then save the *contents* of the R console window to show to your outside expert. In addition to the lines and results leading to the problem, including ALL messages that R produces, you must include the line (and its results)
sessionInfo()
in the material you show the expert.

Screenshots are not a good way to capture information. The informative way to get the contents of the console window is by highlighting the entire window (including the off-screen part) and saving it in a text file. Show your text listing to an R expert.

A.1.4 XLConnect: All Operating Systems

The XLConnect package lets you read MS Excel files directly from R on any operating system. You may use an R statement similar to the following

```
library(XLConnect)
WB <-  ## pathname of file with some additional information
  loadWorkbook(
## "c:/Users/rmh/MyWorkbook.xlsx"  ## rmh pathname in Windows
    "~/MyWorkbook.xlsx"            ## rmh pathname in Macintosh
  )
mydata <- readWorksheet(WB, sheet="Sheet1", region="A1:D11")
```

If you get an error from library(XLConnect) of the form
Error : .onLoad failed in loadNamespace() for 'rJava'
then you need to install java on your machine from http://java.com. The java installer will ask you if you want to install ask as your default search provider. You may deselect both checkboxes to retain your preferred search provider.

A.2 Installing **R**—Updating

R is under constant development with new releases every few months. At this writing (August 2015) the current release is R-3.2.2 (2015-08-14).

See the FAQ for general update information, and in addition the Windows FAQ or MacOs X FAQ for those operating systems. Links to all three FAQs are available at http://cran.r-project.org/faqs.html. The FAQ files are included in the documentation placed on your machine during installation.

The update.packages mechanism works for packages on CRAN. It does not work for packages downloaded from elsewhere. Specifically, Windows users with RExcel installed will need to update the RExcel packages by reinstalling them as described in Section D.1.2.

A.3 Using **R**

A.3.1 Starting the R Console

In this book our primary access to the R language is from the command line. On most computer systems clicking the R icon **R** on the desktop will start an R session in a console window. Other options are to begin within an Emacs window with M-x R, to start an R-Studio session, to start R at the Unix or MS-DOS command line prompt, or to start R through one of several other R-aware editors.

With any of these, the R offers a prompt "> ", the user types a command, R responds to the user's command and then offers another prompt "> ". A very simple command-line interaction is shown in Table A.1.

Table A.1 Very simple command-line interaction.

```
> ## Simple R session
> 3 + 4
[1] 7

> pnorm(c(-1.96, -1.645, -0.6745, 0, 0.6745, 1.645, 1.96))
[1] 0.02500 0.04998 0.25000 0.50000 0.75000 0.95002 0.97500

>
```

A.3.2 Making the Functions in the HH Package Available to the Current Session

At the R prompt, enter

```
library(HH)
```

This will load the **HH** package and several others. All **HH** functions are now available to you at the command line.

A.3.3 Access HH Datasets

All **HH** datasets are accessed with the R data() function. The first six observations are displayed with the head() function, for example

```
data(fat)
head(fat)
```

A.3.4 Learning the R Language

R is a dialect of the S language. The easiest way to learn it is from the manuals that are distributed with R in the doc/manual directory; you can find the pathname with the R call

```
system.file("../../doc/manual")
```

or

```
WindowsPath(system.file("../../doc/manual"))
```

Open the manual directory with your computer's tools, and then read the pdf or html files. Start with the R-intro and the R-FAQ files. With the Windows **Rgui**, you can access the manuals from the menu item Help > Manuals (in PDF). From the Macintosh **R.app**, you can access the manuals from the menu item R Help.

A Note on Notation: Slashes

Inside R, on any computer system, pathnames always use only the forward slashes "/".

In Linux and Macintosh, operating system pathnames use only the forward slashes "/".

In Windows, operating system pathnames—at the MS-DOS prompt shell CMD, in the Windows icon Properties windows, and in the **Windows Explorer** file-name entry bar—are written with backslashes "\". The **HH** package provides a convenience function WindowsPath to convert pathnames from the forward slash notation to the backslash notation.

A.3.5 Duplicating All HH Examples

Script files containing R code for all examples in the book are available for you to use to duplicate the examples (table and figures) in the book, or to use as templates for your own analyses. You may open these files in an R-aware editor.

See the discussion in Section B.2 for more details.

A.3.5.1 Linux and Macintosh

The script files for the second edition of this book are in the directory

```
HHscriptnames()
```

The script files for the first edition of this book are in the directory

```
HHscriptnames(edition=1)
```

A.3.5.2 Windows.

The script files for the second edition of this book are in the directory

```
WindowsPath(HHscriptnames())
```

The script files for the first edition of this book are in the directory

```
WindowsPath(HHscriptnames(edition=1))
```

A.3.6 Learning the Functions in R

Help on any function is available. For example, to learn about the ancovaplot function, type

```
?ancovaplot
```

To see a function in action, you can run the examples on the help page. You can do them one at a time by manually copying the code from the example section of a help page and pasting it into the R console. You can do them all together with the example function, for example

```
example("ancovaplot")
```

Some functions have demonstration scripts available, for example,

```
demo("ancova")
```

The list of demos available for a specific package is available by giving the package name

```
demo(package="HH")
```

The list of all demos available for currently loaded packages is available by

```
demo()
```

See ?demo for more information on the demo function.

The demo and example functions have optional arguments ask=FALSE and echo=TRUE. The default value ask=TRUE means the user has to press the ENTER key every time a new picture is ready to be drawn. The default echo=FALSE often has the effect that only the last of a series of **lattice** or **ggplot2** graphs will be displayed. See FAQ 7.22:

> 7.22 Why do lattice/trellis graphics not work?
>
> The most likely reason is that you forgot to tell R to display the graph. **lattice** functions such as xyplot() create a graph object, but do not display it (the same is true of **ggplot2** graphics, and **trellis** graphics in S-Plus). The print() method for the graph object produces the actual display. When you use these functions interactively at the command line, the result is automatically printed, but in source() or inside your own functions you will need an explicit print() statement.

A.3.7 Learning the lattice Functions in R

One of the best places to learn the **lattice** functions is the original trellis documentation: the S-Plus Users Manual (Becker et al., 1996a) and a descriptive paper with examples (Becker et al., 1996b). Both are available for download.

A.3.8 Graphs in an Interactive Session

We frequently find during an interactive session that we wish to back up and compare our current graph with previous graphs.

For R on the Macintosh using the quartz device, the most recent 16 figures are available by COMMAND-LEFTARROW and COMMAND-RIGHTARROW.

For R on **Windows** using the `windows` device, previous graphs are available if you turn on the graphical history of your device. This can be done with the mouse (by clicking **History** > **Recording** on the device menu) or by entering the R command

```
options(graphics.record = TRUE)
```

You can now navigate between graphs with the PGUP and PGDN keys.

A.4 S/R Language Style

S is a language. R is a dialect of S. Languages have standard styles in which they are written. When a language is displayed without paying attention to the style, it looks unattractive and may be illegible. It may also give valid statements that are not what the author intended. Read what you turn in before turning it in.

The basic style conventions are simple. They are also self-evident after they have been pointed out. Look at the examples in the book's code files in the directory

```
HHscriptnames()
```

and in the R manuals.

1. Use the courier font for computer listings.
   ```
   This is courier.
   ```
   This is Times Roman.
   Notice that displaying program output in a font other than one for which it was designed destroys the alignment and makes the output illegible. We illustrate the illegibility of improper font choice in Table A.2 by displaying the first few lines of the `data(tv)` dataset from Chapter 4 in both correct and incorrect fonts.

Table A.2 Correct and incorrect alignment of computer listings. The columns in the correct example are properly aligned. The concept of columns isn't even visible in the incorrect example.

Courier (with correct alignment of columns)	Times Roman (alignment is lost)

```> tv[1:5, 1:3]```
```   life.exp ppl.per.tv ppl.per.phys```
``` Argentina   70.5   4.0   370```
```Bangladesh   53.5   315.0   6166```
```   Brazil   65.0   4.0   684```
```   Canada   76.5   1.7   449```
```   China   70.0   8.0   643```
	> tv[1:5, 1:3]
	life.exp ppl.per.tv ppl.per.phys
	Argentina 70.5 4.0 370
	Bangladesh 53.5 315.0 6166
	Brazil 65.0 4.0 684
	Canada 76.5 1.7 449
	China 70.0 8.0 643

2. Use sensible spacing to distinguish the words and symbols visually. This convention allows people to read the program.

---

bad: `abc<-def`   no space surrounding the `<-`

---

good: `abc <- def`

---

3. Use sensible indentation to display the structure of long statements. Additional arguments on continuation lines are most easily parsed by people when they are aligned with the parentheses that define their depth in the set of nested parentheses.

---

bad: `names(tv) <- c("life.exp","ppl.per.tv","ppl.per.phys",`
       `"fem.life.exp","male.life.exp")`

---

good: `names(tv) <- c("life.exp",`
      `                "ppl.per.tv",`
      `                "ppl.per.phys",`
      `                "fem.life.exp",`
      `                "male.life.exp")`

Use Emacs (or other R-aware editor) to help with indentation. For example, open up a new file `tmp.r` in Emacs (or another editor) and type the above bad example—in two lines with the indentation exactly as displayed. Emacs in ESS [S] mode and other R-aware editors will automatically indent it correctly.

4. Use a page width in the Commands window that your word processor and printer supports. We recommend

   `options(width=80)`

   if you work with the natural width of 8.5in×11in paper ("letter" paper in the US. The rest of the world uses "A4" paper at 210cm×297cm) with 10-pt type. If you use a word processor that insists on folding lines at some shorter width (72 characters is a common—and inappropriate—default folding width), you must either take control of your word processor, or tell R to use a shorter width. Table A.3 shows a fragment from an `anova` output with two different width settings for the word processor.

Table A.3 Legible and illegible printings of the same table. The illegible table was inappropriately folded by an out-of-control word processor. You, the user, must take control of folding widths.

Legible:

```
> anova(fat2.lm)
Analysis of Variance Table

Response: bodyfat

Terms added sequentially (first to last)
            Df Sum of Sq  Mean Sq  F Value       Pr(F)
  abdomin    1  2440.500 2440.500 101.1718 0.00000000
   biceps    1   209.317  209.317   8.6773 0.00513392
Residuals  44  1061.382   24.122
```

Illegible (folded at 31 characters):

```
> anova(fat2.lm)
Analysis of Variance Table

Response: bodyfat

Terms added sequentially (first
 to last)
            Df Sum of Sq  Mean Sq
  F Value       Pr(F)
  abdomin    1  2440.500 2440.500
101.1718 0.00000000
   biceps    1   209.317  209.317
  8.6773 0.00513392
Residuals  44  1061.382   24.122
```

5. Reserved names. R has functions with the single-letter names c, s, and t. R also has many functions whose names are commonly used statistical terms, for example: mean, median, resid, fitted, data. If you inadvertently name an object with one of the names used by the system, your object might mask the system object and strange errors would ensue. Do not use system names for your variables. You can check with the statement

```
conflicts(detail=TRUE)
```

A.5 Getting Help While Learning and Using R

Although this section is written in terms of the R email list, its recommendations apply to all situations, in particular, to getting help from your instructor while reading this book and learning R.

R has an email help list. The archives are available and can be searched. Queries sent to the help list will be forwarded to several thousand people world-wide and will be archived. For basic information on the list, read the note that is appended to the bottom of EVERY email on the R-help list, and follow its links:

```
R-help@r-project.org mailing list -- To UNSUBSCRIBE and more,
see https://stat.ethz.ch/mailman/listinfo/r-help
PLEASE do read the posting guide
http://www.R-project.org/posting-guide.html and
provide commented, minimal, self-contained, reproducible code.
```

R-help is a plain text email list. Posting in HTML mangles your code, making it hard to read. Please send your question in plain text and make the code reproducible. There are two helpful sites on reproducible code

```
http://adv-r.had.co.nz/Reproducibility.html
```

```
http://stackoverflow.com/questions/5963269/
how-to-make-a-great-r-reproducible-example
```

When outside help is needed, save the *contents* of the R console window to show to your outside expert. In addition to the lines and results leading to the problem, including ALL messages that R produces, you must include the line (and its results)
```
sessionInfo()
```
in the material you show the expert.

Screenshots are not a good way to capture information. The informative way to get the contents of the console window is by highlighting the text of the entire window (including the off-screen part) and saving it in a text file. Show your text listing to an R expert.

A.6 R Inexplicable Error Messages—Some Debugging Hints

In general, weird and inexplicable errors mean that there are masked function names. That's the easy part. The trick is to find which name. The name conflict is frequently inside a function that has been called by the function that you called directly. The general method, which we usually won't need, is to trace the action of the function you called, and all the functions it called in turn. See ?trace, ?recover, ?browser, and ?debugger for help on using these functions.

One method we will use is to find all occurrences of our names that might mask system functions. R provides two functions that help us. See ?find and ?conflicts for further detail.

find: Returns a vector of names, or positions of databases and/or frames that contain an object.

- This example is problem because the user has used a standard function name "data" for a different purpose

```
> args(data)
function (..., list = character(), package = NULL,
    lib.loc = NULL, verbose = getOption("verbose"),
    envir = .GlobalEnv)
NULL
> data <- data.frame(a=1:3, b=4:6)
> data
  a b
1 1 4
2 2 5
3 3 6
> args(data)
NULL
> find("data")
[1] ".GlobalEnv"     "package:utils"
> rm(data)
> args(data)
function (..., list = character(), package = NULL,
    lib.loc = NULL, verbose = getOption("verbose"),
    envir = .GlobalEnv)
NULL
>
```

conflicts: This function checks a specified portion of the search list for items that appear more than once.

- The only items we need to worry about are the ones that appear in our working directory.

```
> data <- data.frame(a=1:3, b=4:6)
> conflicts(detail=TRUE)
$.GlobalEnv
[1] "data"

$'package:utils'
[1] "data"

$'package:methods'
[1] "body<-"    "kronecker"

$'package:base'
[1] "body<-"    "kronecker"

> rm(data)
```

Once we have found those names we must assign their value to some other variable name and then remove them from the working directory. In the above example, we have used the system name "data" for one of our variable names. We must assign the value to a name without conflict, and then remove the conflicting name.

```
> data <- data.frame(a=1:3, b=4:6)
> find("data")
[1] ".GlobalEnv"    "package:utils"
> CountingData <- data
> rm(data)
> find("data")
[1] "package:utils"
>
```

Appendix B

HH

Every graph and table in this book is an example of the types of graphs and analytic tables that readers can produce for their own data using functions in either base R or the **HH** package (Heiberger, 2015). Please see Section A.1 for details on installing **HH** and additional packages.

When you see a graph or table you need, open the script file for that chapter and use the code there on your data. For example, the MMC plot (Mean–mean Multiple Comparisons plot) is described in Chapter 7, and the first MMC plot in that Chapter is Figure 7.3. Therefore you can enter at the R prompt:

```
HHscriptnames(7)
```

and discover the pathname for the script file. Open that file in your favorite R-aware editor. Start at the top and enter chunks (that is what a set of code lines is called in these files which were created directly from the manuscript using the R function Stangle) from the top until the figure you are looking for appears. That gives the sequence of code you will need to apply to your own data and model.

B.1 Contents of the HH Package

The **HH** package contains several types of items.

1. Functions for the types of graphs illustrated in the book. Most of the graphical functions in **HH** are built on the trellis objects in the **lattice** package.

2. R scripts for all figures and tables in the book.

3. R data objects for all datasets used in the book that are not part of base R.

4. Additional R functions, some analysis and some utility, that I like to use.

© Springer Science+Business Media New York 2015
R.M. Heiberger, B. Holland, *Statistical Analysis and Data Display*,
Springer Texts in Statistics, DOI 10.1007/978-1-4939-2122-5

B.2 R Scripts for all Figures and Tables in the Book

Files containing R scripts for all figures and tables in the book, both Second and First Editions, are included with the **HH** package. The details of pathnames to the script files differ by computer operating systems, and often by individual computer.

To duplicate the figures and tables in the book, open the appropriate script file in an R-aware editor. Highlight and send over to the R console one chunk at a time. Each script file is consistent within itself. Code chunks later in a script will frequently depend on the earlier chunks within the same script having already been executed.

Second Edition: The R function HHscriptnames displays the full pathnames of Second Edition script files for your computer. For example, HHscriptnames(7) displays the full pathname for Chapter 7. The pathname for Chapter 7 relative to the **HH** package is HH/scripts/hh2/mcomp.r. Valid values for the chapternumbers argument for the Second Edition are c(1:18, LETTERS[1:15]).

First Edition: First Edition script file pathnames are similar, for example, the relative path for Chapter 7 is HH/scripts/hh1/Ch07-mcomp.r. The full pathnames of First Edition files for your computer are displayed by the R statement HHscriptnames(7, edition=1). Valid values for the chapternumbers argument for the First Edition are c(1:18).

The next few subsections show sample full pathnames of Second Edition script filenames for several different operating systems.

B.2.1 Macintosh

On Macintosh the full pathname will appear as something like

```
> HHscriptnames(7)
7 "/Library/Frameworks/R.framework/Versions/3.2/Resources/
                        library/HH/scripts/hh2/mcomp.R"
```

B.2.2 Linux

On Linux, the full pathname will appear something like

```
> HHscriptnames(7)
7 "/home/rmh/R/x86_64-unknown-linux-gnu-library/3.2/HH/
                        scripts/hh2/mcomp.R"
```

B.2.3 Windows

On Windows the full pathname will appear as something like

```
> HHscriptnames(7)
7 "C:/Users/rmh/Documents/R/win-library/3.2/HH/scripts/
                                                  hh2/mcomp.R"
```

You might prefer it to appear with Windows-style path separators (with the escaped backslash that looks like a double backslash)

```
> WindowsPath(HHscriptnames(7), display=FALSE)
7 "C:\\Users\\rmh\\Documents\\R\\win-library\\3.2\\HH\
                                      \scripts\\hh2\\mcomp.R"
```

or unquoted and with the single backslash

```
> WindowsPath(HHscriptnames(7))
7 C:\Users\rmh\Documents\R\win-library\3.2\HH\scripts\
                                              hh2\mcomp.R
```

Some of these variants will work with Windows Explorer (depending on which version of Windows) or your favorite editor, and some won't.

B.3 Functions in the HH Package

There are many functions in the **HH** package, and in the rest of R, that you will need to learn about. The easiest way is to use the documentation that is included with R. For example, to learn about the linear regression function lm (for "Linear Model"), just ask R with the simple "?" command:

```
?lm
```

and a window will open with the description. Try it now.

B.4 HH and S+

Package version HH_2.1-29 of 2009-05-27 (Heiberger, 2009) is still available at CSAN. This version of the package is appropriate for the First Edition of the book. It includes very little of the material developed after the publication of the First Edition. Once I started using the features of R's **latticeExtra** package it no longer made sense to continue compatibility with S-Plus. **HH_2.1-29** doesn't have data(); instead it has a datasets subdirectory.

Appendix C

Rcmdr: R Commander

The R Commander, released as the package **Rcmdr** (Fox, 2005; John Fox et al., 2015), is a platform-independent basic-statistics GUI (graphical user interface) for R, based on the **tcltk** package (part of R).

We illustrate how to use it by reconstructing the two panels of Figure 9.5 directly from the **Rcmdr** menu. We load **Rcmdr** indirectly by explicitly loading our package **RcmdrPlugin.HH** (Heiberger and with contributions from Burt Holland, 2015). We then use the menu to bring in the hardness dataset, compute the quadratic model, display the squared residuals for the quadratic model (duplicating the left panel of Figure 9.5), display the squared residuals for the linear model (duplicating the right panel of Figure 9.5). The linear model was fit implicitly and its summary is not automatically printed. We show the summary of the quadratic model.

Figures C.1–C.14 illustrate all the steps summarized above.

© Springer Science+Business Media New York 2015
R.M. Heiberger, B. Holland, *Statistical Analysis and Data Display*,
Springer Texts in Statistics, DOI 10.1007/978-1-4939-2122-5

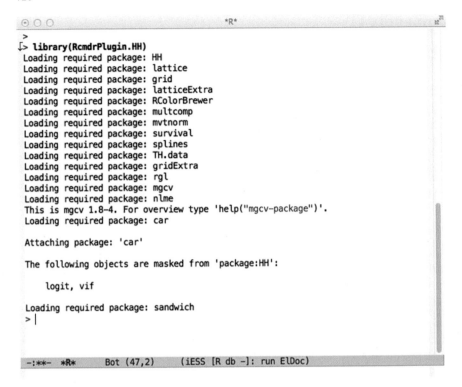

Fig. C.1 From the *R* buffer (or the R Console) enter library(RcmdrPlugin.HH). This also loads HH and Rcmdr and several other packages. It also open the Rcmdr window shown in Figure C.2.

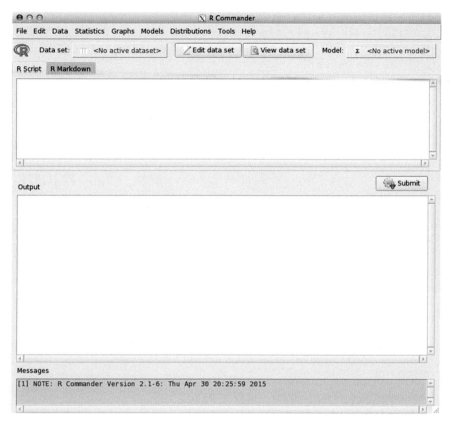

Fig. C.2 The Rcmdr window as it appears when first opened. It shows a menu bar, a tool bar, and three subwindows.

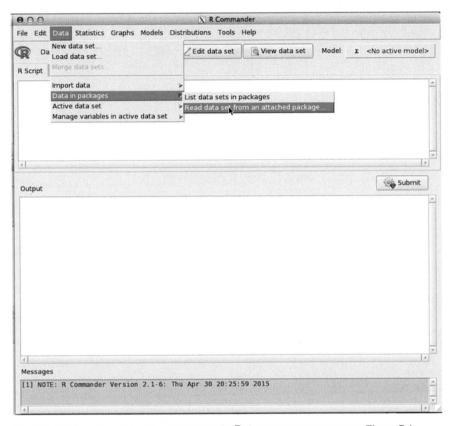

Fig. C.3 We bring in a dataset by clicking on the Data menu sequence to open Figure C.4.

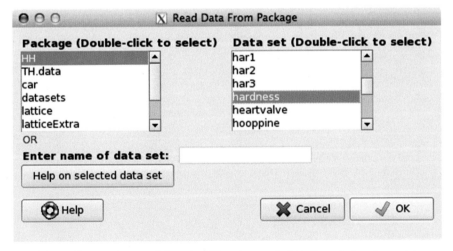

Fig. C.4 In this menu box we click on the **HH** package and within it, on the hardness data.

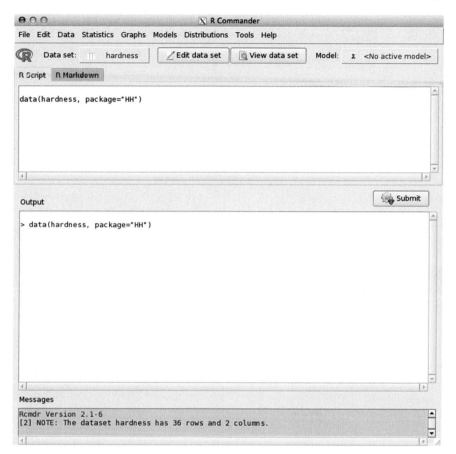

Fig. C.5 The Data set: item in the tool bar now shows hardness as the active dataset. The R Script subwindow show the R command that was generated by the menu sequence. The Output subwindow shows the transcript of the R session where the command was executed. The Messages subwindow give information on the dataset that was opened.

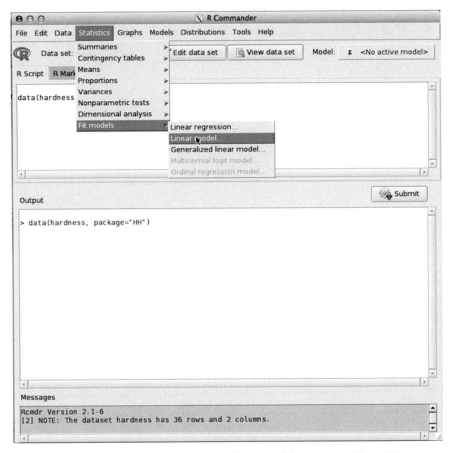

Fig. C.6 From the Statistics menu we open the Linear model menu box in Figure C.7.

Fig. C.7 Specify the linear model. The user can enter the model by typing or by clicking on the menu items.

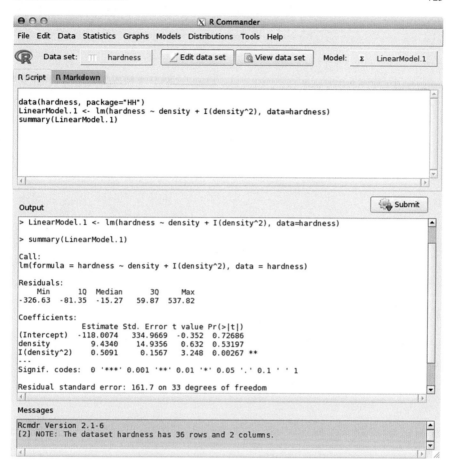

Fig. C.8 The Linear model menu item wrote the R commands shown in the R Script sub-window and executed them in the Output window. The model is stored in the R "lm" object named LinearModel.1. The Model: item in the tool bar now shows LinearModel.1 as the active model.

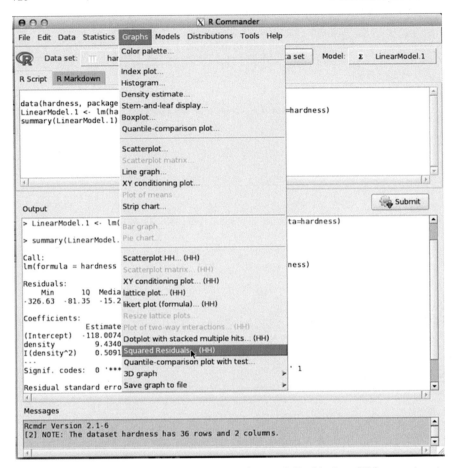

Fig. C.9 Use the Graphs menu to get to the Squared Residuals...(HH) menu box in Figure C.10.

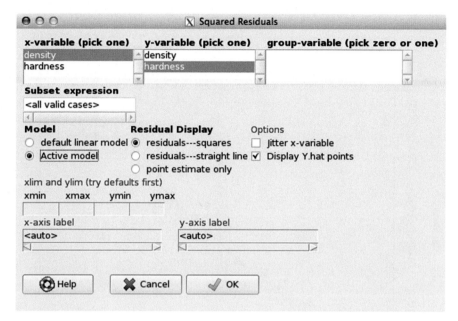

Fig. C.10 Specify the x-variable, the y-variable, and the active model (`LinearModel.1`) to get Figure C.11.

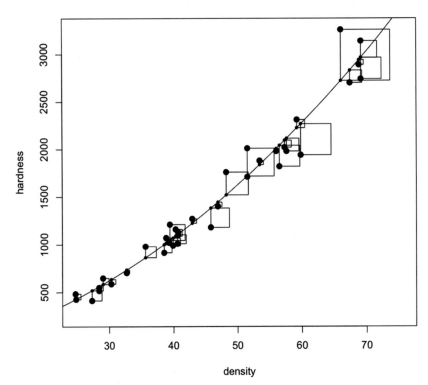

Fig. C.11 This is the squared residuals for the quadratic model (duplicating the right panel of Figure 9.5).

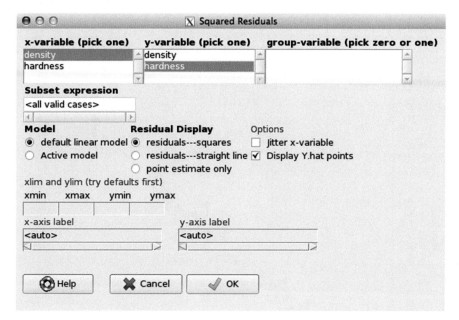

Fig. C.12 We repeated Figure C.9 to get the menu box again. This time we took the default linear model to get Figure C.13.

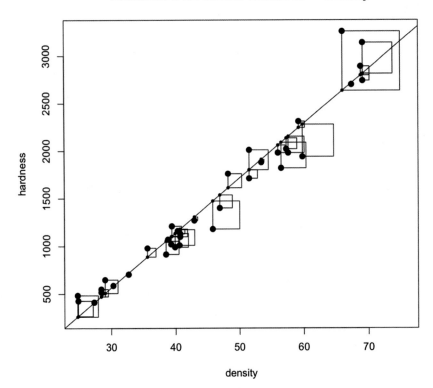

Fig. C.13 This is the squared residuals for the linear model (duplicating the left panel of Figure 9.5).

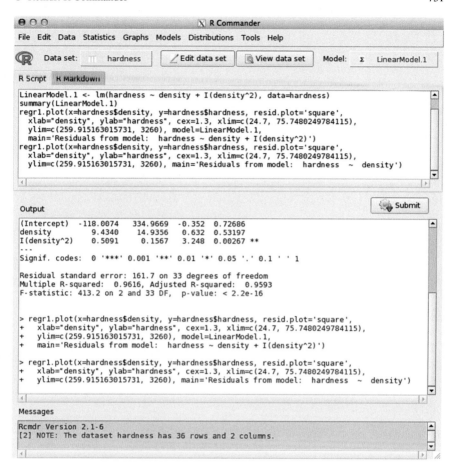

Fig. C.14 The Rcmdr window now shows the summary for the quadratic regression and the specifications for the two graphs in Figures C.11 and C.13.

Appendix D

RExcel: Embedding R inside **Excel** on **Windows**

If you are running on a **Windows** machine and have access to MS **Excel**, then we recommend that you install **RExcel** (Baier and Neuwirth, 2007; Neuwirth, 2014). **RExcel** is free of charge for "single user non-commercial use" with 32-bit **Excel**. Any other use will require a license. Please see the license at rcom.univie.ac.at for details.

    RExcel seamlessly integrates the entire set of **R**'s statistical and graphical methods into **Excel**, allowing students to focus on statistical methods and concepts and minimizing the distraction of learning a new programming language. Data can be transferred between **R** and **Excel** "the **Excel** way" by selecting worksheet ranges and using **Excel** menus. **RExcel** has embedded the **Rcmdr** menu into the **Excel** ribbon. Thus **R**'s basic statistical functions and selected advanced methods are available from an **Excel** menu. Almost all R functions can be used as worksheet functions in Excel. Results of the computations and statistical graphics can be returned back into **Excel** worksheet ranges. **RExcel** allows the use of **Excel** scroll bars and check boxes to create and animate R graphics as an interactive analysis tool.

    See Heiberger and Neuwirth (2009) for the book *R through Excel: A Spreadsheet Interface for Statistics, Data Analysis, and Graphics*. This book is designed as a computational supplement for any Statistics course.

    RExcel works with **Excel** only on **Windows**. **Excel** on the **Macintosh** uses a completely different interprocess communications system.

© Springer Science+Business Media New York 2015

R.M. Heiberger, B. Holland, *Statistical Analysis and Data Display*,

Springer Texts in Statistics, DOI 10.1007/978-1-4939-2122-5

D.1 Installing **RExcel** for **Windows**

You must have MS Excel (2007, 2010, or 2013) installed on your Windows machine. You need to purchase Excel separately. Excel 2013 is the current version (in early 2015). RExcel is free of charge for "single user non-commercial use" with 32-bit Excel. Any other use will require a license.

D.1.1 Install R

Begin by installing R and the necessary packages as described in Section A.1.

D.1.2 Install Two R Packages Needed by RExcel

You will also need two more R packages that must be installed as computer Administrator.

Start R as Administrator (on Windows 7 and 8 you need to right-click the R icon and click the "Run as administrator" item). In R, run the following commands (again, you must have started R as Administrator to do this)

```
## Tell Windows that R should have the same access to the outside
## internet that is granted to Internet Explorer.
setInternet2()
install.packages(c("rscproxy","rcom"),
                 repos="http://rcom.univie.ac.at/download",
                 type="binary",
                 lib=.Library)
library(rcom)
comRegisterRegistry()
```

Close R with q("no"). If it asks
    Save workspace image? [y/n/c]:
answer with n.

D.1.3 Install *RExcel* and Related Software

Go to `http://rcom.univie.ac.at` and click on the Download tab. Download and execute the following four files (or newer releases if available)

- `statconnDCOM3.6-OB2_Noncommercial`

- `RExcel 3.2.15`

- `RthroughExcelWorkbooksInstaller_1.2-10.exe`

- `SWord 1.0-1B1 Noncommercial` SWord (Baier, 2014) is an add-in package for MSword that makes it possible to embed R code in a MSword document. The R code will be automatically executed and the output from the R code will be included within the MSword document. SWord is free for non-commercial use. Any other use will require a license. SWord is a separate program from RExcel and is not required for RExcel.

These installer `.exe` files will ask for administrator approval, as they write in the `Program Files` directory and write to the Windows registry as part of the installation. Once they are installed, they run in normal user mode.

D.1.4 Install Rcmdr to Work with *RExcel*

In order for RExcel to place the **Rcmdr** menu on the Excel ribbon, it is necessary that **Rcmdr** be installed in the `C:/Program Files/R/R-x.y.z/library` directory and not in the `C:/Users/rmh/*/R/win-library/x.y` directory. If **Rcmdr** is installed in the user directory, it must be removed before reinstalling it as Administrator. Remove it with the `remove.packages` function using

        remove.packages(c("Rcmdr", "RcmdrMisc"))

Then see the installation details in Section A.1.2.2.

D.1.5 Additional Information on Installing *RExcel*

Additional RExcel installation information is available in the Wiki page at

        `http://rcom.univie.ac.at`

D.2 Using RExcel

D.2.1 Automatic Recalculation of an R Function

RExcel places R inside the Excel automatic recalculation model. Figure D.1 by
Heiberger and Neuwirth was originally presented in Robbins et al. (2009) using
Excel 2007. We reproduce it here with Excel 2013.

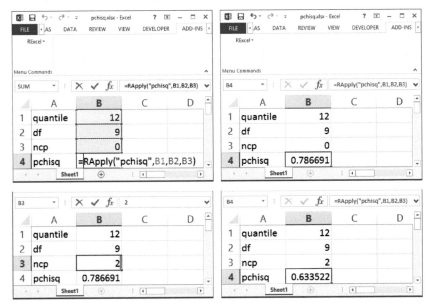

Fig. D.1 Any R function can be used in Excel with the RExcel worksheet function RApply. The
formula =RApply("pchisq", B1, B2, B3) computes the value of the noncentral distribution
function for the quantile-value, the degrees of freedom, and the noncentrality parameter in cells
B1, B2, and B3, respectively, and returns its value into cell B4. When the value of one of the
arguments, in this example the noncentrality parameter in cell B3, is changed, the value of the
cumulative distribution is automatically updated by Excel to the appropriate new value.

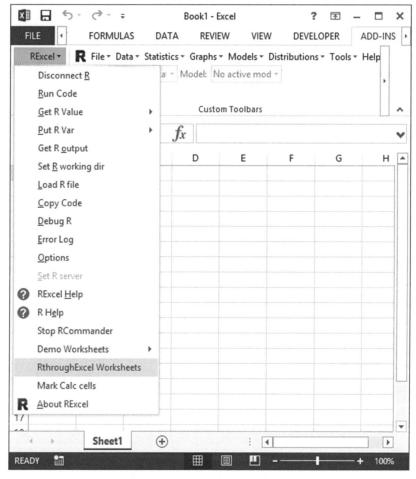

Fig. D.2 Retrieve the StudentData into Excel. From the RExcel Add-In tab, click RthroughExcel Worksheets. This brings up the BookFilesTOC worksheet in Figure D.3.

D.2.2 Transferring Data To/From R and Excel

Datasets can be transferred in either direction to/from Excel from/to R. In Figures D.2–D.4 we bring in a dataset from an Excel worksheet, transfer it to R, and make it the active dataset for use with **Rcmdr**.

The StudentData was collected by Erich Neuwirth for over ten years from students in his classes at the University of Vienna. The StudentData dataset is included with RExcel.

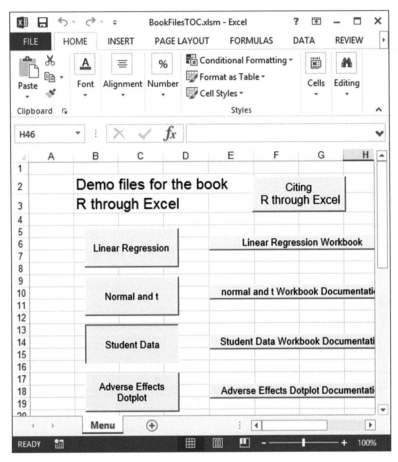

Fig. D.3 Click the StudentData button to bring up the StudentData worksheet in Figure D.4.

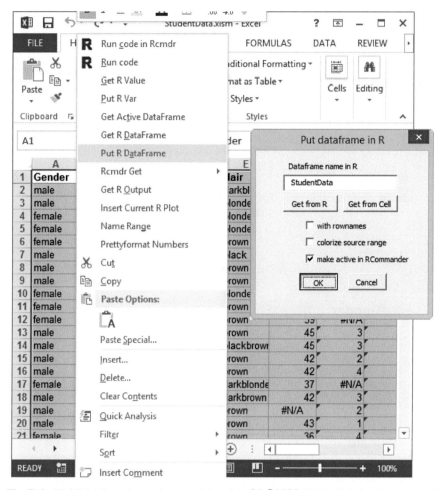

Fig. D.4 Highlight the entire region containing data A1:Q1126. Right-click for the menu and select Put R DataFrame. Click OK on the "Put dataframe in R" Dialog box. This places the dataset name StudentData into the **Rcmdr**'s active Dataset box in Figure D.5.

D.2.3 Control of a lattice Plot from an *Excel*/*Rcmdr* Menu

The example in Figures D.5–D.8 is originally from Heiberger and Neuwirth (2009) using Excel 2007. We reproduce it here with Excel 2013. We made it the active dataset for **Rcmdr** in Figures D.2–D.4.

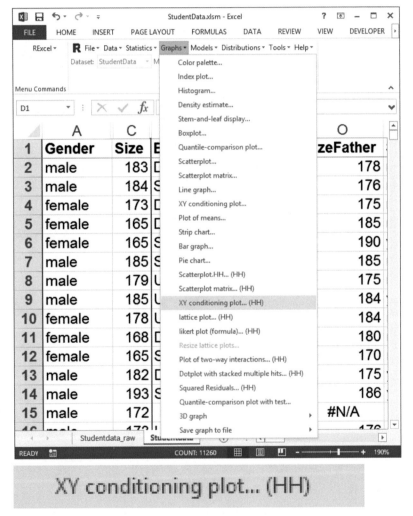

Fig. D.5 RExcel has placed the **Rcmdr** menu onto the Excel ribbon. Click the Graphs tab to get the menu and then click XY conditioning plot... (HH) to get the Dialog box in Figure D.6. The (HH) in the menu item means the function was added to the **Rcmdr** menu by our **RcmdrPlu-gin.HH** package (Heiberger and with contributions from Burt Holland, 2015).

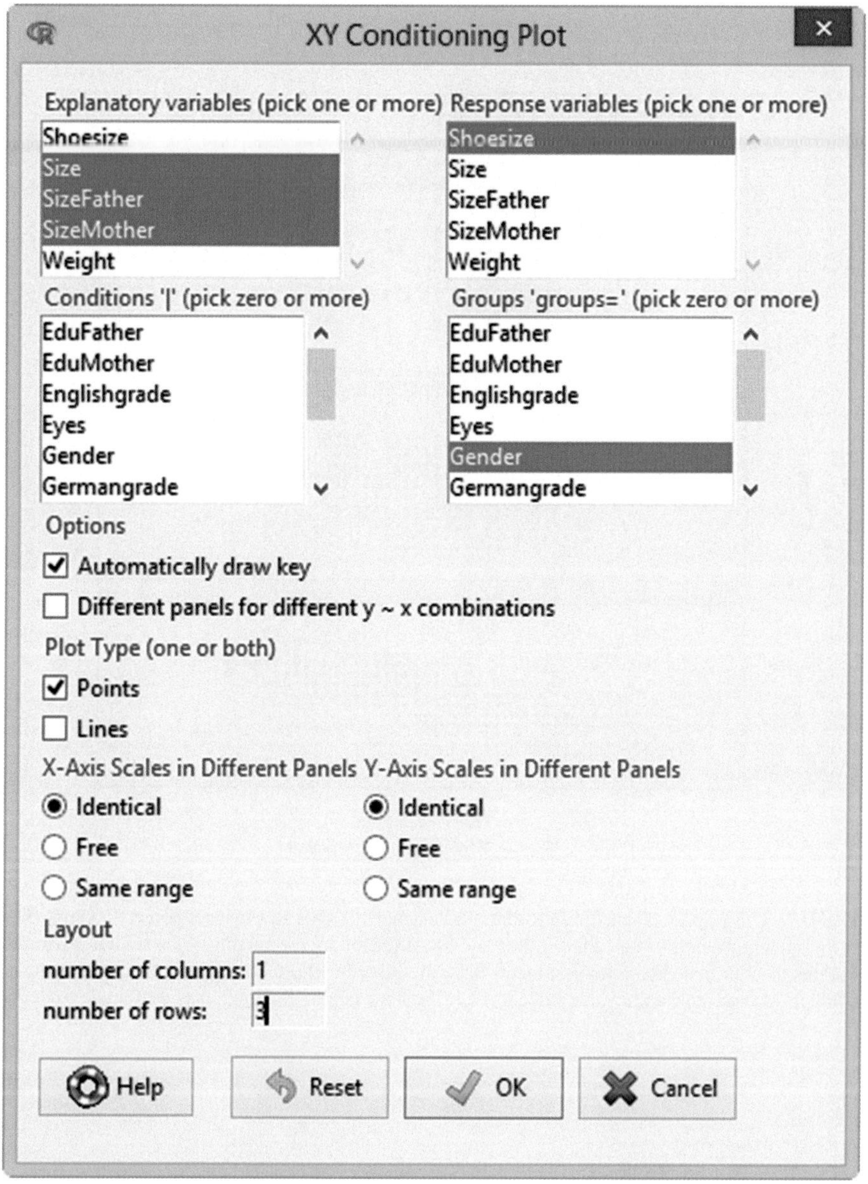

Fig. D.6 The user fills in the Dialog box to specify the graph in Figure D.7 and generate the R commands displayed in Figure D.8.

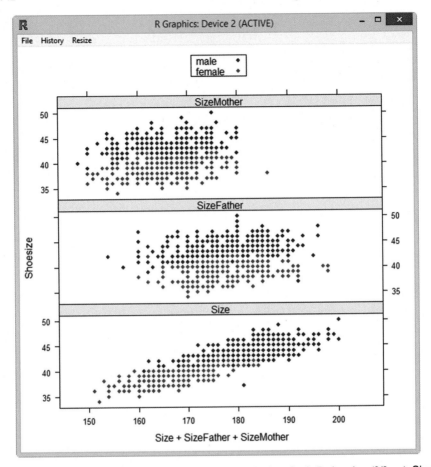

Fig. D.7 The graph is displayed. Shoesize is the student's shoe size in Paris points (2/3 cm). Size is the student's height in cm. SizeFather and SizeMother are the heights of the student's parents. Fathers of both male and female students have the same height distribution as the male students. Mothers of both male and female students have the same height distribution as the female students.

```
R Commander                                                    −  □  ✕

R Script   R Markdown

xyplot(Shoesize ~ Size + SizeFather + SizeMother, layout=c(1, 3),
   groups=Gender, type="p", pch=16, auto.key=list(border=TRUE),
   par.settings=simpleTheme(pch=16), scales=list(x=list(relation='same'),
   y=list(relation='same')), data=StudentData)
```

Fig. D.8 The generated R code is displayed.

Appendix E

Shiny: Web-Based Access to R Functions

Shiny (Chang et al., 2015; RStudio, 2015) is an R package that provides an R language interface for writing interactive web applications. Apps built with **shiny** place the power of R behind an interactive webpage that can be used by non-programmers. A full tutorial and gallery are available at the Shiny web site.

We have animated several of the graphs in the **HH** package using **shiny**.

© Springer Science+Business Media New York 2015
R.M. Heiberger, B. Holland, *Statistical Analysis and Data Display*,
Springer Texts in Statistics, DOI 10.1007/978-1-4939-2122-5

E.1 NTplot

The NTplot function shows significance levels and power for the normal or *t*-distributions. Figure E.1, an interactive version of the top panel of the middle section of Figure 3.20, is specified with

```
NTplot(mean0=8, mean1=8.411, sd=2, n=64, cex.prob=1.3,
       shiny=TRUE)
```

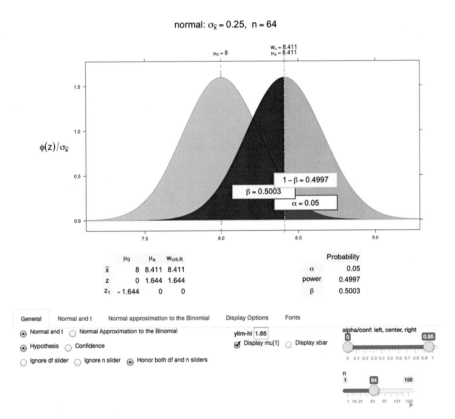

Fig. E.1 This is an interactive version of Figure 3.20 constructed with the **shiny** package. Adjusting the n slider at the bottom right can produce all three columns of Figure 3.20. Clicking the ▶ button below the slider will dynamically move through all values of *n* from 1 through 150, including the three that are displayed in Figure 3.20. There are additional controls on the Normal and t, Display Options, and Fonts tabs that will show the power and beta panels.

E.2 `bivariateNormal`

Figure E.2 is an interactive version of Figure 3.9 showing the bivariate normal density in 3D space with various correlations and various viewpoints.

```
shiny::runApp(system.file("shiny/bivariateNormal",
                          package="HH"))
```

Bivariate Normal Density

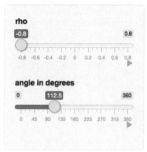

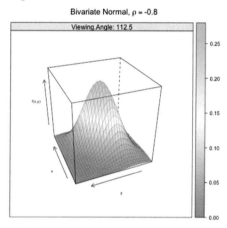

Fig. E.2 This is an interactive version of Figure 3.9 constructed with the **shiny** package. Adjusting the rho slider changes the correlation between x and y. Adjusting the angle in degrees slider rotates the figure through all the viewpoint angles shown in Figure 3.10. Both sliders can be made dynamic by clicking their ▶ buttons.

E.3 `bivariateNormalScatterplot`

Figure E.3 is a dynamic version of Figure 3.8 specified with

```
shiny::runApp(system.file("shiny/bivariateNormalScatterplot",
                          package="HH"))
```

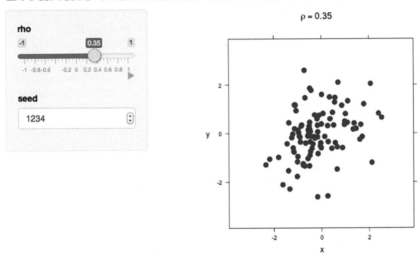

Fig. E.3 This is an interactive version of Figure 3.8 constructed with the **shiny** package. Adjusting the rho slider changes the correlation between x and y. By clicking the ► button, the figure will transition through the panels of Figure 3.8.

E.4 `PopulationPyramid`

Figure E.4 is an interactive version of Figure 15.19 showing the population pyramid for the United States annually for the years 1900–1979 specified with

```
shiny::runApp(system.file("shiny/PopulationPyramid",
                          package="HH"))
```

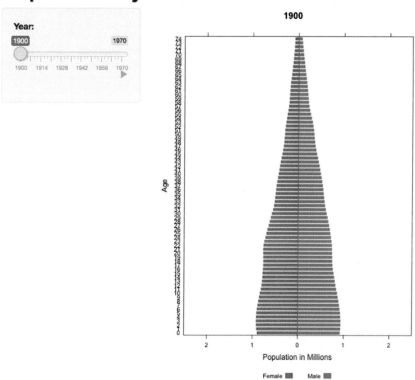

Fig. E.4 This is an interactive version of Figure 15.19 constructed with the **shiny** package. Adjusting the Year slider changes the year in the range 1900–1970. By clicking the ▶ button, the figure will dynamically transition through the panels of Figure 15.19.

Appendix F

R Packages

The R program as supplied by R-Core on the CRAN (Comprehensive R Archive Network) page (CRAN, 2015) consists of the base program and about 30 required and recommended packages. Everything else is a contributed package. There are about 6500 contributed packages (April 2015). Our **HH** is a contributed package.

F.1 What Is a Package?

R packages are extensions to R.

Each package is a collection of functions and datasets designed to work together for a specific purpose. The **HH** package is designed to provide computing support for the techniques discussed in this book . The **lattice** package (a recommended package) is designed to provide `xyplot` and related graphics functions. Most of the graphics functions in **HH** are built on the functions in **lattice**.

Packages consist at a minimum of functions written in R, datasets that can be brought into the R working directory, and documentation for all functions and datasets. Some packages include subroutines written in Fortran or C.

F.2 Installing and Loading R Packages

The base and recommended packages are installed on your computer when you download and install R itself. All other packages must be explicitly installed (downloaded from CRAN and placed into an R-determined location on your computer).

© Springer Science+Business Media New York 2015   749
R.M. Heiberger, B. Holland, *Statistical Analysis and Data Display*,
Springer Texts in Statistics, DOI 10.1007/978-1-4939-2122-5

The packages available at CRAN are most easily installed into your computer with a command of the form

```
install.packages("MyPackage")
```

The R GUIs usually have a menu item that constructs this statement for you.

The functions in an installed package are not automatically available when you start an R session. It is necessary to load them into the current session, usually with the library function. Most examples in this book require you to enter

```
library(HH)
```

(once per R session) before doing anything else. The **HH** package loads **lattice** and several additional packages.

The list of R packages installed on your computer is seen with the R command

```
library()
```

The list of R packages loaded into your current R session is seen with the R command

```
search()
```

F.3 Where Are the Packages on Your Computer?

Once a package has been installed on your computer it is kept in a directory (called a "library" in R terminology). The base and recommended packages are stored under R itself in directory

```
system.file("..")
```

Files inside the installed packages are in an internal format and cannot be read by an editor directly from the file system.

Contributed packages will usually be installed in a directory in your user space.

In Windows that might be something like

```
C:/Users/yourloginname/Documents/R/win-library/3.2
```

or

```
C:/Users/yourloginname/AppData/Roaming/R/win-library/3.2
```

On Macintosh it will be something like

```
/Users/yourloginname/Library/R/3.2/library
```

You can find out where the installed packages are stored by loading one and then entering

```
searchpaths()
```

(searchpaths() is similar to search() but with the full pathname included in the output, not just the package name). For example,

```
library(HH)
searchpaths()
```

F.4 Structure of an R Package

The package developer writes individual source files. The R build system (see the *Writing R Extensions* manual) has procedures for checking coherency and then for building the source package and the binaries.

The packages at CRAN are available in three formats. The source packages (what the package designer wrote and what you should read when you want to read the code) are stored as `packagename.tar.gz` files. The binary packages for **Windows** are stored as `packagename.zip` files. The binary packages for **Macintosh** are stored as `packagename.tgz` files.

F.5 Writing and Building Your Own Package

At some point in your analysis project you will have accumulated several of your own functions that you quite frequently use. At that point it will be time to collect them into a package.

We do not say much here about designing and writing functions in this book. Begin with *An Introduction to R* in file
```
system.file("../../doc/manual/R-intro.pdf")
```
It includes a chapter "Writing your own functions".

Nor do we say much here about building a package. The official reference is the *Writing R Extensions* manual that also comes with R in file
```
system.file("../../doc/manual/R-exts.pdf")
```
When you are ready, begin by looking at the help file
```
?package.skeleton
```
to see how the pieces fit together.

It will help to have someone work with you the first time you build a package. You build the package with the operating system command
```
R CMD check YourPackageName
```
and then install it on your own machine with the command
```
R CMD INSTALL --build YourPackageName
```
The checking process is very thorough, and gets more thorough at every release of R. Understanding how to respond to the messages from the check is the specific place where it will help to have someone already familiar with package building.

F.6 Building Your Own Package with **Windows**

The MS Windows operating system does not include many programs that are central to building R packages. You will need to download and install the most recent Rtools from CRAN. See Section A.1.2.4 for download information.

You will need to include Rtools in your PATH environment variable to enable the R CMD check packagename command to work. See "Appendix D The Windows toolset" in *R Installation and Administration* manual at

```
system.file("../../doc/manual/R-admin.pdf")
```

Appendix G

Computational Precision and Floating-Point Arithmetic

Computers use *floating point* arithmetic. The floating point system is not identical to the real-number system that we (teachers and students) know well, having studied it from kindergarten onward. In this section we show several examples to illustrate and emphasize the distinction.

The principal characteristic of real numbers is that we can have as many digits as we wish. The principal characteristic of floating point numbers is that we are limited in the number of digits that we can work with. In double-precision IEEE 754 arithmetic, we are limited to exactly 53 binary digits (approximately 16 decimal digits)

The consequences of the use of floating point numbers are pervasive, and present even with numbers we normally think of as far from the boundaries. For detailed information please see FAQ 7.31 in file

```
system.file("../../doc/FAQ")
```

The help menus in **Rgui** in Windows and **R.app** on Macintosh have direct links to the FAQ file.

G.1 Examples

Let us start by looking at two simple examples that require basic familiarity with floating point arithmetic.

1. Why is .9 not recognized to be the same as (.3 + .6)?
   Table G.1 shows that .9 is not perceived to have the same value as .3 + .6, when calculated in floating point (that is, when calculated by a computer). The difference between the two values is not 0, but is instead a number on the order of

© Springer Science+Business Media New York 2015
R.M. Heiberger, B. Holland, *Statistical Analysis and Data Display*,
Springer Texts in Statistics, DOI 10.1007/978-1-4939-2122-5

machine epsilon (the smallest number ϵ such that $1 + \epsilon > 1$). In R, the standard mathematical comparison operators recognize the difference. There is a function all.equal which tests for *near equality*. See ?all.equal for details.

Table G.1 Calculations showing that the floating point numbers .9 and .3 + .6 are not stored the same inside the computer. R comparison operators recognize the numbers as different.

```
> c(.9, (.3 + .6))
[1] 0.9 0.9

> .9 == (.3 + .6)
[1] FALSE

> .9 - (.3 + .6)
[1] 1.11e-16

> identical(.9, (.3 + .6))
[1] FALSE

> all.equal(.9, (.3 + .6))
[1] TRUE
```

Table G.2 $\left(\sqrt{2} \right)^2 \neq 2$ in floating point arithmetic inside the computer.

```
> c(2, sqrt(2)^2)
[1] 2 2

> sqrt(2)^2
[1] 2

> 2 == sqrt(2)^2
[1] FALSE

> 2 - sqrt(2)^2
[1] -4.441e-16

> identical(2, sqrt(2)^2)
[1] FALSE

> all.equal(2, sqrt(2)^2)
[1] TRUE
```

2. Why is $\left(\sqrt{2}\right)^2$ not recognized to be the same as 2?

Table G.2 shows that the difference between the two values $\left(\sqrt{2}\right)^2$ and 2 is not 0, but is instead a number on the order of machine epsilon (the smallest number ϵ such that $1 + \epsilon > 1$).

We will pursue these examples further in Section G.7, but first we need to introduce floating point numbers—the number system used inside the computer.

G.2 Floating Point Numbers in the IEEE 754 Floating-Point Standard

The number system we are most familiar with is the infinite-precision base-10 system. Any number can be represented as the infinite sum

$$\pm (a_0 \times 10^0 + a_1 \times 10^{-1} + a_2 \times 10^{-2} + \ldots) \times 10^p$$

where p can be any positive integer, and the values a_i are digits selected from decimal digits $\{0, 1, 2, 3, 4, 5, 6, 7, 8, 9\}$. For example, the decimal number 3.3125 is expressed as

$$3.3125 = (3 \times 10^0 + 3 \times 10^{-1} + 1 \times 10^{-2} + 2 \times 10^{-3} + 5 \times 10^{-4}) \times 10^0$$
$$= 3.3125 \times 1$$

In this example, there are 4 decimal digits after the radix point. There is no limit to the number of digits that could have been specified. For decimal numbers the term *decimal point* is usually used in preference to the more general term *radix point*.

Floating point arithmetic in computers uses a finite-precision base-2 (binary) system for representation of numbers. Most computers today use the 53-bit IEEE 754 system, with numbers represented by the finite sum

$$\pm (a_0 \times 2^0 + a_1 \times 2^{-1} + a_2 \times 2^{-2} + \ldots + a_{52} \times 2^{-}52) \times 2^p$$

where p is an integer in the range -1022 to 1023 (expressed as decimal numbers), the values a_i are digits selected from $\{0, 1\}$, and the subscripts and powers i are decimal numbers selected from $\{0, 1, \ldots, 52\}$. The decimal number 3.125_{10} is 11.0101_2 in binary.

$$3.3125_{10} = 11.0101_2 = (1 \times 2^0 + 1 \times 2^{-1} + 0 \times 2^{-2} + 1 \times 2^{-3} + 0 \times 2^{-4} + 1 \times 2^{-5}) \times 2_{10}$$
$$= 1.10101_2 \times 2_{10}$$

This example (in the normalized form $1.10101_2 \times 2_{10}$) has five binary digits (bits) after the radix point (binary point in this case). There is a maximum of 52 binary positions after the binary point.

Strings of 0 and 1 are hard for people to read. They are usually collected into units of 4 bits (called a byte).

The IEEE 754 standard requires the base $\beta = 2$ number system with $p = 53$ base-2 digits. Except for 0, the numbers in internal representation are always *normalized* with the leading bit always 1. Since it is always 1, there is no need to store it and only 52 bits are actually needed for 53-bit precision. A string of 0 and 1 is difficult for humans to read. Therefore every set of 4 bits is represented as a single hexadecimal digit, from the set {0 1 2 3 4 5 6 7 8 9 a b c d e f}, representing the decimal values {0 1 2 3 4 5 6 7 8 9 10 11 12 13 14 15}. The 52 stored bits can be displayed with 13 hex digits. Since the base is $\beta = 2$, the exponent of an IEEE 754 floating point number must be a power of 2. The double-precision computer numbers contain 64 bits, allocated 52 for the significant, 1 for the sign, and 11 for the exponent. The 11 bits for the exponent can express $2^{11} = 2048$ unique values. These are assigned to range from -2^{-1022} to 2^{1023}, with the remaining 2 exponent values used for special cases (zero and the *special quantities* NaN and ∞).

The number 3.3125_{10} is represented in hexadecimal (base-16) notation as

$$3.3125_{10} = 3.50_{16} = (1 \times 16^0 + a_{16} \times 16^{-1} + 8_{16} \times 16^{-2}) \times 2_{10}$$
$$= 1.a8_{16} \times 2_{10}$$
$$= 1.10101000_2 \times 2_{10}$$

There are two hex digits after the radix point (binary point, not hex point because the normalization is by powers of 2_{10} not powers of 16_{10}).

The R function sprintf is used to specify the printed format of numbers. The letter a in format sprintf("%+13.13a", x) tells R to print the numbers in hexadecimal notation. The "13"s say to use 13 hexadecimal digits after the binary point. See ?sprintf for more detail on the formatting specifications used by the sprintf function. Several numbers, simple decimals, and simple multiples of powers of 1/2 are shown in Table G.3 in both decimal and binary notation.

G.3 Multiple Precision Floating Point

The R package **Rmpfr** allows the construction and use of arbitrary precision floating point numbers. It was designed, and is usually used, for higher-precision arithmetic—situations where the 53-bit double-precision numbers are not precise

Table G.3 Numbers, some simple integers divided by 10, and some fractions constructed as multiples of powers of 1/2. The $i/10$ decimal numbers are stored as repeating binaries in the hexadecimal notation until they run out of digits. There are only 52 bits (binary digits) after the binary point. For decimal input 0.1 we see that the repeating hex digit is "9" until the last position where it is rounded up to "a".

```
> nums <- c(0, .0625, .1, .3, .3125, .5, .6, (.3 + .6), .9,  1)

> data.frame("decimal-2"=nums,
+            "decimal-17"=format(nums, digits=17),
+            hexadecimal=sprintf("%+13.13a", nums))
    decimal.2          decimal.17             hexadecimal
1     0.0000 0.00000000000000000 +0x0.0000000000000p+0
2     0.0625 0.06250000000000000 +0x1.0000000000000p-4
3     0.1000 0.10000000000000001 +0x1.999999999999ap-4
4     0.3000 0.29999999999999999 +0x1.3333333333333p-2
5     0.3125 0.31250000000000000 +0x1.4000000000000p-2
6     0.5000 0.50000000000000000 +0x1.0000000000000p-1
7     0.6000 0.59999999999999998 +0x1.3333333333333p-1
8     0.9000 0.89999999999999991 +0x1.ccccccccccccccp-1
9     0.9000 0.90000000000000002 +0x1.ccccccccccccdp-1
10    1.0000 1.00000000000000000 +0x1.0000000000000p+0
```

enough. In this Appendix we use it for lower-precision arithmetic—four or five significant digits. In this way it will be much easier to illustrate how the behavior of floating point numbers differs from the behavior of real numbers.

G.4 Binary Format

It is often easier to see the details of the numerical behavior when numbers are displayed in binary, not in the hex format of sprintf("%+13.13a", x). The **Rmpfr** package includes a binary display format for numbers. The formatBin function uses sprintf to construct a hex display format and then modifies it by replacing each hex character with its 4-bit expansion as shown in Table G.4.

Optionally (with argument scientific=FALSE), all binary numbers can be formatted to show aligned radix points. There is also a formatHex function which is essentially a wrapper for sprintf. Both functions are used in the examples in this Appendix. Table G.5 illustrates both functions, including the optional scientific argument, with a 4-bit arithmetic example.

Table G.4 Four-bit expansions for the sixteen hex digits. We show both lowercase [a:f] and uppercase [A:F] for the hex digits.

```
> Rmpfr:::HextoBin
            1        2        3        4        5        6        7
    "0000"  "0001"  "0010"  "0011"  "0100"  "0101"  "0110"  "0111"
            8        9        A        B        C        D        E        F
    "1000"  "1001"  "1010"  "1011"  "1100"  "1101"  "1110"  "1111"
            a        b        c        d        e        f
    "1010"  "1011"  "1100"  "1101"  "1110"  "1111"
```

G.5 Round to Even

The IEEE 754 standard calls for "rounding ties to even". The explanation here is from `help(mpfr, package="Rmpfr")`:

> The *round to nearest* ("N") mode, the default here, works as in the IEEE 754 standard: in case the number to be rounded lies exactly in the middle of two representable numbers, it is rounded to the one with the least significant bit set to zero. For example, the number 5/2, which is represented by (10.1) in binary, is rounded to (10.0)=2 with a precision of two bits, and not to (11.0)=3. This rule avoids the *drift* phenomenon mentioned by Knuth in volume 2 of *The Art of Computer Programming* (Section 4.2.2).

G.6 Base-10, 2-Digit Arithmetic

Hex numbers are hard to fathom the first time they are seen. We therefore look at a simple example of finite-precision arithmetic with 2 significant decimal digits.

Calculate the sum of squares of three numbers in 2-digit base-10 arithmetic. For concreteness, use the example

$$2^2 + 11^2 + 15^2$$

This requires rounding to 2 significant digits at *every* intermediate step. The steps are easy. Putting your head around the steps is hard.

We rewrite the expression as a fully parenthesized algebraic expression, so we don't need to worry about precedence of operators at this step.

$$((2^2) + (11^2)) + (15^2)$$

Now we can evaluate the parenthesized groups from the inside out.

Table G.5 Integers from 0 to 39 stored as 4-bit **mpfr** numbers. The numbers from 17 to 39 are rounded to four significant bits. All numbers in the "16" and "24" columns are multiples of 2, and all numbers in the "32" columns are multiples of 4. The numbers are displayed in decimal, hex, binary, and binary with aligned radix points. To interpret the aligned binary numbers, replace the "_" placeholder character with a zero.

```
> library(Rmpfr)

> FourBits <- mpfr(matrix(0:39, 8, 5), precBits=4)

> dimnames(FourBits) <- list(0:7, c(0,8,16,24,32))

> FourBits
'mpfrMatrix' of dim(.) =  (8, 5) of precision  4   bits
      0    8    16    24    32
0 0.00 8.00 16.0 24.0 32.0
1 1.00 9.00 16.0 24.0 32.0
2 2.00 10.0 18.0 26.0 32.0
3 3.00 11.0 20.0 28.0 36.0
4 4.00 12.0 20.0 28.0 36.0
5 5.00 13.0 20.0 28.0 36.0
6 6.00 14.0 22.0 30.0 40.0
7 7.00 15.0 24.0 32.0 40.0

> formatHex(FourBits)
   0        8         16        24        32
0 +0x0.0p+0 +0x1.0p+3 +0x1.0p+4 +0x1.8p+4 +0x1.0p+5
1 +0x1.0p+0 +0x1.2p+3 +0x1.0p+4 +0x1.8p+4 +0x1.0p+5
2 +0x1.0p+1 +0x1.4p+3 +0x1.2p+4 +0x1.ap+4 +0x1.0p+5
3 +0x1.8p+1 +0x1.6p+3 +0x1.4p+4 +0x1.cp+4 +0x1.2p+5
4 +0x1.0p+2 +0x1.8p+3 +0x1.4p+4 +0x1.cp+4 +0x1.2p+5
5 +0x1.4p+2 +0x1.ap+3 +0x1.4p+4 +0x1.cp+4 +0x1.2p+5
6 +0x1.8p+2 +0x1.cp+3 +0x1.6p+4 +0x1.ep+4 +0x1.4p+5
7 +0x1.cp+2 +0x1.ep+3 +0x1.8p+4 +0x1.0p+5 +0x1.4p+5

> formatBin(FourBits)
   0           8           16          24          32
0 +0b0.000p+0 +0b1.000p+3 +0b1.000p+4 +0b1.100p+4 +0b1.000p+5
1 +0b1.000p+0 +0b1.001p+3 +0b1.000p+4 +0b1.100p+4 +0b1.000p+5
2 +0b1.000p+1 +0b1.010p+3 +0b1.001p+4 +0b1.101p+4 +0b1.000p+5
3 +0b1.100p+1 +0b1.011p+3 +0b1.010p+4 +0b1.110p+4 +0b1.001p+5
4 +0b1.000p+2 +0b1.100p+3 +0b1.010p+4 +0b1.110p+4 +0b1.001p+5
5 +0b1.010p+2 +0b1.101p+3 +0b1.010p+4 +0b1.110p+4 +0b1.001p+5
6 +0b1.100p+2 +0b1.110p+3 +0b1.011p+4 +0b1.111p+4 +0b1.010p+5
7 +0b1.110p+2 +0b1.111p+3 +0b1.100p+4 +0b1.000p+5 +0b1.010p+5

> formatBin(FourBits, scientific=FALSE)
   0             8             16            24            32
0 +0b_____0.000  +0b__1000.___ +0b_1000_.___ +0b_1100_.___ +0b1000__.___
1 +0b_____1.000  +0b__1001.___ +0b_1000_.___ +0b_1100_.___ +0b1000__.___
2 +0b____10.00_  +0b__1010.___ +0b_1001_.___ +0b_1101_.___ +0b1000__.___
3 +0b____11.00_  +0b__1011.___ +0b_1010_.___ +0b_1110_.___ +0b1001__.___
4 +0b___100.0__  +0b__1100.___ +0b_1010_.___ +0b_1110_.___ +0b1001__.___
5 +0b___101.0__  +0b__1101.___ +0b_1010_.___ +0b_1110_.___ +0b1001__.___
6 +0b___110.0__  +0b__1110.___ +0b_1011_.___ +0b_1111_.___ +0b1010__.___
7 +0b___111.0__  +0b__1111.___ +0b_1100_.___ +0b1000__.___ +0b1010__.___
```

```
( (2²)  +  (11²) )  +  (15²) ## parenthesized expression
( (4)   +  (121) )  +  (225) ## square each term
( 4     +  120   )  +   220  ## round each term to two significant decimal digits
(     124        )  +   220  ## calculate the intermediate sum
(     120        )  +   220  ## round the intermediate sum to two decimal digits
              340           ## sum the terms
```

Compare this to the full precision arithmetic

```
( (2²)  +  (11²) )  +  (15²) ## parenthesized expression
( 4     +  121   )  +  225   ## square each term
(     125        )  +  225   ## calculate the intermediate sum
              350           ## sum the terms
```

We see immediately that two-decimal-digit rounding at each stage gives an answer that is not the same as the one from familiar arithmetic with real numbers.

G.7 Why Is .9 Not Recognized to Be the Same as (.3 + .6)?

We can now continue with the first example from Section G.1. The floating point binary representation of 0.3 and the floating point representation of 0.6 must be aligned on the binary point before the addition. When the numbers are aligned by shifting the smaller number right one position, the last bit of the smaller number has nowhere to go and is lost. The sum is therefore one bit too small compared to the floating point binary representation of 0.9. Details are in Table G.6.

G.8 Why Is $\left(\sqrt{2} \right)^2$ Not Recognized to Be the Same as 2?

We continue with the second example from Section G.1. The binary representation inside the machine of the two numbers $\left(\sqrt{2} \right)^2$ and 2 is not identical. We see in Table G.7 that they differ by one bit in the 53rd binary digit.

G.9 `zapsmall` to Round Small Values to Zero for Display

R provides a function that rounds small values (those close to the machine epsilon) to zero. We use this function for printing of many tables where we wish to interpret numbers close to machine epsilon as if they were zero. See Table G.8 for an example.

Table G.6 Now let's add 0.3 and 0.6 in hex:

```
0.3  +0x1.3333333333333p-2 = +0x0.9999999999999p-1  aligned binary (see below)
0.6  +0x1.3333333333333p-1 = +0x1.3333333333333p-1
-------------------------    --------------------
0.9  add of aligned binary   +0x1.ccccccccccccccp-1
0.9  convert from decimal     +0x1.cccccccccccccdp-1
```

We need to align binary points for addition. The shift is calculated by converting hex to binary, shifting one bit to the right to get the same p−1 exponent, regrouping four bits into hex characters, and allowing the last bit to fall off:

$$1.0011\ 0011\ 0011\ \ldots\ 0011\ \times 2^{-2}\ \rightarrow\ .1001\ 1001\ 1001\ \ldots\ 1001\ |\ 1 \times 2^{-1}$$

```
> nums369 <- c(.3, .6, .3+.6, 9)

> nums369df <-
+ data.frame("decimal-2"=nums369,
+            "decimal-17"=format(nums369, digits=17),
+            hexadecimal=sprintf("%+13.13a", nums369))

> nums369df[3,1] <- "0.3 + 0.6"

> nums369df
  decimal.2        decimal.17           hexadecimal
1       0.3 0.29999999999999999 +0x1.3333333333333p-2
2       0.6 0.59999999999999998 +0x1.3333333333333p-1
3 0.3 + 0.6 0.89999999999999991 +0x1.ccccccccccccccp-1
4         9 9.00000000000000000 +0x1.2000000000000p+3
```

Table G.7 The binary representation of the two numbers $\left(\sqrt{2}\right)^2$ and 2 is not identical. They differ by one bit in the 53rd binary digit.

```
> sprintf("%+13.13a", c(2, sqrt(2)^2))
[1] "+0x1.0000000000000p+1" "+0x1.0000000000001p+1"
```

Table G.8 We frequently wish to interpret numbers that are very different in magnitude as if the smaller one is effectively zero. The display function zapsmall provides that capability.

```
> c(100, 1e-10)
[1] 1e+02 1e-10

> zapsmall(c(100, 1e-10))
[1] 100   0
```

G.10 Apparent Violation of Elementary Factoring

We show a simple example of disastrous cancellation (loss of high-order digits), where the floating point statement

$$a^2 - b^2 \neq (a + b) \times (a - b)$$

is an inequality, not an equation, for some surprising values of a and b. Table G.9 shows two examples, a decimal example for which the equality holds so we can use our intuition to see what is happening, and a hex example at the boundary of rounding so we can see precisely how the equality fails.

Table G.9 Two examples comparing $a^2 - b^2$ to $(a + b) \times (a - b)$. On the top, the numbers are decimal $a = 101$ and $b=102$ and the equality holds on a machine using IEEE 754 floating point arithmetic. On the bottom, the numbers are hexadecimal $a = $ 0x8000001 and $b = $ 0x8000002 and the equality fails to hold on a machine using IEEE 754 floating point arithmetic. The outlined $\boxed{0}$ in the decimal column for a^2 with x=+0x8000000 would have been a $\boxed{1}$ if we had 54-bit arithmetic. Since we have only 53 bits available to store numbers, the 54th bit was rounded to 0 by the Round to Even rule (see Section G.5). The marker $\updownarrow$ in the hex column for a^2 with x=+0x8000000 shows that one more hex digit would be needed to indicate the squared value precisely.

	Decimal 100 =	Hex +0x64
x	100	+0x1.9000000000000p+6
a <- x+1	101	+0x1.9400000000000p+6
b <- x+2	102	+0x1.9800000000000p+6
a^2	10201	+0x1.3ec8000000000p+13
b^2	10404	+0x1.4520000000000p+13
b^2 - a^2	203	+0x1.9600000000000p+7
(b+a) * (b-a)	203	+0x1.9600000000000p+7

	Decimal 134217728 =	Hex +0x8000000
x	134217728	+0x1.0000000000000p+27
a <- x+1	134217729	+0x1.0000002000000p+27
b <- x+2	134217730	+0x1.0000004000000p+27
a^2	18014398777917440	+0x1.0000004000000p+54
b^2	18014399046352900	+0x1.0000008000001p+54
b^2 - a^2	268435460	+0x1.0000004000000p+28
(b+a) * (b-a)	268435459	+0x1.0000003000000p+28

G.11 Variance Calculations

Once we understand disastrous cancellation, we can study algorithms for the calculation of variance. Compare the two common formulas for calculating sample variance, the two-pass formula and the disastrous one-pass formula.

Two-pass formula   One-pass formula

$$\left(\sum_{i=1}^{n}(x_i - \bar{x})^2\right)/(n-1) \qquad \left(\left(\sum_{i=1}^{n}x_i^2\right) - n\bar{x}^2\right)/(n-1)$$

Table G.10 shows the calculation of the variance by both formulas. For $x = (1, 2, 3)$, var$(x) = 1$ by both formulas. For $x = (k+1, k+2, k+3)$, var$(x) = 1$ by both formulas for $k \leq 10^7$. For $k = 10^8$, the one-pass formula gives 0. The one-pass formula is often shown in introductory books with the name "machine formula". The "machine" it is referring to is the desk calculator, not the digital computer. The one-pass formula gives valid answers for numbers with only a few significant figures (about half the number of digits for machine precision), and therefore does not belong in a general algorithm. The name "one-pass" is reflective of the older computation technology where scalars, not the vector, were the fundamental data unit. See Section G.14 where we show the one-pass formula written with an explicit loop on scalars.

We can show what is happening in these two algorithms by looking at the binary display of the numbers. We do so in Section G.12 with presentations in Tables G.11 and G.12. Table G.11 shows what happens for double precision arithmetic (53 significant bits, approximately 16 significant decimal digits). Table G.12 shows the same behavior with 5-bit arithmetic (approximately 1.5 significant decimal digits).

G.12 Variance Calculations at the Precision Boundary

Table G.11 shows the calculation of the sample variance for three sequential numbers at the boundary of precision of 53-bit floating point numbers. The numbers in column "15" fit within 53 bits and their variance is the variance of $k + (1, 2, 3)$ which is 1. The numbers $10^{16} + (1, 2, 3)$ in column "16" in Table G.11 require 54 bits for precise representation. They are therefore rounded to $10^{16} + c(0, 2, 4)$ to fit within the capabilities of 53-bit floating point numbers. The variance of the numbers in column "16" is calculated as the variance of $k + (0, 2, 4)$ which is 4. When we place the numbers into a 54-bit representation (not possible with the standard 53-bit floating point), the calculated variance is the anticipated 1.

Table G.12 shows the calculation of the sample variance for three sequential numbers at the boundary of precision of 5-bit floating point numbers. The numbers {33, 34, 35} on the left side need six significant bits to be represented precisely.

Table G.10 The one-pass formula fails at $x = (10^8 + 1, 10^8 + 2, 10^8 + 3)$ (about half as many significant digits as machine precision). The two-pass formula is stable to the limit of machine precision. The calculated value at the boundary of machine precision for the two-pass formula is the correctly calculated floating point value. Please see Section G.12 and Tables G.11 and G.12 for the explanation.

```
> varone <- function(x) {               > vartwo <- function(x) {
+    n <- length(x)                      +    n <- length(x)
+    xbar <- mean(x)                     +    xbar <- mean(x)
+    (sum(x^2) - n*xbar^2) / (n-1)       +    sum((x-xbar)^2) / (n-1)
+ }                                      + }

> x <- 1:3                              > x <- 1:3

> varone(x)                             > vartwo(x)
[1] 1                                   [1] 1

> varone(x+10^7)                        > vartwo(x+10^7)
[1] 1                                   [1] 1

> ## half machine precision            > ## half machine precision
> varone(x+10^8)                        > vartwo(x+10^8)
[1] 0                                   [1] 1

> varone(x+10^15)                       > vartwo(x+10^15)
[1] 0                                   [1] 1

> ## boundary of machine precision     > ## boundary of machine precision
> ##                                    > ## See next table.
> varone(x+10^16)                       > vartwo(x+10^16)
[1] 0                                   [1] 4

> varone(x+10^17)                       > vartwo(x+10^17)
[1] 0                                   [1] 0
```

Following the Round to Even rule, they are rounded to {32, 34, 36} in the five-bit representation on the right side. The easiest way to see the rounding is in the `scientific=FALSE` binary presentation (the last section on both sides of the Table G.12, repeated in Table G.13).

There is also a third formula, with additional protection against cancellation, called the "corrected two-pass algorithm".

$$y = x - \bar{x}$$
$$\sum(y - \bar{y})^2 / (n - 1)$$

We define a function in Table G.14 and illustrate its use in a very tight boundary case (the 54-bit column "16" of Table G.11) in Table G.15.

Table G.11 Variance of numbers at the boundary of 53-bit double precision arithmetic. Column "15" fits within 53 bits and the variance is calculated as 1. Column "16" requires 54 bits for precise representation. With 53-bit floating point arithmetic the variance is calculated as 4. With the extended precision to 54 bits, the variance is calculated as 1.

```
> x <- 1:3; p <- 15:16

> xx <- t(outer(10^p, x, '+')); dimnames(xx) <- list(x, p)

> print(xx, digits=17)
                15                16
1 1000000000000001 10000000000000000
2 1000000000000002 10000000000000002
3 1000000000000003 10000000000000004

> formatHex(xx)
    15                  16
1 +0x1.c6bf526340008p+49 +0x1.1c37937e08000p+53
2 +0x1.c6bf526340010p+49 +0x1.1c37937e08001p+53
3 +0x1.c6bf526340018p+49 +0x1.1c37937e08002p+53

> var(xx[,"15"])
[1] 1

> var(xx[,"16"])
[1] 4

> x54 <- mpfr(1:3, 54)

> xx54 <- t(outer(10^p, x54, '+')); dimnames(xx54) <- list(x, p)

> xx54
'mpfrMatrix' of dim(.) =  (3, 2) of precision  54   bits
                15                16
1 1000000000000001.00 10000000000000001.0
2 1000000000000002.00 10000000000000002.0
3 1000000000000003.00 10000000000000003.0

> vartwo(xx54[,"16"] - mean(xx54[,"16"]))
1 'mpfr' number of precision  54    bits
[1] 1

> ## hex for 54-bit numbers is not currently available from R.
>

> ## We manually constructed it here.
> formatHex(xx54)  ## We manually constructed this
    15                  16
1 +0x1.c6bf5263400080p+49 +0x1.1c37937e080008p+53
2 +0x1.c6bf5263400100p+49 +0x1.1c37937e080010p+53
3 +0x1.c6bf5263400180p+49 +0x1.1c37937e080018p+53
```

Table G.12 Numbers with six and five significant bits. The six-bit integers 33 and 35 on the left side cannot be expressed precisely with only five significant bits. They are rounded to 32 and 36 in the five-bit representation on the right side. The variance of the numbers {33, 34, 35} is 1. The variance of the numbers {32, 34, 36} is 4. Please see Section G.12 for the discussion of this table. Please see Table G.13 for additional details.

```
> y <- 1:3; q <- 3:5; yy <- t(outer(2^q, y, '+'))
> yy6 <- mpfr(yy, 6); dimnames(yy6) <- list(y, q)
> yy6
'mpfrMatrix' of dim(.) = (3, 3) of precision 6   bits
     3    4    5
1 9.00 17.0 33.0
2 10.0 18.0 34.0
3 11.0 19.0 35.0
> vartwo(yy6[,"5"])
1 'mpfr' number of precision 53   bits
[1] 1
> formatBin(yy6)
     3              4              5
1 +0b1.00100p+3 +0b1.00010p+4 +0b1.00001p+5
2 +0b1.01000p+3 +0b1.00100p+4 +0b1.00010p+5
3 +0b1.01100p+3 +0b1.00110p+4 +0b1.00011p+5
> formatBin(yy6, scientific=FALSE)
     3              4              5
1 +0b_1001.00 +0b_10001.0_ +0b100001.--
2 +0b_1010.00 +0b_10010.0_ +0b100010.--
3 +0b_1011.00 +0b_10011.0_ +0b100011.--
```

```
> y <- 1:3; q <- 3:5; yy <- t(outer(2^q, y, '+'))
> yy5 <- mpfr(yy, 5); dimnames(yy5) <- list(y, q)
> yy5
'mpfrMatrix' of dim(.) = (3, 3) of precision 5   bits
     3    4    5
1 9.00 17.0 32.0
2 10.0 18.0 34.0
3 11.0 19.0 36.0
> vartwo(yy5[,"5"])
1 'mpfr' number of precision 53   bits
[1] 4
> formatBin(yy5)
     3             4             5
1 +0b1.0010p+3 +0b1.0001p+4 +0b1.0000p+5
2 +0b1.0100p+3 +0b1.0010p+4 +0b1.0001p+5
3 +0b1.0110p+3 +0b1.0011p+4 +0b1.0010p+5
> formatBin(yy5, scientific=FALSE)
     3            4            5
1 +0b_1001.0 +0b_10001.- +0b10000_.--
2 +0b_1010.0 +0b_10010.- +0b10001_.--
3 +0b_1011.0 +0b_10011.- +0b10010_.--
```

Table G.13 This table focuses on the last displays in Table G.12. The last three bits in the 6-bit display of 33 (001) show a 1 in the "1" position. The number is rounded to the nearest even number (000) in the "2" digit (with a 0 in the "1" position) and truncated to (00_) in the 5-bit display. The last three bits (011) of 35 are rounded to the nearest even number (100) in the "2" digit and truncated to (10_) In both values, the resulting "2" digit is (0). The last three bits (010) of 34 already have a (0) in the "1" digit and therefore no rounding is needed. The sample variance of {33, 34, 35} is 1. When those numbers are rounded to five-bit binary, they become {32, 34, 36}. The sample variance of {32, 34, 36} is 4.

6-bit binary		rounded to 5-bit binary		
decimal	binary	displayed as 6-bit	truncated to 5-bit	equivalent decimal
33	+0b100001.	+0b100000.	+0b10000_.	32
34	+0b100010.	+0b100010.	+0b10001_.	34
35	+0b100011.	+0b100100.	+0b10010_.	36

Table G.14 The corrected two-pass algorithm centers the data by subtracting the mean, and then uses the two-pass algorithm on the centered data. It helps in some boundary conditions, for example the one shown in Table G.15.

```
> vartwoC <- function(x) {
+    vartwo(x-mean(x))
+ }

> x <- 1:3

> vartwoC(x)
[1] 1

> vartwoC(x+10^7)
[1] 1

> ## half machine precision
> vartwoC(x+10^8)
[1] 1

> vartwoC(x+10^15)
[1] 1

> ## boundary of machine precision
> ##
> vartwoC(x+10^16)
[1] 4

> vartwoC(x+10^17)
[1] 0
```

Table G.15 `vartwo` doesn't work for some problems on the boundary, such as this example at the boundary of 54-bit arithmetic. Summing the numbers effectively required one more significant binary digit. Since there are no more digits available, the data was rounded and the variance is not what our real-number intuition led us to expect. `vartwoC` does work.

```
> vartwo(xx54[,"15"])
1 'mpfr' number of precision   54    bits
[1] 1


> ## wrong answer.  numbers were shifted one binary position.
> vartwo(xx54[,"16"])
1 'mpfr' number of precision   54    bits
[1] 2.5


> ## vartwoC protects against that problem and gets the right answer.
> vartwoC(xx54[,"16"])
1 'mpfr' number of precision   54    bits
[1] 1


> sum(xx54[1:2,"16"])
1 'mpfr' number of precision   54    bits
[1] 20000000000000004


> ## Adding the first two numbers effectively doubled the numbers which
> ## means the significant bits were shifted one more place to the left.
> ## The first value was rounded up. Looking at just the last three bytes
> ## (where the last three bits are guaranteed 0):
> ##     +0x008p53 + 0x010p53 -> +0x010p53 + 0x010p53 -> +0x020p53
>
> sum((xx54)[1:3,"16"])       ## too high
1 'mpfr' number of precision   54    bits
[1] 30000000000000008


> sum((xx54)[3:1,"16"])       ## too low
1 'mpfr' number of precision   54    bits
[1] 30000000000000004


> sum((xx54)[c(1,3,2),"16"]) ## just right
1 'mpfr' number of precision   54    bits
[1] 30000000000000006
```

G.13 Can the Answer to the Calculation be Represented?

Chan et al. (1983) discuss various strategies needed to make sure that the fundamental goal of numerical analysis is achieved:

> If the input values can be represented by the computer, and if the answer can be represented by the computer, then the calculation should get the right answer.

It is very easy to construct an easy sum-of-squares problem for which naive calculations cannot get the right answer. The programmer's task is to get the right answer.

The Pythagorean Theorem tells us that $z = \sqrt{x^2 + y^2}$ will be an integer for several well known sets of triples $\{x, y, z\}$. The triple $\{3, 4, 5\}$ is probably the best known. The triple $\{3k, 4k, 5k\}$ for any k is also a triple which works. Table G.16 shows an example of k for which naive calculation fails, and for which Mod (modulus), one of R's base function, works. The goal is to understand how Mod is written.

Table G.16 The naive square-root-of-the-sum-of-squares algorithm gets the right answer in two of the three cases shown here. Mod gets the right answer in all three cases.

```
> x <- 3; y <- 4

> sqrt(x^2 + y^2)
[1] 5

> Mod(x + 1i*y)
[1] 5

> x <- 3e100; y <- 4e100

> sqrt(x^2 + y^2)
[1] 5e+100

> Mod(x + y*1i)
[1] 5e+100

> x <- 3e305; y <- 4e305

> sqrt(x^2 + y^2)
[1] Inf

> Mod(x + y*1i)
[1] 5e+305
```

The problem is that squaring arguments with a large exponent causes floating point overflow, which R interprets as infinite. The repair, shown in the MyMod function in Table G.17, is to rescale the numbers to a smaller exponent and then take the square root of the sum of squares. At the end the exponent is restored.

Table G.17 The MyMod function rescales the numbers and gets the right answer in all three cases. Only the third case is shown here.

```
> MyMod <- function(x, y) {
+    XYmax <- max(abs(c(x, y)))
+    xx <- x/XYmax
+    yy <- y/XYmax
+
+    result <- sqrt(xx^2 + yy^2)
+
+    result * XYmax
+ }

> x^2
[1] Inf

> y^2
[1] Inf

> MyMod(x, y)
[1] 5e+305
```

G.14 Explicit Loops

Desk calculator technology had different technical goals than digital computer technology. Entering the data manually multiple times was expensive and to be avoided. Table G.18 shows a scalar version of the one-pass algorithm for calculating sample variance. The explicit loop uses each entered value x[i] twice before going on to the next value. The term *one-pass* refers to the single entry of the data value for both accumulations $\left(\sum(x_i) \text{ and } \sum(x_i^2)\right)$ on the scalars in the vector x. Explicit scalar loops are exceedingly slow in R and are normally avoided. When we use vectorized operations (as in statements such as sum(x^2)) the looping is implicit at the user level and is done at machine speeds inside R.

We timed the scalar version of the one-pass algorithm along with the vectorized one-pass and two-pass algorithms. The explicitly looped one-pass algorithm varoneScalar is much slower than the vectorized algorithms. The vectorized onepass and twopass algorithms are about equally fast. Only the twopass algorithm gives the correct calculation to the precision of the computer.

Table G.18 The one-pass formula written as a scalar loop. This made sense for desk calculators because the user keyed in each number exactly once. It is about the most inefficient way to write code for R. In this example it is about 60 times slower than the vectorized algorithms.

```
> varoneScalar <- function(x) {
+    ## This is a pedagogical example.
+    ## Do not use this as a model for writing code.
+    n <- length(x)
+    sumx <- 0
+    sumx2 <- 0
+    for (i in 1:n) {
+      sumx <- sumx + x[i]
+      sumx2 <- sumx2 + x[i]^2
+    }
+    (sumx2 - (sumx^2)/n) / (n-1)
+ }

> x <- 1:3

> varoneScalar(x)
[1] 1

> varoneScalar(x+10^7)
[1] 1

> ## half machine precision
> varoneScalar(x+10^8)
[1] 0

> xx <- rnorm(1000)

> ## explicit loops are much slower in R
> system.time(for (j in 1:1000) varoneScalar(xx))
   user  system elapsed
  1.249   0.309   1.483

> system.time(for (j in 1:1000) varone(xx))
   user  system elapsed
  0.020   0.002   0.021

> system.time(for (j in 1:1000) vartwo(xx))
   user  system elapsed
  0.021   0.001   0.022
```

Appendix H

Other Statistical Software

The statistical analyses described in this book can be calculated with other software than R.

Readers are welcome to work the examples and exercises in this book using other software. All datasets used in either the text or exercises are available in ASCII characters in csv (comma-separated-values) format in the zip file

http://astro.ocis.temple.edu/~rmh/HH2/HH2datasets.zip

The reader must be aware of several issues when using these datasets.

1. All data sets.

   The first row of the csv file contains variable names. There is one fewer name than columns of data with the convention that the initial unnamed column is the row number or row name. Missing observations are coded NA. Factors are stored as character strings. When converting them back to factors, verify that the factor levels are ordered correctly. Names (row names, column names, level names for factors, and values of character variables) may include blanks and other non-alphanumeric characters.

2. Time Series datasets.

   co2, elnino, employM16, nottem, ozone are stored one row per year in twelve columns named with the month names. Missing observations are coded NA. When converting this back to a time series in any software system, verify that the months are identified as a factor in the correct calendar order. Factor levels may include blanks and other non-alphanumeric characters.

   product, tser.mystery.X, tser.mystery.Y, tser.mystery.Z, tsq are stored as a single column named x. Read the problem description for any other information.

© Springer Science+Business Media New York 2015

R.M. Heiberger, B. Holland, *Statistical Analysis and Data Display*,

Springer Texts in Statistics, DOI 10.1007/978-1-4939-2122-5

3. Very Long character strings.

   SFF8121 contains very long character strings with embedded newline characters and with embedded commas. It is the only `csv` file in this set that has been saved with double quotes around all character strings.

Appendix I

Mathematics Preliminaries

A certain degree of mathematical maturity is a prerequisite for understanding the material in this book. Many chapters in this book require a basic understanding of these areas of mathematics:

- algebra

- differential calculus

- matrix algebra, with special attention devoted to quadratic forms, eigenvalues and eigenvectors, transformations of coordinate systems, and ellipsoids in matrix notation

- combinations and permutations

- floating point arithmetic

This appendix provides a brief review of these topics at a level comparable to the book's exposition of statistics.

I.1 Algebra Review

We begin with some topics in algebra, focusing on the case of two dimensions. The labels x and y are given to the horizontal and vertical dimensions, respectively.

© Springer Science+Business Media New York 2015
R.M. Heiberger, B. Holland, *Statistical Analysis and Data Display*,
Springer Texts in Statistics, DOI 10.1007/978-1-4939-2122-5

I.1.1 Line

The general equation of a straight line is given by $y = a + bx$, where a and b are constants with $b \neq \pm\infty$. The line in Figure I.1 intersects the y-axis at $y = a$ and the x-axis at $x = -a/b$, and has slope b.

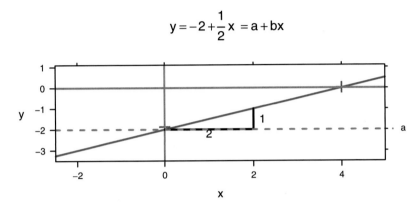

$$y = -2 + \frac{1}{2}x = a + bx$$

Fig. I.1 Straight line $y = 2 + 1/2x$ with intercepts and slope indicated.

I.1.2 Parabola

The general equation of a *parabola* having a vertical axis is the quadratic equation $y = ax^2 + bx + c$ for constants a, b, c with $a \neq 0$. The graph of the parabola opens upward if $a > 0$ and attains a minimum when $x = -b/2a$. The graph opens downward if $a < 0$ and attains a maximum when $x = -b/2a$. The quantity $d = b^2 - 4ac$ is called the *discriminant*. The parabola intersects the horizontal axis at

$$x = \frac{-b \pm \sqrt{d}}{2a} \tag{I.1}$$

The number of intersections, or real roots, is 2, 1, or 0 according to whether $d >, =, < 0$. The parabola intersects the y-axis at $y = c$. Equation (I.1) is referred to as the quadratic formula for solving the equation $ax^2 + bx + c = 0$. We illustrate a parabola opening upward in Figure I.2.

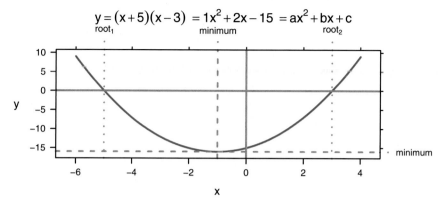

Fig. I.2 Parabola with x-roots and minimum indicated.

I.1.3 Ellipse

The equation

$$\frac{x^2}{a^2} + \frac{y^2}{b^2} = 1$$

represents an *ellipse* centered at $(0, 0)$ with major and minor axes parallel to the coordinate system. When $a > b$, the semimajor axis has length a and the semiminor axis has length b. We graph a sample ellipse, with $a = 3$ and $b = 2$, in Figure I.3.

An ellipse centered at (μ_x, μ_y) is obtained by replacing x and y in the above with $x - \mu_x$ and $y - \mu_y$, respectively. Ellipses having axes nonparallel to the coordinate system are important in statistics and will be discussed in Section I.4.13 as an example of the use of matrix notation.

I.1.4 Simultaneous Equations

A common algebraic problem is the determination of the solution to two (or more) simultaneous equations. In the case of two linear equations, the number of solutions may be 0, 1, or ∞. There are no solutions if the equations are contradictory, such as $x + y = 8$ and $x + y = 9$; there are an infinite number of solutions if one equation is indistinct from the other, for example $x + y = 8$ and $2x + 2y = 16$. When there is a unique solution, several approaches exist for finding it. One of these is adding a carefully chosen multiple of one equation to the other equation so as to result in an easily solved new linear equation involving just one variable. For example, suppose

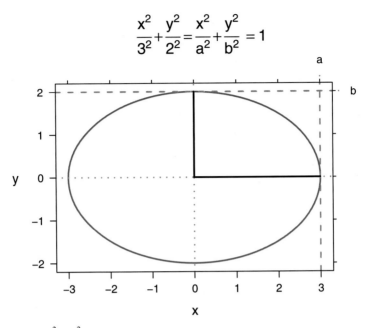

$$\frac{x^2}{3^2}+\frac{y^2}{2^2}=\frac{x^2}{a^2}+\frac{y^2}{b^2}=1$$

Fig. I.3 Ellipse $\dfrac{x^2}{3^2} + \dfrac{y^2}{2^2} = 1$

the two equations are $x+y = 8$ and $2x-3y = 1$. Adding three times the first equation to the second yields $5x + 0y = 25$, which implies $x = 5$ and then $y = 3$. We illustrate in Figure I.4.

I.1.5 Exponential and Logarithm Functions

Two additional elementary functions are the exponential function and logarithmic function. The exponential function $y = c_1 e^{c_2 x}$ (where c_1, c_2 are nonzero constants) has the property that the rate of change in y in response to a change in x is proportional to the current value of y. The logarithmic function $y = c \ln(x)$, $x > 0$ is useful in situations when successive changes in x are geometrical (i.e., proportional to the current value of x). The exponential function and the natural (to the base e) logarithm are inverse to each other. We illustrate both in Figure I.5.

$$x + y = 8, \quad 2x - 3y = 1$$

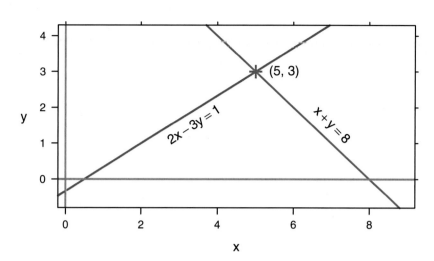

Fig. I.4 Solution of simultaneous equations at the intersection of the two straight lines described by the equations.

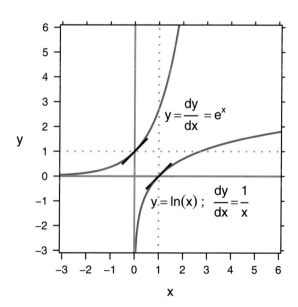

Fig. I.5 Exponential and logarithmic functions with derivatives at x=0.

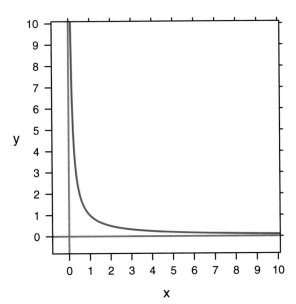

Fig. I.6 Asymptotes of the hyperbola $y = 1/x$ at the horizontal axis as $x \to \infty$ and the vertical axis as $x \to 0^+$.

I.1.6 Asymptote

An asymptote is a straight line that is gradually approached by a curved line. This concept is used to describe the ultimate behavior of the curved line. For example in Figure I.6, the graph of $y = \frac{1}{x}$ has the horizontal axis as its asymptote as $x \to \infty$ and the vertical axis as its asymptote as $x \to 0^+$.

I.2 Elementary Differential Calculus

If $y = f(x)$ expresses the functional relationship between x and y, the *derivative* of y with respect to x, denoted $\frac{dy}{dx}$ or $D_x y$ or $f'(x)$, is a new function of x. For each value of x, $f'(x)$ gives the relative amount that y changes in response to a small change in x. For example, if $y = f(x) = x^2$, then it can be shown that $f'(x) = 2x$. When $x = 3$, a small increase in x will beget a sixfold increase in y because $f'(3) = 2(3) = 6$. Graphically, $f'(x_0)$ is the slope of the straight line tangent to $f(x)$ at $x = x_0$.

If $f(x) = a_0 + a_1 x + a_2 x^2 + \ldots + a_m x^m$, an mth-degree polynomial, then $f'(x) = a_1 + 2a_2 x + 3a_3 x^2 + \ldots + m a_m x^{m-1}$. This rule can be used to differentiate (i.e., find the derivative of) many common functions. If $f(x)$ can be expressed as the product of two functions, say $f(x) = g(x)\, h(x)$, then its derivative is given by the *product rule*

$f'(x) = g(x) h'(x) + g'(x) h(x)$. This can be used, for example, to find the derivative of $(3x^2 + 4x + 5)(-8x^2 + 7x - 6)$ without first multiplying the quadratics.

The most important application of $f'(x)$ is in finding relative extrema (i.e., maxima or minima) of $f(x)$. A *necessary* condition for x_0 to be an extremum of $f(x)$ is that $f'(x_0) = 0$. This follows from the interpretation of the derivative as a tangent slope. Additional investigation is needed to confirm that such an x_0 corresponds to either a maximum or minimum, and then to determine which of the two it is. For example, if $f(x) = x^3 - 3x$, then $f'(x) = 3x^2 - 3$. Setting $f'(x) = 0$, we find $x = \pm 1$. $x = 1$ corresponds to a local minimum and $x = -1$ corresponds to a local maximum. As another example, consider $f(x) = x^3$. While for this function, $f'(0) = 0$, $x = 0$ is neither a relative minimum nor a relative maximum of $f(x)$.

Example of an Optimization Problem

A rectangular cardboard poster is to have 150 square inches for printed matter. It is to have a 3-inch margin at the top and bottom and a 2-inch margin on each side. Find the dimensions of the poster so that the amount of cardboard used is minimized.

Solution I.1. Let the vertical dimension of the printed matter be x. Then for the printed area to be 150, the horizontal dimension of the printed matter is $\frac{150}{x}$. Then allowing for the margins, the vertical and horizontal dimensions of the poster are $x + 6$ and $\frac{150}{x} + 4$. The product of these dimensions is the area of the poster that we seek to minimize: $a(x) = 174 + 4x + \frac{900}{x}$. Taking the derivative and setting it equal to zero gives $a'(x) = 4 - 900x^{-2} = 0$, which leads to the positive solution $x = 15$. Thus the minimum-sized poster with required printed area is 21 inches high by 14 inches wide, and its printed area is 15 inches high by 10 inches wide.

I.3 An Application of Differential Calculus

We introduce Newton's method for solutions of an equation of the form $f(x) = 0$. A common application is the need to solve $f'(x) = 0$ to find extrema, as discussed in Section I.2. Many equations of this type are readily solvable by successively moving toward isolation of a lone x on one side of the equation, or via a specialized technique such as the quadratic formula. In other situations one must employ one of the number of numerical techniques designed for this purpose, one of which is Newton's method.

Newton's method has the disadvantage of requiring knowledge of the derivative $f'(x)$, but it will often converge to a solution within a small number of iterations. As with all procedures for dealing with this problem, one must start with a first approximation x_0, and if this is "too far" from the solution x^*, the procedure may fail to converge.

The idea behind Newton's method is not difficult to understand. It is based on the equation of the tangent line to $f(x)$ at $x = x_0$. If this tangent line intersects the x-axis at $x = x_1$, then

$$f'(x_0) = \frac{f(x_0)}{x_0 - x_1} \quad \rightarrow \quad x_1 = x_0 - \frac{f(x_0)}{f'(x_0)}$$

The iteration then proceeds with

$$x_2 = x_1 - \frac{f(x_1)}{f'(x_1)}$$

and so on.

An Illustration of Newton's Method

Consider solving $f(x) = x^3 - x - 5 = 0$, $f'(x) = 3x^2 - 1$, and let $x_0 = 2$. You can verify with Figure I.7 the following sequence:

i	x_i	$f(x_i)$
0	2.00000000	$1.00 10^0$
1	1.90909091	$4.88 10^{-2}$
2	1.90417486	$1.38 10^{-4}$
3	1.90416086	$1.14 10^{-9}$

I.4 Topics in Matrix Algebra

We next provide an overview of selected topics from matrix algebra that are useful in applied statistics. Not only do matrices (the plural of matrix) allow for a more concise notation than scalar algebra, but they are an indispensable tool for communicating statistical findings. Additional material on matrix algebra is contained in the appendices of most books dealing with regression analysis, linear models, or multivariate analysis.

A matrix is a rectangular array consisting of r rows and c columns. A *vector* is a special type of matrix, having either r or c equal to 1. Data files are often arranged as a matrix, such that each variable is one column and each observation is one row. Systems of linear equations may be succinctly written in matrix notation.

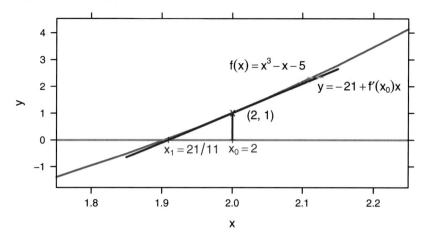

Fig. I.7 An illustration of Newton's Method.

Multivariate analysis involves probability distributions of random vectors rather than scalar random variables. Each component of a random vector is a scalar random variable. The variances and covariances of the components of a random vector are arranged in a variance–covariance matrix. (Such a symmetric matrix V, also called covariance matrix or dispersion matrix, has the variances on the main diagonal and the covariance between variables i and j in its row i column j position. See Section 3.3.5.) The multivariate normal distribution of the random k-vector x with mean vector μ and covariance matrix V, with notation

$$x \sim N(\mu, V) \tag{I.2}$$

and probability density function

$$f(x) = \frac{1}{(2\pi)^{k/2} \, |V|^{1/2}} \, e^{(x-\mu)'V^{-1}(x-\mu)/2} \tag{I.3}$$

is the most important distribution used in multivariate analysis. We give examples of the bivariate (multivariate with $k = 2$) normal in Sections 3.3.5 and J.4.2.

Matrices may be used to translate from one coordinate system to another. Orthogonal matrices perform rigid rotations in order to facilitate meaningful interpretation of results.

I.4.1 Elementary Operations

sum: If two matrices are of the same size, their sum is the new matrix of this size
comprised of the element-wise sums. The difference of two matrices of the same
size is defined similarly.

transpose: If A is an $r \times c$ matrix, then its *transpose*, denoted A', is the $c \times r$ matrix
having columns identical to the rows of A and conversely.

inner product: The *inner product* of two vectors of the same size, say $u =
(u_1, u_2, \ldots, u_k)'$ and $v = (v_1, v_2, \ldots, v_k)'$, is written $u'v = v'u = \sum_{i=1}^{k} u_i v_i$, i.e.,
the sum of the products of the corresponding elements of the two vectors. The
inner product of a vector with itself yields the sum of squares of the elements of
this vector. A vector u is said to be normalized if $u'u = 1$, i.e., if its (Euclidean)
length equals 1. Two vectors u and v are said to be orthogonal if their inner product is zero: $u'v = 0$.

matrix product: The matrix product AB of an $r \times c$ matrix A with an $m \times n$ matrix
B is defined when $c = m$. The element in row i and column j of AB is calculated
as the inner product of the ith row of A and the jth column of B. The condition
$c = m$ assures that the vectors forming the inner products are of the same size.
Matrix addition has the mathematical properties of commutativity and associativity. Matrix multiplication has only associativity. When AB is defined, BA may
have different dimensions from AB or even be undefined.

Matrix multiplication is used when expressing systems of linear equations in
matrix notation. Earlier we considered the system $x + y = 8$ and $2x - 3y = 1$. If
we define the 2×1 vectors $X = \begin{pmatrix} x \\ y \end{pmatrix}$ and $c = \begin{pmatrix} 8 \\ 1 \end{pmatrix}$ and the 2×2 matrix $A = \begin{pmatrix} 1 & 1 \\ 2 & -3 \end{pmatrix}$,
then this system can be written $AX = c$. In this context, the matrix A is referred
to as the *coefficient* matrix.

transpose: The transpose of the product of two matrices is the product of their
transposes in reverse order: $(AB)' = B'A'$.

square: A matrix is said to be square if it has the same number of rows as columns.

identity: The $n \times n$ identity matrix has ones on its main diagonal positions (those
where the row number equals the column number) and zeros everywhere else.
Thus, for example, the 3×3 identity matrix is

$$I_3 = \begin{pmatrix} 1 & 0 & 0 \\ 0 & 1 & 0 \\ 0 & 0 & 1 \end{pmatrix}$$

The identity matrix plays the same role in matrix notation as does the number 1
in scalar notation: If I denotes an identity matrix such that the indicated multiplication is defined, then $AI = A$ and $IA = A$.

J_n: We define J_n to be the $n \times n$ matrix having all entries equal to 1.

symmetric: A square matrix A is said to be a symmetric matrix if $A = A'$. This means that row number i corresponds to column number i for all i. Let a_{ij} denote the element in row i and column j of matrix A. Then if A is square, its *trace* is the sum of its main diagonal elements: $\text{trace}(A) = \sum_i a_{ii}$.

operation count: Numerical analysts count the number of multiplications in an algorithm as an indicator of the costliness of the algorithm. A vector inner product $\sum_{i=1}^{n} a_i b_i$, for example, takes n multiplications to complete. There are other operations (indexing, adding) that must also be performed. Rather than report them explicitly, we say instead that the amount of computation is proportional to the number of multiplications. We indicate the proportionality by saying the operation count is $O(n)$ (read as "big 'O' of n"). Similarly, the operation count for matrix multiplication is proportional to n^3 and is reported as $O(n^3)$.

determinant: The *determinant* of a square matrix A, denoted $|A|$, is a scalar calculated from the elements of A. In the 2×2 case where

$$A = \begin{pmatrix} a_{11} & a_{12} \\ a_{21} & a_{22} \end{pmatrix}$$

the determinant is $|A| = a_{11}a_{22} - a_{21}a_{12}$. If A is a square coefficient matrix of a system of linear equations (thus implying that the system has the same number of equations as unknowns), then the system has a unique solution if and only if $|A| \neq 0$. The determinant has useful mathematical properties, but is totally impractical from a computational standpoint. It is almost never needed as an intermediate calculation. There is almost always a cheaper way to calculate the final answer.

nonsingular: If $|A| \neq 0$, then A is said to be a nonsingular matrix. There is no vector v other than $v = 0$ such that $Av \equiv 0$.

inverse: A nonsingular matrix has associated with it a unique *inverse*, denoted A^{-1}. The inverse has the property that $AA^{-1} = A^{-1}A = I$. The unique solution to a system of linear equations $AX = c$ is then $X = A^{-1}c$. For example, the inverse of

$$A = \begin{pmatrix} a_{11} & a_{12} \\ a_{21} & a_{22} \end{pmatrix} \quad \text{is} \quad \frac{1}{|A|} \begin{pmatrix} a_{22} & -a_{12} \\ -a_{21} & a_{11} \end{pmatrix}$$

provided that $|A| \neq 0$. Note that this is a mathematical identity. It is not to be interpreted as an algorithm for calculation of the inverse. As an algorithm it is very expensive, requiring $O(2n^3)$ arithmetic operations for an $n \times n$ matrix A. An efficient algorithm requires only $O(n^3)$ operations.

singular: If $|A| = 0$, then A is said to be a singular matrix. There exists at least one vector v other than $v = 0$ such that $Av \equiv 0$. A singular matrix does not have an inverse.

idempotent: A square matrix A is said to be an idempotent matrix if $AA = A$, i.e., A equals the product of A with itself. A simple example is

$$\begin{pmatrix} .5 & -.5 \\ -.5 & .5 \end{pmatrix}$$

I.4.2 Linear Independence

A matrix X consists of a set of column vectors

$$\underset{n\times(1+p)}{X} = [\mathbf{1}X_1X_2\ldots X_p] = [X_0X_1X_2\ldots X_p]$$

The columns are numbered $0, 1, \ldots, p$.

The matrix X is said to have linearly dependent columns if there exists a nonzero $(1 + p)$-vector ℓ such that

$$X\ell = 0$$

or, equivalently,

$$\underset{n\times1}{\left(\sum_j \ell_j X_j\right)} = \underset{n\times1}{(0)}$$

The matrix X is said to have linearly independent columns if no such vector ℓ exists. For example, the matrix

$$\underset{4\times(1+4)}{X} = \begin{pmatrix} 1 & 1 & 0 & 0 & 0 \\ 1 & 0 & 1 & 0 & 0 \\ 1 & 0 & 0 & 1 & 0 \\ 1 & 0 & 0 & 0 & 1 \end{pmatrix} \tag{I.4}$$

has linearly dependent columns because there exists a vector $\ell = (-1\,1\,1\,1\,1)'$ such that $X\ell = 0$.

The matrix

$$\underset{4\times(1+3)}{X_{(,-1)}} = \begin{pmatrix} 1 & 0 & 0 & 0 \\ 1 & 1 & 0 & 0 \\ 1 & 0 & 1 & 0 \\ 1 & 0 & 0 & 1 \end{pmatrix} \tag{I.5}$$

has linearly independent columns because there exists no nonzero vector ℓ such that $X\ell = 0$.

The *rank* of a matrix is the number of linearly independent columns it contains. Both matrices above, X and $X_{(,-1)}$ in Equations (I.4) and (I.5), have rank 4. For any matrix X, rank(X) = rank$(X'X)$. A *full-rank* matrix is one whose rank is equal to the minimum of its row and column dimensions, that is,

$$\text{rank}\left(\underset{r \times c}{A}\right) = \min(r, c)$$

I.4.3 Rank

The rank of a matrix A is the maximum number of linearly independent rows (equivalently, the maximum number of linearly independent columns) the matrix has. For example, the matrix

$$A = \begin{pmatrix} 1 & 1 & 2 \\ 1 & 2 & 3 \\ 1 & 3 & 4 \end{pmatrix}$$

has rank 2 since columns 1 and 2 add to column 3, but any two of the columns are linearly independent of one another (i.e., are not proportional to one another).

I.4.4 Quadratic Forms

A *quadratic form* is a scalar resulting from the matrix product $x'Ax$, where x is a $k \times 1$ vector and A is a $k \times k$ symmetric matrix. The matrix A is termed the matrix of the quadratic form. Complicated sums of squares and products as often occur in an *analysis of variance* can be written as quadratic forms. If x has a standardized multivariate normal distribution $x \sim N(0, I)$ [see Equation (I.2)], then $x'Ax$ has a χ^2-distribution with v degrees of freedom if and only if A is an idempotent matrix with rank v. For example, the numerator of the usual univariate sample variance, $\sum_{i=1}^{n}(x_i - \bar{x})^2$, can be written as $x'Ax$ where $x = (x_1, x_2, \ldots, x_n)'$ and $A = I_n - \frac{1}{n}J_n$. It can be shown that this matrix A has rank $n - 1$, the degrees of freedom associated with the sample variance.

If x is a random vector with expected value μ and covariance matrix V, then the expected value of the quadratic form $x'Ax$ is

$$E(x'Ax) = \mu'A\mu + \text{trace}(AV)$$

Result (I.6) does not require that x has a multivariate normal distribution.

A square symmetric matrix A is said to be a *positive definite* (abbreviated p.d.) matrix if $x'Ax > 0$ for all vectors x other than a vector of zeros. The matrix associated with the quadratic form representation of a sum of squares is always p.d.

I.4.5 Orthogonal Transformations

A matrix M is said to be *orthogonal* if $M'M = I$. An example is

$$M = \begin{pmatrix} 1/\sqrt{3} & 1/\sqrt{2} & 1/\sqrt{6} \\ 1/\sqrt{3} & -1/\sqrt{2} & 1/\sqrt{6} \\ 1/\sqrt{3} & 0 & -2/\sqrt{6} \end{pmatrix}$$

A transformation based on an orthogonal matrix is called an *orthogonal transformation*. Such transformations are rotations that preserve relative distances: $y = Mx \rightarrow y'y = x'M'Mx = x'x$. Orthogonal transformations are frequently encountered in statistics. A common use of them is to transform from a correlated set of random variables x to an uncorrelated set y.

The columns of an orthogonal matrix are said to be *orthonormal*. The columns are orthogonal to each other, that is $M'_{.j}M_{.j'} = 0$ for $j \neq j'$. In addition, the columns have been scaled so that $M'_{.j}M_{.j} = 1$.

I.4.6 Orthogonal Basis

If $\text{rank}\begin{pmatrix} A \\ {}_{r \times c} \end{pmatrix} = p < \min(r, c)$ and $c \leq r$, then any set of p linearly independent columns of A constitutes a *basis* for A. The set of all vectors that can be expressed as a linear combination Xv of the columns of A is called the column space of A, denoted $C(A)$. Therefore, $C(A)$ is completely specified by any basis of A. We say that the columns of the basis *span* the column space of A.

An *orthogonal* basis for A is a basis for A with the property that any two vectors comprising it are orthogonal. Starting from an arbitrary basis for A, algorithms are available for constructing an orthogonal basis for A. We show one algorithm, the Gram–Schmidt process, in Section I.4.8.

A *basis* set of column vectors for a matrix X is a set of column vectors U_i that span the same linear space as the original columns X_i. An orthogonal basis is a set of column vectors that are mutually orthogonal, that is $U'_i U_j = 0$ for $i \neq j$. An orthonormal basis is an orthogonal basis whose columns have been rescaled to have norm $\| U_i \| = \sqrt{U'_i U_i} = 1$.

I.4.7 Matrix Factorization—QR

Any rectangular matrix $\underset{n\times m}{X}$ can be factored into the product of an matrix with orthogonal columns $\underset{n\times m}{Q}$ and an upper triangular $\underset{m\times m}{R}$

$$\underset{n\times m}{X} = \underset{n\times m}{Q} \; \underset{m\times m}{R}$$

The columns of Q span the same column space as the columns of the original matrix X. This means that any linear combination of the columns of X, say Xv, can be constructed as a linear combination of the columns of Q. Specifically, using the associative law, $Xv = (QR)v = Q(Rv)$.

The numerically efficient R function qr is the computational heart of the linear models and analysis of variance functions. The intermediate matrices Q and R are usually not explicitly produced. If you wish to see them, use the qr.Q and qr.R functions. An example showing the QR factorization using the qr function is in Table I.1.

An expository R function illustrating the construction of the QR factorization is in Section I.4.8.

I.4.8 Modified Gram–Schmidt (MGS) Algorithm

There are many algorithms available to construct the matrix factorization. We show one, the Modified Gram–Schmidt (MGS) algorithm Bjork (1967). "Modified" means that the entire presentation is in terms of the columns Q_i of the matrix under construction. The MGS algorithm is numerically stable when calculated in finite precision. The original Gram–Schmidt (GS) algorithm, which constructs Q in terms of the columns X_i of the original matrix, is not numerically stable and should not be used for computation.

Let $X_{n\times m} = [X_1 X_2 \ldots X_m]$. The results of the factorization will be stored in $Q_{n\times m}$ and $R_{m\times m}$. The columns of X and Q and both the rows and columns of R are numbered $1, \ldots, m$.

Table I.1 Illustration of *QR* algorithm to factor *X* into the product of an orthogonal matrix and an upper triangular matrix.

```
> X <- matrix(c(1,3,6,4,2,3,8,6,4,5,3,2), 4, 3)

> X
     [,1] [,2] [,3]
[1,]    1    2    4
[2,]    3    3    5
[3,]    6    8    3
[4,]    4    6    2

> crossprod(X)
     [,1] [,2] [,3]
[1,]   62   83   45
[2,]   83  113   59
[3,]   45   59   54


> ## use the efficient calculation
> X.qr <- qr(X)

> qr.Q(X.qr) ## display q
          [,1]      [,2]     [,3]
[1,] -0.127  0.48139  0.8188
[2,] -0.381 -0.73969  0.4755
[3,] -0.762 -0.02348 -0.3038
[4,] -0.508  0.46965 -0.1057

> qr.R(X.qr) ## display r
         [,1]     [,2]     [,3]
[1,] -7.874 -10.541 -5.7150
[2,]  0.000   1.374 -0.9041
[3,]  0.000   0.000  4.5301

> zapsmall(crossprod(qr.Q(X.qr)))  ## identity
     [,1] [,2] [,3]
[1,]    1    0    0
[2,]    0    1    0
[3,]    0    0    1

> crossprod(qr.R(X.qr))      ## reproduce crossprod(X)
     [,1] [,2] [,3]
[1,]   62   83   45
[2,]   83  113   59
[3,]   45   59   54

> qr.X(X.qr) ## reproduce X
     [,1] [,2] [,3]
[1,]    1    2    4
[2,]    3    3    5
[3,]    6    8    3
[4,]    4    6    2
```

We will construct Q and R in steps.

1. Initialize R to 0.

$$R \leftarrow \mathbf{0}$$

2. Initialize Q to X.

$$Q \leftarrow X$$

3. Initialize the column counter.

$$i \leftarrow 1$$

4. Normalize column Q_i.

$$r_{i,i} \leftarrow \sqrt{Q_i' Q_i}$$
$$Q_i \leftarrow Q_i / r_{i,i}$$

If $i = m$, we are done.

5. For each of the remaining columns Q_j, $j = i + 1, \ldots, m$, find the component of Q_j orthogonal to Q_i by

$$r_{i,j} \leftarrow Q_i' Q_j$$
$$Q_j \leftarrow Q_j - Q_i r_{i,j}$$

6. Update the column counter.

$$i \leftarrow i + 1$$

7. Repeat steps 4–6 until completion.

An R version of this expository algorithm is in Table I.2. An example using the expository function is in Table I.3.

Table I.2   An expository algorithm for the Modified Gram–Schmidt Algorithm. The function is a direct translation of the pseudo-code in Section I.4.8.

```
## modified Gram-Schmidt orthogonalization

mgs <- function(x) {
  ## modified Gram-Schmidt orthogonalization

  ## this is an expository algorithm
  ## this is not an efficient computing algorithm

  ## q[,j] is the normalized residual from the least squares fit of
  ## x[,j] on the preceding normalized columns q[,1:(j-1)]

  n <- nrow(x)
  m <- ncol(x)

  q <- x
  r <- matrix(0, m, m)

  for (i in 1:m) {
    r[i,i] <- sqrt(sum(q[,i]^2)) ## length of q[,i]
    q[,i] <- q[,i] / r[i,i]       ## normalize q[,i]

    if (i < m) {   ## if we still have columns to go
      for (j in (i+1):m) {
        r[i,j] <- sum(q[,i] * q[,j]) ## length of projection of q[,j] on q[,i]
        q[,j] <- q[,j] - q[,i] * r[i,j] ## remove projection of q[,j] on q[,i]
      }
    }
  }
  list(q=q, r=r)
}
```

Table I.3 Illustration of the expository algorithm for the Modified Gram–Schmidt Algorithm shown in Table I.2. These are the same values (up to multiplication by -1) as calculated by the `qr` function in Table I.1.

```
X <- matrix(c(1,3,0,4,2,3,8,6,4,5,3,2), 4, 3)
X
crossprod(X)

## use the expository function defined in the previous chunk

X.mgs <- mgs(X)
X.mgs  ## q is orthogonal, r is upper triangular
## These are identical to the results of qr(X)
## up to the sign of the columns of q and the rows of r.

zapsmall(crossprod(X.mgs$q))  ## identity
crossprod(X.mgs$r)    ## reproduces crossprod(X)

X.mgs$q %*% X.mgs$r    ## reproduces X
```

I.4.9 Matrix Factorization—Cholesky

Any square positive definite matrix $\underset{m \times m}{S}$ can be factored into the product of an upper triangle matrix $\underset{n \times m}{R}$ and its transpose

$$S = R'R$$

When S has been constructed as the cross product $S = X'X$ of a rectangular matrix $\underset{n \times m}{X}$, then the upper triangular matrix R is the same matrix we get from the QR factorization.

$$S = X'X = (QR)'(QR) = R'(Q'Q)R = R'R$$

The numerically efficient R function `chol` is illustrated in Table I.4.

I.4.10 Orthogonal Polynomials

Consider the k-vector $v = (v_1, v_2, \ldots, v_k)'$, where $v_1 < v_2 < \ldots < v_k$. Construct a matrix $V = [v^0, v^1, v^2, \ldots, v^{k-1}]$, where we use the notation $v^j = (v_1^j, v_2^j, \ldots, v_k^j)'$

Table I.4 Illustration of Cholesky factorization of a square positive definite matrix into an upper triangular factor and its transpose.

```
> X <- matrix(c(1,3,6,4,2,3,8,6,4,5,3,2), 4, 3)

> M <- crossprod(X)

> M
      [,1] [,2] [,3]
[1,]   62   83   45
[2,]   83  113   59
[3,]   45   59   54

> chol(M)
        [,1]     [,2]      [,3]
[1,] 7.874 10.541   5.7150
[2,] 0.000  1.374  -0.9041
[3,] 0.000  0.000   4.5301

> crossprod(chol(M)) ## reproduce M
      [,1] [,2] [,3]
[1,]   62   83   45
[2,]   83  113   59
[3,]   45   59   54
```

An orthogonal basis Q constructed from the matrix V is called a set of orthogonal polynomials. In the analysis of variance and related techniques, we often construct dummy variables for ordered factors from a set of contrasts that are orthogonal polynomials. See Figure 10.4 and the surrounding discussion in Section 10.4 for an illustration.

I.4.11 Projection Matrices

Given any matrix $\underset{n\times m}{X} = \underset{n\times m}{Q} \; \underset{m\times m}{R}$ the matrix $P_X = X(X'X)^{-1}X' = QQ'$ is a *projection matrix* that projects an *n*-vector y onto the space spanned by the columns of X, that is, the product $P_X y$ is in the column space $C(X)$. If X has m columns and rank $r \le m$, then the eigenvalues of P_X consist of r 1s and $m - r$ 0s. See Table I.5.

Table I.5 Projection of a 3-vector onto the space of its first two coordinates.

```
> X <- matrix(c(3,1,0, 1,2,0, 0,0,0), 3, 3)

> P <- cbind(qr.Q(qr(X))[, 1:2], 0)

> P
          [,1]     [,2] [,3]
[1,] -0.9487  0.3162    0
[2,] -0.3162 -0.9487    0
[3,]  0.0000  0.0000    0

> crossprod(P)
     [,1] [,2] [,3]
[1,]   1    0    0
[2,]   0    1    0
[3,]   0    0    0

> y <- matrix(1:3)

> y
     [,1]
[1,]   1
[2,]   2
[3,]   3

> P %*% y
         [,1]
[1,] -0.3162
[2,] -2.2136
[3,]  0.0000

> sqrt(sum(y[1:2,]^2))
[1] 2.236

> sqrt(sum((P %*% y)^2))
[1] 2.236
```

I.4.12 Geometry of Matrices

We provide some details of the application to two-dimensional geometry. Each two-dimensional vector represents a point; alternatively, a directed line segment from the origin to this point. A 2×2 matrix postmultiplied by a vector transforms this point to another point. Consider the orthogonal matrix

$$M = \begin{pmatrix} \cos(\theta) & \sin(\theta) \\ -\sin(\theta) & \cos(\theta) \end{pmatrix}$$

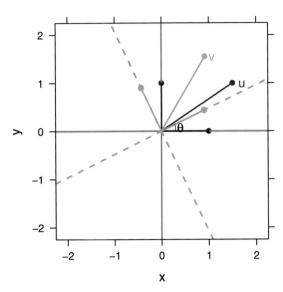

Fig. I.8 Rotation of a vector u and its coordinate system into v by angle θ.

and let u be a 2×1 vector representing a point in two dimensions. Then $v = Mu$ produces a new point v which is where u appears in the new coordinate system formed by rotating the old one θ degrees around the origin. See Figure I.8.

If x and y are each two-dimensional vectors and θ is the angle between them, then $\cos(\theta) = x'y / \sqrt{x'x\, y'y} = \text{corr}(x, y)$, the *correlation* between these two vectors. Note that if the vectors are orthogonal so that $x'y = 0$, then $\cos(\theta) = 0$ and $\theta = 90°$.

I.4.13 Eigenvalues and Eigenvectors

Next we study the concepts of eigenvectors and eigenvalues of an $n \times n$ symmetric matrix V.

If $V\xi = \lambda\xi$, where ξ is an $n \times 1$ vector and λ is a scalar, then λ is said to be an *eigenvalue* of V with corresponding *eigenvector* ξ. Without loss of generality, we can take the eigenvector to be normalized. Geometrically, the matrix V transforms its eigenvectors into multiples of themselves. Any two distinct eigenvectors are orthogonal: $\xi_i'\xi_j = 0$, $i \neq j$. V can be written as its *spectral decomposition* $V = \sum_i \lambda_i \xi_i \xi_i'$. V can be written as its *eigenvalue factorization* $V = \Xi\Lambda\Xi'$.

A matrix is nonsingular if and only if it has only nonzero eigenvalues. A matrix is positive definite if and only if all of its eigenvalues are positive. The eigenvalues of V^{-1} are the reciprocals of the eigenvalues of V. The determinant of a matrix equals the product of its eigenvalues $|V| = \prod \lambda_i$, and calculating the eigenvalues is normally the most efficient way to calculate the determinant. The trace of a matrix equals the sum of its eigenvalues $\text{trace}(V) = \sum \lambda_i$.

Consider the problem of choosing x to maximize $x'Vx$ subject to $x'x = 1$. The maximum value is the largest eigenvalue of V and is attained when x is the eigenvector corresponding to this eigenvalue. Similarly, $x'Vx$ is minimized (subject to $x'x = 1$) at the value of the smallest eigenvalue of V, occurring at the corresponding eigenvector.

Here is an example of hand calculation of eigenvalues and eigenvectors. Consider the 2×2 matrix

$$V = \begin{pmatrix} 2 & 1 \\ 1 & 1 \end{pmatrix}$$

Its 2 eigenvalues are the 2 scalar solutions λ to the equation $|V - \lambda I| = 0$. Taking the determinant leads to $(2 - \lambda)(1 - \lambda) - 1 = 0 \implies \lambda^2 - 3\lambda + 1 = 0 \implies \lambda = (3 \pm \sqrt{5})/2$, which expands to ≈ 2.618 or ≈ 0.382. (Note that we are explicit about the approximation to 3 decimal digits. Our usual practice is *not* to round answers.) The eigenvector $\xi = \begin{pmatrix} \xi_{11} \\ \xi_{21} \end{pmatrix}$ corresponding to $\lambda \approx 2.618$ is the solution to the equation $V\xi \approx 2.618\xi$. This implies that $2\xi_{11} + \xi_{21} \approx 2.618\xi_{11}$, and coupled with the normalization restriction $\xi_{11}^2 + \xi_{21}^2 = 1$ we find that $\xi_{11} \approx .8507$ and $\xi_{21} \approx .5257$. The eigenvector corresponding to the other eigenvalue is found similarly.

As a geometric application of eigenvalues, consider the ellipse having equation $(x - \mu)' V^{-1}(x - \mu) = b$. In statistics, this ellipse based on the inverse V^{-1} is the form of a confidence ellipse for μ. Let $\lambda_1 < \lambda_2$ be the eigenvalues of V with corresponding normalized eigenvectors $\xi_1 = \begin{pmatrix} \xi_{11} \\ \xi_{21} \end{pmatrix}$, $\xi_2 = \begin{pmatrix} \xi_{12} \\ \xi_{22} \end{pmatrix}$. Then the semimajor axis of this ellipse has length $\sqrt{\lambda_2 b}$ and the semiminor axis has length $\sqrt{\lambda_1 b}$. The angle between the extension of the semimajor axis and the horizontal axis is $\arctan\left(\frac{\xi_{12}}{\xi_{22}}\right)$.

Continuing the example, we calculate the eigenvalues of $V = \begin{pmatrix} 2 & 1 \\ 1 & 1 \end{pmatrix}$ in Table I.6 and graph the ellipse $x'Vx = x'\begin{pmatrix} 2 & 1 \\ 1 & 1 \end{pmatrix}x = 1$ in Figure I.9.

I.4.14 Singular Value Decomposition

Let M be an arbitrary $r \times c$ matrix, U the $r \times r$ matrix containing the eigenvectors of MM', and W the $c \times c$ matrix containing the eigenvectors of $M'M$. Let Δ be the $r \times c$ matrix having $\delta_{ij} = 0$ ($i \neq j$). If $r \geq c$ (the usual case in statistical applications), define $\delta_{ii} = $ the square root of the eigenvalue of $M'M$ corresponding to the eigenvector in the ith column of W. If $r < c$, define $\delta_{ii} = $ the square root of the eigenvalue of MM' corresponding to the eigenvector in the ith column of U. Then the *singular value decomposition* of M is $M = U\Delta W$. Note that the number of

Table I.6 Eigenvalues and eigenvectors of $V = \begin{pmatrix} 2 & 1 \\ 1 & 1 \end{pmatrix}$.

```
> V <- matrix(c(2, 1, 1, 1), 2, 2)

> V
     [,1] [,2]
[1,]    2    1
[2,]    1    1

> eV <- eigen(V)

> eV
$values
[1] 2.618 0.382

$vectors
          [,1]     [,2]
[1,] -0.8507   0.5257
[2,] -0.5257  -0.8507


> sqrt(eV$val)  ## semimajor and semiminor axis lengths
[1] 1.618 0.618

> ## angle of axes in radians
> atan(c(eV$vec[2,1]/eV$vec[1,1], eV$vec[2,2]/eV$vec[1,2]))
[1]   0.5536 -1.0172

> ## = -pi/2 ## right angle
> diff(atan(c(eV$vec[2,1]/eV$vec[1,1], eV$vec[2,2]/eV$vec[1,2])))
[1] -1.571
```

nonzero diagonal values in Δ is min(r, c). We show a numerical example in standard mathematical notation in Table I.7. We show the same example calculated with the svd function in Tables I.8 and I.9.

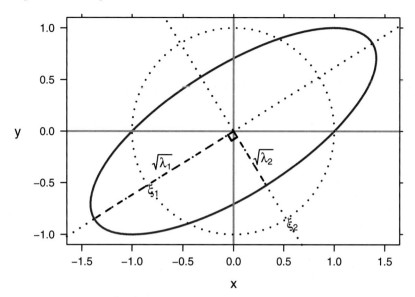

Fig. I.9 Ellipse $x'Vx = x'\begin{pmatrix} 2 & 1 \\ 1 & 1 \end{pmatrix}x = 1$. We show the unit circle with the normalized eigenvectors, the ellipse, and the semimajor and semiminor axes whose lengths are the eigenvalues multiplied by the square root of the eigenvectors.

Table I.7 Matrix multiplication of the components of the Singular Value Decomposition of the matrix M in Table I.9).

$$M = U\Delta V' = \begin{bmatrix} 1 & 2 & 4 \\ 3 & 3 & 5 \\ 6 & 8 & 3 \\ 4 & 6 & 2 \end{bmatrix} =$$

$$\begin{bmatrix} -0.2561 & 0.6098 & -0.6935 \\ -0.4102 & 0.6334 & 0.5907 \\ -0.7125 & -0.3674 & 0.1755 \\ -0.5084 & -0.3034 & -0.3733 \end{bmatrix} \begin{bmatrix} 14.481 & 0.000 & 0.000 \\ 0.000 & 4.324 & 0.000 \\ 0.000 & 0.000 & 0.783 \end{bmatrix} \begin{bmatrix} -0.5383 & -0.7246 & -0.4302 \\ -0.2099 & -0.3791 & 0.9012 \\ 0.8162 & -0.5755 & -0.0520 \end{bmatrix}$$

Table I.8 Illustration of singular value decomposition, part I (to be continued in Table I.9).

```
> M <- matrix(c(1,3,6,4,2,3,8,6,4,5,3,2), 4, 3)

> M
     [,1] [,2] [,3]
[1,]   1    2    4
[2,]   3    3    5
[3,]   6    8    3
[4,]   4    6    2

> M.svd <- svd(M)

> M.svd
$d
[1] 14.4806  4.3243  0.7825

$u
         [,1]     [,2]     [,3]
[1,] -0.2561  0.6098 -0.6935
[2,] -0.4102  0.6334  0.5907
[3,] -0.7125 -0.3674  0.1755
[4,] -0.5084 -0.3034 -0.3733

$v
         [,1]     [,2]      [,3]
[1,] -0.5383 -0.2099  0.81617
[2,] -0.7246 -0.3791 -0.57547
[3,] -0.4302  0.9012 -0.05198


> zapsmall(crossprod(M.svd$u))
     [,1] [,2] [,3]
[1,]   1    0    0
[2,]   0    1    0
[3,]   0    0    1

> zapsmall(crossprod(M.svd$v))
     [,1] [,2] [,3]
[1,]   1    0    0
[2,]   0    1    0
[3,]   0    0    1

> M.svd$u %*% diag(M.svd$d) %*% t(M.svd$v)
     [,1] [,2] [,3]
[1,]   1    2    4
[2,]   3    3    5
[3,]   6    8    3
[4,]   4    6    2
```

Table I.9 Illustration of singular value decomposition, part II (continued from Table I.8). Relation between singular value decomposition and Eigenvalue decomposition.

```
> eigen(tcrossprod(M))
$values
[1]   2.097e+02   1.870e+01   6.123e-01  -5.978e-15

$vectors
          [,1]      [,2]      [,3]      [,4]
[1,]  -0.2561    0.6098    0.6935   -0.2857
[2,]  -0.4102    0.6334   -0.5907    0.2857
[3,]  -0.7125   -0.3674   -0.1755   -0.5714
[4,]  -0.5084   -0.3034    0.3733    0.7143


> eigen(crossprod(M))
$values
[1]  209.6877   18.7000     0.6123

$vectors
          [,1]      [,2]      [,3]
[1,]  -0.5383   -0.2099   0.81617
[2,]  -0.7246   -0.3791  -0.57547
[3,]  -0.4302    0.9012  -0.05198


> M.svd$d^2
[1]  209.6877   18.7000     0.6123
```

I.4.15 Generalized Inverse

For any rectangular matrix $\underset{n \times m}{X} = U\Delta W'$, the Moore–Penrose generalized inverse is defined as

$$X^- = W\Delta^{-1}U'$$

Since $\Delta = \mathrm{diag}(\delta_i)$ is a diagonal matrix, its inverse is $\Delta^{-1} = \mathrm{diag}(\delta_i^{-1})$. The definition is extended to the situation when $\mathrm{rank}(X) < \min(n, m)$ by using $0^{-1} = 0$. See Table I.10 for an example.

When $\mathrm{rank}(X) = m = n$, hence the inverse exists, the generalized inverse is equal to the inverse.

Table I.10 Illustration of the generalized inverse.

```
> M <- matrix(c(1,3,6,4,2,3,8,6,4,5,3,2), 4, 3)

> M
     [,1] [,2] [,3]
[1,]    1    2    4
[2,]    3    3    5
[3,]    6    8    3
[4,]    4    6    2

> library(MASS)

> Mi <- ginv(M)

> Mi
        [,1]     [,2]     [,3]     [,4]
[1,] -0.7434   0.6006  0.22741 -0.35569
[2,]  0.4694  -0.4694 -0.06122  0.32653
[3,]  0.1808   0.1050 -0.06706 -0.02332

> zapsmall(eigen(M %*% Mi)$value)
[1] 1 1 1 0

> zapsmall(Mi %*% M)
     [,1] [,2] [,3]
[1,]    1    0    0
[2,]    0    1    0
[3,]    0    0    1

> M.svd$v %*% diag(1/M.svd$d) %*% t(M.svd$u)
        [,1]     [,2]     [,3]     [,4]
[1,] -0.7434   0.6006  0.22741 -0.35569
[2,]  0.4694  -0.4694 -0.06122  0.32653
[3,]  0.1808   0.1050 -0.06706 -0.02332
```

I.4.16 Solving Linear Equations

There are three cases.

I.4.16.1 $n = m = \text{rank}(X)$

Given a matrix $\underset{n \times m}{X}$ and an n-vector y, the solution β of the linear equation

$$y = X\beta$$

is uniquely defined by

$$\beta = X^{-1}y$$

when X is invertible, that is when $n = m = \text{rank}(X)$.

I.4.16.2 $n > m = \text{rank}(X)$

When $n > m = \text{rank}(X)$, the linear equation is said to be overdetermined. Some form of arbitrary constraint is needed to find a solution. The most frequently used technique is least-squares, a technique in which $\hat{\beta}$ is chosen to minimize the norm of the residual vector.

$$\min_{\beta} \| y - X\beta \|^2 = \sum_{i=1}^{n} \left(y_i - \sum_{j=1}^{m} x_{ij}\beta_j \right)^2$$

The solution $\hat{\beta}$ is found by solving the related linear equations

$$X'y = X'X\hat{\beta}$$

The solution is often expressed as

$$\hat{\beta} = (X'X)^{-1}(X'y) = X^- y$$

This is a definition, not an efficient computing algorithm. The primary efficient algorithm in R is the QR algorithm as defined in qr and related functions, including lm.

I.4.16.3 $m > p = \mathrm{rank}(X)$

When $m > p = \mathrm{rank}(X)$, there are an infinite number of solutions to the linear equation. The singular value decomposition $X = U\Delta W'$ will have $m - p$ zero values along the diagonal of Δ. Let β_0 be one solution. Then

$$\beta_\gamma = \beta_0 + W\begin{pmatrix} 0 \\ \gamma \end{pmatrix}$$

where $\mathbf{0}$ is a vector of p zeros and γ is any vector of length $m - p$, is also a solution.

I.5 Combinations and Permutations

I.5.1 Factorial

For a positive integer n, the notation $n!$, read "n factorial", is used to indicate the product of all the integers from 1 through n:

$$n! = n \times (n - 1) \times \ldots \times 1 = n \times (n - 1)!$$

The factorial of zero, $0!$, is separately defined to equal 1.

  Thus

n	n	$=$	n	$=$	$n((n-1)!)$
0	0	$=$	1	$=$	1
1	1	$=$	1	$=$	1×1
2	2	$=$	2	$=$	2×1
3	3	$=$	6	$=$	3×2
4	4	$=$	24	$=$	4×6
5	5	$=$	120	$=$	5×24
$\vdots$	$\vdots$	$=$	$\vdots$	$=$	$\vdots$

I.5.2 Permutations

The notation $_nP_p$, read "n permute p", indicates the number of ways to select p distinct items from n possible items where two different orderings of the same p items are considered to be distinct. Equivalently, $_nP_p$ is the number of distinct ways of *arranging* p items from n possible items:

$$_nP_p = \frac{n!}{(n-p)!}$$

For example,

$$_5P_3 = \frac{5!}{(5-3)!} = \frac{5 \times 4 \times 3 \times 2 \times 1}{2 \times 1} = 5 \times 4 \times 3 = 60$$

I.5.3 Combinations

The notation $_nC_p$ or $\binom{n}{p}$, read "n choose p", indicates the number of ways to select p distinct items from n possible items, where two different orderings of the same p items are considered to be the same selection. Equivalently, $_nC_p$ is the number of distinct ways of *choosing* p items from n possible items:

$$\binom{n}{p} =_n C_p = \frac{n!}{p! \times (n-p)!} = \frac{_nP_p}{p!}$$

For example,

$$\binom{5}{3} = \frac{_5C_3}{2!} = \frac{5!}{3!(5-3)!} = \frac{5 \times 4 \times 3 \times 2 \times 1}{(3 \times 2 \times 1)(2 \times 1)} = \frac{5 \times 4}{2 \times 1} = 10$$

I.6 Exercises

Exercise I.1.

Start from the matrix in Equation (I.6)

$$A = \begin{pmatrix} 1 & 1 & 2 \\ 1 & 2 & 3 \\ 1 & 3 & 4 \end{pmatrix}$$

Give an example of a basis for A. Then give an example of a vector in $C(A)$ and also a vector not in $C(A)$. Give an example of an orthogonal basis of A, demonstrating that it is orthogonal.

Verify that Equation (I.6) defines a family of solutions to the set of linear equations with $p = \text{rank}(X) < m$.

Appendix J

Probability Distributions

We list, with some discussion, several common probability distributions. We illustrate 21 distributions, 20 distributions in the R **stats** package and one in the **HH** package, for which all three functions (d* for density, p* for probability, and q* for quantile) are available. We also include the Studentized Range distribution for which only the p* and q* functions are available, and the discrete multinomial and continuous multivariate normal. The d* functions give the density $f(x)$ for continuous distributions or the discrete density $f(i)$ for discrete distributions. The p* functions give the cumulative distribution, the probability that an observation is less than or equal to the value x

$$F(x) = P(X \le x) = \begin{cases} \displaystyle\int_{-\infty}^{x} f(x)\, dx & \text{for continuous distributions} \\[2ex] \displaystyle\sum_{i=-\infty}^{x} f(i) & \text{for discrete distributions} \end{cases}$$

The q* functions give the quantiles $F^{-1}(p)$, that is the inverse of the probability function $F(x)$.

In the example illustrations all three functions (d*, p*, and q*) are shown and evaluated at sample X and for specific values of the parameters. For distributions with finite support, the entire domain of x is shown. For distributions with infinite support, the domain of x showing most of the probability is shown.

For the continuous distributions, we show the plot of the density function. The darker color shows the probability (area) to the left of x, and the lighter color shows the probability (area) to the right of x. d*(X) gives the height of the density function at X, p*(X) gives the probability (area) to the left of X, and q*(p) recovers X from the probability p.

© Springer Science+Business Media New York 2015
R.M. Heiberger, B. Holland, *Statistical Analysis and Data Display*,
Springer Texts in Statistics, DOI 10.1007/978-1-4939-2122-5

For the discrete distributions, we show the plot of the discrete density function. The darkest color shows the probability at X, the intermediate color shows the probability strictly left of X, and the lightest color shows the probability to the right of X. d*(X) gives the probability at X, p*(X) gives the probability to the left of and including X, and q*(p) recovers X from the probability p.

We list the continuous central distributions in Section J.1, the continuous noncentral distributions in Section J.2, and the discrete distributions in Section J.3. Within each section the distributions are ordered alphabetically.

J.1 Continuous Central Distributions

J.1.1 Beta

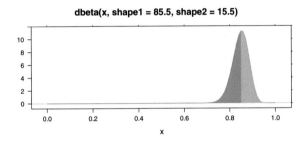

dbeta(x, shape1 = 85.5, shape2 = 15.5)

```
>    dbeta(.85, shape1=85.5, shape2=15.5)
[1] 11.22

>    pbeta(.85, shape1=85.5, shape2=15.5)
[1] 0.5131

>    qbeta(0.5131489, shape1=85.5, shape2=15.5)
[1] 0.85
```

This two-parameter distribution is often used to model phenomena restricted to the range $(0, 1)$, for example sample proportions. It is used in Section 5.1.2 to construct alternative one-sided confidence intervals on a population proportion.

J.1.2 Cauchy

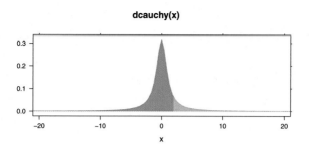

```
>   dcauchy(1.96)
[1] 0.06574

>   pcauchy(1.96)
[1] 0.8498

>   qcauchy(0.8498286)
[1] 1.96
```

The Cauchy distribution is the same as the *t*-distribution with 1 degree of freedom. It's special feature is that it does not have a finite population mean.

J.1.3 Chi-Square

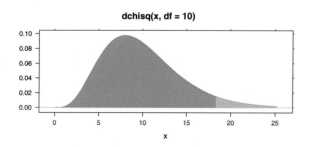

```
>   dchisq(18.31, df=10)
[1] 0.01547

>   pchisq(18.31, df=10)
[1] 0.95

>   qchisq(0.9500458, df=10)
[1] 18.31
```

The central χ^2 distribution with k degrees of freedom is the distribution of the sum of squares of k independent standard normal r.v.'s. If $k > 2$, a χ^2 r.v. has a unimodal, positively skewed PDF starting at zero and asymptotically tapering to the horizontal axis for large values. The mean of this distribution is k, and k is also approximately its median if k is large. The r.v. $[(n-1)s^2]/\sigma^2$, where s^2 is the variance of a normal sample, has a χ^2 distribution with $n - 1$ degrees of freedom.

This distribution is used in inferences about the variance (or s.d.) of a single population and as the approximate distribution of many nonparametric test statistics, including goodness-of-it tests and tests for association in contingency tables.

J.1.4 Exponential

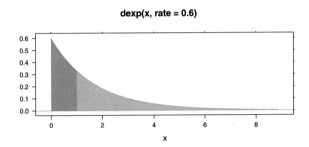

dexp(x, rate = 0.6)

```
>   dexp(1, rate=.6)
[1] 0.3293

>   pexp(1, rate=.6)
[1] 0.4512

>   qexp(0.4511884, rate=.6)
[1] 1
```

μ is both the mean and standard deviation of this distribution. R parameterizes the exponential distribution with the rate $1/\mu$, the reciprocal of the mean μ. Times between successive Poisson events with mean rate of occurrence μ have the exponential distribution. The exponential distribution is the only distribution with the "lack of memory" or "lack of deterioration" property, which states that the probability that an exponential random variable exceeds $t_1 + t_2$ given that it exceeds t_1 equals the probability that it exceeds t_2.

J.1.5 F

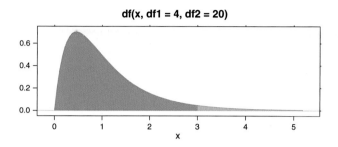

df(x, df1 = 4, df2 = 20)

```
>    df(3, df1=4, df2=20)
[1] 0.0469

>    pf(3, df1=4, df2=20)
[1] 0.9568

>    qf(0.956799, df1=4, df2=20)
[1] 3
```

The F distribution is related to the χ^2 distribution. If U_i, for $i = \{1, 2\}$, is a χ^2 r.v. with ν_i degrees of freedom, and if U_1 and U_2 are independent, then $F = U_1/U_2$ has an F distribution with ν_1 and ν_2 df. This distribution is extensively used in problems involving the comparison of variances of two normal populations or comparisons of means of two or more normal populations.

J.1.6 Gamma

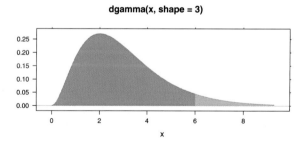

dgamma(x, shape = 3)

```
>    dgamma(6, shape=3)
[1] 0.04462

>    pgamma(6, shape=3)
[1] 0.938

>    qgamma(0.9380312, shape=3)
[1] 6
```

From ?dgamma:

The Gamma distribution with parameters 'shape' = a and 'scale' = s has density
$$f(x) = 1/(s^a \Gamma(a)) x^{(a-1)} e^{-(x/s)}$$
for $x \geq 0$, $a > 0$ and $s > 0$. (Here $\Gamma(a)$ is the function implemented by R's gamma() and defined in its help. Note that $a = 0$ corresponds to the trivial distribution with all mass at point 0.)

The mean and variance are $E(X) = a \times s$ and $\text{Var}(X) = a \times s^2$.

The special case of the gamma distribution with shape=1 is the exponential distribution.

J.1.7 Log Normal

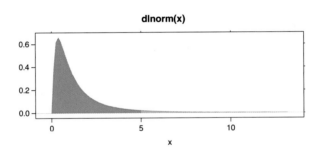

```
>    dlnorm(5)
[1] 0.02185

>    plnorm(5)
[1] 0.9462

>    qlnorm(0.9462397)
[1] 5
```

A r.v. X is said to have a *lognormal* distribution with parameters μ and σ if $Y = \ln(X)$ is $N(\mu, \sigma^2)$; i.e., if Y is normal, then e^Y is lognormal. This is a positively skewed unimodal distribution defined for $x > 0$. It is commonly used as a good approximation for positively skewed data, such as a distribution of income.

J.1.8 Logistic

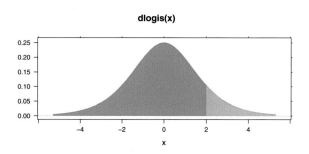

dlogis(x)

```
> dlogis(2)
[1] 0.105

> plogis(2)
[1] 0.8808

> qlogis(0.8807971)
[1] 2
```

From ?dlogis:

The Logistic distribution with `location=m` and `scale=s` has distribution function
$$F(x) = 1/\left(1 + \exp\left(-(x - m)/s\right)\right)$$
and density
$$f(x) = (1/s)\ \exp\left((x - m)/s\right)\ \left(1 + \exp\left((x - m)/s\right)\right)^{-2}.$$
It is a long-tailed distribution with mean m and variance $(\pi^2/3)s^2$.
   `qlogis(p)` is the same as the well known *logit* function, $\text{logit}(p) = \log(p/(1 - p))$, the log odds function, and `plogis(x)` has consequently been called the *inverse logit*.
   The distribution function is a rescaled hyperbolic tangent, $\text{plogis}(x) = (1 + \tanh(x/2))/2$, and it is called a *sigmoid function* in contexts such as neural networks.

J.1.9 Normal

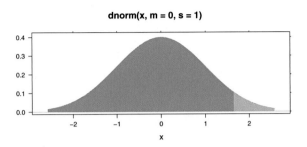

dnorm(x, m = 0, s = 1)

```
>    dnorm(1.645, m=0, s=1)
[1] 0.1031

>    pnorm(1.645, m=0, s=1)
[1] 0.95

>    qnorm(0.95, m=0, s=1)
[1] 1.645
```

This distribution was introduced in Section 3.4.2. If Z is standard normal $N(0, 1)$, the standard normal density ϕ and cumulative distribution Φ functions are

$$\phi(z) = \frac{1}{\sqrt{2\pi}} \exp\left(\frac{z^2}{2}\right)$$

$$\Phi(Z) = \int_{-\infty}^{Z} \phi(z)\, dz$$

The general density, for random variable x with mean μ and variance σ^2, is

$$f(x \mid \mu, \sigma^2) = \frac{1}{\sigma}\, \phi\left(\frac{x-\mu}{\sigma}\right) = \frac{1}{\sqrt{2\pi\sigma^2}} \exp\left(\frac{(x-\mu)^2}{2\sigma^2}\right)$$

The term *probit* is an alternate notation for the inverse function Φ^{-1}.

$$q = \Phi^{-1}(p) = \text{probit}(P)$$

is the inverse function such that $p = \Phi(q)$.

J.1.10 Studentized Range Distribution

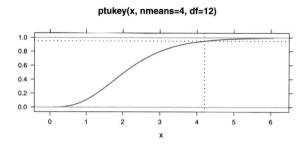

ptukey(x, nmeans=4, df=12)

```
>    ptukey(4.199, nmeans=4, df=12)
[1] 0.95

>    qtukey(0.95, nmeans=4, df=12)
[1] 4.199
```

This distribution is used in the Tukey multiple comparisons procedure discussed in Section 6.3. Let $\bar{y}_{(1)}$ and $\bar{y}_{(a)}$ denote the smallest and largest means of samples of size n drawn from a populations having a common variance σ^2, and let $s = \sqrt{MS_{Res}}$ be the estimate of σ calculated from the ANOVA table, for example, Table 6.2. Then the random variable

$$Q = \frac{\bar{y}_{(a)} - \bar{y}_{(1)}}{s/\sqrt{n}}$$

has a Studentized range distribution with parameters a and $df_{Res} = a(n - 1)$. The Studentized range distribution is defined on the domain $0 \le q < \infty$. R provides the ptukey and qtukey functions, but not the density function dtukey. We therefore show the cumulative probability function instead of the density for the Studentized range distribution.

J.1.11 (Student's) T

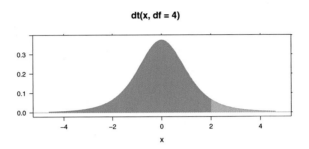

```
>    dt(2, df=4)
[1] 0.06629

>    pt(2, df=4)
[1] 0.9419

>    qt(0.9419417, df=4)
[1] 2
```

This distribution was introduced in Section 3.4.3.

J.1.12 Uniform

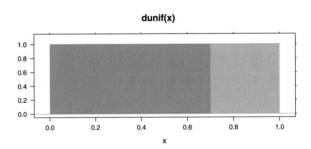

```
>    dunif(.7)
[1] 1

>    punif(.7)
[1] 0.7

>    qunif(.7)
[1] 0.7
```

All real numbers between *a* and *b* are equally likely. The standard case has $a = 0$ and $b = 1$. Hypothesis tests work by mapping an appropriate null distribution to the uniform (with the appropriate p* function in R).

J.1.13 Weibull

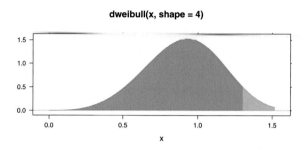

dweibull(x, shape = 4)

```
>    dweibull(1.3, shape=4)
[1] 0.5052

>    pweibull(1.3, shape=4)
[1] 0.9425

>    qweibull(0.9425075, shape=4)
[1] 1.3
```

From dweibull:

The Weibull distribution with 'shape' parameter a and 'scale' parameter b has density given by
$$f(x) = (a/b)\,(x/b)^{(a-1)}\,\exp(-(x/b)^a)$$
for $x > 0$. The cumulative distribution function is $F(x) = 1 - \exp(-(x/b)^a)$ on $x > 0$, the mean is $E(X) = b\Gamma(1 + 1/a)$, and the $\text{Var}(X) = b^2\left(\Gamma(1 + 2/a) - (\Gamma(1 + 1/a))^2\right)$.

J.2 Noncentral Continuous Probability Distributions

In hypothesis testing, except in special cases such as testing with the normal distribution, one deals with a central distribution when the null hypothesis is true and an analogous noncentral distribution when the null hypothesis is false. Thus calculations of probabilities under the alternative hypothesis, as are required when doing Type II error analysis and when constructing O.C. (beta curves) and power curves, necessitate the use of noncentral distributions.

The forms of the t, chi-square, and F distributions we've considered thus far have all been central distributions. For example, if $\bar{X}$ is the mean and s the standard deviation of a random sample of size n from a normal population with mean μ, then $t = (\bar{x} - \mu)/(s/\sqrt{n})$ has a central t distribution with $n - 1$ df. Suppose, however, that the population mean is instead μ_1, different from μ. Then t above is said to have a noncentral t distribution with $n - 1$ df and a *noncentrality parameter* proportional to

$((\mu - \mu_1)/\sigma)^2$. (If $\mu = \mu_1$, so that the noncentrality parameter is zero, the noncentral t distribution reduces to the central t distribution.)

A noncentral chi-square (χ^2) r.v. is a sum of squares of independent normal r.v.'s each with s.d. 1 but at least some of which have a nonzero mean. A noncentral F r.v. is the ratio of a noncentral chi-square r.v. to a central chi-square r.v., where the two chi-squares are independent.

For tests using the t, chi-square, or F distribution, the power of the test (protection against Type II errors) is an increasing function of the noncentrality parameter.

Noncentral distributions are specified with one more parameter than their corresponding central distribution. Consequently, tabulations of their cumulative distribution function appear much less frequently than those for central distributions. Fewer statistical software packages include them. There is no noncentral normal distribution. The distribution under the alternative hypothesis is just an ordinary normal distribution with a shifted mean.

The R functions for noncentral t, chi-square, and F CDFs are the same as those for the corresponding central distribution with the addition of an argument for the noncentrality parameter ncp. The noncentrality parameter defaults to zero (hence to a central distribution) if it is not specified.

The figures in Sections J.2.1, J.2.2, and J.2.3 show the noncentral distribution along with the corresponding central distribution. This way it is possible to see that a positive noncentrality parameter shifts the mode to the right.

J.2.1 Chi-Square: Noncentral

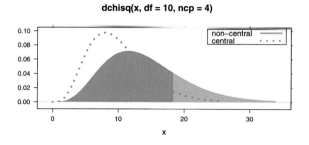

```
>   dchisq(18.31, df=10, ncp=4)
[1] 0.0408

>   pchisq(18.31, df=10, ncp=4)
[1] 0.7852

>   qchisq(0.7852264, df=10, ncp=4)
[1] 18.31
```

See discussion in Section 14.8.2 and example in Figure D.1.

J.2.2 T: Noncentral

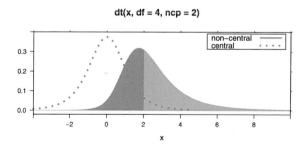

```
>   dt(2, df=4, ncp=2)
[1] 0.3082

>   pt(2, df=4, ncp=2)
[1] 0.4557

>   qt(0.455672, df=4, ncp=2)
[1] 2
```

See examples in Figures 3.24, 5.2, and 5.10.

J.2.3 F: Noncentral

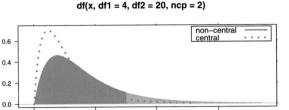

df(x, df1 = 4, df2 = 20, ncp = 2)

```
>    df(3, df1=4, df2=20, ncp=2)
[1] 0.1062

>    pf(3, df1=4, df2=20, ncp=2)
[1] 0.871

>    qf(0.8710256, df1=4, df2=20, ncp=2)
[1] 3
```

See example in Section 6.5 and discussion in Section 14.8.2.

J.3 Discrete Distributions

Discrete distributions are defined to have nonzero values on a set of integers.

$$F(x) = P(X \le x) = \sum_{i=-\infty}^{x} f(i)$$

The inverse functions (the q* functions in R) are sensitive to the precision of the numerical representation.

Computers use finite precision floating point arithmetic, precise to 53 significant binary digits (bits)—approximately 17 decimal digits. They do not use the real number system that we are familiar with. Simple decimal repeating fractions such as these are not stored precisely with finite precision machine arithmetic. All of the individual values in this example are automatically rounded to 53 bits when they are entered into the computer. None of them are exactly represented inside the computer. See Appendix G for more on the floating point arithmetic used in computers. In several of the discrete distribution examples here, it has been necessary to display the *p* values to 17 decimal digits in order to get the desired answer from the q* functions.

Here is a simple example, the 6-level discrete uniform (one fair die), to illustrate the problem. In this example, rounding produces several results from the qdiscunif function that are one unit off (either too large or too small).

			$F(i)$	
			rounded to	machine precision, rounded to 53 bits,
i	$f(i)$	fraction	4 decimal digits	≈ 17 decimal digits
1	1/6	1/6	0.1667	0.16666666666666666
2	1/6	2/6	0.3333	0.33333333333333331
3	1/6	3/6	0.5000	0.50000000000000000
4	1/6	4/6	0.6667	0.66666666666666663
5	1/6	5/6	0.8333	0.83333333333333337
6	1/6	6/6	1.0000	1.00000000000000000

```
> ## this is printing precision, not internal representation
> old.digits <- options(digits=7)

> ddiscunif(1:6, 6)
[1] 0.1666667 0.1666667 0.1666667 0.1666667 0.1666667 0.1666667

> pdiscunif(1:6, 6)
[1] 0.1666667 0.3333333 0.5000000 0.6666667 0.8333333 1.0000000

> qdiscunif(pdiscunif(1:6, 6), 6)
[1] 1 2 3 4 5 6

> round(pdiscunif(1:6, 6), 4) ## rounded to four decimal digits
[1] 0.1667 0.3333 0.5000 0.6667 0.8333 1.0000

> ## inverse after rounding to four decimal digits
> qdiscunif(round(pdiscunif(1:6, 6), 4), 6)
[1] 1 1 3 4 4 6

> options(old.digits)
```

J.3.1 Discrete Uniform

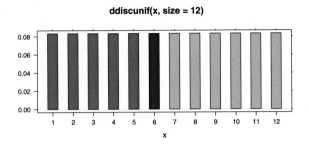

ddiscunif(x, size = 12)

```
>   ddiscunif(6, size=12)
[1] 0.08333

>   pdiscunif(6, size=12)
[1] 0.5

>   qdiscunif(.5, size=12)
[1] 6
```

In the discrete uniform distribution, the integers from 1 to n are equally likely. The population mean is $(n + 1)/2$. The population variance is $(n^2 - 1)/12$. The distribution function is

J.3.2 Binomial

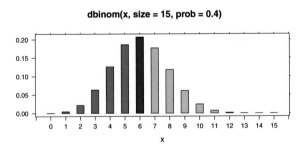

dbinom(x, size = 15, prob = 0.4)

```
> ## probability of exactly 6 Heads
> dbinom(6, size=15, prob=.4)
[1] 0.2066

> ## probability of 6 or fewer Heads
> ## extra precision is needed
> print(pbinom(6, size=15, prob=.4), digits=17)
[1] 0.60981315570892769

> ## q, the number for which the probability of seeing
> ## q or fewer Heads is 0.60981315570892769
> qbinom(0.60981315570892769, size=15, prob=.4)
[1] 6
```

The binomial distribution was introduced in Section 3.4.1. If X has a binomial distribution with parameters n (n=size, number of coins tossed simultaneously) and p (p=prob, probability of one coin landing Heads on one toss), then the binomial distribution gives the probability of observing exactly X heads.

J.3.3 Geometric

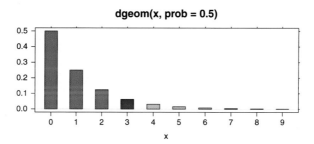

dgeom(x, prob = 0.5)

```
>    dgeom(3, prob=.5)
[1] 0.0625

>    pgeom(3, prob=.5)
[1] 0.9375

>    qgeom(0.9375, prob=.5)
[1] 3
```

From ?dgeom:

The geometric distribution with 'prob' = p has density
$$p(x) = p(1 - p)^x$$
for $x = 0, 1, 2, ..., 0 < p \leq 1$.
    The quantile is defined as the smallest value x such that $F(x) \geq p$, where F is the distribution function.

J.3.4 Hypergeometric

dhyper(x, m = 5, n = 6, k = 7)

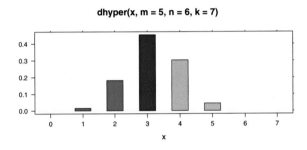

```
>    dhyper(3, m=5, n=6, k=7)
[1] 0.4545

>    print(phyper(3, m=5, n=6, k=7), digits=17)
[1] 0.65151515151515149

>    qhyper(0.65151515151515149, m=5, n=6, k=7)
[1] 3
```

    The hypergeometric distribution is used in Chapter 15. We sample *n* items *without* replacement from a population of *N* items comprised of *M* successes and *N − M* failures. Then the number of successes *X* observed in the population is said to have a hypergeometric distribution with parameters *N*, *M*, and *n*.

J.3.5 Negative Binomial

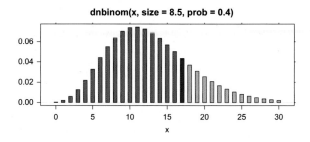

```
>   dnbinom(17, size=8.5, prob=.4)
[1] 0.04338

>   print(pnbinom(17, size=8.5, prob=.4), digits=17)
[1] 0.81209497223034977

>   qnbinom(0.81209497223034977, size=8.5, prob=.4)
[1] 17
```

The negative binomial distribution with size $= n$ and prob $= p$ has density

$$\frac{\Gamma(x+n)}{\Gamma(n)\,x!}\, p^n (1-p)^x$$

for $x = 0, 1, 2, \ldots, n > 0$, and $0 < p \le 1$.

This represents the number of failures which occur in a sequence of Bernoulli trials before a target number of successes is reached. The mean is $\mu = n(1-p)/p$ and variance is $n(1-p)/p^2$.

J.3.6 Poisson

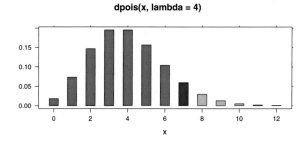

```
>    dpois(7, lambda=4)
[1] 0.05954

>    print(ppois(7, lambda=4), digits=17)
[1] 0.94886638420715264

>    qpois(0.94886638420715264, lambda=4)
[1] 7
```

Let the random variable X be the number of occurrences of some event that are observed in a unit of time, volume, area, etc., and let λ be the mean number of occurrences of the event per unit, assumed to be constant throughout the process that generates the occurrences. Suppose that the occurrence(s) of the event in any one unit are independent of the occurrence(s) of the event in any other nonoverlapping unit. Then X has a Poisson distribution with parameter λ.

J.3.7 Signed Rank

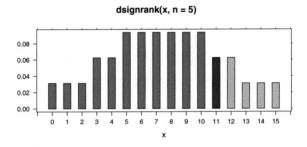

```
>    dsignrank(11, n=5)
[1] 0.0625

>    psignrank(11, n=5)
[1] 0.8438

>    qsignrank(0.84375, n=5)
[1] 11
```

See the discussion in Section 16.3.2.

J.3.8 Wilcoxon

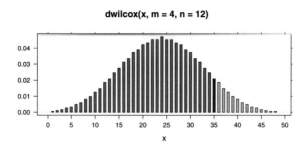

dwilcox(x, m = 4, n = 12)

```
>    dwilcox(35, m=4, n=12)
[1] 0.02088

>    print(pwilcox(35, m=4, n=12), digits=17)
[1] 0.9148351648351648

>    qwilcox(0.9148351648351648, m=4, n=12)
[1] 35
```

See Table 16.7 for an example of a rank sum test.

J.4 Multivariate Distributions

J.4.1 Multinomial

dmultinom(x, prob = c(1,2,5))

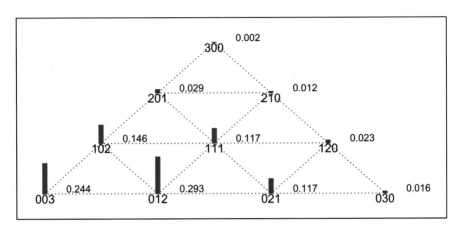

```
> ## This example is based on ?dmultinom in R
> ## all possible outcomes of Multinom(N = 3, K = 3)
> X <- t(as.matrix(expand.grid(0:3, 0:3)))

> X <- X[, colSums(X) <= 3]

> X <- rbind(X, 3:3 - colSums(X))

> dimnames(X) <- list(letters[1:3], apply(X, 2, paste, collapse=""))

> Y <- round(apply(X, 2, function(x) dmultinom(x, prob = c(1,2,5))), 3)

> rbind(X, Y)
     003    102    201    300    012    111    210    021    120    030
a 0.000  1.000  2.000  3.000  0.000  1.000  2.000  0.000  1.000  0.000
b 0.000  0.000  0.000  0.000  1.000  1.000  1.000  2.000  2.000  3.000
c 3.000  2.000  1.000  0.000  2.000  1.000  0.000  1.000  0.000  0.000
Y 0.244  0.146  0.029  0.002  0.293  0.117  0.012  0.117  0.023  0.016
```

   The (discrete) *multinomial* distribution is a generalization of the binomial distribution to the case of $k > 2$ categories. Suppose there are n independent trials, each of which can result in just one of k possible categories such that p_j is the probability of resulting in the jth of these k categories. (Hence $p_1 + p_2 + \ldots + p_k = 1$.) Let X_j be the number of occurrences in category j. Then the vector $(X_1, X_2, ..., X_k)$ is said to have a multinomial distribution with parameters $n, p_1, p_2, \ldots, p_k$. Its PMF is

$$P(X_j = x_j \mid j = 1, \ldots, k) = \frac{n! \, p_1^{x_1} p_2^{x_2} \cdots p_k^{x_k}}{x_1! x_2! \ldots x_k!}, \quad x_1 + x_2 + \ldots x_k = n$$

If a proportion p_j of a population of customers prefers product number $j, j = 1, \ldots, k$, among k products, then the multinomial distribution provides the probability of observing any particular configuration of preferences among a random sample of n customers.

J.4.2 Multivariate Normal

Bivariate Normal: dmvnorm(c(−1, −1))

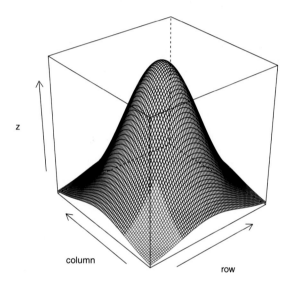

```
> dmvnorm(c(-1, -1))
[1] 0.05855

> pmvnorm(upper=c(-1, -1))[1]
[1] 0.02517

> qmvnorm(0.02517, mean=c(0,0))$quantile
[1] -1
```

See Sections 3.3.5 and I.4 for examples.

Appendix K

Working Style

Working style in a computer environment depends on two interrelated concepts: which programs you use and how you use them.

For statistical analysis we are recommending R as the primary computational tool. There are other very good programs, and it is imperative that you develop a working understanding of several of them. See Appendix H for information on how to use the datasets discussed in this book with other software.

An excellent text editor is an indispensable tool for the statistical analyst. The editor is the single program in which we spend most of our time. We use it for looking at raw data, for writing commands in the statistical languages we use, for reading the output tables produced by our statistical programs, for writing reports, and for reading and writing correspondence about our studies to our clients, consultants, supervisors, and subordinates. We discuss our requirements for a text editor in Section K.1. Our personal choice of editor is Emacs (Free Software Foundation, 2015), which we discuss in Appendix M. There are other excellent editors which satisfy our requirements.

We discuss in Section K.2 the types of interaction that are possible with R. We recommend in Section K.3 working with files of R commands. We discuss in Section K.4 our recommendations for organization of the files within the operating system's directory structure.

K.1 Text Editor

As we indicated in the Preface on page ix, our goal in teaching statistical languages is to make the student aware of the capabilities of the language for describing data and their analyses. The language is approached through the text editor.

© Springer Science+Business Media New York 2015
R.M. Heiberger, B. Holland, *Statistical Analysis and Data Display*,
Springer Texts in Statistics, DOI 10.1007/978-1-4939-2122-5

We distinguish between the concepts of *text editing* (discussed in this Appendix) and *word processing* (discussed in Appendix O). Text editing is moving characters around on the screen with the expectation that they will stay where you put them. This is critical when writing computer programs where the physical placement of lines and characters on the page is part of what the computer program interprets. In the R language the two layouts of the same characters in Table K.1 have completely different interpretations.

Word processing is moving sentences, paragraphs, sections, figures, and cross-references around. A word processor can be used as a text editor by manually turning off many of the word processing features.

Table K.1 Two different interpretations of the same characters {"3", "+", "4"} that depend on their placement on separate lines. If the first set of input lines were reformatted according to English language paragraph formatting rules it would be interpreted as if it were the second set, which has a completely different meaning. This is the simplest possible example of why we make a distinction between text editing and word processing.

	R Input	
Two lines		One line
3		3 + 4
+ 4		

	R Output	
> 3		> 3 + 4
[1] 3		[1] 7
> + 4		
[1] 4		

K.1.1 Requirements for an Editor

These are the requirements we place on any text editor that is to be used for interacting with a computing language:

1. Permit easy modification of computing instructions and facilitate their resubmission for processing

2. Be able to operate on output as well as input

3. Be able to cut, paste, and rearrange text; to search documents for strings of text; to work with rectangular regions of text

4. Be aware of the language, for example, to illustrate the syntax with appropriate indentation and with color and font highlighting, to detect syntactic errors in input code, and to check the spelling of statistical keywords

5. Handle multiple files simultaneously

6. Interact cleanly with the presentation documents (reports and correspondence) based on the statistical results

7. Check spelling of words in natural languages

8. Permit placement of graphics within text.

9. Permit placement of mathematical expressions in the text.

10. Work with Unicode to give access to all character sets in all human languages.

K.1.2 Choice of Editor

Our preference for an editor is Emacs with **ESS**, which we discuss in Appendix M. In Section M.5 we show how Emacs satisfies the requirements in Section K.1.1.

There are many other options for an editor, usually within the context of an Integrated Development Environment (IDE), a software application that provides comprehensive facilities to computer programmers for software development. An IDE normally consists of a source code editor, build-automation tools, and a debugger. For more information on IDEs see Wikipedia (2015).

For an annotated list, with links to their webpages, of other editors and IDEs that work with R, see Grosjean (2012).

We discuss word processors such as MS Word (Microsoft, 2015) in Section O.1. In Section O.1.1 we show how MS Word satisfies some of the requirements in Section K.1.

K.2 Types of interaction with R

1. CLI Command Line Interface: The user types R statements and the system responds with requested output. Examples: R in a shell/terminal/CMD window, R in the *R* buffer in Emacs, R in the Console window in the Rgui.exe for Windows, the R.app for Macintosh, and the **JGR** package on all operating systems.

2. Menu or Dialog Box: Dropdown lists of command names, often with default settings for arguments. This allows point and click access to many R functions with standard settings of options. The **Rcmdr** (R Commander) package described in Appendix C is such a system.

3. Spreadsheet. The RExcel system described in Appendix D allows access to all R functions from within the automatic recalculation mode of Excel for Windows. The **Rcmdr** interface is incorporated into the Excel menu ribbon.

4. Web-based interface. Technology for embedding R applications within an html page to provide interactive access to R functions for non-programmers. See Appendix E for a discussion of the **shiny** system.

5. Document based interface. The end user writes a document (book, paper, report) in a standard document writing system (LaTeX or Word, for example) with embedded R code. Three examples are Sweave (Leisch and R-core, 2014), knitr (Xie, 2015), and SWord (Baier, 2014). The second edition of HH is written using Sweave (see help(Sweave, package="utils")). All R code, leading to all graphs and tables in the book, is included in the LaTeX source files for the chapters. The code files for the **HH** package, located with the HHscriptnames() function, were pulled from the LaTeX files by the Stangle function which is part of the Sweave framework.

6. GUI Graphical User Interface: anything other than the command line interface.

K.3 Script File

Our personal working style (and the one we recommend to our readers) is to write a file of commands (a *script file* with extension .R or .r) that specify the analysis, run the commands and review the graphs and tables so produced, and then correct and augment the analysis specifications. We construct the command file interactively. Initially we write the code to read the data into the program and to prepare several graphs and the initial tables. If there are errors (in typing or in programming logic), we correct them *in the file* and rerun the commands. As we progress and gain insight into what the data say, we add to the code to produce additional graphs and tables. When we have finished, we have a file of commands that could be run as a batch job to produce the complete output. We also have a collection of graphs and tables that can be stored in additional files.

K.4 Directory Structure

When we have many files, we need to organize them within the directory structure of our computer's operating system.

K.4.1 Directory Structure of This Book

This book is written in LaTeX (Lamport, 1994). Our organizational structure uses the directory structure provided by the computer's operating systems. For this book we have a main directory (hh2) and subdirectories for code (containing .R files), transcripts (.Rout) files, and figures (.pdf) files. The main directory contains all the .tex files (one per chapter plus a master file), our LaTeX style file (hh2.sty), and several other support files. We have a work directory parallel to the hh2 directory for experiments with R code, for correspondence with our publisher, and for correspondence with the people who were kind enough to read and comment on drafts of this book.

The R functions that we developed for the book, and all datasets used in the book, are collected into the **HH** package (Heiberger, 2015) distributed through CRAN.

The master copy of the R scripts for all figures and tables in the second edition is included in the *.tex source files for the individual chapters. We use the **Sweave** functions included in R to execute the code directly from the *.tex files. When we are ready to distribute the code, we pull the R code from the *.tex files with the Stangle function (part of R's **Sweave** system) and place them into the **HH** package. The script files in the **HH** package for the book can be located by the package user with the HHscriptnames() function.

We hope that readers of our book, and more generally users of R, design and collect their own functions into a personal package (it doesn't have to be distributed through CRAN). Once you have more than a few functions that are part of your working pattern, maintaining them in a package is much simpler than inventing your own idiosyncratic way of keeping track of what you have. We say a few words about building a package in Appendix F and refer you to the R document

    system.file("../../doc/manual/R-exts.pdf")

for complete information.

K.4.2 Directory Structure for Users of This Book

It is critical to realize that your work is yours. Your work is not part of the computer's operating system, nor of the installed software. The operating system, the editor, R, and other software are kept in system directories. In order to protect the integrity of the system you will usually not have write access to those locations.

You need a home directory, the one where you keep all your personal subdirectories and files. Your operating system automatically creates a HOME directory for you. R can find its name with the command Sys.getenv("HOME"). Everything you do should be in a subdirectory of your home directory.

For example, each time we teach a course, we create a directory for that course. For last year's course based on this book, we used the directory 8003.f14 directly under the HOME directory. We have a subdirectory of the course directory for the syllabus, for each class session, and for student records. We keep handouts, R scripts, and transcripts of the class R session as files within each class session's directory.

K.4.3 Other User Directories

We recommend a separate directory for each project. It will make your life much easier a year from now when you try to find something.

Appendix L

Writing Style

Reports, including homework exercises, based on computer printout must be typed correctly. We recommend LATEX (the standard required by many statistics and mathematics journals, and the typesetting package with which we wrote this book). We do accept other word processing software. Whichever software you use, you must use it correctly. We discuss in this appendix some of the fundamentals about good technical writing.

L.1 Typographic Style

Specific style issues that you must be aware of are

1. Fonts: Computer listings in R (and S-Plus and SAS and many other statistical software systems) are designed for monowidth fonts (such as Courier) where all characters (including spaces) are equally wide. These listings are unreadable in Times Roman or any other *proportional font*. English text looks best in a proportional font (where, for example, the letter "M" is wider than the letter "i") such as Times Roman. Table L.1 shows the difference with simple alphabetic text.

   Table L.2 shows an example of the issue for computer listings. The Courier rendition is consistent with the design of the output by the program designer. The Times Roman is exactly the same text dropped into an environment that is incorrectly attempting to space it in accordance with English language typesetting rules. **In our classes, we return UNREAD any papers that use a proportional font for computer listings.**

2. Alignment: Numbers in a column are to be aligned on decimal points. Alignment makes it possible to visually compare printed numbers in columns. There are two

© Springer Science+Business Media New York 2015   837
R.M. Heiberger, B. Holland, *Statistical Analysis and Data Display*,
Springer Texts in Statistics, DOI 10.1007/978-1-4939-2122-5

Table L.1 We placed a vertical rule after four characters on each line in both Courier and Times Roman. The Courier rules are aligned because all characters in Courier are exactly the same width. The Times Roman rules are not aligned because each character in Times Roman has its own width.

Courier	Times Roman
Wide| and narrow.	Wide| and narrow.
Lett|ers in each row align	Lett|ers in each row do not align
with| letters in the previous row.	with| letters in the previous row.

Table L.2 R output displayed correctly in a monowidth font, and incorrectly in a proportional font. The same text, the output from an R `summary.data.frame` call, is displayed in both fonts. The unequal width of the characters in the Times Roman font destroys the vertical alignment that is necessary for interpretation of this listing.

Courier (correct spacing)	Times Roman (incorrect spacing)
``` > summary(ex0221) weight code Min.:23.20 1:35 1st Qu.:24.75 2:24 Median:25.70 Mean:25.79 3rd Qu.:26.50 Max.:31.10 ```	> summary(ex0221) weight code Min.:23.20 1:35 1st Qu.:24.75 2:24 Median:25.70 Mean:25.79 3rd Qu.:26.50 Max.:31.10

Table L.3 Alignment of decimal points in the numbers on the left makes it easy to compare the magnitudes of the numbers. Centering of the numbers on the right, and therefore non-alignment of decimal points, makes it difficult to compare the magnitudes of the numbers.

Correct	Wrong
123.45	123.45
12.34	12.34
−4.32	−4.32
0.12	0.12

reasons for getting it wrong. One is carelessness. The other is blind copying from a source that gets it wrong. We show an example of both alignments in Table L.3.

3. Minus signs and dashes: There are four distinct concepts that have four different typographical symbols in well-designed fonts. On typewriters all four are

usually displayed with the symbol "-" that appears on the hyphen key (next to the number 0). You are expected to know the difference and to use the symbols correctly.

Table L.4 shows an example of the correct usage of all four symbols and the keys in LATEX and MS Word.

Table L.4 Correct usage of all four dash-like symbols (- – – —) and the keys to generate them in LATEX and MS Word.

Symbol	Use	Example	LATEX	MS Word
- hyphen	compound word	*t*-test	-	-
– en dash	range	100–120	--	ctrl-num –
– minus	negation	−12	$-$	Insert-menu/symbol.../ –
— em dash	apposition	punctuation—like this	---	alt-ctrl-num –

The misuse of dashes that touches my (rmh) hot button the most is misuse of hyphen when minus is meant, for example

correct	WRONG
+12.2	+12.2
−12.2	-12.2

In this wrong usage, the "+" and "-" are not aligned and consequently the decimal point is not aligned.

4. Right margins and folding: Table L.5 intentionally misuses formatting to illustrate how bad it can look. This usually occurs when the R window width is wider than your word processor is willing to work with. Verify that you picked an options()$width consistent with your word processor. You can make the width narrower in R by using the R command

    ```
    options(width=72)
    ```

5. Quotation marks: Quotation marks in Times Roman are directional. This is "Times Roman correct" (with a left-sided symbol on the left and a right-sided symbol on the right). This is "Times Roman incorrect" (with the right-sided symbol on both sides). In the typewriter font recognized by R, quotation marks are vertical, not directional. and are the same on both sides. In typewriter font, this is "typewriter correct" (same non-directional symbol on both sides).

Table L.5 Intentional misuse of formatting. Never turn in anything that looks like these bad examples. The top section has lines of text that extend beyond the right margin of the page. The middle section is an R table that has been arbitrarily folded at 49 mono-spaced letter widths. The bottom section retains the lines, but places them in a proportional font and ignores the alignment of columns.

Do not allow the right margins of your work to run off the edge of the page. It is hard to read text that isn't visible.

Do not allow lines to be arbitrarily folded in a way that destroys the formatting.

This is particularly a problem if you copy output from the R Console window to an editor in a word-processing mode. Word processors (as distinct from text editors) by default enforce English-language conventions (such as maximum line length and proportional font) on code and output text that is designed for column alignment and a fixed-width font. Use an editor, such as Emacs with ESS, that is aware of the R formatting conventions. Most word processors do have an option to set sections in fixed-width Courier font and to give the write control of margins.

Folding
makes
this table
impossible to
read.

```
                              Sum of
Source              DF        Squares    Mean Squa
re     F Value    Pr > F

Model               2         2649.816730    1324.9083
65       54.92     <.0001
Error               44        1061.382419       24.1223
28
Corrected Total     46        3711.199149
```

Column
alignment
ignored.
Table is
unreadable.

Sum of
Source DF Squares Mean Square F Value Pr > F

Model 2 2649.816730 1324.908365 54.92 <.0001
Error 44 1061.382419 24.122328
Corrected Total 46 3711.199149

L.2 Graphical Presentation Style

Graphs designed for someone to read must be legible. Legibility includes the items listed in this section and in Chapter 4 (some of which are repeated here).

L.2.1 Resolution

Figure L.1 shows the same graph drawn on a vector graphics device (pdf() in this example) and a bitmap device (png() in this example).

Vector graphics devices define objects to be drawn in terms of geometrical primitives such as points, lines, curves, and shapes or polygons. Vector graphics can be magnified without pixalation. Current vector graphics formats are pdf, ps, and wmf.

By contrast bitmap (or raster) graphics devices define objects in terms of the set of pixels used in one specific magnification. Further magnification gives larger discrete dots and not smooth objects. Current bitmap formats are png, bmp, tif, and gif.

Figure L.1 panels a and c are drawn with a vector graphics device. They are clear and crisp at any magnification (two magnifications are shown here). Figure L.1 panels b and d are drawn with a bitmapped graphics device. Panel b is not clear, and the magified version in panel d is granular and fuzzy.

L.2.2 Aspect Ratio

The graphs are initially drawn for the device size they see at the time they are drawn. The *aspect ratio* (the ratio of the width to height in graphic units) is set at that time. Plotting symbols and text are positioned to look right at that initial magnification with that size device. The graphs in Figure L.1 honor the aspect ratio. Both the x and y dimensions are scaled identically in those panels.

Changing the aspect ratio after the graph has been drawn interferes with the message of the graph. It is most evident when both axes use the same units, but is visible even when the units are different.

The graph in Figure L.2 does not honor the aspect ratio, and the graph becomes very hard to read. The width is stretched to twice its initial size and the height is left at the original height. As a consequence, the circles used as plotting symbols are stretched to ellipses. The font used for the labels is stretched to visually change the shapes of the letters. About half of the zero characters look like circles. The

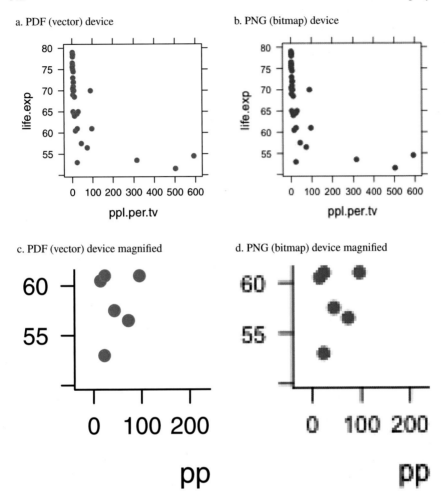

Fig. L.1 Panels a and b were drawn with the same command addressed to different devices. Panel a uses the `pdf` vector device and panel b uses the `png` bitmap device. Even at low magnification the difference between the two images is clear. The circle glyphs, the text, and the lines are crisp on the vector device. The circle glyphs, the text, and the lines are granular on the bitmap device. Panels c and d are the lower left corners of panels a and b magnified by a scale of 2. The vector display is just as crisp at this magnification. The bitmap display is more granular and fuzzy. Not even the straight lines are clear in panel d.

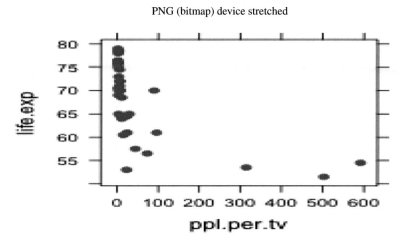

Fig. L.2 The figure here is the same plot shown in Figure L.1 panel b, this time with its *x*-dimension stretched. The circular glyphs are now ellipses. The thin numeral "0" are now circles. The vertical straight lines are even fuzzier than before.

other half look different because the pixel break points are placed differently on the underlying letter shapes.

It is possible to damage the aspect ratio even with a good typographic system (I intentionally did so in Figure L.2 using LATEX). Under normal circumstances in LATEX the aspect ratio is retained.

It is too easy to damage the aspect ratio with a drag-and-drop editing system by stretching an image at the top, bottom, or sides of an image. It is also easy to maintain the aspect ratio when controlling the size of an image, by stretching only at the corners, and never stretching the top, bottom, or sides of an image.

L.2.3 Other Features

Other features to be aware of were discussed and illustrated in Chapter 4 and are summarized here.

1. In scatterplot matrices, a NW–SE main diagonal has a consequence of multiple axes of symmetry that interfere with easy reading. The (SW–NE) main diagonal (the defaul in `splom` has a single axis of symmetry both within and between panels. See the examples and discussion in Section 4.7 for more detail.

2. The panels in a scatterplot matrix should be square to emphasize the equivalence of rows and columns. The intention of the scatterplot matrix is the comparison of

y ˜ x with x ˜ y, and maintaining the same scale in both orientations facilitates the comparison of reflected panels. Compare Figures 4.10 and 4.11.

3. Choice of plotting symbols, open circles, closed circles, other symbols (triangles, squares, etc), letters. See the itemized discussion in Section 4.1.

4. Choice of colors. See Section 4.10 for discussion of color vision.

5. Location of ticks and labels inside or outside the panels. In scatterplot matrices, placing ticks and labels inside the main diagonal makes the frees more surface area for the panels themselves. Compare Figures 4.10 and 4.11.

6. Space between panels. Compare Figures 4.10 and 4.11.

L.3 English Writing Style

1. Check spelling

   - Select the right homophone (words that sound alike): brake vs break. Both spellings will be accepted by a spell-checking program.

   - Learn to spell technical words correctly. The following words seem to be particularly liable to misspelling:

     - separate: The fourth letter is "a".

     - correlation: The letter "r" is doubled.

     - collinear: The letter "l" is doubled.

     - stationary: not moving

     - stationery: writing paper and envelopes

     - symmetric: The letter "m" is doubled.

     - asymmetric: The letter "s" is single.

     - Tukey: John W. Tukey

     - turkey: a bird

   - "p-value" is preferred (with p in math italic). "P-value" is not OK (with "P" in uppercase roman).

   - Spell people's names correctly and with proper capitalization (John W. Tukey, Dennis Cook).

2. Punctuation.

   "." ":" "," ";" always touch the preceding character. They always have a space after them.

L.4 Programming Style and Common Errors

1. Data entry: Use real variable names. The default variable names "X_1" and "X_2" carry no information. Variable names like "height" and "weight" carry information.

2. Data entry: probably you don't need to do that step. Don't reenter by hand the numbers that you already have in machine-readable form.

3. Use `dump("varname","")` to get ASCII versions of R variables that can be read back into R with no loss (of digits, labeling, attributes). The output from the dump can be copied into email and copied back from email. The output from the simpler print commands will frequently get garbled in email. See Table L.6 for an illustration where the original class of two of the columns is lost when we neglected to use the `dump` function.

4. The R function `splom()` for scatterplot matrices by default gives easy-to-read plots with a single axis of symmetry over the entire set of square panels. The R `pairs()` function (for all pairwise two-variable graphs) by default gives many conflicting axes of symmetry and rectangular panels. See Figure 4.12 and the accompanying discussion.

5. Analyze the experiment given you. Don't ignore the block factor. Usually the block sum of squares is removed first, before looking at the treatment effects. In R, this is done by placing the block factor first in the model formula, for example, in Section 12.9 we use `aov(plasma ~ id + time)` so the sequential analysis of variance table will read

```
id
time
Residuals
```

This way, in non-balanced designs, the sequential sum of squares for the treatment factor (`time` in this example) is properly adjusted for the blocking factor `id`.

6. We normally recommend the use of the R command language, not a menu system, when you are learning the techniques. You will get

- much better-looking output
- more control
- the ability to reproduce what you did

7. When GUI point-and-click operations have been used to construct preliminary graphical (or tabular) views of the data, the commands corresponding to these operations are frequently displayed. **Rcmdr** for example displays the generated commands in its R Script window (see Figure C.14 for an example). These commands can then be used as components in the construction of more complex commands needed to produce highly customized graphs.

Table L.6 We construct a data.frame and then display it twice, once by typing the variable name, the second time by using the dump function. When we re-enter the typed text back to R, we lose the structure of the data.frame. When we re-enter the dumped structure, we retain the original structure.

```
> tmp <- data.frame(aa=1:3, bb=factor(4:6), cc=letters[7:9],
+                        dd=factor(LETTERS[10:12]), stringsAsFactors=FALSE)

> str(tmp)
'data.frame': 3 obs. of  4 variables:
 $ aa: int  1 2 3
 $ bb: Factor w/ 3 levels "4","5","6": 1 2 3
 $ cc: chr  "g" "h" "i"
 $ dd: Factor w/ 3 levels "J","K","L": 1 2 3

> tmp
  aa bb cc dd
1  1  4  g  J
2  2  5  h  K
3  3  6  i  L

> dump("tmp", "")
tmp <-
structure(list(aa = 1:3, bb = structure(1:3, .Label = c("4",
"5", "6"), class = "factor"), cc = c("g", "h", "i"),
dd = structure(1:3, .Label = c("J",
"K", "L"), class = "factor")), .Names = c("aa", "bb", "cc", "dd"
), row.names = c(NA, -3L), class = "data.frame")

> tmp <- read.table(text="
+    aa bb cc dd
+ 1  1  4  g  J
+ 2  2  5  h  K
+ 3  3  6  i  L
+ ", header=TRUE)

> sapply(tmp, class)
        aa        bb        cc        dd
"integer" "integer"  "factor"  "factor"

> tmp <-
+ structure(list(aa = 1:3, bb = structure(1:3, .Label = c("4",
+ "5", "6"), class = "factor"), cc = c("g", "h", "i"),
+ dd = structure(1:3, .Label = c("J",
+ "K", "L"), class = "factor")), .Names = c("aa", "bb", "cc", "dd"
+ ), row.names = c(NA, -3L), class = "data.frame")

> sapply(tmp, class)
          aa          bb          cc          dd
   "integer"    "factor" "character"    "factor"
```

8. Store results of an R function call in a variable to permit easy extraction of various displays from the results. For example,

```
mydata <- data.frame(x=1:6, y=c(1,4,2,3,6,2))

my.lm <- lm( y ~ x , data=mydata)

summary(my.lm) ## summary() method on lm argument
old.mfrow <- par(mfrow=c(2,2)) ## four panels on the graphics device
plot(my.lm)     ## plot() method on lm argument
par(old.mfrow)                  ## restore previous arrangement
coef(my.lm)     ## coef() method on lm argument
anova(my.lm)    ## anova() method on lm argument
resid(my.lm)    ## resid() method on lm argument
predict(my.lm) ## predict() method on lm argument
```

9. Analysis of Variance *requires* that the classification factor be declared as a factor. Otherwise you will get a nonsense analysis. The wrong degrees of freedom for a treatment effect is usually the indicator that you forgot the factor(treatment) command in R.

10. The degrees of freedom for a problem always comes from the Residual or ERROR line of the ANOVA table. In multiple-stratum models, the Residual line in each stratum provides the comparison value (denominator Mean Square and degrees of freedom) for effects in that stratum.

11. Please use par(mfrow=c(2,2)) (as illustrated above in item 8) for plotting the results of an lm() or aov(). That way the plot uses only one piece of paper, not four.

12. All comparable graphs must be on the same scale—on the same axes is often better. See Section 17.6.2, especially Figure 17.19, for an example of comparable scaling in the top panels and noncomparable scaling in the bottom panels. See Figure 4.9 for comparable scaling in Panels a and b and noncomparable scaling in Panels c, d, and e.

L.5 Presentation of Results

This list is designed for our classroom setting. It is more generally applicable.

1. Use the minus sign "−4" in numbers. We do not accept hyphens "-4". See Table L.4.

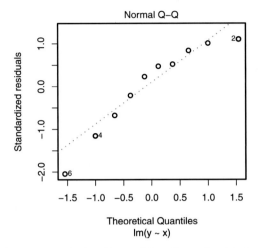

Fig. L.3 `qqplot` of a regression of noise. The usual three "largest" residuals are labeled, and none of the labeled residuals are big.

2. For multiple comparisons we can use the MMC (Section 7.2) default of Tukey pairwise comparisons. The `mmc` and `mmcplot` functions in the **HH** package are built on the `glht` function in the **multcomp** package.

3. We don't do multiple comparisons of blocks. We know there are differences. That is why we chose to use that factor as a block, not as a treatment. See Section 12.6.

4. Write an experiment description that tells the reader how to reproduce the experiment.

5. Distinguish between "bigger than the others" and "big". The `plot.lm()` labels the three biggest points. It doesn't care if they are significantly big. In Figure L.3, for example, the `qqplot` of a regression of random noise, the usual three "largest" residuals are labeled, and none of the labeled residuals are big.

6. `summary.lm(...)` doesn't usually provide interesting information in designed experiments. `summary.aov(..., split=list())` is frequently interesting. The ANOVA table in Table 12.13, for example, shows the partition of the 10-df "strain nested within combination" into two easily interpretable 5-df sums of squares, "strain within clover" and "strain within clover+alfalfa". It is easy to interpret these partitioned sums of squares along with the interaction means at the bottom of Table 12.12 and in the upper-left panel of Figure 12.12.

   Had we used `summary.lm` we would have gotten information in Table 12.15 on the regression coefficients for the dummy variables, and we would need to see the dummy variables in Table 12.14 for the coefficients themselves to make any sense.

7. Always state the conclusions in the context of the problem.

8. Please do not turn in impossible or illegal formatting. For example, the line breaks in the middle of words or character strings in Table L.7 are unacceptable.

Table L.7 Impossible formatting in R output.

```
## line break in the middle of a word
Residual standard error: 0.4811 on 10 degrees of fre
dom

## wrong: line break in string
plot(y ~ x, main="abc
def")

plot(y ~ x,                    ### correct
     main="abc def")
```

9. Do not turn in lists or tables of data values in homework assignments. We know the data values. In the classroom situation we gave them to you. You may show a few observations to verify that you have read the data correctly. For example:

```
data(gunload)
head(gunload)
```

10. On the other hand, plots of the data are very interesting. We usually expect to see appropriate plots of the data (scatterplot matrix, interaction plots, sets of boxplots) and of the analysis.

11. We do not want a cover page for homework. This is our personal style. A class-full of essentially empty cover pages weighs too much and wastes paper.

12. Use spacing for legibility, for example:

```
abc<--5+3        is hard to read.
abc <- -5 + 3    is easy to read.
```

13. When you copy output, particularly by mouse from a document in a monowidth font to one with a default proportionally spaced font, make sure you keep the original spacing and indentation.

14. Short, complete answers are best.

Appendix M

Accessing R Through a Powerful Editor —With Emacs and ESS as the Example

This Appendix is a discussion of the use of a powerful editor and programming environment. We use Emacs terminology and examples because we are most familiar with it—we use Emacs with ESS as our primary editing environment. One of us (RMH) is a coauthor of ESS.

Much of the discussion applies with only small changes to use of many of the other high-quality editors. See Grosjean (2012) for an annotated list (with links) of other editors that are used in programming R.

Emacs (Stallman, 2015) is a mature, powerful, and easily extensible text editing system freely available under the GNU General Public License for a large number of platforms, including Linux, Macintosh, and Windows. Emacs shares some features with word processors and, more importantly, shares many characteristics with operating systems. Most importantly, Emacs can interact with and control other programs either as subprocesses or as cooperating processes.

The name "Emacs" was originally chosen as an acronym for *Editor MACroS*. Richard M. Stallman got a MacArthur genius award in 1990 for the development of Emacs. Emacs comes from the Free Software Foundation, also known as the GNU project (GNU is Not Unix).

Emacs provides facilities that go beyond simple insertion and deletion: viewing two or more files at once (see Figures M.1 and M.2); editing formatted text; visual comparison of two similar files (Figure M.1); and navigation in units of characters, words, lines, sentences, paragraphs, and pages. Emacs knows the syntax of each programming language. It can provide automatic indentation of programs. It can highlight with fonts or colors specified syntactic characteristics. Emacs is extensible using a dialect of Lisp (Chassell, 1999; Graham, 1996). This means that new functions, with user interaction, can be written for common and repeated text editing tasks.

ESS (Mächler et al., 2015; Rossini et al., 2004) extends Emacs to provide a functional, easily extensible, and uniform interface for multiple statistical packages.

© Springer Science+Business Media New York 2015   851
R.M. Heiberger, B. Holland, *Statistical Analysis and Data Display*,
Springer Texts in Statistics, DOI 10.1007/978-1-4939-2122-5

One of us (RMH) is a coauthor of **ESS**. Several of the other coauthors are members of R-Core, the primary authors of R itself. Currently **ESS** works with R, S+, SAS, Stata, OpenBUGS/JAGS, and Julia. The online documentation includes an introduction in file `ESS/ess/doc/intro.pdf` (an early version of Rossini et al. (2004)). Online help is available from within Emacs and **ESS**.

M.1 Emacs Features

M.1.1 Text Editing

Most programming and documentation tasks fall under the realm of text editing. This work is enhanced by features such as contextual highlighting and recognition of special reserved words appropriate to the programming language in use. In addition, editor behaviors such as folding, outlining, and bookmarks can assist with maneuvering around a file. We discuss in Appendix K the set of capabilities we expect a text editing program to have. Emacs automatically detects mismatched parentheses and other types of common syntax and typing mistakes.

Typesetting and word processing, which focus on the presentation of a document, are tasks that are not pure text editing. Emacs shares many features with word processing programs and cooperates with document preparation systems such as LATEX (discussed in Section N) and HTML (discussed in Appendix E).

We strongly recommend that students in our graduate statistics classes use Emacs as their primary text editor. The primary reason for this recommendation is that Emacs is the first general editor we know of that fully understands the `syntax` and formatting rules for the statistical language R that we use in our courses. Other editing systems designed to work with R are described and linked to in the webpage provided by Grosjean (2012). Emacs has many other advantages (listed above), as evidenced by Richard Stallman having won a MacArthur award in 1992 for developing Emacs.

M.1.2 File Comparison

Visual file comparisons are one of the most powerful capabilities provided by Emacs. Emacs' `ediff` function builds on the standard Unix `diff` command and is therefore immediately available for Linux and Macintosh users. Windows users must first install `Rtools` and place `Rtools` in the PATH (see Section F.6).

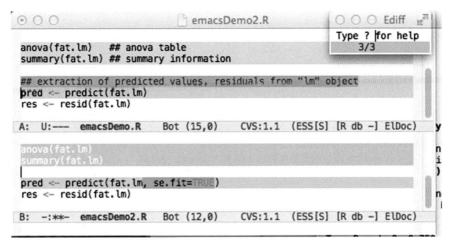

Fig. M.1 `ediff` of two similar files. All mismatches between the two files are detected and highlighted. The `ediff` control frame shows that we are currently in the third of three detected differences between the two files. The matching sections of the third chunk are highlighted in light pink in the top buffer and light green in the bottom buffer. The mismatching sections of the third chunk are highlighted in darker pink in the top buffer and darker green in the bottom buffer.

For a simple example, let us compare the current version of our R script for a homework exercise with the first version we started yesterday. Figure M.1 shows the comparison.

M.1.3 Buffers

Emacs can work with many multiple files simultaneously. It brings each into a *buffer*. A buffer is copy of a file within the Emacs editor. Any editing changes made to the contents of a buffer are temporary until the buffer is saved back into the file system. A buffer can hold a file, a directory listing, an interactive session with the operating system, or an interactive instance of another running program. Emacs allows you to open and edit an unlimited number of files and login sessions simultaneously, running each in its own buffer. The files or login sessions can be local or on another computer, anywhere in the world. You can run simultaneous multiple sessions. The size of a buffer is limited only by the size of the computer. One of us (RMH) normally has several buffers visible and frequently has hundreds of open buffers (several chapters, their code files, their transcript files, the console buffer for R, help files, directory listings on remote computers, handouts for classes, and a listing of the currently open buffers).

M.1.4 Shell Mode

Emacs includes a *shell mode* in which a terminal interaction runs inside an Emacs buffer. The Unix terminology for the program that runs an interactive command line session is a "shell". There are several commonly used shell programs: sh is the original and most fundamental shell program. Other Unix shell programs are csh and bash. The MS-DOS prompt window (c:/Windows/System32/cmd.exe) is the native shell program in MS Windows. We usually use the sh included in Rtools as our shell under MS Windows.

A terminal interaction running inside an Emacs buffer is much more powerful than one run in an ordinary terminal emulator window. The entire live login session inside an Emacs buffer is just another editable buffer (with full search capability). The only distinction is that both you and the computer program you are working with can write to the buffer. This is exceedingly important because it means nothing ever rolls off the top of the screen and gets lost. Just roll it back. The session can be saved to a file and then is subject to automatic backup to protect you from system crash or loss of connection to a remote machine.

M.1.5 Controlling Other Programs

A shell running in an Emacs buffer is normally used to run another program (R for example). Frequently we can drop the intermediate step and have Emacs run the other program directly. **ESS** provides that capability for R. The advantage of running R directly through Emacs is that it becomes possible to design the interactivity that allows a buffer containing R code to send that code directly to the running R process.

ESS (see Section M.2) builds on shell mode to provide modes for interacting with statistical processes. The terminal interaction can be local (on the same computer on which Emacs is running) or remote (anywhere else).

M.2 ESS

Figure M.2 is a screenshot showing an interactive Emacs session. Emacs is a powerful program, hence the figure is quite dense. We discuss many of its components in the following subsections.

The discussion here is based on Rossini et al. (2004). **ESS** provides:

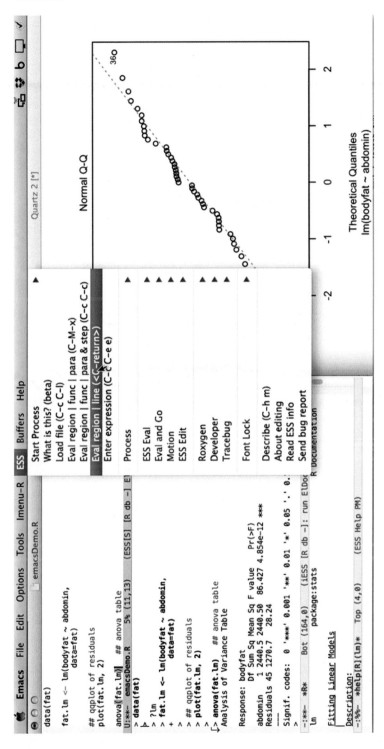

Fig. M.2 Interaction with R process through Emacs with ESS. An Emacs frame with three buffers is shown on the left. The top buffer shows the R file emacsDemo.R that we are executing one line at a time. The cursor is at the end of the line anova(fat.lm), with the closing right paren and its matching left paren highlighted together. The middle buffer *R* shows the R process with the printed result from the anova(fat.lm) command. The down arrow ⇩ in the left margin shows the beginning of the output from the most recently executed line. The bottom buffer *help[R](lm)* shows the help file from the R command ?lm. In the center is the **ESS** menu. On the right is the graph drawn by the plot(fat.lm, 2) statement. More detail on this figure is presented in Section M.2.

M.2.1 Syntactic Indentation and Color/Font-Based Source Code Highlighting

The **ESS** interface includes a description of the syntax and grammar of each statistical language it knows about. This gives **ESS** the ability to edit the programming language code, often more smoothly than with editors distributed with the languages. The process of programming code is enhanced as **ESS** provides the user with a clear presentation of the code with syntax highlighting to denote assignment, reserved words, strings, and comments. The upper-left buffer (labeled emacsDemo.R) in Figure M.2 contains an R script file. The mode-line for the buffer is in the "mode-line" face indicating that this is the active buffer. The mode-lines for the other, inactive, buffers in this figure are in the lighter-colored "mode-line-inactive" face. We can tell from the ** in the mode-line that the file is still under construction. Those two symbols would be replaced by -- once the file is saved.

In order to illustrate syntactic indentation, we show the statement fat.lm <- lm(bodyfat ~ abdomin, data=fat) on two lines in Figure M.2. When we entered the <RET> character after the comma, **Emacs** automatically indented the second line of the statement and placed the cursor aligned with the first character after the open paren.

The assignment arrow, a keyword in the language, is detected and colored in the "constant" face. Comments indicated by "##" are colored in the "comment" face. The cursor, represented by a blinking "|", is redundantly located by the (11,13) (indicating row and column in the file) in the mode-line. In this snapshot the cursor immediately follows a closing paren ")", hence both the closing paren and its matching opening paren "(" are highlighted in the paren-match face. Mismatched parens, as in Figure M.3, are shown in the paren-mismatch face.

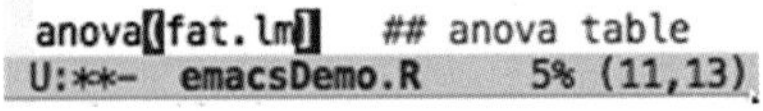

Fig. M.3 Mismatched parens are shown in a glaring color to help protect the user from many types of typographical errors.

M.2.2 Partial Code Evaluation

Emacs/ESS can send individual lines, entire function definitions, marked regions, and whole edited buffers from the window in which the code is displayed for editing to the statistical language/program for execution. **Emacs/ESS** sends the code directly to the running program and receives the printed output back from the program in an editable **Emacs** buffer (the *R* buffer). This is a major improvement over cut-and-paste as it does not require switching buffers or windows.

We show this twice in Figure M.2. The dropdown menu is set to show the various options on which subset of the buffer will be sent over. The menu also shows the key-stroke combinations that can be typed directly and thereby avoid using the menu at all. The cursor is on the line anova(fat.lm) and that line was sent over by this command.

The *R* buffer has just received the command (the hooked down arrow ∫ in the left margin shows the beginning of the output from the most recently executed line). The command and its output are displayed. In addition to receiving and executing lines sent over from the script file, the *R* buffer is also an ordinary R console and the user can type directly into the *R* buffer.

ESS facilitates the editing of files of R scripts by providing a means for loading and error-checking of small sections of code (as illustrated in Figure M.2). This allows for source-level debugging of batch files.

M.2.3 Object Name Completion

In addition, for languages in the S family (S developed at Bell Labs, S+ from TIBCO, and R) **ESS** provides object-name completion of user- and system-defined functions and data, and filename completion for files in the computer.

M.2.4 Process Interaction

ESS builds on **Emacs**'s facilities to interface directly with the running R process. The output of the package goes directly to the editable *R* text buffer in **Emacs**. **Emacs** has historically referred to processes under its control as "inferior", accounting for the name inferior **ESS** (iESS) shown in the mode-line of the *R* buffer in Figure M.2. This mode allows for command-line editing and for recalling and searching the history of previously entered commands. Filename completion, and object-name and function-name completion are available. Transcripts are easily recorded and can be edited into an ideal activity log, which can then be saved. **ESS** intercepts calls to the internal help system for R and displays the help files inside an **Emacs** buffer (see the bottom left buffer in Figure M.2).

M.2.5 Interacting with Statistical Programs on Remote Computers

ESS provides the facility to edit and run programs on remote machines in the same session and with the same simplicity as if they were running on the local machine.

M.2.6 Transcript Editing and Reuse

Once a transcript log is generated, perhaps by saving an iESS buffer, transcript mode assists with reuse of part or all of the entered commands. It permits editing and re-evaluating the commands directly from the saved transcript. This is useful for demonstration of techniques as well as for reconstruction of data analyses. There currently exist functions within **ESS** for "cleaning" transcripts from the R, S+, and SAS languages back to source code by finding all input lines and isolating them into an input file.

By default transcript files (files with extensions *.Rout *.rt *.Sout *.st) open into read-only buffers. The intent is to protect the history of your analysis sequence. If you need to make them writable, use C-x C-q.

M.2.7 Help File Editing (R)

ESS also provides an interface for writing help files for R functions and packages (see Appendix F). It provides the ability to view and execute embedded R source code directly from the help file in the same manner as **ESS** normally handles code from a source file. ESS Help mode provides syntax highlighting and the ability to submit code from the help file to a running R process.

M.3 Learning Emacs

There are several ways to learn Emacs. Whichever you choose, it will give you access to the most comprehensive editing system.

M.3.1 GUI (Graphical User Interface)

Emacs provides a GUI (graphical user interface) with dropdown menus (shown in Figure M.2) or with toolbar icons (not shown here). An excellent guide to menu-based Emacs usage, with the delightful title "Emacs has no learning curve: Emacs and **ESS**", is available in Johnson (2015). Detailed help on Emacs is available from the Help dropdown menu item.

M.3.2 Keyboard Interface

Emacs, in its present incarnation, goes back to 1984. A precursor goes back to 1976. One of us (RMH) started using Emacs about 1990. Emacs was originally designed in a keyboard-only environment, long before mice and multi-windowed "desktop" environments. We still prefer the keyboard as it is faster (once learned) and has more capabilities than the menu-based GUI options. The dropdown menus themselves show the keyboard equivalents. For keyboard-based usage use one or more of

1. Tutorial. Enter "C-h t". Then read the file and follow the instructions. You are working with your own private copy of the TUTORIAL file, so you can practice the keystrokes as suggested.

2. Manual. The manual is online in the hyperlinked Info system. It can be accessed from the menu or by entering "C-h i" to bring up the Info: buffer.

3. Help. You can search within the dropdown Help menu. Or, to find help apropos a topic, for example to answer the question "How do I save my current editing buffer to a file?", enter "C-h a save RET" and get a list of all commands with save as part of their name. You probably want the command save-buffer and will see that you can use that command by typing "C-x C-s" or by using the FILES pull-down menu.

4. Reference Card to Emacs keystroke commands. The Emacs reference card is in the directory accessed by "C-h r C-x d ../etc/refcards". The English card is in file refcard.pdf. Several other languages are also available as files with names **-refcard.pdf. Reference cards for other Free Software Foundation programs are also in the same directory.

M.4 Nuisances with **Windows** and **Emacs**

When R has been started by an ordinary user, as opposed to Administrator, then installed packages are placed into a personal library. The R GUI under **Windows** uses the directory
```
"C:/Users/loginname/Documents/R/win-library/x.y"
```
as the location of your personal library. The *R* buffer inside **Emacs** uses
```
"C:/Users/loginname/AppData/Roaming/R/win-library/x.y"
```
as the location of your personal library. The notation x.y must be replaced by the first two digits of your current version of R. For example, with R-3.3.0, "x.y" would become "3.3". Neither will automatically see packages that were installed into the directory used by the other. You must tell it about the other directory.

You can tell the R GUI to use the additional directory with the statement
```
.libPaths(
    "C:/Users/loginname/AppData/Roaming/R/win-library/x.y")
```

You can tell *R* to use the additional directory with the statement
```
.libPaths(
    "C:/Users/loginname/Documents/R/win-library/x.y")
```

M.5 Requirements

Emacs satisfies the requirements detailed in Section K.1.1.

1. **ESS** provides full interaction between the commands file, the statistical process, and the transcript.

2. The statistical process runs in an **Emacs** buffer and is therefore fully searchable and editable.

3. Cut and paste is standard.

4. **Emacs** comes with modes specialized for all standard computing languages (**C**, **C++**, **Fortran**, **VBA**). **ESS** provides the modes for R and S+ (and for **SAS** and several others). Each mode by default highlights all keywords in its language, is aware of recommended indentation patterns and other formatting issues, and has communication with its associated running program.

5. **Emacs** handles multiple files as part of its basic design.

6. **Emacs** has a LaTeX mode, and therefore provides editing access to the best mathematical typesetting system currently available anywhere.

7. **Emacs** has several text modes.

8. Several spell-check programs works with **Emacs**; we use ispell.

9. **Emacs** permits embedding of graphics directly into the **Emacs** buffer.

10. Graphics in PDF, PostScript, and bitmap formats can be embedded into LaTeX documents.

11. **Emacs** includes encoding (Unicode) for all human languages. Enter C-H h to display the HELLO file, which lists many languages and characters.

Appendix N
LaTeX

We used LaTeX (Lamport, 1994) as the document preparation system for writing this book. LaTeX knows all the intricacies of mathematical typesetting and makes it very easy to include figures and tables into the manuscript. LaTeX is the standard required by many statistics and mathematics journals. LaTeX can adapt to any standard style, such as those used by book publishers and journals, or you can write your own.

The LaTeX document preparation system is written as a collection of macros for Donald Knuth's TeX program (Knuth, 1984). LaTeX adds to TeX a collection of commands that simplify typesetting by letting the user concentrate on the structure of the text.

TeX is a sophisticated program designed to produce high-quality typesetting, especially for mathematical text. The TeX system was developed by Donald E. Knuth at Stanford University. It is now maintained by CTAN, the Comprehensive TeX Archiving Network (Comprehensive TeX Archiving Network, 2002). There are several distributions available. We use MikTeX (Schenk, 2001) on Windows and MacTeX (The TeX Users Group (TUG), 2014) on Macintosh.

The `latex` function in R (Heiberger and Harrell, 1994; Frank E Harrell et al., 2014) may be used to prepare formatted typeset tables for a LaTeX document. Several tables in this book (8.1, 9.1, 12.4, 15.5, and 15.8) were prepared in this way.

N.1 Organization Using LaTeX

There are several ways to approach LaTeX. The way we used for this book was to write each chapter in its own file, and then combine them into book using the style file provided by our publisher Springer.

© Springer Science+Business Media New York 2015
R.M. Heiberger, B. Holland, *Statistical Analysis and Data Display*,
Springer Texts in Statistics, DOI 10.1007/978-1-4939-2122-5

N.2 Setting Equations

This is equation 8.1 as we typed it:

```
\begin{eqnarray*}
y_i = \beta_0 + \beta_1 x_i   + \epsilon_i \cond{for $i=1,
      \dots,n$}
\end{eqnarray*}
```

This is how it appears when typeset:

$$y_i = \beta_0 + \beta_1 x_i + \epsilon_i \quad \text{for } i = 1, \dots, n$$

N.3 Coordination with R

The Second Edition of this book was written using the `Sweave` and `Stangle` functions from the **utils** package. All R code is included within the LATEX source for the book. `Stangle` reads the LATEX input files, isolates from the input files the actual code that produced the tables and figures, and collects the code into the script files that are included with the **HH** package. See `help(Sweave, package="utils")` for details on writing using `Sweave`.

N.4 Global Changes: Specification of Fonts

In LATEX, it is very easy to make global changes to a document style. For example, placing programs and transcripts into a `verbatim` environment takes care of the font. Should we later decide that we would like a different monowidth font, say "Adobe Courier" instead of the default "Computer Modern Typewriter", we include the single statement `\renewcommand{\ttdefault}{pcr}` and then immediately ALL instances of the typewriter font will change. We illustrate in Table N.1.

Compare this simple change of a single specification in LATEX to the more labor-intensive way of accomplishing the same task in a visual formatting systems (MS Word, for example). In a visual formatting system, programs and transcripts must be individually highlighted and then explicitly placed into Courier. Changing ALL instances of the typewriter font requires manually highlighting and changing EACH of them individually, one at a time.

Table N.1 Switching the typewriter font (LaTeX command \tt) in LaTeX from the default "Computer Modern Typewriter" to "Adobe Courier" and back.

```
This is the default Computer Modern Typewriter font.

%% switch to Adobe Courier
\renewcommand{\ttdefault}{pcr}

This is Adobe Courier font.

%% return to Computer Modern Typewriter
\renewcommand{\ttdefault}{cmtt}

And back to the Computer Modern Typewriter font.
```

Appendix O

Word Processors and Spreadsheets

Word processing is moving sentences, paragraphs, sections, figures, and cross-references around. Most word processors can be used as a text editor by manually turning off many of the word processing features.

Spreadsheet software is used to operate on tables of numbers.

Microsoft **Word** and Microsoft **Excel** are the most prevalent word processor and spreadsheet software systems. Most of what we say here applies to all such systems.

O.1 Microsoft Word

Microsoft **Word** is probably the most prevalent text editor and word processor today.

MS **Word** is configured by default as a word processor. It can be used as a text editor by changing those default settings. The most critical features are the font and the paragraph reflow. Courier (or another monowidth font in which all letters are equally wide) should be used for program writing or for summary reports in which your displayed output from R is included. The software output from R is designed to look right (alignment and spacing) with monowidth fonts. It looks terrible, to the point of illegibility, in proportional fonts. See examples in Section A.4 and Appendix L. Other word processor features to turn off are spell checking and syntax checking, both of which are designed to make sense with English (or another natural language) but not with programming languages.

© Springer Science+Business Media New York 2015
R.M. Heiberger, B. Holland, *Statistical Analysis and Data Display*,
Springer Texts in Statistics, DOI 10.1007/978-1-4939-2122-5

If you are using an editor that thinks it knows more than you, be very sure that the `*.r`, `*.Rout`, and `*.rt` files are displayed in Courier and fit within the margin settings of the word processor.

O.1.1 Editing Requirements

MS Word satisfies some of the requirements detailed in Section K.1.1.

1. MS Word can edit the commands file. It does not interact directly with the running statistical process; manual cut-and-paste is required.

2. The output from the statistical process is in a window independent of MS Word. The output can be picked up and pasted into an MS Word window.

3. Cut and paste is standard.

4. When checking is on, MS Word will by default inappropriately check computer programs for English syntax and spelling.

5. MS Word handles multiple files as part of its basic design.

6. Reports can be written and output text can be embedded into them;

7. MS Word has syntax and spell-checking facilities limited to the natural language (English in our case).

8. Graphics can be pasted directly into an MS Word document.

9. Mathematical formulas can be entered in an MS Word document.

10. SWord and other connections between R and MS Word.

11. MS Word works with Unicode, providing access to all character sets in all human languages. On the Insert tab, click Symbols and then More Symbols. In the font box, find Arial Unicode MS.

O.1.2 SWord

SWord (Baier, 2014) is an add-in package for MS Word that makes it possible to embed R code in an MS Word document. The R code will be automatically executed and the output from the R code will be included within the MS Word document. SWord is free for non-commercial use. Any other use will require a license. Please see Section D.1.3 for installation information.

O.2 Microsoft Excel

Microsoft Excel is probably the most prevalent spreadsheet program today.

We believe MS Excel is well suited for two tasks: as a small-scale database management system and as a way of organizing calculations. We do not recommend Excel for the actual calculations.

O.2.1 Database Management

R (and S+ and SAS) can read and write Excel files. Since many people within an organization collect and distribute their data in Excel spreadsheets, this is a very important feature.

O.2.2 Organizing Calculations

R can be connected directly to Excel on Windows via RExcel (Baier and Neuwirth, 2007) using DCOM, Microsoft's protocol for exchanging information across programs. See Appendix D for further details and for download information. Used in this way, Excel can be used similarly to the ways that Emacs and MS Word are used. Excel can be used to control R, for example by putting R commands inside Excel cells and making them subject to automatic recalculation. Or R can be in control, and use Excel as one of its subroutines.

O.2.3 Excel as a Statistical Calculator

We believe Excel is usually a poor choice for statistical computations because:

1. As of this writing (Excel 2013 for Windows and Excel 2011 for Macintosh) at least some of the built-in statistical functions do not include even basic numerical protection.

   Table O.1 and Figure O.1 show the calculation of the variance of three numbers by R's var function and Excel's VAR function. Figure O.1 shows a simple set of variance calculations where Excel reveals an algorithmic error. We show the erroneous number as a hex number in Table O.2. For comparison, Table O.1 shows the correct calculations by R.

B9	: ⊗ ⊘ (⚬ *fx*	=VAR(10^A9+1,10^A9+2,10^A9+3)
	A	**B**
1	K	VAR(10^K+1,10^K+2,10^K+3)
2	0	1
3	1	1
4	2	1
5	15	1
6	16	4
7	17	0
8	26	0
9	27	28334198897217900000000
10	28	0

Fig. O.1 Calculation of the variance by Excel for the sequence $(10^k) + 1, (10^k) + 1, (10^k) + 1$ for several values of k. The real-valued result for all values of k is exactly 1. The 2.833×10^{22} value shown for $k = 27$ is the indication of an erroneous algorithm in Excel. We see this for Excel 2013 on Windows and Excel 2011 on Macintosh. The floating point result for $k = 0 : 15$ is exactly 1 as shown. The correct floating point result for all $k >= 17$ is exactly 0. The floating point value 4 for $k = 16$ is correct. We illustrate the correct floating point behavior in Section G.12. Table O.2 shows the hexadecimal display for the erroneous value.

Table O.1 Calculation of the variance by R for the sequence $(10^k) + 1, (10^k) + 1, (10^k) + 1$ for several values of k. The real-valued result for all values of k is exactly 1. R gets the correct floating point variance for all values of k. The floating point result for $k = 0 : 15$ is exactly 1. The floating point value 4 for $k = 16$ is correct. The floating point result for $k >= 17$ is exactly 0. We illustrate the correct floating point behavior in Section G.12.

```
> k <- c(0, 1, 2, 15, 16, 17, 26, 27, 28)

> cbind(k=k, var=apply(cbind(10^k + 1, 10^k + 2, 10^k + 3), 1, var))
        k var
 [1,]   0   1
 [2,]   1   1
 [3,]   2   1
 [4,]  15   1
 [5,]  16   4
 [6,]  17   0
 [7,]  26   0
 [8,]  27   0
 [9,]  28   0
```

Table O.2 The first statement shows the hex display of the number Excel shows in Figure O.1 as the result of var($10^{27} + 1, 10^{27} + 2, 10^{27} + 3$). It is very close to a pretty hex number suggesting that there is a numeric overflow somewhere in the algorithm. The second statement shows the hex display of the three input numbers. They are all identical to machine precision.

```
> sprintf("%+13.13a", 28334198897217900000000)
[1] "+0x1.8000000000007p+74"

> as.matrix(sprintf("%+13.13a", c(10^27 + 1, 10^27 + 2, 10^27 + 3)))
      [,1]
[1,] "+0x1.9d971e4fe8402p+89"
[2,] "+0x1.9d971e4fe8402p+89"
[3,] "+0x1.9d971e4fe8402p+89"
```

Earlier versions of MS Excel got different wrong answers for the variance of three numbers. Most notably MS Excel 2002 gets the right answer for 10^7 and the wrong answer for 10^8, suggesting that it was using the numerically unstable one-pass algorithm. Later releases got different sets of wrong answers.

2. Most add-in packages are not standard and are not powerful. If add-ins are used along with an introductory textbook, they will most likely be limited in capability to the level of the text. They are unlikely to be available on computers in a work situation.

   RExcel is an exception. It uses R from the Excel interface and therefore has all the power and generality of R.

References

A.A. Adish, S.A. Esrey, T.W. Gyorkos, J. Jean-Baptiste, A. Rojhani, Effect of consumption of food cooked in iron pots on iron status and growth of young children: a randomised trial. Lancet **353**(9154), 712–716 (1999)

A. Agresti, *Categorical Data Analysis* (Wiley, New York, 1990)

A. Agresti, B. Caffo, Simple and effective confidence intervals for proportions and differences of proportions result from adding two successes and two failures. Am. Stat. **54**(4), 280–288 (2000)

Albuquerque Board of Realtors (1993). URL: `http://lib.stat.cmu.edu/DASL/Stories/homeprice.html`

American Statistical Association, Career Center (2015). URL: `http://www.amstat.org/careers`

O. Amit, R.M. Heiberger, P.W. Lane, Graphical approaches to the analysis of safety data from clinical trials. Pharm. Stat. **7**(1), 20–35 (2008). URL: `http://www3.interscience.wiley.com/journal/114129388/abstract`

A.H. Anderson, E.B. Jensen, G. Schou, Two-way analysis of variance with correlated errors. Int. Stat. Rev. **49**, 153–167 (1981)

R.L. Anderson, T.A. Bancroft, *Statistical Theory in Research* (McGraw-Hill, New York,1952)

V.L. Anderson, R.A. McLean, *Design of Experiments* (Marcel Dekker, New York,1974)

D.F. Andrews, A.M. Herzberg, *Data: A Collection of Problems from Many Fields for the Student and Research Worker* (Springer, New York,1985). URL: `http://lib.stat.cmu.edu/datasets/Andrews/`

E. Anionwu, D. Watford, M. Brozovic, B. Kirkwood, Sickle cell disease in a British urban community. Br. Med. J. **282**, 283–286 (1981)

© Springer Science+Business Media New York 2015
R.M. Heiberger, B. Holland, *Statistical Analysis and Data Display*,
Springer Texts in Statistics, DOI 10.1007/978-1-4939-2122-5

P.K. Asabere, F.E. Huffman, Negative and positive impacts of golf course proximity on home prices. Apprais. J. **64**(4), 351–355 (1996)

T. Baier, Sword (2014). URL: http://rcom.univie.ac.at

T. Baier, E. Neuwirth, Excel :: Com :: R. Comput. Stand. **22**(1), 91–108 (2007)

M.K. Barnett, F.C. Mead, A 2^4 factorial experiment in four blocks of eight. Appl. Stat. **5**, 122–131 (1956)

R.A. Becker, J.M. Chambers, A.R. Wilks, *The S Language; A Programming Environment for Data Analysis and Graphics* (Wadsworth & Brooks/Cole, Pacific Grove, 1988)

R.A. Becker, W.S. Cleveland, S-PLUS Trellis Graphics User's Manual (1996a). URL: http://www.stat.purdue.edu/~wsc/papers/trellis.user.pdf

Becker, R.A., W.S. Cleveland, M.-J. Shyu, S.P. Kaluzny, A tour of trellis graphics (1996b). URL: http://www2.research.att.com/areas/stat/doc/95. 12.color.ps

Y.Y.M. Bishop, S.E. Fienberg, P.W. Holland, *Discrete Multivariate Analysis* (MIT, Cambridge,1975)

A. Bjork, Solving least squares problems by Gram–Schmidt orthogonalization. BIT **7**, 1–21 (1967)

C.I. Bliss, *Statistics in Biology* (McGraw-Hill, New York, 1967)

C.R. Blyth, On Simpson's paradox and the sure-thing principle. J. Am. Stat. Assoc. **67**, 364–366 (1972)

B.L. Bowerman, R.T. O'Connell, *Linear Statistical Models* (Duxbury, Belmont, 1990)

G.E.P. Box, D.R. Cox, An analysis of transformations. J. R. Stat. Soc. B **26**, 211–252 (1964)

G.E.P. Box, W.G. Hunter, J.S. Hunter, *Statistics for Experimenters* (Wiley, New York, 1978)

G.E.P. Box, G.M. Jenkins, *Time Series Analysis: Forecasting and Control*, revised edn. (Holden-Day, San Francisco, 1976)

R.G. Braungart, Family status, socialization and student politics: a multivariate analysis. Am. J. Sociol. **77**, 108–130 (1971)

C. Brewer, Colorbrewer (2002). URL: http://colorbrewer.org

L. Brochard, J. Mancebo, M. Wysocki, F. Lofaso, G. Conti, A. Rauss, G. Simonneau, S. Benito, A. Gasparetto, F. Lemaire, D. Isabey, A. Harf, Noninvasive ventilation for acute exacerbations of chronic pulmonary disease. N. Engl. J. Med. **333**(13), 817–822 (1995)

D.G. Brooks, S.S. Carroll, W.A. Verdini, Characterizing the domain of a regression model. Am. Stat. **42**, 187–190 (1988)

B.W. Brown Jr., Prediction analyses for binary data, in *Biostatistics Casebook*, ed. by R.G. Miller Jr., B. Efron, D.W. Brown Jr., L.E. Moses (Wiley, New York, 1980)

M.B. Brown, A.B. Forsyth, Robust tests for equality of variances. J. Am. Stat. Assoc. **69**, 364–367 (1974)

Bureau of the Census, *Statistical Abstract of the United States* (U.S. Department of Commerce, Washington, DC, 2001)

T.T. Cai, One-sided confidence intervals in discrete distributions. J. Stat. Plan. Inference **131**, 63–88 (2003). URL: http://www-stat.wharton.upenn.edu/~tcai/paper/1sidedCI.pdf

E. Cameron, L. Pauling, Supplemental ascorbate in the supportive treatment of cancer: re-evaluation of prolongation of survival times in terminal human cancer. Proc. Natl. Acad. Sci. USA **75**, 4538–4542 (1978)

J.M. Chambers, W.S. Cleveland, B. Kleiner, P.A. Tukey, *Graphical Methods for Data Analysis* (Wadsworth, Belmont, 1983)

T.F.C. Chan, G.H. Golub, R.J. LeVeque, Algorithms for computing the sample variance: analysis and recommendations.Am. Stat. **37**(3), 242–247 (1983)

W. Chang, J. Cheng, J.J. Allaire, Y. Xie, J. McPherson, *shiny: Web Application Framework for R. R Package Version 0.11.1* (2015). URL: http://CRAN.R-project.org/package=shiny

R. Chassell, *Programming in Emacs Lisp: An Introduction*. Free Software Foundation, 2nd edn. (1999) URL: https://www.gnu.org/software/emacs/manual/html_node/eintr/

T.W. Chin, E. Hall, C. Gravelle, J. Speers, The influence of Salk vaccination on the epidemic pattern and spread of the virus in the community. Am. J. Hyg. **73**, 67–94 (1961)

S. Chu, Diamond ring pricing using linear regression. J. Stat. Educ. **4** (1996). URL: http://www.amstat.org/publications/jse/v4n3/datasets.chu.html

W.S. Cleveland, *Visualizing Data* (Hobart Press, Summit, 1993)

W.S. Cleveland, R. McGill, Graphical perception: theory, experimentation, and application to the development of graphical methods. J. Am. Stat. Assoc. **79**, 531–554 (1984)

W.G. Cochran, G.M. Cox, *Experimental Designs*, 2nd edn. (Wiley, New York, 1957)

D.R. Collett, *Modelling Binary Data* (Chapman & Hall, London, 1991)

Comprehensive TeX Archiving Network CTAN (2002). URL: ftp://metalab.unc.edu/pub/packages/TeX/index.html

W.J. Conover, M.E. Johnson, M.M. Johnson, A comparative study of tests for homogeneity of variances, with applications to the outer continental shelf bidding data. Technometrics **23**, 351–361 (1981)

Consumer Reports, Hot dogs, Consumer Reports (1986), pp. 366–367. URL: http://lib.stat.cmu.edu/DASL/Stories/Hotdogs.html

R.D. Cook, S. Weisberg, *Applied Regression Including Computing and Graphics* (Wiley, New York, 1999)

T. Cook, *Convictions for Drunkenness* (New Society, 1971)

CRAN, *Comprehensive R Archive Network* (2015). URL: http://CRAN.R-project.org

Cytel Software Corporation, *Logxact Statistical Software: Release 5* (2004). URL: http://www.cytel.com/LogXact/logxact_brochure.pdf

S.R. Dalal, E.B. Fowlkes, B. Hoadley, Risk analysis of the space shuttle: pre-challenger prediction of failure. J. Am. Stat. Assoc. **84**, 945–957 (1989)

C. Darwin, *The Effect of Cross- and Self-fertilization in the Vegetable Kingdom*, 2nd edn. (John Murray, London, 1876)

Data Archive, J. Stat. Educ. (1997). URL: http://www.amstat.org/publications/jse/jse_data_archive.html

O.L. Davies, *Design and Analysis of Industrial Experiments* (Oliver and Boyd, London, 1954)

O.L. Davies, P.L. Goldsmith (eds.), *Statistical Methods in Research and Production*, 4th edn. (Oliver and Boyd, London, 1972)

M.M. Desu, D. Raghavarao, *Nonparametric Statistical Methods for Complete and Censored Data*, 1st edn. (Chapman & Hall, Boca Raton, 2003)

R. Dougherty, A. Wade (2006).URL: http://vischeck.com

D. Edwards, J.J. Berry, The efficiency of simulation-based multiple comparisons. Biometrics **43**, 913–928 (1987)

M.E. Ellis, K.R. Neal, A.K. Webb, Is smoking a risk factor for pneumonia for patients with chickenpox? Br. Med. J. **294**, 1002 (1987)

J.D. Emerson, M.A. Stoto, Transforming data, in *Understanding Robust and Exploratory Data Analysis*, ed. by D.C. Hoaglin, F. Mosteller, J.W. Tukey (Wiley, New York, 1983)

L.W. Erdman, Studies to determine if antibiosis occurs among Rhizobia: 1. Between *Rhizobium meliloti* and *Rhizobium trifolii*. J. Am. Soc. Agron. **38**, 251–258 (1946)

J.L. Fleiss, *Statistical Methods for Rates and Proportions*, 2nd edn. (Wiley, New York, 1981)

Forbes Magazine (1993). URL: `http://lib.stat.cmu.edu/DASL/Datafiles/ceodat.html`

R.S. Fouts, Acquisition and testing of gestural signs in four young chimpanzees. Science **180**, 978–980 (1973)

J. Fox, *Regression Diagnostics: An Introduction* (Sage, Thousand Oaks, 1991)

J. Fox, The R Commander: a basic statistics graphical user interface to R. J. Stat. Softw. **14**(9), 1–42 (2005). URL: `http://www.jstatsoft.org/v14/i09`

E.H. Frank Jr., with contributions from Charles Dupont, and many others, *Hmisc: Harrell Miscellaneous. R Package Version 3.14-6* (2014). URL: `http://CRAN.R-project.org/package=Hmisc`

Free Software Foundation, Emacs (2015). URL: `http://www.gnu.org/software/emacs/`

R.J. Freund, R.C. Littell, *SAS System for Regression* (SAS Institute, Inc., Cary, 1991)

J.H. Goodnight, Tests of hypotheses in fixed effects linear models. Technical Report R-101 (SAS Institute, Cary, 1978)

P. Graham, *ANSI Common Lisp* (Prentice Hall, Upper Saddle River, 1996)

M.J. Greenacre, *Theory and Applications of Correspondence Analysis* (Academic, New York, 1984)

P. Grosjean, Ide/script editors. Annotated list with links to many editing environments for R. (2012). URL: `http://www.sciviews.org/_rgui/`

R.F. Gunst, R.L. Mason, *Regression Analysis and Its Application: A Data-Oriented Approach* (Marcel Dekker, New York, 1980)

L.C. Hamilton, Saving water: a causal model of household conservation. Sociol. Perspect. **26**(4), 355–374 (1983)

L.C. Hamilton, *Regression with Graphics* (Brooks-Cole, Belmont, 1992)

D.J. Hand, F. Daly, A.D. Lunn, K.J. McConway, E. Ostrowski, *A Handbook of Small Data Sets* (Chapman and Hall, London, 1994)

D.D. Harrison, D.E. Harrison, T.J. Janik, R. Cailliet, J.R. Ferrantelli, J.W. Hass, B. Holland, Modeling of the sagittal cervical spine as a method to discriminate hypo-lordosis: results of elliptical and circular modeling in 72 asymptomatic subjects, 52 acute neck pain subjects, and 70 chronic neck pain subjects. *Spine* **29**(22):2485–2492 (2004)

D.E. Harrison, R. Cailliet, D.D. Harrison, T.J. Janik, B. Holland, Changes in sagittal lumbar configuration with a new method of extension traction combined with spinal manipulation and its clinical significance: non-randomized clinical control trial. Arch. Phys. Med. Rehabil. **83**(11), 1585–1591 (2002)

R.M. Heavenrich, J.D. Murrell, K.H. Hellman, *Light Duty Automotive Technology and Fuel Economy Trends through 1991* (U.S. Environmental Protection Agency, Ann Arbor, 1991)

R.M. Heiberger, *Computation for the Analysis of Designed Experiments* (Wiley, New York, 1989)

R.M. Heiberger, *HH: Statistical Analysis and Data Display: Heiberger and Holland. Spotfire S+ Package Version 2.1-29* (2009). URL: `http://csan.insightful.com/PackageDetails.aspx?Package=HH`

R.M. Heiberger, *HH: Statistical Analysis and Data Display: Heiberger and Holland. R Package Version 3.1-15* (2015). URL: `http://CRAN.R-project.org/package=HH`

R.M. Heiberger, F.E. Harrell Jr., Design of object-oriented functions in S for screen display, interface and control of other programs (SAS and LATEX), and S programming, in *Computing Science and Statistics*, vol. 26 (1994), pp. 367–371. The software is available in both S-Plus and R in `library(hmisc)`. It is included in the S-Plus distribution. It may be downloaded for R from the `contrib` page of the R Development Core Team (2004) website

R.M. Heiberger, B. Holland, *Statistical Analysis and Data Display: An Intermediate Course with Examples in S-Plus, R, and SAS*, 1st edn. (Springer, New York, 2004)

R.M. Heiberger, B. Holland, Mean–mean multiple comparison displays for families of linear contrasts. J. Comput. Graph. Stat. **14**(4), 937–955 (2006)

R.M. Heiberger, E. Neuwirth, *R through Excel: A Spreadsheet Interface for Statistics, Data Analysis, and Graphics* (Springer, New York, 2009). URL: `http://www.springer.com/978-1-4419-0051-7`

R.M. Heiberger, N.B. Robbins, Design of diverging stacked bar charts for Likert scales and other applications. J. Stat. Softw. **57**(5), 1–32 (2014). URL: `http://www.jstatsoft.org/v57/i05/`

R.M. Heiberger, P. Teles, Displays for direct comparison of ARIMA models. Am. Stat. **56**, 131–138, 258–260 (2002)

R.M. Heiberger, with contributions from H. Burt, *RcmdrPlugin.HH: Rcmdr Support for the HH Package. R Package Version 1.1-42* (2015). URL: `http://CRAN.R-project.org/package=RcmdrPlugin.HH`

C.R. Hicks, K.V. Turner Jr., Fundamental *Concepts in Design of Experiments*, 5th edn. (Oxford, New York, 1999)

J.E. Higgens, G.G. Koch, Variable selection and generalized chi-square analysis of categorical data applied to a large cross-sectional occupational health survey. Int. Stat. Rev. **45**, 51–62 (1977)

D.C. Hoaglin, F. Mosteller, J.W. Tukey (eds.), *Understanding Robust and Exploratory Data Analysis* (Wiley, New York, 1983)

Y. Hochberg, A sharper Bonferroni procedure for multiple tests of significance. Biometrika **75**, 800–803 (1988)

Y. Hochberg, A.C. Tamhane, *Multiple Comparison Procedures* (Wiley, NewYork, 1987)

M. Holzer, Mild therapeutic hypothermia to improve the neurologic outcome after cardiac arrest. N. Engl. J. Med. **346**(8), 549–556 (2002)

D.W. Hosmer, S. Lemeshow, *Applied Logistic Regression*, 2nd edn. (Wiley, New York, 2000)

J. Hsu, M. Peruggia, Graphical representations of Tukey's multiple comparison method. J. Comput. Graph. Stat. **3**, 143–161 (1994)

R. Ihaka, P. Murrell, K. Hornik, J.C. Fisher, A. Zeileis, *Colorspace: Color Space Manipulation. R Package Version 1.2-4* (2013). URL: http://CRAN.R-project.org/package=colorspace

R.L. Iman, *A Data-Based Approach to Statistics* (Duxbury, Belmont, 1994)

Insightful Corp., S-Plus Statistical Software: Release 6.1 (2002). URL: http://www.insightful.com.

F. John et al., *R Commander. R Package Version 2.1-6* (2015). URL: http://CRAN.R-project.org/package=Rcmdr

N.L. Johnson, F.C. Leone, *Statistics and Experimental Design in Engineering and the Physical Sciences*, vol. 2 (Wiley, New York, 1967)

P.E. Johnson, Emacs has no learning curve: Emacs and ess (2015). URL: http://pj.freefaculty.org/guides/Rcourse/emacs-ess/emacs-ess.pdf

P.O. Johnson, F. Tsao, Factorial design and covariance in the study of individual educational development. Psychometrika **10**, 133–162 (1945)

R.W. Johnson, Fitting percentage of body fat to simple body measurements. J. Stat. Educ. **4**(1) (1996). URL: http://www.amstat.org/publications/jse/archive.htm

D.E. Knuth, *The TeXbook* (Addison-Wesley, Reading, 1984)

L. Krantz, *1999–2000 Jobs Rated Almanac: The Best and Worst Jobs—250 in All—Ranked by More Than a Dozen Vital Factors Including Salary, Stress, Benefits and More* (St. Martins Press, New York, 1999). URL: http://www.hallmaps.com/almanacs_yearbooks/29.shtml

L. Lamport, *LaTeX: A Document Preparation System: User's Guide and Reference Manual* (Addison-Wesley, Boston, 1994)

M. Lavine, Problems in extrapolation illustrated with space shuttle o-ring data. J. Am. Stat. Assoc. **86**, 919–921 (1991)

A.J. Lea, New observations on distribution of neoplasms of female breast in certain European countries. Br. Med. J. **1**, 488–490 (1965)

E.T. Lee, *Statistical Methods for Survival Data Analysis* (Lifetime Learning Publications, Belmont, 1980)

E. Lehmann, *Nonparametrics—Statistical Methods Based on Ranks*, revised first edition (Prentice Hall, New York, 1998)

F. Leisch, R-core, Sweave: automatic generation of reports (2014). URL: `https://stat.ethz.ch/R-manual/R-devel/library/utils/doc/Sweave.pdf`

A.Y. Lewin, M.F. Shakun, *Policy Sciences, Methodology and Cases* (Pergammon, Oxford, 1976). URL: `http://lib.stat.cmu.edu/DASL/Stories/airpollutionfilters.html`

R. Likert, A technique for the measurement of attitudes. Arch. Psychol. **140**(55), 1–55 (1932)

R.J.A. Little, D.B. Rubin, *Statistical Analysis with Missing Data*, 2nd edn. (Wiley, New York, 2002)

J. Longley, An appraisal of least squares programs for the electronic computer from the point of view of the user. J. Am. Stat. Assoc. **62**, 819–841 (1967)

A. Luo, T. Keyes, Second set of results in from the career track member survey, in *Amstat News* (American Statistical Association, Arlington, 2005), pp. 14–15

M. Mächler, S. Eglen, R.M. Heiberger, K. Hornik, S.P. Luque, H. Redesting, A.J. Rossini, R. Sparapani, V. Spinu, ESS (Emacs Speaks Statistics) (2015). URL: `http://ESS.R-project.org/`

P. McCullagh, J.A. Nelder, *Generalized Linear Models* (Chapman and Hall, London, 1983)

C.R. Mehta, N.R. Patel, P. Senchaudhuri, Efficient Monte Carlo methods for conditional logistic regression. J. Am. Stat. Assoc. **95**(449), 99–108 (2000)

W.M. Mendenhall, J.T. Parsons, S.P. Stringer, N.J. Cassissi, R.R. Million, T2 oral tongue carcinoma treated with radiotherapy: analysis of local control and complications. Radiother. Oncol. **16**, 275–282 (1989)

D. Meyer, A. Zeileis, K. Hornik, The strucplot framework: visualizing multi-way contingency tables with vcd. J. Stat. Softw. **17**(3), 1–48 (2006). URL: `http://www.jstatsoft.org/v17/i03/`

D. Meyer, A. Zeileis, K. Hornik, *vcd: Visualizing Categorical Data. R Package Version 1.2-13* (2012). URL: `http://CRAN.R-project.org/package=vcd`

Microsoft, Inc., Word (2015). URL: `https://products.office.com/en-us/Word`

G.A. Milliken, D.E. Johnson, *Analysis of Messy Data*, vol. I (Wadsworth, Belmont, 1984)

D.C. Montgomery, *Design and Analysis of Experiments*, 4th edn. (Wiley, New York, 1997)

D.C. Montgomery, *Design and Analysis of Experiments*, 5th edn. (Wiley, New York, 2001)

D.S. Moore, G.P. McCabe, *Introduction to the Practice of Statistics* (Freeman, New York, 1989)

F. Mosteller, J.W. Tukey, *Data Analysis and Regression* (Addison-Wesley, Reading, 1977)

J.D. Murray, G. Dunn, P. Williams, A. Tarnopolsky, Factors affecting the consumption of psychotropic drugs. Psychol. Med. **11**, 551–560 (1981)

P. Murrell, *R Graphics*, 2nd edn. (CRC, Boca Raton, 2011). URL: http://www.taylorandfrancis.com/books/details/9781439831762/

R.H. Myers, *Classical and Modern Regression with Applications*, Chap. 5 (PWS-Kent, Boston, 1990), p. 218

S.C. Narula, J.T. Wellington, Prediction, linear regression and the minimum sum of errors. Technometrics **19**, 185–190 (1977)

U.S. Navy, *Procedures and Analyses for Staffing Standards Development: Data/Regression Analysis Handbook* (Navy Manpower and Material Analysis Center, San Diego, 1979)

J.A. Nelder, A reformulation of linear models. J. R. Stat. Soc. **140**(1), 48–77 (1977). doi:10.2307/2344517

J. Neter, M.H. Kutner, C.J. Nachtsheim, W. Wasserman, *Applied Linear Statistical Models*, 4th edn. (Irwin, Homewood, 1996)

E. Neuwirth, *RColorBrewer: ColorBrewer Palettes. R Package Version 1.0-5* (2011). URL: http://CRAN.R-project.org/package=RColorBrewer

E. Neuwirth, RExcel (2014). URL: http://rcom.univie.ac.at

New Zealand Ministry of Research Science and Technology, Staying in science (2006). URL: http://www.morst.govt.nz/Documents/publications/researchreports/Staying-in-Science-summary.pdf

D.F. Nicholls, The analysis of time series—the time domain approach. Aust. J. Stat. **21**, 93–120 (1979)

NIST, National Institute of Standards and Technology, Statistical Engineering Division (2002). URL: http://www.itl.nist.gov/div898/software/dataplot.html/datasets.htm

NIST, National Institute of Standards and Technology, Data set of southern oscillations, in *NIST/SEMATECH e-Handbook of Statistical Methods* (2005). URL: http://www.itl.nist.gov/div898/handbook/pmc/section4/pmc4412.htm

Olympic Committee, Salt Lake City 2002 Winter Olympics (2001). URL: http://www.saltlake2002.com

R.L. Ott, *An Introduction to Statistical Methods and Data Analysis*, 4th edn. (Duxbury, Belmont, 1993)

S.C. Pearce, *The Agricultural Field Experiment* (Wiley, New York, 1983)

K. Penrose, A. Nelson, A. Fisher, Generalized body composition prediction equation for men using simple measurement techniques (abstract). Med. Sci. Sports Exerc. **17**(2), 189 (1985)

D.H. Peterson (ed.), *Aspects of Climate Variability in the Pacific and the Western Americas*. Number 55 in Geophysical Monograph (American Geophysical Union, Washington, DC, 1990)

R.G. Peterson, *Design and Analysis of Experiments* (Marcel Dekker, New York/Basel, 1985)

R Core Team, *R: A Language and Environment for Statistical Computing* (R Foundation for Statistical Computing, Vienna, 2015). URL: http://www.R-project.org/

R Development Core Team, *R: A Language and Environment for Statistical Computing* (R Foundation for Statistical Computing, Vienna, 2004). URL: http://www.R-project.org. ISBN 3-900051-00-3

W. Rasband, Imagej 1.46t (2015). URL: http://rsb.info.nih.gov/ij

N.B. Robbins, *Creating More Effective Graphs* (Chart House, Ramsey, 2005, reissued 2013) [Originally Wiley-Interscience]

N.B. Robbins, R.M. Heiberger, E. Neuwirth, C. Ritter, Professional statistics and graphics accessible from excel, in *Presented at the Joint Statistics Meetings* (American Statistical Association, Washington, DC, 2009). URL: http://rcom.univie.ac.at/papers/handoutJSM2009.pdf

W.S. Robinson, Ecological correlations and the behavior of individuals. Am. Sociol. Rev. **15**, 351–357 (1950)

A.J. Rossini, R.M. Heiberger, R.A. Sparapani, M. Mächler, K. Hornik, Emacs Speaks Statistics (ESS): a multiplatform, multipackage development environment for statistical analysis. J. Comput. Graph. Stat. **13**(1), 247–261 (2004). URL: http://dx.doi.org/10.1198/1061860042985

A.J. Rossman, Televisions, physicians, and life expectancy. J. Stat. Educ. (1994). URL: http://www.amstat.org/publications/jse/archive.htm

RStudio, Shiny: a web application framework for r (2015). URL: http://shiny.rstudio.com

D. Sarkar, *Lattice: Multivariate Data Visualization with R* (Springer, New York, 2008). URL: http://lmdvr.r-forge.r-project.org. ISBN 978-0-387-75968-5

D. Sarkar, *lattice: Lattice Graphics. R Package Version 0.20-29* (2014). URL: http://CRAN.R-project.org/package=lattice

D. Sarkar, F. Andrews, *latticeExtra: Extra Graphical Utilities Based on Lattice. R Package Version 0.6-26* (2013). URL: http://CRAN.R-project.org/package=latticeExtra

S. Sarkar, Some probability inequalities for ordered MTP_2 random variables: a proof of the Simes conjecture. Ann. Stat. **26**, 494–504 (1998)

SAS Institute, Inc., The four types of estimable functions, in *SAS/STAT User's Guide* (SAS Institute, Inc., Cary, 1999)

C. Schenk, MikTeX (2001). URL: ftp://metalab.unc.edu/pub/packages/TeX/systems/win32/miktex

S.R. Searle, *Linear Models* (Wiley, New York, 1971)

H.C. Selvin, Durkheim's suicide: further thoughts on a methodological classic, in *Émile Durkheim*, ed. by R.A. Nisbet (Prentice Hall, Englewood Cliffs, NJ, 1965), pp. 113–136

R.T. Senie, P.P. Rosen, M.L. Lesser, D.W. Kinne, Breast self-examinations and medical examination relating to breast cancer stage. Am. J. Public Health **71**, 583–590 (1981)

N. Shaw, *Manual of Meteorology*, vol. 1 (Cambridge University Press, Cambridge, 1942)

W.J. Shih, S. Weisberg, Assessing influence in multiple linear regression with incomplete data.Technometrics **28**, 231–240 (1986)

J. Simpson, A. Olsen, J.C. Eden, A Bayesian analysis of a multiplicative treatment effect in weather modification. Technometrics **17**, 161–166 (1975)

W. Smith, L. Gonick, *The Cartoon Guide to Statistics* (HarperCollins, New York, 1993)

G.W. Snedecor, W.G. Cochran, *Statistical Methods*, 7th edn. (Iowa State University Press, Ames, 1980)

R.R. Sokal, F.J. Rohlf, *Biometry*, 2nd edn. (W.H. Freeman, New York, 1981)

R.M. Stallman, Emacs (2015). URL: http://www.gnu.org/software/emacs/

R.G.D. Steel, J.H. Torrie, *Principles and Procedures of Statistics*, 1st edn. (McGraw-Hill, Auckland, 1960)

P.H. Sulzberger, The effects of temperature on the strength of wood, plywood and glued joints. Technical Report (Aeronautical Research Consultative Committee, Australia, Department of Supply, 1953)

N. Teasdale, C. Bard, J. LaRue, M. Fleury, On the cognitive penetrability of posture control. Exp. Aging Res. **19**, 1–13 (1993)

The TeX Users Group (TUG), The MacTeX-2014 Distribution (2014). URL: http://tug.org/mactex/

TIBCO Software Inc., *TIBCO Spotfire S+: Release 8.2* (2010). URL: `https://edelivery.tibco.com/storefront/eval/tibco-spotfire-s-/prod10222.html`

TIBCO Software Inc., *The TIBCO Enterprise Runtime for R Engine* (2014). URL: `http://spotfire.tibco.com/discover-spotfire/what-does-spotfire-do/predictive-analytics/tibco-enterprise-runtime-for-r-terr`

R. Till, *Statistical Methods for the Earth Scientist* (Macmillan, London, 1974)

E.R. Tufte, *The Visual Display of Quantitative Information*, 2nd edn. (Graphics Press, Cheshire, 2001)

J.W. Tukey, One degree of freedom for nonadditivity. Biometrics **5**(3), 232–242 (1949)

W. Vandaele, Participation in illegitimate activities: Erlich revisited, in *Deterrence and Incapacitation*, ed. by A. Blumstein, J. Cohen, D. Nagin (National Academy of Sciences, Washington, DC, 1978), pp. 270–335

P.K. VanVliet, J.M. Gupta, Tham-v-sodium bicarbonate in idiopathic respiratory distress syndrome. Arch. Dis. Child. **48**, 249–255 (1973)

W.N. Venables, Exegeses on linear models, in *Proceedings of the S-PLUS Users Conference*, Washington, DC (1998). URL: `http://www.stats.ox.ac.uk/pub/MASS3/Exegeses.pdf`

W.N. Venables, B.D. Ripley, *Modern Applied Statistics with S-PLUS*, 2nd edn. (Springer, New York, 1997)

H. Wainer, *Graphical Tales of Fate and Deception from Napoleon Bonaparte to Ross Perot* (Copernicus Books, New York, 1997)

W.W.S. Wei, *Time Series Analysis, Univariate and Multivariate Methods* (Addison-Wesley, Reading, 1990)

A.M. Weindling, F.M. Bamford, R.A. Whittall, Health of juvenile delinquents. Br. Med. J. **292**, 447 (1986)

S. Weisberg, *Applied Linear Regression*, 2nd edn. (Wiley, New York, 1985)

I. Westbrooke, Simpson's paradox: an example in a New Zealand survey of jury composition. Chance **11**, 40–42 (1998)

P.H. Westfall, D. Rom, Bootstrap step-down testing with multivariate location shift data. Unpublished (1990)

H. Wickham, *Ggplot2: Elegant Graphics for Data Analysis* (Springer, New York, 2009). URL: `http://had.co.nz/ggplot2/book`

Wikipedia, *Integrated Development Environment* (2015). URL: `http://en.wikipedia.org/wiki/Integrated_development_environment`

L. Wilkinson, *The Grammar of Graphics* (Springer, New York, 1999)

A.F. Williams, Teenage passengers in motor vehicle crashes: a summary of current research. Technical report (Insurance Institute for Highway Safety, Arlington, 2001)

E.J. Williams, *Regression Analysis* (Wiley, New York, 1959)

N. Woods, P. Fletcher, A. Hughes, *Statistics in Language Studies* (Cambridge University Press, Cambridge, 1986)

World Almanac and Book of Facts, *World Almanac and Book of Facts*, 2002 edn. (World Almanac Books, New York, 2001)

E.L. Wynder, A. Naravvette, G.E. Arostegui, J.L. Llambes, Study of environmental factors in cancer of the respiratory tract in Cuba. J. Natl. Cancer Inst. **20**, 665–673 (1958)

Y. Xie (2015). URL: `http://cran.r-project.org/web/packages/knitr/knitr.pdf`

F. Yates, The analysis of multiple classifications with unequal numbers in the different classes. J. Am. Stat. Assoc. **29**, 51–56 (1934)

F. Yates, *The Design and Analysis of Factorial Experiments* (Imperial Bureau of Soil Science, Harpenden, 1937)

Index of Datasets

© Springer Science+Business Media New York 2015
R.M. Heiberger, B. Holland, *Statistical Analysis and Data Display*,
Springer Texts in Statistics, DOI 10.1007/978-1-4939-2122-5

Index

Printed in the United States
By Bookmasters